Nikolay I. Kolev

Multiphase Flow Dynamics 1

Nikolay I. Kolev

Multiphase Flow Dynamics 1
Fundamentals

2nd ed.

With 114 Figures and CD-ROM

Dr.-Ing. habil. Nikolay I. Kolev
Framatome ANP GmbH
P.O. Box 3220
91050 Erlangen
Germany

ISBN 3-540-22106-9 **Springer Berlin Heidelberg New York**

Library of Congress Control Number: 2004111217

This work is subject to copyright. All rights are reserved, whether the whole or part of the material is concerned, specifically the rights of translation, reprinting, reuse of illustrations, recitation, broadcasting, reproduction on microfilm or in other ways, and storage in data banks. Duplication of this publication or parts thereof is permitted only under the provisions of the German Copyright Law of September 9, 1965, in its current version, and permission for use must always be obtained from Springer-Verlag. Violations are liable to prosecution under German Copyright Law.

Springer is a part of Springer Science+Business Media
springeronline.com

© Springer-Verlag Berlin Heidelberg 2002, 2005
Printed in Germany

The use of general descriptive names, registered names, trademarks, etc. in this publication does not imply, even in the absence of a specific statement, that such names are exempt from the relevant protective laws and regulations and therefore free for general use.

Typesetting: Digital data supplied by author.
Production: PTP-Berlin Protago-TeX-Production GmbH, Germany
Cover-Design: medionet AG, Berlin
Printed on acid-free paper   62/3020 Yu - 5 4 3 2 1 0

To Iva, Rali and Sonja with love!

Silence, Feb. 2004, Nikolay Ivanov Kolev, 36× 48cm oil on linen

A FEW WORDS ABOUT THE SECOND EXTENDED EDITION

The reader will find in the second edition the information already included in the first one improved and extended in several places. Chapter 3 of Volume I has been completely rewritten. It now contains the next step in the generalization of the theory of the equations of states for arbitrary real mixtures. Now with one and the same formalism a mixture of miscible and immiscible components in arbitrary solid, liquid or gaseous states mixed and/or dissolved can be treated. This is a powerful method towards creating a universal flow analyzer. Chapter 6 has been extended with cases including details of modeling of combustion and detonation of hydrogen by taking into account the equilibrium dissociation. In Chapter 9, dealing with detonation during melt-water interaction, additional introductory information is given for the detonation of hydrogen in closed pipes taking into account the dissociation of the generated steam. A new Chapter 11 is inserted between the former Chapters 10 and 11 giving the mathematical tools for computing eigenvalues and eigenvectors and for determination of the type of systems of partial differential equations. The procedure for transformation of a hyperbolic system into canonical form is also provided. Then the relations between eigenvalues and critical flow and between eigenvalues and propagation velocity of small perturbation are briefly defined. This is in fact a translation of one chapter of my first book published in German by Springer in 1986. In Chapter 12 about the numerical solution methods, the variation of the volume-porosity with time is systematically incorporated into the numerical formalism. Appendix 2 of Volume I contains some additional information about orthogonal grid generation. Chapter 26 of Volume II, which is included in the accompanying CD, contains some additional experiments and movies documenting the performance of the method for fast pressure wave propagation in 2D geometry and interesting acoustical problems of melt-water interaction. Of course misprints and some layout deficiencies have also been removed as is usual for a second edition of such voluminous material.

The form of the second improved and extended edition has been reached after I received many communications from all over the world from colleagues and friends commenting on different aspects of the two volumes or requesting additional information. I thank all of you who have contributed in this way to improving the two volumes.

Erlangen, February 2004   Nikolay Ivanov Kolev

INTRODUCTION

Multi-phase flows are not only part of our natural environment such as rainy or snowy winds, tornadoes, typhoons, air and water pollution, volcanic activities etc., but also are working processes in a variety of conventional and nuclear power plants, combustion engines, propulsion systems, flows inside the human body, oil and gas production and transport, chemical industry, biological industry, process technology in the metallurgical industry or in food production etc. The list is by far not exhaustive. For instance everything to do with phase changes is associated with multi-phase flows. The industrial use of multi-phase systems requires methods for predicting their behavior. This explains the "explosion" of scientific publications in this field in the last 30 years. Some countries, such as Japan, have declared this field to be of strategic importance for future technological development.

Probably the first known systematic study on two-phase flow was done during the Second World War by the Soviet scientist *Teletov* [12] and published in 1958 as "On the problem of fluid dynamics of two-phase mixtures". Two books that appeared in Russia and the USA in 1969 by *Mamaev* et al. [7] and by *Wallis* [13] played an important role in educating a generation of scientists in this discipline including me. Both books contain valuable information mainly for steady state flows in pipes. *Hewitt* and *Hall-Taylor* published in 1974 "Annular two-phase flow" [5]. The book also considers steady state pipe flows. The usefulness of the idea of a three-fluid description of two-phase flows was clearly demonstrated on annular flows with entrainment and deposition. *Ishii* [6] published in 1975 the book "Thermo-fluid dynamic theory of two-phase flow", which contained a rigorous derivation of *time*-averaged conservation equations for the so called two-fluid separated and diffusion momentum equations models. This book founded the basics for new measurement methods appearing on the market later. *R. Nigmatulin* published "Fundamentals of mechanics of heterogeneous media" [8] in Russian in 1978. The book mainly considers one-dimensional two-phase flows. Interesting particular wave dynamics solutions are obtaining for specific sets of assumptions for dispersed systems. The book was extended mainly with mechanical interaction constitutive relations and translated into English in 1991 [9]. The next important book [2] for two-phase steam-water flow in turbines was published by *Deich* and *Philipoff* in 1981 in Russian. Again mainly steady state, one-dimensional flows are considered. *Delhaye* et al. published in the same year "Thermohydraulics of two-phase systems for industrial design and nuclear engineering" [3]. The book contains the main ideas of local volume averaging, and considers mainly many steady state one-dimensional flows. One year later, in 1982, *Hetsroni* edited the "Handbook of multi phase systems" [4] containing the state of the art of constitutive interfacial relationships for practical use. The book is still a valuable source of empirical information for different disciplines dealing with multi-phase flows. In the monograph "Interfacial transport phenomena" [10] published by *Slattery* in 1990 complete, rigorous derivations of the local volume-averaged two-fluid conservation equations are presented together with a variety of aspects of the fundamentals of the interfacial processes based on his long years of work. *Slattery*'s first edition appeared in 1978. Some aspects of the heat and mass transfer theory of two-phase flow are now included in modern text books such us "Thermodynamics" by *Baer* and "Technical thermodynamics" by *Stephan* and *Mayinger*, see [1] and [11].

It is noticeable that none of the above mentioned books is devoted in particular to *numerical methods* of solution of the fundamental systems of partial differential equations describing multi-phase flows. Analytical methods still do not exist. I published in 1986 the book "Transient two-phase flows" with Springer-Verlag in German, discussing several engineering methods and practical examples for integrating systems of partial differential equations describing two- and three-fluid flows in pipes.

Since 1984 I have worked intensively on creating numerical algorithms for describing complicated multi-phase multi-component flows in pipe networks and complex *three-dimensional* geometries mainly for nuclear safety applications. Note that the mathematical description of multi-dimensional two- and multi-phase flows is a scientific discipline with considerable activity in the last 30 years. In addition thousands of scientists have collected for years experimental information in this field. But there is still a lack of a systematic presentation of the theory and practice of *numerical multi-phase fluid dynamics*. This book is intended to fill this gap.

Numerical multi-phase fluid dynamics is the science of the derivation and the numerical integration of the conservation equations reflecting the mass momentum and energy conservation for multi-phase processes in nature and technology at different scales in time and space. The emphasis of this book is on the generic links within computational predictive models between

- fundamentals,
- numerical methods,
- empirical information for the constitutive interfacial phenomena, and
- comparison with experimental data at different levels of complexity.

The reader will realize how strong the mutual influence of the four model constituencies is. There are still many attempts to attack these problems using single-phase fluid mechanics by simply extending existing single-phase computer codes with additional fields and linking with differential terms outside of the code without increasing the strength of the feedback in the numerical integration methods. The success of this approach in describing low concentration suspensions and dispersed systems without strong thermal interactions should not confuse the engineer about the real limitations of this method.

This monograph can be considered also as a handbook on numerical modeling of three strongly interacting fluids with dynamic fragmentation and coalescence representing multi-phase multi-component systems. Some aspects of the author's ideas, such us the three-fluid entropy concept with dynamic fragmentation and coalescence for describing multi-phase, multi-component flows by local volume-averaged and time-averaged conservation equations, have been published previously in separate papers but are collected here in a single context for the first time. An important contribution of this book to the state of the art is also the rigorous thermodynamic treatment of multi-phase systems, consisting of different mixtures. It is also the first time of publishing the basics of the boundary fitted description of multi-phase flows and an appropriate numerical method for integrating them with proven convergence. It is well known in engineering practice that "the devil is hidden in the details". This book gives many hints and details on how to design computational methods for multi-phase flow analysis and demonstrates the power of the method in the attached compact disc and in the last chapter in Volume 2 by presenting successful comparisons between predictions and experimental data or analytical benchmarks for a class of problems with a complexity not known in the multi-phase literature up to now. It starts with the single-phase U-tube problem and ends with explosive interaction between molten melt and cold water in complicated 3D geometry and condensation shocks in complicated pipe networks containing acoustically interacting valves and other components.

Erlangen, Spring 2002   Nikolay Ivanov Kolev

References

1. Baer HD (1996) Thermodynamik, Springer, Berlin Heidelberg New York
2. Deich ME, Philipoff GA (1981) Gas dynamics of two phase flows. Energoisdat, Moscow (in Russian)
3. Delhaye JM, Giot M, Reithmuller ML (1981) Thermohydraulics of two-phase systems for industrial design and nuclear engineering, Hemisphere, New York, McGraw Hill, New York
4. Hetstroni G (1982) Handbook of multi phase systems. Hemisphere, Washington, McGraw-Hill, New York
5. Hewitt GF and Hall-Taylor NS (1974) Annular two-phase flow, Pergamon, Oxford
6. Ishii M (1975) Thermo-fluid dynamic theory of two-phase flow, Eyrolles, Paris
7. Mamaev WA, Odicharia GS, Semeonov NI, Tociging AA (1969) Gidrodinamika gasogidkostnych smesey w trubach, Moskva
8. Nigmatulin RI (1978) Fundamentals of mechanics of heterogeneous media, Nauka, Moscow, 336 pp (in Russian)

9. Nigmatulin RI (1991) Dynamics of multi-phase media, revised and augmented rdition, Hemisphere, New York
10. Slattery JC (1990) Interfacial transport phenomena, Springer, Berlin Heidelberg New York
11. Stephan K and Mayinger F (1998) Technische Thermodynamik, Bd.1, Springer, 15. Auflage
12. Teletov SG (1958) On the problem of fluid dynamics of two-phase mixtures, I. Hydrodynamic and energy equations, Bulletin of the Moscow University, no 2 p 15
13. Wallis GB (1969) One-dimensional two-phase flow, McGraw-Hill, New York

SUMMARY

This monograph contains theory, methods and practical experience for describing complex transient multi-phase processes in arbitrary geometrical configurations. It is intended to help applied scientists and practicing engineers to understand better natural and industrial processes containing dynamic evolutions of complex multi-phase flows. The book is also intended to be a useful source of information for students in the high semesters and in PhD programs.

This monograph consists of two volumes:

Vol. 1 Fundamentals (14 Chapters and 2 Appendixes), 746 pages + CD-ROM
Vol. 2 Mechanical and thermal interactions (26 Chapters), 690 pages

In Volume 1 the concept of three-fluid modeling is introduced. Each of the fields consists of multi-components grouped into an inert and a non-inert components group. Each field has its own velocity in space and its own temperature allowing mechanical and thermodynamic non-equilibrium among the fields. The idea of dynamic fragmentation and coalescence is introduced. Using the *Slattery-Whitaker* local spatial averaging theorem and the *Leibnitz* rule, the local volume-averaged mass, momentum and energy conservation equations are rigorously derived for heterogeneous porous structures. Successively a time averaging is performed. A discussion is provided on particle size spectra and averaging, cutting off the lower part of the spectrum due to mass transfer, the effect of the averaging on the effective velocity difference etc. In the derivation of the momentum equations special attention is paid to rearranging the pressure surface integrals in order to demonstrate the physical meaning of the originating source terms in the averaged systems and their link to hyperbolicity. The Reynolds stress concept is introduced for multi-phase flows. Before deriving the energy conservation in Chapter 5, I provide a Chapter 3 in which it is shown how to generate thermodynamic properties and the substantial derivatives for different kinds of mixtures by knowing the properties of the particular constituents. This chapter provides the necessary information to understand the entropy concept which is presented in Chapter 5. In the author's experience understanding the complex energy conservation for multi-phase systems and especially the entropy concept is very difficult for most students and practicing engineers. That is why Chapter 4 is provided as an introduction, showing the variety of notation of the energy conservation principle for single-phase multi-component flows. The local volume-averaged and time-averaged energy conservation equation is derived in Chapter 5 in different notation forms in terms of specific internal energy, specific enthalpy, specific entropy, and temperatures. The introduction of the entropy principle for such complex systems is given in detail in order to enable the practical use of the entropy concept. The useful "conservation of volume" equation is also derived. Examples for better understanding are given for the simple cases of lumped parameters – Chapter 6, infinite heat exchange without interfacial mass transfer, discharge of gas from a volume, injection of inert gas in a closed volume initially filled with inert gas, heat input in a gas in a closed volume, steam injection in a steam-air mixture, chemical reaction in a gas mixture in a closed volume and hydrogen combustion in an inert atmosphere. The exergy for a multi-phase, multi-component system is introduced in Chapter 7 and discussed for the example of judging the efficiency of a heat pump. Simplification of the resulting system of PDEs to the case of one-dimensional flow is presented in Chapter 8. Some interesting aspects of the fluid structure coupling such as pipe deformation due to temporal pressure change in the flow, and forces acting on the internal pipe walls are discussed. The idea of algebraic slip is presented. From the system thus obtained the next step of simplification leads to the system of ordinary differential equations describing the critical multi-phase, multi-component flow by means of three velocity fields. Modeling of valves and pumps is discussed in the context of modeling of networks consisting of pipes, valves, pumps and other different components. Another case of simplification of the theory of multiphase flows is presented in Chapter 9, where the theory of continuum sound waves and discontinuous shock waves for melt-water interaction is presented. In order to easily understand it, the corresponding theory for single- and two-phase flows is reviewed as an introduction. Finally an interesting application for the interaction of molten uranium and aluminum oxides with water, as well of the interaction of molten iron with water is presented. Chapter 10 is devoted to the derivation of the conservation equations for multi-phase multi-component multi-velocity field flow in general curvilinear coordinate systems. For a better understanding of the mathematical basics used in this chapter two appendixes are provided: Appendix 1 in which a brief introduction to vector analysis is given and Appendix 2 in which the basics of the coordinate transformation theory are summarized. Chapter 11 describes numerical solution methods for different multi-phase flow problems. The first order donor-cell method is presented in detail by discretizing the governing equations, creating a strong interfacial velocity coupling, and strong pressure-velocity coupling. Different approximations for the pressure equations are derived and three different solution methods are discussed in detail. One of them is based on the Newton iterations for minimizing the residuals by using the conjugate gradients. A method for temperature inversion is presented. Several details are given enabling scientists and engineers to use this chapter for their own computer code development, such us integration procedure (implicit method), time step and accuracy control. Finally some high-order discretization schemes for convection-diffusion terms such us space exponential scheme and other high-order up-winding schemas are presented. Different analytical derivations are provided in Appendixes 11.1 to 11.8 including the analytical derivatives of the residual error of each equation with respect to the dependent variables. Some important basic definitions that are required for describing a pipe networks are introduced. Chapter 12 presents a numerical solution method for multi-phase flow problems in multiple blocks of curvilinear coordinate systems generalizing in fact the experience gained by Chapter 11. Several important details of how to derive explicit pressure equations are provided. The advantage of using orthogonal grids also is easily derived from this chapter. This completes the basics of the multi-phase, multi-component flow dynamics.

Chapter 13 provides the mathematical tools for determination of the type of a system of partial differential equations. The procedures for computing eigenvalues and eigenvectors and the algorithm for transferring a hyperbolic system into canonical form are given. The relations between eigenvalues and critical flow and eigenvalues and propagation velocity of small perturbations are also defined.

Chapter 14 provides several numerical simulations as illustrations of the power of the methods presented in this monograph. A compact disc is attached that contains movies corresponding to particular cases discussed in this chapter. The movies can be played with any tool capable of accepting *avi*- or animated *gif*-files.

Volume 2 is devoted to the so-called closure laws: the important constitutive relations for mechanical and thermal interactions. The structure of the volume has the character of a state-of-the-art review and a selection of the best available approaches for describing interfacial processes. In many cases the original contribution of the author is incorporated into the overall presentation. The most important aspects of the presentation are that it stems from the author's long years of experience of developing computer codes. The emphasis is on the practical use of these relationships: either as stand alone estimation methods or within a framework of computer codes.

NOMENCLATURE

Latin

A	cross section, m^2
A	surface vector
a	speed of sound, m/s
a_{lw}	surface of the field l wetting the wall w per unit flow volume $\sum_{l=1}^{l_{max}} Vol_l$ belonging to control volume Vol (local volume interface area density of the structure w), m^{-1}
$a_{l\sigma}$	surface of the velocity field l contacting the neighboring fields per unit flow volume $\sum_{l=1}^{l_{max}} Vol_l$ belonging to control volume Vol (local volume interface area density of the velocity field l), m^{-1}
a_l	total surface of the velocity field l per unit flow volume $\sum_{l=1}^{l_{max}} Vol_l$ belonging to control volume Vol (local volume interface area density of the velocity field l), m^{-1}
Cu_i	Courant criterion corresponding to each eigenvalue, *dimensionless*
C_{il}	mass concentration of the inert component i in the velocity field l
c	coefficients, *dimensionless*
C_m	mass concentration of the component m in the velocity field, *dimensionless*
C_i	mass concentration of the component i in the velocity field, *dimensionless*
c_p	specific heat at constant pressure, $J/(kgK)$
c^{vm}	virtual mass force coefficient, *dimensionless*
c^d	drag force coefficient, *dimensionless*
c^L	lift force coefficient, *dimensionless*
D_{hy}	hydraulic diameter (4 times cross-sectional area / perimeter), m
D_{3E}	diameter of the entrained droplets, m
D_{ld}	size of the bubbles produced after one nucleation cycle on the solid structure, bubble departure diameter, m

D_{1dm}	size of bubbles produced after one nucleation cycle on the inert solid particles of field $m = 2, 3$
D_{1ch}	critical size for homogeneous nucleation, m
D_{1cd}	critical size in presence of dissolved gases, m
D'_l	most probable particle size, m
D_l	characteristic length of the velocity field l, particle size in case of fragmented field, m
D^l_{il}	coefficient of molecular diffusion for species i into the field l, m^2/s
D^t_{il}	coefficient of turbulent diffusion, m^2/s
D^*_{il}	total diffusion coefficient, m^2/s
DC_{il}	right-hand side of the non-conservative conservation equation for the inert component, $kg/(sm^3)$
D	diffusivity, m^2/s
d	total differential
E	total energy, J
e	specific internal energy, J/kg
$F(\xi)$	function introduced first in Eq. (42) Chapter 2
$F, f(...$	function of (...
f	force per unit flow volume, N/m^3
f	fraction of entrained melt or water in the detonation theory
F_{lw}	surfaces separating the velocity field l from the neighboring structure within Vol, m^2
$F_{l\sigma}$	surfaces separating the velocity field l from the neighboring velocity field within Vol, m^2
F	surface defining the control volume Vol, m^2
f_{im}	frequency of the nuclei generated from one activated seed on the particle belonging to the donor velocity field m, s^{-1}
f_{lw}	frequency of the bubble generation from one activated seed on the channel wall, s^{-1}
$f_{l,coal}$	coalescence frequency, s^{-1}
g	acceleration due to gravity, m/s^2
H	height, m
h	specific enthalpy, J/kg
h_i	eigenvectors corresponding to each eigenvalue
I	unit matrix, *dimensionless*
i	unit vector along the x-axis
J	matrix, *Jacobian*
j	unit vector along the y-axis

k	unit vector along the *k*-axis
k	cell number
k	kinetic energy of turbulent pulsation, m^2/s^2
k_{il}^T	coefficient of thermo-diffusion, *dimensionless*
k_{il}^p	coefficient of baro-diffusion, *dimensionless*
L	length, *m*
M_i	kg-mole mass of the species *i*, *kg/mole*
m	total mass, *kg*
$\mathbf{n}_{\Delta V}$	unit vector pointing along $\Delta \mathbf{V}_{ml}$, *dimensionless*
n	unit vector pointing outwards from the control volume *Vol*, *dimensionless*
$\mathbf{n}_{le}$	unit surface vector pointing outwards from the control volume *Vol*
$\mathbf{n}_{l\sigma}$	unit interface vector pointing outwards from the velocity field *l*
n_{il}	number of the particle from species *i* per unit flow volume, m^{-3}
n_l	number of particles of field *i* per unit flow volume, particle number density of the velocity field *l*, m^{-3}
$\dot{n}_{coal}$	number of particles disappearing due to coalescence per unit time and unit volume, m^{-3}
$\dot{n}_{l,kin}$	particle production rate due to nucleation during evaporation or condensation, $1/(m^3 s)$
n_{lw}'''	number of the activated seeds on unit area of the wall, m^{-2}
$\dot{n}_{lh}$	number of the nuclei generated by homogeneous nucleation in the donor velocity field per unit time and unit volume of the flow, $1/(m^3 s)$
$\dot{n}_{l,dis}$	number of the nuclei generated from dissolved gases in the donor velocity field per unit time and unit volume of the flow, $1/(m^3 s)$
$\dot{n}_{l,sp}$	number of particles of the velocity field *l* arising due to hydrodynamic disintegration per unit time and unit volume of the flow, $1/(m^3 s)$
P	probability
P	irreversibly dissipated power from the viscous forces due to deformation of the local volume and time average velocities in the space, W/kg
Per	perimeter, *m*
p_{li}	*l* = 1: partial pressure inside the velocity field *l*
	l = 2,3: pressure of the velocity field *l*
p	pressure, *Pa*
$\dot{q}'''$	thermal power per unit flow volume introduced into the fluid, W/m^3
$\dot{q}_{\sigma l}'''$	*l* = 1,2,3. Thermal power per unit flow volume introduced from the interface into the velocity field *l*, W/m^3

$\dot{q}'''_{w\sigma l}$	thermal power per unit flow volume introduced from the structure interface into the velocity field l, W/m^3
R	mean radius of the interface curvature, m
$\mathbf{r}(x,y,z)$	position vector, m
R	(with indexes) gas constant, $J/(kgK)$
$\mathbf{s}$	arc length vector, m
S	total entropy, J/K
s	specific entropy, $J/(kgK)$
Sc^t	turbulent *Schmidt* number, *dimensionless*
T	temperature, K
T_l	temperature of the velocity field l, K
T	shear stress tensor, N/m^2
t	unit tangent vector
U	dependent variables vector
Vol	control volume, m^3
$Vol^{1/3}$	size of the control volume, m
Vol_l	volume available for the field l inside the control volume, m^3
$\sum_{l=1}^{l_{max}} Vol_l$	volume available for the flow inside the control volume, m^3
V	instantaneous fluid velocity with components, u, v, w in r, θ, and z direction, m/s
$\mathbf{V}_l^\tau$	instantaneous field velocity with components, $u_l^\vartheta, v_l^\tau, w_l^\tau$ in r, θ, and z direction, m/s
$\mathbf{V}_l$	time-averaged velocity, m/s
$\mathbf{V}_l'$	pulsation component of the instantaneous velocity field, m/s
$\Delta \mathbf{V}_{lm}$	$\mathbf{V}_l - \mathbf{V}_m$, velocity difference, disperse phase l, continuous phase m carrying l, m/s
$\delta_i V_l^\tau$	diffusion velocity, m/s
$\mathbf{V}_{l\sigma}^\tau$	interface velocity vector, m/s
$\mathbf{V}_l^\tau \gamma$	instantaneous vector with components, $u_l^\vartheta \gamma_r, v_l^\tau \gamma_\theta, w_l^\tau \gamma_z$ in r, θ, and z directions, m/s
v	specific volume, m^3/kg
x	mass fraction, *dimensionless*
y	distance between the bottom of the pipe and the center of mass of the liquid, m
×	vector product

Greek

α_l	part of $\gamma_v Vol$ available to the velocity field l, local instantaneous volume fraction of the velocity field l, *dimensionless*
α_{il}	the same as α_l in the case of gas mixtures; in the case of mixtures consisting of liquid and macroscopic solid particles, the part of $\gamma_v Vol$ available to the inert component i of the velocity field l, local instantaneous volume fraction of the inert component i of the velocity field l, *dimensionless*
$\alpha_{l,\max}$	≈ 0.62, limit for the closest possible packing of particles, *dimensionless*
γ_v	the part of $dVol$ available for the flow, volumetric porosity, *dimensionless*
γ	surface permeability, *dimensionless*
$\vec{\gamma}$	directional surface permeability with components $\gamma_r, \gamma_\theta, \gamma_z$, *dimensionless*
Δ	finite difference
δ	small deviation with respect to a given value
δ_l	= 1 for continuous field;
	= 0 for disperse field, *dimensionless*
∂	partial differential
ε	dissipation rate for kinetic energy from turbulent fluctuation, power irreversibly dissipated by the viscous forces due to turbulent fluctuations, W/kg
η	dynamic viscosity, $kg/(ms)$
θ	θ-coordinate in the cylindrical or spherical coordinate systems, *rad*
κ	= 0 for Cartesian coordinates,
	= 1 for cylindrical coordinates
κ	isentropic exponent
κ_l	curvature of the surface of the velocity field l, m
λ	thermal conductivity, $W/(mK)$
λ	eigenvalue
μ_l^τ	local volume-averaged mass transferred into the velocity field l per unit time and unit mixture flow volume, local volume-averaged instantaneous mass source density of the velocity field l, $kg/(m^3 s)$
μ_l	time average of μ_l^τ, $kg/(m^3 s)$
μ_{wl}	mass transport from exterior source into the velocity field l, $kg/(m^3 s)$
μ_{il}^τ	local volume-averaged inert mass from species i transferred into the velocity field l per unit time and unit mixture flow volume, local volume-

	averaged instantaneous mass source density of the inert component i of the velocity field l, $kg/(m^3 s)$
μ_{il}	time average of μ_{il}^τ, $kg/(m^3 s)$
μ_{iml}^τ	local volume-averaged instantaneous mass source density of the inert component i of the velocity field l due to mass transfer from field m, $kg/(m^3 s)$
μ_{iml}	time average of μ_{iml}^τ, $kg/(m^3 s)$
μ_{ilm}^τ	local volume-averaged instantaneous mass source density of the inert component i of the velocity field l due to mass transfer from field l into velocity field m, $kg/(m^3 s)$
μ_{ilm}	time average of μ_{ilm}^τ, $kg/(m^3 s)$
ν	kinematic viscosity, m^2/s
ν_l^t	coefficient of turbulent kinematic viscosity, m^2/s
ξ	angle between $\mathbf{n}_{l\sigma}$ and $\Delta \mathbf{V}_{lm}$, rad
ρ	density, kg/m^3
ρ	instantaneous density, density; without indexes, mixture density, kg/m^3
ρ_l	instantaneous field density, kg/m^3
ρ_{il}	instantaneous inert component density of the velocity field l, kg/m^3
$\langle \rho_l \rangle^l$	intrinsic local volume-averaged phase density, kg/m^3
$(\rho w)_{23}$	entrainment mass flow rate, $kg/(m^2 s)$
$(\rho w)_{32}$	deposition mass flow rate, $kg/(m^2 s)$
$(\rho_l \mathbf{V}_l^\tau)^{le}$	local intrinsic surface mass flow rate, $kg/(m^2 s)$
σ, σ_{12}	surface tension between phases 1 and 2, N/m
τ	time, s
φ	angle giving the projection of the position of the surface point in the plane normal to $\Delta \mathbf{V}_{lm}$, rad
$\chi_l^{m\sigma}$	the product of the effective heat transfer coefficient and the interfacial area density, $W/(m^3 K)$. The subscript l denotes inside the velocity field l. The superscript $m\sigma$ denotes location at the interface σ dividing field m from field l. The superscript is only used if the interfacial heat transfer is associated with mass transfer. If there is heat transfer only, the linearized interaction coefficient is assigned the subscript ml only, indicating the interface at which the heat transfer takes place.

Subscripts

c	continuous
d	disperse
lm	from l to m or l acting on m
w	region "outside of the flow"
e	entrances and exits for control volume Vol
l	velocity field l, intrinsic field average
i	inert components inside the field l, non-condensable gases in the gas field $l = 1$, or microscopic particles in water in field 2 or 3
i	corresponding to the eigenvalue λ_i in Chapter 4
M	non-inert component
m	mixture of entrained coolant and entrained melt debris that is in thermal and mechanical equilibrium behind the shock front
ml	from m into l
iml	from im into il
max	maximum number of points
n	inert component
0	at the beginning of the time step
E	entrainment
$coal$	coalescence
sp	splitting, fragmentation
σ	interface
τ	old time level
$\tau + \Delta\tau$	new time level
*	initial
0	reference conditions
p,v,s	at constant p,v,s, respectively
L	left
R	right
1	vapor or in front of the shock wave
2	water or behind the shock wave
3	melt
4	entrained coolant behind the front – entrained coolant
5	micro-particles after the thermal interaction – entrained melt

Superscripts

$'$	time fluctuation
'	saturated steam
''	saturated liquid
'''	saturated solid phase
A	air
d	drag

e	heterogeneous
i	component (either gas or solid particles) of the velocity field
i_{max}	maximum for the number of the components inside the velocity field
L	lift
l	intrinsic field average
le	intrinsic surface average
$l\sigma$	averaged over the surface of the sphere
m	component
n	normal
n	old iteration
$n+1$	new iteration
t	turbulent, tangential
vm	virtual mass
τ	temporal, instantaneous
$\overline{}$	averaging sign

Operators

$\nabla \cdot$	divergence
∇	gradient
∇_n	normal component of the gradient
∇_t	tangential component of the gradient
∇_l	surface gradient operator, $1/m$
∇^2	*Laplacian*
$\langle \ \rangle$	local volume average
$\langle \ \rangle^l$	local intrinsic volume average
$\langle \ \rangle^{le}$	local intrinsic surface average

Nomenclature required for coordinate transformations

(x, y, z) coordinates of a Cartesian, left oriented coordinate system (*Euclidean* space). Another notation which is simultaneously used is x_i $(i = 1, 2, 3)$: x_1, x_2, x_3

(ξ, η, ζ) coordinates of the curvilinear coordinate system called transformed coordinate system. Another notation which is simultaneously used is ξ^i $(i = 1, 2, 3)$: ξ^1, ξ^2, ξ^3

$\mathbf{V}_{cs}$ the velocity of the curvilinear coordinate system

$\sqrt{g}$ Jacobian determinant or *Jacobian* of the coordinate transformation
$x = f(\xi,\eta,\zeta)$, $y = g(\xi,\eta,\zeta)$, $z = h(\xi,\eta,\zeta)$

a_{ij} elements of the *Jacobian* determinant

a^{ij} elements of the determinant transferring the partial derivatives with respect to the transformed coordinates into partial derivatives with respect to the physical coordinates. The second superscript indicates the Cartesian components of the contravariant vectors

$(\mathbf{a}_1, \mathbf{a}_2, \mathbf{a}_3)$ covariant base vectors of the curvilinear coordinate system tangent vectors to the three curvilinear coordinate lines represented by (ξ,η,ζ)

$(\mathbf{a}^1, \mathbf{a}^2, \mathbf{a}^3)$ contravariant base vectors, normal to a coordinate surface on which the coordinates ξ, η and ζ are constant, respectively

g_{ij} covariant metric tensor (symmetric)

g^{ij} contravariant metric tensor (symmetric)

$(\mathbf{e}^1, \mathbf{e}^2, \mathbf{e}^3)$ unit vectors normal to a coordinate surface on which the coordinates ξ, η and ζ are constant, respectively

V^i $= \mathbf{a}^i \cdot \mathbf{V}$, contravariant components of the vector $\mathbf{V}$

V_i $= \mathbf{a}_i \cdot \mathbf{V}$, covariant components of the vector $\mathbf{V}$

$(\gamma_\xi, \gamma_\eta, \gamma_\zeta)$ permeabilities of coordinate surfaces on which the coordinates ξ, η and ζ are constant, respectively

Greek

A, α	Alpha	I, ι	Iota	Σ, σ	Sigma
B, β	Beta	K, κ	Kappa	T, τ	Tau
Γ, γ	Gamma	Λ, λ	Lambda	Φ, φ	Phi
Δ, δ	Delta	M, μ	Mu	X, χ	Chi
E, ε	Epsilon	N, ν	Nu	Υ, υ	Ypsilon
Z, ζ	Zeta	Ξ, ξ	Xi	Ψ, ψ	Psi
H, η	Eta	O, o	Omikron	Ω, ω	Omega
Θ, ϑ	Theta	Π, π	Pi		
		P, ρ	Rho		

Table of Contents

1 Mass conservation 1
   1.1 Introduction 1
   1.2 Basic definitions 2
   1.3 Non-structured and structured fields 10
   1.4 *Slattery* and *Whitaker's* local spatial averaging theorem 10
   1.5 General transport equation (*Leibnitz* rule) 13
   1.6 Local volume-averaged mass conservation equation 14
   1.7 Time average 18
   1.8 Local volume-averaged component conservation equations 20
   1.9 Local volume- and time-averaged conservation equations 22
   1.10 Conservation equations for the number density of particles 27
   1.11 Implication of the assumption of mono-dispersity in a cell 33
      1.11.1 Particle size spectrum and averaging 33
      1.11.2 Cutting of the lower part of the spectrum due to mass transfer 35
      1.11.3 The effect of the averaging on the effective velocity difference 37
   1.12 Stratified structure 38
   1.13 Final remarks and conclusions 39
   References 41

2 Momentums conservation 45
   2.1. Introduction 45
   2.2. Local volume-averaged momentum equations 46
      2.2.1 Single-phase momentum equations 46
      2.2.2 Interface force balance (momentum jump condition) 46
      2.2.3 Local volume averaging of the single-phase momentum equation 54
   2.3 Rearrangement of the surface integrals 56
   2.4 Local volume average and time average 61
   2.5 Viscous and *Reynolds* stresses 62
   2.6 Non-equal bulk and boundary layer pressures 65
      2.6.1 Continuous interface 65
      2.6.2 Dispersed interface 82
   2.7 Working form for dispersed and continuous phase 92
   2.8 General working form for dispersed and continuous phases 97
   2.9 Some practical simplifications 99
   2.10 Conclusion 103
   Appendix 2.1 104

Appendix 2.2 105
Appendix 2.3 106
References 109

3 Derivatives for the equations of state.. 115
3.1 Introduction 115
3.2 Multi-component mixtures of miscible and non-miscible components ... 117
    3.2.1 Computation of partial pressures for known mass concentrations, system pressure and temperature 119
    3.2.2 Partial derivatives of the equation of state $\rho = \rho\left(p,T,C_{2...i_{max}}\right)$ 126
    3.2.3 Partial derivatives in the equation of state $T = T\left(\varphi,p,C_{2...i_{max}}\right)$, where $\varphi = s,h,e$ 132
    3.2.4 Chemical potential 142
    3.2.5 Partial derivatives in the equation of state $\rho = \rho\left(p,\varphi,C_{2...i_{max}}\right)$, where $\varphi = s,h,e$ 152
3.3 Mixture of liquid and microscopic solid particles of different chemical substances 155
    3.3.1 Partial derivatives in the equation of state $\rho = \rho\left(p,T,C_{2...i_{max}}\right)$ 155
    3.3.2 Partial derivatives in the equation of state $T = T\left(p,\varphi,C_{2...i_{max}}\right)$ where $\varphi = h,e,s$ 156
3.4 Single-component equilibrium fluid 157
    3.4.1 Superheated vapor 158
    3.4.2 Reconstruction of equation of state by using a limited amount of data available 159
    3.4.3 Vapor-liquid mixture in thermodynamic equilibrium 167
    3.4.4 Liquid-solid mixture in thermodynamic equilibrium 168
    3.4.5 Solid phase 168
Appendix 3.1 Application of the theory to steam-air mixtures 169
Appendix 3.2 Useful references for computing properties of single constituents 170
References 173

4 On the variety of notations of the energy conservation for single-phase flow 177
4.1 Introduction 178
4.2 Mass and momentum conservation, energy conservation 178
4.3 Simple notation of the energy conservation equation 179
4.4 The entropy 180
4.5 Equation of state 181
4.6 Variety of notation of the energy conservation principle 182
    4.6.1 Temperature 182
    4.6.2 Specific enthalpy 182

4.7 Summary of different notations 183
4.8 The equivalence of the canonical forms 184
4.9 Equivalence of the analytical solutions 188
4.10 Equivalence of the numerical solutions? 188
    4.10.1 Explicit first order method of characteristics 188
    4.10.2 The perfect gas shock tube: benchmark for numerical methods 190
4.11 Interpenetrating fluids 199
4.12 Summary of different notations for interpenetrating fluids 205
Appendix 4.1 Analytical solution of the shock tube problem 207
Appendix 4.2 Achievable accuracy of the donor-cell method for
single-phase flows 211
References 214

5 First and second laws of the thermodynamics 217
5.1 Introduction 217
5.2 Instantaneous local volume average energy equations 220
5.3 *Dalton* and *Fick*'s laws, center of mass mixture velocity, caloric
mixture properties 228
5.4 Enthalpy equation 230
5.5 Internal energy equation 235
5.6 Entropy equation 235
5.7 Local volume- and time-averaged entropy equation 238
5.8 Local volume- and time-averaged internal energy equation 244
5.9 Local volume- and time-averaged specific enthalpy equation 246
5.10 Non-conservative and semi-conservative forms of the entropy
equation 248
5.11 Comments on the source terms in the mixture entropy equation 250
5.12 Viscous dissipation 256
5.13 Temperature equation 260
5.14 Second law of the thermodynamics 264
5.15 Mixture volume conservation equation 265
5.16 Linearized form of the source term for the temperature equation 271
5.17 Interface conditions 279
5.18 Lumped parameter volumes 281
5.19 Final remarks 282
References 282

6 Some simple applications of the mass and energy conservation 285
6.1 Infinite heat exchange without interfacial mass transfer 285
6.2 Discharge of gas from a volume 287
6.3 Injection of inert gas in a closed volume initially filled with inert gas 290
6.4 Heat input in a gas in a closed volume 291
6.5 Steam injection in a steam-air mixture 292
6.6 Chemical reaction in a gas mixture in a closed volume 295

6.7 Hydrogen combustion in an inert atmosphere 297
  6.7.1 Simple introduction to combustion kinetics 297
  6.7.2 Ignition temperature and ignition concentration limits 299
  6.7.3 Detonability concentration limits 301
  6.7.4 The heat release due to combustion 301
  6.7.5 Equilibrium dissociation 303
  6.7.6 Source terms of the energy conservation of the gas phase 308
  6.7.7 Temperature and pressure changes in a closed control volume; adiabatic temperature of the burned gases 310
References 315

7 Exergy of multi-phase multi-component systems 317
7.1 Introduction 317
7.2 The pseudo-exergy equation for single-fluid systems 317
7.3 The fundamental exergy equation 319
  7.3.1 The exergy definition in accordance with *Reynolds* and *Perkins* 319
  7.3.2 The exergy definition in accordance with *Gouy* (l'énergie utilisable, 1889) 320
  7.3.3 The exergy definition appropriate for estimation of the volume change work 322
  7.3.4 The exergy definition appropriate for estimation of the technical work 323
7.4 Some interesting consequences of the fundamental exergy equation 323
7.5 Judging the efficiency of a heat pump as an example of application of the exergy 325
7.6 Three-fluid multi-component systems 327
7.7 Practical relevance 330
References 331

8 One-dimensional three-fluid flows 333
8.1 Summary of the local volume- and time-averaged conservation equations 333
8.2 Treatment of the field pressure gradient forces 337
  8.2.1 Dispersed flows 337
  8.2.2 Stratified flow 337
8.3 Pipe deformation due to temporal pressure change in the flow 338
8.4 Some simple cases 339
8.5 Slip model – transient flow 347
8.6 Slip model – steady state. Critical mass flow rate 352
8.7 Forces acting on the pipes due to the flow – theoretical basics 359
8.8 Relief valves 367
  8.8.1 Introduction 367
  8.8.2 Valve characteristics, model formulation 367
  8.8.3 Analytical solution 372
  8.8.4 Fitting the piecewise solution on two known position – time points 374
  8.8.5 Fitting the piecewise solution on known velocity and position for a given time 376

8.8.6 Idealized valve characteristics 377
8.8.7 Recommendations for the application of the model in system
computer codes 380
8.8.8 Some illustrations of the valve performance model 383
8.8.9 Nomenclature for Section 8.8 389
8.9 Pump model 391
8.9.1 Variables defining the pump behavior 391
8.9.2 Theoretical basics 394
8.9.3 *Suter* diagram 402
8.9.4 Computational procedure 410
8.9.5 Centrifugal pump drive model 411
8.9.6 Extension of the theory to multi-phase flow 411
References 416

9 Detonation waves caused by chemical reactions or by melt-coolant interactions 419
9.1 Introduction 419
9.2 Single-phase theory 421
9.2.1 Continuum sound waves (*Laplace*) 421
9.2.2 Discontinuum shock waves (*Rankine-Hugoniot*) 422
9.2.3 The *Landau* and *Liftshitz* analytical solution for detonation in perfect gases 427
9.2.4 Numerical solution for detonation in closed pipes 431
9.3 Multi-phase flow 434
9.3.1 Continuum sound waves 434
9.3.2 Discontinuum shock waves 436
9.3.3 Melt-coolant interaction detonations 438
9.3.4 Similarity to and differences from the *Yuen* and *Theofanous* formalism 444
9.3.5 Numerical solution method 444
9.4 Detonation waves in water mixed with different molten materials 445
9.4.1 UO_2 water system 446
9.4.2 Efficiencies 451
9.4.3 The maximum coolant entrainment ratio 454
9.5 Conclusions 455
9.6 Practical significance 458
Appendix 9.1 Specific heat capacity at constant pressure for urania and alumina 458
References 459

10 Conservation equations in general curvilinear coordinate systems 463
10.1 Introduction 463
10.2 Field mass conservation equations 464

10.3 Mass conservation equations for components inside the field – conservative form 467
10.4 Field mass conservation equations for components inside the field – non-conservative form 470
10.5. Particles number conservation equations for each velocity field 470
10.6 Field entropy conservation equations – conservative form 471
10.7 Field entropy conservation equations – non-conservative form 472
10.8 Irreversible power dissipation caused by the viscous forces 473
10.9 The non-conservative entropy equation in terms of temperature and pressure 475
10.10 The volume conservation equation 477
10.11 The momentum equations 479
10.12 The flux concept, conservative and semi-conservative forms 487
    10.12.1 Mass conservation equation 488
    10.12.2 Entropy equation 489
    10.12.3 Temperature equation 490
    10.12.4 Momentum conservation in the x-direction 491
    10.12.5 Momentum conservation in the y-direction 492
    10.12.6 Momentum conservation in the z-direction 493
10.13 Concluding remarks 495
References 495

11 Type of the system of PDEs 497
11.1 Eigenvalues, eigenvectors, canonical form 497
11.2 Physical interpretation 500
    11.2.1 Eigenvalues and propagation velocity of perturbations 500
    11.2.2 Eigenvalues and propagation velocity of harmonic oscillations 501
    11.2.3 Eigenvalues and critical flow 502
References 503

12 Numerical solution methods for multi-phase flow problems 505
12.1 Introduction 505
12.2 Formulation of the mathematical problem 505
12.3 Space discretization and location of the discrete variables 508
12.4 Discretization of the mass conservation equations 512
12.5 First order donor-cell finite difference approximations 514
12.6 Discretization of the concentration equations 516
12.7 Discretization of the entropy equation 518
12.8 Discretization of the temperature equation 519
12.9. Physical significance of the necessary convergence condition 522
12.10. Implicit discretization of momentum equations 524
12.11 Pressure equations for IVA2 and IVA3 computer codes 531
12.12 A *Newton*-type iteration method for multi-phase flows 535
12.13 Integration procedure: implicit method 545
12.14 Time step and accuracy control 547
12.15 High order discretization schemes for convection-diffusion terms 548

12.15.1 Space exponential scheme 548
12.15.2 High order upwinding 551
12.15.3 Constrained interpolation profile (CIP) method 554
12.16 Pipe networks: some basic definitions 561
  12.16.1 Pipes 561
  12.16.2 Axis in the space 563
  12.16.3 Diameters of pipe sections 564
  12.16.4 Reductions 565
  12.16.5 Elbows 565
  12.16.6 Creating a library of pipes 566
  12.16.7 Sub system network 567
  12.16.8 Discretization of pipes 568
  12.16.9 Knots 568
  Appendix 12.1 Definitions applicable to discretization of the mass
  conservation equations 570
  Appendix 12.2 Discretization of the concentration equations 573
  Appendix 12.3 Harmonic averaged diffusion coefficients 576
  Appendix 12.4. Discretized radial momentum equation 578
  Appendix 12.5 The $\bar{a}$ coefficients for Eq. (12.46) 584
  Appendix 12.6 Discretization of the angular momentum equation 584
  Appendix 12.7 Discretization of the axial momentum equation 586
  Appendix 12.8 Analytical derivatives for the residual error of each
  equation with respect to the dependent variables 588
  Appendix 12.9 Simple introduction to iterative methods for solution of
  algebraic systems 592
References 593

13 Numerical methods for multi-phase flow in curvilinear coordinate systems 599
13.1 Introduction 599
13.2 Nodes, grids, meshes, topology - some basic definitions 601
13.3 Formulation of the mathematical problem 602
13.4 Discretization of the mass conservation equations 604
  13.4.1 Integration over a finite time step and finite control volume 604
  13.4.2 The donor-cell concept 607
  13.4.3 Two methods for computing the finite difference
  approximations of the contravariant vectors at the cell center 610
  13.4.4 Discretization of the diffusion terms 612
13.5 Discretization of the entropy equation 618
13.6 Discretization of the temperature equation 619
13.7 Discretization of the particle number density equation 619
13.8 Discretization of the x momentum equation 620
13.9 Discretization of the y momentum equation 622
13.10. Discretization of the z momentum equation 623
13.11 Pressure-velocity coupling 624

13.12 Staggered x momentum equation.. 629
Appendix 13.1 Harmonic averaged diffusion coefficients 641
Appendix 13.2 Off-diagonal viscous diffusion terms of the x momentum
equation 643
Appendix 13.3 Off-diagonal viscous diffusion terms of the y momentum
equation 647
Appendix 13.4 Off-diagonal viscous diffusion terms of the z momentum
equation 650
References 653

Appendix 1 Brief introduction to vector analysis 657

Appendix 2 Basics of the coordinate transformation theory 687

Index 747

Chapter 14 of Volume 1 and Chapter 26 of volume 2 are available in pdf format on the CD-ROM attached to Volume 1. The system requirements are Windows 98 and higher. Both pdf files contain links to computer animations. To see the animations, one double clicks on the active links contained inside the pdf documents. The animations are then displayed in an internet browser, such Microsoft Internet Explorer or Netscape. Alternatively, gif-file animations are also provided.

14 Visual demonstration of the method.. 1
    14.1 Melt-water interactions... 2
        14.1.1 Cases 1 to 4 2
        14.1.2 Cases 5, 6 and 7.. 8
        14.1.3 Cases 8 to 10 13
        14.1.4 Cases 11 and 12.. 24
        14.1.5 Case 13 27
        14.1.6 Case 14 28
    14.2 Pipe networks 31
        14.2.1 Case 15 31
    14.3 3D steam-water interaction.. 33
        14.3.1 Case 16 33
    14.4 Three dimensional steam-water interaction in presence of non-
    condensable gases.. 34
        14.4.1 Case 17 34
    14.5 Three dimensional steam production in boiling water reactor............... 36
        14.5.1 Case 18 36
    References 38

26 Validation of multi-phase flow models 1
    26.1 Introduction 3
    26.2 Material relocation – gravitational waves (2D) 10
        26.2.1 U-tube benchmarks 10

26.2.2 Gravitational 2D waves 15
26.3 Steady state single-phase nozzle flow... 16
26.4 Pressure waves – single phase 17
   26.4.1 Gas in a shock tube.. 17
   26.4.2 Water in a shock tube 21
   26.4.3 Pressure wave propagation in a cylinder vessel with free surface (2D) 22
26.5 2D: N_2 explosion in space filled previously with air 26
26.6 2D: N_2 explosion in space with internals filled previously with water 28
26.7 Film entrainment in pipe flow.. 32
26.8 Water flashing in nozzle flow 34
26.9 Pipe blow-down with flashing 38
   26.9.1 Single pipe... 38
   26.9.2 Complex pipe network 42
26.10 Boiling, critical heat flux, post-critical heat flux heat transfer............... 43
26.11 Film boiling 50
26.12 Behavior of clouds of cold and very hot spheres in water 52
26.13 Experiments with dynamic fragmentation and coalescence................. 57
   26.13.1 L14 experiment... 57
   26.13.2 L20 and L24 experiments.. 61
   26.13.3 Uncertainty in the prediction of non-explosive melt-water interactions 62
   26.13.4 Conclusions 63
26.14 L28, L31 experiment.. 64
26.15 PREMIX-13 experiment 69
26.16 PREMIX 17 and 18 experiments 75
26.17 RIT and IKE experiments 88
26.18 Assessment for detonation analysis 89
26.19 Examples of 3D capabilities 90
   26.19.1 Case 1. Rigid body steady rotation problem.............................. 90
   26.19.2 Case 2. Pure radial symmetric flow.. 92
   26.19.3 Case 3. Radial-azimuthal symmetric flow.................................. 94
   26.19.4 Case 4. Small-break loss of coolant.. 96
   26.19.5 Case 5. Asymmetric steam-water interaction in a vessel [65]........ 98
   26.19.6 Case 6. Melt relocation in a pressure vessel 102
26.20 General conclusions 104
References.. 104

1 Mass conservation

*"...all changes in the nature happen so that
the mass lost by one body is added to other...".*

1748, *M. Lomonosov's* letter to *L. Euler*

The heterogeneous porous-media formulation proposed first by Sha, Chao and Soo is applied with some modifications to derive mass conservation equations for multi-phase flows conditionally divided into 3 velocity fields. In addition, the method is extended by incorporating the concept of dynamic particle fragmentation and coalescence. The link between kinetic generation of particles and the mass source terms is specified. The link between mass source terms and the change in bubble/droplet size due to evaporation or condensation is presented for local volume- and time- averaged source terms. The concept of mono-dispersity is discussed and a method is proposed for computation of the disappearance of particles due to evaporation and condensation.

1.1 Introduction

The creation of computer codes for modeling multi-phase flows in industrial facilities is very complicated, time-consuming and expensive. That is why the fundamentals on which such codes are based are subject to continuous review in order to incorporate the state of the art of knowledge into the current version of the code in question. One of the important elements of the codes is the system of partial differential equations governing the flow. The understanding of each particular term in these equations is very important for the application. From the large number of formulations of the conservation equations for multi-phase flows, see [47, 12, 4, 15] and the references given there, the heterogeneous porous-media formulation presented by *Sha, Chao* and *Soo* [45] is to date the most effective concept available for deriving multi-phase equations applicable to flows in complicated technical facilities. For technical structures, the introduction of the local volume porosity and directional permeability is a convenient formalism to describe the real distances between the flow volumes. Geometry simplifications that lead to distortion of the modeled acoustic process characteristics of the systems are not necessary. This is the reason behind the use of the heterogeneous porous-media formulation in the IVA computer codes development in the author's work in the last 15 years.

This work follows, with some modification, the method proposed by *Sha, Chao* and *Soo* to derive mass conservation equations for each velocity field. In addition, this method is extended in two ways, viz. by also including inert components in each velocity field additionally, and by including the concept of dynamic particle fragmentation and coalescence. This Chapter presents a improved version of the work published in [52].

1.2 Basic definitions

Consider the idealized flow patterns of the three-phase flows shown in Fig. 1.1. What do all these flow patterns have in common? How can they be described by a single mathematical formalism that allows transition from any flow pattern to any other flow pattern? How can the complicated interactions between the participating components be described as simply as possible? In the discussion that follows a number of basic definitions are introduced or reiterated so as to permit successful treatment of such flows using a unified algorithm.

Zemansky [54] p. 566 defines "a phase as a system or a portion of the system composed of any number of chemical constituents satisfying the requirements (a) that it is homogeneous and (b) that it has a definite boundary". In this sense, consideration of a multi-phase flow system assumes the consistency described below:

1) A homogeneous gas phase having a definite boundary. The gas phase can be continuous (non-structured) or dispersed (structured). The gas phase in the gas/droplet flow is an example of a non-structured gas phase. The bubbles in the bubble/liquid flow are an example of a structured phase. Besides the steam, the gas phase contains $i_{1\max} - 1$ groups of species of inert components. Inert means that these components do not experiences changes in the state of aggregate during the process of consideration. The gas components occupy the entire gas volume in accordance with the *Dalton*'s law, and therefore possess the definite boundary of the gas phase it selves. The gas phase velocity field will be denoted as no.1 and $l = 1$ assigned to this. Velocity field no.1 has the temperature T_1.

2) A mixture of liquid water and a number of species $i_{2\max} - 1$ of microscopic solid particles. The particles are homogeneous and have a definite boundary surrounded by liquid only. The water has its own definite outer boundary. An example of such a mixture is a liquid film carrying $i_{2\max} - 1$ groups of radioactive solid species. This means that the mixture of water with $i_{2\max} - 1$ groups of species is a

mixture of $i_{2\max}$ phases. The mixture velocity field defined here will be denoted as no. 2. Velocity field no. 2 can likewise be non-structured or structured. Velocity field no. 2 has the temperature T_2.

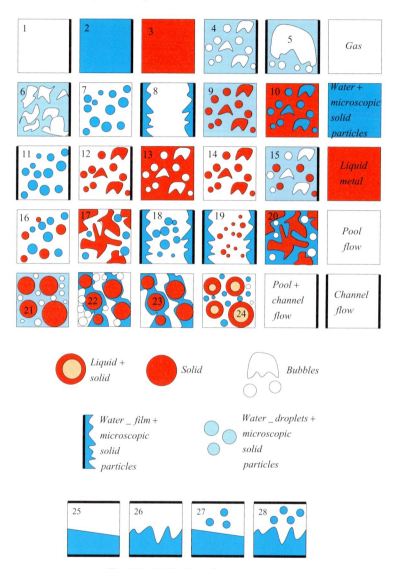

Fig. 1.1. Multi-phase flow patterns

3) A mixture of the type of velocity field no. 2 of liquid with $i_{3\max} - 1$ groups of species of microscopic solid particles. The mixture can be non-structured or structured; this is denoted as velocity field no. 3. Velocity field no. 3 has the temperature T_3. In the event that only one inert component occupies velocity field no. 3, this will be allowed to be macroscopic, either a liquid, a liquid-solid mixture in homogenous equilibrium or solid particles only. This allows versatile application of the concept.

The entire flow under consideration consists of $i_{2\max} + i_{3\max} + 1$ phases and is conditionally divided into three velocity fields. In order to avoid the need to write indices to the indices, *il* will be written in place of i_l.

An example demonstrating the necessity for the use of three velocity fields is the modeling of a mixture consisting of gas, water, and a liquid metal, whose densities have the approximate ratios $1 : 10^3 : 10^4$. In a transient, these three fields will have considerably differing velocities and temperatures.

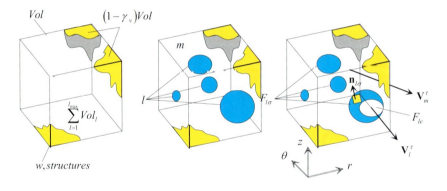

Fig. 1.2. Control volume for definition of the mass conservation equation, partially occupied by structure and two different velocity fields

Turning our attention again to the flow patterns depicted in Fig. 1.1, and keeping in mind that in reality these change their characteristic sizes chaotically, it is obvious that it is not possible to determine the details of the thermodynamic and flow parameters for each component of each velocity field. Consequently, some type of averaging must be implemented. Following *Anderson* and *Jackson, Slattery*, and *Whitaker* [1, 46, 50], a control volume is allocated to every point in space, the thermodynamic and flow properties averaged for each velocity field are assigned to the center of the control volume. Since there is a control volume associated with every point in space, a field of average values can be generated for all thermodynamic and flow properties. The field of the average properties is therefore smooth, and space derivatives of these averaged properties exist. Consider the control volume *Vol* occupied by a non-movable structure in addition to the three velocity fields, see Figs. 1.2 and 1.3. While the individual *l*-field volumes Vol_l may be functions of time and space, the control volume *Vol* is not.

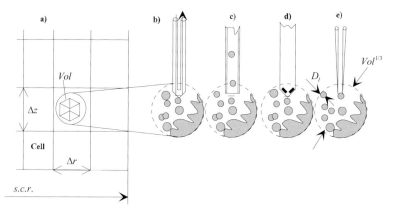

Fig. 1.3. Comparison of possible scales of local volume averaging, scale of the measuring devices, scale of the computational region (s.c.r.) and the global flow dimensions

Velocity field *l* is taken to have the characteristic length D_l (e.g. bubble size, droplet size), which is much larger than the molecular free path, Fig. 1.3. The size of the computational region of interest here is much larger than the size of the local structure D_l and larger than the spatial changes in the flow parameters of interest. The choice of the size of the control volume, which is of order of $Vol^{1/3}$ has a major impact on the meaning of the averaged values. From the various approaches possible, there are two which are meaningful.

i) The size of the control volume is larger than the characteristic configuration length D_l of the field *l*, i.e.,

molecular free path $<< D_l << Vol^{1/3} <<$ *size of the computational region* (1.1)

This approach is useful for fine dispersed flows.

ii) The size of the control volume is comparable to the characteristic configuration length D_l of the field *l* , i.e.,

molecular free path $<< D_l \approx Vol^{1/3} <<$ *size of the computational region* (1.2)

This approach is meaningful for direct simulation or for simulation of flow pattern with stratification, for example.

The geometry of the non-movable structures inside *Vol* is characterized by the volumetric porosity γ_v, which is defined as the ratio of the volume occupied by the flow mixture inside the control volume, $\sum_{l=1}^{3} Vol_l$, and the control volume *Vol*:

$$\gamma_v = \frac{\sum_{l=1}^{3} Vol_l}{Vol}. \tag{1.3}$$

Consequently the part of *Vol* occupied by structures is $1-\gamma_v$. Inside the volume available for the flow, $\sum_{l=1}^{3} Vol_l$, it is assumed that three velocity fields coexist. The instantaneous geometry of the velocity field is defined in a similar way as the non-movable structure. The local instantaneous volume fraction of the velocity field *l*, α_l, is defined as the part of $\gamma_v Vol$ occupied by the velocity field *l*:

$$\alpha_l = \frac{Vol_l}{\sum_{l=1}^{3} Vol_l} = \frac{Vol_l}{\gamma_v Vol}. \tag{1.4}$$

In general, this varies with time and location. By definition, we have

$$\sum_{l=1}^{3} \alpha_l = 1. \tag{1.4a}$$

On the basis of the definitions introduced above, the part of the control volume *Vol* occupied by the velocity field *l* is

$$Vol_l = \alpha_l \gamma_v Vol. \tag{1.5}$$

The discussion here is restricted to the right-handed Cartesian and cylindrical coordinates designated with r, θ, and z. The instantaneous field velocity $\mathbf{V}_l^\tau (r, \theta, z, \tau)$ is defined through the entire region occupied by the velocity field *l*. The components of the velocity vectors along the axes are

$$\mathbf{V}_l^\tau = \left(u_l^\tau, v_l^\tau, w_l^\tau \right). \tag{1.6}$$

The control volume *Vol* is bounded by a surface *F*. The velocity field within the volume Vol_l is bounded by a surface F_l. Smooth closed surfaces in space are orientable. The orientation of the surface with respect to the field *l* is given with the unit vector $\mathbf{n}_l$ attached to the *l*-surface and pointing outwards of the field *l*. The interface belongs to both neighboring phases. The surfaces F, F_l are defined as a scalars. The surface F_l has the following three constituents:

(a) F_{le}, the control volume entrances and exits crossing the field l, being also part of the control volume surface F.

(b) $F_{l\sigma}$, the interface between the field l and the surrounding field or fields m.

(c) F_{lw}, the interface between the solid structure, wall, and the field.

Consider the small part of the surface F, ΔF. The part of this surface occupied by the flow is $\sum_{l=1}^{3} \Delta F_l$. The ratio

$$\gamma = \frac{1}{\Delta F} \sum_{l=1}^{3} \Delta F_{le} \qquad (1.7)$$

is also known as *surface permeability*. We consider the surface permeability as ratio of scalars. Here

$$dF_{le} < \Delta F_{le} < F_{le}, \qquad (1.8)$$

and

$$dF < \Delta F < F. \qquad (1.9)$$

Note the difference between surface permeability and the "permeability coefficients" in standard use for description of the pressure drop in porous media. Also note that the values of γ_v at the center and at the surface of the control volume may be different, because they belong to different averaging volumes. This definition of γ_v is sufficient to describe the isotropic porosity that is found in homogeneous porous bodies. For most of the technical structures forming channels for multi-phase flow it is extremely convenient to introduce the surface permeability γ in addition to the volumetric porosity γ_v in order to describe obstacles to the flow inside the space of interest, i.e., non-isotropic porosity (heterogeneous porous body). As far I know, *Gentry* et al. [8] introduced for first time this concept in 1966 using the term volume and surface flow fractions for volumetric porosity and surface permeabilities. Valuable comments for the practical use of the concept are given by *Hirt* in [14]. Note that inclined surfaces in orthogonal structured grids can also be modeled by this method by specifying correctly the part of the boundary cells that is occupied by flow and the part of the surfaces that are open for the flow. γ-values are likewise permitted to be prescribed functions of time in addition to functions of space. Additionally, the vector

$$\vec{\gamma} = \mathbf{n}_{l,x}\,\gamma + \mathbf{n}_{l,y}\,\gamma + \mathbf{n}_{l,z}\,\gamma = \vec{\gamma}_x + \vec{\gamma}_y + \vec{\gamma}_z \qquad (1.10)$$

is referred to as the directional surface permeability.

The counterpart of γ for each velocity field is then introduced, field surface fraction α_l^e (called some times heterogeneous volume fraction), which is the part of the surface $\gamma\Delta F$ crossing the field l:

$$\alpha_l^e = \Delta F_{le} / (\gamma\Delta F). \qquad (1.11)$$

This means that

$$\sum_{l=1}^{3}\alpha_l^e = 1. \qquad (1.11a)$$

The impact of the choice of the size for the control volume can now be clearly understood by examining the relationship between the heterogeneous volume fraction and the local volume fraction. A large size for the control volume compared to the characteristic configuration length, assumption (i), leads to

$$\alpha_l^e = \alpha_l. \qquad (1.12)$$

This assumption is reasonable for dispersed flows. If the size of the control volume is comparable with the characteristic configuration length of the field l, assumption (ii), the following results

$$\alpha_l^e \neq \alpha_l. \qquad (1.13)$$

In both cases the local volume averaging is mathematically permissible.

Consider any scalar property of field l designated with φ_l. Inside the space occupied by the field l this property is considered to be smooth. In general the averages of the property φ_l given below are required for the derivation of local volume average equations.

1) The *local volume average* is defined by

1.2 Basic definitions

$$\langle \varphi_l \rangle := \frac{1}{Vol} \int_{Vol_l} \varphi_l dVol .$$ (1.14)

Note that, when the property φ_l is constant, the local volume average given by Eq. (1.14) does not equal this constant.

2) The *intrinsic field average* is defined by

$$\langle \varphi_l \rangle^l := \frac{1}{Vol_l} \int_{Vol_l} \varphi_l dVol .$$ (1.15)

These two average properties are obviously related by

$$\langle \varphi_l \rangle = \alpha_l \gamma_v \langle \varphi_l \rangle^l .$$ (1.16)

The averages defined by Eqs. (1.14) and (1.15) are applicable not only for scalars but also for vectors and tensors.

The macroscopic density $\langle \rho_l \rangle^l$ is the intrinsic field average over the space occupied by velocity field l. Outside the velocity field l the density $\langle \rho_l \rangle^l$ is not defined. The macroscopic density, referred to simply as density in the following, obeys the law expressed by the macroscopic equation of state. The equation of state describes the interdependence between density, the intrinsic field-averaged pressure, and the intrinsic field-averaged temperature, $\langle \rho_l \rangle^l = f\left(\langle p_l \rangle^l, \langle T_l \rangle^l\right)$ frequently denoted in short as $\rho_l = \rho_l(p_l, T_l)$ in the sense of intrinsic volume average.

3) The *intrinsic surface average* is defined by

$$\langle \varphi_l \rangle^{le} := \frac{1}{\Delta F_{le}} \int_{\Delta F_{le}} \varphi_l \, \mathbf{n}_l dF .$$ (1.17)

The result of the surface averaging of a scalar is a vector. Therefore $\langle \varphi_l \rangle^{le}$ is a vector. The product of the scalar property φ_l and $\mathbf{V}_l^{\tau}$, $\varphi_l \mathbf{V}_l^{\tau}$, is referred to as the flux of φ_l along the flow direction. This is a vector. The scalar product of the three-dimensional vector field $\varphi_l \mathbf{V}_l^{\tau}$ across the oriented surface dF in the direction $\mathbf{n}_l$, $\varphi_l \mathbf{V}_l^{\tau} \cdot \mathbf{n}_l dF$, is the flow of property φ_l normal through the surface dF. The vector form of Eq. (1.17) is

$$\left\langle \varphi_l \mathbf{V}_l^\tau \right\rangle^{le} := \frac{1}{\Delta F_{le}} \int_{\Delta F_{le}} \varphi_l \mathbf{V}_l^\tau \cdot \mathbf{n}_l dF \qquad (1.18)$$

Note that the intrinsic surface average of a scalar is a vector and the intrinsic surface average of a vector is a scalar. The notation $\langle \ \rangle$, $\langle \ \rangle^l$, and $\langle \ \rangle^{le}$ should be considered as operators defined by Eqs. (1.14, 1.15, 1.17). Here I adapted the notation used by *Whitaker* [53].

For practical applications the following definition of the weighted average velocity is used

$$\left\langle \mathbf{V}_l^\tau \right\rangle^{le,\rho} = \left\langle \rho_l \mathbf{V}_l^\tau \right\rangle^{le} / \left\langle \rho_l \right\rangle^l. \qquad (1.19)$$

1.3 Non-structured and structured fields

As already mentioned, within the averaging volume the fields are essentially allowed to be (a) non-structured or (b) structured. For the non-structured fields the spatial variation of the properties within the continuous volume is smooth, and therefore space differentiating of the properties is allowed inside the space occupied by the field. Consequently the *Gauss-Ostrogradskii* theorem for transformation of surface integrals into volume integrals is applicable. This is not the case for a structured fluid, where the properties within the continuous fragment are smooth, but the field volume inside the control volume consists of several fragments. In this case simple extension of the *Gauss-Ostrogradskii* theorem for transfer of surface integral into volume integral over the fragmented volume occupied by the field is not possible unless one extends the definition of the properties in the sense of distributions, see the discussion by *Gray* and *Lee* in [9], or uses some type of special treatment as described in the next section.

1.4 *Slattery* and *Whitaker's* local spatial averaging theorem

In the following text the operator $\nabla \langle \ \rangle$ is used for the *gradient* of a scalar which is a vector, and $\nabla \cdot \langle \ \rangle$ for the *divergence* of a vector which is a scalar. The following mathematical tools are used to derive local volume-averaged field conservation equations for the property φ_l.

a) The *spatial averaging theorem*, Eq. (22) in [53],

$$\nabla \langle \varphi_l \rangle = \nabla \cdot \left(\alpha_l \gamma_v \langle \varphi_l \rangle^l \right) = \frac{1}{Vol} \int_{F_{le}} \varphi_l \, \mathbf{n}_l dF . \qquad (1.20)$$

Here $\nabla \langle \varphi_l \rangle$ is a vector. The vector form of the spatial averaging theorem is given by Eq. (23) in [53].

$$\nabla \cdot \langle \varphi_l \mathbf{V}_l^\tau \rangle = \nabla \cdot \left(\alpha_l \gamma_v \langle \varphi_l \mathbf{V}_l^\tau \rangle^l \right) = \frac{1}{Vol} \int_{F_{le}} \varphi_l \mathbf{V}_l^\tau \cdot \mathbf{n}_l dF , \qquad (1.21)$$

Here $\nabla \cdot \langle \varphi_l \mathbf{V}_l^\tau \rangle$ is a scalar. For the validation of Eqs. (1.20) and (1.21) see Refs. [1, 46, 50, 53].

Useful consequences of Eq. (1.20) are obtained for $\varphi_l = 1$:

$$\nabla \langle \alpha_l \gamma_v \rangle = \frac{1}{Vol} \int_{F_{le}} \mathbf{n}_l dF = \frac{1}{Vol} \int_F \alpha_l^e \gamma \, \mathbf{n}_l dF' = \nabla \left(\alpha_l^e \gamma \right) . \qquad (1.22)$$

Taking into account Eqs. (1.4a) and (1.11a) the sum of the above equation for all three fields yields

$$\nabla \langle \gamma_v \rangle = \frac{1}{Vol} \int_F \gamma \, \mathbf{n}_l dF' = \nabla \gamma . \qquad (1.23)$$

It follows from Eq. (1.21) that

$$\nabla \cdot \langle \varphi_l \mathbf{V}_l^\tau \rangle = \frac{1}{Vol} \int_{F_{le}} \varphi_l \mathbf{V}_l^\tau \cdot \mathbf{n}_l dF = \frac{1}{Vol} \int_F \alpha_l^e \gamma \langle \varphi_l \mathbf{V}_l^\tau \rangle^{el} \cdot \mathbf{n}_l dF' . \qquad (1.24)$$

By introducing local surface averaging, Eq. (1.18), the last surface integral can be transformed into the divergence for a smooth vector field

$$\nabla \cdot \langle \varphi_l \mathbf{V}_l^\tau \rangle = \nabla \cdot \left(\alpha_l^e \gamma \langle \varphi_l \mathbf{V}_l^\tau \rangle^{el} \right) . \qquad (1.25)$$

This equation is similar to Eq. (2.26) obtained in Ref. [45], where, however, the surface averaging was performed over the entire surface F of the control volume rather than over local ΔF which is a part of F.

Note: Comparing Eq. (1.25) with Eq. (1.21) we realize something very interesting

$$\nabla \cdot \left\langle \varphi_l \mathbf{V}_l^{\tau} \right\rangle = \nabla \cdot \left(\alpha_l \gamma_v \left\langle \varphi_l \mathbf{V}_l^{\tau} \right\rangle^l \right) = \nabla \cdot \left(\alpha_l^e \gamma \left\langle \varphi_l \mathbf{V}_l^{\tau} \right\rangle^{el} \right). \qquad (1.25a)$$

> This means, that one can work either with local volume averaged fluxes $\left\langle \varphi_l \mathbf{V}_l^{\tau} \right\rangle^l$ or with local surface averaged fluxes $\left\langle \varphi_l \mathbf{V}_l^{\tau} \right\rangle^{el}$. In the first case, one has to use in the divergence expression the local volume fractions α_l and γ_v, and in the second case the local surface fractions α_l^e and γ.

There are many literature sources where this difference is not clearly made.

b) The *Gauss-Ostrogradskii* theorem is applied to the volume Vol_l. The resulting expression is divided by Vol. The result is

$$\frac{1}{Vol} \int_{Vol_l} \nabla \varphi_l \, dVol = \frac{1}{Vol} \int_{F_{le}} \varphi_l \cdot \mathbf{n}_l dF + \frac{1}{Vol} \int_{F_{l\sigma}+F_{lw}} \varphi_l \cdot \mathbf{n}_l dF. \qquad (1.26)$$

Replacing the first integral of the RHS of Eq. (1.26) with Eq. (1.20) one obtains

$$\left\langle \nabla \varphi_l \right\rangle = \nabla \left\langle \varphi_l \right\rangle + \frac{1}{Vol} \int_{F_{l\sigma}+F_{lw}} \varphi_l \cdot \mathbf{n}_l dF. \qquad (1.27)$$

This equation expresses the volume average of derivatives in the form of derivatives of volume average and a surface integral. While the derivatives of the non-average property φ_l may be a non-smooth function of location, the local volume-averaged property $\left\langle \varphi_l \right\rangle$ is a smooth function of location and can be differentiated. This is an extremely interesting consequence of the *Slattery-Whitaker's spatial averaging theorem*, which allows one to write local volume average differential conservation equations for non-structured as well for structured fluids. For one-dimensional flow Eq. (1.27) reduces to Eq. (3) by *Delhaye* in [4], p. 160 published in 1981.

The analogous form for a vector is

$$\left\langle \nabla \cdot \left(\varphi_l \mathbf{V}_l^{\tau} \right) \right\rangle = \nabla \cdot \left\langle \varphi_l \mathbf{V}_l^{\tau} \right\rangle + \frac{1}{Vol} \int_{F_{l\sigma}+F_{lw}} \varphi_l \mathbf{V}_l^{\tau} \cdot \mathbf{n}_l dF. \qquad (1.28)$$

A different route to this result was given in Ref. [9] by extending the definition of the field properties in the sense of distributions.

Useful consequences of Eq. (1.27) are obtained by setting $\varphi_l = 1$

$$\nabla \left(\alpha_l \gamma_v \right) = -\frac{1}{Vol} \int_{F_{l\sigma}+F_{lw}} \mathbf{n}_l dF = \nabla \left(\alpha_l^e \gamma \right) \qquad (1.29)$$

- compare with Eq. (1.22). Summing over all the fields, and keeping in mind Eq. (1.4a) that $\sum_{l=1}^{3} \alpha_l = 1$ and $\mathbf{n}_w = -\mathbf{n}_l$, the following is obtained

$$\nabla \gamma_v = -\frac{1}{Vol} \sum_{l=1}^{3} \int_{F_{l\sigma}} \mathbf{n}_l dF - \frac{1}{Vol} \sum_{l=1}^{3} \int_{F_{lw}} \mathbf{n}_l dF = \frac{1}{Vol} \int_{F_w} \mathbf{n}_w dF . \qquad (1.30)$$

While the first integral is a sum of repeating terms with alternating signs and therefore is equal to zero, the second integral will be zero only in the case of immersed structure inclusions. Otherwise the second integral is not equal to zero.

1.5 General transport equation (*Leibnitz* rule)

The general transport equation applied to the field volume Vol_l is used in the form

$$\frac{d}{d\tau} \int_{Vol_l} \varphi_l \, dVol = \int_{Vol_l} \frac{\partial \varphi_l}{\partial \tau} dVol + \int_{F_{l\sigma}+F_{lw}+F_{le}} \varphi_l \mathbf{V}_{l\sigma}^\tau \cdot \mathbf{n}_l \, dF \qquad (1.31)$$

using Eq. (1.14), divided by the constant Vol and rearranged,

$$\left\langle \frac{\partial \varphi_l}{\partial \tau} \right\rangle = \frac{d}{d\tau} \langle \varphi_l \rangle - \frac{1}{Vol} \int_{F_{l\sigma}+F_{lw}+F_{le}} \varphi_l \mathbf{V}_{l\sigma}^\tau \cdot \mathbf{n}_l dF . \qquad (1.32)$$

$\mathbf{V}_{l\sigma}^\tau$ is the instantaneous interface velocity. Replacing the total differential

$$\frac{d}{d\tau} \langle \varphi_l \rangle = \frac{\partial}{\partial \tau} \langle \varphi_l \rangle + \nabla \cdot \langle \varphi_l \mathbf{V}_l^\tau \rangle ,$$

and using Eq. (1.21) we see that the components at the control volume surfaces cancels and therefore

$$\boxed{\left\langle \frac{\partial \varphi_l}{\partial \tau} \right\rangle = \frac{\partial}{\partial \tau} \langle \varphi_l \rangle - \frac{1}{Vol} \int_{F_{l\sigma}+F_{lw}} \varphi_l \mathbf{V}_{l\sigma}^{\tau} \cdot \mathbf{n}_l dF \; .} \qquad (1.32a)$$

Since $\langle \varphi_l \rangle = \alpha_l \gamma_v \langle \varphi_l \rangle^l$, Eq. (1.16), we finally obtain

$$\boxed{\left\langle \frac{\partial \varphi_l}{\partial \tau} \right\rangle = \frac{\partial}{\partial \tau} \left(\alpha_l \gamma_v \langle \varphi_l \rangle^l \right) - \frac{1}{Vol} \int_{F_{l\sigma}+F_{lw}} \varphi_l \mathbf{V}_{l\sigma}^{\tau} \cdot \mathbf{n}_l dF \; .} \qquad (1.32b)$$

A useful consequence is obtained by setting $\varphi_l = 1$:

$$\frac{\partial}{\partial \tau} \left(\alpha_l \gamma_v \right) = \frac{1}{Vol} \int_{F_{l\sigma}+F_{lw}} \mathbf{V}_{l\sigma}^{\tau} \cdot \mathbf{n}_l dF \; . \qquad (1.33)$$

1.6 Local volume-averaged mass conservation equation

The principle of conservation of mass for the control volume *Vol* can be expressed verbally as follows:

> The change in the mass of the velocity field *l* within *Vol* with time equals the net mass flow of the field *l* through the surface *F* and through the interface of the velocity field $F_{l\sigma}$.

The conservation equation for the property φ_l valid inside Vol_l is the classical one:

$$\frac{\partial \varphi_l}{\partial \tau} + \nabla \cdot \left(\varphi_l \mathbf{V}_l^{\tau} \right) = 0 \; . \qquad (1.34)$$

Performing volume averaging,

$$\left\langle \frac{\partial \varphi_l}{\partial \tau} \right\rangle + \left\langle \nabla \cdot \left(\varphi_l \mathbf{V}_l^{\tau} \right) \right\rangle = 0 \; , \qquad (1.35)$$

and using Eqs. (1.32a) and (1.28), the local average of the derivatives is replaced by derivatives of the local average and additional terms. The result is

$$\frac{\partial}{\partial \tau}\langle \varphi_l \rangle + \nabla \cdot \langle \varphi_l \mathbf{V}_l^\tau \rangle = -\frac{1}{Vol} \int_{F_{l\sigma}+F_{lw}} \varphi_l (\mathbf{V}_l^\tau - \mathbf{V}_{l\sigma}^\tau) \cdot \mathbf{n}_l dF, \qquad (1.36)$$

or, replacing the volume average with its equivalents,

$$\frac{\partial}{\partial \tau}\left(\alpha_l \gamma_v \langle \varphi_l \rangle^l\right) + \nabla \cdot \left(\alpha_l \gamma \langle \varphi_l \mathbf{V}_l^\tau \rangle^{le}\right) = -\frac{1}{Vol} \int_{F_{l\sigma}+F_{lw}} \varphi_l (\mathbf{V}_l^\tau - \mathbf{V}_{l\sigma}^\tau) \cdot \mathbf{n}_l dF, \quad (1.37)$$

valid for non-structured as well for structured velocity fields. Finally, the mass conservation equation is easily obtained by setting $\varphi_l = \rho_l$ and using the definition of the weighted average velocity, Eq. (1.19). The result is

$$\frac{\partial}{\partial \tau}\left(\alpha_l \langle \rho_l \rangle^l \gamma_v\right) + \nabla \cdot \left(\alpha_l \langle \rho_l \rangle^l \langle \mathbf{V}_l^\tau \rangle^{le} \gamma\right) = \gamma_v \mu_l^\tau, \qquad (1.38)$$

which is Eq. (3.9) or (5.7) of [45] with γ_v incorporated into the time derivatives. Leaving γ_v under the time derivatives allows modeling of structure with varying γ_v over time, e.g. deformable structures that depend on local pressure differences or structures deformed by actions independent of the flow parameter but governing the flow. Here

$$\mu_l^\tau = -\frac{1}{\gamma_v Vol} \int_{F_{l\sigma}+F_{lw}} \rho_l (\mathbf{V}_l^\tau - \mathbf{V}_{l\sigma}^\tau) \cdot \mathbf{n}_l dF$$

$$= -\frac{F_{l\sigma}}{\sum_{l=1}^{3} Vol_l} \frac{1}{F_{l\sigma}} \int_{F_{l\sigma}} \rho_l (\mathbf{V}_l^\tau - \mathbf{V}_{l\sigma}^\tau) \cdot \mathbf{n}_{l\sigma} dF - \frac{F_{lw}}{\sum_{l=1}^{3} Vol_l} \frac{1}{F_{lw}} \int_{F_{lw}} \rho_l (\mathbf{V}_l^\tau - \mathbf{V}_{lw}^\tau) \cdot \mathbf{n}_{lw} dF$$

$$(1.39)$$

is the local volume-averaged mass transferred into the velocity field l per unit time and unit mixture flow volume $\sum_{l=1}^{3} Vol_l$. This term is referred to as local volume-averaged instantaneous mass source density in $kg/(m^3 s)$ for the velocity field l. Note the difference between the mass source per unit volume of flow, μ_l^τ, and the mass source per unit of the control volume Vol, $\gamma_v \mu_l^\tau$. The ratio of the interface surface to the mixture volume $F_{l\sigma} / \sum_{l=1}^{3} Vol_l$ is referred in the literature as *interfacial area density*. It is customary in two-phase flow theory to incorporate local volume interface density of the velocity field l in the form of the interface per unit flow volume $\sum_{l=1}^{3} Vol_l$ belonging to the control volume Vol:

$$a_{l\sigma} = F_{l\sigma} / \sum_{l=1}^{3} Vol_l \,. \tag{1.40}$$

This is an important dependent variable. In general, it varies with time and location. The means for describing this variable are discussed in detail in Sections 1.10 and 1.11. In addition, the local volume interface density of the structure w is defined as follows:

$$a_{lw} = F_{lw} / \sum_{l=1}^{3} Vol_l = 4 / D_{hy} \,, \tag{1.41}$$

where D_{hy} is the hydraulic diameter (4 times cross-sectional area divided by perimeter).

For numerical integration it should not be forgotten that the volume fraction at the surfaces of the discretization volume may differ from the local volume fraction of the cells as already mentioned, simply because these are associated with different points in space. Setting these volume fractions to be equal, as is usually implemented in the widely used donor-cell method, introduces non-physical diffusion. This kind of diffusion is in no way associated with numerical diffusion. It influences the modeling of disperse flow pattern less than does free surface modeling. The performance of the numerical model can be improved even for the case of free surfaces in the control volume by extension of the numerical technique developed by *Hirt* and *Nichols* [13] (1981) to cover multi-phase flows.

If the interface is immaterial and consequently does not accumulate mass we have,

$$\rho_l \left(\mathbf{V}_l^\tau - \mathbf{V}_{l\sigma}^\tau \right) \cdot \mathbf{n}_l + \rho_m \left(\mathbf{V}_m^\tau - \mathbf{V}_{m\sigma}^\tau \right) \cdot \mathbf{n}_m = 0 \,. \tag{1.42}$$

This is the instantaneous mass balance of the interface. Note that $\mathbf{n}_l = -\mathbf{n}_m$, and that the contact discontinuity velocity is common for the both sides of the interface,

$$\mathbf{V}_{l\sigma}^\tau = \mathbf{V}_{m\sigma}^\tau = \mathbf{V}_{lm}^\tau \,. \tag{1.43}$$

Interfaces at which we have

$$\mathbf{V}_l^\tau \cdot \mathbf{n}_l = \mathbf{V}_{ml}^\tau \cdot \mathbf{n}_m \qquad (1.44)$$

are called *impermeable*. Such surface may be solid-fluid or fluid-fluid interface [52], p. 307. Surfaces at which

$$\mathbf{V}_l^\tau = \mathbf{V}_{ml}^\tau = \mathbf{0} \qquad (1.45)$$

are called *impermeable fixed* surfaces [52], p. 307. It is convenient to write the mass flow rates perpendicular to the interface in the following form

$$(\rho w)_{lm} = \rho_l (\mathbf{V}_l^\tau - \mathbf{V}_{lm}^\tau) \cdot \mathbf{n}_l \qquad (1.46)$$

$$(\rho w)_{ml} = \rho_m (\mathbf{V}_m^\tau - \mathbf{V}_{lm}^\tau) \cdot \mathbf{n}_m . \qquad (1.47)$$

Then the Eq. (1.42) reads

$$(\rho w)_{lm} = -(\rho w)_{ml} . \qquad (1.48)$$

Consequently the relations between the velocity components normal to the interface and the density are

$$(\mathbf{V}_l^\tau - \mathbf{V}_m^\tau) \cdot \mathbf{n}_l = \left(\frac{1}{\rho_l} - \frac{1}{\rho_m} \right) (\rho w)_{lm} , \qquad (1.49)$$

for the case of mass transfer across the interface and

$$V_{lm}^{n,\tau} = \frac{\rho_l V_l^{n,\tau} - \rho_m V_m^{n,\tau}}{\rho_l - \rho_m} , \qquad (1.50)$$

$$V_l^{n,\tau} - V_{lm}^{n,\tau} = \frac{\rho_m}{\rho_l - \rho_m} \left(V_m^{n,\tau} - V_l^{n,\tau} \right) , \qquad (1.51)$$

$$V_m^{n,\tau} - V_{lm}^{n,\tau} = \frac{\rho_l}{\rho_l - \rho_m} \left(V_m^{n,\tau} - V_l^{n,\tau} \right) , \qquad (1.52)$$

for the "shock wave discontinuity" without mass transfer. The volume average mass balance of the surface $F_{l\sigma}$ common to the fields m and l is

$$\frac{1}{Vol} \int_{F_{lm}^{\tau}} \left[\rho_l (\mathbf{V}_l^{\tau} - \mathbf{V}_{lm}^{\tau}) - \rho_m (\mathbf{V}_m^{\tau} - \mathbf{V}_{lm}^{\tau}) \right] \cdot \mathbf{n}_m dF = 0 . \tag{1.53}$$

1.7 Time average

Splitting time dependent variables, e.g. Φ^{τ} and Ψ^{τ}, into their mean Φ, Ψ and fluctuation parts Φ', Ψ'

$$\Phi^{\tau} = \Phi + \Phi', \quad \Psi^{\tau} = \Psi + \Psi', \tag{1.54}$$

and time averaging them in the following way

$$\Phi = \frac{1}{2\Delta\tau} \int_{-\Delta\tau}^{\Delta\tau} \Phi^{\tau} d\tau , \tag{1.55}$$

is known in the literature as a *Reynolds* averaging. Here $\Delta\tau$ is a time scale large relative to the time scale of turbulent fluctuations, and small relative to the time scale to which we wish to resolve. This averaging process has the following properties

$$\overline{\Phi^{\tau}} = \Phi, \quad \overline{\Psi^{\tau}} = \Psi, \tag{1.56}$$

$$\overline{a\Phi^{\tau} + b\Psi^{\tau}} = a\Phi + b\Psi , \tag{1.57}$$

$$\overline{\Phi^{\tau}\Psi^{\tau}} = \Phi\Psi + \overline{\Phi'\Psi'} , \tag{1.58}$$

$$\overline{\frac{\partial \Phi^{\tau}}{\partial \tau}} = \frac{\partial \Phi}{\partial \tau}, \quad \overline{\nabla \Phi^{\tau}} = \nabla \Phi . \tag{1.59}$$

The instantaneous surface-averaged velocity of the field l, $\left\langle \mathbf{V}_l^{\tau} \right\rangle^{le}$, can be expressed as the sum of the surface-averaged velocity which is subsequently time averaged,

$$V_l := \overline{\left\langle \mathbf{V}_l^{\tau} \right\rangle^{le}} \tag{1.60}$$

and a pulsation component V_l',

$$\left\langle \mathbf{V}_l^\tau \right\rangle^{le} = V_l + V_l' \qquad (1.61)$$

as proposed by *Reynolds*. Substituting for the appropriate terms in the mass conservation equation, performing time averaging, and dropping any averaging signs for the sake of simple notation leads to

$$\frac{\partial}{\partial \tau}(\alpha_l \rho_l \gamma_v) + \nabla \cdot (\alpha_l \rho_l \mathbf{V}_l \gamma) = \gamma_v \mu_l . \qquad (1.62)$$

For convenience of numerical integration, the source term μ_l is split up into a sum of pairs of non-negative terms:

$$\mu_l = \sum_{m=1}^{l_{\max},w} (\mu_{ml} - \mu_{lm}) . \qquad (1.63)$$

Thus, source terms with two subscripts are non-negative. Two successive subscripts denote the direction of the mass transfer. For instance, ml denotes the transferred mass per unit time and unit volume of the flow from velocity field m into l. As a consequence of this definition, source terms with two identical subscripts are equal to zero $\mu_{mm} = 0$. w is used to denote the region outside of the flow. μ_{wl} denotes mass transport from an exterior source into the velocity field l.

If an interface is immaterial and consequently does not accumulate mass, the following equation holds at this interface:

$$\sum_{l=1}^{3} \mu_l = 0 . \qquad (1.64)$$

For $\gamma_v = \gamma = 1$, the Eq. (1.62) has been successfully used in thousands of publications in two-phase flow literature, tacitly in the majority of cases in the sense of a local volume average and successively time-averaged equation. *There is no doubt as to its validity* for velocities far below the velocity of light. If used incorrectly, however, this can give rise to non-solvable numerical problems. An example for usage that can lead to such problems is the use of the left-hand side in the sense of local volume and time average and the right-hand side in the sense of instantaneous sources. For processes with intense mass transfer (condensation shocks, flushing etc.) this is incorrect. The right-hand side must be the local volume average, with this successively time averaged over the current time step.

After some rearrangements the Eq. (1.62) can be written in the following form:

$$\frac{\partial \alpha_l}{\partial \tau} + \nabla \cdot \left(\frac{\gamma \mathbf{V}_l}{\gamma_v} \alpha_l \right) = \frac{\mu_l}{\rho_l} - \alpha_l \left(\frac{\partial \ln \gamma_v \rho_l}{\partial \tau} + \frac{\gamma \mathbf{V}_l}{\gamma_v} \cdot \nabla \ln \gamma_v \rho_l \right).$$

We see that the effective convection velocity of the volumetric fraction inside a complex structure is not the velocity itself but the expression $\gamma \mathbf{V}_l / \gamma_v$.

1.8 Local volume-averaged component conservation equations

The definition given previously should now be recalled, i.e. that each velocity field consists of one non-inert component and several inert components. Each component is designated with i_l. As mentioned in Section 1.2. in order to avoid complicated indices, the designation *il* will be used in the text which follows.

The local volume-averaged instantaneous mass conservation equation for the microscopic component *il* in the velocity field *l* can be expressed as follows:

> Inside the part of the control volume filled with the component *il*, the net mass flow of the component *il* must equal the rate of increase in mass for the component *il*, or mathematically:

$$\frac{\partial}{\partial \tau} \left(\alpha_{il} \langle \rho_{il} \rangle^l \gamma_v \right) + \nabla \cdot \left(\alpha_{il} \langle \rho_{il} \rangle^l \langle \mathbf{V}_{il}^\tau \rangle^{le} \gamma \right)$$

$$= -\frac{1}{Vol} \int_{F_{l\sigma}+F_{lw}} \alpha_{il} \rho_{il} (\mathbf{V}_{il}^\tau - \mathbf{V}_{il\sigma}^\tau) \cdot \mathbf{n}_l dF = \gamma_v \mu_{il}^\tau. \tag{1.65}$$

For a gas mixture it follows from *Dalton*'s law that $\alpha_{il} = \alpha_l$, whereas for mixtures consisting of liquid and macroscopic solid particles $\alpha_{il} \neq \alpha_l$. The instantaneous mass concentration of the component *i* in *l* is now defined by:

$$C_{il}^\tau = \alpha_{il} \langle \rho_{il} \rangle^l / \left(\alpha_l \langle \rho_l \rangle^l \right). \tag{1.66}$$

The center of mass (c.m.) velocity is given by intrinsic surface-averaged field velocity $\langle \mathbf{V}_l^\tau \rangle^{le}$ introduced previously:

$$\alpha_l \langle \rho_l \rangle^l \langle \mathbf{V}_l^\tau \rangle^{le} = \sum_{i=1}^{i_{max}} \alpha_{il} \langle \rho_{il} \rangle^l \langle \mathbf{V}_{il}^\tau \rangle^{le} = \alpha_l \langle \rho_l \rangle^l \sum_{i=1}^{i_{max}} C_{il}^\tau \langle \mathbf{V}_{il}^\tau \rangle^{le}. \tag{1.67}$$

1.8 Local volume-averaged component conservation equations

Consequently

$$\langle V_l^\tau \rangle^{le} = \sum_{i=1}^{i_{max}} C_{il}^\tau \langle V_{il}^\tau \rangle^{le}. \qquad (1.68)$$

It is convenient for description of transport of the microscopic component *il* in the velocity field *l* to replace the velocity component $\langle V_{il}^\tau \rangle^{le}$ by the sum of the center of mass velocity for the particular field $\langle V_l^\tau \rangle^{le}$ and the deviation from the c. m. velocity or the so-called *diffusion velocity* of the inert component $\delta_i \langle V_l^\tau \rangle^{le}$, this yielding the following

$$\langle V_{il}^\tau \rangle^{le} = \langle V_l^\tau \rangle^{le} + \left(\delta \langle V_l^\tau \rangle^{le}\right)_i. \qquad (1.69)$$

Fick [7] (1855) noticed that the mass flow rate of the inert component with respect to the total mass flow rate of the continuous mixture including the inert component is proportional to the gradient of the concentration of this inert component

$$\alpha_{il} \langle \rho_{il} \rangle^l \left(\delta \langle V_l^\tau \rangle^{le}\right)_i = -\alpha_l \langle \rho_l \rangle^l \delta_l D_{il}^l \nabla C_{il}^\tau. \qquad (1.70)$$

The coefficient of proportionality, D_{il}^l, is known as the coefficient of molecular diffusion. The diffusion mass flow rate is directed from regions with higher concentration to regions with lower concentration, with this reflected by the minus sign in the assumption made by *Fick* (which has subsequently come to be known as *Fick*'s law), because many processes in nature and industrial equipment are successfully described mathematically by the above approach - so-called diffusion processes. Here $\delta_l = 1$ for a continuous field and $\delta_l = 0$ for a disperse field. Substituting in Eq. (1.65) yields

$$\frac{\partial}{\partial \tau}\left(\alpha_l \langle \rho_l \rangle^l C_{il}^\tau \gamma_v\right) + \nabla \cdot \left(\alpha_l \langle \rho_l \rangle^l \langle V_l^\tau \rangle^{le} C_{il}^\tau \gamma\right) - \nabla \cdot \left(\alpha_l \langle \rho_l \rangle^l \delta_l D_{il}^l \nabla C_{il}^\tau \gamma\right) = \gamma_v \mu_{il}^\tau. \qquad (1.71)$$

Molecular diffusion has microscopic character, as it is caused by molecular interactions. The general expression (1.70) is actually

$$\alpha_{il} \langle \rho_{il} \rangle^l \left(\delta \langle V_l^\tau \rangle^{le}\right)_i = -\alpha_l \langle \rho_l \rangle^l \delta_l D_{il}^l \left(\nabla C_{il}^\tau + \frac{k_{il}^T}{T_l}\nabla T_l + \frac{k_{il}^p}{p}\nabla p\right). \qquad (1.72)$$

k_{il}^T and k_{il}^p are the dimensionless coefficients of thermo-diffusion and baro-diffusion, which are functions of the component concentrations [10]. They tend to zero in the limiting cases of pure substances. Note that

$$\sum_{i=1}^{i_{\max}} D_{il}^l \nabla C_{il}^\tau = 0 , \qquad (1.73)$$

$$\sum_{i=1}^{i_{\max}} k_{il}^T = 0 , \qquad (1.74)$$

$$\sum_{i=1}^{i_{\max}} k_{il}^p = 0 . \qquad (1.75)$$

The effect of the baro-diffusion is usually negligible. The thermo-diffusion may be substantial if the components gave a quite different molecular mass (e.g. mixture of hydrogen and freon), large temperature gradients, and concentrations around $1/i_{\max}$. Usually thermo-diffusion is also neglected in practical computations [10]. The special theoretical treatment and experimental experience of how to determine the molecular diffusion coefficients in multi-component mixtures is a science in its own right. This topic is beyond the scope of this book. In this context, it should merely be only noted that in line with the thermodynamics of irreversible processes, the thermal diffusivity and the diffusion coefficients influence each other. The interested reader can find useful information in *Reid* et al. [41].

1.9 Local volume- and time-averaged conservation equations

As already mentioned, the c. m. velocity of the velocity field can be expressed as time-averaged c. m. velocity and a pulsation component as proposed by *Reynolds*, see Eq. (1.61). The same can be performed for the concentrations,

$$C_{il}^\tau = C_{il} + C_{il}' , \qquad (1.76)$$

where

$$C_{il} := \overline{C_{il}^\tau} . \qquad (1.77)$$

Substituting appropriately in Eq. (1.71) and performing time averaging, and omitting averaging signs

$$\overline{\alpha_l \langle \rho_l \rangle^l C_{il}^\tau} = \overline{\alpha_l \langle \rho_l \rangle^l} \overline{C_{il}^\tau} \equiv \alpha_l \rho_l C_{il}, \qquad (1.78)$$

$$\overline{\alpha_l \langle \rho_l \rangle^l C_{il}'} = 0, \qquad (1.79)$$

$$\overline{\alpha_l \langle \rho_l \rangle^l C_{il} \mathbf{V}_l} = \overline{\alpha_l \langle \rho_l \rangle^l} \overline{C_{il} \mathbf{V}_l} \equiv \alpha_l \rho_l C_{il} \mathbf{V}_l, \qquad (1.80)$$

$$\overline{\alpha_l \langle \rho_l \rangle^l C_{il} \mathbf{V}_l'} = 0, \qquad (1.81)$$

$$\overline{\alpha_l \langle \rho_l \rangle^l C_{il}' \mathbf{V}_l} = 0, \qquad (1.82)$$

$$\overline{\alpha_l \langle \rho_l \rangle^l \delta_l D_{il}^l \nabla C_{il}} = \overline{\alpha_l \langle \rho_l \rangle^l} \overline{\delta_l D_{il}^l \nabla C_{il}} \equiv \alpha_l \rho_l \delta_l D_{il}^l \nabla C_{il}, \qquad (1.83)$$

$$\overline{\alpha_l \langle \rho_l \rangle^l \delta_l D_{il}^l \nabla C_{il}'} = 0, \qquad (1.84)$$

$$\overline{\mu_{il}^\tau} = \mu_{il}, \qquad (1.85)$$

the following is obtained:

$$\frac{\partial}{\partial \tau}\left(\alpha_l \rho_l C_{il} \gamma_v\right) + \nabla \cdot \left(\alpha_l \rho_l C_{il} \mathbf{V}_l \gamma\right) + \nabla \cdot \left(\alpha_l \rho_l \overline{C_{il}' \mathbf{V}_l'} \gamma\right) - \nabla \cdot \left(\alpha_l \rho_l \delta_l D_{il}^l \nabla C_{il} \gamma\right) = \gamma_v \mu_{il}. \qquad (1.86)$$

For the sake of simplicity, averaging signs are omitted except in the turbulent diffusion term, which will be discussed next. The diffusion can also have macroscopic character, being caused by the macroscopic strokes between large eddies with dimensions considerably larger than the molecular dimensions - turbulent diffusion. In a mixture at rest the molecular strokes are the only mechanism driving diffusion. In real flows both mechanisms are observed. The higher the velocity of the flow, the higher the effect of turbulent diffusion. O. Reynolds assumed

$$\overline{C_{il}' \mathbf{V}_l'} = -D_{il}^t \nabla C_{il}, \qquad (1.87)$$

where the coefficient of turbulent diffusion D_{il}^t is proportional to the coefficient of turbulent kinematic viscosity (this is not valid for turbulence of electro-conductive liquids in a strong magnetic field):

$$D_{il}^t = v_l^t / Sc^t, \qquad (1.88)$$

where the proportionality is determined by the turbulent *Schmidt* number (e.g. $Sc^t = 0.77$ if no other information is available). This coefficient is not a thermodynamic property of the material as is the molecular coefficient, but forms a property of the flow. Thus, having for the total diffusion coefficient

$$D_{il}^* = \delta_l D_{il}^l + D_{il}^t, \qquad (1.89)$$

in a final step, the local volume and time average mass conservation equation valid for each species *il* inside each velocity field *l* is obtained:

$$\frac{\partial}{\partial \tau}(\alpha_l \rho_l C_{il} \gamma_v) + \nabla \cdot \left[\alpha_l \rho_l \gamma \left(\mathbf{V}_l C_{il} - D_{il}^* \nabla C_{il} \right) \right]$$

$$= \gamma_v \sum_{m=1}^{3,w} (\mu_{iml} - \mu_{ilm}) \equiv \gamma_v \mu_{il}. \qquad (1.90)$$

For $\alpha_l = 1$, $\gamma_v = \gamma = 1$ this is the well known concentration equation from single-phase flow dynamics. There is no doubt to its general validity as previously mentioned for velocities much smaller then the light velocity. Eq. (1.90) has been successfully used in IVA codes with diffusion terms neglected, see [18, 20, 21, 22, 24, 25, 27, 28] and with diffusion terms [36].

If the surface is immaterial and does not accumulate mass from the inert component, then the mass jump condition is $\sum_{m=1}^{3} \mu_{il}$.

Note that there is no net mass diffusively transported across a cross section perpendicular to the strongest concentration gradient. Mathematically it is expressed as follows

$$\sum_{i=1}^{i_{max}} \left(D_{il}^* \nabla C_{il} \right) = 0, \qquad (1.91)$$

or

$$D_{1l}^* \nabla C_{1l} = -\sum_{i=2}^{i_{max}} \left(D_{il}^* \nabla C_{il} \right). \qquad (1.92)$$

For mono-disperse particles, i.e., particles with constant particle size D_{il} inside the control volume Eq. (1.90) can be divided by the mass of the single particle in order to obtain an alternative form

$$\frac{\partial}{\partial \tau}(n_{il}\gamma_v) + \nabla \cdot \left[\gamma \left(\mathbf{V}_l n_{il} - D_{il}^* \nabla n_{il}\right)\right] = \gamma_v \mu_{il} / \left(\rho_{il} \frac{1}{6}\pi D_{il}^3\right). \tag{1.93}$$

Here

$$n_{il} = \alpha_l \rho_l C_{il} / \left(\rho_{il} \frac{1}{6}\pi D_{il}^3\right) \tag{1.94}$$

is the number of the particles of species *i* per unit flow volume.

After expanding the first two terms using the chain rule, and comparing with the local volume and time-averaged mass conservation equation of the velocity field, the non-conservative form as follows is obtained

$$\alpha_l \rho_l \left(\gamma_v \frac{\partial C_{il}}{\partial \tau} + \mathbf{V}_l \gamma \nabla C_{il}\right) - \nabla \left(\alpha_l \rho_l D_{il}^* \gamma \nabla C_{il}\right) = \gamma_v \left(\mu_{il} - C_{il}\mu_l\right). \tag{1.95}$$

For numerical integration it is more convenient to use the so-called semi-conservative form, Eq. (1.64), obtained by splitting the right-hand side into non-negative source and sink terms and then placing the sources at the left-hand side:

$$\alpha_l \rho_l \left(\gamma_v \frac{\partial C_{il}}{\partial \tau} + \mathbf{V}_l \gamma \nabla C_{il}\right) - \nabla \left(\alpha_l \rho_l D_{il}^* \gamma \nabla C_{il}\right) + \gamma_v \mu_l^+ C_{il} = \gamma_v DC_{il}. \tag{1.96}$$

Here

$$\mu_{il}^+ = \sum_{\substack{m=1 \\ m \neq l}}^{3} \mu_{iml}, \tag{1.97}$$

$$DC_{il} = \mu_{il} - C_{il}(\mu_l - \mu_l^+). \tag{1.98}$$

For mass transport among the velocity fields and between the exterior sources and the velocity fields it is assumed that the mass convectively leaving the velocity field has the concentrations of the inert components and the mass entering the field has the concentrations of the donor field.

For the particular case of flow consisting of air-steam and two additional liquid fields with one species of solid particles, as postulated for the IVA codes [18, 20, 21, 22, 24, 25, 27-32, 34-36], the following are obtained

$$DC_{i1} = C_{i1}(\mu_{12} + \mu_{13}), \qquad (1.99)$$

$$DC_{i2} = C_{i2}\mu_{21} + C_{i3}\mu_{32}, \qquad (1.100)$$

$$DC_{i3} = C_{i3}\mu_{31} + C_{i2}\mu_{23}, \qquad (1.101)$$

and

$$\mu_1^+ = \mu_{21} + \mu_{31}, \qquad (1.102)$$

$$\mu_2^+ = \mu_{12} + \mu_{32}, \qquad (1.103)$$

$$\mu_{31}^+ = \mu_{13} + \mu_{23}, \qquad (1.104)$$

or, in abbreviated form

$$\mu_{il}^+ = \mu_l^+. \qquad (1.105)$$

The simplicity of the concentration equation is the reason for the choice of C_{il} as elements of the dependent variables vector describing the flow. As a consequence of this choice, the equation of state for the multi-component mixture, $\rho_l = \rho_l(p, T_l, C_{il}$ for all $i)$, has to be derived from the equation of state for the elementary components of the mixture, $\rho_{il} = \rho_{il}(p_{il}, T_{il})$. This has already been performed by the author and published in [23]. An extended variant of this work is Chapter 3 of this monograph.

Finally let as write the mass jump condition at the interface:

Instant:

$$\alpha_{il}^{\tau\sigma} \rho_{il}^{\sigma} \left(\mathbf{V}_{il}^{\tau} - \mathbf{V}_{lm}^{\tau}\right) \cdot \mathbf{n}_l + \alpha_{im}^{\tau\sigma} \rho_{im}^{\sigma} \left(\mathbf{V}_{im}^{\tau} - \mathbf{V}_{lm}^{\tau}\right) \cdot \mathbf{n}_m = 0. \qquad (1.106)$$

In terms of local volume-averaged parameter:

$$C_{il}^{\tau\sigma} \rho_l \left(\mathbf{V}_l^{\tau} - \mathbf{V}_{lm}^{\tau}\right) \cdot \mathbf{n}_l - \rho_l D_{il}^l \nabla C_{il}^{\tau\sigma} \cdot \mathbf{n}_l$$

$$+ C_{im}^{t\sigma} \rho_m \left(\mathbf{V}_m^\tau - \mathbf{V}_{lm}^\tau \right) \cdot \mathbf{n}_m - \rho_m D_{im}^l \nabla C_{im}^{t\sigma} \cdot \mathbf{n}_m = 0. \qquad (1.107)$$

Local volume and time average:

$$C_{il}^{m\sigma} \rho_l \left(\mathbf{V}_l - \mathbf{V}_{lm} \right) \cdot \mathbf{n}_l - \rho_l D_{il}^* \nabla C_{il}^{m\sigma} \cdot \mathbf{n}_l$$

$$+ C_{im}^{l\sigma} \rho_m \left(\mathbf{V}_m - \mathbf{V}_{lm} \right) \cdot \mathbf{n}_m - \rho_m D_{im}^* \nabla C_{im}^{l\sigma} \cdot \mathbf{n}_m = 0. \qquad (1.108)$$

1.10 Conservation equations for the number density of particles

Modeling of gases and liquids which disintegrate in a finite time from a continuum to a spectrum of particles and vice versa, see Fig. 1.4, requires, in addition to the mass conservation equation, equations balancing the number of particles per unit volume of the mixture. Such particles may originate from nucleation or may be the result of convection from the neighboring cells.

If the flow is dispersed there are several possible reasons for the presence of a spectrum of particle sizes in the local control volume. A practicable simplification is the used concept of mono-dispersity in the control volume. This concept is associated with the assumption that the particles in each control volume are represented by a single local volume and time average particle size. As the volume fraction α_l and the local volume and time average particle number density n_l are known, it is then possible to compute the representative local volume and time average particle diameter

$$D_l = \left(\frac{6 \, \alpha_l}{\pi \, n_l} \right)^{1/3} \qquad (1.109)$$

for any control volume. The approach used to obtain Eq. (1.93) illustrates the philosophy behind the method for obtaining the equation for the conservation of the number of macroscopic particles (droplets or bubbles) per unit flow volume:

$$\frac{\partial}{\partial \tau} (n_l \gamma_v) + \nabla \cdot \left[\left(\mathbf{V}_l n_l - \frac{v_l^t}{Sc^t} \nabla n_l \right) \gamma \right] = \gamma_v \left(\dot{n}_{l,kin} - \dot{n}_{l,coal} + \dot{n}_{l,sp} \right), \qquad (1.110)$$

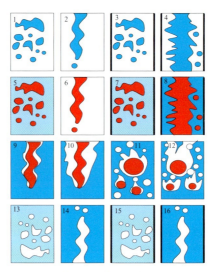

Fig. 1.4. Fragmentation mechanism. (1) Acceleration and turbulence-induced droplet fragmentation in pool flow; (2) jet disintegration in pool; (3) acceleration and turbulence-induced droplet fragmentation in channels; (4) jet fragmentation in channels; (5) droplet in pool; (6) jet in pool; (7) droplet in channel; (8) jet in channel; Liquid metal jet disintegration in liquid: (9) with film boiling - thin film; (10) with film boiling and strong radiation (thick film); Liquid metal droplet disintegration in liquid: (11) with film boiling - thin film; (12) with film boiling and strong radiation (thick film); (13) Bubble fragmentation in pool; (14) Gas jet disintegration in pool; (15) Bubble fragmentation in channels; (16) Gas jet disintegration in channels.

for $\alpha_l > 0$. For simplicity, any averaging signs are omitted. n_l takes values between zero and approximately 10^{12} per cubic meter ($10^{12}\ m^{-3}$ is presumably the upper limit for nucleation during evaporation or condensation). $n_l = 0$ for an existing velocity field here will be associated with a continuous fluid. $0 < n_l Vol_{cel} < 1$, where Vol_{cel} is the volume under consideration, can be interpreted as continuous fluid (jet or free surface) with excited interface. $n_l Vol_{cel} \to 0$ means a stable surface, and $n_l Vol_{cel} \to 1$ means an unstable interface. $n_l Vol_{cel} > 1$ means that there is no longer a continuous velocity field l. Here $-v_l^t / Sc^t$ is the diffusive flux for particle number density resulting from the fact that the particles are in random motion in the presence of a particle density gradient. In accordance with *Batchelor* [2],

$$-\frac{v_l^t}{Sc^t} \approx const\ D_l \overline{(\mathbf{V}_l' \cdot \mathbf{V}_l')}^{1/2} = const\ D_l \left(\Delta \mathbf{V}_{lm} \cdot \Delta \mathbf{V}_{lm}\right)^{1/2} \left[H(\alpha_l)\right]^{1/2}, \quad (1.111)$$

where the constant is approximately unity,

$$H(\alpha_l) \approx \frac{\alpha_l}{\alpha_{l,\max}}\left(1 - \frac{\alpha_l}{\alpha_{l,\max}}\right), \tag{1.112}$$

and $\alpha_{l,\max} \approx 0.62$ is the limit for the closest possible packing of particles. $\overline{(\mathbf{V}_l' \cdot \mathbf{V}_l')}$ is the mean square of the velocity fluctuations and $(\Delta \mathbf{V}_{lm} \cdot \Delta \mathbf{V}_{lm})^{1/2}$ is the magnitude of the relative velocity. This choice meets the requirements for disappearance of velocity fluctuations at the limit $\alpha_l \to 0$ and at the limit for the closest possible packing $\alpha_l \to \alpha_{l,\max}$.

For the locally mono-disperse system there is a unique relationship between volume, concentration, particle number density and the local particle length scale or the interfacial area density a_d (interfacial area F_d divided by the mixture flow volume $\sum_{l=1}^{l_{\max}} Vol_l$),

$$a_d = F_d \Big/ \sum_{l=1}^{l_{\max}} Vol_l = n_d F_{d,\text{single particle}} = \alpha_d\, F_{d,\text{single particle}}\big/ Vol_{d,\text{single particle}}$$

$$= \alpha_d \frac{6}{D_d} = 6\left(\frac{\pi}{6} n_d\right)^{1/3} \alpha_d^{2/3}. \tag{1.113}$$

Therefore the use of one of the variables n_d or D_d or a_d is equivalent and only the simplicity of the conservation equation dictates to prefer n_d.

The production terms in the right-hand side of Eq. (1.109) can be classified as kinetic and non-kinetic terms. The kinetic terms are

$$\dot{n}_{l,kin} = f_{lw} \frac{4}{D_{hy}} n''_{lw} + \alpha_m f_{im} \sum_{i=1}^{l_{\max}} n_{im} + \alpha_m \left(\dot{n}_{l,h} + \dot{n}_{l,dis}\right). \tag{1.114}$$

In the case of bubble flow, $l = 1$, $m = 2$. For this case, the following then apply: $f_{lw} \dfrac{4}{D_{hy}} n''_{lw}$ is the number of bubbles generated at the wall per unit time and unit volume of the flow (wall cavity nucleation rate). f_{lw} is the frequency of bubble generation at one activated seed on the channel wall. n''_{lw} is the number of activated seeds per unit area of the wall.

The term $\alpha_m f_{im} \sum_{i=1}^{i_{max}} n_{im}$ gives the number of bubbles generated from the solid particles homogeneously mixed with the second velocity field per unit time and unit volume of the flow. f_{im} is the frequency of bubbles generated from one activated seed on the particle belonging to the second velocity field. $\dot{n}_{l,h} + \dot{n}_{l,dis}$ is the bulk liquid nucleation rate consisting of the number of bubbles generated by homogeneous nucleation in the second velocity field per unit time and unit volume of the flow, $\dot{n}_{l,h}$, and the number of bubbles generated from the dissolved gases in the second velocity field per unit time and unit volume of the flow, $\dot{n}_{l,dis}$.

The non-kinetic *mechanical* terms are

$$\dot{n}_{l,coal} = n_l f_{l,coal} / 2, \qquad (1.115)$$

this denoting the number of the bubbles that disappear due to *coalescence* per unit time and unit volume of the flow, and

$$\dot{n}_{l,sp} = n_l f_{l,sp}, \qquad (1.116)$$

which is the number of the bubbles arising due to hydrodynamic bubble disintegration per unit time and unit volume of the flow. $f_{l,coal}$ is the *coalescence frequency of two colliding particles*. $f_{l,sp}$ is the *fragmentation frequency of single particle*. The coalescence frequency of a single particle is defined as a product of the *collision frequency per single particle* $f_{d,col}$ and the *coalescence probability* of two colliding particles $P_{d,coal}$. Different causes for the collisions are associated with different collision frequencies $f_{d,col}^s$, $f_{d,col}^{no}$ and $f_{d,col}^o$. Similarly, different causes for collisions are associated with different coalescence probabilities $P_{d,coal}^{no}$, $P_{d,coal}^o$. $P_{d,coal}^o$ is the probability of oscillatory coalescence (due to turbulent oscillations). $P_{d,coal}^{no}$ is the probability of non- oscillatory coalescence (due to non-uniform velocity field in space). Thus the final expression takes the form

$$f_{d,coal} = \left(f_{d,col}^s + f_{d,col}^{no} \right) P_{d,coal}^{no} + f_{d,col}^o P_{d,coal}^o. \qquad (1.117)$$

$f_{d,col}^o$ is the frequency of collision due to turbulence fluctuation of the particles. It depends on the fluctuation component of the velocity of the particles, V_d',

$$f_{d,col}^o = f\left(\Delta V_d', \alpha_d, ..., ... \right). \qquad (1.118)$$

The superscript o is used to remember that this is an oscillatory frequency. $f_{d,col}^{s}$ and $f_{d,col}^{no}$ are the frequencies of collision due to the convective motion. The splitting in two components is due to the use of the concept of mono-dispersity. The computational collisions, $f_{d,col}^{no}$, are caused only by the mean relative velocity between the field with averaged particle size and the surrounding continuum ΔV_{dd}^{no}. Correct mathematical averaging of the velocity difference gives an additional component ΔV_{dd}^{s} associated with $f_{d,col}^{s}$. $f_{d,col}^{s}$ is zero for real mono-disperse systems. Thus

$$f_{d,col}^{no} = f\left(\Delta V_{dd}^{no}, \alpha_d, ..., ...\right), \qquad (1.119)$$

and

$$f_{d,col}^{s} = f\left(\Delta V_{dd}^{s}, \alpha_d, ..., ...\right). \qquad (1.120)$$

The superscript no is used to remember that this is a non-oscillatory frequency. The superscript s is used to remember that this frequency is associated with a spectrum of a particle sizes.

In a similar way, the source terms for the conservation equation for droplet number density, $l = 3$ are defined.

For channel flow it is assumed here that the second velocity field is continuous and the third disperse. In this case, besides the fragmentation and coalescence that also exist in pool flow, the entrainment and deposition will influence not only the mass but also the particle number density balance

$$\mu_{23} - \mu_{32} = \frac{4}{D_h}(1-\alpha_2)^{1/2}\left[(\rho w)_{23} - (\rho w)_{32}\right], \qquad (1.121)$$

$$\dot{n}_{23} - \dot{n}_{32} = \frac{6}{\pi}\frac{4}{D_h}(1-\alpha_2)^{1/2}\left[(\rho w)_{23}/(\rho_2 D_{3E}^3) - (\rho w)_{32}/(\rho_3 D_3^3)\right], \qquad (1.122)$$

where $(\rho w)_{23}$ and $(\rho w)_{32}$ are entrainment and deposition mass flow rate and D_{3E} is the diameter of the entrained droplets.

The large variety of phenomena leading to fragmentation and coalescence are discussed by the author in Refs. [26] and more recently [35]. In Volume II of this monograph the reader will find the current status of empirical knowledge in this field and the way in which this can be used to compute the fragmentation and coalescence production rates.

It is appropriate now to highlight an interesting potential of this concept. Multiplying n_l with $\sum_{l=1}^{3} Vol_l$ gives the number of particles in a single control volume or in a single discretization volume during numerical integration. If there is more than one particle in the cell,

$$n_l \sum_{l=1}^{3} Vol_l > 1, \qquad (1.123)$$

the field is dispersed. Consequently, the mechanisms governing fragmentation and coalescence are the mechanisms for dispersed field l surrounded by m, e.g., droplets or bubble fragmentation and coalescence. If

$$n_l \sum_{l=1}^{3} Vol_l \leq 1, \qquad (1.124)$$

the field is continuous. This means that the mechanisms controlling potential fragmentation are the mechanisms known for continuous field l surrounded by the field m. An example is jet fragmentation, where both participating fields at the jet region are continuous. In this case there is no coalescence. Thus, natural transition from continuum to disperse and vice versa can be modeled. In addition, a very important memory effect of the multi-phase structure is taken into account in this modeling approach. An example for the use of this approach was given in Ref. [27], for interaction and fragmentation of gas, molten metal and water during a transient.

Equation (1.109) without diffusion and non-kinetic production terms was successfully used by *Kocamustafaogulari* and *Ishii* [17] (1983) for the modeling of single-component boiling systems. A comparison of non-equilibrium model predictions with experimental data for flashing in *Laval* nozzles performed by the author [19] (1985) showed that an additional differential equation for description of the particle number density is necessary to obtain a more accurate prediction than the widespread approach of assuming an almost arbitrary number density within the range of 10^9 to $10^{13} cm^{-3}$. *Riznic* and *Ishii* [42] (1989) applied the *Kocamustafaogulari* and *Ishii* approach with success to the modeling of single-component flushing systems. *Deich* and *Philippov* [5] analyzed the pressure and temperature distribution inside the eddies of subcooled steam and came to the conclusion that eddies with appropriate dimensions serve as a nuclei for condensation, this providing further substantiation for use of the method discussed above.

The derived particle number density conservation equation can be used to compute the surface energy associated with the interface. Multiplying the bubble number density, n_l, by the surface energy of a single bubble, $\pi D_1^2 \sigma_{12}$, the surface energy per unit mixture volume is then obtained. The conservation equation for this energy is similar to Eq. (1.109). It is interesting to note that the surface energy is transported by the convection and diffusion of the *discrete* velocity fields but that the terms supplying the change in this energy (kinetic origination, collision, splitting) result from the energy of the surrounding *continuous* liquid.

1.11 Implication of the assumption of mono-dispersity in a cell

1.11.1 Particle size spectrum and averaging

The result of the initial, boundary conditions, convection and local coalescence, fragmentation entrainment and deposition is a spectrum of particle size. In this Section we will demonstrate what influence the replacement of the spectrum of particle sizes with a single particle size may have on heat and mass transfer processes and how it can be incorporate approximately in the analysis.

For the illustration of this we use the observed in 1938 *Nukiama - Tanasawa* distribution [39]

$$\frac{P(D_{di})}{D'_d} = 4 \left(\frac{D_{di}}{D'_d} \right)^2 e^{-2(D_{di}/D'_d)}. \qquad (1.125)$$

Here $P(D_{di})$ is the probability that a particle size is between D_{di} and $D_{di} + \delta D_{di}$. D'_d is the most probable particle size, the size where the probability distribution function has its maximum value. The particle sizes may take values between zero and a maximum value

$$0 < D_{di} < D_{d,\max}. \qquad (1.126)$$

Thus, if we know D'_d and $D_{d,\max}$, the particle distribution is uniquely characterized. *MacVean* (see *Wallis* [49] (1969)) found that a great deal of data could be correlated by assuming that

$$D'_d = D_d / 2, \qquad (1.127)$$

where D_d is the volume-averaged particle size. The relationship between D_d and the maximum particle size, $D_{d,\max}$, is reported as

$$D_{d,\max} \approx (2.04 \text{ to } 3.13)\, D_d \qquad (1.128)$$

(see *Pilch* et al. [40] (1981) and *Kataoka* et al. [16] (1983), among others. Other distributions give slightly different results e.g. *Kolomentzev* and *Dushkin* [37] (1985)

$$\frac{P(D_{di})}{D'_d} = 8\left(\frac{D_{di}}{D'_d}\right)^2 e^{-4(D_{di}/D'_d)^2} \qquad (1.129)$$

came to $D_{d,\max} \approx 1.33\, D_d$. Other distributions are reported by *Rosin* and *Rammler* [43] (1933), *Griffith* [11] (1943) and *Mugele* and *Evans* [38] (1951).

In contrast to equation (1.113) for mono-disperse spherical particles the interfacial area density in the general case is

$$a_d = F_d \Big/ \sum_{l=1}^{l_{\max}} Vol_l = n_d F_{d,\text{single particle}} = \alpha_d\, F_{d,\text{single particle}} \Big/ Vol_{d,\text{single particle}} =$$

$$\frac{6\alpha_d}{D_d^{sm}} = \frac{6\alpha_d}{D_d}\frac{D_d}{D_d^{sm}}. \qquad (1.130)$$

Here

$$D_d^{sm} = \left[\int_0^{D_{d,\max}} D_d^3 P(D_d)\, dD_d \Big/ \int_0^{D_{d,\max}} D_d^2 P(D_d)\, dD_d\right] \qquad (1.131)$$

is the so called *Sauter* mean diameter (see *Sauter* [44] (1929)). A droplet with diameter equal to the *Sauter* mean diameter has the same surface to volume ratio as the entire spray. Using the *Nukiama-Tanasawa* distribution, Eq. (1.125), and Eqs. (1.127) and (1.128), after evaluating Eq. (1.131) analytically we obtain for the *Sauter* mean diameter

$$D_d^{sm} \approx (1.154 \text{ to } 1.238)\, D_d \quad (D_d^{sm} < 1.25 D_d),$$

or

$$D_d / D_d^{sm} \approx 0.867 \text{ to } 0.807 > 0.8, \qquad (1.132)$$

which is perfectly confirmed by the measurements summarized by *Faeth* in [6] - $D_d^{sm}/D_d \approx 1.2$. This is a very useful result, allowing use of the volume fraction

α_d and the particle number density n_d in cases where the assumption of mono-dispersity does not hold.

1.11.2 Cutting of the lower part of the spectrum due to mass transfer

Use of the particle number density conservation equations for each of the three velocity fields as already implemented in the IVA computer code series since 1985 is an extremely practical method of modeling the scale of the field. A number of important features of the concept of mono-dispersity will now be discussed.

The time-averaged source terms associated with the change in the aggregate state for the mass conservation equations are

$$\mu_l = \frac{1}{\Delta \tau} \int_0^{\Delta \tau} \frac{d}{d\tau}\left(n_l \rho_l \frac{\pi}{6} D_l^3\right) d\tau = \left[\Delta\left(n_l \rho_l \frac{\pi}{6} D_l^3\right)\right] / \Delta \tau$$

$$= \rho_{l0} \frac{\pi}{6} \left[D_{ld}^3 f_{lw} \frac{4}{D_{hy}} n_{lw}'' + D_{ldm}^3 \alpha_m f_{im} \sum_{i=1}^{i_{max}} n_{im} + \alpha_m \left(D_{l,h}^3 \dot{n}_{l,h} + D_{l,dis}^3 \dot{n}_{l,dis} \right) \right]$$

$$+ \alpha_{l0} \rho_{l0} \left[\rho_l D_l^3 / \left(\rho_{l0} D_{l0}^3 \right) - 1 \right] / \Delta \tau . \qquad (1.133)$$

Here ρ_{l0}, n_{l0} and D_{l0} are density, particle number density, and particle size at the beginning of the time step $\Delta \tau$. D_{ld} is the size with which the bubbles are produced after one nucleation cycle on the solid structure, i.e., the particle departure diameter. D_{ldm} is the size with which the bubbles are produced after one nucleation cycle on the inert solid particles of the field $m = 2$. If the nucleation takes place in the bulk of the donor field, the size is equal to the smallest stable size $D_{l,h}$, or in case of nucleation on dissolved gases, $D_{l,dis}$. More information for modeling of particulate processes is provided in Volume II of this monograph.

The first term in the above equation gives the time and local volume-averaged mass production due to nucleation; the second term gives the time and local volume average of mass production due to mass change for existing particles.

It is important to note that the second term describes changes in the particle size due to evaporation and condensation. In accordance with the concept of mono-dispersity, the mass changes leading to a decrease in the local volume and time average particle size also cause the disappearance of those particles in the spectrum having sizes smaller than the size change. This statement, together with the assumption made for the form of the spectrum, provides the basics for computation of the averaged number of disappearing particles. Suppose the particle size distribution obeys the *Nukiama - Tanasawa* [39] law

$$P(D_l) = 4(D_l / D_l')^2 \exp\left[-2(D_l / D_l')\right]. \tag{1.134}$$

Suppose the form of the distribution remains unchanged during the time step. Having in mind that $\mu_{lm} = -\alpha_{l0}\rho_{l0}\left[\rho_l D_l^3 / (\rho_{l0} D_{l0}^3) - 1\right] / \Delta \tau$ the volume-averaged diameter changes within this time by

$$\Delta D_l = D_{l0} - D_l = D_{l0}\left(1 - D_l / D_{l0}\right) = D_{l0}\left\{1 - \left[\left(1 - \frac{\mu_{lm}\Delta\tau}{\alpha_{l0}\rho_{l0}}\right)\frac{\rho_{l0}}{\rho_l}\right]^{1/3}\right\}$$

$$\approx D_{l0}\left\{1 - \left[1 - \frac{\mu_{lm}\Delta\tau}{\alpha_{l0}\rho_{l0}}\right]^{1/3}\right\}. \tag{1.135}$$

and all particles having sizes $0 \le D_l \le 2\Delta D_l$, namely,

$$n_l \int_0^{2\Delta D_l} P(D_l) dD_l = 2n_l\left\{1 - \exp\left(-4\frac{\Delta D_l}{D_l'}\right)\left[1 + 4\frac{\Delta D_l}{D_l'}\left(1 + 2\frac{\Delta D_l}{D_l'}\right)\right]\right\}, \tag{1.136}$$

disappear. The averaged particle sink per unit time and unit mixture volume is consequently

$$\dot{n}_{l,spectrum\_cut} = (n_l / \Delta\tau) 2\left\{1 - \exp\left(-4\frac{\Delta D_l}{D_l'}\right)\left[1 + 4\frac{\Delta D_l}{D_l'}\left(1 + 2\frac{\Delta D_l}{D_l'}\right)\right]\right\}, \tag{1.137}$$

for $\Delta D_l \ge D_l'$. The latter condition means that the averaged particle size is ΔD_l, with particle size ranging approximately within the interval $2\Delta D_l$. It was found that for shock condensation of bubbles the inclusion of Eq. (1.137) for prediction of a reduction in the number density for disappearance of steam volume fractions is very important [29-31, 34, 36]. The same is true for the disappearance of water droplets in highly superheated gases due to evaporation. The latter can be the case in steam explosion analysis where gas is superheated due to previous contact with melt, with this followed by water droplet entry into the superheated gas regions. Combustion processes and spray cooling of hot gas jets are further examples.

Equation (1.137) can be written in the following form

$$\dot{n}_{l,spectrum\_cut} = n_l f_{l,spectrum\_cut}, \tag{1.138}$$

where

$$f_{l,spectrum\_cut} = \frac{2}{\Delta\tau}\left\{1-\exp\left(-4\frac{\Delta D_l}{D_l'}\right)\left[1+4\frac{\Delta D_l}{D_l'}\left(1+2\frac{\Delta D_l}{D_l'}\right)\right]\right\}, \tag{1.139}$$

is the frequency of disappearance of particles due to the spectrum cutting.

1.11.3 The effect of the averaging on the effective velocity difference

Consider i groups of particles with different diameters having a concentration per unit volume called particle number density

$$n_{di} = n_d P(D_{di}) \qquad i = 1, I \tag{1.140}$$

that is dependent on the diameter. Here $P(D_{di})$ is the probability of a particle to have a size between D_{di} and $D_{di} + \delta D_{di}$. We assume that the continuous velocity field possesses a constant velocity w_c. The difference of the continuum and the particle velocity is approximately described by the following equations

$$\left(1+b_d c_d^{vm}\right)\frac{\partial \Delta w_{cd}}{\partial \tau} + \left(\frac{1}{\rho_c} - \frac{1}{\rho_d}\right)\nabla p + b_d \frac{3}{4}\frac{c_{cd}^d}{D_d}|\Delta w_{cd}|\Delta w_{cd} = 0, \tag{1.141}$$

$$\left(1+b_{di} c_{cdi}^{vm}\right)\frac{\partial \Delta w_{cdi}}{\partial \tau} + \left(\frac{1}{\rho_c} - \frac{1}{\rho_d}\right)\nabla p + b_{di} \frac{3}{4}\frac{c_{cdi}^d}{D_{di}}|\Delta w_{cdi}|\Delta w_{cdi} = 0, \tag{1.142}$$

where $b_d = (\alpha_c \rho_c + \alpha_d \rho_d)/(\rho_c \rho_d)$, and $b_{di} = (\alpha_c \rho_c + \alpha_{di} \rho_d)/(\rho_c \rho_d)$. In this case we see that each particular group possesses a different relative velocity with respect to the continuum and consequently the groups are moving relatively to each other. Therefore *collision in real nature is an inevitable phenomenon*. For steady flow we have from Eqs. (1.141) and (1.142)

$$\Delta w_{cdi} = \Delta w_{cd}\left(\frac{b_d}{b_{di}}\frac{c_d^d}{c_{cdi}^d}\frac{D_{di}}{D_d}\right)^{1/2} = \Delta w_{cd} f_i. \tag{1.143}$$

Obviously there is a resulting average velocity

$$\Delta V_{dd} = \frac{1}{n_d}\sum_i n_{di}\left|\Delta w_{cd} - \Delta w_{cdi}\right| = \left|\Delta w_{cd}\right|\frac{1}{n_d}\sum_i n_{di}\left|1-f_i\right| = \left|\Delta w_{cd}\right|\sum_i P(D_{di})\left|1-f_i\right|,$$
(1.144)

which is the

$$\frac{1}{n_d}\sum_i n_{di}\left|1-f_i\right| \qquad (1.145)$$

part of the average relative velocity of the droplets with respect to the continuum. For mono-dispersed particles

$$f_1 = f_2 = \ldots = f_i = 1 \qquad (1.146)$$

and

$$\Delta V_{dd} = 0. \qquad (1.147)$$

The expression

$$\frac{1}{n_d}\sum_i n_{di}\left|1-f_i\right| = \int_0^{D_{d,\max}/D_d} P(D_{di})\left|1-f_i\right| d(D_{di}/D_d), \qquad (1.148)$$

can be numerically estimated using known distribution functions e.g. the *Nukiama - Tanasava* distribution [9] (1939) as given by Eq. (1.125). Furthermore

$$f_i = \left(\frac{b*+1}{b*+\alpha_{di}/\alpha_d}\frac{c_d^d}{c_{cdi}^d}\frac{D_{di}}{D_d}\right)^{1/2}, \qquad (1.149)$$

where $b* = \alpha_c \rho_c / \alpha_d \rho_d$ can be expressed as function of D_{di}/D_d having in mind that $\alpha_{di} = (\pi D_{di}^3/6)n_{di}$ and $\alpha_d = (\pi D_d^3/6)n_d$ and using Eq. (1.140),

$$\alpha_{di}/\alpha_d = (D_{di}/D_d)^3 P(D_{di}). \qquad (1.150)$$

1.12 Stratified structure

If the selected scale of discretization δ is smaller than the characteristic length of the fields,

$$\delta|\kappa_l| < 1, \tag{1.151}$$

the local control volume contains a surface dividing for instance field l from field m. The unit vector $\mathbf{n}_l$ pointing outwards from the field l is an important local characteristic of this surface. It is defined from the spatial distribution of the volume fraction of the field l as follows

$$\mathbf{n}_l = -\frac{\nabla(\alpha_l \gamma)}{|\nabla(\alpha_l \gamma)|}. \tag{1.152}$$

The curvature of the surface is conveniently computed following the derivation of *Brackbill* et al. [3],

$$\kappa_l = \nabla \cdot \mathbf{n}_l, \tag{1.153}$$

and is defined positive if the center of the curvature is in field l.

1.13 Final remarks and conclusions

The heterogeneous porous-media formulation proposed first by *Sha, Chao* and *Soo* was applied with some modifications to derive mass conservation equations for multi-phase multi-component flows conditionally divided into three velocity fields. Application of the *Slattery-Whitaker* local volume average for continuous and dispersed fields provides the tool needed for derivation of mass conservation equations valid for continuous as well as for dispersed fields. In addition, the method was extended by including the concept of dynamic particle fragmentation and coalescence. The result of this derivation yields the following three local volume and time average equations applicable for each velocity field. Here, the equations are written in the scalar form for the most frequently used Cartesian and cylindrical coordinate systems to simplify their direct use by the reader for his particular application.

$$\frac{\partial}{\partial \tau}(\alpha_l \rho_l \gamma_v) + \frac{1}{r^\kappa}\frac{\partial}{\partial r}(r^\kappa \alpha_l \rho_l u_l \gamma_r) + \frac{1}{r^\kappa}\frac{\partial}{\partial \theta}(\alpha_l \rho_l v_l \gamma_\theta) + \frac{\partial}{\partial z}(\alpha_l \rho_l w_l \gamma_z)$$

$$= \gamma_v \sum_{m=1}^{l_{\max},w}(\mu_{ml} - \mu_{lm}), \tag{1.154}$$

$$\frac{\partial}{\partial \tau}(\alpha_l \rho_l C_{il} \gamma_v) + \frac{1}{r^\kappa}\frac{\partial}{\partial r}\left[r^\kappa \alpha_l \rho_l \left(u_l C_{il} - D_{il}^* \frac{\partial C_{il}}{\partial r}\right)\gamma_r\right]$$

$$+\frac{1}{r^{\kappa}}\frac{\partial}{\partial\theta}\left[\alpha_{l}\rho_{l}\left(v_{l}C_{il}-D_{il}^{*}\frac{1}{r^{\kappa}}\frac{\partial C_{il}}{\partial\theta}\right)\gamma_{\theta}\right]+\frac{\partial}{\partial z}\left[\alpha_{l}\rho_{l}\left(w_{l}C_{il}-D_{il}^{*}\frac{\partial C_{il}}{\partial z}\right)\gamma_{z}\right]$$

$$=\gamma_{v}\sum_{m=1}^{l_{\max},w}\left(\mu_{iml}-\mu_{ilm}\right) \quad \text{for } \alpha_{l}\geq 0, \qquad (1.155)$$

$$\frac{\partial}{\partial\tau}\left(n_{l}\gamma_{v}\right)+\frac{1}{r^{\kappa}}\frac{\partial}{\partial r}\left[r^{\kappa}\left(u_{l}n_{l}-\frac{v_{l}^{t}}{Sc_{l}^{t}}\frac{\partial n_{l}}{\partial r}\right)\gamma_{r}\right]+\frac{1}{r^{\kappa}}\frac{\partial}{\partial\theta}\left[\left(v_{l}n_{l}-\frac{v_{l}^{t}}{Sc_{l}^{t}}\frac{1}{r^{\kappa}}\frac{\partial n_{l}}{\partial\theta}\right)\gamma_{\theta}\right]$$

$$+\frac{\partial}{\partial z}\left[\left(w_{l}n_{l}-\frac{v_{l}^{t}}{Sc_{l}^{t}}\frac{\partial n_{l}}{\partial z}\right)\gamma_{z}\right]=\gamma_{v}\left(\dot{n}_{l,kin}-\dot{n}_{l,coal}+\dot{n}_{l,sp}\right)$$

$$\equiv \gamma_{v}n_{l}\left\{\frac{1}{n_{l}}\dot{n}_{l,kin}+f_{l,sp}-\frac{1}{2}\left[\left(f_{d,col}^{s}+f_{d,col}^{no}\right)P_{d,coal}^{no}+f_{d,col}^{o}P_{d,coal}^{o}\right]-f_{l,spectrum\_cut}\right\}$$

for $\alpha_{l}\geq 0$. (1.156)

For flow in a pipe the entrainment and deposition terms have to be taken into account. The link between kinetic generation of particles and the mass source terms is specified. The link between mass sources and the change in the bubble/droplet size due to evaporation or condensation was presented for local volume and time-averaged source terms. The concept of mono-dispersity is discussed and a method is proposed for computation of the disappearance of particles due to evaporation and condensation.

The following conclusion can be drawn from this derivation:

1. For numerical integration, the size of discretization for the above local volume and time average equations is allowed to be smaller than, equal to, or larger than the characteristic length scale D_l of the velocity field, unless other mathematical constraints associated with the numerical method used or concept used for the computation of the fragmentation and coalescence sources are imposed. The characteristic length scale D_l is understood to be: $D_l = 2/|\kappa_l|$, for continuum, and D_l = particle size, for disperse field.

2. The size of the spatial discretization for the resulting equations should not be an order of magnitude larger than the characteristic length scale of the fragmentation. This ensures that important local events will not be obscured as a result of averaging of the properties in the finite volume. For example, insufficiently

fine spatial discretization on modeling of melt-water interaction may lead to non-prediction of steam explosions for situations where explosions have been observed in experiments because of averaging over large volume having regions with intensive fragmentation and others without such fragmentation.

3. The use of the concept of mono-dispersity is associated with the effect of disappearance of part of the particles during a given time step in the case of coincident mass transfer that reduces the local volume fraction of the phase.

4. Kinetic particle source terms are directly related to mass source terms in the mass conservation equation.

5. Non-kinetic particle source terms such as fragmentation or coalescence influence average particle size and, with the size, the amount of evaporation or condensation. This means that the non-kinetic particle source terms indirectly influence the mass of the field. These terms introduce important *time delays* compared to the approach where the quasi-static particle size under local conditions is used. This is of major significance in the analysis of severe accidents such as steam explosion or processes associated with condensation shocks.

References

1. Anderson TB, Jackson R (1967) A fluid mechanical description of fluidized beds, Ind. Eng. Fundam., vol 6 p 527
2. Batchelor GK (1970) The stress system in a suspension of force-free particles, J. Fluid Mech., vol 42 pp 545-570
3. Brackbill JU, Kothe DB, Zemach C (1992) A continuum method for modeling surface tension, J. Comput. Phys., vol 100 pp 335-354
4. Delhaye JM, Giot M, Reithmuller ML (1981) Thermohydraulics of two-phase systems for industrial design and nuclear engineering, Hemisphere Publishing Corporation, New York, McGraw Hill Book Company, New York
5. Deich ME, Philipoff G A (1981) Gas dynamics of two phase flows. (In Russian) Energoisdat, Moscow
6. Faeth GM, 1995, Spray combustion: A Review, Proc. of The 2nd International Conference on Multiphase Flow '95 Kyoto, April 3-7, 1995, Kyoto, Japan
7. Fick A (1855) Über Diffusion. Ann. der Physik, 94 p 59
8. Gentry RA, Martin RE and Daly BJ (1966) An Eulerian differencing method for unsteady compressible flow problems, J. Comp. Physics, vol 1 p 87
9. Gray WG, Lee PCY (1977) On the theorems for local volume averaging of multi-phase system, Int. J. Multi-Phase Flow, vol 3 pp 222-340
10. Grigorieva VA and Zorina VM (eds) (1988) Handbook of thermal engineering, thermal engineering experiment, in Russian, 2d Edition, Moskva, Atomisdat, vol 2,
11. Griffith L (1943) A theory of the size distribution of particles in a comminuted system, Canadian J. Research, vol 21 A no. 6 pp 57-64

12. Hetstrony G (1982) Handbook of multi phase systems. Hemisphere Publ. Corp., Washington et al., McGraw-Hill Book Company, New York ...
13. Hirt CW, Nichols BD (1981) Volume of fluid (VOF) method for dynamics of free boundaries, J. Comput. Phys., vol. 39 pp 201 – 225
14. Hirt CW (1993) Volume-fraction techniques: powerful tools for wind engineering, Journal of Wind Engineering and Industrial Aerodynmics, vol 46&47 pp 327-338
15. Ishii M (1975) Thermo-fluid dynamic theory of two-phase flow, Eyrolles, Paris
16. Kataoka I, Ishii M, Mishima K (June 1983) Transactions of the ASME, vol 105 pp 230-238
17. Kocamustafaogulari G, Ishii M (1983) Interfacial area and nucleation site density in boiling systems, Int. J. Heat Mass transfer, vol 26 no 9 pp 1377-1387
18. Kolev NI (March 1985) Transiente Drephasen Dreikomponenten Strömung, Teil 1: Formulierung des Differentialgleichungssystems, KfK Report 3910
19. Kolev NI (August 1985) Transiente Dreiphasen Dreikomponenten Stroemung, Teil 2: Eindimensionales Schlupfmodell Vergleich Theorie-Experiment, KfK Report 3926
20. Kolev NI (1986)Transiente Dreiphasen Dreikomponenten Strömung, Teil 3: 3D-Dreifluid-Diffusionsmodell, KfK Report 4080
21. Kolev N I (1986) Transient three-dimensional three-phase three-component nonequilibrium flow in porous bodies described by three-velocity fields, Kernenergie 29 pp 383-392
22. Kolev NI (August 1987) A three field-diffusion model of three-phase, three-component flow for the tansient 3D-computer code IVA2/001. Nuclear Technology, vol 78 pp 95-131
23. Kolev NI (1990) Derivatives for the state equations of multi-component mixtures for universal multi-component flow models, Nuclear Science and Engineering, vol 108 pp 74-87
24. Kolev NI (Sept. 1991) A three-field model of transient 3D multi-phase, three-component flow for the computer code IVA3, Part 1: Theoretical basics: Conservation and state equations, numerics, KfK Report 4948
25. Kolev NI (1991) IVA3: A transient 3D three-phase, three-component flow analyzer, Proc. of the Int. Top. Meeting on Safety of Thermal Reactors, Portland, Oregon, July 21-25, 1991, pp 171-180. Presented at the 7th Meeting of the IAHR Working Group on Advanced Nuclear Reactor Thermal-Hydraulics, Kernforschungszentrum Karlsruhe, August 27 to 29 1991
26. Kolev NI (1993) Fragmentation and coalescence dynamics in multi-phase flows, Experimental Thermal and Fluid Science, vol 6 pp 211-251
27. Kolev NI (1993) The code IVA3 for modelling of transient three-phase flows in complicated 3D geometry, Kerntechnik, vol 58 no 3 pp 147-156
28. Kolev NI (1993) IVA3 NW: Computer code for modeling of transient three phase flow in complicated 3D geometry connected with industrial networks, Proc. of the Sixth Int. Top. Meeting on Nuclear Reactor Therrmmal Hydraulics, Oct. 5-8, 1993, Grenoble, France
29. Kolev NI (1993) Berechnung der Fluiddynamischen Vorgänge bei einem Sperrwasserkühlerrohrbruch, Projekt KKW Emsland, Siemens KWU Report R232/93/0002
30. Kolev NI (1993) IVA3-NW A three phase flow network analyzer. Input description, Siemens KWU Report R232/93/E0041
31. Kolev NI (1993) IVA3-NW Components: Relief Valves, Pumps, Heat Structures, Siemens KWU Report R232/93/E0050

32. Kolev NI (1994) The influence of the mutual bubble interaction on the bubble departure diameter, Experimental Thermal and Fluid Science, 8 pp 167-174
33. Kolev NI (1994) The code IVA4: Modeling of mass conservation in multi-phase multi-component flows in heterogeneous porous media, Kerntechnik, vol 59 no 4-5 pp 226-237
34. Kolev NI (1996) Three fluid modeling with dynamic fragmentation and coalescence fiction or daily practice? 7th FARO Experts Group Meeting Ispra, October 15-16, 1996; Proceedings of OECD/CSNI Workshop on Transient thermal-hydraulic and neutronic codes requirements, Annapolis, MD, U.S.A., 5th-8th November 1996; 4th World Conference on Experimental Heat Transfer, Fluid Mechanics and Thermodynamics, ExHFT 4, Brussels, June 2-6, 1997; ASME Fluids Engineering Conference & Exhibition, The Hyatt Regency Vancouver, Vancouver, British Columbia, CANADA June 22-26, 1997, Invited Paper; Proceedings of 1997 International Seminar on Vapor Explosions and Explosive Eruptions (AMIGO-IMI), May 22-24, Aoba Kinen Kaikan of Tohoku University, Sendai-City, Japan
35. Kolev NI (1998) Can melt-water interaction jeopardize the containment integrity of the EPR? Part 3: Fragmentation and coalescence dynamics in multi-phase flows, KWU NA-T/1998/E083a, Project EPR
36. Kolev NI (1999) Verification of IVA5 computer code for melt-water interaction analysis, Part 1: Single phase flow, Part 2: Two-phase flow, three-phase flow with cold and hot solid spheres, Part 3: Three-phase flow with dynamic fragmentation and coalescence, Part 4: Three-phase flow with dynamic fragmentation and coalescence – alumna experiments, CD Proceedings of the Ninth International Topical Meeting on Nuclear Reactor Thermal Hydraulics (NURETH-9), San Francisco, California, October 3-8,1999, Log. Nr. 315
37. Kolomentzev AI, Dushkin AL (1985) Vlianie teplovoj i dinamiceskoj neravnovesnosty faz na pokasatel adiabatj v dwuchfasnijh sredach, TE 8 pp 53-55
38. Mugele R A, Evans H D (151) Droplet size distribution in sprays, Ing. Eng. Chem., vol.43 pp 1317-1324
39. Nukiama S, Tanasawa Y (1931) Trans. Soc. Mech. Engrs. (Japan), vol 4 no 14 p.86
40. Pilch M, Erdman CA, Reynolds AB (August 1981) Acceleration induced fragmentation of liquid drops, Charlottesville, VA: Department of Nucl. Eng., University of Virginia, NUREG/CR-2247
41. Reid RC, Prausnitz JM, Sherwood TK (1982) The properties of gases and liquids, Third Edition, McGraw-Hill Book Company, New York
42. Riznic J R, Ishii M (1989) Bubble number density and vapor generation in flushing flow, Int. J. Heat Mass Transfer, vol 32 no 10 pp 1821-1833
43. Rosin P, Rammler E, (1933) Laws governing the fineness of powdered coal, J. Inst. Fuel, vol 7 pp 29-36
44. Sauter J (1929) NACA Rept. TM-518
45. Sha T, Chao BT, Soo SL (1984) Porous-media formulation for multi-phase flow with heat transfer, Nuclear Engineering and Design, vol 82 pp 93-106
46. Slattery JC (1967) Flow of viscoelastic fluids through porous media, AIChE Journal, vol 13 p 1066
47. Slattery JC (1990) Interfacial transport phenomena, Springer, Berlin Heidelberg New York
48. Teletov SG (1958) On the problem of fluid dynamics of two-phase mixtures, I. Hydrodynamic and energy equations, Bulletin of the Moscow University, No 2, p 15

49. Wallis GB (1969) One-dimensional two-phase flow, McGraw-Hil, New York 1969
50. Whitaker S (1967) Diffusion and dispersion in porous media, AIChE Journal, vol 13 p 420
51. Whitaker S (1969) Advances in theory of fluid motion in porous media, Ind. Eng. Chem., vol 61 no 12 pp 14-28
52. Whitaker S (1977) Experimental principles of heat transfer, Pergamon Press Inc., New York, 1977
53. Whitaker S (1985) A simple geometrical derivation of the spatial averaging theorem, Chemical Engineering Education, Chemical Engineering Education, pp 18-21 pp 50-52
54. Zemansky MW (1968) Heat and thermodynamics, McGraw-Hill Book Company, Fifth Edition

2 Momentums conservation

The time rate for change in momentum of a body equals the net force exerted on it.

Isaac Newton, Philosophiae naturalis principia, 1687

The heterogeneous porous body concept in conjunction with the local volume and time averaging are used to derive rigorously the momentum equations for multi-phase flows conditionally divided into three velocity fields. All interfacial integrals are suitably transformed in order to enable practical application. Some minor simplifications are introduced in the general equation finally obtained, and working equations for each of the three velocity fields are recommended for general use in multi-phase fluid dynamic analysis.

2.1. Introduction

As already mentioned in Chapter 1, the heterogeneous porous-media formulation presented by *Sha* et al. [57] is to date the most effective concept for deriving multi-phase equations applicable to flows in complicated technical facilities - for comparison see also the Refs. [57, 59, 19, 14]. In the present chapter this concept is used to derive momentum equations for multi-phase flows in heterogeneous porous structures as defined in the previous Chapter [1]. The flow is conditionally divided into three velocity fields. In addition, and going beyond the concept in [57], the local volume average equations obtained are time-averaged. Each of the interfacial terms obtained is rearranged appropriately to permit its practical evaluation. This yields a working form which is applicable to a large variety of problems. This Chapter is an improved and extended version of the work published in [38].

The strategy we follow is: We first apply the momentum equations for each of the velocity fields excluding the interfaces by replacing their actions by forces. Then, we write a force balance at the interfaces considering them as immaterial and therefore inertialess. This interfacial force balance links the momentum equations valid for the both sides of the interface.

2.2. Local volume-averaged momentum equations

2.2.1 Single-phase momentum equations

> The time rate for change in momentum of a body relative to an inertial frame of reference equals the net force exerted on it (Newton).

Applied to a single continuum this principle results in Euler's first law of continuum mechanics [66]. This principle applied to each velocity field within the control volume except the interface yields the well-known local instantaneous momentum equation

$$\frac{\partial}{\partial \tau}\left(\rho_l \mathbf{V}_l^r\right) + \nabla \cdot \left(\rho_l \mathbf{V}_l^r \mathbf{V}_l^r - \mathbf{T}_l^r\right) + \nabla p_l^r + \rho_l \mathbf{g} = 0, \qquad (2.1)$$

which is valid only inside the velocity field l excluding the interface. Here the positive velocity direction gives the negative force direction - a commonly used definition. The total stress tensor is split into $p_l^r \mathbf{I}$ and $\mathbf{T}_l^r$. p_l^r is the static pressure inside field l, $\mathbf{I}$ is the unit matrix, and $\mathbf{T}_l^r$ is the shear stress tensor. $\mathbf{V}_l^r \mathbf{V}_l^r$ is the dyadic product of two vectors $\mathbf{V}_l^r$ and $\mathbf{V}_l^r$ - see Appendix 1. It is a second order tensor. $\mathbf{g}$ is the vector of the gravitational acceleration. Equation (2.1) is the generally accepted momentums balance for single-phase for velocities much smaller then the velocity of light.

2.2.2 Interface force balance (momentum jump condition)

Next we abstract a volume around the interface having thickness ε converging to zero. Decoupling mechanically the control volume from the adjacent fields requires replacing the action of the forces on the volume by equivalent forces. In this case the *Cauchy*'s lemma holds:

> The stress vectors acting upon opposite sides of the same surface at a given point are equal in magnitude and opposite in direction [60] p.32.

Characterization of the interface, surface tension: Consider two fluids with different densities, fluid m and fluid l as presented on Fig. 2.1.

2.2. Local volume-averaged momentum equations 47

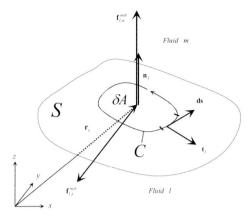

Fig. 2.1. Surface force - geometry definitions.

Fluid m is a gas and fluid l is a liquid, $\rho_l > \rho_m$. A three-dimensional surface S, described by the position vector $\mathbf{r}_S(x,y,z)$, separates both fluids. We call such fluids *imiscible*. The unit normal vector of the liquid interface $\mathbf{n}_l$ is pointing outside of the liquid l and is an important local characteristic of this surface. It is defined from the spatial distribution of the volume fraction of the field l as follows

$$\mathbf{n}_l = -\frac{\nabla(\alpha_l \gamma)}{|\nabla(\alpha_l \gamma)|}. \tag{2.2}$$

Due to different molecular attraction forces at the two sides of the surface a resulting attraction force with special properties arises at the surface. The force exists only at the surface and acts at the denser fluid l. This force is called *surface force*. The surface force per unit mixture volume is denoted with $\mathbf{f}_l^{m\sigma}$. The subscript l indicates that the force acts at the field l and the subscript $m\sigma$ indicates that the surface is an interface with filed m. We extract from the surface S an infinitesimal part δA around the point $\mathbf{r}_S(x,y,z)$ so that

$$\delta \mathbf{A} = \mathbf{n}_l \delta A. \tag{2.3}$$

The closed curve C contains δA. The closed curve C is a oriented counter clock wise. Consider the infinitesimal directed line element $\mathbf{ds}$ called *arc length vector*. The unit tangent vector to the surface S at C that is perpendicular to $\mathbf{ds}$ is $\mathbf{t}$. In this case the relation holds

$$\mathbf{t}ds = \mathbf{ds} \times \mathbf{n}_l. \tag{2.4}$$

The surface force exerted on the surface δA by the surface outside of δA across the directed line element **ds** is equal to $\sigma_{lm} tds$. Here σ_{lm} is a material property being a force tangential to S per unit length called *surface tension*. It may vary with the surface properties like temperature, concentration of the impurities of the liquid etc. The net surface force on the element δA is then obtained by summing all forces $\sigma_{lm} tds$ exerted on each element of arc length ds, $\oint_C \sigma_{lm} tds$. Using Eq. (2.4) gives

$$\oint_C \sigma_{lm} tds = \oint_C \sigma_{lm} \mathbf{ds} \times \mathbf{n}_l . \qquad (2.5)$$

The *Stokes* theorem, see [64] for derivation, allows one to transfer the integral over a closed curve to an integral over the surface closed by this curve

$$\oint_C \mathbf{ds} \times \mathbf{n}_l \sigma_{lm} = \iint_S (\mathbf{n}_l \times \nabla) \times \mathbf{n}_l \sigma_{lm} dA . \qquad (2.6)$$

We see that for the *infinitesimal* surface δA the *surface force per unit interface* is

$$(\mathbf{n}_l \times \nabla) \times \mathbf{n}_l \sigma_{lm} = \sigma_{lm} \left[(\mathbf{n}_l \times \nabla) \times \mathbf{n}_l \right] + (\mathbf{n}_l \times \nabla \sigma_{lm}) \times \mathbf{n}_l . \qquad (2.7)$$

This force can be split into a normal and a tangential component by splitting the gradient into a sum of normal and tangential components

$$\nabla = \nabla_n + \nabla_t , \qquad (2.8)$$

where

$$\nabla_n = (\nabla \cdot \mathbf{n}_l) \mathbf{n}_l \qquad (2.9)$$

and

$$\nabla_t = \nabla - \nabla_n . \qquad (2.10)$$

Using this splitting and after some mathematical manipulation *Brackbill* et al. [7] simplified Eq. (2.7) and obtained finally the very important result,

$$(\mathbf{n}_l \times \nabla) \times \mathbf{n}_l \sigma = -\mathbf{n}_l \sigma_{lm} (\nabla \cdot \mathbf{n}_l) + \nabla_t \sigma_{lm} = \mathbf{n}_l \sigma_{lm} \kappa_l + \nabla_t \sigma_{lm} , \qquad (2.11)$$

where

2.2. Local volume-averaged momentum equations

$$\kappa_l = -(\nabla \cdot \mathbf{n}_l) = \nabla \cdot \left[\frac{\nabla(\alpha_l \gamma)}{|\nabla(\alpha_l \gamma)|} \right] \quad (2.12)$$

is the curvature of the interface defined only by the gradient of the interface unit vector. Equation (2.12) is used in Eq. (5.4), p.23 in [15]. Note that the mathematical definition of curvature is the sum of the two principal curvatures which are magnitudes of two vectors. This sum is always positive. The expression resulting from Eq. (2.12) defines curvature with sign, which means - with its orientation. The curvature is positive if the center of the curvature is in the fluid m. In other words the positive curvature κ_l is oriented along the normal vector $\mathbf{n}_l$.

Let us as an example estimate the curvature of a liquid layer in stratified flow between two horizontal planes with a gap equal to H. The interface is described by the curve $z^* = \frac{z}{H} = \alpha_2(x)$ being in the plane $y = const$. We use as coordinates (x, y, z^*). The gradient of the liquid volume fraction is then

$$\nabla \alpha_2(x) = \frac{\partial \alpha_2(x)}{\partial x}\mathbf{i} + \frac{\partial \alpha_2(x)}{\partial z^*}\mathbf{k} = \frac{\partial \alpha_2(x)}{\partial x}\mathbf{i} + \frac{\partial z^*}{\partial z^*}\mathbf{k} = \frac{\partial \alpha_2(x)}{\partial x}\mathbf{i} + \mathbf{k}.$$

The magnitude of the gradient is then

$$|\nabla \alpha_2(x)| = \left\{ 1 + \left[\frac{\partial \alpha_2(x)}{\partial x}\right]^2 \right\}^{1/2}.$$

The normal vector is

$$\mathbf{n}_2 = -\frac{\nabla \alpha_2(x)}{|\nabla \alpha_2(x)|} = -\left(\frac{\frac{\partial \alpha_2(x)}{\partial x}}{\left\{1 + \left[\frac{\partial \alpha_2(x)}{\partial x}\right]^2\right\}^{1/2}}\mathbf{i} + \frac{1}{\left\{1 + \left[\frac{\partial \alpha_2(x)}{\partial x}\right]^2\right\}^{1/2}}\mathbf{k} \right)$$

The curvature in accordance with Eq. (2.12) is then

$$\kappa_l = -(\nabla \cdot \mathbf{n}_l) = \nabla \cdot \left[\frac{\nabla(\alpha_l \gamma)}{|\nabla(\alpha_l \gamma)|} \right]$$

$$= \nabla \cdot \left(\frac{\dfrac{\partial \alpha_2(x)}{\partial x}}{\left\{1 + \left[\dfrac{\partial \alpha_2(x)}{\partial x}\right]^2\right\}^{1/2}} \mathbf{i} + \frac{1}{\left\{1 + \left[\dfrac{\partial \alpha_2(x)}{\partial x}\right]^2\right\}^{1/2}} \mathbf{k} \right)$$

$$= \frac{\partial}{\partial x} \frac{\dfrac{\partial \alpha_2(x)}{\partial x}}{\left\{1 + \left[\dfrac{\partial \alpha_2(x)}{\partial x}\right]^2\right\}^{1/2}} + \frac{\partial}{\partial z^*} \frac{1}{\left\{1 + \left[\dfrac{\partial \alpha_2(x)}{\partial x}\right]^2\right\}^{1/2}}.$$

Having in mind that

$$\frac{\partial}{\partial z^*} \frac{1}{\left\{1 + \left[\dfrac{\partial \alpha_2(x)}{\partial x}\right]^2\right\}^{1/2}} = -\frac{\dfrac{\partial \alpha_2(x)}{\partial x} \dfrac{\partial^2 \alpha_2(x)}{\partial x \partial z^*}}{\left\{1 + \left[\dfrac{\partial \alpha_2(x)}{\partial x}\right]^2\right\}^{3/2}} = 0$$

one finally obtains the well-known expression

$$\kappa_2 = \frac{\partial^2 \alpha_2(x)}{\partial x^2} \bigg/ \left\{1 + \left[\frac{\partial \alpha_2(x)}{\partial x}\right]^2\right\}^{3/2}.$$

The higher pressure is in the fluid medium on the concave side of the interface, since surface force in case of constant surface tension is a net normal force directed toward the center of curvature of the interface.

Note, that the term $\sigma_{lm}\kappa_l$ becomes important for

$$1/|\kappa_l| < 0.001m.$$ (2.13)

The term $\nabla_t \sigma_{lm}$ can be expressed as

$$\nabla_t \sigma_{lm} = (\nabla_t T_l)\frac{\partial \sigma_{lm}}{\partial T_l} + \sum_{i=1}^{i_{max}} (\nabla_t C_{il})\frac{\partial \sigma_{lm}}{\partial C_{il}}.$$ (2.14)

These terms describes the well-known *Marangoni* effect.

The grid density for computational analysis can be judged by comparing the size of the control volume Δx with the curvature. Obvious, if and only if the volume is small enough,

$$\Delta x |\kappa_l| < 1, \qquad (2.15)$$

the curvature can be resolved by the computational analysis.

No velocity variation across the interface: Consider the case for which there is no velocity variation across the interface (e.g. stagnant fluids) - Fig. 2.2.

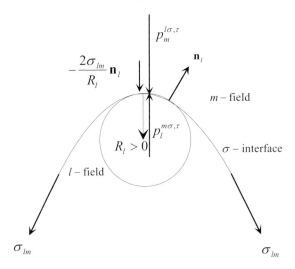

Fig. 2.2. Interface force equilibrium without mass transfer and Marangoni effect (Laplace)

The interface pressure $p_l^{m\sigma,\tau}$ is the normal force per unit surface acting on field l. This force acts *inside* the field l in the immediate vicinity of the interface and in the opposite direction on the interface control volume. It is *different* from the surface force exerted by the pressure of the neighboring velocity field, $p_m^{l\sigma,\tau}$. If there are no other forces except pressure and surface tension we have

$$p_l^{m\sigma,\tau}\mathbf{n}_l + p_m^{l\sigma,\tau}\mathbf{n}_m + \nabla_l \sigma_{lm} + \sigma_{lm}\kappa_l \mathbf{n}_l = 0. \qquad (2.16)$$

This equation is known as the *Laplace* equation. If in addition we have viscous forces acting on the two sides the momentum balance is

$$\left(\mathbf{T}_l^{m\sigma,\tau} - p_l^{m\sigma,\tau}\mathbf{I}\right)\cdot\mathbf{n}_l + \left(\mathbf{T}_m^{l\sigma,\tau} - p_m^{l\sigma,\tau}\mathbf{I}\right)\cdot\mathbf{n}_m - \nabla_l \sigma_{lm} - \sigma_{lm}\kappa_l \mathbf{n}_l = 0. \qquad (2.17)$$

Velocity variation across the interface: If mass is transferred from one to the other fluid for whatever reason, the interface moves in space not only convectively but also controlled by the amount of mass transferred between the fields – Fig. 2.3. In this case the interface velocity $\mathbf{V}_{lm}^\tau$ is not equal to the neighboring field velocities. The mass flow rate

$$(\rho w)_{lm} = \rho_l \left(\mathbf{V}_l^\tau - \mathbf{V}_{lm}^\tau \right) \cdot \mathbf{n}_l$$

enters the interface control volume and exerts the force $\rho_l \mathbf{V}_l^\tau \left(\mathbf{V}_l^\tau - \mathbf{V}_{lm}^\tau \right) \cdot \mathbf{n}_l$ per unit surface on it. Note that this force has the same direction as the pressure force inside the field *l*. Similarly, we have a reactive force $\rho_m \mathbf{V}_m^\tau \left(\mathbf{V}_m^\tau - \mathbf{V}_{lm}^\tau \right) \cdot \mathbf{n}_m$ exerted per unit surface on the control volume by the leaving mass flow rate. Assuming that the control volume moves with the normal component of the interface velocity, $\mathbf{V}_{lm}^\tau \cdot \mathbf{n}_l$, we obtain the following force balance

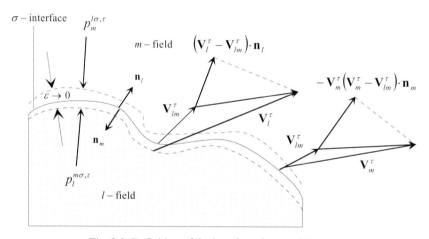

Fig. 2.3. Definition of the interface characteristics

$$-\rho_l \mathbf{V}_l^\tau \left(\mathbf{V}_l^\tau - \mathbf{V}_{lm}^\tau \right) \cdot \mathbf{n}_l + \left(\mathbf{T}_l^{m\sigma,\tau} - p_l^{m\sigma,\tau} \mathbf{I} \right) \cdot \mathbf{n}_l$$

$$-\rho_m \mathbf{V}_m^\tau \left(\mathbf{V}_m^\tau - \mathbf{V}_{m\sigma}^\tau \right) \cdot \mathbf{n}_m + \left(-p_{m\sigma}^\tau \mathbf{I} + \mathbf{T}_{m\sigma}^\tau \right) \cdot \mathbf{n}_m - \nabla_l \sigma_{lm} - \sigma_{lm} \kappa_l \mathbf{n}_l = 0 \, . \quad (2.18)$$

This is the general form of the interfacial momentum jump condition. It is convenient to rewrite the above equation by using the mass jump condition at the interface

2.2. Local volume-averaged momentum equations

$$\rho_l(\mathbf{V}_l^\tau - \mathbf{V}_{lm}^\tau)\cdot\mathbf{n}_l + \rho_m(\mathbf{V}_m^\tau - \mathbf{V}_{lm}^\tau)\cdot\mathbf{n}_m = 0, \qquad (2.19)$$

which is Eq. (1.42). The result is

$$\left[\rho_l(\mathbf{V}_l^\tau - \mathbf{V}_{lm}^\tau)(\mathbf{V}_m^\tau - \mathbf{V}_l^\tau) + \left(p_m^{l\sigma,\tau} - p_l^{m\sigma,\tau}\right)\mathbf{I} + \mathbf{T}_l^{m\sigma,\tau} - \mathbf{T}_m^{l\sigma,\tau} - \sigma_{lm}\kappa_l\right]\cdot\mathbf{n}_l - \nabla_l\sigma_{lm} = 0 \qquad (2.20)$$

or

$$(\rho w)_{lm}\left(\mathbf{V}_m^\tau - \mathbf{V}_l^\tau\right) + \left(p_m^{l\sigma,\tau} - p_l^{m\sigma,\tau}\right)\mathbf{n}_l + \left[\mathbf{T}_l^{m\sigma,\tau} - \mathbf{T}_m^{l\sigma,\tau} - \sigma_{lm}\kappa_l\right]\cdot\mathbf{n}_l - \nabla_l\sigma_{lm} = 0. \qquad (2.21)$$

The projection of this force to the normal direction is obtained by scalar multiplication of the above equation with the unit vector $\mathbf{n}_l$. The result is

$$(\rho w)_{lm}\left(\mathbf{V}_m^\tau - \mathbf{V}_l^\tau\right)\cdot\mathbf{n}_l + p_m^{l\sigma,\tau} - p_l^{m\sigma,\tau} - \sigma_{lm}\kappa_l + \left[\left(\mathbf{T}_l^{m\sigma,\tau} - \mathbf{T}_m^{l\sigma,\tau}\right)\cdot\mathbf{n}_l\right]\cdot\mathbf{n}_l$$

$$-\left(\nabla_l\sigma_{lm}\right)\cdot\mathbf{n}_l = 0. \qquad (2.22)$$

Using the mass conservation at the interface we finally have an important force balance normal to the interface

$$(\rho w)_{lm}^2\left(\frac{1}{\rho_m} - \frac{1}{\rho_l}\right) + p_m^{l\sigma,\tau} - p_l^{m\sigma,\tau} - \sigma_{lm}\kappa_l + \left[\left(\mathbf{T}_l^{m\sigma,\tau} - \mathbf{T}_m^{l\sigma,\tau}\right)\cdot\mathbf{n}_l\right]\cdot\mathbf{n}_l$$

$$-\left(\nabla_l\sigma_{lm}\right)\cdot\mathbf{n}_l = 0. \qquad (2.23)$$

> Neglecting all the forces except those caused by pressure and interfacial mass transfer results in the surprising conclusion that during the mass transfer the pressure in the denser fluid is always larger than the pressure in the lighter fluid independently on the direction of the mass transfer - Delhaye [13], p. 52, Eq. (2.64).

For the limiting case of no interfacial mass transfer and dominance of the pressure difference the velocity of the interface can be expressed as a function of the pressure difference and the velocities in the bulk of the fields.

$$V_{lm}^{n,\tau} = V_l^{n,\tau} - \frac{p_m^{l\sigma,\tau} - p_l^{m\sigma,\tau}}{\rho_l\left(V_m^{n,\tau} - V_l^{n,\tau}\right)}. \qquad (2.24)$$

This velocity is called contact discontinuity velocity. Replacing the discontinuity velocity with Eq. (1.42) we obtain

$$\left(V_l^{n,\tau} - V_m^{n,\tau}\right)^2 = \frac{\rho_l - \rho_m}{\rho_l \rho_m}\left(p_l^{m\sigma,\tau} - p_m^{l\sigma,\tau}\right). \quad (2.25)$$

For the case $\rho_l \gg \rho_m$ we have the expected result that the pressure difference equals the stagnation pressure at the side of the lighter medium

$$p_l^{m\sigma,\tau} - p_m^{l\sigma,\tau} = \rho_m \left(V_l^{n,\tau} - V_m^{n,\tau}\right)^2. \quad (2.26)$$

2.2.3 Local volume averaging of the single-phase momentum equation

The aim here is to average Eq. (2.1) over the total control volume - see Fig. 2.4.

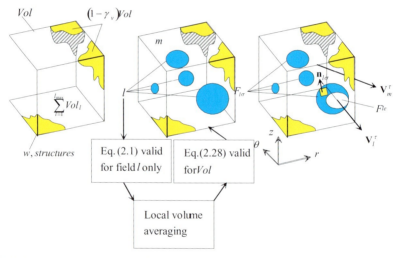

Fig. 2.4. Definition regions for single-phase instantaneous momentum balance and the local volume average balance

The mathematical tools used to derive local volume-averaged field conservation equations for the property being any scalar, vectorial, or tensorial function of time and location [1,58,69,71,18,70] are once again *Slattery - Whitaker*'s spatial averaging theorem, together with the *Gauss - Ostrogradskii* theorem, and the general transport equation (*Leibnitz* rule). Applying the local volume average to Eq. (2.1), the following is obtained

2.2. Local volume-averaged momentum equations

$$\left\langle \frac{\partial}{\partial \tau}(\rho_l \mathbf{V}_l^{\tau}) \right\rangle + \left\langle \nabla \cdot (\rho_l \mathbf{V}_l^{\tau} \mathbf{V}_l^{\tau}) \right\rangle - \left\langle \nabla \cdot (\mathbf{T}_l^{\tau}) \right\rangle + \left\langle \nabla p_l^{\tau} \right\rangle + \left\langle \rho_l \mathbf{g} \right\rangle = 0, \quad (2.27)$$

or using Eqs. (1.28), (1.32) and (1.28) or Ref. [37]

$$\frac{\partial}{\partial \tau}\left\langle \rho_l \mathbf{V}_l^{\tau} \right\rangle + \nabla \cdot \left\langle \rho_l \mathbf{V}_l^{\tau} \mathbf{V}_l^{\tau} \right\rangle + \nabla \cdot \left\langle p_l^{\tau} \mathbf{I} - \mathbf{T}_l^{\tau} \right\rangle + \frac{1}{Vol} \int_{F_{l\sigma}+F_{lw}} \rho_l \mathbf{V}_l^{\tau}(\mathbf{V}_l^{\tau} - \mathbf{V}_l^{\sigma,\tau}) \cdot \mathbf{n}_l dF$$

$$- \frac{1}{Vol} \int_{F_{l\sigma}+F_{lw}} \left(-p_l^{\tau} \mathbf{I} + \mathbf{T}_l^{\tau} \right) \cdot \mathbf{n}_l dF + \mathbf{g}\rho_l = 0 \quad (2.28)$$

The volume average momentum equation can be rewritten using the *weighted average*, see Eq. (1.19) or Ref. [37]

$$\left\langle \mathbf{V}_l^{\tau} \right\rangle^{le,\rho} = \left\langle \rho_l \mathbf{V}_l^{\tau} \right\rangle^{le} / \left\langle \rho_l \right\rangle^{l}, \quad (2.29)$$

and

$$\left\langle \mathbf{V}_l^{\tau} \right\rangle^{le} \left\langle \mathbf{V}_l^{\tau} \right\rangle^{le} = \left\langle \rho_l \mathbf{V}_l^{\tau} \mathbf{V}_l^{\tau} \right\rangle^{le} / \left\langle \rho_l \right\rangle^{l}. \quad (2.30)$$

The result is

$$\frac{\partial}{\partial \tau}\left(\alpha_l \gamma_v \left\langle \rho_l \right\rangle^{l} \left\langle \mathbf{V}_l^{\tau} \right\rangle^{le} \right) + \nabla \cdot \left(\alpha_l^e \gamma \left\langle \rho_l \right\rangle^{l} \left\langle \mathbf{V}_l^{\tau} \right\rangle^{le} \left\langle \mathbf{V}_l^{\tau} \right\rangle^{le} \right) - \nabla \cdot \left(\alpha_l^e \gamma \left\langle \mathbf{T}_l^{\tau} \right\rangle^{le} \right)$$

$$+ \nabla \cdot \left(\alpha_l^e \gamma \left\langle p_l^{\tau} \right\rangle^{le} \right) + \alpha_l \gamma_v \left\langle \rho_l \right\rangle^{l} \mathbf{g} - \frac{1}{Vol}\int_{F_{l\sigma}}\left[-p_l^{m\sigma,\tau} \mathbf{I} + \mathbf{T}_l^{m\sigma,\tau} - \rho_l \mathbf{V}_l^{\tau}(\mathbf{V}_l^{\tau} - \mathbf{V}_{lm}^{\tau}) \right] \cdot \mathbf{n}_l dF$$

$$- \frac{1}{Vol}\int_{F_{lw}}\left[-p_l^{w\sigma,\tau} \mathbf{I} + \mathbf{T}_l^{w\sigma,\tau} - \rho_l \mathbf{V}_l^{\tau}(\mathbf{V}_l^{\tau} - \mathbf{V}_{lw}^{\tau}) \right] \cdot \mathbf{n}_l dF = 0. \quad (2.31)$$

We assume that the weighted average of products can be replaced by the products of the average. This should be borne in mind when constructing a numerical algorithm for solving the final system and selecting the size of the finite volume so as to be not so large as to violate the validity of this assumption.

Note the differences between Eq. (2.31) and the final result obtained in Ref. [23], Eq. (3.16):

(a) the directional permeability is used here instead of the volumetric porosity in the pressure gradient term, and

(b) as for the mass conservation equation in Ref. [37], the volumetric porosity is kept below the time differential since it can be a function of time in a number of interesting applications.

For the case $\gamma_v = \gamma = 1$ and one-dimensional flow, Eq. (2.31) reduces to Eq. (15) derived by *Delhaye* in Ref. [19], p. 163.

Equation (2.31), the rigorously derived local volume average momentum equation, is not amenable to direct use in computational models without further transformation. In order to facilitate its practical use

(i) the integral expression has to be evaluated, and

(ii) the time averaging must be performed subsequently.

These steps, which go beyond Ref. [57], are performed in Sections 2.3 and 2.4.

2.3 Rearrangement of the surface integrals

The expression under the surface integral is replaced by its equivalent from the momentum jump condition, Eq. (2.18):

$$-\frac{1}{Vol}\int_{F_{l\sigma}}\left[-\rho_l \mathbf{V}_l^\tau\left(\mathbf{V}_l^\tau - \mathbf{V}_{lm}^\tau\right) + \mathbf{T}_l^{m\sigma,\tau} - p_l^{m\sigma,\tau}\mathbf{I}\right]\cdot\mathbf{n}_l dF$$

$$= -\frac{1}{Vol}\int_{F_{l\sigma}}\left\{\left[-\rho_m \mathbf{V}_m^\tau\left(\mathbf{V}_m^\tau - \mathbf{V}_{m\sigma}^\tau\right) + \mathbf{T}_m^{l\sigma,\tau} - p_m^{l\sigma,\tau}\mathbf{I} + \sigma_{lm}\kappa_l\right]\cdot\mathbf{n}_l + \nabla_l\sigma_{lm}\right\} dF \text{ . (2.32)}$$

Note that there are no surface force terms in Eq. (2.31). Equation (2.32) reflects the action of the surface forces and the action of the stresses caused by the surrounding field m on l. Note also that if the momentum equations are written for two neighboring fields the interface forces will appear in both equation with opposite sign only if the equations are applied to a common dividing surface. Otherwise, the exchange terms in question will be non-symmetric.

The *intrinsic surface-averaged* field pressure at the entrances and exits of the control volume crossing the field m is $\langle p_m^\tau \rangle^{me}$. We call it bulk pressure inside the velocity field m at this particular surface. The interfacial m-side pressure $p_m^{l\sigma,\tau}$ can be expressed as the sum of the intrinsic averaged pressure, $\langle p_m^\tau \rangle^{me}$, which is not a function of the position at the interface inside the control volume and can be taken outside of the integral sign, and a pressure difference $\Delta p_m^{l\sigma,\tau}$, which is a function of the position at the interface in the control volume

$$p_m^{l\sigma,\tau} = \langle p_m^\tau \rangle^{me} + \Delta p_m^{l\sigma,\tau}. \qquad (2.33)$$

The same is performed for all other fields. Similarly, the surface pressure of the field structure interface is expressed as

$$p_l^{w\sigma,\tau} = \langle p_l^\tau \rangle^{le} + \Delta p_l^{w\sigma,\tau}. \qquad (2.34)$$

The interfacial pressure differs from the bulk pressures of the corresponding fields $m\sigma$, and lw, $\Delta p_m^{l\sigma,\tau} \neq 0$ and $\Delta p_l^{w\sigma,\tau} \neq 0$. This occurs because obstacles to the continuous phase can cause local velocity decreases or increases at the interface velocity boundary layer, resulting in increased or decreased pressure relative to bulk pressure. In order to estimate the surface integrals the exact dependence of the pressure as a function of the position at the interface inside the control volume has to be elaborated for each idealized flow pattern and form of the structure. The same is valid for the viscous shear stresses $\mathbf{T}_m^{l\sigma,\tau}$ and $\mathbf{T}_l^{w\sigma,\tau}$. Note, that in order to estimate the interfacial pressure integrals for practical use of the momentum equations the time averaging must be performed first to make it admissible to use real pressure distributions measured experimentally on bodies in turbulent flow.

Substituting Eqs. (2.29) to (2.31) into Eq. (2.28) we obtain

$$\frac{\partial}{\partial \tau}\left(\alpha_l \gamma_v \langle \rho_l \rangle^l \langle \mathbf{V}_l^\tau \rangle^{le}\right) + \nabla \cdot \left(\alpha_l \gamma \langle \rho_l \rangle^l \langle \mathbf{V}_l^\tau \rangle^{le} \langle \mathbf{V}_l^\tau \rangle^{le}\right) - \nabla \cdot \left(\alpha_l \gamma \langle \mathbf{T}_l^\tau \rangle^{le}\right)$$

$$+ \nabla \left(\alpha_l^e \gamma \langle p_l^\tau \rangle^{le}\right) + \alpha_l \gamma_v \langle \rho_l \rangle^l \mathbf{g} + \frac{1}{Vol}\int_{F_{l\sigma}}\langle p_m^\tau \rangle^{me} \mathbf{n}_l dF + \frac{1}{Vol}\int_{F_{lw}}\langle p_l^\tau \rangle^{le} \mathbf{n}_l dF$$

$$- \frac{1}{Vol}\int_{F_{l\sigma}}\left\{\left[-\Delta p_m^{l\sigma,\tau}\mathbf{I} + \mathbf{T}_m^{l\sigma,\tau} - \rho_m \mathbf{V}_m^\tau\left(\mathbf{V}_m^\tau - \mathbf{V}_{lm}^\tau\right) + \sigma_{lm}\kappa_l\right]\cdot\mathbf{n}_l + \nabla_t \sigma_{lm}\right\}dF$$

$$- \frac{1}{Vol}\int_{F_{lw}}\left[-\Delta p_l^{w\sigma,\tau}\mathbf{I} + \mathbf{T}_l^{w\sigma,\tau} - \rho_l \mathbf{V}_l^\tau\left(\mathbf{V}_l^\tau - \mathbf{V}_{lw}^\tau\right)\right]\cdot\mathbf{n}_l dF = 0. \qquad (2.35)$$

Keeping in mind that $\langle p_m^\tau \rangle^{me}$ and $\langle p_l^\tau \rangle^{le}$ are not functions of the interface position inside the control volume, and making use of Eq. (1.32) (see also [37]), the bulk pressure integrals can then be rewritten as follows:

$$\langle p_m^\tau \rangle^{me} \frac{1}{Vol} \int_{F_{l\sigma}} \mathbf{n}_l dF + \langle p_l^\tau \rangle^{le} \frac{1}{Vol} \int_{F_{lw}} \mathbf{n}_{lw} dF$$

$$= -\langle p_m^\tau \rangle^{me} \nabla(\alpha_l^e \gamma) + \left(\langle p_l^\tau \rangle^{le} - \langle p_m^\tau \rangle^{me}\right) \frac{1}{Vol} \int_{F_{lw}} \mathbf{n}_l dF . \tag{2.36}$$

Rearranging the surface tension integrals: The surface force per unit volume of the mixture is in fact the local volume average of the surface tension force

$$\mathbf{f}_l^{m\sigma} = -\frac{1}{Vol} \iint_{F_{l\sigma}} (\mathbf{n}_l \sigma_{lm} \kappa + \nabla_t \sigma_{lm}) dF . \tag{2.37}$$

Note that the orientation of this force is defined with respect the coordinate system given in Fig. 2.1. Using Eq. (1.29) we have

$$\mathbf{f}_l^{m\sigma} = -\sigma_{lm} \kappa_l \frac{1}{Vol} \iint_{F_{l\sigma}} \mathbf{n}_l \sigma_{lm} \kappa dF - \frac{1}{Vol} \iint_{F_{l\sigma}} (\nabla_t \sigma_{lm}) dF$$

$$= \sigma_{lm} \kappa_l \left[\nabla(\alpha_l^e \gamma) + \frac{1}{Vol} \int_{F_{lw}} \mathbf{n}_l dF \right] - \frac{1}{Vol} \iint_{F_{l\sigma}} (\nabla_t \sigma_{lm}) dF , \tag{2.38}$$

with normal

$$\mathbf{f}_{l,n}^{m\sigma} = \sigma_{lm} \kappa_l \left[\nabla(\alpha_l^e \gamma) + \frac{1}{Vol} \int_{F_{lw}} \mathbf{n}_l dF \right] = \sigma_{lm} \nabla \cdot \left[\frac{\nabla(\alpha_l \gamma)}{|\nabla(\alpha_l \gamma)|} \right] \left[\nabla(\alpha_l^e \gamma) + \frac{1}{Vol} \int_{F_{lw}} \mathbf{n}_l dF \right]$$

for $\alpha_l > 0$ (2.39)

and tangential

$$\mathbf{f}_{l,t}^{m\sigma} = -\frac{1}{Vol} \iint_{F_{l\sigma}} (\nabla_t \sigma_{lm}) dF \approx -a_{lm} \nabla_t \sigma_{lm} \tag{2.40}$$

surface force components per unit volume of the mixture, respectively. Here a_{lm} is the interfacial area density. The local volume averaged surface force is called some times in the literature *continuum surface force* or abbreviated CSF, see *Brackbill* et al. [7].

Note that if the surface tension is a constant in space there is no resulting tangential force component. At plane surfaces the curvature is zero and therefore

2.3 Rearrangement of the surface integrals

there is no normal force acting at the liquid. If the liquid consists of clouds of spheres the local surface force creates only a difference in pressures inside and outside the sphere, but there is not a net force influencing the total movement either of a single droplet or of the cloud of the droplets in the space due to this force. We express this fact by multiplying the surface force by the function δ_1 being one for continuum and zero for disperse field.

Using Eqs. (2.33) and (2.35) the pressure and the surface tension terms can be rearranged as follows

$$\nabla \cdot \left(\alpha_i^e \gamma \langle p_i^\tau \rangle^{le} \right) + \frac{1}{Vol} \int_{F_{l\sigma}} \langle p_m^\tau \rangle^{me} \mathbf{n}_l dF + \frac{1}{Vol} \int_{F_{lw}} \langle p_i^\tau \rangle^{le} \mathbf{n}_l dF$$

$$- \frac{1}{Vol} \iint_{F_{l\sigma}} (\mathbf{n}_l \sigma_{lm} \kappa + \nabla_t \sigma_{lm}) dF$$

$$= \nabla \cdot \left(\alpha_i^e \gamma \langle p_i^\tau \rangle^{le} \right) - \langle p_m^\tau \rangle^{me} \nabla (\alpha_i^e \gamma) + \left(\langle p_i^\tau \rangle^{le} - \langle p_m^\tau \rangle^{me} \right) \frac{1}{Vol} \int_{F_{lw}} \mathbf{n}_l dF$$

$$+ \delta_l \sigma_{lm} \kappa_l \left[\nabla (\alpha_i^e \gamma) + \frac{1}{Vol} \int_{F_{lw}} \mathbf{n}_l dF \right] - \delta_l \frac{1}{Vol} \iint_{F_{l\sigma}} (\nabla_t \sigma_{lm}) dF$$

$$= \alpha_i^e \gamma \nabla \cdot \langle p_i^\tau \rangle^{le} + \left(\langle p_i^\tau \rangle^{le} - \langle p_m^\tau \rangle^{me} + \delta_l \sigma_{lm} \kappa_l \right) \left(\nabla (\alpha_i^e \gamma) + \frac{1}{Vol} \int_{F_{lw}} \mathbf{n}_l dF \right)$$

$$- \delta_l \frac{1}{Vol} \iint_{F_{l\sigma}} (\nabla_t \sigma_{lm}) dF . \qquad (2.41)$$

Note that

$$\alpha_i^e \gamma \nabla \cdot \langle p_i^\tau \rangle^{le}$$

stands in Eq. (2.38) for

$$\nabla \cdot \left(\alpha_i^e \gamma \langle p_i^\tau \rangle^{le} \right) - \langle p_i^\tau \rangle^{le} \nabla \cdot (\alpha_i^e \gamma).$$

Rearranging the integrals defining interfacial momentum transfer due to mass transfer: Again using the mass jump condition at the interface, which is Eq. (1.42)

$$\rho_m V_m^\tau \left(V_m^\tau - V_{lm}^\tau \right) = \rho_l V_m^\tau \left(V_l^\tau - V_{lm}^\tau \right), \qquad (2.42)$$

the surface integral is rearranged as follows:

$$-\frac{1}{Vol} \int_{F_{l\sigma}} \left[-\rho_m V_m^\tau \left(V_m^\tau - V_{lm}^\tau \right) \right] \cdot \mathbf{n}_l dF = \frac{1}{Vol} \int_{F_{l\sigma}} \left[\rho_l V_m^\tau \left(V_l^\tau - V_{lm}^\tau \right) \right] \cdot \mathbf{n}_l dF . \quad (2.43)$$

For practical applications the mass source term is split into non-negative components as already explained in Chapter 1 (see also [37]). In addition, it is assumed that the mass emitted from a field has the velocity of the donor field. As a result, the local volume-averaged interfacial forces related to mass transfer through the interfaces can then be rewritten as follows:

a) The components related to mass injection into, or suction from, the field through the field-structure interface are replaced by the "donor" hypothesis

$$-\frac{1}{Vol} \int_{F_{lw}} \left[\rho_l V_m^\tau \left(V_l^\tau - V_{lw}^\tau \right) \right] \cdot \mathbf{n}_l dF = \gamma_v \left(\mu_{wl}^\tau \left\langle V_w^\tau \right\rangle^{we} - \mu_{lw}^\tau \left\langle V_l^\tau \right\rangle^{le} \right). \quad (2.44)$$

b) The components due to evaporation, condensation, entrainment and deposition are replaced by the "donor" hypothesis

$$-\frac{1}{Vol} \int_{F_{l\sigma}} \left[\rho_l V_l^\tau \left(V_l^\tau - V_{l\sigma}^\tau \right) \right] \cdot \mathbf{n}_l dF = \gamma_v \sum_{m=1}^{3} \left(\mu_{ml}^\tau \left\langle V_m^\tau \right\rangle^{me} - \mu_{lm}^\tau \left\langle V_l^\tau \right\rangle^{le} \right). \quad (2.45)$$

The sum of all interface mass transfer components is then

$$\gamma_v \sum_{m=1}^{3,w} \left(\mu_{ml}^\tau \left\langle V_m^\tau \right\rangle^{me} - \mu_{lm}^\tau \left\langle V_l^\tau \right\rangle^{le} \right). \qquad (2.46)$$

Thus, the final form obtained for the local volume average momentum equation is as follows,

$$\frac{\partial}{\partial \tau} \left(\alpha_l \gamma_v \left\langle \rho_l \right\rangle^l \left\langle V_l^\tau \right\rangle^{le} \right) + \nabla \cdot \left(\alpha_l^e \gamma \left\langle \rho_l \right\rangle^l \left\langle V_l^\tau \right\rangle^{le} \left\langle V_l^\tau \right\rangle^{le} \right) - \nabla \cdot \left(\alpha_l^e \gamma \left\langle \mathbf{T}_l^\tau \right\rangle^{le} \right) + \alpha_l \gamma_v \left\langle \rho_l \right\rangle^l \mathbf{g}$$

$$+ \alpha_l^e \gamma \nabla \left\langle p_l^\tau \right\rangle^{le} + \left(\left\langle p_l^\tau \right\rangle^{le} - \left\langle p_m^\tau \right\rangle^{me} + \delta_l \sigma_{lm} \kappa_l \right) \left(\nabla \left(\alpha_l^e \gamma \right) + \frac{1}{Vol} \int_{F_{lw}} \mathbf{n}_l dF \right)$$

$$-\delta_l \frac{1}{Vol} \int_{F_{l\sigma}} (\nabla_t \sigma_{lm}) dF + \frac{1}{Vol} \int_{F_{l\sigma}} \left(\Delta p_m^{l\sigma,\tau} \mathbf{I} - \mathbf{T}_m^{l\sigma,\tau} \right) \cdot \mathbf{n}_l dF$$

$$+ \frac{1}{Vol} \int_{F_{lw}} \left(\Delta p_l^{w\sigma,\tau} \mathbf{I} - \mathbf{T}_l^{w\sigma,\tau} \right) \cdot \mathbf{n}_l dF = \gamma_v \sum_{m=1}^{3,w} \left(\mu_{ml}^\tau \left\langle \mathbf{V}_m^\tau \right\rangle^{me} - \mu_{lm}^\tau \left\langle \mathbf{V}_l^\tau \right\rangle^{le} \right), \quad (2.47)$$

and is independent of whether the field is structured or non-structured. Before we continue with the estimation of the remaining integrals, we will perform first a time averaging of Eq. (2.47).

2.4 Local volume average and time average

The instantaneous surface-averaged velocity of the field l, $\left\langle \mathbf{V}_l^\tau \right\rangle^{le}$, can be expressed as the sum of the surface-averaged velocity which is subsequently time-averaged,

$$V_l = \overline{\left\langle \mathbf{V}_l^\tau \right\rangle^{le}} \quad (2.48)$$

and a pulsation component V_l',

$$\left\langle \mathbf{V}_l^\tau \right\rangle^{le} = V_l + V_l', \quad (2.49)$$

as proposed by *Reynolds*. The fluctuation of the velocity is the predominant fluctuation component relative to, say, the fluctuation of α_l or ρ_l. Introduction of Eq. (2.49) in the momentum conservation equation and the time averaging yields

$$\frac{\partial}{\partial \tau}\left(\alpha_l \rho_l \mathbf{V}_l \gamma_v\right) + \nabla \cdot \left(\alpha_l^e \rho_l \mathbf{V}_l \mathbf{V}_l \gamma\right) + \nabla \cdot \left[\alpha_l^e \gamma \left(\rho_l \overline{\mathbf{V}_l' \mathbf{V}_l'} - \delta_l \mathbf{T}_l\right)\right] + \alpha_l^e \gamma \nabla p_l + \alpha_l \gamma_v \rho_l \mathbf{g}$$

$$+ \left(p_l - p_m + \delta_l \sigma_{lm} \kappa_l\right)\left(\nabla(\alpha_l^e \gamma) + \frac{1}{Vol}\int_{F_{lw}} \mathbf{n}_l dF\right) - \delta_l \frac{1}{Vol}\int_{F_{l\sigma}} (\nabla_t \sigma_{lm}) dF$$

$$+ \frac{1}{Vol}\int_{F_{l\sigma}} \left(\Delta p_m^{l\sigma} \mathbf{I} - \mathbf{T}_m^{l\sigma}\right) \cdot \mathbf{n}_l dF + \frac{1}{Vol}\int_{F_{lw}} \left(\Delta p_l^{w\sigma} \mathbf{I} - \mathbf{T}_l^{w\sigma}\right) \cdot \mathbf{n}_l dF$$

$$= \gamma_v \sum_{\substack{k=1 \\ k \neq l}}^{3,w} \left(\mu_{kl} \mathbf{V}_k - \mu_{lk} \mathbf{V}_l \right) \quad (2.50\mathrm{a})$$

see also Appendix 2.1. It is evident from Eq. (2.50a) that the products of the pulsation velocity components, called *Reynolds* stresses, act on the flow introducing additional macroscopic cohesion inside the velocity field. Equation (2.50a) is applied on the field l including the surface up to the m-side interface. For dispersed flows it is convenient to have also the momentum equation of the continuum in a primitive form without using the momentum jump condition. In this case the time average of Eq. (2.31) for field m after introducing Eq. (2.33) is

$$\frac{\partial}{\partial \tau}\left(\alpha_m \rho_m \mathbf{V}_m \gamma_v \right) + \nabla \cdot \left(\alpha_m^e \rho_m \mathbf{V}_m \mathbf{V}_m \gamma \right) + \nabla \cdot \left[\alpha_m^e \gamma \left(\rho_m \overline{\mathbf{V}_m' \mathbf{V}_m'} - \delta_m \mathbf{T}_m \right) \right] + \alpha_m \gamma_v \rho_m \mathbf{g}$$

$$+ \alpha_m^e \nabla p_m - \frac{1}{Vol} \int_{F_{m\sigma}} \left(\Delta p_m^{l\sigma} \mathbf{I} - \mathbf{T}_m^{l\sigma} \right) \cdot \mathbf{n}_l dF + \frac{1}{Vol} \int_{F_{mw}} \left(\Delta p_m^{w\sigma} \mathbf{I} - \mathbf{T}_m^{w\sigma} \right) \cdot \mathbf{n}_m dF$$

$$= \gamma_v \sum_{\substack{k=1 \\ k \neq m}}^{3,w} \left(\mu_{km} \mathbf{V}_k - \mu_{mk} \mathbf{V}_m \right). \quad (2.50\mathrm{b})$$

Comparing Eq. (2.50a) with Eq. (2.50b) we realize that the term

$$-\frac{1}{Vol} \int_{F_{m\sigma}} \left(\Delta p_m^{l\sigma} \mathbf{I} - \mathbf{T}_m^{l\sigma} \right) \cdot \mathbf{n}_l dF$$

appears in the both equations with opposite sign. For practical computation we recommend the use of a couple of equations having common interface in order to easily control the momentum conservation at the selected common interface.

2.5 Viscous and *Reynolds* stresses

The solid-body rotation and translation of the fluid element does not cause any deformation and therefore no internal viscous stresses in the fluid. Only the deformation of the fluid element causes viscous stress resisting this deformation. For estimation of the relationship between deformation and viscous stress, the heuristic approach proposed by *Helmholtz* and *Stokes* can be used for the continuous, intrinsic isotropic, non-structured field, see *Schlichting* [56] p. 58. The background conditions behind this approach will now be recalled: (a) the field is a continuum, (b) small velocity changes are considered, (c) only the linear part of the *Taylor* series is taken into account, and (d) linear dependence between stresses and velocity deformations (*Newtonian* continuum). The mathematical notation for this hypothesis is

$$\mathbf{T} = \eta \left\{ \left[\nabla \mathbf{V} + (\nabla \mathbf{V})^T \right] - \frac{2}{3} (\nabla \cdot \mathbf{V}) \mathbf{I} \right\}, \tag{2.51}$$

where **T** is the second order tensor for the momentum flux, $\nabla \mathbf{V}$ is the dyadic product of the *Nabla* operator and the velocity vector (a second order tensor), ^{T} designates the transposed tensor, and $\nabla \cdot \mathbf{V}$ is the divergence of the velocity vector. Equation (2.51) contains the *Stokes* result for the relation of the bulk viscosity equal to -2/3 dynamic viscosity, see *Schliching*, [56] p.60. From the mechanical equilibrium condition for all angular momenta around an axis for vanishing dimensions of the control volume, one obtains the symmetry of the components of the viscous stress tensor, see *Schlichting* [56], p. 50 for *Cartesian* coordinates.

The search for a quantitative estimation of the Reynolds stresses for multi-phase flows is in its initial stage. A possible step in the right direction, in analogy to single-phase turbulence, is the use of the *Boussinesq* hypothesis (1877) for the turbulent eddy viscosity inside the velocity field, namely

$$-\rho_l \overline{\mathbf{V}_l' \mathbf{V}_l'} = \eta^t \left\{ \left[\nabla \mathbf{V} + (\nabla \mathbf{V})^T \right] - \frac{2}{3} (\nabla \cdot \mathbf{V} + k) \mathbf{I} \right\}. \tag{2.52}$$

where

$$k = \frac{1}{2} \left(\overline{u'u'} + \overline{v'v'} + \overline{w'w'} \right) \tag{2.53}$$

is the kinetic energy of the turbulent fluctuations of the velocity component, see *Schlichting* [56]. The *Boussinesq* hypothesis does not solve the closure problem of the turbulence description of the velocity field. It simply transfers it to the search for a formalism for estimating the so-called turbulent dynamic viscosity coefficient η^t.

It is important to emphasize that, in spite of the fact that several processes in single-phase fluid dynamics can be successfully described by the *Helmholtz-Stokes* and by the *Boussinesq* hypotheses; these hypotheses have never been derived from experiments or proven by abstract arguments. This limitation of the hypotheses should be borne in mind when they are applied.

While the heuristic approach proposed by *Helmholtz* and *Stokes*, Eq. (2.51), is valid only for the continuous part of each velocity field, the *Bousinesq* hypothesis is useful for continuous and disperse velocity fields. This behavior is again described here by introducing for each velocity field the multiplier δ_l in Eq. (2.50), where $\delta_l = 0$ for dispersed field and $\delta_l = 1$ for continuous non- structured fields.

For a single field, $\delta_l = 1$, the description of the viscous and Reynolds stresses reduces to the widely accepted expression.

One can then substitutes for the viscous and turbulent shear stresses and define the turbulent pressure as

$$p'_l = \eta'_l \frac{2}{3} k_l = \rho_l v'_l \frac{1}{3}(u'u' + v'v' + w'w'), \qquad (2.54)$$

to obtain

$$\nabla \cdot \left[\alpha_l^e \gamma \left(\rho_l \overline{\mathbf{V}'_l \mathbf{V}'}_l - \delta_l \mathbf{T}_l \right) \right]$$

$$= -\nabla \left(\alpha_l^e \gamma p'_l \right) - \nabla \cdot \left\{ \alpha_l^e \rho_l \gamma v_l^* \left[\nabla \mathbf{V}_l + (\nabla \mathbf{V}_l)^T - \frac{2}{3} (\nabla \cdot \mathbf{V}_l) \mathbf{I} \right] \right\}, \qquad (2.55)$$

where

$$v_l^* = v'_l + \delta_l v_l. \qquad (2.56)$$

The term $-\nabla \left(\alpha_l^e \gamma \, p'_l \right)$ is usually absorbed from the pressure term $\alpha_l^e \gamma \, \nabla p_l$. The term

$$-p'_l \nabla \left(\alpha_l^e \gamma \right) = \frac{1}{Vol} \int_{F_{l\sigma}} p'_l \mathbf{n}_l dF + \frac{1}{Vol} \int_{F_{lw}} p'_l \mathbf{n}_l dF \qquad (2.57)$$

is usually included in the pressure differences between the bulk pressure and boundary layer pressure. Hence, from here on, p'_l will no longer appear in the notation.

It is interesting to note that from the kinetic theory for two colliding particles plus their added mass the following is obtained

$$\nabla \left(\alpha_l^e \gamma \, p'_l \right) \approx \nabla \left[\alpha_l^e \gamma \left(\rho_l + c_l^{vm} \rho_m \right) \frac{1}{3} \mathbf{V}_l'^2 \right], \qquad (2.58)$$

which seems to be a more realistic estimate because the oscillations of the particles are more accurately taken into account.

2.6 Non-equal bulk and boundary layer pressures

2.6.1 Continuous interface

2.6.1.1 3D flows

Examples for the existence of continuous interfaces are the stratified pool, see Fig. 2.5, and annular pipe flow. The treatment of the interface depends very much on the numerical method used. If the numerical method is able to resolve the interface itself and the two attached boundary layers (see for instance [20] for more information), the interface momentum jump condition is the only information which is needed to close the mathematical description of the interface. If this is not the case, special treatment of the processes at the interface is necessary. In this Section we discuss some possibilities.

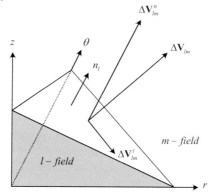

Fig. 2.5. Continuous interface

Consider a liquid (l) – gas (m) flow without mass transfer. The computational cells are so large that the surface is at best represented by piecewise planes at which the surface tension is neglected. The compressibility of the gas is much larger then the compressibility of the liquid. In this case we can assume that there is almost no difference between the bulk and liquid side interface pressure

$$\Delta p_l^{m\sigma} = 0 . \tag{2.59}$$

The pressure change across the gas side boundary layer is then approximated by the stagnation pressure

$$\Delta p_m^{l\sigma} = \rho_m \left(V_l^n - V_m^n\right)^2 = \rho_m \left|\Delta \mathbf{V}_{lm}^n\right|^2 sign\left(V_l^n - V_m^n\right), \quad (2.60)$$

- see Fig. 2.6. The normal velocity difference required in the above expression can be obtained by splitting the relative velocity vector at the interface $\Delta \mathbf{V}_{lm}$ into a component which is parallel to $\mathbf{n}_l$

$$\Delta \mathbf{V}_{lm}^n = proj_{\mathbf{n}_l} \Delta \mathbf{V}_{lm} = \left(\frac{\Delta \mathbf{V}_{lm} \cdot \mathbf{n}_l}{\mathbf{n}_l \cdot \mathbf{n}_l}\right) \mathbf{n}_l = \left(\Delta \mathbf{V}_{lm} \cdot \mathbf{n}_l\right) \mathbf{n}_l, \quad (2.61)$$

with a magnitude

$$\left|\Delta \mathbf{V}_{lm}^n\right| = \left|\Delta \mathbf{V}_{lm} \cdot \mathbf{n}_l\right| = \sqrt{\left[n_{lx}\left(u_l - u_m\right)\right]^2 + \left[n_{ly}\left(v_l - v_m\right)\right]^2 + \left[n_{lz}\left(w_l - w_m\right)\right]^2} \quad (2.62)$$

and a component orthogonal to $\mathbf{n}_l$,

$$\Delta \mathbf{V}_{lm}^t = \Delta \mathbf{V}_{lm} - proj_{\mathbf{e}_l} \Delta \mathbf{V}_{lm} = \Delta \mathbf{V}_{lm} - \left(\Delta \mathbf{V}_{lm} \cdot \mathbf{n}_l\right) \mathbf{n}_l$$

$$= \left[\Delta u_{lm} - n_{lx}\left(\Delta \mathbf{V}_{lm} \cdot \mathbf{n}_l\right)\right] \mathbf{i} + \left[\Delta v_{lm} - n_{ly}\left(\Delta \mathbf{V}_{lm} \cdot \mathbf{n}_l\right)\right] \mathbf{j} + \left[\Delta w_{lm} - n_{lz}\left(\Delta \mathbf{V}_{lm} \cdot \mathbf{n}_l\right)\right] \mathbf{k}$$
$$(2.63)$$

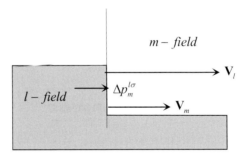

Fig. 2.6. Stagnation pressure in stratified flow

In a similar way, the stagnation pressure difference at the field-structure interface can be estimated,

$$\Delta p_l^{w\sigma} = \rho_l \left(V_l^n - V_w^n\right)^2 sign\left(V_l^n - V_w^n\right). \quad (2.64)$$

For the case of $\Delta p_m^{l\sigma} \approx const$ within *Vol* the following can be written

$$\frac{1}{Vol}\int_{F_{l\sigma}}\Delta p_m^{l\sigma}\mathbf{n}_l dF=\Delta p_m^{l\sigma}\frac{1}{Vol}\int_{F_{l\sigma}}\mathbf{n}_l dF=-\Delta p_m^{l\sigma}\left[\nabla\left(\alpha_l^e\gamma\right)+\frac{1}{Vol}\int_{F_{lw}}\mathbf{n}_l dF\right].$$
(2.65)

For $\mathbf{V}_l^n > \mathbf{V}_m^n$ and decreasing $\alpha_l^e\gamma$ in space, this force resists the field l. If there is no difference in the average normal velocities at the interface the above term is zero.

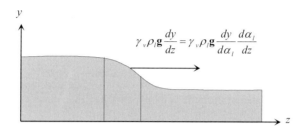

Fig. 2.7. Geodesic pressure force

If the tangential average velocity difference differs from zero, there is a tangential viscous shear force

$$-\frac{1}{Vol}\int_{F_{l\sigma}}\mathbf{T}_{m\sigma}\cdot\mathbf{n}_{l\sigma}dF=c_{ml}\left|\Delta\mathbf{V}_{lm}^t\right|\Delta\mathbf{V}_{lm}^t.$$
(2.66)

Here c_{ml} has to be computed using empirical correlation in case of the large scale of the cells not resolving the details of the boundary layer. For a wavy surface, the interface share coefficient should be increased by a component for form drag caused by the non-uniform pressure distribution, which results in additional tangential force. Similarly, the wall viscous shear stress is

$$-\frac{1}{Vol}\int_{F_{lw}}\mathbf{T}_{lw}\cdot\mathbf{n}_{lw}dF=c_{wl}\left|\mathbf{V}_l^t\right|\mathbf{V}_l^t,$$
(2.67)

where like in the previous case, c_{wl} has to be computed using empirical correlation. Note that in case of stratified rectangular duct flow in the z direction - see Fig. 2.7, the change in the liquid thickness causes a lateral geodesic pressure force

$$\gamma_v\rho_l\mathbf{g}\frac{dy}{dz}=\gamma_v\rho_l\mathbf{g}\frac{dy}{d\alpha_l}\frac{d\alpha_l}{dz}.$$
(2.68)

Here y is the distance between the bottom of the duct and the center of mass of the liquid. In three-dimensional models this force automatically arises due to differences in the local bulk pressure having the geodesic pressure as a component. This force should be taken into account in one-dimensional models. If this force is neglected, the one-dimensional model will not be able to predict water flow in a horizontal pipe with negligible gas-induced shear. In the next section we consider this problem in more detail.

2.6.1.2 Stratified flow in horizontal or inclined rectangular channels

Geometrical characteristics: Stratified flow may exists in regions with such relative velocities between the liquid and the gas which does not cause instabilities leading to slugging.

Some important geometrical characteristics are specified here – compare with Fig. 2.8. The perimeter of the pipe is then

$$Per_{1w} = 2(a + H), \qquad (2.69)$$

and the wetted perimeters for the gas and the liquid parts are

$$Per_{1w} = a + 2(H - \delta_{2F}) = a + 2\alpha_1 H, \qquad (2.70)$$

$$Per_{2w} = a + 2\delta_{1F} = a + 2\alpha_2 H. \qquad (2.71)$$

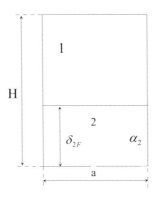

Fig. 2.8. Definition of the geometrical characteristics of the stratified flow

The gas-liquid interface median is then a, and the liquid level

$$\delta_{2F} = \alpha_2 H. \qquad (2.72)$$

2.6 Non-equal bulk and boundary layer pressures

The hydraulic diameters for the gas and the liquid for computation of the pressure drop due to friction with the wall are therefore

$$D_{h1} = 4\alpha_1 F / Per_{1w} = \frac{4\alpha_1 aH}{a + 2H\alpha_1}, \qquad (2.73)$$

$$D_{h2} = 4\alpha_2 F / Per_{2w} = \frac{4\alpha_2 aH}{a + 2H\alpha_2}, \qquad (2.74)$$

and the corresponding *Reynolds* numbers

$$\text{Re}_1 = \frac{\alpha_1 \rho_1 w_1}{\eta_1} \frac{4\alpha_1 aH}{a + 2H\alpha_1}, \qquad (2.75)$$

$$\text{Re}_2 = \frac{\alpha_2 \rho_2 w_2}{\eta_2} \frac{4\alpha_2 aH}{a + 2H\alpha_2}. \qquad (2.76)$$

Here F is the channel cross section and Per_{1w} and Per_{2w} are the perimeters wet by gas and film, respectively. If one considers the core of the flow the hydraulic diameter for computation of the friction pressure loss component gas-liquid is then

$$D_{h12} = 4\alpha_1 F / (Per_{1w} + a) = \frac{2\alpha_1 aH}{a + H\alpha_1} \qquad (2.77)$$

and the corresponding *Reynolds* number

$$\text{Re}_1 = \frac{\alpha_1 \rho_1 |w_1 - w_2|}{\eta_1} \frac{2\alpha_1 aH}{a + H\alpha_1}. \qquad (2.78)$$

The gas-wall, liquid-wall and gas-liquid interfacial area densities are

$$a_{1w} = \frac{Per_{1w}}{F} = \frac{a + 2\alpha_1 H}{aH}, \qquad (2.79)$$

$$a_{2w} = \frac{Per_{2w}}{F} = \frac{a + 2\alpha_2 H}{aH}, \qquad (2.80)$$

$$a_{12} = \frac{a}{F} = \frac{1}{H}. \qquad (2.81)$$

For the estimation of flow pattern transition criterion the following expression is some times required

$$\frac{d\alpha_2}{d\delta_{2F}} = \frac{1}{H}. \tag{2.82}$$

Using the geometric characteristics and the Reynolds numbers the interfacial interaction coefficients can be computing by means of empirical correlations as discussed in Volume II of this monograph.

Gravitational (hydrostatic) pressure variation across the flow cross section of horizontal pipe: In stratified flow the gravitation is a dominant force. The cross section averaged gas gravity pressure difference with respect to the interface is

$$\Delta p_1^{2\sigma} = \rho_1 g \alpha_1 \frac{H}{2}. \tag{2.83}$$

The cross section averaged liquid gravity pressure difference with respect to the interface is

$$\Delta p_2^{1\sigma} = -\rho_2 g \alpha_2 \frac{H}{2}. \tag{2.84}$$

The cross section averaged field pressures in terms of the interfacial pressure are then

$$p_1 = p_1^{2\sigma} - \Delta p_1^{2\sigma} = p_1^{2\sigma} - \rho_1 g \alpha_1 \frac{H}{2}, \tag{2.85}$$

$$p_2 = p_2^{1\sigma} - \Delta p_2^{1\sigma} = p_2^{1\sigma} + \rho_2 g \alpha_2 \frac{H}{2}, \tag{2.86}$$

recalling the definition Eqs. (2.33) and (2.34). The averaged pressure in the cross section can be expressed as the cross section weighted averaged pressures inside the fields

$$p = \alpha_1 p_1 + \alpha_2 p_2 = \alpha_1 p_1^{2\sigma} + \alpha_2 p_2^{1\sigma} + g \frac{H}{2}\left(\alpha_2^2 \rho_2 - \alpha_1^2 \rho_1\right) \tag{2.87}$$

Neglecting the surface tension we obtain

$$p \approx p_\sigma + g \frac{H}{2}\left(\alpha_2^2 \rho_2 - \alpha_1^2 \rho_1\right), \tag{2.88}$$

or

2.6 Non-equal bulk and boundary layer pressures

$$p_\sigma \approx p - g\frac{H}{2}\left(\alpha_2^2 \rho_2 - \alpha_1^2 \rho_1\right). \tag{2.89}$$

The averaged pressure in the gas and in the liquid phase can be then expressed as a function of the system averaged pressure and geometrical characteristics by replacing p_σ in Eqs. (2.85) and (2.86):

$$p_1 = p - \frac{H}{2}g\alpha_2\rho \quad \text{for} \quad \alpha_1 > 0 \tag{2.90}$$

$$p_2 = p + \frac{H}{2}g\alpha_1\rho \quad \text{for} \quad \alpha_2 > 0 \tag{2.91}$$

where $\rho = \alpha_1\rho_1 + \alpha_2\rho_2$ is the homogeneous mixture density. The check $\alpha_1 p_1 + \alpha_2 p_2 = p$ proves the correctness of the computation. The difference between both averaged pressures is then

$$\Delta p_{21} = p_2 - p_1 = \frac{H}{2}g\rho \quad \text{for} \quad \alpha_1 > 0 \text{ and } \alpha_2 > 0 \tag{2.92}$$

Delhaye [13], p. 89, therefore

$$p_1 = p - \alpha_2 \Delta p_{21}, \tag{2.93}$$

$$p_2 = p + \alpha_1 \Delta p_{21}, \tag{2.94}$$

and consequently

$$\frac{\partial p_1}{\partial z} = \frac{\partial p}{\partial z} - \Delta p_{21}\frac{\partial \alpha_2}{\partial z} - \alpha_2\frac{\partial \Delta p_{21}}{\partial z}, \tag{2.95}$$

$$\frac{\partial p_2}{\partial z} = \frac{\partial p}{\partial z} + \Delta p_{21}\frac{\partial \alpha_1}{\partial z} + \alpha_1\frac{\partial \Delta p_{21}}{\partial z}. \tag{2.96}$$

Now we can write the specific form of the following general terms of the momentum equation

$$\ldots + \alpha_1^e \gamma_z \frac{\partial p_1}{\partial z} + \left(p_l - p_m - \delta_l \sigma_{lm} \kappa_l - \Delta p_m^{l\sigma}\right)\left[\frac{\partial}{\partial z}\left(\alpha_1^e \gamma_z\right) - \delta_l\frac{\partial \gamma_z}{\partial z}\right]$$

$$- \Delta p_l^{w\sigma}\delta_l \frac{\partial \gamma_z}{\partial z} + \gamma_v\alpha_l\rho_l g \cos\varphi\ldots \tag{2.97}$$

Note that both fields are continuous and therefore $\delta_l = 1$ and the effect of the surface tension is neglected

$$... + \alpha_1 \gamma_z \frac{\partial p_1}{\partial z} - \gamma_z \left(\Delta p_{21} + \Delta p_2^{1\sigma}\right) \frac{\partial \alpha_1}{\partial z} - \Delta p_1^{w\sigma} \frac{\partial \gamma_z}{\partial z} ... \quad (2.98)$$

$$... + \alpha_2 \gamma_z \frac{\partial p_2}{\partial z} + \gamma_z \left(\Delta p_{21} - \Delta p_1^{2\sigma}\right) \frac{\partial \alpha_2}{\partial z} - \Delta p_2^{w\sigma} \frac{\partial \gamma_z}{\partial z} ... \quad (2.99)$$

Substituting the field pressures we finally obtain

$$... + \gamma_z \left[\alpha_1 \frac{\partial p}{\partial z} - \alpha_1 \alpha_2 \frac{\partial \Delta p_{21}}{\partial z} - \left(\alpha_2 \Delta p_{21} + \Delta p_2^{1\sigma}\right) \frac{\partial \alpha_1}{\partial z}\right] - \Delta p_1^{w\sigma} \frac{\partial \gamma_z}{\partial z} ... \quad (2.100)$$

$$... + \gamma_z \left[\alpha_2 \frac{\partial p}{\partial z} + \alpha_1 \alpha_2 \frac{\partial \Delta p_{21}}{\partial z} - \left(\alpha_1 \Delta p_{21} - \Delta p_1^{2\sigma}\right) \frac{\partial \alpha_1}{\partial z}\right] - \Delta p_2^{w\sigma} \frac{\partial \gamma_z}{\partial z} ... \quad (2.101)$$

Assuming that the change of the densities contributes much less to the change of

$$\frac{\partial \Delta p_{21}}{\partial z} \approx \frac{\partial \Delta p_{21}}{\partial \alpha_1} \frac{\partial \alpha_1}{\partial z} = g \frac{H}{2} (\rho_1 - \rho_2) \frac{\partial \alpha_1}{\partial z} \quad (2.102)$$

then the change of the local volume fraction becomes

$$... + \gamma_z \left[\alpha_1 \frac{\partial p}{\partial z} - gH\alpha_1\alpha_2 (\rho_1 - \rho_2) \frac{\partial \alpha_1}{\partial z}\right] - \Delta p_1^{w\sigma} \frac{\partial \gamma_z}{\partial z} ... \quad (2.103)$$

$$... + \gamma_z \left[\alpha_2 \frac{\partial p}{\partial z} + gH\alpha_1\alpha_2 (\rho_1 - \rho_2) \frac{\partial \alpha_1}{\partial z}\right] - \Delta p_2^{w\sigma} \frac{\partial \gamma_z}{\partial z} ... \quad (2.104)$$

As a plausibility check note that for $\alpha_1 \to 0$ or $\alpha_2 \to 0$ the term containing the derivative of the volume fraction of velocity field 1 converges to zero and the momentum equations take the expected form. The sum of the two momentum equations gives

$$... + \gamma_z \frac{\partial p}{\partial z} - \left(\Delta p_1^{w\sigma} + \Delta p_2^{w\sigma}\right) \frac{\partial \gamma_z}{\partial z} ... \quad (2.105)$$

2.6 Non-equal bulk and boundary layer pressures

Note that the gravitational force is already taken into account in Eqs. (2.103) and (2.104) and there is no need for an additional term $\gamma_v \alpha_1 \rho_1 g \cos\varphi$. Stability criteria for the stratified flow can be obtained from the eigen values analysis. For simplicity assuming incompressible flow the mass and momentum equations of stratified flow in a straight pipe with constant cross section are

$$\alpha_1 \frac{\partial w_1}{\partial z} + \alpha_2 \frac{\partial w_2}{\partial z} + (w_1 - w_2)\frac{\partial \alpha_1}{\partial z} = 0 \qquad (2.106)$$

$$\frac{\partial \alpha_1}{\partial \tau} + \alpha_1 \frac{\partial w_1}{\partial z} + w_1 \frac{\partial \alpha_1}{\partial z} = 0 \qquad (2.107)$$

$$\frac{\partial w_1}{\partial \tau} + w_1 \frac{\partial w_1}{\partial z} + \frac{1}{\rho_1}\frac{\partial p}{\partial z} + gH\frac{\alpha_2(\rho_2 - \rho_1)}{\rho_1}\frac{\partial \alpha_1}{\partial z} = 0 \qquad (2.108)$$

$$\frac{\partial w_2}{\partial \tau} + w_2 \frac{\partial w_2}{\partial z} + \frac{1}{\rho_2}\frac{\partial p}{\partial z} - gH\frac{\alpha_1(\rho_2 - \rho_1)}{\rho_2}\frac{\partial \alpha_1}{\partial z} = 0 \qquad (2.109)$$

or in matrix notation

$$\begin{pmatrix} 0 & 0 & 0 & 0 \\ 0 & 1 & 0 & 0 \\ 0 & 0 & 1 & 0 \\ 0 & 0 & 0 & 1 \end{pmatrix} \frac{\partial}{\partial \tau}\begin{pmatrix} p \\ \alpha_1 \\ w_1 \\ w_2 \end{pmatrix} + \begin{pmatrix} 0 & (w_1 - w_2) & \alpha_1 & \alpha_2 \\ 0 & w_1 & \alpha_1 & 0 \\ \frac{1}{\rho_1} & gH\frac{\alpha_2(\rho_2-\rho_1)}{\rho_1} & w_1 & 0 \\ \frac{1}{\rho_2} & -gH\frac{\alpha_1(\rho_2-\rho_1)}{\rho_2} & 0 & w_2 \end{pmatrix}\frac{\partial}{\partial z}\begin{pmatrix} p \\ \alpha_1 \\ w_1 \\ w_2 \end{pmatrix} = 0.$$

(2.110)

If you are not familiar with the analysis of the type of a system of partial differenttial equations by first computing the eigenvalues, eigenvectors and canonical forms it is recommended first to read Section 11 before continuing here.

The eigenvalues are defined by the characteristics equations

2 Momentums conservation

$$\begin{pmatrix} 0 & (w_1 - w_2) & \alpha_1 & \alpha_2 \\ 0 & w_1 - \lambda & \alpha_1 & 0 \\ \dfrac{1}{\rho_1} & gH\dfrac{\alpha_2(\rho_2 - \rho_1)}{\rho_1} & w_1 - \lambda & 0 \\ \dfrac{1}{\rho_2} & -gH\dfrac{\alpha_1(\rho_2 - \rho_1)}{\rho_2} & 0 & w_2 - \lambda \end{pmatrix} = 0 \qquad (2.111)$$

or

$$\frac{\alpha_1}{\rho_1}(w_1 - w_2)(w_2 - \lambda) - \frac{\alpha_1}{\rho_1}(w_1 - \lambda)(w_2 - \lambda) - \frac{\alpha_2}{\rho_2}(w_1 - \lambda)^2$$

$$+ gH\frac{\alpha_1\alpha_2(\rho_2 - \rho_1)}{\rho_1\rho_2} = 0 \qquad (2.112)$$

or

$$\left(\frac{\alpha_1}{\rho_1} + \frac{\alpha_2}{\rho_2}\right)\lambda^2 - 2\left(\frac{\alpha_1}{\rho_1}w_2 + \frac{\alpha_2}{\rho_2}w_1\right)\lambda + \frac{\alpha_1}{\rho_1}w_2^2 + \frac{\alpha_2}{\rho_2}w_1^2 - gH\frac{\alpha_1\alpha_2(\rho_2 - \rho_1)}{\rho_1\rho_2} = 0. \qquad (2.113)$$

This equation is in fact consistent with the gravity long wave theory of *Milne-Thomson* [49] 1955. There are two eigen values

$$\lambda_{1,2} = \frac{\dfrac{\alpha_1}{\rho_1}w_2 + \dfrac{\alpha_2}{\rho_2}w_1 \pm \sqrt{\left(\dfrac{\alpha_1}{\rho_1}w_2 + \dfrac{\alpha_2}{\rho_2}w_1\right)^2 - \left(\dfrac{\alpha_1}{\rho_1} + \dfrac{\alpha_2}{\rho_2}\right)\left(\dfrac{\alpha_1}{\rho_1}w_2^2 + \dfrac{\alpha_2}{\rho_2}w_1^2 - gH\dfrac{\alpha_1\alpha_2(\rho_2 - \rho_1)}{\rho_1\rho_2}\right)}}{\left(\dfrac{\alpha_1}{\rho_1} + \dfrac{\alpha_2}{\rho_2}\right)}$$

(2.114)

or after rearranging

$$\lambda_{1,2} = \frac{\frac{\alpha_1}{\rho_1} w_2 + \frac{\alpha_2}{\rho_2} w_1 \pm \sqrt{\left[gH(\rho_2 - \rho_1)\left(\frac{\alpha_1}{\rho_1} + \frac{\alpha_2}{\rho_2}\right) - (w_1 - w_2)^2\right]\frac{\alpha_1 \alpha_2}{\rho_1 \rho_2}}}{\left(\frac{\alpha_1}{\rho_1} + \frac{\alpha_2}{\rho_2}\right)},$$

(2.115)

which are real and different from each other if

$$(w_1 - w_2)^2 < gH(\rho_2 - \rho_1)\left(\frac{\alpha_1}{\rho_1} + \frac{\alpha_2}{\rho_2}\right).$$ (2.116)

In fact this is the *Kelvin-Helmholtz* stability criterion. If the above condition is satisfied the system describing the flow is hyperbolic. Violating the above conditions results in nature in other flow patterns different from the stratified one. Condition (2.116) is equivalent to Eq. (2.216) derived by *Delhaye* [13], p. 90 in 1981. *Brauner* and *Maron* included in 1992 the surface tension effect in their stability analysis and obtained

$$(w_1 - w_2)^2 < H\left(\frac{\alpha_1}{\rho_1} + \frac{\alpha_2}{\rho_2}\right)\left[g(\rho_2 - \rho_1) + \sigma_{12} k^2\right]$$

– see Eq. (28a) in [8] – which for neglected surface tension results in Eq. (2.116). Here k is the real wave number.

2.6.1.3 Stratified flow in horizontal or inclined pipes

Geometrical characteristics: The geometric flow characteristics for round pipes are non-linearly dependent on the liquid level, which makes the computation somewhat more complicated.

Some important geometrical characteristics are specified here – compare with Fig. 2.9. The angle with the origin of the pipe axis defined between the upwards oriented vertical and the liquid-gas-wall triple point is defined as a function of the liquid volume fraction by the equation

$$f(\theta) = -(1 - \alpha_2)\pi + \theta - \sin\theta \cos\theta = 0.$$ (2.117)

The derivative

$$\frac{d\theta}{d\alpha_1} = \frac{\pi}{2\sin^2\theta}$$ (2.118)

will be used later. Having in mind that

$$\frac{df}{d\theta} = 2\sin^2\theta \qquad (2.119)$$

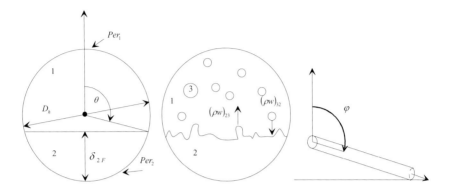

Fig. 2.9. Definition of the geometrical characteristics of the stratified flow.

the solution with respect to the angle can be obtained by using the *Newton* iteration method as follows

$$\theta = \theta_0 - \frac{f_0}{df/d\theta} = \theta_0 + \frac{(1-\alpha_2)\pi - \theta_0 + \sin\theta_0 \cos\theta_0}{2\sin^2\theta_0} \qquad (2.120)$$

where index 0 stays for the previous guess. The iteration starts with an initial value of $\pi/2$ [76]. *Biberg* proposed in 1999 [77] accurate direct approximation

$$\theta = \pi\alpha_2 + \left(\frac{3\pi}{2}\right)^{1/3}\left[1 - 2\alpha_2 + \alpha_2^{1/3} - (1-\alpha_2)^{1/3}\right] \qquad (2.120b)$$

with an error less then $\pm 0.002 rad$ or

$$\theta = \pi\alpha_2 + \left(\frac{3\pi}{2}\right)^{1/3}\left[1 - 2\alpha_2 + \alpha_2^{1/3} - (1-\alpha_2)^{1/3}\right]$$

$$-\frac{1}{200}\alpha_2(1-\alpha_2)(1-2\alpha_2)\left\{1 + 4\left[\alpha_2^2 + (1-\alpha_2)^2\right]\right\}, \qquad (2.120c)$$

with an error less then $\pm 0.00005 rad$. The perimeter of the pipe is then

$$Per_{1w} = \pi D_h, \qquad (2.121)$$

and the wetted perimeters for the gas and the liquid parts are

$$Per_{1w} = \theta D_h,\qquad(2.122)$$

$$Per_{2w} = (\pi - \theta) D_h.\qquad(2.123)$$

The gas-liquid interface median is then

$$b = D_h \sin(\pi - \theta) = D_h \sin\theta,\qquad(2.124)$$

and the liquid level

$$\delta_{2F} = \frac{1}{2} D_h (1 + \cos\theta).\qquad(2.125)$$

The hydraulic diameters for the gas and the liquid for computation of the pressure drop due to friction with the wall are therefore

$$D_{h1} = 4\alpha_1 F / Per_{1w} = \frac{\pi}{\theta}\alpha_1 D_h\qquad(2.126)$$

$$D_{h2} = 4\alpha_2 F / Per_{2w} = \frac{\pi}{\pi - \theta}\alpha_2 D_h\qquad(2.127)$$

and the corresponding *Reynolds* numbers

$$\mathrm{Re}_1 = \frac{\alpha_1 \rho_1 w_1 D_h}{\eta_1}\frac{\pi}{\theta},\qquad(2.128)$$

$$\mathrm{Re}_2 = \frac{\alpha_2 \rho_2 w_2 D_h}{\eta_2}\frac{\pi}{\pi - \theta}.\qquad(2.129)$$

Here F is the channel cross section and Per_1 and Per_2 are the wet perimeters of gas and film, respectively.

If one considers the core of the flow the hydraulic diameter for computation of the friction pressure loss component at the gas-liquid interface is then

$$D_{h12} = 4\alpha_1 F / (Per_{1w} + b) = \frac{\pi}{\theta + \sin\theta}\alpha_1 D_h\qquad(2.130)$$

and the corresponding *Reynolds* number

$$\text{Re}_1 = \frac{\alpha_1 \rho_1 |w_1 - w_2| D_h}{\eta_1} \frac{\pi}{\theta + \sin\theta}. \tag{2.131}$$

The gas-wall, liquid-wall and gas-liquid interfacial area densities are

$$a_{1w} = \frac{Per_{1w}}{F} = \frac{4\alpha_1}{D_{h1}} = \frac{\theta}{\pi} \frac{4}{D_h}, \tag{2.132}$$

$$a_{2w} = \frac{Per_{2w}}{F} = \frac{4\alpha_2}{D_{h2}} = \frac{\pi - \theta}{\pi} \frac{4}{D_h}, \tag{2.133}$$

$$a_{12} = \frac{b}{F} = \frac{\sin\theta}{\pi} \frac{4}{D_h}. \tag{2.134}$$

Geffraye et al. [37] approximated this relation for a smooth interface with

$$a_{12} \cong \frac{8}{\pi D_h} \sqrt{\alpha_2 (1 - \alpha_2)}. \tag{2.135}$$

For the estimation of flow pattern transition criterion the following expression is some times required

$$\frac{d\alpha_2}{d\delta_{2F}} = \frac{4}{D_h} \frac{\sin\theta}{\pi}. \tag{2.136}$$

Gravitational (hydrostatic) pressure variation across the flow cross section of a horizontal pipe: In stratified flow the gravitation is a dominant force. The cross section averaged gas gravity pressure difference with respect to the interface is

$$\Delta p_1^{2\sigma} = \rho_1 g D_h \left(\frac{\sin^3 \theta}{3\pi \alpha_1} - \frac{1}{2} \cos\theta \right). \tag{2.137}$$

The cross section averaged liquid gravity pressure difference with respect to the bottom of the pipe is

$$\Delta p_2^{1\sigma} = -\rho_2 g D_h \left(\frac{\sin^3 \theta}{3\pi \alpha_2} + \frac{1}{2} \cos\theta \right). \tag{2.138}$$

With respect to the interfacial pressure we have

2.6 Non-equal bulk and boundary layer pressures

$$p_1 = p_1^{2\sigma} - \Delta p_1^{2\sigma} = p_1^{2\sigma} - \rho_1 g D_h \left(\frac{\sin^3 \theta}{3\pi \alpha_1} - \frac{1}{2}\cos\theta \right), \qquad (2.139)$$

$$p_2 = p_2^{1\sigma} - \Delta p_2^{1\sigma} = p_2^{1\sigma} + \rho_2 g D_h \left(\frac{\sin^3 \theta}{3\pi \alpha_2} + \frac{1}{2}\cos\theta \right), \qquad (2.140)$$

which are in fact Eqs. (55) and (60) in *Ransom* et al. [54], p. 30. The averaged pressure in the cross section can be expressed as the cross section weighted averaged pressures inside the fields

$$p = \alpha_1 p_1 + \alpha_2 p_2 = \alpha_1 p_1^{2\sigma} + \alpha_2 p_2^{1\sigma} + gD_h \left[\frac{\sin^3 \theta}{3\pi}(\rho_2 - \rho_1) + (\alpha_2 \rho_2 + \alpha_1 \rho_1)\frac{1}{2}\cos\theta \right]. \qquad (2.141)$$

Neglecting the surface tension we obtain

$$p \approx p_\sigma + gD_h \left[\frac{\sin^3 \theta}{3\pi}(\rho_2 - \rho_1) + (\alpha_2 \rho_2 + \alpha_1 \rho_1)\frac{1}{2}\cos\theta \right], \qquad (2.142)$$

or

$$p_\sigma \approx p - gD_h \left[\frac{\sin^3 \theta}{3\pi}(\rho_2 - \rho_1) + (\alpha_2 \rho_2 + \alpha_1 \rho_1)\frac{1}{2}\cos\theta \right]. \qquad (2.143)$$

The averaged pressure in the gas and in the liquid phase can then be expressed as a function of the system averaged pressure and geometrical characteristics:

$$p_1 = p - gD_h \left[\frac{\sin^3 \theta}{3\pi \alpha_1}(\alpha_1 \rho_2 + \alpha_2 \rho_1) + \alpha_2 (\rho_2 - \rho_1)\frac{1}{2}\cos\theta \right] \quad \text{for} \quad \alpha_1 > 0 \qquad (2.144)$$

$$p_2 = p + gD_h \left[\frac{\sin^3 \theta}{3\pi \alpha_2}(\alpha_1 \rho_2 + \alpha_2 \rho_1) + \alpha_1 (\rho_2 - \rho_1)\frac{1}{2}\cos\theta \right] \quad \text{for} \quad \alpha_2 > 0 \qquad (2.145)$$

The check $\alpha_1 p_1 + \alpha_2 p_2 = p$ proves the correctness of the computation. The difference between both averaged pressures is then

$$\Delta p_{21} = p_2 - p_1 = gD_h \left[\frac{\sin^3 \theta}{3\pi} \left(\frac{\rho_2}{\alpha_2} + \frac{\rho_1}{\alpha_1} \right) + (\rho_2 - \rho_1) \frac{1}{2} \cos \theta \right]$$

for $\alpha_1 > 0$ and $\alpha_2 > 0$ (2.146)

and therefore Eqs. (2.93-2.96) are valid also for a circular pipe. Assuming that the change of the densities contributes much less to the change of Δp_{21} then the change of the local volume fraction results in

$$\frac{\partial \Delta p_{21}}{\partial \alpha_1} \approx \frac{\partial \Delta p_{21}}{\partial \alpha_1} \frac{\partial \alpha_1}{\partial z}$$

$$= gD_h \left[\frac{\sin^3 \theta}{3\pi} \left(\frac{\rho_2}{\alpha_2^2} - \frac{\rho_1}{\alpha_1^2} \right) + \left(\frac{\rho_2}{\alpha_2} + \frac{\rho_1}{\alpha_1} \right) \frac{1}{2} \cos \theta - (\rho_2 - \rho_1) \frac{\pi}{4 \sin \theta} \right] \frac{\partial \alpha_1}{\partial z}.$$
(2.147)

This equation was obtained by taking into account that θ is also an implicit function of α_1 through Eq. (2.117) and using Eq. (2.118). Using Eqs. (2.156), (2.146), (2.137) and (2.138) and substituting into the momentum equations (2.100) and (2.101) we finally obtain

$$... + \gamma_z \left[\alpha_1 \frac{\partial p}{\partial z} + \alpha_1 \alpha_2 (\rho_2 - \rho_1) \frac{\pi g D_h}{4 \sin \theta} \frac{\partial \alpha_1}{\partial z} \right] - \Delta p_1^{w\sigma} \frac{\partial \gamma_z}{\partial z} ... \quad (2.148)$$

$$... + \gamma_z \left[\alpha_2 \frac{\partial p}{\partial z} - \alpha_1 \alpha_2 (\rho_2 - \rho_1) \frac{\pi g D_h}{4 \sin \theta} \frac{\partial \alpha_1}{\partial z} \right] - \Delta p_2^{w\sigma} \frac{\partial \gamma_z}{\partial z} ... \quad (2.149)$$

As plausibility check note that for $\alpha_1 \to 0$ the term containing the derivative of the volume fraction of velocity field 1 converges to zero and the momentum equation for field 2 takes the expected form. The sum of the two equations is then

$$... + \gamma_z \frac{\partial p}{\partial z} - \left(\Delta p_1^{w\sigma} + \Delta p_2^{w\sigma} \right) \frac{\partial \gamma_z}{\partial z} ... \quad (2.150)$$

Comparing with the momentum equations for the rectangular channel we find instead of H the length $\frac{\pi D_h}{4 \sin \theta} = \frac{\pi D_h^2}{4b}$, which is the height of the rectangular channel having the same cross section as the pipe and base equal to the gas-liquid median from Eq. (2.124). Therefore the stability condition for stratified flow in a pipe is

$$(w_1 - w_2)^2 < g \frac{\pi D_h}{4\sin\theta}(\rho_2 - \rho_1)\left(\frac{\alpha_1}{\rho_1} + \frac{\alpha_2}{\rho_2}\right), \qquad (2.151a)$$

or in an alternative form

$$(w_1 - w_2)^2 < g(\rho_2 - \rho_1)\left(\frac{\alpha_1}{\rho_1} + \frac{\alpha_2}{\rho_2}\right)\frac{d\delta_{2F}}{d\alpha_2}, \qquad (2.151b)$$

which is a generalized *Kelvin-Helmholtz* stability criterion valid for pipes with arbitrary cross section. Sunstituting Eq. (2.82) in the above equation we obtain Eq. (2.116) for rectangular channels. This result was obtained by *Barnea* and *Titel* in 1994 – see Eq. (26) in [2] for the inviscid case. It is identical with Eq. (6.9) p.313 obtained by *de Crecy* [11] in 1986.

Dividing the momentum equations by $\alpha_1\rho_1$ and $\alpha_2\rho_2$ respectively, and subtracting the second from the first we obtain

$$\ldots + \gamma_z \left[\left(\frac{1}{\rho_1} - \frac{1}{\rho_2}\right)\frac{\partial p}{\partial z} + \frac{\rho}{\rho_1\rho_2}(\rho_2 - \rho_1)\frac{\pi g D_h}{4\sin\theta}\frac{\partial \alpha_1}{\partial z}\right] - \left(\frac{\Delta p_1^{w\sigma}}{\alpha_1\rho_1} - \frac{\Delta p_2^{w\sigma}}{\alpha_2\rho_2}\right)\frac{\partial \gamma_z}{\partial z} \ldots$$
$$(2.152)$$

The second term in the first brackets is exactly Eq. (58) obtained by *Ransom* et al. [54], p.30.

Note the general notation of the coefficients of $\partial\alpha_1/\partial z$ in Eqs. (2.103, 2.104) and Eqs. (2.148, 2.149)

$$\alpha_1\alpha_2(\rho_2 - \rho_1)g\frac{F}{b}\frac{\partial \alpha_1}{\partial z},$$

where F is the channel cross section and b is the gas-liquid interface median. This result was obtained by *de Crecy* in 1986 – see in [11] Eq. (6.3) p.312.

Teletov and *Mamaev* et al., see in [48] 1969, presented analytical solutions for stratified flow between two parallel plates and stratified flow in circular tubes, respectively. The gas and liquid phase are considered incompressible. No heat and mass transfer was considered. The pressure gradient is assumed constant and the flow was considered to be stationary and fully developed. The velocities at the wall are assumed to be zero and the velocity at the interface was assumed to be equal for both phases. The solution for the circular tube is found after introducing a bipolar coordinate transformation and integration, and is expressed as cross section averaged velocities as a function of the flow parameter in the form

$$w_1 = \frac{\pi R^4}{8\eta_1}\left[\frac{dp}{dz} - \rho_1 g \cos(\mathbf{g},z)\right]\varphi_1(\alpha_1,\eta_2/\eta_1),$$

$$w_2 = \frac{\pi R^4}{8\eta_2}\left[\frac{dp}{dz} - \rho_2 g \cos(\mathbf{g},z)\right]\varphi_2(\alpha_1,\eta_2/\eta_1).$$

$\varphi_1(\alpha_1,\eta_2/\eta_1)$ and $\varphi_2(\alpha_1,\eta_2/\eta_1)$ are complicated integral functions. For practical use they are presented in graphical form.

2.6.2 Dispersed interface

2.6.2.1 General

In this section we provide a guide for derivation of a constitutive relation for mechanical interaction between a dispersed field l and the surrounding continuum m. An example for such flow is bubbly flow. In other words we discuss a possible simplification of the surface integrals in Eq. (2.50). For a dispersed phase l, the viscous shear at the interface is negligible for non-*Stokes* flows:

$$-\frac{1}{Vol}\int_{F_{l\sigma}} \mathbf{T}_l^{m\sigma} \cdot \mathbf{n}_l dF \approx 0. \tag{2.153}$$

The viscous effects also in the continuum at the interface are neglected,

$$-\frac{1}{Vol}\int_{F_{m\sigma}} \mathbf{T}_m^{l\sigma} \cdot \mathbf{n}_m dF \approx 0. \tag{2.154}$$

Note that a 1 cm bubble in water having a relative velocity of 10 *cm/s* has a *Reynolds* number of about 100. For Reynolds numbers less than 24 the viscous effect in the continuum is important. For larger Reynolds number which is often the case in the nature the viscous effects can be neglected. The force of the *Marangoni* effect,

$$\delta_l \frac{1}{Vol} \int_{F_{l\sigma}} (\nabla_t \sigma_{lm}) dF = 0 \tag{2.155}$$

can be neglected for the majority of the macroscopic processes. If the dispersed phase is assumed to have no wall contact, $F_{lw} = 0$, the following results

$$\frac{1}{Vol} \int_{F_{lw}} \left(\Delta p_l^{w\sigma} \mathbf{I} - \mathbf{T}_l^{w\sigma} \right) \cdot \mathbf{n}_l dF = 0. \qquad (2.156)$$

The difference between the bubble bulk pressure and the bubble interface pressure is also negligible,

$$+ \frac{1}{Vol} \int_{F_{m\sigma}} \Delta p_l^{m\sigma} \mathbf{n}_m dF \approx 0. \qquad (2.157)$$

The momentum equation for the dispersed field is

$$\frac{\partial}{\partial \tau} \left(\alpha_l \rho_l \mathbf{V}_l \gamma_v \right) + \nabla \cdot \left(\alpha_l^e \rho_l \mathbf{V}_l \mathbf{V}_l \gamma \right) + \nabla \cdot \left[\alpha_l^e \left(\rho_l \overline{\mathbf{V}_l' \mathbf{V}_l'} \right) \gamma \right] + \alpha_l^e \gamma \nabla p_l + \alpha_l \gamma_v \rho_l \mathbf{g}$$

$$+ \left(p_l - p_m + \delta_l \sigma_{lm} \kappa_l \right) \nabla \left(\alpha_l^e \gamma \right) + \frac{1}{Vol} \int_{F_{l\sigma}} \Delta p_m^{l\sigma} \mathbf{n}_l dF = \gamma_v \sum_{\substack{k=1 \\ k \ne l}}^{3,w} \left(\mu_{kl} \mathbf{V}_k - \mu_{lk} \mathbf{V}_l \right).$$

(2.158)

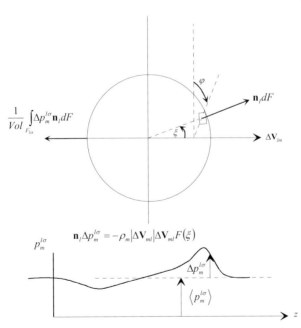

Fig. 2.10. Difference between the bulk pressure and interfacial pressure inside the velocity field m for steady state flow

The dispersed equation for the continuum (2.50b) is

$$\frac{\partial}{\partial \tau}\left(\alpha_m \rho_m \mathbf{V}_m \gamma_v\right) + \nabla \cdot \left(\alpha_m^e \rho_m \mathbf{V}_m \mathbf{V}_m \gamma\right) + \nabla \cdot \left[\alpha_m^e \gamma \left(\rho_m \overline{\mathbf{V}'_m \mathbf{V}'_m} - \delta_m \mathbf{T}_m\right)\right] + \alpha_m \gamma_v \rho_m \mathbf{g}$$

$$+ \alpha_m^e \nabla p_m - \frac{1}{Vol} \int_{F_{m\sigma}} \Delta p_m^{l\sigma} \mathbf{n}_l dF + \frac{1}{Vol} \int_{F_{mw}} \left(\Delta p_m^{w\sigma} \mathbf{I} - \mathbf{T}_m^{w\sigma}\right) \cdot \mathbf{n}_m dF$$

$$= \gamma_v \sum_{\substack{k=1 \\ k \neq m}}^{3,w} \left(\mu_{km} \mathbf{V}_k - \mu_{mk} \mathbf{V}_m\right) \qquad (2.159)$$

The term in the momentum equation for the dispersed field, which remains to be estimated is

$$\frac{1}{Vol} \int_{F_{l\sigma}} \Delta p_m^{l\sigma} \mathbf{n}_l dF .$$

For this, information is needed about the pressure distribution over the surface of a single particle - see Fig. 2.10. For illustration of the estimation of this integral we assume a family of mono-disperse spheres. For dispersed flow with a small concentration of the dispersed phase, the flow about each sphere can be considered as unaffected by its neighbors. The interface pressure distribution along the surface can be generally represented by the following expression

$$\Delta p_m^{l\sigma} \mathbf{n}_l = -\frac{1}{2} \rho_m R_l \left[\frac{\partial}{\partial \tau} \Delta \mathbf{V}_{ml} + \left(\mathbf{V}_l \cdot \nabla\right) \Delta \mathbf{V}_{ml}\right] \cos \xi + \rho_m \left|\Delta \mathbf{V}_{ml}\right|^2 F(\xi)$$

$$(2.160)$$

see, for example, *Stuhmiller* [63] (1977). Here a spherical coordinate system is used with the main axis along $\Delta \mathbf{V}_{lm}$. The polar angle ξ is measured with respect to the direction of $\Delta \mathbf{V}_{lm}$. The azimuthal angle is φ. The force per unit surface is split into a component parallel to $\Delta \mathbf{V}_{lm}$ and a component perpendicular to $\Delta \mathbf{V}_{lm}$. The integration is then performed. Note that Eq. (2.160) does not depend on φ and therefore the perpendicular component for the symmetric body is **0**. Note also that the pressure distribution in Eq. (2.160) does not take into account the spatial variation of the continuum velocity. The later will give rise even for a symmetric body a force component perpendicular to the $\Delta \mathbf{V}_{lm}$. For the integration we need the following relations. The differential surface element of the sphere (rotational body) is $dF = R_l^2 \sin \xi \, d\xi \, d\varphi$. The projection of the interfacial pressure force on direction $\Delta \mathbf{V}_{lm}$ for a single particle is therefore

$$\frac{1}{Vol}\int_{F_{l\sigma}}\Delta p_{m}^{l\sigma}\mathbf{n}_{l}\cos\xi dF = R_{l}^{2}\int_{0}^{2\pi}\int_{0}^{\pi}\Delta p_{m}^{l\sigma}\mathbf{n}_{l}\cos\xi\sin\xi\,d\xi\,d\varphi. \tag{2.161}$$

and on the plane normal to $\Delta \mathbf{V}_{lm}$

$$\frac{1}{Vol}\int_{F_{l\sigma}}\Delta p_{m}^{l\sigma}\mathbf{n}_{l}\sin\xi dF = R_{l}^{2}\int_{0}^{2\pi}\int_{0}^{\pi}\Delta p_{m}^{l\sigma}\mathbf{n}_{l}\sin^{2}\xi\,d\xi\,d\varphi. \tag{2.162}$$

For our case of a rotational body the above integral gives **0**. The collective force acting on the cloud of $n_l \sum_{l=1}^{3} Vol_l$ spheres in the control volume Vol per unit control volume in the axial direction is therefore

$$\frac{\sum_{l=1}^{3}Vol_{l}}{Vol}n_{l}R_{l}^{2}\int_{0}^{2\pi}\int_{0}^{\pi}\Delta p_{m}^{l\sigma}\mathbf{n}_{l}\cos\xi\sin\xi\,d\xi\,d\varphi = \gamma_{v}n_{l}R_{l}^{2}\int_{0}^{2\pi}\int_{0}^{\pi}\Delta p_{m}^{l\sigma}\mathbf{n}_{l}\cos\xi\sin\xi\,d\xi\,d\varphi$$
$$\tag{2.163}$$

and in the plane normal to $\Delta \mathbf{V}_{ml}$

$$\frac{\sum_{l=1}^{3}Vol_{l}}{Vol}n_{l}R_{l}^{2}\int_{0}^{2\pi}\int_{0}^{\pi}\Delta p_{m}^{l\sigma}\mathbf{n}_{l}\sin^{2}\xi\,d\xi\,d\varphi = \gamma_{v}n_{l}R_{l}^{2}\int_{0}^{2\pi}\int_{0}^{\pi}\Delta p_{m}^{l\sigma}\mathbf{n}_{l}\sin^{2}\xi\,d\xi\,d\varphi. \tag{2.164}$$

Estimation of the integrals (2.163) and (2.164) provides a practical approach for computing the interfacial forces. What remains after the integration and some rearrangements given in the next section is

$$\frac{1}{Vol}\int_{F_{l\sigma}}\Delta p_{m}^{l\sigma}\mathbf{n}_{l}dF = -\gamma_{v}\alpha_{l}\frac{1}{2}\rho_{m}\left[\frac{\partial}{\partial\tau}\Delta\mathbf{V}_{ml} + (\mathbf{V}_{l}\cdot\nabla)\Delta\mathbf{V}_{ml}\right]$$

$$+0.37 c_{ml}^{d}\rho_{m}\left|\Delta\mathbf{V}_{ml}\right|^{2}\nabla\left(\alpha_{l}^{e}\gamma\right) - \gamma_{v}\alpha_{l}\rho_{m}\frac{1}{D_{l}}\frac{3}{4}c_{ml}^{d}\left|\Delta\mathbf{V}_{ml}\right|\Delta\mathbf{V}_{ml}$$

$$=\gamma_{v}\left(\mathbf{f}_{l}^{vm}+\mathbf{f}_{l}^{d}\right)+\Delta p_{m}^{l\sigma*}\frac{1}{Vol}\int_{F_{l}^{m\sigma}}\mathbf{n}_{l}dF. \tag{2.165}$$

This result for one-dimensional flow was obtained by *Stuhmiller* [63] (1977). If one takes into account in Eq. (2.160) the non-isotropy of the continuum velocity field we obtain the general form

$$\frac{1}{Vol} \int_{F_{l\sigma}} \Delta p_m^{l\sigma} \mathbf{n}_l dF = \gamma_v \left(\mathbf{f}_l^{vm} + \mathbf{f}_l^{d} + \mathbf{f}_l^{L} \right) + \Delta p_m^{l\sigma*} \frac{1}{Vol} \int_{F_l^{m\sigma}} \mathbf{n}_l dF . \qquad (2.166)$$

The force components $\mathbf{f}_l^{vm}$, $\mathbf{f}_l^{d}$, $\mathbf{f}_l^{L}$, and $\Delta p_m^{l\sigma*} \frac{1}{Vol} \int_{F_l^{m\sigma}} \mathbf{n}_l dF$ are called virtual mass force, drag force, lift force, and stagnation pressure force, respectively. The detailed discussion toof them is given below. Empirical information how to compute these forces is given in Volume II.

The pressure distribution around a particle may influence the local mass transfer. In case of strong thermodynamic non-equilibrium the larger pressure difference across the interface at the stagnation point may lead to lower evaporation compared to the rear point. It may lead to a reactive resulting force at the droplet, which manifests itself as an effective drag reduction. A strong condensation may lead to the opposite effect. Although such arguments may sound reasonable one should be careful because there is no accurate theoretical or experimental treatment of this problem.

2.6.2.2 Virtual mass force

Consider the integral defined by Eq. (2.161) taken over the first term of Eq. (2.160)

$$\gamma_v \mathbf{f}_l^{vm} = -\gamma_v n_l \frac{1}{2} \rho_m \left[\frac{\partial}{\partial \tau} \Delta \mathbf{V}_{ml} + (\mathbf{V}_l \cdot \nabla) \Delta \mathbf{V}_{ml} \right] R_l^3 \int_0^{2\pi} d\varphi \int_0^{\pi} \cos^2 \xi \sin \xi d\xi$$

$$= -\gamma_v n_l \frac{4}{3} \pi R_l^3 \frac{1}{2} \rho_m \left[\frac{\partial}{\partial \tau} \Delta \mathbf{V}_{ml} + (\mathbf{V}_l \cdot \nabla) \Delta \mathbf{V}_{ml} \right]$$

$$= -\gamma_v \alpha_l \frac{1}{2} \rho_m \left[\frac{\partial}{\partial \tau} \Delta \mathbf{V}_{ml} + (\mathbf{V}_l \cdot \nabla) \Delta \mathbf{V}_{ml} \right]. \qquad (2.167)$$

The force $\mathbf{f}_l^{vm}$ is the *virtual mass force* per unit mixture volume. Here the virtual mass coefficient is $c_{ml}^{vm} = 1/2$. The general form of the virtual mass force with the accuracy of an empirical coefficient was proposed first by *Prandtl* [53] (1952), *Lamb* [46] (1952) and *Milne-Thomson* [49] (1968) in the same form

$$\mathbf{f}_d^{vm} = -\alpha_d \rho_c c_d^{vm} \left[\frac{\partial}{\partial \tau} \Delta \mathbf{V}_{cd} + (\mathbf{V}_d \cdot \nabla) \Delta \mathbf{V}_{cd} \right], \qquad (2.168)$$

where the subscripts c and d mean continuous and disperse, respectively. The scalar force components in *Cartesian* and in *cylindrical* coordinates are

$$f_{d,r}^{vm} = -\alpha_d \rho_c c_d^{vm} \left\{ \begin{array}{l} \dfrac{\partial}{\partial \tau}(u_c - u_d) + u_d \dfrac{\partial}{\partial r}(u_c - u_d) \\[6pt] +v_d \dfrac{1}{r^\kappa} \dfrac{\partial}{\partial \theta}(u_c - u_d) + w_d \dfrac{\partial}{\partial z}(u_c - u_d) \end{array} \right\}, \qquad (2.169)$$

$$f_{d,\theta}^{vm} = -\alpha_d \rho_c c_d^{vm} \left\{ \begin{array}{l} \dfrac{\partial}{\partial \tau}(v_c - v_d) + u_d \dfrac{\partial}{\partial r}(v_c - v_d) \\[6pt] +v_d \dfrac{1}{r^\kappa} \dfrac{\partial}{\partial \theta}(v_c - v_d) + w_d \dfrac{\partial}{\partial z}(v_c - v_d) \end{array} \right\}, \qquad (2.170)$$

$$f_{d,z}^{vm} = -\alpha_d \rho_c c_d^{vm} \left\{ \begin{array}{l} \dfrac{\partial}{\partial \tau}(w_c - w_d) + u_d \dfrac{\partial}{\partial r}(w_c - w_d) \\[6pt] +v_d \dfrac{1}{r^\kappa} \dfrac{\partial}{\partial \theta}(w_c - w_d) + w_d \dfrac{\partial}{\partial z}(w_c - w_d) \end{array} \right\}. \qquad (2.171)$$

The virtual mass force is experienced by the body as if it had an additional mass during its translation relative to the continuum. This explains the other name used for this force, *added mass force*. For larger particle concentrations, c_d^{vm} is a function of α_l. Expressions for practical computation of the virtual mass coefficient for dispersed fields with larger concentration can be found in Refs. [5,6,9,10,24,44,46,49,50,52,53,55,67,72,73,74,75]. These references represent the state of the art in this field. *Winatabe* et al. [72] (1990) proposed the use of only the transient part of the virtual mass force for the case of strong transients.

2.6.2.3 Form drag and stagnation pressure force

For inviscid (ideal) potential flow we have in Eq. (2.160)

$$F(\xi) = \frac{1}{8}\left(9\cos^2 \xi - 5\right), \qquad (2.172)$$

see *Lamb* [47]. This profile does not give any resulting force component (*d'Alembert's* paradox).

In the nature $F(\xi)$ gives an non-symmetric profile as indicated in Fig. 2.10. For an idealized non-symmetric fore-aft profile, *Nigmatulin* [51] (1979) estimated the integrals of Eqs. (2.160) and (2.161) for the second term of the right-hand side of Eq. (2.159) for bubbles in bubble-liquid flows. *Biesheavel* and *van Wijngarden* [5] (1984) computed analytically the coefficients for spherical bubbles for *Nigmatulin*'s derivation.

A more general approach was proposed by *Hwang* and *Schen* [22] (1992). For the general case, $F(\xi)$ is determined experimentally, see *Schlichting* [56] (1959), p.21, Fig. 1.11. *Hwang* and *Schen* [22] (1992) provided a method for computing the pressure distribution around a sphere for Reynolds numbers greater than 3000, where

$$F(\xi) = \frac{1}{4}\left\{\frac{2\lambda^2-1}{\lambda^2+4} + \frac{9}{2}\frac{e^{-\lambda\xi}}{\lambda^2+4}\left[2\lambda\sin(2\xi) + \cos(2\xi)\right]\right\}, \qquad (2.173)$$

see Eq. (12) in [22]. When the drag coefficient c_{ml}^d is known and the equation

$$c_{ml}^d = 45\frac{1-e^{-\pi\lambda}}{(\lambda^2+4)(\lambda^2+16)} \qquad (2.173)$$

is solved for the smaller real root of λ, there is a method for estimating the integrals analytically. The reader will find in [22] the final result of *Hwang* and *Schen*'s derivation, characterized by anisotropic forces.

Before *Hwang* and *Schen* had published their work, *Stuhmiller* [63] (1977) had already rewritten the term $F(\xi)$ as follows

$$F(\xi) = \langle F(\xi)\rangle^{1\sigma} + F(\xi) - \langle F(\xi)\rangle^{1\sigma}, \qquad (2.175)$$

where $\langle F(\xi)\rangle^{1\sigma}$ represents the surface average over a single sphere. For a *turbulent pressure distribution* around a sphere as given by *Schlichting* (1959), the interface average of the function $F(\xi)$ is

$$\langle F(\xi)\rangle^{1\sigma} = -0.37 c_{ml}^d, \qquad (2.176)$$

where c_{ml}^d is the form drag coefficient for single particles. In the literature some times $\langle F(\xi) \rangle^{l\sigma}$ is set to the constant value ¼ - see for instance *Lamb* [47]. This means that the integral over the second term of Eq. (2.160) can be split into two parts. The first part can be estimated directly exploiting the fact that $\langle F(\xi) \rangle^{l\sigma}$ does not depend on the position at the interface. The result is

$$\rho_m |\Delta \mathbf{V}_{ml}|^2 \int_{F_{l\sigma}^*} \langle F(\xi) \rangle^{l\sigma} \mathbf{n}_l dF = -\rho_m |\Delta \mathbf{V}_{ml}|^2 \langle F(\xi) \rangle^{l\sigma} \left[\nabla(\alpha_l^e \gamma) + \frac{1}{Vol} \int_{F_{lw}} \mathbf{n}_l dF \right]$$

$$= 0.37 c_{ml}^d \rho_m |\Delta \mathbf{V}_{ml}|^2 \left[\nabla(\alpha_l^e \gamma) + \frac{1}{Vol} \int_{F_{lw}} \mathbf{n}_l dF \right] = 0.37 c_{ml}^d \rho_m |\Delta \mathbf{V}_{ml}|^2 \nabla(\alpha_l^e \gamma),$$

(2.177)

this resulting in an effective stagnation pressure difference

$$\Delta p_m^{l\sigma*} = -0.37 c_{ml}^d \rho_m |\Delta \mathbf{V}_{ml}|^2 \qquad (2.178)$$

similar to that discussed for stratified flow, Eq. (2.64). To derive Eq. (2.177), Eq. (29) from Ref. [37] is also used together with the fact that there is no contact between the dispersed field and the wall, $F_{lw} = 0$.

The second part is the net force experienced by the particle due to non-uniform pressure distribution around the particle, the so-called form drag force:

$$\gamma_v \mathbf{f}_l^d = -\gamma_v n_l \rho_m |\Delta \mathbf{V}_{ml}| \Delta \mathbf{V}_{ml} R_l^2 \int_0^{2\pi} d\varphi \int_0^{\pi} \left[F(\xi) - \langle F(\xi) \rangle^{l\sigma} \right] \cos \xi \sin \xi d\xi$$

$$= -\gamma_v n_l \rho_m |\Delta \mathbf{V}_{ml}| \Delta \mathbf{V}_{ml} c_{ml}^d \frac{1}{2} \pi R_l^2 = -\gamma_v \left[\alpha_l / \left(\frac{4}{3} \pi R_l^3 \right) \right] \rho_m |\Delta \mathbf{V}_{ml}| \Delta \mathbf{V}_{ml} c_{ml}^d \frac{1}{2} \pi R_l^2$$

$$= -\gamma_v \alpha_l \rho_m \frac{1}{D_l} \frac{3}{4} c_{ml}^d |\Delta \mathbf{V}_{ml}| \Delta \mathbf{V}_{ml}. \qquad (2.179)$$

The form drag force per unit mixture volume is therefore

$$\mathbf{f}_l^d = -\alpha_l \rho_m \frac{1}{D_l} \frac{3}{4} c_{ml}^d |\Delta \mathbf{V}_{ml}| \Delta \mathbf{V}_{ml}. \qquad (2.180)$$

For larger volume fractions α_l one should take into account the dependence of the drag force on the volume fraction - also see Refs. [74, 24], for example. For drag forces in two-phase flows the study by *Ishii* and *Mishima* [24] (1984) is recommended.

2.6.2.4 Lift force

The force component perpendicular to the relative velocity direction is called lateral or *lift force*. The lift force is zero for bodies exposed to symmetrical flow

$$\mathbf{f}_{ml}^L = -\alpha_l \rho_m c_{ml}^L (\mathbf{V}_l - \mathbf{V}_m) \times (\nabla \times \mathbf{V}_m)$$

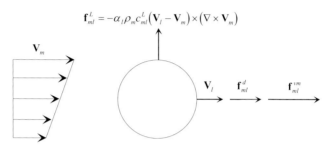

Fig. 2.11. Drag, virtual mass, and lift forces acting simultaneously on the field *l*

$$\mathbf{f}_l^L = 0. \qquad (2.181)$$

A body exposed to asymmetrical flow experiences a lateral force - see Fig. 2.11. The general form of the lateral lift force for inviscid flows is given by *Drew* and *Lahey* [16] (1987)

$$\mathbf{f}_{cd}^L = -\alpha_d \rho_c c_{cd}^L (\mathbf{V}_d - \mathbf{V}_c) \times (\nabla \times \mathbf{V}_c). \qquad (2.182)$$

The scalar components for *Cartesian* and *cylindrical* coordinates are

$$f_{cd,r}^L = -\alpha_d \rho_c c_{cd}^L \left\{ (v_d - v_c) \left[\frac{1}{r^\kappa} \frac{\partial}{\partial r} (r^\kappa v_c) - \frac{1}{r^\kappa} \frac{\partial u_c}{\partial \theta} \right] - (w_d - w_c) \left(\frac{\partial u_c}{\partial z} - \frac{\partial w_c}{\partial r} \right) \right\}, \qquad (2.183)$$

$$f_{cd,\theta}^L = -\alpha_d \rho_c c_{cd}^L \left\{ (w_d - w_c) \left[\frac{1}{r^\kappa} \frac{\partial w_c}{\partial \theta} - \frac{\partial v_c}{\partial z} \right] - (u_d - u_c) \left(\frac{1}{r^\kappa} \frac{\partial}{\partial r} (r^\kappa v_c) - \frac{1}{r^\kappa} \frac{\partial u_c}{\partial \theta} \right) \right\}, \qquad (2.184)$$

$$f_{cd,z}^L = -\alpha_d \rho_c c_{cd}^L \left\{ (u_d - u_c) \left[\frac{\partial u_c}{\partial z} - \frac{\partial w_c}{\partial r} \right] - (v_d - v_c) \left(\frac{1}{r^\kappa} \frac{\partial w_c}{\partial \theta} - \frac{\partial v_c}{\partial z} \right) \right\}. \qquad (2.185)$$

The lift coefficient must be derived experimentally. The reader will find information on modeling of the lift force in Refs. [12,62,4,61,21,68,16,45,17,3].

2.6.2.5 Interfacial structure forces

The continuous field interacts with the wall structures. Careful estimation of the surface integral

$$\frac{1}{Vol} \int_{F_{mw}} \Delta p_m^{w\sigma} \mathbf{n}_m dF$$

is required, especially in the case of variable geometry of the structure in space. The discussion of flow on immersed bodies given in Sections 2.6.2.2 through 2.6.2.4 is also valid for the case of a porous solid structure with a characteristic size of D_w. This means that the stagnation pressure force, form drag, virtual mass force and lift force must also be incorporated,

$$\frac{1}{Vol} \int_{F_{mw}} \left(\Delta p_m^{w\sigma} \mathbf{I} - \mathbf{T}_m^{w\sigma} \right) \cdot \mathbf{n}_m dF = \gamma_v \left(f_{wm}^d + f_{wm}^{vm} + f_{wm}^L \right) + \Delta p_m^{w\sigma*} \frac{1}{Vol} \int_{F_{mw}} \mathbf{n}_m dF .$$
(2.186)

For a continuous velocity field wetting the total structure, $F_{mw} = F_w$ and therefore

$$\frac{1}{Vol} \int_{F_{mw}} \mathbf{n}_m dF = \frac{1}{Vol} \int_{F_w} \mathbf{n}_m dF = \nabla (1-\gamma) = -\nabla \gamma .$$
(2.187)

which is in fact Eq. (1.29) with $\mathbf{n}_l = -\mathbf{n}_w$ and $\alpha_l^e = 1$. Consequently

$$\frac{1}{Vol} \int_{F_{mw}} \left(\Delta p_m^{w\sigma} \mathbf{I} - \mathbf{T}_m^{w\sigma} \right) \cdot \mathbf{n}_m dF = \gamma_v \left(f_{wm}^d + f_{wm}^{vm} + f_{wm}^L \right) - \Delta p_m^{w\sigma*} \nabla \gamma .$$
(2.188)

Here the following applies for a disperse structure (flow through porous media)

$$\mathbf{f}_{wm}^{vm} = \rho_m c_w^{vm} \left[\frac{\partial}{\partial \tau} \Delta \mathbf{V}_{mw} + (\mathbf{V}_w \cdot \nabla) \Delta \mathbf{V}_{mw} \right] = \overline{c_w^{vm}} \left[\frac{\partial}{\partial \tau} \Delta \mathbf{V}_{mw} + (\mathbf{V}_w \cdot \nabla) \Delta \mathbf{V}_{mw} \right],$$
(2.189)

$$\mathbf{f}_{wm}^d = \rho_m \frac{1}{D_w} \frac{3}{4} c_{mw}^d |\Delta \mathbf{V}_{mw}| \Delta \mathbf{V}_{mw} = \overline{c_{mw}^d} |\Delta \mathbf{V}_{mw}| \Delta \mathbf{V}_{mw} ,$$
(2.190)

$$\mathbf{f}_{wm}^L = \rho_m c_{mw}^L \left(\mathbf{V}_w - \mathbf{V}_m\right) \times \left(\nabla \times \mathbf{V}_m\right) = \overline{c_{mw}^L \left(\mathbf{V}_w - \mathbf{V}_m\right) \times \left(\nabla \times \mathbf{V}_m\right)}. \quad (2.191)$$

For the case of a wall at rest we have

$$\mathbf{f}_{wm}^{vm} = \rho_m c_w^{vm} \frac{\partial \mathbf{V}_m}{\partial \tau} = \overline{c_w^{vm} \frac{\partial \mathbf{V}_m}{\partial \tau}}, \quad (2.192)$$

$$\mathbf{f}_{wm}^{d} = \rho_m \frac{1}{D_w} \frac{3}{4} c_{mw}^d \left|\mathbf{V}_m\right| \mathbf{V}_m = \overline{c_{mw}^d \left|\mathbf{V}_m\right| \mathbf{V}_m}, \quad (2.193)$$

$$\mathbf{f}_{wm}^L = -\rho_m c_{mw}^L \mathbf{V}_m \times \left(\nabla \times \mathbf{V}_m\right) = -\overline{c_{mw}^L \mathbf{V}_m \times \left(\nabla \times \mathbf{V}_m\right)}. \quad (2.194)$$

The shear (friction) force for channels is usually incorporated into $\overline{c}_{mw}^d$. The same is performed for the drag resulting from local changes in the flow cross-section for the specific flow direction.

2.7 Working form for dispersed and continuous phase

Thus the momentum equation of a dispersed velocity field takes the form

$$\frac{\partial}{\partial \tau}\left(\alpha_l \rho_l \mathbf{V}_l \gamma_v\right) + \nabla \cdot \left(\alpha_l^e \rho_l \mathbf{V}_l \mathbf{V}_l \gamma\right) + \nabla \cdot \left[\alpha_l^e \left(\rho_l \overline{\mathbf{V}_l' \mathbf{V}_l'}\right)\gamma\right] + \alpha_l^e \gamma \nabla p_l + \alpha_l \gamma_v \rho_l \mathbf{g}$$

$$+ \left(p_l - p_m + \delta_l \sigma_{lm}\kappa_l\right)\nabla\left(\alpha_l^e \gamma\right) - \Delta p_m^{l\sigma*}\nabla\left(\alpha_l^e \gamma\right)$$

$$-\gamma_v \alpha_l \rho_m \left\{ \begin{array}{l} c_{ml}^{vm}\left[\dfrac{\partial}{\partial \tau}\Delta \mathbf{V}_{ml} + \left(\mathbf{V}_l \cdot \nabla\right)\Delta \mathbf{V}_{ml}\right] \\ \\ -c_{ml}^L \Delta \mathbf{V}_{ml} \times \left(\nabla \times \mathbf{V}_m\right) + \dfrac{1}{D_l}\dfrac{3}{4}c_{ml}^d \left|\Delta \mathbf{V}_{ml}\right|\Delta \mathbf{V}_{ml} \end{array} \right\}$$

$$= \gamma_v \sum_{\substack{k=1 \\ k \ne l}}^{3,w}\left(\mu_{kl}\mathbf{V}_k - \mu_{lk}\mathbf{V}_l\right). \quad (2.195)$$

The momentum equation of the continuous phase takes the following form so far

$$\frac{\partial}{\partial \tau}\left(\alpha_m \rho_m \mathbf{V}_m \gamma_v\right) + \nabla \cdot \left(\alpha_m^e \rho_m \mathbf{V}_m \mathbf{V}_m \gamma\right) + \nabla \cdot \left[\alpha_m^e \gamma \left(\rho_m \overline{\mathbf{V}_m' \mathbf{V}_m'} - \mathbf{T}_m\right)\right] + \alpha_m^e \gamma \nabla p_m$$

$$+\alpha_m \gamma_v \rho_m \mathbf{g} + \Delta p_m^{l\sigma*} \nabla (\alpha_l^e \gamma) - \Delta p_m^{w\sigma*} \nabla \gamma$$

$$+\gamma_v \alpha_l \rho_m \left\{ \begin{array}{l} c_{ml}^{vm} \left[\dfrac{\partial}{\partial \tau} \Delta \mathbf{V}_{ml} + (\mathbf{V}_l \cdot \nabla) \Delta \mathbf{V}_{ml} \right] \\[2ex] -c_{ml}^L \Delta \mathbf{V}_{ml} \times (\nabla \times \mathbf{V}_m) + \dfrac{1}{D_l} \dfrac{3}{4} c_{ml}^d |\Delta \mathbf{V}_{ml}| \Delta \mathbf{V}_{ml} \end{array} \right\}$$

$$+\gamma_v \rho_m \left[c_w^{vm} \dfrac{1}{2} \dfrac{\partial \mathbf{V}_m}{\partial \tau} - c_{mw}^L \mathbf{V}_m \times (\nabla \times \mathbf{V}_m) + \dfrac{1}{D_w} \dfrac{3}{4} c_{mw}^d |\mathbf{V}_m| \mathbf{V}_m \right]$$

$$= \gamma_v \sum_{\substack{k=1 \\ k \neq m}}^{3,w} (\mu_{km} \mathbf{V}_k - \mu_{mk} \mathbf{V}_m). \qquad (2.196)$$

The relation between the two bulk pressures is given by the momentum jump condition. With the assumptions made in Section 2.6.2.1 the momentum jump condition, Eq. (2.23), reduces to

$$(\rho w)_{ml}^2 \left(\dfrac{1}{\rho_l} - \dfrac{1}{\rho_m} \right) + p_l^{m\sigma,\tau} - p_m^{l\sigma,\tau} - \sigma_{ml} \kappa_m = 0. \qquad (2.197)$$

We exchange the subscripts l and m because the surface tension is assumed to belong to the liquid phase. After time averaging we obtain

$$(\rho w)_{ml}^2 \left(\dfrac{1}{\rho_l} - \dfrac{1}{\rho_m} \right) + p_l^{m\sigma} - p_m^{l\sigma} + \sigma_{ml} \kappa_l = 0, \qquad (2.198)$$

or in terms of bulk pressure

$$p_l - p_m + \sigma_{ml} \kappa_l = \Delta p_m^{l\sigma} + (\rho w)_{ml}^2 \left(\dfrac{1}{\rho_m} - \dfrac{1}{\rho_l} \right). \qquad (2.199)$$

Actually the pressure difference $\Delta p_m^{l\sigma}$ varies over the surface and some surface-averaged value

$$\frac{1}{F_{l\sigma}}\left|\int_{F_{l\sigma}}\Delta p_m^{l\sigma}\mathbf{n}_l dF\right| = \frac{Vol}{F_{l\sigma}}\frac{1}{Vol}\left|\int_{F_{l\sigma}}\Delta p_m^{l\sigma}\mathbf{n}_l dF\right| \qquad (2.200)$$

may be used. There is no experience in this field and future investigations are necessary. Approximations are thinkable for predominant surface tension and low mass transfer

$$p_l - p_m + \sigma_{ml}\kappa_l \approx 0, \qquad (2.201)$$

or for predominant mass transfer

$$p_l - p_m + \sigma_{ml}\kappa_l \approx (\rho w)_{ml}^2 \left(\frac{1}{\rho_m} - \frac{1}{\rho_l}\right). \qquad (2.202)$$

Note that for spheres $\kappa_l = 1/R_l + 1/R_l = 2/R_l$. Remember that if the radius is inside the field the curvature is negative.

Now let us as a practical illustration of the application of the theory analyze the eigen values for bubble flow without mass transfer in a one-dimensional horizontal channel with constant cross section assuming non-compressible phases and neglecting the diffusion terms. The governing system simplifies then to

$$p = p_2, \qquad (2.203)$$

$$p_1 = p + \sigma_{12}\kappa_2, \qquad (2.204)$$

$$\alpha_1 \frac{\partial w_1}{\partial z} + \alpha_2 \frac{\partial w_2}{\partial z} + (w_1 - w_2)\frac{\partial \alpha_1}{\partial z} = 0 \qquad (2.205)$$

$$\frac{\partial \alpha_1}{\partial \tau} + \alpha_1 \frac{\partial w_1}{\partial z} + w_1 \frac{\partial \alpha_1}{\partial z} = 0 \qquad (2.206)$$

$$\alpha_1\left(\rho_1 + \rho_2 c_{21}^{vm}\right)\left(\frac{\partial w_1}{\partial \tau} + w_1 \frac{\partial w_1}{\partial z}\right) - \alpha_1\rho_2 c_{21}^{vm}\left(\frac{\partial w_2}{\partial \tau} + w_1 \frac{\partial w_2}{\partial z}\right) + \alpha_1 \frac{\partial p}{\partial z} - \Delta p_2^{l\sigma*}\frac{\partial \alpha_1}{\partial z}$$

$$= \alpha_1\rho_2 \frac{1}{D_1}\frac{3}{4}c_{21}^d|\Delta w_{21}|\Delta w_{21}, \qquad (2.207)$$

$$-\alpha_1\rho_2 c_{21}^{vm}\left(\frac{\partial w_1}{\partial \tau} + w_1 \frac{\partial w_1}{\partial z}\right) + \left(\alpha_2 + \alpha_1 c_{21}^{vm} + c_w^{vm}\right)\rho_2 \frac{\partial w_2}{\partial \tau} + \left(\alpha_2 w_2 + \alpha_1 c_{21}^{vm} w_1\right)\rho_2 \frac{\partial w_2}{\partial z}$$

$$+\alpha_2 \frac{\partial p}{\partial z} + \Delta p_2^{1\sigma*} \frac{\partial \alpha_1}{\partial z} = -\frac{3}{4}\rho_2\left(\alpha_1 \frac{1}{D_1} c_{21}^d |\Delta w_{21}|\Delta w_{21} + \frac{1}{D_w} c_{2w}^d |w_2| w_2\right)$$

(2.208)

In matrix notation we have

$$\begin{pmatrix} 0 & 0 & 0 & 0 \\ 0 & 1 & 0 & 0 \\ 0 & 0 & \alpha_1(\rho_1 + \rho_2 c_{21}^{vm}) & -\alpha_1\rho_2 c_{21}^{vm} \\ 0 & 0 & -\alpha_1\rho_2 c_{21}^{vm} & (\alpha_2 + \alpha_1 c_{21}^{vm} + c_w^{vm})\rho_2 \end{pmatrix} \frac{\partial}{\partial \tau} \begin{pmatrix} p \\ \alpha_1 \\ w_1 \\ w_2 \end{pmatrix}$$

$$+ \begin{pmatrix} 0 & w_1 - w_2 & \alpha_1 & \alpha_2 \\ 0 & w_1 & \alpha_1 & 0 \\ \alpha_1 & -\Delta p_2^{1\sigma*} & \alpha_1(\rho_1 + \rho_2 c_{21}^{vm})w_1 & -\alpha_1\rho_2 c_{21}^{vm} w_1 \\ \alpha_2 & \Delta p_2^{1\sigma*} & -\alpha_1\rho_2 c_{21}^{vm} w_1 & (\alpha_2 w_2 + \alpha_1 c_{21}^{vm} w_1)\rho_2 \end{pmatrix} \frac{\partial}{\partial z} \begin{pmatrix} p \\ \alpha_1 \\ w_1 \\ w_2 \end{pmatrix}$$

$$= \begin{pmatrix} 0 \\ 0 \\ \alpha_1 \rho_2 \frac{1}{D_1} \frac{3}{4} c_{21}^d |\Delta w_{21}|\Delta w_{21} \\ -\frac{3}{4}\rho_2 \left(\alpha_1 \frac{1}{D_1} c_{21}^d |\Delta w_{21}|\Delta w_{21} + \frac{1}{D_w} c_{2w}^d |w_2| w_2\right) \end{pmatrix}$$

(2.209)

If you are not familiar with the analysis of the type of a system of partial differenttial equations by first computing the eigenvalues, eigenvectors and canonical forms it is recommended first to read Section 11 before continuing here.

The characteristic equation for determining the eigenvalues is then

$$\begin{vmatrix} 0 & w_1 - w_2 & \alpha_1 & \alpha_2 \\ 0 & w_1 - \lambda & \alpha_1 & 0 \\ \alpha_1 & -\Delta p_2^{1\sigma*} & \alpha_1(\rho_1 + \rho_2 c_{21}^{vm})(w_1 - \lambda) & -\alpha_1\rho_2 c_{21}^{vm}(w_1 - \lambda) \\ \alpha_2 & \Delta p_2^{1\sigma*} & -\alpha_1\rho_2 c_{21}^{vm}(w_1 - \lambda) & \left[\alpha_2 w_2 + \alpha_1 c_{21}^{vm} w_1 - \lambda(\alpha_2 + \alpha_1 c_{21}^{vm} + c_w^{vm})\right]\rho_2 \end{vmatrix} = 0.$$

(2.210)

or

$$\alpha_2 \left(c_{21}^{vm} + \alpha_2 \frac{\rho_1}{\rho_2} \right)(w_1 - \lambda)^2 + \alpha_1 (w_2 - \lambda)\left[\alpha_2 w_2 + \alpha_1 c_{21}^{vm} w_1 - \lambda \left(\alpha_2 + \alpha_1 c_{21}^{vm} + c_w^{vm} \right) \right]$$

$$-\alpha_1 \alpha_2 (w_1 - w_2) c_{21}^{vm} (w_1 - \lambda) + \alpha_2 \frac{\Delta p_2^{1\sigma*}}{\rho_2} = 0 \qquad (2.211)$$

or

$$a\lambda^2 - 2b\lambda + c = 0 \qquad (2.212)$$

where

$$a = \alpha_2 \left(\alpha_1 - \alpha_2 \frac{\rho_1}{\rho_2} \right) + \left(\alpha_1^2 - \alpha_2 \right) c_{21}^{vm} + \alpha_1 c_w^{vm} > 0, \qquad (2.213)$$

$$b = \frac{1}{2}\left\{ \left[\alpha_1 w_2 + \left(\alpha_1^2 - 2\alpha_2 - \alpha_1 \alpha_2 \right) w_1 \right] c_{21}^{vm} + \alpha_1 \left(2\alpha_2 + c_w^{vm} \right) w_2 - 2w_1 \alpha_2^2 \frac{\rho_1}{\rho_2} \right\},$$

(2.214)

$$c = \left(\alpha_1 w_1 w_2 + \alpha_2^2 w_1^2 \right) c_{21}^{vm} + \alpha_1 \alpha_2 w_2^2 + \alpha_2^2 \frac{\rho_1}{\rho_2} w_1^2 + \alpha_2 \frac{\Delta p_2^{1\sigma*}}{\rho_2}, \qquad (2.215)$$

with two real solutions

$$\lambda_{1,2} = \frac{b \pm \sqrt{b^2 - ac}}{a} \qquad (2.216)$$

for

$$b^2 > ac \qquad (2.217)$$

which is in fact the stability criterion for bubbly flow.

2.8 General working form for dispersed and continuous phases

In this section we write a single equation valid for both disperse and continuous phases. First we compare the terms in the two equations

$$\mathbf{f}_{d\alpha} = \left(p_d - p_c + \delta_d \sigma_{dc} \kappa_d - \Delta p_c^{d\sigma*} \right) \nabla \left(\alpha_d^e \gamma \right), \qquad (2.218)$$

and

$$\mathbf{f}_{c\alpha} = \Delta p_c^{d\sigma*} \nabla \left(\alpha_d^e \gamma \right), \qquad (2.219)$$

and realize that in both cases the force is a function of the gradient of the disperse volume fraction multiplied by different multipliers. With this notation we have

$$\boxed{\begin{aligned}&\frac{\partial}{\partial \tau}\left(\alpha_l \rho_l \mathbf{V}_l \gamma_v\right)+\nabla \cdot \left(\alpha_l^e \rho_l \gamma \left\{\mathbf{V}_l \mathbf{V}_l - v_l^* \left[\nabla \mathbf{V}_l + (\nabla \mathbf{V}_l)^T - \frac{2}{3}(\nabla \cdot \mathbf{V}_l)\mathbf{I}\right]\right\}\right)\\ &+\alpha_l^e \nabla p_l + \alpha_l \rho_l \mathbf{g}\gamma_v - \Delta p_l^{w\sigma*}\nabla\gamma + \mathbf{f}_{l\alpha}\\ &-\gamma_v \sum_{\substack{m=1 \\ m\neq l}}^{3}\left\{\begin{aligned}&\overline{c}_{ml}^{d}\left|\Delta \mathbf{V}_{ml}\right|\Delta \mathbf{V}_{ml}\\ &+\overline{c}_{ml}^{L}\left(\mathbf{V}_l - \mathbf{V}_m\right)\times\left(\nabla \times \mathbf{V}_m\right)+\overline{c}_{ml}^{vm}\left[\frac{\partial}{\partial \tau}\Delta\mathbf{V}_{ml}+\left(\mathbf{V}_l \cdot \nabla\right)\Delta\mathbf{V}_{ml}\right]\end{aligned}\right\}\\ &+\gamma_v\left[\overline{c}_{lw}^{d}\left|\mathbf{V}_l\right|\mathbf{V}_l + \rho_l c_{lw}^{vm}\frac{\partial \mathbf{V}_l}{\partial \tau}-\rho_l c_{lw}^{L}\mathbf{V}_l \times(\nabla \times \mathbf{V}_l)\right]=\gamma_v \sum_{m=1}^{3,w}\left(\mu_{ml}\mathbf{V}_m - \mu_{lm}\mathbf{V}_l\right).\end{aligned}}$$

(2.220)

Note that

$$\Delta p_d^{w\sigma*} = 0, \qquad (2.221)$$

$$\Delta p_c^{w\sigma*} \neq 0. \qquad (2.222)$$

Similarly

$$c_{dw}^{d}, c_{dw}^{vm}, c_{dw}^{L} = 0, \qquad (2.223)$$

and

$$c_{cw}^d, c_{cw}^{vm}, c_{cw}^L \neq 0, \qquad (2.224)$$

if the continuum is in a contact with the wall. For easy programming, a couple of simple drag, lift and virtual mass coefficients combined as follows is introduced for each field

$$\overline{c}_{ml}^d = \overline{c}_{lm}^d = \frac{3}{4}\left(\alpha_m \rho_l c_{lm}^d / D_m + \alpha_l \rho_m c_{ml}^d / D_l \right), \qquad (2.225)$$

$$\overline{c}_{ml}^L = \overline{c}_{lm}^L = \alpha_m \rho_l c_{lm}^L + \alpha_l \rho_m c_{ml}^L, \qquad (2.226)$$

$$\overline{c}_{ml}^{vm} = \overline{c}_{lm}^{vm} = \alpha_m \rho_l c_{lm}^{vm} + \alpha_l \rho_m c_{ml}^{vm}. \qquad (2.227)$$

If field l is disperse surrounded by the continuous m-field, the coefficient c_{ml} is not equal to zero and c_{lm} is equal to zero and vice versa. With other words if the second subscript refers to a disperse field the local size of dispersion is positive and the coefficients are not equal to zero. For application in computer codes the following general notation is recommended

$$\mathbf{f}_l = \sum_{\substack{m=1 \\ m \neq l}}^{3,w} \overline{c}_{ml}^d \mathbf{f}_{ml} \qquad (2.228)$$

to take into account the fact that a control volume may contain two dispersed fields carried by one continuous field. This approach together with implicit discretization of the momentum equations, their strong coupling through a special numerical procedure and comparison with data for three-phase bubble flow was presented by *Kolev* et al. in Ref. [30]. In Volume II the reader will find additional information on practical computation of drag forces in multi-phase flows.

Equation (2.220) is the rigorously derived local volume and time average momentum balance for multi-phase flows in heterogeneous porous structures conditionally divided into three velocity fields.

The non-conservative form of the momentum conservation equation in component notation is given in Appendix 2.3. In the same appendix some interesting single-phase analytical solutions are given. They can be used as benchmarks for testing the accuracy of the numerical solution methods.

2.9 Some practical simplifications

Equation (2.220) has been used since 1984 in the IVA1 to IVA6 computer codes [25-29,31-35,40,41,43] with the following simplifications:

$$p_l - p_m + \frac{2\sigma_{lm}}{R_l} - \Delta p_m^{l\sigma*} \approx 0, \tag{2.229}$$

$$p_l \approx p_m = p, \tag{2.230}$$

$$v_l' \approx 0. \tag{2.231}$$

Assumption (2.229) is quite close to the local volume and time average interfacial jump condition at the interface and therefore does not lead to any problems for slow interfacial mass transfer.

Assumption (2.230) leads to the so-called *single-pressure model*. It should be emphasized that the most important interfacial pressure forces, which are considerably larger in magnitude than the error introduced by the single-pressure assumption, have already been taken into account. This assumption likewise does not lead to any problems. In this type of single-pressure model the hyperbolicy is preserved thanks to the stabilizing viscous, drag and virtual mass terms. Neglect of the viscous, drag and virtual mass terms leads to unphysical models.

Assumption (2.231) was dictated by the lack of knowledge. When information for v_l' becomes available, this can easily be included, as the viscous terms have already been taken into account.

The resulting simplified form of Eq. (2.220) is therefore

$$\frac{\partial}{\partial \tau}\left(\alpha_l \rho_l \mathbf{V}_l \gamma_v\right) + \nabla \cdot \left(\alpha_l^e \rho_l \gamma \left\{ \mathbf{V}_l \mathbf{V}_l - v_l^* \left[\nabla \mathbf{V}_l + \left(\nabla \mathbf{V}_l\right)^T - \frac{2}{3}\left(\nabla \cdot \mathbf{V}_l\right)\mathbf{I}\right]\right\}\right)$$

$$+ \alpha_l^e \gamma \nabla p + \alpha_l \rho_l \mathbf{g} \gamma_v$$

$$-\gamma_v \sum_{\substack{m=1 \\ m \neq l}}^{3} \left\{ \begin{array}{l} \overline{c}_{ml}^{d} |\Delta \mathbf{V}_{ml}| \cdot \Delta \mathbf{V}_{ml} \\ + \overline{c}_{ml}^{L} (\mathbf{V}_l - \mathbf{V}_m) \times (\nabla \times \mathbf{V}_m) + \overline{c}_{ml}^{vm} \left[\frac{\partial}{\partial \tau} \Delta \mathbf{V}_{ml} + (\mathbf{V}_l \cdot \nabla) \Delta \mathbf{V}_{ml} \right] \end{array} \right\}$$

$$+\gamma_v \left[\overline{c}_{lw}^{d} |\mathbf{V}_l| \cdot \mathbf{V}_l + \rho_l c_{lw}^{vm} \frac{\partial \mathbf{V}_l}{\partial \tau} - \rho_l c_{lw}^{L} \mathbf{V}_l \times (\nabla \times \mathbf{V}_l) \right]$$

$$= \gamma_v \sum_{m=1}^{3,w} (\mu_{ml} \mathbf{V}_m - \mu_{lm} \mathbf{V}_l). \tag{2.232}$$

The form of Eq. (2.232) is some times called *conservative* in order to distinguish it from the *non-conservative* form. The non-conservative form is derived by applying the chain rule to the first two terms and inserting the mass-conservation equation (1.45). The resulting equation,

$$\alpha_l \rho_l \left[\gamma_v \frac{\partial \mathbf{V}_l}{\partial \tau} + (\mathbf{V}_l \gamma \nabla) \mathbf{V}_l \right] - \nabla \cdot \left\{ \alpha_l^e \rho_l \gamma v_l^* \left[\nabla \mathbf{V}_l + (\nabla \mathbf{V}_l)^T - \frac{2}{3} (\nabla \cdot \mathbf{V}_l) \mathbf{I} \right] \right\}$$

$$+\alpha_l^e \gamma \nabla p + \alpha_l \rho_l \mathbf{g} \gamma_v$$

$$-\gamma_v \sum_{\substack{m=1 \\ m \neq l}}^{3} \left\{ \begin{array}{l} \overline{c}_{ml}^{d} |\Delta \mathbf{V}_{ml}| \cdot \Delta \mathbf{V}_{ml} \\ + \overline{c}_{ml}^{L} (\mathbf{V}_l - \mathbf{V}_m) \times (\nabla \times \mathbf{V}_m) + \overline{c}_{ml}^{vm} \left[\frac{\partial}{\partial \tau} \Delta \mathbf{V}_{ml} + (\mathbf{V}_l \cdot \nabla) \Delta \mathbf{V}_{ml} \right] \end{array} \right\}$$

$$+\gamma_v \left[\overline{c}_{lw}^{d} |\mathbf{V}_l| \cdot \mathbf{V}_l + \rho_l c_{lw}^{vm} \frac{\partial \mathbf{V}_l}{\partial \tau} - \rho_l c_{lw}^{L} \mathbf{V}_l \times (\nabla \times \mathbf{V}_l) \right]$$

$$= \gamma_v \left\{ \sum_{m=1}^{3} [\mu_{ml} (\mathbf{V}_m - \mathbf{V}_l)] + \mu_{wl} (\mathbf{V}_{wl} - \mathbf{V}_l) + \mu_{lw} (\mathbf{V}_{lw} - \mathbf{V}_l) \right\}, \tag{2.233}$$

contains some extremely interesting information, namely

(a) mass sinks of the velocity field *l* have no influence on the velocity change (an exception is the controlled flow suction from the structure through the structure interface), and

2.9 Some practical simplifications

(b) mass sources from a donor field whose velocities differ from the velocity of the receiving field influence the velocity change.

To facilitate direct use of the vector equation (2.333), we give its scalar components for the most frequently used cylindrical, $\kappa = 1$, and Cartesian, $\kappa = 0$, coordinate systems.

r direction:

$$\frac{\partial}{\partial \tau}(\alpha_l \rho_l u_l \gamma_v) + \frac{1}{r^\kappa}\frac{\partial}{\partial r}\left[r^\kappa \alpha_l \rho_l \left(u_l u_l - v_l^* \frac{\partial u_l}{\partial r}\right)\gamma_r\right]$$

$$+ \frac{1}{r^\kappa}\frac{\partial}{\partial \theta}\left[\alpha_l \rho_l \left(v_l u_l - v_l^* \frac{1}{r^\kappa}\frac{\partial u_l}{\partial \theta}\right)\gamma_\theta\right]$$

$$- \frac{\kappa}{r^\kappa}\alpha_l \rho_l \left[v_l v_l - \frac{2}{r^\kappa}v_l^*\left(\frac{\partial v_l}{\partial \theta} + u_l\right)\right]\gamma_\theta + \frac{\partial}{\partial z}\left[\alpha_l \rho_l\left(w_l u_l - v_l^*\frac{\partial u_l}{\partial z}\right)\gamma_z\right]$$

$$+\alpha_l \gamma_r \frac{\partial p}{\partial r} + (\alpha_l \rho_l g_r + f_{lu})\gamma_v = \gamma_v \sum_{m=1}^{3,w}(\mu_{ml} u_{ml} - \mu_{lm} u_l) + f_{vlu} \qquad (2.234)$$

where

$$f_{vlu} = \frac{1}{r^\kappa}\frac{\partial}{\partial r}\left(r^\kappa \alpha_l \rho_l v_l^* \frac{\partial u_l}{\partial r}\gamma_r\right) + \frac{1}{r^\kappa}\frac{\partial}{\partial \theta}\left[\alpha_l \rho_l v_l^* r^\kappa \frac{\partial}{\partial r}\left(\frac{v_l}{r^\kappa}\right)\gamma_\theta\right]$$

$$+ \frac{\partial}{\partial z}\left(\alpha_l \rho_l v_l^* \frac{\partial w_l}{\partial z}\gamma_z\right) - \frac{2}{3}\frac{1}{r^\kappa}\left\{\frac{\partial}{\partial r}\left[r^\kappa (\alpha_l \rho_l v_l^* \nabla \cdot \mathbf{V}_l)\gamma_r\right] - (\alpha_l \rho_l v_l^* \nabla \cdot \mathbf{V}_l)\gamma_\theta\right\};$$

$$(2.235)$$

θ direction

$$\frac{\partial}{\partial \tau}(\alpha_l \rho_l v_l \gamma_v) + \frac{1}{r^\kappa}\frac{\partial}{\partial r}\left[r^\kappa \alpha_l \rho_l \left(u_l v_l - v_l^* \frac{\partial v_l}{\partial r}\right)\gamma_r\right]$$

$$+ \frac{1}{r^\kappa}\frac{\partial}{\partial \theta}\left[\alpha_l \rho_l \left(v_l v_l - v_l^* \frac{1}{r^\kappa}\frac{\partial v_l}{\partial \theta}\right)\gamma_\theta\right]$$

$$+\frac{\kappa}{r^\kappa}\alpha_l\rho_l\left[v_lu_l-r^\kappa v_l^*\frac{\partial}{\partial r}\left(\frac{v_l}{r^\kappa}\right)-v_l^*\frac{1}{r^\kappa}\frac{\partial u_l}{\partial\theta}\right]\gamma_\theta+\frac{\partial}{\partial z}\left[\alpha_l\rho_l\left(w_lv_l-v_l^*\frac{\partial v_l}{\partial z}\right)\gamma_z\right]$$

$$+\alpha_l\gamma_\theta\frac{1}{r^\kappa}\frac{\partial p}{\partial\theta}+\left(\alpha_l\rho_lg_\theta+f_{lv}\right)\gamma_v=\gamma_v\sum_{m=1}^{3,w}\left(\mu_{ml}v_{ml}-\mu_{lm}v_l\right)+f_{vlv} \qquad (2.236)$$

where

$$f_{vlv}=\frac{1}{r^\kappa}\frac{\partial}{\partial r}\left[\alpha_l\rho_lv_l^*\left(\frac{\partial u_l}{\partial\theta}-v_l\right)\gamma_r\right]+\frac{1}{r^\kappa}\frac{\partial}{\partial\theta}\left[\alpha_l\rho_lv_l^*\frac{1}{r^\kappa}\left(\frac{\partial v_l}{\partial\theta}+2u_l\right)\gamma_\theta\right]$$

$$+\frac{\partial}{\partial z}\left(\alpha_l\rho_lv_l^*\frac{1}{r^\kappa}\frac{\partial w_l}{\partial\theta}\gamma_z\right)-\frac{2}{3}\frac{1}{r^\kappa}\frac{\partial}{\partial\theta}\left[\left(\alpha_l\rho_lv_l^*\nabla\cdot\mathbf{V}_l\right)\gamma_\theta\right]; \qquad (2.237)$$

z direction

$$\frac{\partial}{\partial\tau}\left(\alpha_l\rho_lw_l\gamma_v\right)+\frac{1}{r^\kappa}\frac{\partial}{\partial r}\left[r^\kappa\alpha_l\rho_l\left(u_lw_l-v_l^*\frac{\partial w_l}{\partial r}\right)\gamma_r\right]$$

$$+\frac{1}{r^\kappa}\frac{\partial}{\partial\theta}\left[\alpha_l\rho_l\left(v_lw_l-v_l^*\frac{1}{r^\kappa}\frac{\partial w_l}{\partial\theta}\right)\gamma_\theta\right]+\frac{\partial}{\partial z}\left[\alpha_l\rho_l\left(w_lw_l-v_l^*\frac{\partial w_l}{\partial z}\right)\gamma_z\right]$$

$$+\alpha_l\gamma_z\frac{\partial p}{\partial z}+\left(\alpha_l\rho_lg_z+f_{lw}\right)\gamma_v=\gamma_v\sum_{m=1}^{3,w}\left(\mu_{ml}w_{ml}-\mu_{lm}w_l\right)+f_{vlw} , \qquad (2.238)$$

where

$$f_{vlw}=\frac{1}{r^\kappa}\frac{\partial}{\partial r}\left(r^\kappa\alpha_l\rho_lv_l^*\frac{\partial u_l}{\partial z}\gamma_r\right)+\frac{1}{r^\kappa}\frac{\partial}{\partial\theta}\left(\alpha_l\rho_lv_l^*\frac{\partial v_l}{\partial z}\gamma_\theta\right)+\frac{\partial}{\partial z}\left(\alpha_l\rho_lv_l^*\frac{\partial w_l}{\partial z}\gamma_z\right)$$

$$-\frac{2}{3}\frac{\partial}{\partial z}\left[\left(\alpha_l\rho_lv_l^*\nabla\cdot\mathbf{V}_l\right)\gamma_z\right]. \qquad (2.239)$$

Here

$$v_l^*\nabla\cdot\mathbf{V}_l=v_l^*\left[\frac{1}{r^\kappa}\frac{\partial}{\partial r}\left(r^\kappa u_l\right)+\frac{1}{r^\kappa}\frac{\partial v_l}{\partial\theta}+\frac{\partial w_l}{\partial z}\right]. \qquad (2.240)$$

Note that all interfacial forces are designated with $\mathbf{f}_l$ with components f_{lu}, f_{lv}, f_{lw}.

The viscous terms have been rearranged in order to obtain a *convection-diffusion form* for the left hand side of the momentum equations. The residual terms are pooled into the momentum source terms $\mathbf{f}_v$. This notation is justified for two reasons. First, for a single-phase flow in a pool (unrestricted flows) and a constant density, the source terms $\mathbf{f}_v$ are equal to zero, which *intuitively leads to the idea that the main viscous influence is outside the $\mathbf{f}_v$ source terms*. This reasoning led some authors to derive an explicit discretization for the $\mathbf{f}_v$ source terms for single-phase flow applications, see *Trent* and *Eyler* [65] (1983), or even to their neglect. Second, methods with known mathematical properties for the discretization of convection-diffusion equations have already been developed, and these can be applied directly.

Note that for Cartesian coordinates the convective components contain spatial derivatives. In the case of cylindrical coordinates the convective part contains in addition two components not containing spatial derivatives. The component

$$centrifugal\ force = -\frac{\kappa}{r^\kappa}\alpha_l \rho_l v_l v_{l\theta} \qquad (2.241)$$

is known in the literature as centrufagal forces. It gives the effective force component in the r direction resulting from fluid motion in the θ direction. The component

$$Coriolis\ force = \frac{\kappa}{r^\kappa}\alpha_l \rho_l v_l u_l \gamma_\theta \qquad (2.242)$$

Is known in the literature as *Coriolis* force. It is an effective force component in the θ direction when there is flow in both the r and θ directions. This results in the components of the viscous stress tensor, $\tau_{l,\theta\theta}$ and $\tau_{l,r\theta}$ corresponding to these forces and acting in the opposite directions.

2.10 Conclusion

The positive experience with Eq. (2.334) to (2.340) in the development of the IVA code [25, 36, 41, 43] allows one to recommend these equations for general use. One should keep in mind for application purposes that both sides of the equations are local volume and time averages.

As next step in this direction, a rigorous formulation of the equations reflecting the second law of thermodynamics for this multi-phase multi-component system has been successfully derived, which is an important new result of this development. The result of this derivation has formed the subject of a Chapter 4, see also [39], and a comment to this publication [42]. Understanding of the local volume and time average momentum equations is a prerequisite for understanding the second law of thermodynamics and its extremely interesting application to yield a simple description of this highly complicated system.

Appendix 2.1

Substituting in the momentum conservation equation and performing time averaging,

$$\overline{\langle p_l^\tau \rangle^{le}} = p_l,$$

$$\overline{\langle p_m^\tau \rangle^{me}} = p_m,$$

$$\overline{\alpha_l \gamma_v \langle \rho_l \rangle^l \mathbf{V}_l} = \alpha_l \gamma_v \rho_l \mathbf{V}_l,$$

$$\overline{\alpha_l \gamma_v \langle \rho_l \rangle^l \mathbf{V}_l'} = 0,$$

$$\overline{\alpha_l^e \gamma \langle \rho_l \rangle^l \mathbf{V}_l \mathbf{V}_l} = \alpha_l^e \gamma \rho_l \mathbf{V}_l \mathbf{V}_l,$$

$$\overline{\alpha_l^e \gamma \langle \rho_l \rangle^l \mathbf{V}_l' \mathbf{V}_l} = 0,$$

$$\overline{\alpha_l^e \gamma \langle \rho_l \rangle^l \mathbf{V}_l' \mathbf{V}_l'} = \alpha_l^e \gamma \rho_l \overline{\mathbf{V}_l' \mathbf{V}_l'},$$

$$\overline{\nabla \cdot \left(\alpha_l^e \gamma \langle \mathbf{T}_l^\tau \rangle^{le} \right)} = \nabla \cdot \left(\alpha_l^e \gamma \mathbf{T}_l \right),$$

$$\overline{\alpha_l^e \gamma \nabla \cdot \langle p_l^\tau \rangle^{le}} = \alpha_l^e \gamma \nabla \cdot p_l,$$

$$\overline{\alpha_l \gamma_v \langle \rho_l \rangle^l \mathbf{g}} = \alpha_l \gamma_v \rho_l \mathbf{g},$$

$$\sum_{m=1}^{3} \overline{\left(\mu_{ml}^\tau \langle \mathbf{V}_m^\tau \rangle^{me} - \mu_{lm}^\tau \langle \mathbf{V}_l^\tau \rangle^{le} \right)} \approx \sum_{m=1}^{3} \left(\mu_{ml} \mathbf{V}_m - \mu_{lm} \mathbf{V}_l \right),$$

and

$$+\left(p_l - p_m + \delta_l \sigma_{lm} \kappa_l\right)\left(\nabla\left(\alpha_l^e \gamma\right) + \frac{1}{Vol}\int_{F_{lw}} \mathbf{n}_l dF\right) - \delta_l \frac{1}{Vol}\int_{F_{l\sigma}} (\nabla_t \sigma_{lm}) dF$$

$$+ \frac{1}{Vol}\int_{F_{l\sigma}} \left(\Delta p_m^{l\sigma} \mathbf{I} - \mathbf{T}_m^{l\sigma}\right)\cdot \mathbf{n}_l dF + \frac{1}{Vol}\int_{F_{lw}} \left(\Delta p_l^{w\sigma} \mathbf{I} - \mathbf{T}_l^{w\sigma}\right)\cdot \mathbf{n}_l dF$$

$$= \gamma_v \sum_{m=1}^{3,w}\left(\mu_{ml}\mathbf{V}_m - \mu_{lm}\mathbf{V}_l\right)$$

$$\int_{F_{l\sigma}}\left\{\left(-\Delta p_m^{l\sigma}\mathbf{I} + \mathbf{T}_m^{l\sigma}\right)\cdot \mathbf{n}_l - \nabla_t \sigma_{lm}\right\}dF + \int_{F_{lw}}\left(-\Delta p_l^{w\sigma}\mathbf{I} + \mathbf{T}_l^{w\sigma}\right)\cdot \mathbf{n}_l dF$$

$$= \int_{F_{l\sigma}}\left\{\left(-\Delta p_m^{l\sigma}\mathbf{I} + \mathbf{T}_m^{l\sigma}\right)\cdot \mathbf{n}_l - \nabla_t \sigma_{lm}\right\}dF + \int_{F_{lw}}\left(-\Delta p_l^{w\sigma}\mathbf{I} + \mathbf{T}_l^{w\sigma}\right)dF$$

one obtains Eqn. (2.50a).

Appendix 2.2

The normal velocity difference can be obtained by splitting the relative velocity vector at the interface $\Delta \mathbf{V}_{lm}$ into a component which is parallel to $\mathbf{n}_l$

$$\Delta \mathbf{V}_{lm}^n = proj_{\mathbf{n}_l} \Delta \mathbf{V}_{lm} = \left(\frac{\Delta \mathbf{V}_{lm} \cdot \mathbf{n}_l}{\mathbf{n}_l \cdot \mathbf{n}_l}\right)\mathbf{n}_l = \left(\Delta \mathbf{V}_{lm}\cdot \mathbf{n}_l\right)\mathbf{n}_l,$$

with a magnitude

$$\left|\Delta \mathbf{V}_{lm}^n\right| = \left|\Delta \mathbf{V}_{lm}\cdot \mathbf{n}_l\right| = \sqrt{\left[n_{lx}(u_l - u_m)\right]^2 + \left[n_{ly}(v_l - v_m)\right]^2 + \left[n_{lz}(w_l - w_m)\right]^2}$$

and a component orthogonal to $\mathbf{n}_l$,

$$\Delta \mathbf{V}_{lm}^t = \Delta \mathbf{V}_{lm} - proj_{\mathbf{e}_l}\Delta \mathbf{V}_{lm} = \Delta \mathbf{V}_{lm} - \left(\Delta \mathbf{V}_{lm}\cdot \mathbf{n}_l\right)\mathbf{n}_l$$

$$= \left[\Delta u_{lm} - n_{lx}\left(\Delta \mathbf{V}_{lm}\cdot \mathbf{n}_l\right)\right]\mathbf{i} + \left[\Delta v_{lm} - n_{ly}\left(\Delta \mathbf{V}_{lm}\cdot \mathbf{n}_l\right)\right]\mathbf{j} + \left[\Delta w_{lm} - n_{lz}\left(\Delta \mathbf{V}_{lm}\cdot \mathbf{n}_l\right)\right]\mathbf{k}$$

Appendix 2.3

The non-conservative form of Eqs. (2.195) is

r direction:

$$\alpha_l \rho_l \left(\gamma_v \frac{\partial u_l}{\partial \tau} + \gamma_r \frac{1}{2} \frac{\partial u_l^2}{\partial r} + v_l \gamma_\theta \frac{1}{r^\kappa} \frac{\partial u_l}{\partial \theta} + w_l \gamma_z \frac{\partial u_l}{\partial z} \right)$$

$$-\frac{1}{r^\kappa} \frac{\partial}{\partial r} \left(r^\kappa \alpha_l \rho_l v_l^* \frac{\partial u_l}{\partial r} \gamma_r \right) - \frac{1}{r^\kappa} \frac{\partial}{\partial \theta} \left(\alpha_l \rho_l v_l^* \frac{1}{r^\kappa} \frac{\partial u_l}{\partial \theta} \gamma_\theta \right) - \frac{\partial}{\partial z} \left(\alpha_l \rho_l v_l^* \frac{\partial u_l}{\partial z} \gamma_z \right)$$

$$-\frac{\kappa}{r^\kappa} \alpha_l \rho_l \left[v_l v_l - \frac{2}{r^\kappa} v_l^* \left(\frac{\partial v_l}{\partial \theta} + u_l \right) \right] \gamma_\theta + \alpha_l \gamma_r \frac{\partial p}{\partial r} + \left(\alpha_l \rho_l g_r + f_{lu} \right) \gamma_v$$

$$= \gamma_v \left\{ \sum_{m=1}^{3} \left[\mu_{ml} \left(u_m - u_l \right) \right] + \mu_{wl} \left(u_{wl} - u_l \right) - \mu_{lw} \left(u_{lw} - u_l \right) \right\} + f_{vlu}.$$

θ direction:

$$\alpha_l \rho_l \left(\gamma_v \frac{\partial v_l}{\partial \tau} + u_l \gamma_r \frac{\partial v_l}{\partial r} + \gamma_\theta \frac{1}{2} \frac{1}{r^\kappa} \frac{\partial v_l^2}{\partial \theta} + w_l \gamma_z \frac{\partial v_l}{\partial z} \right)$$

$$-\frac{1}{r^\kappa} \frac{\partial}{\partial r} \left(r^\kappa \alpha_l \rho_l v_l^* \frac{\partial v_l}{\partial r} \gamma_r \right) - \frac{1}{r^\kappa} \frac{\partial}{\partial \theta} \left(\alpha_l \rho_l v_l^* \frac{1}{r^\kappa} \frac{\partial v_l}{\partial \theta} \gamma_\theta \right) - \frac{\partial}{\partial z} \left(\alpha_l \rho_l v_l^* \frac{\partial v_l}{\partial z} \gamma_z \right)$$

$$+\frac{\kappa}{r^\kappa} \alpha_l \rho_l \left[v_l u_l - r^\kappa v_l^* \frac{\partial}{\partial r} \left(\frac{v_l}{r^\kappa} \right) - v_l^* \frac{1}{r^\kappa} \frac{\partial u_l}{\partial \theta} \right] \gamma_\theta + \alpha_l \gamma_\theta \frac{1}{r^\kappa} \frac{\partial p}{\partial \theta} + \left(\alpha_l \rho_l g_\theta + f_{lv} \right) \gamma_v.$$

$$= \gamma_v \left\{ \sum_{m=1}^{3} \left[\mu_{ml} \left(v_m - v_l \right) \right] + \mu_{wl} \left(v_{wl} - v_l \right) - \mu_{lw} \left(v_{lw} - v_l \right) \right\} + f_{vlv}$$

z direction:

$$\alpha_l \rho_l \left(\gamma_v \frac{\partial w_l}{\partial \tau} + u_l \gamma_r \frac{\partial w_l}{\partial r} + v_l \gamma_\theta \frac{1}{r^\kappa} \frac{\partial w_l}{\partial \theta} + \gamma_z \frac{1}{2} \frac{\partial w_l^2}{\partial z} \right)$$

$$-\frac{1}{r^\kappa} \frac{\partial}{\partial r} \left(r^\kappa \alpha_l \rho_l v_l^* \frac{\partial w_l}{\partial r} \gamma_r \right) - \frac{1}{r^\kappa} \frac{\partial}{\partial \theta} \left(\alpha_l \rho_l v_l^* \frac{1}{r^\kappa} \frac{\partial w_l}{\partial \theta} \gamma_\theta \right) - \frac{\partial}{\partial z} \left(\alpha_l \rho_l v_l^* \frac{\partial w_l}{\partial z} \gamma_z \right)$$

$$+\alpha_1 \gamma_z \frac{\partial p}{\partial z} + (\alpha_1 \rho_1 g_z + f_{lw})\gamma_v$$

$$= \gamma_v \left\{ \sum_{m=1}^{3} \left[\mu_{ml}(w_m - w_l) \right] + \mu_{wl}(w_{wl} - w_l) - \mu_{lw}(w_{lw} - w_l) \right\} + f_{vlw}.$$

Some simple single-phase test cases: For testing numerical solutions it is important to provide a set of simple benchmarks having analytical solutions. Some of them are presented here.

Rigid body steady rotation problem: This test problem presents a hollow cylinder with symmetric flow in the azimuthal direction, see Fig. A2.3-1. No axial and radial flow exists. The mass conservation equation gives $\frac{\partial v}{\partial \theta} = 0$. The r direction momentum equation simplifies to $\rho \frac{v^2}{r} = \frac{\partial p}{\partial r}$, and the θ direction momentum equation gives $\frac{\partial p}{\partial \theta} = 0$. For constant rotational frequency ω, $(v(r)=r\omega)$, the analytical solution of the radial momentum equation is $p - p_0 = \frac{1}{2}\rho\omega^2 (r^2 - r_0^2)$

or $p - p_0 = \frac{1}{2}\rho [v(r)]^2 \left[1 - \left(\frac{r_0}{r} \right)^2 \right]$.

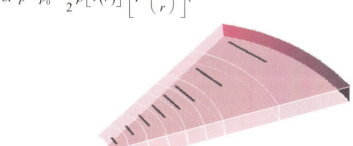

Fig. A2.3-1 Geometry of the test problem *rigid body steady rotation*

Pure radial symmetric flow: This test problem presents a hollow cylinder with symmetric flow in the radial direction, see Fig. A2.3-2.

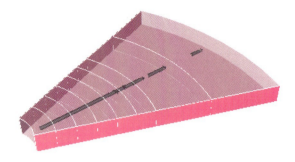

Fig. A2.3-2 Geometry of the test problem *pure radial symmetric flow*.

No axial and azimuthal flow exists. The mass conservation equation gives $\frac{\partial}{\partial r}(ru) = 0$. The r direction momentum equation simplifies to $\rho \frac{1}{2}\frac{\partial u^2}{\partial r} = -\frac{\partial p}{\partial r}$, and the θ direction momentum equation gives $\frac{\partial p}{\partial \theta} = 0$. From the mass conservation we have $u = u_0 \frac{r_0}{r}$. The analytical solution of the radial momentum equation is the well-known *Bernulli* equation $p - p_0 = -\frac{1}{2}\rho(u^2 - u_0^2)$ or

$$p - p_0 = \frac{1}{2}\rho u_0^2 \left[1 - \left(\frac{r_0}{r}\right)^2\right].$$

Radial-azimuthal symmetric flow: This test problem presents a hollow cylinder with symmetric flow in the radial and azimuthal directions - in fact a superposition of the previous two cases, rigid body steady rotation and pure radial symmetric flow, see Fig. A2.3-3. No axial flow exists. The mass conservation equation gives $\frac{\partial}{\partial r}(ru) = 0$. The r direction momentum equation simplifies to $\rho\left(\frac{1}{2}\frac{\partial u^2}{\partial r} - \frac{v^2}{r}\right) = -\frac{\partial p}{\partial r}$, and the θ-direction momentum equation gives $\rho u\left(\frac{\partial v}{\partial r} + \frac{v}{r}\right) = -\frac{1}{r}\frac{\partial p}{\partial \theta}$. From the mass conservation we have $u = u_0 \frac{r_0}{r}$. From the azimuthal symmetry, $\frac{\partial p}{\partial \theta} = 0$, the θ direction momentum equation gives $\frac{\partial v}{\partial r} + \frac{v}{r} = 0$ or $v = v_0 \frac{r_0}{r}$. Taking into account the both solutions of the mass and of the θ momentum equation the radial momentum equation gives

$$\rho\frac{1}{r}\left(u^2+v^2\right)=\frac{\partial p}{\partial r} \text{ or } \rho\left(u_0^2+v_0^2\right)r_0^2\frac{1}{r^3}=\frac{\partial p}{\partial r} \text{ or }$$

$$p-p_0=\rho\left(u_0^2+v_0^2\right)r_0^2\frac{1}{2}\left(\frac{1}{r_0^2}-\frac{1}{r^2}\right).$$

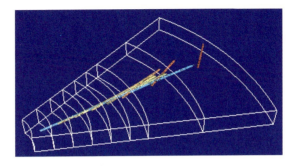

Fig. A2.3-3 Geometry of the test problem *radial-azimuthal symmetric flow*

References

1. Anderson TB, Jackson R (1967) A fluid mechanical description of fluidized beds, Ind. Eng. Fundam., vol 6 pp 527
2. Barnea D, Taitel Y (1994) Interfacial and structural stability, Int. J. Multiphase Flow, vol 20 Suppl pp 387-414
3. Bataille J, Lance M, Marie JL (1990) Bubble turbulent shear flows, ASME Winter Annular Meeting, Dallas, Nov. 1990
4. Bernemann K, Steiff A, Weinspach PM (1991) Zum Einfluss von längsangeströmten Rohrbündeln auf die großräumige Flüssigkeitsströmung in Blasensäulen, Chem. Ing. Tech., vol 63 no 1 pp 76-77
5. Biesheuvel A, van Wijngaarden L (1984) Two-phase flow equations for a dilute dispersion of gas bubbles in liquid, J. Fluid Mechanics, 168 pp 301-318
6. Biesheuvel A, Spollstra S (1989) The added mass coefficient of dispersion of gas bubbles in liquid, Int. J. Multiphase Flow, no. 15 pp 911-924
7. Brackbill JU, Kothe DB, Zemach C (1992) A continuum method for modeling surface tension, Journal of Computational Physics, vol 100 pp 335-354
8. Brauner N, Maron DM (1992) Stability analysis of stratified liquid-liquid flow, Int. J. Multiphase Flow, vol 18 no 1 pp 103-121
9. Cook TL, Harlow FH (1983) VORT: A computer code for bubble two-phase flow. Los Alamos National Laboratory documents LA-10021-MS
10. Cook TL, Harlow FH (1984) Virtual mass in multi-phase flow, Int. J. Multiphase Flow, vol. 10 no 6 pp 691-696
11. de Crecy F (1986) Modeling of stratified two-phase flow in pipes, pumps and other devices, Int. J. Multiphase Flow, vol 12 no 3 pp 307-323

12. Deich ME, Philipoff GA (1981) Gas dynamics of two phase flows, Energoisdat, Moskva
13. Delhaye JM (1981) Basic equations for two-phase flow, in Bergles AE et al (eds) Two-phase flow and heat transfer in power and process industries, Hemisphere Publishing Corporation, McGraw-Hill Book Company, New York
14. Delhaye JM, Giot M, Riethmuller ML (1981) Thermohydraulics of two-phase systems for industrial design and nuclear engineering, Hemisphere Publishing Corporation, New York, McGraw Hill Book Company, New York
15. Drazin PG, Reid WH (1981) Hydrodynamic Stability, Cambridge Univ. Press, Cambridge, UK
16. Drew DA, Lahey RT Jr (1987) The virtual mass and lift force on a sphere in rotating and straining flow, Int. J. Multiphase Flow, vol. 13 no 1, pp 113-121
17. Erichhorn R, Small S (1969) Experiments on the lift and drag of spheres suspended in a poiseuille flow, J. Fluid Mech., vol 20-3 pp 513
18. Gray WG, Lee PCY (1977) On the theorems for local volume averaging of multi-phase system, Int. J. Multi-Phase Flow, vol 3 pp 222-340
19. Hetstrony G (1982) Handbook of multi-phase systems. Hemisphere Publ. Corp., Washington et al., McGraw-Hill Book Company, New York et al.
20. Hirt CW, Nichols BD (1981) Volume of fluid (VOF) method for dynamics of free boundaries, J. of Comp. Physics, vol 39 p 201-225
21. Ho BP, Leal LG (1976) Internal migration of rigid spheres in two-dimensional unidirectional flows, J. Fluid Mech., vol 78 no 2 p 385
22. Hwang GJ, Schen HH (Sept. 21-24, 1992) Tensorial solid phase pressure from hydrodynamic interaction in fluid-solid flows. Proc. of the Fifth International Topical Meeting On Reactor Thermal Hydraulics, NURETH-5, Salt Lake City, UT, USA, IV pp 966-971
23. Ishii M (1975) Thermo-fluid dynamic theory of two-phase flow, Eyrolles, Paris
24. Ishii M, Michima K (1984) Two-fluid model and hydrodynamic constitutive relations, NSE 82 pp 107-126
25. Kolev NI (March 1985) Transiente Dreiphasen-Dreikomponenten Strömung, Teil 1: Formulierung des Differentialgleichungssystems, KfK 3910
26. Kolev NI (1986) Transiente Dreiphasen-Dreikomponenten Strömung, Teil 3: 3D-Dreifluid-Diffusionsmodell, KfK 4080
27. Kolev NI (1986) Transient three-dimensional three-phase three-component non equilibrium flow in porous bodies described by three-velocity fields, Kernenergie vol 29 no 10 pp 383-392
28. Kolev NI (Aug. 1987), A three field-diffusion model of three-phase, three-component Flow for the transient 3D-computer code IVA2/001. Nuclear Technology, vol 78 pp 95-131
29. Kolev NI (1991) IVA3: A transient 3D three-phase, three-component flow analyzer, Proc. of the Int. Top. Meeting on Safety of Thermal Reactors, Portland, Oregon, July 21-25, 1991, pp 171-180. The same paper was presented to the 7th Meeting of the IAHR Working Group on Advanced Nuclear Reactor Thermal - Hydraulics, Kernforschungszentrum Karlsruhe, August 27 to 29, 1991
30. Kolev NI, Tomiyama A, Sakaguchi T (Sept. 1991) Modeling of the mechanical interaction between the velocity fields in three-phase flow, Experimental Thermal and Fluid Science, vol 4 no 5 pp 525-545

31. Kolev NI (Sept. 1991) A three-field model of transient 3D multi-phase, three-component flow for the computer code IVA3, Part 1: Theoretical basics: conservation and state equations, Numerics. KfK 4948, Kernforschungszentrum Karlsruhe
32. Kolev NI (1993) The code IVA3 for modeling of transient three-phase flows in complicated 3D geometry, Kerntechnik, vol 58 no 3 pp 147 - 156
33. Kolev NI (1993) IVA3 NW: Computer code for modeling of transient three-phase flow in complicated 3D geometry connected with industrial networks, Proc. of the Sixth Int. Top. Meeting on Nuclear Reactor Thermal Hydraulics, Oct. 5-8, 1993, Grenoble, France
34. Kolev NI (1993) Berechnung der Fluiddynamischen Vorgänge bei einem Sperrwasser-Kühlerrohrbruch, Projekt KKW Emsland, Siemens KWU Report R232/93/0002
35. Kolev NI (1993) IVA3-NW A three phase flow network analyzer. Input description, Siemens KWU Report R232/93/E0041
36. Kolev NI (1993) IVA3-NW components: relief valves, pumps, heat structures, Siemens KWU Report R232/93/E0050
37. Kolev NI (1994) IVA4: Modeling of mass conservation in multi-phase multi-component flows in heterogeneous porous media. Kerntechnik, vol 59 no 4-5 pp 226-237
38. Kolev NI (1994) The code IVA4: Modelling of momentum conservation in multi-phase multi-component flows in heterogeneous porous media, Kerntechnik, vol 59 no 6 pp 249-258
39. Kolev NI (1995) The code IVA4: Second law of thermodynamics for multi phase flows in heterogeneous porous media, Kerntechnik, vol 60 no 1, pp 1-39
40. Kolev NI (1994) The influence of the mutual bubble interaction on the bubble departure diameter, Experimental Thermal and Fluid Science, vol 8 pp 167-174
41. Kolev NI (1996) Three Fluid Modeling With Dynamic Fragmentation and Coalescence Fiction or Daily practice? 7th FARO Experts Group Meeting Ispra, October 15-16, 1996; Proceedings of OECD/CSNI Workshop on Transient thermal-hydraulic and neutronic codes requirements, Annapolis, MD, U.S.A., 5th-8th November 1996; 4th World Conference on Experimental Heat Transfer, Fluid Mechanics and Thermodynamics, ExHFT 4, Brussels, June 2-6, 1997; ASME Fluids Engineering Conference & Exhibition, The Hyatt Regency Vancouver, Vancouver, British Columbia, CANADA June 22-26, 1997, Invited Paper; Proceedings of 1997 International Seminar on Vapor Explosions and Explosive Eruptions (AMIGO-IMI), May 22-24, Aoba Kinen Kaikan of Tohoku University, Sendai-City, Japan.
42. Kolev NI (1997) Comments on the entropy concept, Kerntechnik, vol 62 no 1 pp 67-70
43. Kolev N I (1999) Verification of IVA5 computer code for melt-water interaction analysis, Part 1: Single phase flow, Part 2: Two-phase flow, three-phase flow with cold and hot solid spheres, Part 3: Three-phase flow with dynamic fragmentation and coalescence, Part 4: Three-phase flow with dynamic fragmentation and coalescence – alumna experiments, CD Proceedings of the Ninth International Topical Meeting on Nuclear Reactor Thermal Hydraulics (NURETH-9), San Francisco, California, October 3-8,1999, Log. Nr. 315
44. Lahey RT Jr (Jan. 1991) Void wave propagation phenomena in two-phase flow, AIChE Journal, vol 31 no 1 pp 123-135
45. Lahey RT Jr (1990) The analysis of phase separation and phase distribution phenomena using two-fluid models, NED 122 pp 17-40
46. Lamb MA (1945) Hydrodynamics, Cambridge University Press, Cambridge

47. Lamb H (1945) Hydrodynamics, Dover, New York
48. Mamaev WA, Odicharia G S, Semeonov N I, Tociging A A (1969) Gidrodinamika gasogidkostnych smesey w trubach, Moskva
49. Milne-Thomson LM (1968) Theoretical Hydrodynamics, MacMillan & Co. Ltd., London
50. Mokeyev GY (1977) Effect of particle concentration on their drag induced mass, Fluid. Mech. Sov. Res., vol 6 p 161
51. Nigmatulin RT (1979) Spatial averaging in the mechanics of heterogeneous and dispersed systems, Int. J. of Multiphase Flow, vol. 5 pp. 353-389
52. No HC, Kazimi MS (1985) Effects of virtual mass of the mathematical characteristics and numerical stability of the two-fluid model, NSE 89 pp. 197-206
53. Prandtl L (1952) Essentials of Fluid Dynamics, Blackie & Son, Glasgow pp. 342
54. Ransom VH et al. (March 1987) RELAP5/MOD2 Code manual, vol 1: Code structure, system models, and solution methods, NUREG/CR-4312, EGG-2396, rev 1
55. Ruggles AE et al (1988) An investigation of the propagation of pressure perturbation in bubbly air/water flows, Trans. ASME J. Heat Transfer , vol 110 pp 494-499
56. Schlichting H (1959) Boundary layer theory, Mc Graw-Hill, New York
57. Sha T, Chao BT, Soo SL (1984) Porous-media formulation for multi-phase flow with heat transfer, Nuclear Engineering and Design, vol 82 pp 93-106
58. Slattery JC (1967) Flow of visco-elastic fluids through porous media, AIChE Journal, vol 13 pp 1066
59. Slattery JC (1990) Interfacial transport phenomena, Springer-Verlag, Berlin Heidelberg New York
60. Slattery JC (1999) Advanced transport phenomena, Cambridge University Press
61. Soo SL, Tung, S K (1972) J. Powder Techn., vol 6 p 283
62. Staffman PG (1965) The lift on a small sphere in a slow shear flow, J. Fluid Mech., vol 22, Part 2 pp 385-400
63. Stuhmiller JH (1977) The influence of the interfacial pressure forces on the character of the two-phase flow model, Proc. of the 1977 ASME Symp. on Computational Techniques for Non-Equilibrium Two-Phase Phenomena, pp 118-124
64. Thomas GB, Jr., Finney RL, Weir MD (1998) Calculus and analytic geometry, 9th Edition, Addison-Wesley Publishing Company, Reading, MA
65. Trent DS, Eyler LL (1983) Application of the TEMPTEST computer code for simulating hydrogen distribution in model containment structures, PNL-SA-10781, DE 83 002725
66. Truesdell C (1968) Essays in the history of mechanics, Springer-Verlag, New York
67. van Wijngaarden L (1976) Hydrodynamic interaction between gas bubbles in liquid, J. Fluid Mech., vol 77 pp 27-44
68. Vasseur P, Cox RG (1976) The lateral migration of spherical particles in two-dimensional shear flows, J. Fluid Mech., vol 78 Part 2 pp 385-413
69. Whitaker S (1967) Diffusion and dispersion in porous media, AIChE Journal, vol 13 pp 420
70. Whitaker S (1969) Advances in theory of fluid motion in porous media, Ind. Engrg. Chem., vol 61 no 12 pp 14-28
71. Whitaker S (1985) A Simple geometrical derivation of the spatial averaging theorem, Chemical Engineering Education, pp 18-21, pp 50-52

72. Winatabe T, Hirano M, Tanabe F, Kamo H (1990) The effect of the virtual mass force term on the numerical stability and efficiency of the system calculations, Nuclear Engineering and design, vol 120 pp 181-192
73. Wallis GB (1969) One-dimensional two-phase flow, McGraw-Hill, New York
74. Zuber N (1964) On the dispersed two-phase flow in the laminar flow regime, Chem. Eng. Science, vol 49 pp 897-917
75. Zwick SA (1906) J. Math. Phys., vol 4 p 289
76. Kolev NI (1977) Two-phase two-component flow (air-water steam-water) among the safety compartments of the nuclear power plants with water cooled nuclear reactors during lose of coolant accidents, PhD Thesis, Technical University Dresden
77. Biberg D (December 1999) An explicit approximation for the wetted angle in two-phase stratified pipe flow, The Canadian Journal of Chemical Engineering, vol 77 pp 1221-1224

3 Derivatives for the equations of state

Derivation of partial derivatives for isothermal multi-component mixtures needed for development of universal models for multi-phase multi-component flows is presented. The equations of state and the derivative approximations as functions of temperature and pressure for the elementary mixture constituents are assumed to be known. The so called universal fluid is introduced consisting of an arbitrary number of miscible and non-miscible components. This fluid model describes in its limiting cases gas, or gas mixture, or liquid, or solution of liquids, or gas-liquid solutions with an arbitrary number of gaseous and liquid components, or gas-liquid solutions containing immiscible liquid or solid particles. In addition, one component liquid-gas and solid-liquid equilibrium mixtures are considered.

3.1 Introduction

Numerical modeling of complicated physical phenomena such as multi-component multi-phase flows is a powerful tool supplementing experiments and enabling optimum design of complicated technical facilities. The wide range of computer codes developed over the past 30 years for the description of multi-dimensional single-, two- and multiphase flows inevitably leads to the step of developing a universal flow analyzer. Such a computer code should model transient and steady-state three-dimensional flows in a complicated geometry with arbitrary internals. The flow should be described by multi-velocity fields, each of them consisting of an arbitrary number of chemical components.

The local volume- and time-averaged fundamental equations are derived by applying the so-called instantaneous equations inside the velocity field and averaging these over space and time. But even the instantaneous conservation equations for mass, momentum and energy are averages too in the sense that the motions of the individual molecules are averaged. This leads to loss of information on the thermodynamic behavior of the system. The information lost must be provided by the state and transport equations which already incorporate the averaging procedure by virtue of their derivation.

For a mathematical description of the flow, time and three space coordinates are generally used as independent variables, with a set of time- and space-dependent variables, e.g. p and T as dependent variables. Besides the velocities that describe the flow, there are dependent variables that describe the thermodynamic state of the particular velocity field. This group of variables has to consist of mutually independent variables, e.g. (p,T), or (p,h), or (p,s), or (p,ρ), etc., see for example, *Kestin* [11] (1979). Some valuable references are also given in Appendix 3.2. All other properties of the velocity field are a unique function of these mutually independent variables. Analytical or tabular approximations of the equation of state and of the transport properties as a function of pressure (p) and temperature (T) are in general use. As already mentioned, the set of dependent variables can be transformed from (p,T) to another set, e.g. (p,h) or (p,s), and the system of PDEs can be rearranged in terms of the chosen set. From the large variety of systems that result, only one has the remarkable quality of being the simplest. Interestingly enough, it is just this system which is obtained if the specific entropy of the velocity fields is used as an element of the dependent variable vector. This will be demonstrated in Chapters 4 and 5. An interesting dilemma thus arises: on the one hand, the desire to integrate the simplest possible system of PDEs in order to save computer time, to reduce errors during code development, and to concentrate on other related models incorporated into the code; on the other, the desire to use an existing library of analytical approximations for the equation of state and for the transport equation as a function of pressure and temperature. These two tendencies are not contradictory. A solution is easily obtained if the equation of state is used in the form

$$T = T(p,s), \quad dT = \left(\frac{\partial T}{\partial p}\right)_s dp + \left(\frac{\partial T}{\partial s}\right)_p ds \qquad (3.1)$$

and

$$\rho = \rho(p,s), \quad d\rho = \left(\frac{\partial \rho}{\partial p}\right)_s dp + \left(\frac{\partial \rho}{\partial s}\right)_p ds \qquad (3.2)$$

where the derivatives $(\partial T/\partial p)_s, (\partial T/\partial s)_p, (\partial \rho/\partial p)_s$, and $(\partial \rho/\partial s)_p$ are functions of (p,T). The derivatives $(\partial \rho/\partial p)_s$ and $(\partial \rho/\partial s)_p$ are needed for development of the very important link between the density change and the change in the other dependent variables describing the flow. This link is, at the same time, the link between the mass, momentum, and energy-conservation equations. This construct for the solution methods is called the *entropy concept*.

The results presented can be used in the framework of the entropy concept, or in any other concepts for development of solution methods in the field of multi-phase fluid dynamics. The expressions used in the IVA computer code family [18-20, 22] are the simple cases for two components of the general expressions presented in this chapter for the *n*-component. Complete implementation of the theory for computational analyses was performed by this author in [21, 23, 24].

3.2 Multi-component mixtures of miscible and non-miscible components

Every fluid that seams to be pure in nature is in fact a mixture of substances. The idealization pure applied to a fluid is very helpful in science. In many cases replacing mixtures of one predominant fluid with some traces of other substances is a very good approximation to work with. But there are a lot of other applications where even these small "impurities" substantially control processes. Examples are the initiation of phase transitions in metastable liquids and gases – the so called heterogeneous nucleation process starting at proffered "impurities" such as fine gas bubbles or solid particles. Another example is the solution and dissolution process of gases. In some technological applications such processes may lead to explosive gas mixtures which under some circumstances may ignite and destroy the facility.

The results obtained in [12, 13] (1986) for mixtures consisting of one real and one ideal gas were generalized in 1991 in [17] for mixtures with an unlimited number of chemical gas components. Chapter 3.2 in [24] was an extended version of [12, 13]. Here we extend the theory once again for mixtures consisting simultaneously of miscible and non-miscible chemical constituents.

Consider a mixture of $i_{max} = m_{max} + n_{max}$ components, where m_{max} is the number of the <u>m</u>iscible components and n_{max} is the number of the <u>n</u>on-miscible components.

Examples of miscible components are:

a) Gases inside a gas mixture;
b) Gases dissolved in liquid;
c) Liquids dissolved in liquid.

In this case each component occupies the total volume occupied by the mixture itself. The volumetric fraction of each miscible component α_i in the multiphase flow is therefore equal to the volume fraction of the corresponding velocity field, α, and the definition of the density of the mixture $\alpha\rho = \sum_{i=1}^{i_{max}} \alpha_i \rho_i$ simplifies to $\rho = \sum_{i=1}^{i_{max}} \rho_i$. As a result, the definition of the mass concentrations simplifies to $C_i = \alpha_i \rho_i / (\alpha\rho) = \rho_i / \rho$. The total pressure p is then the sum of the partial pressures of the miscible components, $p = \sum_{i=1}^{m_{max}} p_i$. This relation expresses the law of *Dalton*.

Examples of non-miscible components are:

a) Mixtures of non-miscible liquids;
b) Mixture of solid particles and liquid.

As already said, in a mixture consisting of miscible components, each component occupies the total volume occupied by the mixture itself. This is not the case for mixtures consisting of non-miscible components. The volume fraction occupied by each species differs from the volume fraction occupied by the mixture itself. The definition of the mixture density is then $\left(\sum_{i=1}^{i_{max}} \alpha_i \right) \rho = \sum_{i=1}^{i_{max}} \alpha_i \rho_i$, and the definition of the mass concentrations therefore

$$C_i = \alpha_i \rho_i / \left(\sum_{i=1}^{i_{max}} \alpha_i \rho_i \right) = \alpha_i \rho_i / \left[\left(\sum_{i=1}^{i_{max}} \alpha_i \right) \rho \right]$$

does not simplify, as does that for gas mixtures.

The non-miscible components experience the same total pressure which is equal to the total mixture pressure $p_n = p$, where $n = m_{max} + 1, m_{max} + n_{max}$.

Examples of mixture of miscible and non-miscible fluids are in fact all fluids in the nature such us:

a) Liquid water containing dissolved gases and impurities such us solid particles of different chemical components;
b) Lava consisting of several molten components containing dissolved gases and solid particles from other then the molten species;
c) Air containing microscopic impurities such us dust.

There are many examples of dramatic events happening in nature and in technology due to the release of the dissolved components inside mixtures:

a) Volcanic explosions initiated first by pressure release and followed by a dramatic release of the dissolved gases;
b) Choked flow in propulsion machines caused by gas release due to a pressure drop;
c) Hydrogen and oxygen release and accumulation in nuclear power plant pipelines that could be ignited and cause explosive damage etc.

3.2.1 Computation of partial pressures for known mass concentrations, system pressure and temperature

Task definition. The mass concentration of each constituent, designated C_i is

$$C_i := \rho_i / \sum_{m=1}^{i_{max}} \rho_m = \rho_i / \rho, \qquad (3.3)$$

where

$$\frac{1}{\rho} = \frac{1 - \sum_{n=m_{max}+1}^{i_{max}} C_n}{\sum_{m=1}^{m_{max}} \rho_m(p_m, T)} + \sum_{n=m_{max}+1}^{i_{max}} \frac{C_n}{\rho_n(p, T)}, \qquad (3.4)$$

is the mixture density. Equations (3.3) and (3.4) are valid for mixture consisting simultaneously of both miscible and non-miscible components. The system pressure, temperature of the mixture and mass concentration of each constituent,

$$p, T, C_{1,2,\ldots,i_{max}-1}, C_{i_{max}} = 1 - \sum_{k=1}^{i_{max}-1} C_k$$

are, by definition, known. The partial pressure of each particular miscible component is sought.

Solution. We start with the $m_{max} - 1$ definition equations for the mass concentrations,

$$\rho_i(p_i, T) = C_i \rho, \quad i = 1, m_{max} - 1, \qquad (3.5)$$

and with *Dalton*'s law valid for the miscible components (i.e., the sum of the partial pressures = system pressure)

$$p = \sum_{m=1}^{m_{max}} p_m. \qquad (3.6)$$

In the general case of real fluids, Eqs. (3.5) and (3.6) are a system of non-linear algebraic equations for the unknowns $p_m, m = 1, m_{max}$, which can be solved numerically by iterations. Before showing this solution we will first give the well-known solution for which the miscible components are perfect fluids. It can be used as a first approximation for the numerical solution.

3 Derivatives for the equations of state

The solution if the miscible components are perfect fluids. For miscible fluids assumed to be *perfect* we have

$$\rho_i = \frac{p_i}{R_i T}. \qquad (3.7)$$

This simplifies the system (3.5), (3.6) considerably. We first substitute the mixture density in Eq. (3.5) using (3.4). Then we substitute each of the m densities of the miscible fluids using (3.7) in the resulting equation. After multiplying both sides of the resulting equation by T we finally obtain

$$\frac{p_i}{R_i} = C_i \left[\frac{1 - \sum_{n=m_{max}+1}^{i_{max}} C_n}{\sum_{i=m}^{m_{max}} \frac{P_m}{R_m}} + \frac{1}{T} \sum_{n=m_{max}+1}^{i_{max}} \frac{C_n}{P_n(p,T)} \right]^{-1}, \quad i = 1, m_{max} - 1. \qquad (3.8)$$

Now the system (3.6) and (3.8) can be solved analytically for the unknown partial pressures in the following way. Select the component, m_{max}, whose pressure $p_{m_{max}}$ must be calculated first. Add the $m_{max} - 1$ equations (3.8)

$$\sum_{k=1}^{m_{max}-1} p_k = \sum_{k=1}^{m_{max}-1} R_k C_k \left[\frac{1 - \sum_{n=m_{max}+1}^{i_{max}} C_n}{\sum_{i=m}^{m_{max}} \frac{P_m}{R_m}} + \frac{1}{T} \sum_{n=m_{max}+1}^{i_{max}} \frac{C_n}{P_n(p,T)} \right]^{-1}, \qquad (3.9)$$

rewrite Eq. (3.6) in the following form

$$p - p_{m_{max}} = \sum_{k=1}^{m_{max}-1} p_k, \qquad (3.10)$$

and replace the sum of the left-hand side of Eq. (3.9) by Eq. (3.10) to obtain

$$p - p_{m_{max}} = \sum_{k=1}^{m_{max}-1} R_k C_k \left[\frac{1 - \sum_{n=m_{max}+1}^{i_{max}} C_n}{\sum_{i=m}^{m_{max}} \frac{P_m}{R_m}} + \frac{1}{T} \sum_{n=m_{max}+1}^{i_{max}} \frac{C_n}{P_n(p,T)} \right]^{-1}. \qquad (3.11)$$

Replace the sum $\sum_{m=1}^{m_{max}} \frac{P_m}{R_m}$ in this equation by the definition equation (3.3) for $C_{m_{max}}$

$$\sum_{m=1}^{m_{max}} \frac{P_m}{R_m} = \frac{P_{m_{max}}}{C_{m_{max}} R_{m_{max}}}, \qquad (3.12)$$

to obtain the quadratic equation

$$\frac{P_{m_{max}}^2}{T} \sum_{n=m_{max}+1}^{i_{max}} \frac{C_n}{P_n(p,T)}$$

$$+ \left[\sum_{k=1}^{m_{max}} C_k R_k - C_{m_{max}} R_{m_{max}} \sum_{n=m_{max}+1}^{i_{max}} C_n - \frac{p}{T} \sum_{n=m_{max}+1}^{i_{max}} \frac{C_n}{P_n(p,T)} \right] P_{m_{max}}$$

$$- \left(1 - \sum_{n=m_{max}+1}^{i_{max}} C_n \right) C_{m_{max}} R_{m_{max}} p = 0. \qquad (3.13)$$

The solution of this equation gives the result we are looking for. Analogously, we compute each of the other partial pressures of the miscible components.

Limiting case for no non-miscible components. For the limiting case of no non-miscible components in the mixture the solution is

$$P_{m_{max}} = \frac{R_{m_{max}} C_{m_{max}}}{\sum_{k=1}^{m_{max}} R_k C_k} p = Y_{m_{max}} p, \qquad (3.14)$$

see Elsner [4] for comparison. Analogously, we compute each of the other partial pressures

$$p_i = Y_i p, \qquad (3.15)$$

where

$$Y_i = \frac{R_i C_i}{\sum_{k=1}^{i_{max}} R_k C_k} = \frac{C_i / M_i}{\sum_{k=1}^{i_{max}} C_k / M_k} \qquad (3.16)$$

is the *molar concentration* of the i-th component of the gas mixture. Here M_i is kg-mole of the constituent i. Note that for a description of transport processes without chemical reactions the use of the mass concentrations is much more convenient than the use of the molar concentrations. For perfect fluids the reverse computation of the mass concentrations if the *kg-mole* concentrations are known is useful

$$C_i = \frac{Y_i M_i}{\sum_{k=1}^{i_{max}} Y_k M_k}. \tag{3.17}$$

The general solution. In reality, the mixture does not consist of perfect fluids but of real fluids. As a result, the solution already obtained can be used as a first approximation for the exact solution of the system (3.5) and (3.6), which then must be obtained by iteration. For this purpose, we use the standard method of *Newton-Raphson*, seeking the zeros for the following functions

$$\varepsilon_i = \rho_i(p_i, T) - C_i \rho, \quad i = 1, m_{max} - 1, \tag{3.18}$$

$$\varepsilon_{m_{max}} = p - \sum_{m=1}^{m_{max}} p_m. \tag{3.19}$$

Small deviations in p_m, δp_m, $m = 1, m_{max} - 1$ lead to a corresponding deviation in ε_m

$$\delta \varepsilon_i = \left(\frac{\partial \rho_i}{\partial p_i}\right)_T \delta p_i - C_i \delta \rho, \quad i = 1, m_{max} - 1, \tag{3.20}$$

and

$$\delta \varepsilon_{m_{max}} = -\sum_{m=1}^{m_{max}} \delta p_m. \tag{3.21}$$

Further to this, the task is reduced to finding such a δp_m, which for a given initial approximation p_m, and

$$\left(\frac{\partial \rho_m}{\partial p_m}\right)_T = f(p_m, T) \tag{3.22}$$

minimizes the residuals of the functions, namely,

$$\delta \varepsilon_m = -\varepsilon_m(p_m, T) \qquad (3.23)$$

or

$$\mathbf{J}\overline{\delta p} = -\mathbf{e} \qquad (3.24)$$

or

$$\overline{\delta p} = -\mathbf{J}^{-1}\mathbf{e}, \qquad (3.25)$$

where the algebraic vector of the pressure increments is

$$\overline{\delta p} = \begin{pmatrix} p_1^{n+1} - p_1^n \\ p_2^{n+1} - p_2^n \\ \ldots \\ p_{m_{\max}}^{n+1} - p_{m_{\max}}^n \end{pmatrix}, \qquad (3.26)$$

and the residuals

$$\mathbf{e} = \begin{pmatrix} \rho_1(p_1, T) - C_1\rho \\ \rho_2(p_2, T) - C_2\rho \\ \ldots \\ p - \sum_{m=1}^{m_{\max}} p_m \end{pmatrix}. \qquad (3.27)$$

Knowing the residuals and the *Jacobian* in the previous iteration step n, the solution improves as follows

$$\overline{p}^{n+1} = \overline{p}^n - (\mathbf{J}^n)^{-1}\overline{\mathbf{e}}. \qquad (3.28)$$

Because p and T are constant the term $\sum_{i=m_{\max}+1}^{i_{\max}} \frac{C_i}{\rho_i(p,T)}$ in the density expression is also a constant. Therefore

$$d\rho = \frac{\left(1 - \sum_{i=m_{max}+1}^{i_{max}} C_i\right)\rho^2}{\left[\sum_{i=1}^{m_{max}} \rho_i(p_i,T)\right]^2} \sum_{i=1}^{m_{max}} d\rho_i(p_i,T), \qquad (3.29)$$

and the *Jacobian* takes the form

$$J_{kj} = \partial \varepsilon_k / \partial p_j = \left\{ \delta_{kj} - C_k \frac{\left(1 - \sum_{i=m_{max}+1}^{i_{max}} C_i\right)\rho^2}{\left[\sum_{i=1}^{m_{max}} \rho_i(p_i,T)\right]^2} \right\} \left(\frac{\partial \rho_k}{\partial p_k}\right)_T \quad \text{for} \quad k = 1,\ldots,m_{max}-1,$$

$$(3.30)$$

$$J_{m_{max} j} = -1. \qquad (3.31)$$

δ_{kj} in Eq. (3.30) is the *Kroneker* delta function ($\delta_{kj} = 0$ for $j = k$, and $= 1$ otherwise). For non-existing non-miscible components Eq. (3.30) reduces to Eq. (3.30) in [24]. If one part of the components is taken to be a perfect gas, the corresponding derivatives in the *Jacobian* are simply

$$\left(\frac{\partial \rho_i}{\partial p_i}\right)_T = \frac{1}{R_i T}. \qquad (3.32)$$

But in the general case

$$\left(\frac{\partial \rho_i}{\partial p_i}\right)_T = \frac{1}{R_i(p_i,T)T}, \qquad (3.33)$$

where

$$R_i(p_i,T) = 1 \Big/ \left[T\left(\frac{\partial \rho_i}{\partial p_i}\right)_T\right] \qquad (3.34)$$

is not a constant, but a function of p_i and T because

$$T\left(\frac{\partial \rho_i}{\partial p_i}\right)_T = f(p_i,T). \qquad (3.35)$$

Using (p,T,ρ_i) instead of (p,T,C_i) as dependent variables we eliminate the need for computing the partial pressure by iterations because we have directly $p_i = p_i(\rho_i, T)$. This approach has its limitation if used with digital computers due to the so called truncation error. Thus below a given *Mach* number

$$Ma_1 < 10^{(n1+n2)/2} \tag{3.36}$$

the density change is then below the truncation error for computers

$$\frac{\Delta \rho_1}{\rho_{10}} \approx 10^{n1} \tag{3.37}$$

and cannot influence the pressure field any more below the value,

$$\frac{\Delta p}{\rho_{10} w_{10}^2} \approx 10^{n2} \tag{3.38}$$

due to $\Delta p \approx a^2 \Delta \rho$, $\tag{3.39}$

$$\frac{\Delta p}{\rho_{10} w_{10}^2} \approx \frac{\Delta \rho_1}{\rho_{10}} / Ma_1^2, \tag{3.40}$$

see *Issa* [10] (1983). More recent discussion of this topic is given in [31] (1999).

Thus, after p, C_i and T are known, the partial pressure and therefore the densities can be calculated as

$$\rho_i = \rho_i(p_i, T) \tag{3.41}$$

as can the partial derivatives for each of the components, $\left(\frac{\partial \rho_i}{\partial p_i}\right)_T$ and $\left(\frac{\partial \rho_i}{\partial T}\right)_{p_i}$.
For a perfect gas the densities and their derivatives can easily be computed from the state equation $\rho_i = \rho_i(p_i, T)$. For real gases, it is assumed that the analytical expression for $\rho_i = \rho_i(p_i, T)$ is known. It is therefore easy to derive analytical expressions to compute the corresponding derivatives. The differential form of the equation of state

$$\rho_i = \rho_i(p_i, T), \tag{3.42}$$

namely,

$$d\rho_i = \left(\frac{\partial \rho_i}{\partial p_i}\right)_T dp_i + \left(\frac{\partial \rho_i}{\partial T}\right)_{p_i} dT, \qquad (3.43)$$

for each of the components is uniquely defined.

3.2.2 Partial derivatives of the equation of state $\rho = \rho\left(p, T, C_{2,\ldots,i_{max}}\right)$

The mixture density

$$\rho = \left[\frac{1 - \sum_{n=m_{max}+1}^{i_{max}} C_n}{\sum_{m=1}^{m_{max}} \rho_m(p_m, T)} + \sum_{n=m_{max}+1}^{i_{max}} \frac{C_n}{\rho_n(p, T)} \right]^{-1} = \rho\left(C_{1,2,\ldots,i_{max}}, p, T\right) \qquad (3.44)$$

is obviously a function of $C_{1,2,\ldots,i_{max}}$, p and T. Note that only $i_{max} - 1$ concentrations are mutually independent due to the fact that

$$\sum_{i=1}^{i_{max}} C_i = 1. \qquad (3.45)$$

Consequently, ρ is simply a function of only $i_{max} - 1$ mutually independent concentrations,

$$\rho = \rho\left(C_{2,\ldots,i_{max}}, p, T\right). \qquad (3.46)$$

Further we solve the following task: Let us assume that $C_{2,\ldots,i_{max}}$, p and T are given, from which, as already shown, all $p_i, \rho_i, \left(\frac{\partial \rho_i}{\partial p_i}\right)_T, \left(\frac{\partial \rho_i}{\partial T}\right)_{p_i}$ and ρ, respectively can be calculated. The expressions defining the partial derivatives for the mixture in the differential form of the equation of state (3.46) are sought:

$$d\rho = \left(\frac{\partial \rho}{\partial p}\right)_{T,all\_C's} dp + \left(\frac{\partial \rho}{\partial T}\right)_{p,all\_C's} dT + \sum_{i=2}^{i_{max}} \left(\frac{\partial \rho}{\partial C_i}\right)_{p,T,all\_C's\_except\_C_i} dC_i \qquad (3.47)$$

3.2 Multi-component mixtures of miscible and non-miscible components

The partial derivatives $\left(\dfrac{\partial p}{\partial \rho}\right)_{T, all\_C's}$, $\left(\dfrac{\partial p}{\partial T}\right)_{p, all\_C's}$ and $\left(\dfrac{\partial p}{\partial C_i}\right)_{p,T, all\_C's\_except\_C_i}$ follow immediately from the differential form of the equation $p = \sum\limits_{m=1}^{m_{max}} p_m$,

$$dp = \sum_{m=1}^{m_{max}} d\left[p_m \left(\rho_m, T\right)\right] = \sum_{m=1}^{m_{max}} d\left[p_m\left(C_m \rho, T\right)\right], \qquad (3.48)$$

where

$$dp_m = \left(\frac{\partial p_m}{\partial \rho_m}\right)_T C_m d\rho + \rho \left(\frac{\partial p_m}{\partial \rho_m}\right)_T dC_m + \left(\frac{\partial p_m}{\partial T}\right)_{\rho_m} dT, \qquad (3.49)$$

or

$$dp = \sum_{m=1}^{m_{max}} d\left[p_m\left(\rho_m, T\right)\right] = \sum_{m=1}^{m_{max}} d\left[p_m\left(C_m \rho, T\right)\right]$$

$$= \sum_{m=1}^{m_{max}} \left[\left(\frac{\partial p_m}{\partial \rho_m}\right)_T (C_m d\rho + \rho dC_m) + \left(\frac{\partial p_m}{\partial T}\right)_{\rho_m} dT\right]$$

$$= \left[\sum_{m=1}^{m_{max}}\left(\frac{\partial p_m}{\partial \rho_i}\right)_T C_m\right] d\rho + \rho \sum_{m=1}^{m_{max}}\left(\frac{\partial p_m}{\partial \rho_m}\right)_T dC_m + \left[\sum_{m=1}^{m_{max}}\left(\frac{\partial p_m}{\partial T}\right)_{\rho_m}\right] dT \qquad (3.50)$$

or after substituting

$$dC_1 = -\sum_{i=2}^{i_{max}} dC_i = -\sum_{i=2}^{m_{max}} dC_i - \sum_{i=m_{max}+1}^{i_{max}} dC_i \qquad (3.51)$$

$$dp = \left[\sum_{m=1}^{m_{max}}\left(\frac{\partial p_m}{\partial \rho_m}\right)_T C_m\right] d\rho + \left[\sum_{m=1}^{m_{max}}\left(\frac{\partial p_m}{\partial T}\right)_{\rho_m}\right] dT$$

$$+ \rho \sum_{m=2}^{m_{max}}\left[\left(\frac{\partial p_m}{\partial \rho_m}\right)_T - \left(\frac{\partial p_1}{\partial \rho_1}\right)_T\right] dC_m - \rho \left(\frac{\partial p_1}{\partial \rho_1}\right)_T \sum_{n=m_{max}+1}^{i_{max}} dC_n \qquad (3.52)$$

namely,

$$\left(\frac{\partial \rho}{\partial p}\right)_{T,all\_C's} = 1 \bigg/ \left[\sum_{m=1}^{m_{max}} \left(\frac{\partial p_m}{\partial \rho_m}\right)_T C_m\right], \qquad (3.53)$$

$$\left(\frac{\partial \rho}{\partial T}\right)_{p,all\_C's} = -\left[\sum_{m=1}^{m_{max}} \left(\frac{\partial p_m}{\partial T}\right)_{\rho_m}\right] \bigg/ \left[\sum_{m=1}^{m_{max}} \left(\frac{\partial p_m}{\partial \rho_m}\right)_T C_m\right], \qquad (3.54)$$

$$\left(\frac{\partial \rho}{\partial C_i}\right)_{p,T,all\_C's\_except\_C_i} = -\rho \left[\left(\frac{\partial p_i}{\partial \rho_i}\right)_T - \left(\frac{\partial p_1}{\partial \rho_1}\right)_T\right] \bigg/ \left[\sum_{m=1}^{m_{max}} \left(\frac{\partial p_m}{\partial \rho_m}\right)_T C_m\right], \qquad (3.55)$$

for $i = 2, m_{max}$, and

$$\left(\frac{\partial \rho}{\partial C_i}\right)_{p,T,all\_C's\_except\_C_i} = \rho \left(\frac{\partial p_1}{\partial \rho_1}\right)_T \bigg/ \left[\sum_{m=1}^{m_{max}} \left(\frac{\partial p_m}{\partial \rho_m}\right)_T C_m\right], \qquad (3.56)$$

for $i = m_{max} + 1, i_{max}$. Here an arbitrary existing component is denoted with 1, $C_1 > 0$. The equations above contain derivatives $\left(\frac{\partial p_i}{\partial \rho_i}\right)_T$, $\left(\frac{\partial p_i}{\partial T}\right)_{\rho_i}$. Only the derivatives $(\partial \rho_i / \partial p_i)_T$ and $(\partial \rho_i / \partial T)_{p_i}$ of the simple components are known, so that the expressions for the mixture derivatives (3.53) through (3.56) in which the known component derivatives explicitly occur are easily obtained by replacing

$$\left(\frac{\partial p_i}{\partial \rho_i}\right)_T = 1 \bigg/ \left(\frac{\partial \rho_i}{\partial p_i}\right)_T, \qquad (3.57)$$

$$\left(\frac{\partial p_i}{\partial T}\right)_{\rho_i} = -\left(\frac{\partial \rho_i}{\partial T}\right)_{p_i} \bigg/ \left(\frac{\partial \rho_i}{\partial p_i}\right)_T. \qquad (3.58)$$

The final result is

$$\boxed{\left(\frac{\partial \rho}{\partial p}\right)_{T,all\_C's} = 1 \bigg/ \left[\sum_{m=1}^{m_{max}} C_m \bigg/ \left(\frac{\partial \rho_m}{\partial p_m}\right)_T\right],} \qquad (3.59)$$

$$\left(\frac{\partial \rho}{\partial T}\right)_{p,all\_C's} = \left(\frac{\partial \rho}{\partial p}\right)_{T,all\_C's} \sum_{m=1}^{m_{max}} \left(\frac{\partial \rho_m}{\partial T}\right)_{p_m} \Bigg/ \left(\frac{\partial \rho_m}{\partial p_m}\right)_T, \quad (3.60)$$

$$\left(\frac{\partial \rho}{\partial C_i}\right)_{p,T,all\_C's\_except\_C_i} = -\rho \left(\frac{\partial \rho}{\partial p}\right)_{T,all\_C's} \left[1\Bigg/\left(\frac{\partial \rho_i}{\partial p_i}\right)_T - 1\Bigg/\left(\frac{\partial \rho_1}{\partial p_1}\right)_T\right], \quad (3.61)$$

for $i = 2, m_{max}$, and

$$\left(\frac{\partial \rho}{\partial C_i}\right)_{p,T,all\_C's\_except\_C_i} = \rho \left(\frac{\partial \rho}{\partial p}\right)_{T,all\_C's} \Bigg/ \left(\frac{\partial \rho_1}{\partial p_1}\right)_T, \quad (3.62)$$

for $i = m_{max} + 1, i_{max}$. With these results Eq. (3.49) can be rewritten as a function of the pressure, temperature and concentrations

$$dp_m = \frac{C_m \left(\frac{\partial p_m}{\partial \rho_m}\right)_T}{\sum_{i=1}^{m_{max}} C_i \left(\frac{\partial p_i}{\partial \rho_i}\right)_T} dp + \frac{1}{\left(\frac{\partial \rho_m}{\partial p_m}\right)_T} \left[C_m \frac{\sum_{i=1}^{m_{max}} \left(\frac{\partial \rho_i}{\partial T}\right)_{p_i} \Bigg/ \left(\frac{\partial \rho_i}{\partial p_i}\right)_T}{\sum_{i=1}^{m_{max}} C_i \left(\frac{\partial p_i}{\partial \rho_i}\right)_T} - \left(\frac{\partial \rho_m}{\partial T}\right)_{p_m} \right] dT$$

$$+ \rho \left(\frac{\partial p_m}{\partial \rho_m}\right)_T \left[dC_m - \frac{C_m}{\sum_{i=1}^{m_{max}} \left(\frac{\partial p_i}{\partial \rho_i}\right)_T C_i} \sum_{i=1}^{m_{max}} \left(\frac{\partial p_i}{\partial \rho_i}\right)_T dC_i \right], \quad (3.63)$$

or

$$dp_m = \left(\frac{\partial p_m}{\partial p}\right)_{T,all\_C's} dp + \left(\frac{\partial p_m}{\partial T}\right)_{p,all\_C's} dT + \sum_{i=2}^{m_{max}} \left(\frac{\partial p_m}{\partial C_i}\right)_{p,T,all\_C's\_except\_C_i} dC_i. \quad (3.64)$$

Therefore

$$\left(\frac{\partial p_m}{\partial p}\right)_{T,all\_C's} = \frac{C_m \big/ \left(\frac{\partial p_m}{\partial p_m}\right)_T}{\sum_{i=1}^{m_{max}} C_i \big/ \left(\frac{\partial p_i}{\partial p_i}\right)_T}, \qquad (3.65)$$

$$\left(\frac{\partial p_m}{\partial T}\right)_{p,all\_C's} = \frac{1}{\left(\frac{\partial p_m}{\partial p_m}\right)_T} \left[C_m \frac{\sum_{i=1}^{m_{max}} \left(\frac{\partial p_i}{\partial T}\right)_{p_i} \big/ \left(\frac{\partial p_i}{\partial p_i}\right)_T}{\sum_{i=1}^{m_{max}} C_i \big/ \left(\frac{\partial p_i}{\partial p_i}\right)_T} - \left(\frac{\partial p_m}{\partial T}\right)_{p_m} \right], \qquad (3.66)$$

$$\left(\frac{\partial p_m}{\partial C_i}\right)_{p,T,all\_C's\_except\_C_i} = \rho \left(\frac{\partial p_m}{\partial p_m}\right)_T \left[\delta_{mi} - C_m \frac{\left(\frac{\partial p_i}{\partial p_i}\right)_T - \left(\frac{\partial p_1}{\partial p_1}\right)_T}{\sum_{i=1}^{m_{max}} C_i \left(\frac{\partial p_i}{\partial p_i}\right)_T} \right]. \qquad (3.67)$$

For the limiting case, where all of the components are taken to be perfect gases, we have

$$\left(\frac{\partial p_i}{\partial p_i}\right)_T = \frac{1}{R_i T}, \qquad (3.68)$$

$$\left(\frac{\partial p_i}{\partial T}\right)_{p_i} = -\frac{p_i}{T}, \qquad (3.69)$$

and

$$\left(\frac{\partial p_1}{\partial p_1}\right)_T = \frac{1}{R_1 T}, \qquad (3.70)$$

$$\left(\frac{\partial p_1}{\partial T}\right)_{p_1} = -\frac{p_1}{T}. \qquad (3.71)$$

Substituting in the equations defining the derivatives this yields

$$\left(\frac{\partial \rho}{\partial p}\right)_{T,all\_C's} = 1 \bigg/ \left(T\sum_{i=1}^{i_{max}} C_i R_i\right) = \frac{1}{RT}, \qquad (3.72)$$

$$\left(\frac{\partial \rho}{\partial T}\right)_{p,all\_C's} = -\left(\sum_{i=1}^{i_{max}} \rho_i R_i\right) \bigg/ \left(T\sum_{i=1}^{i_{max}} C_i R_i\right) = -\rho/T, \qquad (3.73)$$

$$\left(\frac{\partial \rho}{\partial C_i}\right)_{p,T,all\_C's\_except\_C_i} = -\rho(R_i - R_1)/R, \qquad (3.74)$$

where $R = \sum_{i=1}^{i_{max}} C_i R_i$.

Now assume that the gas consists only of one i-th component, $C_1 = 0$, $\rho_1 = 0$, and $C_i = 1$. Using Eq. (3.50), this trivial case yields

$$\left(\frac{\partial \rho}{\partial p}\right)_{T,all\_C's} = \frac{1}{R_i T}, \qquad (3.75)$$

$$\left(\frac{\partial \rho}{\partial T}\right)_{p,all\_C's} = -\rho_i/T, \qquad (3.76)$$

and

$$\left(\frac{\partial \rho}{\partial C_i}\right)_{p,T,all\_C's\_except\_C_i} = 0. \qquad (3.77)$$

The derivatives are not defined in the case of missing miscible components because in this case Eq. (3.6) from which we started our derivation is no longer valid. In this case the Eq. (3.44) simplifies fortunately to

$$\frac{1}{\rho} = \sum_{i=1}^{i_{max}} \frac{C_i}{\rho_i(p,T)} = 1\big/\rho(C_{1,2,\dots,i_{max}}, p, T). \qquad (3.78)$$

The derivatives are then obtained from the differential form

$$d\rho = \sum_{i=1}^{i_{max}} C_i \frac{\rho^2}{\rho_i^2} d\rho_i - \rho^2 \sum_{i=1}^{i_{max}} \frac{1}{\rho_i} dC_i$$

$$= \left[\sum_{i=1}^{i_{max}} C_i \frac{\rho^2}{\rho_i^2}\left(\frac{\partial \rho_i}{\partial p}\right)_T\right] dp + \left[\sum_{i=1}^{i_{max}} C_i \frac{\rho^2}{\rho_i^2}\left(\frac{\partial \rho_i}{\partial T}\right)_p\right] dT - \rho^2 \sum_{i=2}^{i_{max}} \left(\frac{1}{\rho_i} - \frac{1}{\rho_1}\right) dC_i \,. \quad (3.79)$$

The final result is

$$\left(\frac{\partial \rho}{\partial p}\right)_{T, all\_C's} = \sum_{i=1}^{i_{max}} C_i \frac{\rho^2}{\rho_i^2}\left(\frac{\partial \rho_i}{\partial p}\right)_T, \quad (3.80)$$

$$\left(\frac{\partial \rho}{\partial T}\right)_{p, all\_C's} = \sum_{i=1}^{i_{max}} C_i \frac{\rho^2}{\rho_i^2}\left(\frac{\partial \rho_i}{\partial T}\right)_p, \quad (3.81)$$

$$\left(\frac{\partial \rho}{\partial C_i}\right)_{p,T, all\_C's\_except\_C_i} = -\rho^2\left(\frac{1}{\rho_i} - \frac{1}{\rho_1}\right), \quad (3.82)$$

If some of the non-miscible components are considered as non-compressible, their density derivatives are simply set to zero.

3.2.3 Partial derivatives in the equation of state $T = T(\varphi, p, C_{2,\ldots,i_{max}})$, where $\varphi = s, h, e$

The next step is to define the partial derivatives in the linearized equation of state

$$T = T(\varphi, p, C_{2,\ldots,i_{max}}) \quad (3.83)$$

namely

$$dT = \left(\frac{\partial T}{\partial p}\right)_{\varphi, all\_C's} dp + \left(\frac{\partial T}{\partial \varphi}\right)_{p, all\_C's} d\varphi + \sum_{i=2}^{i_{max}} \left(\frac{\partial T}{\partial C_i}\right)_{p,\varphi, all\_C's\_except\_C_i} dC_i, \quad (3.84)$$

where φ may be one of the following variables: specific entropy, specific enthalpy or specific internal energy,

$\varphi = s, h, e$.

This is very important for the construction of a numerical algorithm within the framework of the *entropy, or enthalpy or energy concepts*. Remember that this kind of algorithm describes the behavior of the flow using the specific mixture property φ as elements of the dependent variable vector. As the increments $\delta p, \delta \varphi$ and δC_i are known from the numerical integration of the system of PDEs governing the flow, it is then possible to compute the corresponding increment of the temperature T, and therefore, the particular thermo-physical and transport properties of the velocity field, which depend on T, p and C_i.

Begin with the definition equation for the specific mixture property

$$\rho\varphi = \sum_{i=1}^{i_{max}} \rho_i \varphi_i \qquad (3.85)$$

or

$$\varphi = \sum_{i=1}^{i_{max}} C_i \varphi_i . \qquad (3.86)$$

After differentiating the definition equation for the specific mixture entropy (3.86), the following is obtained

$$d\varphi = \sum_{i=1}^{i_{max}} C_i d\varphi_i + \sum_{i=1}^{i_{max}} \varphi_i dC_i . \qquad (3.87)$$

An arbitrary, but existing component is again denoted with 1, where the mass concentration of this component is uniquely defined by knowing the other concentrations

$$C_1 = 1 - \sum_{i=2}^{i_{max}} C_i \qquad (3.88)$$

and therefore

$$dC_1 = -\sum_{i=2}^{i_{max}} dC_i . \qquad (3.89)$$

Note that this is the only criterion for selecting the above component denoted with 1. Its concentration is not calculated using the differential conservation equation, but merely using Eq. (3.88).

Enthalpy concept $\varphi = h$. In this case Eq. (3.87) is

$$dh = \sum_{i=1}^{i_{max}} C_i dh_i + \sum_{i=1}^{i_{max}} h_i dC_i . \qquad (3.90)$$

Replace in the above equation the differentials of the specific enthalpies using the caloric equations

$$dh_i = c_{pi} dT + \left(\frac{\partial h_i}{\partial p_i}\right)_T dp_i , \qquad (3.91)$$

to obtain

$$dh = \left(\sum_{i=1}^{i_{max}} C_i c_{pi}\right) dT + \sum_{i=1}^{m_{max}} C_i \left(\frac{\partial h_i}{\partial p_i}\right)_T dp_i + \left[\sum_{i=m_{max}+1}^{i_{max}} C_i \left(\frac{\partial h_i}{\partial p}\right)_T\right] dp + \sum_{i=1}^{i_{max}} h_i dC_i . \qquad (3.92)$$

Replace dp_i in the above equation by means of the differentiated equation of state $p_i = p_i(\rho_i, T) = p_i(C_i \rho, T)$, namely

$$dp_i = \left(\frac{\partial p_i}{\partial \rho_i}\right)_T C_i d\rho + \rho \left(\frac{\partial p_i}{\partial \rho_i}\right)_T dC_i + \left(\frac{\partial p_i}{\partial T}\right)_{\rho_i} dT , \qquad (3.93)$$

to obtain

$$dh = \left[\left(\sum_{i=1}^{i_{max}} C_i c_{pi}\right) + \sum_{i=1}^{m_{max}} C_i \left(\frac{\partial h_i}{\partial p_i}\right)_T \left(\frac{\partial p_i}{\partial T}\right)_{\rho_i}\right] dT + \left[\sum_{i=1}^{m_{max}} C_i^2 \left(\frac{\partial h_i}{\partial p_i}\right)_T \left(\frac{\partial p_i}{\partial \rho_i}\right)_T\right] d\rho$$

$$+ \left[\sum_{i=m_{max}+1}^{i_{max}} C_i \left(\frac{\partial h_i}{\partial p}\right)_T\right] dp + \sum_{i=1}^{i_{max}} h_i dC_i + \sum_{i=1}^{m_{max}} \rho C_i \left(\frac{\partial h_i}{\partial p_i}\right)_T \left(\frac{\partial p_i}{\partial \rho_i}\right)_T dC_i . \qquad (3.94)$$

Substituting dC_1 from Eq. (3.89) and $d\rho$ from Eq. (3.47) into Eq. (3.94) yields

$$dh = \left[\sum_{i=1}^{m_{max}} C_i \left\{ c_{pi} + \left(\frac{\partial h_i}{\partial p_i}\right)_T \left[\left(\frac{\partial p_i}{\partial T}\right)_{\rho_i} + C_i \left(\frac{\partial p_i}{\partial \rho_i}\right)_T \left(\frac{\partial \rho}{\partial T}\right)_{p,all\_C's}\right]\right\} + \sum_{i=m_{max}+1}^{i_{max}} C_i c_{pi}\right] dT$$

$$+ \left\{\left[\sum_{i=1}^{m_{max}} C_i^2 \left(\frac{\partial h_i}{\partial p_i}\right)_T \left(\frac{\partial p_i}{\partial \rho_i}\right)_T\right] \left(\frac{\partial \rho}{\partial p}\right)_{T,all\_C's} + \sum_{i=m_{max}+1}^{i_{max}} C_i \left(\frac{\partial h_i}{\partial p}\right)_T\right\} dp$$

$$+ \sum_{i=2}^{i_{max}} \left\{ h_i - h_1 + \left[\sum_{i=1}^{m_{max}} C_i^2 \left(\frac{\partial h_i}{\partial p_i}\right)_T \left(\frac{\partial p_i}{\partial \rho_i}\right)_T\right] \left(\frac{\partial \rho}{\partial C_i}\right)_{p,T,all\_C's\_except\_C_i}\right\} dC_i$$

$$+ \sum_{i=2}^{m_{max}} \rho \left[C_i \left(\frac{\partial h_i}{\partial p_i}\right)_T \left(\frac{\partial p_i}{\partial \rho_i}\right)_T - C_1 \left(\frac{\partial h_1}{\partial p_1}\right)_T \left(\frac{\partial p_1}{\partial \rho_1}\right)_T \right] dC_i, \qquad (3.95)$$

or

$$\boxed{dh = c_p dT + \left(\frac{\partial h}{\partial p}\right)_{T,all\_C's} dp + \sum_{i=2}^{i_{max}} \left(\frac{\partial h}{\partial C_i}\right)_{p,T,all\_C's\_except\_C_i} dC_i \qquad (3.96)}$$

or

$$\boxed{dT = \frac{dh}{c_p} - \frac{1}{c_p}\left(\frac{\partial h}{\partial p}\right)_{T,all\_C's} dp - \frac{1}{c_p} \sum_{i=2}^{i_{max}} \left(\frac{\partial h}{\partial C_i}\right)_{p,T,all\_C's\_except\_C_i} dC_i \qquad (3.97)}$$

where

$$\boxed{\begin{aligned} c_p &= \sum_{i=m_{max}+1}^{i_{max}} C_i c_{pi} + \sum_{i=1}^{m_{max}} C_i \left\{ c_{pi} + \left(\frac{\partial h_i}{\partial p_i}\right)_T \left[\left(\frac{\partial p_i}{\partial T}\right)_{\rho_i} + C_i \left(\frac{\partial p_i}{\partial \rho_i}\right)_T \left(\frac{\partial \rho}{\partial T}\right)_{p,all\_C's}\right]\right\} \\ &= \sum_{i=m_{max}+1}^{i_{max}} C_i c_{pi} + \sum_{i=1}^{m_{max}} C_i \left\{ c_{pi} + \frac{(\partial h_i/\partial p_i)_T}{(\partial \rho_i/\partial p_i)_T} \left[C_i \left(\frac{\partial \rho}{\partial T}\right)_{p,all\_C's} - \left(\frac{\partial \rho_i}{\partial T}\right)_{p_i}\right]\right\}, \quad (3.98) \end{aligned}}$$

$$\left(\frac{\partial h}{\partial p}\right)_{T,all\_C's} = \left(\frac{\partial \rho}{\partial p}\right)_{T,all\_C's} \left[\sum_{i=1}^{m_{max}} C_i^2 \left(\frac{\partial h_i}{\partial p_i}\right)_T \bigg/ \left(\frac{\partial \rho_i}{\partial p_i}\right)_T\right] + \sum_{i=m_{max}+1}^{i_{max}} C_i \left(\frac{\partial h_i}{\partial p}\right)_T , \quad (3.99)$$

$$\left(\frac{\partial h}{\partial C_i}\right)_{p,T,all\_C's\_except\_C_i} = h_i - h_1 + \Delta h_i^{np} \quad (3.100)$$

and

$$\Delta h_i^{np} = \rho \left[C_i \left(\frac{\partial h_i}{\partial p_i}\right)_T \bigg/ \left(\frac{\partial \rho_i}{\partial p_i}\right)_T - C_1 \left(\frac{\partial h_1}{\partial p_1}\right)_T \bigg/ \left(\frac{\partial \rho_1}{\partial p_1}\right)_T\right]$$

$$+ \left[\sum_{i=1}^{m_{max}} C_i^2 \left(\frac{\partial h_i}{\partial p_i}\right)_T \bigg/ \left(\frac{\partial \rho_i}{\partial p_i}\right)_T\right] \left(\frac{\partial \rho}{\partial C_i}\right)_{p,T,all\_C's\_except\_C_i} , \quad (3.101)$$

for $i = 2, m_{max}$ and

$$\Delta h_i^{np} = \left[\sum_{i=1}^{m_{max}} C_i^2 \left(\frac{\partial h_i}{\partial p_i}\right)_T \bigg/ \left(\frac{\partial \rho_i}{\partial p_i}\right)_T\right] \left(\frac{\partial \rho}{\partial C_i}\right)_{p,T,all\_C's\_except\_C_i} \quad (3.102)$$

for $i = m_{max} + 1, i_{max}$.

Important: note that for non-existing miscible components the concentration denoted with "1" in Eq. (3.100) is the first non-miscible component.

Equation (3.100) consists of two parts. In the case of a mixture consisting of perfect fluids, the second part is equal to zero, $\Delta h_{il}^{np} = 0$, because the enthalpies do not depend on the corresponding partial pressures. This also illustrates the meaning of the superscript *np* which stands for non-perfect fluid. From Eq. (3.97) we obtain the analytical expressions for the following derivatives

$$\left(\frac{\partial T}{\partial p}\right)_{h,all\_C's} = -\frac{1}{c_p} \left(\frac{\partial h}{\partial p}\right)_{T,all\_C's} , \quad (3.103)$$

$$\left(\frac{\partial T}{\partial h}\right)_{p,all\_C's} = \frac{1}{c_p}, \qquad (3.104)$$

$$\left(\frac{\partial T}{\partial C_i}\right)_{p,h,all\_C's\_except\_C_i} = -\frac{1}{c_p}\left(\frac{\partial h}{\partial C_i}\right)_{p,T,all\_C's\_except\_C_i}. \qquad (3.105)$$

Checking the above derivation: for a perfect gas mixture, and taking into account that

$$\left(\frac{\partial h_i}{\partial p_i}\right)_T = 0, \qquad (3.106)$$

$$\left(\frac{\partial h_1}{\partial p_1}\right)_T = 0, \qquad (3.107)$$

we obtain

$$c_p = \sum_{i=1}^{i_{max}} C_i c_{pi}, \qquad (3.108)$$

$$\left(\frac{\partial h}{\partial p}\right)_{T,all\_C's} = 0, \qquad (3.109)$$

and

$$\left(\frac{\partial h}{\partial C_i}\right)_{p,T,all\_C's\_except\_C_i} = h_i - h_1, \qquad (3.110)$$

which is the result expected.

Energy concept $\varphi = e$. The derivation already presented for the specific enthalpy is formally identical with the derivation for the specific internal energy. One has to replace formally h with e and write instead c_{pi}, $\left(\frac{\partial e_i}{\partial T}\right)_{p_i,T}$.

Entropy concept $\varphi = s$. Then use the definition equation for the particular specific entropies

$$T\rho_i ds_i = \rho_i dh_i - dp_i \qquad (3.111)$$

or divided by ρ

$$TC_i ds_i = C_i dh_i - \frac{dp_i}{\rho}. \qquad (3.112)$$

Substituting in this equation dh_i using Eq. (3.91), summing the resulting i_{max} equations and comparing the result with Eq. (3.87) we finally obtain

$$Tds = \left(\sum_{i=1}^{i_{max}} C_i dh_i\right) - \frac{dp}{\rho} + T\sum_{i=1}^{i_{max}} s_i dC_i. \qquad (3.113)$$

Note that this is equivalent to the *Gibbs* equation for mixtures

$$Tds = \sum_{i=1}^{i_{max}} d(C_i h_i) - \frac{dp}{\rho} + T\sum_{i=1}^{i_{max}} s_i dC_i - \left(\sum_{i=1}^{i_{max}} h_i dC_i\right) = dh - \frac{dp}{\rho} - \sum_{i=1}^{i_{max}} (h_i - Ts_i) dC_i$$

$$= de + pd\left(\frac{1}{\rho}\right) - \sum_{i=1}^{i_{max}} (h_i - Ts_i) dC_i, \qquad (3.114)$$

in which the $h_i - Ts_i$ is the so called *Gibbs potential* for the single component. The entropy change due to the mixing process is

$$ds_{mixing} = -\frac{1}{T}\sum_{i=1}^{i_{max}} (h_i - Ts_i) dC_i. \qquad (3.115)$$

Substituting dC_1 from Eq. (3.89) into Eq. (3.115) results in

$$ds_{mixing} = -\frac{1}{T}\sum_{i=2}^{i_{max}} \left[h_i - h_1 - T(s_i - s_1)\right] dC_i. \qquad (3.116a)$$

This result follows not from Eq. (3.114) directly but from the comparison of Eq. (3.114) with the energy conservation equation for the mixture. For a mixing process in which there is no total mass change in the system, $\sum_{k=1}^{i_{max}} Y_k M_k = const$, Eq. (3.116a) can be rearranged using Eq. (3.17)

$$ds_{mixing} = -\frac{1}{T}\sum_{i=2}^{i_{max}} C_i \left[h_i - h_1 - T(s_i - s_1) \right] d\ln Y_i .\qquad(3.116b)$$

Remember, that in case of mixing of perfect fluids the molar concentration is equivalent to the ratio of the partial pressure to the total pressure.

Replace the differentials in Eq. (3.114) of the mixture specific enthalpy using the caloric equations (3.96) to obtain

$$ds = \frac{c_p}{T} dT + \frac{1}{\rho T}\left[\rho\left(\frac{\partial h}{\partial p}\right)_{T,all\_C's} - 1\right] dp + \sum_{i=2}^{i_{max}} \left(s_i - s_1 + \frac{\Delta h_i^{np}}{T}\right) dC_i ,\qquad(3.117)$$

or

$$dT = \frac{T}{c_p} ds - \frac{\rho\left(\frac{\partial h}{\partial p}\right)_{T,all\_C's} - 1}{\rho c_p} dp - \frac{T}{c_p}\sum_{i=2}^{i_{max}} \left(\frac{\partial s}{\partial C_i}\right)_{p,T,all\_C's\_except\_C_i} dC_i \qquad(3.118)$$

where

$$\left(\frac{\partial s}{\partial C_i}\right)_{p,T,all\_C's\_except\_C_i} = s_i - s_1 + \Delta s_i^{np} \qquad(3.119)$$

where

$$\Delta s_i^{np} = \frac{\Delta h_i^{np}}{T} . \qquad(3.120)$$

Comparing Eq. (3.100) with Eq. (3.119) we realize that

$$\left(\frac{\partial s}{\partial C_i}\right)_{p,T,all\_C's\_except\_C_i} = \frac{1}{T}\left(\frac{\partial h}{\partial C_i}\right)_{p,T,all\_C's\_except\_C_i} . \qquad(3.121)$$

Like Eq. (3.100) Eq. (3.119) also consists of two parts. In the case of a mixture consisting of perfect fluids, the second part is equal to zero, $\Delta s_{il}^{np} = 0$, because the enthalpies do not depend on the corresponding partial pressures. This again illustrates the meaning of the superscript np which stands for non-perfect fluid. From Eq. (3.118) analytical expressions for the following derivatives are obtained

$$\left(\frac{\partial T}{\partial p}\right)_{s,all\_C's} = \frac{1}{\rho c_p}\left[1-\rho\left(\frac{\partial h}{\partial p}\right)_{T,all\_C's}\right], \qquad (3.122)$$

$$\left(\frac{\partial T}{\partial s}\right)_{p,all\_C's} = \frac{T}{c_p}, \qquad (3.123)$$

$$\left(\frac{\partial T}{\partial C_i}\right)_{p,s,all\_C's\_except\_C_i} = -\frac{T}{c_p}\left(\frac{\partial s}{\partial C_i}\right)_{p,T,all\_C's\_except\_C_i}. \qquad (3.124)$$

Checking the above derivation: for a perfect gas mixture, and taking into account that

$$\left(\frac{\partial h_i}{\partial p_i}\right)_T = 0, \qquad (3.125)$$

$$\left(\frac{\partial h_1}{\partial p_1}\right)_T = 0, \qquad (3.126)$$

we obtain

$$c_p = \sum_{i=1}^{i_{max}} C_i c_{pi}, \qquad (3.127)$$

$$\left(\frac{\partial h}{\partial p}\right)_{T,all\_C's} = 0, \qquad (3.128)$$

and

$$\left(\frac{\partial s}{\partial C_i}\right)_{p,T,all\_C's\_except\_C_i} = s_i - s_1, \qquad (3.129)$$

which is the result expected.

Thus, as already mentioned, after the required derivatives have been computed in Eq. (3.118) in the manner shown, the temperature difference can be easily computed:

$$T - T_a = \left(\frac{\partial T}{\partial p}\right)_{s, all\_C's} (p - p_a) + \left(\frac{\partial T}{\partial s}\right)_{p, all\_C's} (s - s_a)$$

$$+ \sum_{i=2}^{i_{max}} \left(\frac{\partial T}{\partial C_i}\right)_{p, s, all\_C's\_except\_C_i} (C_i - C_{ia}), \qquad (3.130)$$

which corresponds to the increments for the pressure, entropy and concentrations. When the increments δp, δs and δC_i are of considerable magnitude, it is better to take into account their non-linear dependencies in Eq. (3.118),

$$\frac{dT}{T} = \frac{1}{c_p}\left[ds - \sum_{i=2}^{i_{max}} \left(\frac{\partial s}{\partial C_i}\right)_{p,T,all\_C's\_except\_C_i} dC_i\right] + \frac{p}{\rho T c_p}\left[1 - \rho\left(\frac{\partial h}{\partial p}\right)_{T,all\_C's}\right]\frac{dp}{p}$$

(3.131)

in the following way: substitute

$$\Delta s^* = \frac{1}{c_p}\left[s - s_a - \sum_{i=2}^{i_{max}} \left(\frac{\partial s}{\partial C_i}\right)_{p,T,all\_C's\_except\_C_i}(C - C_{ia})\right] \qquad (3.132)$$

$$\bar{R} = \frac{p}{\rho T}\left[1 - \rho\left(\frac{\partial h}{\partial p}\right)_{T,all\_C's}\right] \qquad (3.133)$$

and integrate to obtain

$$\ln(T/T_a) = \Delta s^* + (\bar{R}/c_p)\ln(p/p_a) \qquad (3.134)$$

or

$$T = T_a e^{\Delta s^*}(p/p_a)^{\bar{R}/c_p}, \qquad (3.135)$$

or for a quasi-linear relationship between temperature T and specific enthalpy h

$$h = h_a e^{\Delta s^*} (p/p_a)^{\bar{R}/c_p} . \tag{3.136}$$

This is a canonical equation, as all thermodynamic properties of the gas mixture can be computed as a function of (s, p, C_i) by differentiating. For $\Delta s^* = 0$,

$$T = T_a (p/p_a)^{\bar{R}/c_p} . \tag{3.137}$$

Equation (3.130) is also useful for exact solution of the task $T = ?$ if s, p and $C_{2,3,\ldots,i_{max}}$ are known. This can be performed by iterations (*Newton* method) as follows: the error for the previous step designated with subscript n is

$$\Delta s^n = s - s^n \left(T^n, p, C_{2,3,\ldots,i_{max}} \right) . \tag{3.138}$$

A new temperature T^{n+1} can be computed so as to obtain $\Delta s^{n+1} = 0$,

$$\Delta s^{n+1} - \Delta s^n = -\left(\frac{\partial s}{\partial T} \right)_{p, all\_C's} \left(T^{n+1} - T^n \right) \tag{3.139}$$

or solving with respect to T^{n+1} and using Eq. (3.123)

$$T^{n+1} = T^n \left\{ 1 + \left[s - s^n \left(T^n, p, C_{2,3,\ldots,i_{max}} \right) \right] / c_p \right\} \tag{3.140}$$

or, in more precise form

$$T^{n+1} = T^n \exp\left\{ \left[s - s^n \left(T^n, p, C_{2,3,\ldots,i_{max}} \right) \right] / c_p \right\} \tag{3.141}$$

3.2.4 Chemical potential

3.2.4.1 Gibbs function

Neither the specific entropy, nor the specific enthalpy, nor the specific internal energy can be measured directly. The quantities that can be measured directly are pressure, temperature and concentrations. It is interesting to know whether Eq. (3.114) can be rewritten as a function of measurable variables $f = f(p, T, C's)$. The answer is yes and the result is

$$d(h - Ts) = \frac{dp}{\rho} - sdT + \sum_{i=1}^{i_{max}} (h_i - Ts_i) dC_i . \tag{3.142}$$

3.2 Multi-component mixtures of miscible and non-miscible components

Introducing the so called *free enthalpy* or *Gibbs function*,

$$g = h - Ts \qquad (3.143)$$

Eq. (3.142) takes the remarkable form

$$dg = \frac{dp}{\rho} - sdT + \sum_{i=2}^{i_{max}} \left[h_i - h_1 - T(s_i - s_1) \right] dC_i , \qquad (3.144)$$

with

$$\left(\frac{\partial g}{\partial p} \right)_{T, all\_C's} = \frac{1}{\rho}, \qquad (3.145)$$

$$\left(\frac{\partial g}{\partial T} \right)_{p, all\_C's} = -s, \qquad (3.146)$$

$$\left(\frac{\partial g}{\partial C_i} \right)_{p, s, all\_C's\_except\_C_i} = h_i - h_1 - T(s_i - s_1). \qquad (3.147)$$

3.2.4.2 Definition of the chemical equilibrium

We identify a mixture that is at constant mixture pressure and temperature and that does not change the concentrations of its constituent as a mixture in *chemical equilibrium*.

The remarkable property of Eq. (3.144) is that it provides a quantitative definition of the chemical equilibrium, namely:

$$dg = 0 \quad \text{or} \quad g = const \qquad (3.148)$$

Now consider the following chemical reaction

$$\sum_{i=1}^{i_{max}} n_i Symb_i = 0. \tag{3.149}$$

Here $Symb_i$ is the chemical identification symbol of the i-th species and n_i is the stoichiometric coefficient equal to the number of *kg-moles* of each species that participates in the reaction. Usually $n_i < 0$ for *reactants* that are reducing their mass and $n_i > 0$ for *products* that are increasing their mass in the mixture. An example is $-2H_2 - O_2 + 2H_2O = 0$. During the progress of a *stoichiometric* reaction all concentrations change not arbitrarily but so that the mass change of a single species is always proportional to $n_i M_i$. This is demonstrated as follows. The mass balance of the stoichiometric reaction (3.149) gives

$$\sum_{i=1}^{i_{max}} n_i M_i = 0. \tag{3.150}$$

Applied to our last example: $-2 \times 2kg - 1 \times 32kg + 2 \times 18kg = 0$. Note that $n_i M_i$ is a natural constant belonging to the specific chemical reaction. On other hand, the sum of all mass sources per unit volume of the mixture is equal to zero,

$$\sum_{i=1}^{i_{max}} \mu_i = 0. \tag{3.151}$$

Selecting one arbitrary, but existing, mass source term designated with subscript 1 (usually having the minimal initial concentration if it is going to be consumed) and rearranging we have

$$\sum_{i=1}^{i_{max}} \mu_1 \frac{\mu_i}{\mu_1} = 0. \tag{3.152}$$

Equations (3.150) and (3.151) can then, and only then, be satisfied simultaneously if

$$\mu_i = \frac{n_i M_i}{n_1 M_1} \mu_1. \tag{3.153}$$

Applying this to our example we obtain

$$\mu_{H_2} = \mu_{H_2}, \quad \mu_{O_2} = (32/4)\mu_{H_2} = 8\mu_{H_2}, \quad \mu_{H_2O} = -(36/4)\mu_{H_2} = -9\mu_{H_2}.$$

For the closed volume in which the chemical reaction happens the mass conservation for each species gives

$$dC_i / d\tau = \mu_i / \rho, \qquad (3.154)$$

and therefore

$$\frac{dC_i / d\tau}{dC_i / d\tau} = \frac{\mu_i / \rho}{\mu_i / \rho}, \qquad (3.155)$$

or

$$dC_i = n_i M_i d \frac{C_1}{n_1 M_1} = n_i M_i d\xi, \qquad (3.156)$$

where ξ is sometimes called in the literature the *reaction progress variable*. Substituting in Eq. (3.142) we obtain

$$d(h - Ts) = \frac{dp}{\rho} - s dT + \left[\sum_{i=1}^{i_{max}} (h_i - Ts_i) n_i M_i \right] d\xi. \qquad (3.157)$$

For the case of chemical equilibrium, $dg = 0$, and at constant mixture pressure and temperature, $dp = 0$ and $dT = 0$, Eq. (3.157) reads

$$\sum_{i=1}^{i_{max}} (h_i - Ts_i) n_i M_i = 0. \qquad (3.158)$$

This means that the function $g = g(\xi)$ possesses an extreme $dg = 0$ if Eq. (3.158) is fulfilled for all components. For a single component Eq. (3.142) results in

$$d(h_i - Ts_i) = \frac{dp_i}{\rho_i} - s_i dT. \qquad (3.159)$$

For the case of a constant mixture temperature valid for each species, $dT = 0$, we have

$$d(h_i - Ts_i) = \frac{dp_i}{\rho_i}, \qquad (3.160)$$

or after integration

$$h_i - Ts_i = h_{i0} - Ts_{i0} + \int_{p_0}^{p_i} \frac{dp_i'}{\rho_i}. \tag{3.161}$$

For perfect gases Eq. (3.161) results in

$$h_i - Ts_i = h_{i0} - Ts_{i0} + TR_i \int_{p_0}^{p_i} \frac{dp_i'}{p_i} = h_{i0} - Ts_{i0} + TR_i \ln\left(\frac{p_i}{p_0}\right). \tag{3.162}$$

Substituting (3.162) in Eq. (3.158) we obtain

$$\sum_{i=1}^{i_{max}} (h_{i0} - Ts_{i0}) M_i n_i + T \sum_{i=1}^{i_{max}} n_i M_i R_i \ln\left(\frac{p_i}{p_0}\right) = 0, \tag{3.163}$$

or bearing in mind that

$$R = M_i R_i, \tag{3.164}$$

with $R = 8314$ $J/(kg\text{-}mol\ K)$ being the universal gas constant,

$$\sum_{i=1}^{i_{max}} (h_{i0} - Ts_{i0}) M_i n_i + TR \ln \prod_{i=1}^{i_{max}} \left(\frac{p_i}{p_0}\right)^{n_i} = 0. \tag{3.165}$$

Introducing the so called *pressure equilibrium factor*

$$K_p = \prod_{i=1}^{i_{max}} \left(\frac{p_i}{p_0}\right)^{n_i} = \frac{\prod_{i=1}^{i_{max}} p_i^{n_i}}{p_0^n} = \left(\frac{p}{p_0}\right)^n \prod_{i=1}^{i_{max}} Y_i^{n_i} = \left(\frac{p}{p_0}\right)^n K_y, \tag{3.166}$$

where

$$n = \sum_{i=1}^{i_{max}} n_i, \tag{3.167}$$

$$K_y = \prod_{i=1}^{i_{max}} Y_i^{n_i}, \tag{3.168}$$

3.2 Multi-component mixtures of miscible and non-miscible components 147

and solving Eq. (3.165) with respect to K_p results in the expression well known in chemical thermodynamics defining mathematically the chemical equilibrium

$$K_p = \exp\left[-\frac{1}{TR}\sum_{i=1}^{i_{max}}(h_{i0} - Ts_{i0})n_i M_i\right], \qquad (3.169a)$$

see for comparison in *Warnatz* et al. [36] (2001) p. 42, or

$$K_y = \left(\frac{p_0}{p}\right)^n \exp\left[-\frac{1}{TR}\sum_{i=1}^{i_{max}}(h_{i0} - Ts_{i0})n_i M_i\right]. \qquad (3.169b)$$

Therefore the pressure equilibrium factor *for the particular reaction* (3.149) can be analytically computed from the properties of the constituents. Another reaction generates another expression for the pressure equilibrium constant.

The expression

$$\Delta G_{ch,0} = \sum_{i=1}^{i_{max}}(h_{i0} - Ts_{i0})n_i M_i = \sum_{i=1}^{i_{max}} h_{i0} n_i M_i - T\sum_{i=1}^{i_{max}} s_{i0} n_i M_i, \qquad (3.170)$$

is called the *molar Gibbs energy of the chemical reaction* [36]. The expressions

$$\Delta H_{ch,0} = \sum_{i=1}^{i_{max}} h_{i0} n_i M_i, \qquad (3.171)$$

$$\Delta S_{ch,0} = \sum_{i=1}^{i_{max}} s_{i0} n_i M_i, \qquad (3.172)$$

are called the enthalpy and entropy changes of the mixture due to the chemical reaction, respectively. With this Eq. (3.169b) can also be used in the form

$$K_y = \left(\frac{p_0}{p}\right)^n \exp\left(-\frac{\Delta H_{ch,0}}{TR} + \frac{\Delta S_{ch,0}}{R}\right) = \left(\frac{p_0}{p}\right)^n \exp\left(-\frac{\Delta H_{ch,0}/R}{T} + \sum_{i=1}^{i_{max}} n_i \frac{s_{i0}}{R_i}\right), \qquad (3.173)$$

The choice of the reference pressure varies in the literature. Some authors set it to unity. In this case $K_p = p^n K_y$. If the mixture pressure is selected as the reference pressure we have $K_p = K_y$. In this case $\Delta G_{ch,0} = G(p,T)$. In this case Eq. (3.162) in its integral form $dg_i = TR_i \ln(p_i/p)$ define a dimensionless property $f_i = p_i/p$ called the *fugacity* of component i in the solution of perfect fluids. The general definition of the fugacity of component i is $d \ln f_i = dg_i/(TR_i)$.

3.2.4.3 Partial pressures of perfect fluid compounds in chemical equilibrium

Given the temperature T and the total pressure p of a mixture of j_{max} compounds that may react in a number i_{max} of chemical reactions, then

$$\sum_{j=1}^{j_{max}} n_{ij} Symb_j = 0, \text{ for } i = 1, i_{max}. \tag{3.174}$$

Here n_{ij} is the stoichiometric coefficient, < 0 for reactants and > 0 for products. We look for a solution of j_{max} molar concentrations

$$\mathbf{Y}^T = (Y_1, Y_2, ..., Y_{j_{max}}), \tag{3.175}$$

for which the system is in chemical equilibrium. We know from *Dalton*'s law that the system pressure is the sum of the partial pressures

$$p = \sum_{j=1}^{j_{max}} p_j, \quad Y_i = \frac{p_j}{p}, \quad \sum_{j=1}^{j_{max}} Y_j = 1. \tag{3.176}$$

For each chemical reaction i we have the condition enforcing chemical equilibrium (3.174),

$$K_{y,i} = \prod_{j=1}^{j_{max}} Y_j^{n_j} = \left(\frac{p_0}{p}\right)^{n_i} \exp\left(-\frac{\Delta H_{ch,0,i}/R}{T} + \sum_{j=1}^{j_{max}} n_{ij} \frac{s_{j0}}{R_j}\right), \text{ for } i = 1, i_{max}. \tag{3.177}$$

where

$$n_i = \sum_{j=1}^{j_{max}} n_{ij}. \tag{3.178}$$

Here R is the universal gas constant, R_j is the gas constant of the j-th component,

$$\Delta H_{ch,0,i} = \sum_{j=1}^{j_{max}} h_{0j} n_{ij} M_j \qquad (3.179)$$

and

$$\overline{s}_{0j} = s_{0j}/R_j \qquad (3.180)$$

are the dimensionless reference entropies of the species j. Therefore we have j_{max} unknowns and $1+i_{max}$ equations (3.176), (3.177). The missing $j_{max} - i_{max} - 1$ are obtained from the *proportions* defined by the mass conservation equations

$$\sum_{j=1}^{j_{max}} n_{ij} M_j = 0, \quad \text{for} \quad i = 1, i_{max}, \qquad (3.181)$$

resulting in

$$\frac{dC_j}{n_{ij} M_{ij}} = \frac{dC_{ij*}}{n_{ij*} M_{ij*}}, \quad \text{for} \quad i = 1, i_{max}, \ j = 1, j_{max}, \qquad (3.183)$$

where $ij*$ refers for each i-reaction to the minimum but existing compound j, or

$$\frac{dY_j}{n_{ij}} = \frac{dY_{ij*}}{n_{ij*}}, \quad \text{for} \quad i = 1, i_{max}, \ j = 1, j_{max}. \qquad (3.184)$$

Observe that in a chemical reaction for reactants that completely disappear we have $dY_j = -Y_j$, and for products that do not exist before the reaction but just originate we have $dY_j = Y_j$. Comparing this with the definition of the stoichiometric coefficient for which $n_{ij} < 0$ for reactants and > 0 for products, we have

$$\frac{Y_j}{|n_{ij}|} = \frac{Y_{ij*}}{|n_{ij*}|}, \quad \text{for} \quad i = 1, i_{max}, \ j = 1, j_{max}. \qquad (3.185)$$

Significance. The theory of the chemical equilibrium has very wide-ranging applications that that are particularly important to physical chemists, metallurgists, solid-state physicists, and many engineers. It permits us to predict the equilibrium composition of a mixture of chemicals in an isolated vessel; it tells us how many phases of an alloy can exists together at any particular temperature and pressure etc. Examples of the application of these theoretical results are given in Chapter 6.

3.2.4.4 Phase equilibrium

The idea of chemical equilibrium is extendable to the so called phase equilibrium. If we apply the definitions of the *Gibbs* function, Eq. (3.144), to a mixture of two phases designated with ′ for the liquid and ″ for its vapor we have

$$dg = \frac{dp}{\rho} - sdT + \left[h'' - h' - T(s'' - s')\right]dC' = \frac{dp}{\rho} - sdT - \left[h'' - h' - T(s'' - s')\right]dC''.$$
(3.186)

Because evaporation at a given constant pressure happens for pure liquids at a constant temperature the relation between the entropy change and the enthalpy change is

$$s'' - s' = \frac{h'' - h'}{T}.$$
(3.187)

This relation makes the *Gibbs* function of the mixture independent of the vapor mass concentration change at constant pressure and temperature. Therefore the equilibrium between the liquid and its vapor for any pressure and temperature is defined by

$$dg = 0.$$
(3.188)

Equation (3.187) can be rewritten in the form $h' - Ts' = h'' - Ts''$.

In other words, the equilibrium between the liquid and its vapor is defined by the equality of the specific *Gibbs* functions for each phase

$$g' = g''.$$
(3.189)

If the liquid and the vapor initially in equilibrium are disturbed by dp resulting in dg' and dg'', respectively, but in such a way that the resultant mixture is again in equilibrium we have

$$g' + dg' = g'' + dg'',$$
(3.190)

or after using Eq. (3.189)

$$dg' = dg'',$$
(3.191)

resulting in

$$v'dp - s'dT = v''dp - s''dT \qquad (3.192)$$

and finally

$$\frac{dT}{dp} = \frac{v'' - v'}{s'' - s'}, \qquad (3.193)$$

derived for the first time by *Benoit Pierre Emile Clapeyron* in 1834. The significance of *Clapeyron*'s equation is in the description of the *p-T* line dividing the stable single phase state from the metastable state. This line is called the *saturation line, Clausius*, see in *Elsner* [4] p. 327. Another derivation of Eq. (3.193) is obtained if one visualizes in the *T-s* and *p-v* diagrams the evaporation process as a *Carnot* cycle for infinitesimal values of dT and dp. Equalizing the work computed from both diagrams for the cycle results in Eq. (3.193). Using Eq. (3.187) the above relation reads

$$\boxed{\frac{dT}{dp} = \frac{1}{T}\frac{v'' - v'}{h'' - h'}, \qquad (3.194)}$$

Clapeyron's equations also define the equilibrium state for the case of an equilibrium mixture of liquid and solid. In this case the corresponding properties along the solidification line have to be used.

At low pressure for which the density difference between liquid and vapor is very large Eq.(3.194) simplifies to $dT/dp = v''/(T\Delta h)$. This equation has been known since 1828 in France as the *August* equation. Assuming the vapor behaves as a perfect gas results in $dp/p = (\Delta h/R)dT/T^2$, where R is the vapor gas constant for the specific substance, R. *Clausius* 1850. Assuming that the latent heat of evaporation is a constant, integrating between an initial state 0 and a final state and rearranging results in a useful expression for extrapolation along the saturation line around the known state 0,

$$\frac{p - p_0}{p_0} = \exp\left(\frac{\Delta h}{RT}\frac{T - T_0}{T_0}\right) - 1. \qquad (3.195)$$

A next step in the improvements of Eq. (3.195) is to be assumed that $\Delta h \approx \Delta h_0 + \alpha T$, resulting in $dp/p = (\Delta h_0/RT^2)dT + (\alpha/RT)dT$. After integrating we obtain

$$\ln p = -\frac{\Delta h_0}{RT} + \frac{\alpha}{R}\ln T + \frac{\Delta h_0}{RT_0} + \ln p_0 - \frac{\alpha}{R}\ln T_0, \tag{3.196}$$

or

$$\ln p = A/T + B\ln T + C, \tag{3.197}$$

which is a form empirically proposed by *G. R. Kirchhoff* in 1858. Sometimes only the boiling properties at atmospheric pressure are known. Equation (3.195) allows us to compute the saturation temperature at a pressure different from the atmospheric pressure as long as the pressure remains far below the critical pressure. Equation (3.197) allows the three unknown constants to be fitted on three measured points. In fact the non-linear dependence of the group $(v''-v')/(h''-h')$ on the temperature requires a non-linear approximation of the saturation line of the type $\ln p = f(T)$.

3.2.5 Partial derivatives in the equation of state $\rho = \rho(p,\varphi,C_{2,\dots,i_{max}})$, where $\varphi = s, h, e$

Enthalpy concept $\varphi = h$. Finally, we replace dT in Eq. (3.96) using Eq. (3.47) and substitute

$$\left(\frac{\partial \rho}{\partial p}\right)_{h,all\_C's} = \left(\frac{\partial \rho}{\partial p}\right)_{T,all\_C's} - \left(\frac{\partial \rho}{\partial T}\right)_{p,all\_C's}\frac{1}{c_p}\left(\frac{\partial h}{\partial p}\right)_{T,all\_C's}, \tag{3.198}$$

$$\left(\frac{\partial \rho}{\partial h}\right)_{p,all\_C's} = \frac{1}{c_p}\left(\frac{\partial \rho}{\partial T}\right)_{p,all\_C's}, \tag{3.199}$$

$$\left(\frac{\partial \rho}{\partial C_i}\right)_{p,h,all\_C's\_except\_C_i}$$

$$= \left(\frac{\partial \rho}{\partial C_i}\right)_{p,T,all\_C's\_except\_C_i} - \left(\frac{\partial \rho}{\partial T}\right)_{p,all\_C's}\frac{1}{c_p}\left(\frac{\partial h}{\partial C_i}\right)_{p,T,all\_C's\_except\_C_i}, \tag{3.200}$$

to obtain

3.2 Multi-component mixtures of miscible and non-miscible components

$$d\rho = \left(\frac{\partial \rho}{\partial p}\right)_{h, all\_C's} dp + \left(\frac{\partial \rho}{\partial h}\right)_{p, all\_C's} dh + \sum_{i=2}^{i_{max}} \left(\frac{\partial \rho}{\partial C_i}\right)_{p,h,all\_C's\_except\_C_i} dC_i \quad (3.201)$$

which is the differential form for the equation of state

$$\rho = \rho\left(p, h, C_{2,...,n_{max}}\right). \quad (3.202)$$

Energy concept $\varphi = e$. Again, the derivation already presented for the specific enthalpy is formally identical with the derivation for the specific internal energy. One has to replace formally h with e and write instead c_{pi}, $\left(\frac{\partial e_i}{\partial T}\right)_{p_i, T}$.

Entropy concept $\varphi = s$: We replace dT in Eq. (3.118) using Eq. (3.47) and substitute

$$\left(\frac{\partial \rho}{\partial p}\right)_{s, all\_C's} = \left(\frac{\partial \rho}{\partial p}\right)_{T, all\_C's} - \left(\frac{\partial \rho}{\partial T}\right)_{p, all\_C's} \frac{\rho\left(\frac{\partial h}{\partial p}\right)_{T, all\_C's} - 1}{\rho c_p} = \frac{1}{a^2} = \frac{\rho}{\kappa p},$$

$$(3.203)$$

$$\left(\frac{\partial \rho}{\partial s}\right)_{p, all\_C's} = \left(\frac{\partial \rho}{\partial T}\right)_{p, all\_C's} \frac{T}{c_p}, \quad (3.204)$$

$$\left(\frac{\partial \rho}{\partial C_i}\right)_{p,s,all\_C's\_except\_C_i}$$

$$= \left(\frac{\partial \rho}{\partial C_i}\right)_{p,T,all\_C's\_except\_C_i} - \left(\frac{\partial \rho}{\partial T}\right)_{p,all\_C's} \frac{T}{c_p} \left(\frac{\partial s}{\partial C_i}\right)_{p,T,all\_C's\_except\_C_i}, \quad (3.205)$$

to obtain

$$d\rho = \frac{dp}{a^2} + \left(\frac{\partial \rho}{\partial s}\right)_{p, all\_C's} ds + \sum_{i=2}^{i_{max}} \left(\frac{\partial \rho}{\partial C_i}\right)_{p,s,all\_C's\_except\_C_i} dC_i \quad (3.206)$$

or

$$d\rho = \frac{\rho}{\kappa}\frac{dp}{p} + \left(\frac{\partial \rho}{\partial s}\right)_{p,all\_C's} ds + \sum_{i=2}^{i_{max}} \left(\frac{\partial \rho}{\partial C_i}\right)_{p,s,all\_C's\_except\_C_i} dC_i, \qquad (3.207)$$

which is the differential form for the equation of state

$$\rho = \rho\left(p, s, C_{2,\ldots,n_{max}}\right). \qquad (3.208)$$

The value κ in Eq. (3.207) is defined as

$$\kappa = \frac{\rho}{p} \Bigg/ \left[\left(\frac{\partial \rho}{\partial p}\right)_{T,all\_C's} - \left(\frac{\partial \rho}{\partial T}\right)_{p,all\_C's} \frac{\rho\left(\frac{\partial h}{\partial p}\right)_{T,all\_C's} - 1}{\rho c_p}\right]$$

$$= c_p \Bigg/ \left[\frac{p}{\rho}\left(\frac{\partial \rho}{\partial p}\right)_{T,all\_C's} c_p + \frac{T}{\rho}\left(\frac{\partial \rho}{\partial T}\right)_{p,all\_C's} \overline{R}\right] = \frac{1}{\dfrac{p}{\rho}\left(\dfrac{\partial \rho}{\partial p}\right)_{T,all\_C's}} \frac{c_p}{c_v}, \qquad (3.209)$$

where

$$\overline{R} = \frac{p}{\rho T}\left[1 - \rho\left(\frac{\partial h}{\partial p}\right)_{T,all\_C's}\right], \qquad (3.210)$$

$$c_v = c_p + \frac{\dfrac{T}{\rho}\left(\dfrac{\partial \rho}{\partial T}\right)_{p,all\_C's}}{\dfrac{p}{\rho}\left(\dfrac{\partial \rho}{\partial p}\right)_{T,all\_C's}} \overline{R} = c_p + \frac{\left(\dfrac{\partial \rho}{\partial T}\right)_{p,all\_C's}}{\left(\dfrac{\partial \rho}{\partial p}\right)_{T,all\_C's}} \frac{1}{\rho}\left[1 - \rho\left(\frac{\partial h}{\partial p}\right)_{T,all\_C's}\right], \qquad (3.211)$$

is generally valid for real gases as well for perfect gases. $\overline{R}$ is used to denote the *pseudo gas constant*, which is of course not in fact a constant for real gases. κ is the isentropic exponent for real gas mixtures. Note that κ is not identical to the isentropic exponent for perfect gases in all cases. Only for mixtures of perfect gases does the *pseudo gas constant* $\overline{R}$ reduce to the gas constant for perfect mixtures because $(\partial h/\partial p)_{T,all\_C's} = 0$, and the expression above for the κ coefficient for real gases then reduces to the usual expression defining the isentropic exponent for perfect gas. For proof, we insert Eqs. (3.75), (3.76) into Eq. (3.209). The result is

$$\kappa = \frac{c_p}{c_p - R}. \quad (3.212)$$

An example for application of the general theory to an air-steam mixture is given in Appendix 3.1.

3.3 Mixture of liquid and microscopic solid particles of different chemical substances

This is a limiting case of the theory presented in the previous Section 3.2.

3.3.1 Partial derivatives in the equation of state $\rho = \rho(p, T, C_{2,\ldots,i_{max}})$

For this particular case the derivatives in Eq. (3.47) given by Eqs. (3.80) through (3.82) reduce to

$$\left(\frac{\partial \rho}{\partial p}\right)_{T, all\_C's} = C_1 \frac{\rho^2}{\rho_1^2} \left(\frac{\partial \rho_1}{\partial p}\right)_T, \quad (3.213)$$

$$\left(\frac{\partial \rho}{\partial T}\right)_{p, all\_C's} = C_1 \frac{\rho^2}{\rho_1^2} \left(\frac{\partial \rho_1}{\partial T}\right)_p, \quad (3.214)$$

$$\left(\frac{\partial \rho}{\partial C_i}\right)_{p, T, all\_C's\_except\_C_i} = \rho^2 \left(\frac{1}{\rho_1} - \frac{1}{\rho_i}\right). \quad (3.215)$$

For missing inert components in the liquid, $\sum_{i=2}^{i_{max}} C_i = 0$, and $\rho = \rho_1$, Eqs. (3.213) and (3.214) take values characteristic of the pure liquid. If the mixture consists solely of inert components, $\sum_{i=2}^{i_{max}} C_i = 1$, Eqs. (3.213) and (3.214) are not defined. In this case $\delta\rho = 0$ holds by definition.

Another consequence of Eq. (3.215) is that the concentration change for a given component causes a density change if and only if the density differs from the density of the liquid, $\rho_i \neq \rho_1$. In all other cases, $\rho_i = \rho_1$, the concentration change does not lead to a change in mixture density.

3.3.2 Partial derivatives in the equation of state $T = T(p, \varphi, C_{2...i_{max}})$ where $\varphi = h, e, s$

For this simple case we obtain the following.

Enthalpy concept $\varphi = h$:

$$c_p = \sum_{i=1}^{i_{max}} C_i c_{pi}, \qquad (3.216)$$

$$\left(\frac{\partial h}{\partial p}\right)_{T, all\_C's} = C_1 \left(\frac{\partial h_1}{\partial p}\right)_T. \qquad (3.217)$$

Energy concept $\varphi = e$:

$$\left(\frac{\partial e}{\partial T}\right)_p = \sum_{i=1}^{i_{max}} C_i \left(\frac{\partial e_i}{\partial T}\right)_p, \qquad (3.218)$$

$$\left(\frac{\partial e}{\partial p}\right)_{T, all\_C's} = C_1 \left(\frac{\partial e_1}{\partial p}\right)_T. \qquad (3.219)$$

Entropy concept $\varphi = s$:

$$\left(\frac{\partial s}{\partial C_i}\right)_{p, T, all\_C's\_except\_C_i} = s_i - s_1. \qquad (3.220)$$

If the increments δp, δs and δC_i are known, the temperature increments δT can be computed from Eq. (3.118) in a manner analogous to that used in the preceding section. In the event that the increments δp, δs and δC_i are large in magnitude, it is better to take into account the non-linear character of Eq. (3.118), as already demonstrated using Eqs. (3.131) through (3.141). If $\Delta s^* = 0$, Eq. (3.137) will once again result. For $p = const$ Eq. (3.136) yields

$$T = T_a e^{\Delta s^*}. \qquad (3.221)$$

Usually $\overline{R}$ is significantly smaller for liquids or mixtures of liquids than for gases because of the marked differences in the densities. The temperature change caused by compression or decompression is thus observable after significant changes in pressure. Usually, the entropy change due to the heat and mass transfer causes a change in the liquid temperature $T \approx T_a e^{\Delta s^*}$. The partial derivatives in the equation of state $\rho = \rho(p, s, C_{2,...,n_{\max}})$ are formally identical to those shown in Section 2.4, and are therefore not presented here.

3.4 Single-component equilibrium fluid

The restriction applied that one velocity field is in thermodynamic equilibrium is very strong and is, in fact, not necessary for an adequate description of multi-phase flow behavior. This assumption is, however, used in a number of applications to yield an approximate estimate of the order of magnitude of the processes. For this reason, this section briefly summarizes the derivatives calculated previously (see *Kolev* [13] (1986), for example) for equilibrium mixtures.

Consider the properties of a fluid, assuming that it is in a thermodynamic equilibrium. Additional use will be made here of the following superscripts:

$'$, saturated liquid,

$''$, saturated steam,

$'''$, saturated solid phase.

The properties on the saturation lines are functions of pressure only

$$T' = T'(p), T'' = T''(p), \rho', \rho'', \rho''', s', s'', s''' = f(p).$$

The same is valid for the properties

$$d\rho'/dp, d\rho''/dp, d\rho'''/dp, ds'/dp, ds''/dp, ds'''/dp = f(p).$$

As a result, four regions are distinguished.

3.4.1 Superheated vapor

For $s > s''$ superheated steam exists with properties that are functions of the pressure and of the temperature

$$\rho = \rho(p, T), \tag{3.222}$$

$$d\rho = \left(\frac{\partial \rho}{\partial p}\right)_T dp + \left(\frac{\partial \rho}{\partial T}\right)_p dT, \tag{3.223}$$

$$\left(\frac{\partial \rho}{\partial p}\right)_T, \left(\frac{\partial \rho}{\partial T}\right)_p = f(p, T), \tag{3.224}$$

$$T = T(p, s), \tag{3.225}$$

$$\frac{dT}{T} = \frac{ds}{c_p} + \frac{\bar{R}}{c_p} \frac{dp}{p}, \tag{3.226}$$

$$c_p = c_p(p, T), \tag{3.227}$$

$$\bar{R} = \frac{p}{\rho T}\left[1 - \rho\left(\frac{\partial h}{\partial p}\right)_T\right] = \bar{R}(p, T), \tag{3.228}$$

$$\rho = \rho(p, s), \tag{3.229}$$

$$d\rho = \frac{dp}{a^2} + \left(\frac{\partial \rho}{\partial s}\right)_p ds = \frac{\rho}{\kappa} \frac{dp}{p} + \left(\frac{\partial \rho}{\partial s}\right)_p ds, \tag{3.230}$$

$$\frac{1}{a^2} = \frac{\rho}{\kappa p} = \left(\frac{\partial \rho}{\partial p}\right)_T - \left(\frac{\partial \rho}{\partial T}\right)_p \frac{\rho\left(\frac{\partial h}{\partial p}\right)_T - 1}{\rho c_p}, \tag{3.231}$$

$$\left(\frac{\partial \rho}{\partial s}\right)_p = \left(\frac{\partial \rho}{\partial T}\right)_p \frac{T}{c_p}, \tag{3.232}$$

where

$$c_p = c_p(p, T), \tag{3.233}$$

$$\left(\frac{\partial h}{\partial p}\right)_T = f(p,T). \qquad (3.234)$$

For a perfect gas, the state equations and the derivatives are easily computed. For real gases, it is assumed that analytical expressions for the state equation are known, and therefore analytical expressions for the derivatives required can easily be derived.

3.4.2 Reconstruction of equation of state by using a limited amount of data available

For many new application fields the information regarding thermodynamic data is very limited but order of magnitude analyses are required for the practice. Therefore we present next a few brief examples of how to construct an approximate closed description of the thermodynamic state using only a few data points.

3.4.2.1 Constant thermal expansion coefficient and isothermal compressibility coefficient

The volumetric thermal expansion coefficient defined as

$$\beta = \frac{1}{v}\left(\frac{dv}{dT}\right)_p = -\frac{1}{\rho}\left(\frac{d\rho}{dT}\right)_p, \qquad (3.235)$$

and the isothermal coefficient of compressibility defined as

$$k = -\frac{1}{v}\left(\frac{dv}{dp}\right)_T = \frac{1}{\rho}\left(\frac{d\rho}{dp}\right)_T, \qquad (3.236)$$

are measured experimentally. As an example solid Al_2O_3 has the properties $\beta = 5.0\times 10^{-5} K^{-1}$, $k = 2.8\times 10^{-12} Pa^{-1}$ [26]. Thus the equation of state

$$\rho = \rho(T,p) \qquad (3.237)$$

can be easily constructed starting with

$$\frac{d\rho}{\rho} = \frac{1}{\rho}\left(\frac{\partial \rho}{\partial T}\right)_p dT + \frac{1}{\rho}\left(\frac{\partial \rho}{\partial p}\right)_T dp = -\beta dT + kdp, \qquad (3.238)$$

and integrating with respect to some reference temperature and pressure

$$\rho = \rho_0 \exp\left[-\beta(T-T_0) + k(p-p_0)\right]. \tag{3.239}$$

For construction of the equation of state for specific enthalpy, entropy and internal energy the measured dependence on the specific heat as a function of temperature

$$c_p = c_p(T) \tag{3.240}$$

is necessary. Using the differential relationships

$$dh = c_p dT + \frac{1}{\rho}\left[1 + T\frac{1}{\rho}\left(\frac{\partial \rho}{\partial T}\right)_p\right]dp$$

$$= c_p dT + \frac{(1-\beta T)}{\rho_0}\exp\left[\beta(T-T_0) - k(p-p_0)\right]dp, \tag{3.241}$$

$$ds = \frac{c_p}{T}dT + \frac{1}{\rho^2}\left(\frac{\partial \rho}{\partial T}\right)_p dp = \frac{c_p}{T}dT - \frac{\beta}{\rho_0}\exp\left[\beta(T-T_0) - k(p-p_0)\right]dp, \tag{3.242}$$

$$de = dh - \frac{1}{\rho}dp + \frac{p}{\rho^2}d\rho = \left[c_p + \frac{p}{\rho^2}\left(\frac{\partial \rho}{\partial T}\right)_p\right]dT + \frac{1}{\rho}\left[\frac{T}{\rho}\left(\frac{\partial \rho}{\partial T}\right)_p + \frac{p}{\rho}\left(\frac{\partial \rho}{\partial p}\right)_T\right]dp$$

$$= \left(c_p - \frac{p}{\rho}\beta\right)dT + \frac{1}{\rho}(-\beta T + kp)dp$$

$$= c_p dT - \frac{p\exp\left[-k(p-p_0)\right]}{\rho_0}\beta\exp\left[\beta(T-T_0)\right]dT$$

$$- \frac{\beta T}{\rho_0}\exp\left[\beta(T-T_0)\right]\exp\left[-k(p-p_0)\right]dp$$

$$+ \frac{1}{\rho_0}\exp\left[\beta(T-T_0)\right]\exp\left[-k(p-p_0)\right]kpdp, \tag{3.243}$$

and integrating from some reference pressure and temperature we obtain

$$h = h_0 + \int_{T_0}^{T} c_p(T) dT - \frac{(1-\beta T)}{k\rho_0} \exp[\beta(T-T_0)]\{\exp[-k(p-p_0)]-1\},$$
(3.244)

$$s = s_0 + \int_{T_0}^{T} \frac{c_p(T)}{T} dT + \frac{1}{k\rho_0} \beta \exp[\beta(T-T_0)]\{\exp[-k(p-p_0)]-1\}, \quad (3.245)$$

$$e = e_0 + \int_{T_0}^{T} c_p(T) dT - \frac{p}{\rho_0} \exp[-k(p-p_0)]\{\exp[\beta(T-T_0)]-1\}$$

$$+ \frac{\beta T}{k\rho_0} \exp[\beta(T-T_0)]\{\exp[-k(p-p_0)]-1\}$$

$$+ \frac{1}{\rho_0} \exp[\beta(T-T_0)] \int_{p_0}^{p} \exp[-k(p-p_0)] kp \, dp . \quad (3.246)$$

Many authors enforce $s = 0$ for $T = 0$. This is called in some text books the third law of the thermodynamics [3] as proposed by *Max Plank* in 1912. Note that the solution of the last integral is quite complicated

$$\int_{p_0}^{p} \exp[-k(p-p_0)] kp \, dp = -\frac{\sqrt{\pi}}{2k} \exp\left(\frac{1}{4}k^2 p_0^2\right)\left[erf\left(-kp + \frac{1}{2}kp_0\right) + erf\left(\frac{1}{2}kp_0\right)\right].$$
(3.247)

In addition from the equation of state in the form

$$d\rho = -\beta\rho \frac{T}{c_p} ds + \left\{\rho k - \frac{T}{c_p}\beta^2\right\} dp = \left(\frac{\partial \rho}{\partial s}\right)_p ds + \left(\frac{\partial \rho}{\partial p}\right)_s dp, \quad (3.248)$$

we obtain the derivatives

$$\left(\frac{\partial \rho}{\partial s}\right)_p = -T\beta\rho/c_p, \quad (3.249)$$

$$\left(\frac{\partial \rho}{\partial p}\right)_s = \rho k - T\beta^2/c_p. \quad (3.250)$$

The velocity of sound is therefore

$$a = \frac{1}{\sqrt{\rho k - T\beta^2 / c_p}}. \tag{3.251}$$

The derivatives of the enthalpy, entropy and specific internal energy with respect to pressure and temperature follow immediately from Eqs. (3.229) through (3.243).

$$\left(\frac{\partial h}{\partial T}\right)_p = c_p, \tag{3.252}$$

$$\left(\frac{\partial h}{\partial p}\right)_T = \frac{(1-\beta T)}{\rho_0} \exp\left[\beta(T-T_0) - k(p-p_0)\right], \tag{3.253}$$

$$\left(\frac{\partial s}{\partial T}\right)_p = \frac{c_p}{T}, \tag{3.254}$$

$$\left(\frac{\partial s}{\partial p}\right)_T = -\frac{\beta}{\rho_0} \exp\left[\beta(T-T_0) - k(p-p_0)\right], \tag{3.255}$$

$$\left(\frac{\partial e}{\partial T}\right)_p = c_p - \frac{\beta p}{\rho_0} \exp\left[\beta(T-T_0) - k(p-p_0)\right], \tag{3.256}$$

$$\left(\frac{\partial e}{\partial p}\right)_T = \frac{kp - \beta T}{\rho_0} \exp\left[\beta(T-T_0) - k(p-p_0)\right]. \tag{3.257}$$

3.4.2.2 Properties known only at given a pressure as temperature functions

Frequently in the literature there are measured properties at atmospheric pressure p_0 given as temperature functions only

$$\rho_{p_0} = \rho_{p_0}(T), \tag{3.258}$$

or

$$\left(\frac{\partial \rho}{\partial T}\right)_p = f(T), \tag{3.259}$$

$$c_p = c_p(T). \tag{3.260}$$

In addition the velocity of sound is measured

$$a = a(T). \tag{3.261}$$

For practical application of this information the reconstruction of the properties in consistent data functions is required. Next we construct the analytical form of the density as a function of temperature and pressure. The density derivative with respect to pressure at constant temperature

$$\left(\frac{d\rho}{dp}\right)_T = \frac{1}{a^2} + T\beta^2/c_p = \frac{1}{a^2} + \frac{T}{\rho^2 c_p}\left(\frac{d\rho}{dT}\right)_p^2 = f(T) \tag{3.262}$$

is obviously a function of the temperature only. Therefore the density can be computed by integrating analytically the equation of state

$$d\rho = \left(\frac{\partial \rho}{\partial T}\right)_p dT + \left(\frac{\partial \rho}{\partial p}\right)_T dp. \tag{3.263}$$

The result is

$$\rho = \rho_{p_0} + \left(\frac{\partial \rho}{\partial p}\right)_T (p - p_0). \tag{3.264}$$

Using the differential equations

$$dh = c_p dT + \frac{1}{\rho}\left[1 + T\frac{1}{\rho}\left(\frac{\partial \rho}{\partial T}\right)_p\right]dp = c_p dT + \frac{dp}{\rho} + T\frac{1}{\rho^2}\left(\frac{\partial \rho}{\partial T}\right)_p dp, \tag{3.265}$$

$$ds = \frac{c_p}{T}dT + \frac{1}{\rho^2}\left(\frac{\partial \rho}{\partial T}\right)_p dp, \tag{3.266}$$

$$de = dh - \frac{1}{\rho}dp + \frac{p}{\rho^2}d\rho = \left[c_p + \frac{p}{\rho^2}\left(\frac{\partial \rho}{\partial T}\right)_p\right]dT + \frac{1}{\rho}\left[\frac{T}{\rho}\left(\frac{\partial \rho}{\partial T}\right)_p + \frac{p}{\rho}\left(\frac{\partial \rho}{\partial p}\right)_T\right]dp$$

$$= c_p dT + \frac{p}{\rho^2}\left(\frac{\partial\rho}{\partial T}\right)_p dT + T\left(\frac{\partial\rho}{\partial T}\right)_p \frac{dp}{\left[\rho_{p_0} + \left(\frac{\partial\rho}{\partial p}\right)_T (p-p_0)\right]^2}$$

$$+ \left(\frac{\partial\rho}{\partial p}\right)_T \frac{pdp}{\left[\rho_{p_0} + \left(\frac{\partial\rho}{\partial p}\right)_T (p-p_0)\right]^2}, \qquad (3.267)$$

keeping in mind that

$$\int_{p_0}^{p} \frac{dp}{\rho_{p_0} + \left(\frac{\partial\rho}{\partial p}\right)_T (p-p_0)} = \frac{1}{\left(\frac{\partial\rho}{\partial p}\right)_T} \ln\frac{\rho}{\rho_{p_0}}, \qquad (3.268)$$

$$\int_{p_0}^{p} \frac{dp}{\left[\rho_{p_0} + \left(\frac{\partial\rho}{\partial p}\right)_T (p-p_0)\right]^2} = \frac{p - p_0}{\rho_{p_0}\rho}, \qquad (3.269)$$

$$\left(\frac{\partial\rho}{\partial p}\right)_T \int_{p_0}^{p} \frac{pdp}{\left[\rho_{p_0} + \left(\frac{\partial\rho}{\partial p}\right)_T (p-p_0)\right]^2}$$

$$= \frac{1}{\left(\frac{\partial\rho}{\partial p}\right)_T}\left\{\left[\frac{1}{\rho} - \frac{1}{\rho_{p_0}}\right]\left[\rho_{p_0} - \left(\frac{\partial\rho}{\partial p}\right)_T p_0\right] + \ln\frac{\rho}{\rho_{p_0}}\right\}, \qquad (3.270)$$

and integrating from some reference pressure and temperature we obtain

$$h = h_0 + \int_{T_0}^{T} c_p(T) dT + \frac{1}{\left(\frac{\partial\rho}{\partial p}\right)_T} \ln\frac{\rho}{\rho_{p_0}} + T\left(\frac{\partial\rho}{\partial T}\right)_p \frac{p - p_0}{\rho_{p_0}\rho}, \qquad (3.271)$$

$$s = s_0 + \int_{T_0}^{T} \frac{c_p}{T} dT + \left(\frac{\partial \rho}{\partial T}\right)_p \frac{dp}{\left[\rho_{p_0} + \left(\frac{\partial \rho}{\partial p}\right)_T (p - p_0)\right]^2}$$

$$= s_0 + \int_{T_0}^{T} \frac{c_p}{T} dT + \left(\frac{\partial \rho}{\partial T}\right)_p \frac{p - p_0}{\rho_{p_0} \rho}, \tag{3.272}$$

$$e = e_0 + \int_{T_0}^{T} c_p dT + p \int_{T_0}^{T} \frac{1}{\rho^2} \left(\frac{\partial \rho}{\partial T}\right)_p dT$$

$$+ T \left(\frac{\partial \rho}{\partial T}\right)_p \frac{p - p_0}{\rho_{p_0} \rho} + \frac{1}{\left(\frac{\partial \rho}{\partial p}\right)_T} \left\{ \left(\frac{1}{\rho} - \frac{1}{\rho_{p_0}}\right) \left[\rho_{p_0} - \left(\frac{\partial \rho}{\partial p}\right)_T p_0\right] + \ln \frac{\rho}{\rho_{p_0}} \right\}. \tag{3.273}$$

Again the derivatives are

$$\left(\frac{\partial h}{\partial T}\right)_p = c_p, \tag{3.274}$$

$$\left(\frac{\partial h}{\partial p}\right)_T = \frac{1}{\rho} \left[1 + T \frac{1}{\rho} \left(\frac{\partial \rho}{\partial T}\right)_p\right], \tag{3.275}$$

$$\left(\frac{\partial s}{\partial T}\right)_p = \frac{c_p}{T}, \tag{3.276}$$

$$\left(\frac{\partial s}{\partial p}\right)_T = \frac{1}{\rho^2} \left(\frac{\partial \rho}{\partial T}\right)_p, \tag{3.277}$$

$$\left(\frac{\partial e}{\partial T}\right)_p = \left[c_p + \frac{p}{\rho^2} \left(\frac{\partial \rho}{\partial T}\right)_p\right], \tag{3.278}$$

$$\left(\frac{\partial e}{\partial p}\right)_T = \frac{1}{\rho^2} \left[T \left(\frac{\partial \rho}{\partial T}\right)_p + p \left(\frac{\partial \rho}{\partial p}\right)_T\right]. \tag{3.279}$$

3.4.2.3 Constant density approximations

For an idealized case of constant density we obtain

$$\rho = \rho_0, \qquad (3.280)$$

$$h = h_0 + \int_{T_0}^{T} c_p(T) dT, \qquad (3.281)$$

$$s = s_0 + \int_{T_0}^{T} \frac{c_p}{T} dT, \qquad (3.282)$$

$$e = e_0 + \int_{T_0}^{T} c_p dT. \qquad (3.283)$$

For zero initial values at the reference point we have

$$e \equiv h. \qquad (3.284)$$

3.4.2.4 Perfect gas approximations

For a perfect gas defined by

$$p = RT\rho \qquad (3.285)$$

with constant specific heat at constant pressure the differential Eqs. (3.241, 3.242, 3.243) result in

$$dh = c_p dT + \frac{1}{\rho}\left[1 + T\frac{1}{\rho}\left(\frac{\partial \rho}{\partial T}\right)_p\right] dp = c_p dT, \qquad (3.286)$$

$$ds = c_p \frac{dT}{T} + \frac{1}{\rho^2}\left(\frac{\partial \rho}{\partial T}\right)_p dp = c_p \frac{dT}{T} - R\frac{dp}{p}, \qquad (3.287)$$

$$de = dh - \frac{1}{\rho}dp + \frac{p}{\rho^2}d\rho = \left[c_p + \frac{p}{\rho^2}\left(\frac{\partial \rho}{\partial T}\right)_p\right]dT + \frac{1}{\rho}\left[\frac{T}{\rho}\left(\frac{\partial \rho}{\partial T}\right)_p + \frac{p}{\rho}\left(\frac{\partial \rho}{\partial p}\right)_T\right]dp$$

$$= (c_p - R)dT = c_v dT, \qquad (3.288)$$

or

$$h = c_p(T - T_0),\qquad(3.289)$$

$$e = c_v(T - T_0),\qquad(3.290)$$

$$s = c_p \ln\frac{T}{T_0} - R \ln\frac{p}{p_0}.\qquad(3.291)$$

Modeling multi-phase flows using multi-velocity field models offers the possibility of postulating one of the mixtures described above for each of the fields. The combination of velocity fields and an appropriate number of components in each field make it possible to describe a large variety of flows observed in nature and in the field of engineering.

3.4.3 Vapor-liquid mixture in thermodynamic equilibrium

For $s' < s < s''$ a saturated liquid coexists with its saturated vapor in thermodynamic equilibrium after having enough time for relaxation. The mass concentration of the vapor in the equilibrium mixture is denoted with C''. The mass concentration is used to generate the following definitions for the specific entropy and the density of the mixtures

$$s = C''s'' + (1 - C'')s',\qquad(3.292)$$

$$\frac{1}{\rho} = \frac{C''}{\rho''} + \frac{1 - C''}{\rho'}.\qquad(3.293)$$

As p (and therefore s'', s') and s are known, this permits easy computation of C'' from Eq. (3.292), i.e.,

$$C'' = \frac{s - s'}{s'' - s'}.\qquad(3.294)$$

As C'' and the pressure are known, the density is uniquely defined by Eq. (3.293), i.e.,

$$\rho = \rho(p, s),\qquad(3.295)$$

or

$$d\rho = \frac{dp}{a^2} + \left(\frac{\partial \rho}{\partial s}\right)_p ds. \tag{3.296}$$

The derivatives can be easily obtained after differentiating Eqs. (3.292) and (3.293) and after eliminating dC''. The result is

$$\left(\frac{\partial \rho}{\partial s}\right)_p = -\frac{\rho^2}{\rho'\rho''}\frac{\rho'-\rho''}{s''-s'}, \tag{3.297}$$

$$\frac{1}{a^2} = \frac{\rho}{\kappa p} = \rho^2 \left\{ \frac{C''}{\rho''^2}\frac{d\rho''}{dp} + \frac{1-C''}{\rho'^2}\frac{d\rho'}{dp} - \frac{1}{\rho^2}\left(\frac{\partial \rho}{\partial s}\right)_p \left[C''\frac{ds''}{dp} + (1-C'')\frac{ds'}{dp} \right] \right\}, \tag{3.298}$$

Kolev [13]. As Eq. (3.297) shows, the density change with the entropy for constant pressure depends strongly on ρ, and therefore on the concentration C''.

3.4.4 Liquid-solid mixture in thermodynamic equilibrium

For $s''' < s < s'$, an equilibrium mixture of liquid and solid phases (designated with superscripts $'$ and $'''$ respectively) exists. Replacing the superscript $''$ with $'''$ in the formalism of the previous section the required equation of state and the corresponding derivatives are then obtained.

3.4.5 Solid phase

For $s < s'''$, a solid phase exists. It is assumed here that analytical expressions for the corresponding derivatives are available. Solid phases can be approximately treated as shown in Section 3.4.2.1.

Appendix 3.1 Application of the theory to steam-air mixtures

Consider a mixture consisting of steam and air. This mixture is common in nature. To facilitate easy application, the definitions obtained for the partial derivatives are reduced to those for a two-component mixture consisting of one inert and one non-inert component. Here air is designated with A and steam with S, with air taken as a single ideal gas. The result is

$$\left(\frac{\partial \rho}{\partial p}\right)_{T,all\_C's} = 1 \Big/ \left[C_A R_A T + (1 - C_A) \Big/ \left(\frac{\partial \rho_S}{\partial p_S}\right)_T \right], \qquad \text{(A-1)}$$

$$\left(\frac{\partial \rho}{\partial T}\right)_{p,all\_C's} = \left(\frac{\partial \rho}{\partial p}\right)_{T,all\_C's} \left[-\rho_A R_A + \left(\frac{\partial \rho_S}{\partial T}\right)_{p_S} \Big/ \left(\frac{\partial \rho_S}{\partial p_S}\right)_T \right], \qquad \text{(A-2)}$$

$$\left(\frac{\partial \rho}{\partial C_A}\right)_{p,T,all\_C's\_except\_C_A} = -\rho \left(\frac{\partial \rho}{\partial p}\right)_{T,all\_C's} \left[R_A T - \frac{1}{(\partial \rho_S / \partial p_S)_T} \right], \qquad \text{(A-3)}$$

$$c_p = C_A c_{pA} + (1 - C_A) \left\{ c_{pS} + \frac{(\partial h_S / \partial p_S)_T}{(\partial \rho_S / \partial p_S)_T} \left[(1 - C_A) \left(\frac{\partial \rho}{\partial T}\right)_{p,all\_C's} - \left(\frac{\partial \rho_S}{\partial T}\right)_{p_S} \right] \right\}$$

$$= C_A c_{pA} + (1 - C_A) c_{pS}$$

$$- C_A (1 - C_A) R_A \left[(1 - C_A) \rho + T \left(\frac{\partial \rho_S}{\partial T}\right)_{p_S} \right] \frac{(\partial h_S / \partial p_S)_T}{1 - C_A \left[1 - R_A T (\partial \rho_S / \partial p_S)_T \right]}, \qquad \text{(A-4)}$$

$$\left(\frac{\partial h}{\partial p}\right)_{T,all\_C's} = \left(\frac{\partial \rho}{\partial p}\right)_{T,all\_C's} (1 - C_A)^2 \left(\frac{\partial h_S}{\partial p_S}\right)_T \Big/ \left(\frac{\partial \rho_S}{\partial p_S}\right)_T$$

$$= (1 - C_A)^2 \left(\frac{\partial h_S}{\partial p_S}\right)_T \Big/ \left\{ 1 - C_A \left[1 - R_A T \left(\frac{\partial \rho_S}{\partial p_S}\right)_T \right] \right\}, \qquad \text{(A-5)}$$

$$\left(\frac{\partial s}{\partial C_A}\right)_{p,T,all\_C's\_except\_C_i}$$

$$= s_A - s_S - \frac{\rho}{T}(1-C_A)\left[\left(\frac{\partial h_S}{\partial p_S}\right)_T / \left(\frac{\partial \rho_S}{\partial p_S}\right)_T\right]\left[1-(1-C_A)\frac{1}{\rho}\left(\frac{\partial \rho}{\partial C_A}\right)_{p,T,all\_C's\_except\_C_i}\right]$$

$$= s_A - s_S - (1-C_A)\left(\frac{\partial h_S}{\partial p_S}\right)_T \rho R_A / \left\{1 - C_A\left[1 - R_A T\left(\frac{\partial \rho_S}{\partial p_S}\right)_T\right]\right\}, \qquad \text{(A-6)}$$

and

$$\left(\frac{\partial T}{\partial p}\right)_{s,all\_C's} = \frac{1}{\rho c_p}\left[1 - \rho\left(\frac{\partial h}{\partial p}\right)_{T,all\_C's}\right], \qquad \text{(A-7)}$$

$$\left(\frac{\partial T}{\partial s}\right)_{p,all\_C's} = \frac{T}{c_p}, \qquad \text{(A-8)}$$

$$\left(\frac{\partial T}{\partial C_A}\right)_{p,s,all\_C's\_except\_C_i} = -\frac{T}{c_p}\left(\frac{\partial s}{\partial C_A}\right)_{p,T,all\_C's\_except\_C_A}. \qquad \text{(A-9)}$$

Appendix 3.2 Useful references for computing properties of single constituents

In this appendix a number of references that provide a useful set of approximations for thermodynamic and transport properties of simple constituents are summarized. These approximations are an example of a simple state equation library.

The library discussed below as an example is used in the computer code IVA3 [14-16] (1991). This library consists of a set of analytical approximations for the following *simple* substances: air, water, steam, uranium dioxide in solid, liquid and equilibrium solid/liquid state. Alternatively, the analytical approximations for stainless steel or corium can be used instead of the approximation for uranium dioxide properties.

For air, the *Irvine* and *Liley* [9] (1984) approximations of $\rho, c_p, h = f(T)$ and $s = s(T, p = const)$ are used where the influence of the pressure on the entropy, $-R \ln(p/p_0)$ is added, where p_0 is a reference pressure (e.g. 10^5) and R is the gas constant of air. For steam, the *Irvine* and *Liley* [9] (1984) approximations of

$\rho, c_p, s, h = f(T, p)$ are used. For water the *Rivkin* and *Kremnevskaya* [30] (1977) approximations of $\rho, c_p, h = f(T, p)$ are used. For metastable water, the above analytical properties from *Rivkin* and *Kremnevskaya* [30] (1977) are extrapolated, taking into account the discussion by *Skripov* et al. [32] (1980). The water entropy as a function of temperature and pressure is computed as follows. First the saturation entropy as a function of liquid temperature is computed, $s' = s'[p'(T)]$, using analytical approximations proposed by *Irvine* and *Liley* [9] (1984), with the pressure correction $s = s[s', p'(T) - p]$ then introduced as proposed by *Garland* and *Hand* [7] (1989).

The analytical approximations by *Irvine* and *Liley* [9] (1984) for the steam/water saturation line, $p' = p'(T)$ and $T' = T'(p)$ are used in IVA3. The Clausius-Clapeyron equation for dp'/dT was obtained by taking the first derivative of the analytical approximation with respect to temperature.

The steam/water saturation properties are computed as a function of p and $T'(p)$ using the above approximations.

Analytical approximations for the properties $\rho, c_p, s, h = f(T)$ of solid and liquid uranium dioxide as proposed by *Fischer* [6], *Chawla* et al. [2] and *Fink* et al. [5] (1981-1990) are used.

For the solid stainless steel properties $\rho, c_p, s, h = f(T)$ the analytical approximations proposed by *Chawla* et al. [2] (1981) are recommended. For liquid stainless steel properties $\rho, c_p, s, h = f(T)$ the approximations proposed by *Chawla* et al. [2] and *Touloukian* and *Makita* [33] are recommended.

For the solid/liquid two-phase region for *liquid metal*, the assumptions of thermodynamic equilibrium within the velocity field are used, and the properties then computed as explained in *Kolev* [17] (1991).

The derivatives $(\partial h / \partial p)_T, (\partial \rho / \partial T)_p, (\partial \rho / \partial p)_T$ are easily obtained by differentiating the corresponding analytical approximations.

The transport properties of the *simple* substances are computed as follows. Thermal conductivity and dynamic viscosity of air, and steam $\lambda = \lambda(T)$ - after *Irvine* and *Liley* [9] (1984). The water thermal conductivity $\lambda = \lambda(T, p)$ is computed using the *Rivkin* and *Alexandrov* [29] (1975) approximation and the water dynamic viscosity $\eta = \eta(T,p)$ and surface tension $\sigma = \sigma(T)$ using the TRAC approximation, *Liles* et al. [25] (1981).

The thermal conductivity and dynamic viscosity of the air-steam mixture are computed using the mole weight method of *Wilke* [37] (1950).

The thermal conductivity of solid and liquid uranium dioxide and stainless steel, as well as the dynamic viscosity and surface tension of liquid uranium dioxide and steel, are computed using the *Chawla* et al. [2] (1981) approximations.

Hill and *Miyagawa* [8] proposed in 1997 to use The IAPS Industrial Formulation 1997 for the Thermodynamic Properties of Water and Steam, see in *Wagner* et al. [35], in tabulated form. The use of an equidistant grid on the two independent axes is proposed and storing not only the properties but also their derivatives at a center of the computational cell. Marking this point with the integer coordinates (i,j) the interpolation is made by *Taylor* series expansion

$$f(x,y) = f_{i,j} + (x-x_i)\left(\frac{\partial f}{\partial x}\right)_{y,i,j} + \frac{1}{2}(x-x_i)^2\left(\frac{\partial^2 f}{\partial x^2}\right)_{y,i,j} + (y-y_j)\left(\frac{\partial f}{\partial y}\right)_{x,i,j}$$

$$+ \frac{1}{2}(y-y_j)^2\left(\frac{\partial^2 f}{\partial y^2}\right)_{y,i,j} + (x-x_i)(y-y_j)\left(\frac{\partial^2 f}{\partial x \partial y}\right)_{i,j}$$

In addition to first derivatives, which are usually available as analytical derivatives from the basic functions, the second derivatives have to be computed numerically, e.g. with steps of $0.001K$ and $1000Pa$. Six constants per property and cell center have to be stored. The search of the integer address is quick because of the selected equidistant grid. The reversed functions are also easily obtained by solving analytically the above quadratic equation. Appropriate selected grids contain the critical point at the boundary of a cell (not at the center). Thus, for two such neighboring cells sub- and supercritical approximations are used, respectively. x_i and y_j may be slightly removed from the center if the saturation line crosses the cell. In this case the properties at the stable site of the saturation line are used. The authors also recommended that instead of the functions $v, h, s = f(p,T)$, the more appropriate $pv, h, ps = f(p,T)$ be used, which give finite values for pressures tending to zero.

Reliable source of water and steam approximations is *Wagner* and *Kruse* [35] providing also some inverted approximations for the subcritical region. Some inverted approximations for water and steam are also available in *Meyer-Pittroff* et al. [27]. General approaches for constructing properties for liquids and gases are given in *Reid* et al. [28]. Large data base for caloric and transport properties for pure substances and mixtures is given in *Vargaftik* et al. [34]. The NIST-JANAF thermochemical tables edited by *Chase* [1] are inevitable for computational analysis of reactive flows.

References

1. Chase MW ed (1998) NIST-JANAF Thermochemical Tables, 4th ed Part I, II, American Institute of Physics and American Chemical Society, Woodbury, New York
2. Chawla TC et al (1981) Thermophysical properties of mixed oxide fuel and stainless steel type 316 for use in transition phase analysis, Nuclear Engineering and Design, vol 67 pp 57-74
3. Cordfunke EHP, Konings RJM, Eds (1990) Thermo-chemical data for reactor materials and fission products, North-Holland, Amsterdam
4. Elsner N (1974) Grundlagen der Technischen Thermodynamik, 2. Berichtete Auflage, Akademie-Verlag, Berlin
5. Fink JK, Ghasanov MG, Leibowitz L (May 1982) Properties for reactor safety analysis, ANL-CEN-RSD-82-2
6. Fischer EA (1990) Kernforschungszentrum Karlsruhe GmbH, unpublished report.
7. Garland WJ, Hand BJ (1989) Simple functions for the fast approximation of light water thermodynamic properties, Nuclear Engineering and Design, vol 113 pp 21-34
8. Hill PG, Miyagawa K (April 1997) A tabular Taylor series expansion method for fast calculation of steam properties, Transaction of ASME, Journal of Engineering for Gas Turbines and Power, vol 119 pp 485-491
9. Irvine TF, Liley PE (1984) Steam and gas tables with computer equations, Academic Press, New York
10. Issa RI (1983) Numerical methods for two- and three- dimensional recirculating flows, in Comp. Methods for Turbulent Transonic and Viscous Flow, Ed. J. A. Essers, Hemisphere, Washington, and Springer, Berlin Heidelberg New York, pp 183
11. Kestin J (1979) A course in thermodynamics, vol 1, Hemisphere, Washington
12. Kolev NI (1986) Transient three phase three component non-equilibrium non-homogeneous flow, Nuclear Engineering and Design, vol 91 pp 373-390
13. Kolev NI (1986) Transiente Zweiphasenstromung, Springer, Berlin Heidelberg New York
14. Kolev NI (Sept. 1991) A three-field model of transient 3D multi-phase three-component flow for the computer code IVA3, Part 2: Models for interfacial transport phenomena. Code validation, Kernforschungszentrum Karlsruhe, KfK 4949
15. Kolev NI (Sept. 1991) A three-field model of transient 3D multi-phase three-component flow for the computer code IVA3, Part 1: Theoretical basics: Conservation and state equations, numerics, Kernforschungszentrum Karlsruhe, KfK 4948
16. Kolev NI (Sept. 1991) IVA3: Computer code for modeling of transient three dimensional three phase flow in complicated geometry, Program documentation: Input description, KfK 4950, Kernforschungszentrum Karlsruhe
17. Kolev NI (1991) Derivatives for equations of state of multi-component mixtures for universal multi-component flow models, Nuclear Science and Engineering, vol 108 no 1 pp 74-87

18. Kolev NI (1994) The code IVA4: Modeling of mass conservation in multi-phase multi-component flows in heterogeneous porous media, Kerntechnik, vol 59 no 4-5 pp 226-237
19. Kolev NI (1994) The code IVA4: Modeling of momentum conservation in multi-phase multi-component flows in heterogeneous porous media, Kerntechnik, vol 59 no 6 pp 249-258
20. Kolev NI (1995) The code IVA4: Second law of thermodynamics for multi phase flows in heterogeneous porous media, Kerntechnik, vol 60 no 1 pp 1-39
21. Kolev NI (1996) Three fluid modeling with dynamic fragmentation and coalescence fiction or daily practice? 7th FARO Experts Group Meeting Ispra, October 15-16, 1996; Proceedings of OECD/CSNI Workshop on Transient thermal-hydraulic and neutronic codes requirements, Annapolis, MD, U.S.A., 5th-8th November 1996; 4th World Conference on Experimental Heat Transfer, Fluid Mechanics and Thermodynamics, ExHFT 4, Brussels, June 2-6, 1997; ASME Fluids Engineering Conference & Exhibition, The Hyatt Regency Vancouver, Vancouver, British Columbia, CANADA June 22-26, 1997, Invited Paper; Proceedings of 1997 International Seminar on Vapor Explosions and Explosive Eruptions (AMIGO-IMI), May 22-24, Aoba Kinen Kaikan of Tohoku University, Sendai-City, Japan
22. Kolev NI (1998) On the variety of notation of the energy conservation principle for single phase flow, Kerntechnik, vol 63 no 3 pp 145-156
23. Kolev NI (October 3-8, 1999) Verification of IVA5 computer code for melt-water interaction analysis, Part 1: Single phase flow, Part 2: Two-phase flow, three-phase flow with cold and hot solid spheres, Part 3: Three-phase flow with dynamic fragmentation and coalescence, Part 4: Three-phase flow with dynamic fragmentation and coalescence – alumna experiments, CD Proceedings of the Ninth International Topical Meeting on Nuclear Reactor Thermal Hydraulics (NURETH-9), San Francisco, California
24. Kolev NI (2002) Multi-Phase Flow Dynamics, Vol. 1 Fundamentals + CD, Springer, Berlin, New York, Tokyo, ISBN 3-540-42984-0
25. Liles DR et al (1981) TRAC-PD2 An advanced best-estimate computer program for pressurized water reactor loss-of-coolant accident analysis, NUREG/CR-2054, LA-7709-MS
26. McCahan S, Shepherd JE (January 1993) A thermodynamic model for aluminum-water interaction, Proc. Of the CSNI Specialists Meeting on Fuel-Coolant Interaction, Santa Barbara, California, NUREC/CP-0127
27. Meyer-Pittroff R, Vesper H, Grigul U (May 1969) Einige Umkehrfunktionen und Näherungsgleichungen zur "1967 IFC Formulation for Industrial Use" für Wasser und Wasserdampf, Brennst.-Wärme-Kraft, vol 21 no 5 pp 239
28. Reid RC, Prausnitz JM, Scherwood TK (1982) The properties of gases and liquids, third edition, McGraw-Hill Book Company, New York
29. Rivkin SL, Alexandrov AA (1975) Thermodynamic properties of water and steam, Energia, Moscow, in Russian
30. Rivkin SL, Kremnevskaya EA (1977) Equations of state of water and steam for computer calculations for process and equipment at power stations, Teploenergetika vol 24 no 3 pp 69-73 (in Russian)
31. Sesternhenn J, Müller B, Thomann H (1999) On the calculation problem in calculating compressible low mach number flows, Journal of Computational Physics, vol 151 pp 579-615

32. Skripov VP et al (1980) Thermophysical properties of liquids in meta-stable state, Atomisdat, Moscow, in Russian
33. Touloukian YS, Makita T (1970) Specific heat, non-metallic liquids and gases, IFC/Plenum, New York, vol 6
34. Vargaftik NB, Vonogradov YK and Yargin VS (1996) Handbook of physical properties of liquid and gases, 3d ed., Begell House, New York, Wallingford (UK)
35. Wagner W et al (2000) The IAPS industrial formulation 1997 for the thermodynamic properties of water and steam, Transaction of ASME, Journal of Engineering for Gas Turbines and Power, vol 122 pp 150-182. See also Wagner W and Kruse A (1998) Properties of water and steam, Springer, Berlin Heidelberg
36. Warnatz J, Maas U, Dibble RW (2001) Combustion, 3th Edition, Springer, Berlin Heidelberg New York
37. Wilke CR (1950) J. Chem. Phys., vol 18

4 On the variety of notations of the energy conservation for single-phase flow

The Greek word $\tau\rho o\pi\eta$ stems from the verb $\varepsilon\nu\tau\rho o\pi\varepsilon\iota\eta$ - conversion, transformation, and was used by *R. Clausius* in a sense of "*value representing the sum of all transformations necessary to bring each body or system of bodies to their present state*" (1854 Ann. Phys., 125, p. 390; 1868 Phil. Mag., 35, p. 419; 1875 "Die mechanische Wärmetheorie", vol. 1).

This Chapter recalls the achievements of the classical thermodynamics for describing the thermodynamic behavior of flows. At the end of the 19th century this knowledge had already formed the classical technical thermodynamics.

The classical thermodynamics makes use of different mathematical notation of the first principles. The purpose of this chapter is to remember that all mathematically correct transformations from one notation into the other are simply logical and consistent reflections of the same physics. The use of the specific entropy is not an exception. The importance of the entropy is that it allows one to reflect nature in the simplest mathematical way.

Even being known since 100 years the basic system of partial differential equations is still not analytically solved for the general case. The requirements to solve this system numerically and the virtual possibility to do this using modern computers is the charming characteristic of the science today. The obtained numerical results have always to be critically examined. As we demonstrate here for pressure wave analysis lower order numerical methods predict acceptably only the first few cycles but then the solution degrades destroyed by numerical diffusion. The inability of the lower order numerical methods to solve fluid mechanics equations for initial and boundary conditions defining strong gradients is not evidence that the equations are wrong. That high accuracy solution methods are necessary to satisfy future needs of the industry is beyond question.

4.1 Introduction

Chapter 4 is intended to serve as an introduction to Chapter 5.

The computational fluid mechanics produces a large number of publications in which the mathematical notation of the basic principles and of the thermodynamic relationships are often taken by the authors for self understanding and are rarely explained in detail. One of the prominent examples is the different notation of the energy conservation for flows in the literature. The vector of the dependent variables frequently used may contain either specific energy, or specific enthalpy, or specific entropy, or temperature etc. among other variables. If doing the transformation from one vector into the other correctly the obtained systems of partial differential equation must be completely equivalent to each other. Nevertheless, sometimes the message of the different notations of the first principle is misunderstood by interpreting one of them as "wrong" and other as right. That is why recalling the basics once more seems to be of practical use in order to avoid misunderstandings.

4.2 Mass and momentum conservation, energy conservation

The fluid mechanics emerges as a scientific discipline by first introducing the idea of *control volume*. The flowing continuum can be described mathematically by abstracting a control volume inside the flow and describing what happens in this control volume. If the control volume is stationary to the frame of reference the resulting description of the flow is frequently called *Euler*ian description. If the control volume is stationary to the coordinate system following the trajectory of fluid particles the description is called *Lagrang*ian. *Oswatitsch*'s "Gasdynamik" [14] is my favorite recommendation to start with this topic because of its uncomplicated introduction into the "language of the gas dynamics". Next we write a system of quasi linear partial differential equations describing the behavior of a single-phase flow in an *Euler*ian control volume without derivation.

The idea of conservation of matter was already expressed in ancient philosophy. Much later, in 1748, *M. Lomonosov* wrote in a letter to *L. Euler* "...all changes in the nature happens so that the mass lost by one body is added to other...". Equation (4.1) reflects the principle of conservation of mass per unit volume of the flow probably first mathematically expressed in its present form by *D'Alembert* (1743) in [3]. The second equation is nothing else than the first principle of *Newton* [15] applied on a continuum passing through the control volume probably for the first time by *Euler* (1737, 1755). Again the equation is written per unit flow volume. The third equation reflects the principle of the conservation of energy, *Mayer* (1841-1851), a medical doctor finding that the humans loose more energy in colder regions in the globe then in hotter, which is reflected in the color of the blood due to the different degree of blood oxidation.

$$\frac{\partial \rho}{\partial \tau} + \nabla \cdot (\rho \mathbf{V}) = 0 \tag{4.1}$$

$$\rho \left(\frac{\partial \mathbf{V}}{\partial \tau} + \mathbf{V}\nabla \cdot \mathbf{V} \right) + \nabla \cdot p + \rho \mathbf{g} = 0 \tag{4.2}$$

$$\frac{\partial}{\partial \tau}\left[\rho\left(e + \frac{1}{2}V^2\right)\right] + \nabla \cdot \left[\rho\left(e + p/\rho + \frac{1}{2}V^2\right)\mathbf{V}\right] + \rho \mathbf{g}.\mathbf{V} = 0 \tag{4.3}$$

Here the usual notations of the thermodynamics are used.

Note that these equations are

a) local volume averaged which means that the behavior of a single molecule interacting with its neighbors is not described in detail.

Secondly, they are

b) not time-averaged conservation equation for
c) flow without internal or external heat sources.

That this equation does not reflect all of the complexity of the flows in nature is evident from the fact that

d) the forces resisting the flow due to viscosity are neglected, and consequently
e) the power component of these forces into the energy conservation equation is also neglected.

Nevertheless the system (4.1) through (4.3) reflects several important characteristics observed in real flows and is frequently used in science and practical computations. In other words this system describes well wave dynamics but does not describe any diffusion processes in the flow, like viscous energy dissipation, heat conduction etc.

4.3 Simple notation of the energy conservation equation

The energy conservation equation can be substantially simplified as it will be demonstrated here without any loss of generality. To achieve this first the scalar product of Eq. (4.2) is constructed with the field velocity **V**. The result is a scalar expressing the mechanical energy balance. Subtracting this result from the energy equation results in

$$\frac{\partial}{\partial \tau}(\rho e) + \nabla \cdot (\rho e \mathbf{V}) + p \nabla \cdot \mathbf{V} = 0. \quad (4.4)$$

Obviously, this equation is much more compact than the primitive form of the energy conservation equation (4.3). Additional simplification is reached if differentiating the first two terms and comparing with the mass conservation equation. Dividing by the density results in

$$\frac{\partial e}{\partial \tau} + (\mathbf{V}.\nabla)e + \frac{p}{\rho}\nabla \cdot \mathbf{V} = 0. \quad (4.4a)$$

Note, that the system consisting of Eqs. (4.1, 4.2, 4.3) is completely equivalent to the system consisting of Eqs. (4.1, 4.2, 4.4 or 4.4a).

4.4 The entropy

Equation (4.4) is called frequently in the literature also energy equation. It can be additionally simplified by introducing a new variable. In a closed system the introduced external heat, dq, may be transformed into internal specific energy de, or into expansion work pdv, or into both simultaneously, in a way that no energy is lost $dq = de + pdv$ - energy conservation principle. A unique functional dependence of the type $q = q(e,v)$ cannot be derived. In the mathematics the linear form $dq = de + pdv$ in which dq is not a total differential is called *Pfaff*'s form. The *Gibbs* relation,

$$Tds = de + pdv = de - \frac{p}{\rho^2}d\rho, \quad (4.5)$$

[5] simply says that per unit mass the change of the internal energy de and the expansion work pdv in the control volume, can be expressed as a change of an single variable denoted with s, provided T is the so-called integrating denominator. That T is the integrating denominator was proven 30 years earlier. The integrating denominator itself was found to be the absolute temperature T (*W. Thomson* 1848). *M. Plank* showed that among the many choices for the integrating denominator only the absolute temperature allows to integrate independently from the trajectory of the transformation [16]. The new remarkable variable was called by *R. Clausius* specific *entropy*, see [2]. Specific - because related to unit mass of the fluid. The Greek word for entropy, τροπη, stems from the verb εντροπειη which means conversion or transformation, and is used by *Clausius* in a sense of

"...*value representing the sum of all transformations necessary to bring each body or system of bodies to their present state*". The introduction of the specific entropy allows simple notation of the second principle of the thermodynamics formulated also by *Clausius*. *Gibbs* used Eq. (4.5) as a mathematical expression of the so called equilibrium principle "... *after the entropy of the system reaches the maximum the system is in a state of equilibrium*...". Equation (4.5) has another remarkable property. It is the differential expression of a unique equation of state $s = s(e,v)$ in which the partial derivatives $(\partial e/\partial s)_v = T$ and $(\partial e/\partial v)_s = -p$ are technically measurable properties.

Note that the introduction of the new variable entropy plays also an important role in the statistical mechanics and in the informatics which will not be discussed here.

Thus, making use of the *Gibbs* relation, of the mass conservation equation, and of Eq. (4.5), Eq. (4.4a) is easily transformed into the simplest notation expressing the energy conservation law

$$\frac{\partial s}{\partial \tau} + (\mathbf{V}.\nabla) \cdot s = 0. \quad (4.6)$$

Again note that the system consisting of Eqs. (4.1, 4.2, 4.6) is completely equivalent to the system consisting of Eqs. (4.1, 4.2, 4.4), and consequently to the original system containing Eqs. (4.1, 4.2, 4.3).

4.5 Equation of state

The macroscopic behavior of fluids can be described by describing the behavior of the simple molecule in interaction with others in a control volume [1]. The technical thermodynamics makes use of the so called equation of state in which the local volume- and time-averaged behavior of the fluid is reflected rather than describing the behavior of the single molecule (see the *Rieman's* work from 1859 in which he combines the experimental findings by *Gay-Lussac (1806)*, *Boyle-Mariotte* to one equation of state, or any thermodynamic text book). One example for the equation of state is the dependence of the specific internal energy on the volume-averaged density and on the volume-averaged temperatures, the so called caloric equation of state,

$$e = e(\rho,T). \quad (4.7)$$

Note the useful definition of the specific capacity at constant volume

$$\left(\frac{\partial e}{\partial T}\right)_\rho = \left(\frac{\partial e}{\partial T}\right)_v = c_v \tag{4.8}$$

Perfect fluids are defined by the dependence of the specific internal energy on the temperature only, but not on the density (remember the famous expansion experiments by *Gay-Lussac*). Variety of forms of the equation of state is possible which may lead to a variety of notations of the energy conservation principle.

4.6 Variety of notation of the energy conservation principle

4.6.1 Temperature

Equation (4.4) can be rewritten in terms of temperature using the equation of state (4.7) and the mass conservation equation (4.1). The result is

$$\frac{\partial T}{\partial \tau} + (\mathbf{V}\cdot\nabla)T + \frac{p}{\rho c_v}\left[1 - \frac{\rho^2}{p}\left(\frac{\partial e}{\partial \rho}\right)_T\right]\nabla\cdot\mathbf{V} = 0, \tag{4.9}$$

an equation containing only one time derivative. Note that

$$\frac{p}{T}\left[1 - \frac{\rho^2}{p}\left(\frac{\partial e}{\partial \rho}\right)_T\right] = \left(\frac{\partial p}{\partial T}\right)_v, \tag{4.10}$$

which is an important thermodynamic relationship, see *Reynolds* and *Perkins* [17 p. 266 Eq. 8.54]. Again note that the system consisting of Eqs. (4.1, 4.2, 4.9) is completely equivalent to the system consisting of Eqs. (4.1, 4.2, 4.4), and consequently to the original system containing Eqs. (4.1, 4.2, 4.3).

4.6.2 Specific enthalpy

Another specific variable frequently used in the thermodynamic is the specific enthalpy defined as follows

$$h = e + p/\rho. \tag{4.11}$$

Using the differential form of this definition

$$d(\rho e) + dp = d(\rho h) \tag{4.12}$$

Eq. (4.4) can be written in terms of the specific enthalpy and pressure

$$\frac{\partial}{\partial \tau}(\rho h) + \nabla \cdot (\rho h \mathbf{V}) - \left(\frac{\partial p}{\partial \tau} + \mathbf{V} \cdot \nabla p\right) = 0. \tag{4.13}$$

Again note that the system consisting of Eqs. (4.1, 4.2, 4.13) is completely equivalent to the system consisting of Eqs. (4.1, 4.2, 4.4), and consequently to the original system containing Eqs. (4.1, 4.2, 4.3). Moreover, it is easily shown that using the definition equation for the specific enthalpy (4.11) the *Gibbs* relation (4.5) takes the form

$$T\rho ds = \rho dh - dp \tag{4.14}$$

which immediately transforms Eq. (4.13) into the entropy equation (4.6) again. Equation (4.13) is easily transferred in terms of temperature and pressure by using the following equation of state

$$h = h(T, p) \tag{4.15}$$

or in differential form

$$dh = c_p dT + \left(\frac{\partial h}{\partial p}\right)_T dp. \tag{4.16}$$

The resulting equation is

$$\frac{\partial T}{\partial \tau} + (\mathbf{V} \cdot \nabla) T - \frac{1}{\rho c_p}\left[1 - \rho\left(\frac{\partial h}{\partial p}\right)_T\right]\left[\frac{\partial p}{\partial \tau} + (\mathbf{V} \cdot \nabla) p\right] = 0. \tag{4.17}$$

Again note that the system consisting of Eqs. (4.1, 4.2, 4.17) is completely equivalent to the system consisting of Eqs. (4.1, 4.2, 4.13), and consequently to the original system containing Eqs. (4.1, 4.2, 4.3).

4.7 Summary of different notations

It follows from Sections 4.3 through 4.6 that the system of equations (4.1, 4.2, 4.3) is completely equivalent to one of the systems containing equations (4.1) and (4.2), and one of the following expressions of the first principle

$$\frac{\partial e}{\partial \tau} + (\mathbf{V}\cdot\nabla)e + \frac{p}{\rho}\nabla\cdot\mathbf{V} = 0, \qquad (4.4a)$$

$$\frac{\partial s}{\partial \tau} + (\mathbf{V}\cdot\nabla)s = 0, \qquad (4.6)$$

$$\frac{\partial T}{\partial \tau} + (\mathbf{V}\cdot\nabla)T + \frac{p}{\rho c_v}\left[1 - \frac{\rho^2}{p}\left(\frac{\partial e}{\partial \rho}\right)_T\right]\nabla\cdot\mathbf{V} = 0, \qquad (4.9)$$

$$\frac{\partial h}{\partial \tau} + (\mathbf{V}\cdot\nabla)h - \frac{1}{\rho}\left[\frac{\partial p}{\partial \tau} + (\mathbf{V}\cdot\nabla)p\right] = 0, \qquad (4.13)$$

$$\frac{\partial T}{\partial \tau} + (\mathbf{V}\cdot\nabla)T - \frac{1}{\rho c_p}\left[1 - \rho\left(\frac{\partial h}{\partial p}\right)_T\right]\left[\frac{\partial p}{\partial \tau} + (\mathbf{V}\cdot\nabla)p\right] = 0. \qquad (4.17)$$

These equations are derived without specifying the type of the fluid as perfect or real gas. They are valid for both cases.

4.8 The equivalence of the canonical forms

The system of partial differential equations can be further considerably simplified if they are from a special type called hyperbolic. *Riemann* found that such a system can be transferred to the so called canonical form which consists of ordinary differential equations. Next we will start with two different forms of the system, one using specific entropy and the other using specific internal energy. We then give the eigen values, the eigen vectors and the canonical form of both systems. It will be shown again that the so obtained forms are mathematically identical.

Before performing the analysis of the type of the system we simplify the mass conservation equation as follows. The density in the mass conservation equation is expressed as function of pressure and specific entropy

$$\rho = \rho(p,s) \qquad (4.18)$$

or in differential form

$$d\rho = \frac{dp}{a^2} + \left(\frac{\partial \rho}{\partial s}\right)_p ds. \qquad (4.19)$$

Substituting Eq. (4.19) into Eq. (4.1) and taking into account the entropy Eq. (4.6) the mass conservation equation takes the form of Eq. (4.21). Thus using the following vector of dependent variables

$$U = (p, w, s) \qquad (4.20)$$

the flow is completely described by the system of quasi linear partial differential equations

$$\frac{\partial p}{\partial \tau} + w \frac{\partial p}{\partial z} + \rho a^2 \frac{\partial w}{\partial z} = 0, \qquad (4.21)$$

$$\frac{\partial w}{\partial \tau} + w \frac{\partial w}{\partial z} + \frac{1}{\rho} \frac{\partial p}{\partial z} + g_z = 0, \qquad (4.2)$$

$$\frac{\partial s}{\partial \tau} + w \frac{\partial s}{\partial z} = 0. \qquad (4.6)$$

It is well known that the eigen values of this system defined by

$$\begin{vmatrix} w-\lambda & \rho a^2 & 0 \\ 1/\rho & w-\lambda & 0 \\ 0 & 0 & w-\lambda \end{vmatrix} = 0, \qquad (4.22)$$

are real and different from each other

$$\lambda_{1,2} = w \pm a, \quad \lambda_3 = w. \qquad (4.23, 4.24, 4.25)$$

The eigen vectors of the transposed characteristic matrix

$$\begin{pmatrix} w-\lambda_i & 1/\rho & 0 \\ \rho a^2 & w-\lambda_i & 0 \\ 0 & 0 & w-\lambda_i \end{pmatrix} \begin{pmatrix} h_1 \\ h_2 \\ h_3 \end{pmatrix}_i = 0, \qquad (4.26)$$

are linearly independent

$$h_1 = \begin{pmatrix} 1/(\rho a) \\ 1 \\ 0 \end{pmatrix}, \; h_2 = \begin{pmatrix} -1/(\rho a) \\ 1 \\ 0 \end{pmatrix}, \text{ and } h_3 = \begin{pmatrix} 0 \\ 0 \\ 1 \end{pmatrix}. \qquad (4.27, 4.28, 4.29)$$

Therefore the quasi linear system of partial differential equations is hyperbolic and can be transformed into the very simple and useful canonical form

$$\frac{dz}{d\tau} = w + a, \qquad \frac{1}{\rho a}\frac{dp}{d\tau} + \frac{dw}{d\tau} = -g, \qquad (4.30, 4.31)$$

$$\frac{dz}{d\tau} = w - a, \qquad -\frac{1}{\rho a}\frac{dp}{d\tau} + \frac{dw}{d\tau} = -g, \qquad (4.32, 4.33)$$

$$\frac{dz}{d\tau} = w, \qquad \frac{ds}{d\tau} = 0. \qquad (4.34, 4.35)$$

As far as I know Eqs. (4.30 to 4.35) were obtained for the first time by *Riemann* in 1859 [18]. Remember that each of the ordinary differential equations (4.31, 4.33, 4.35) is valid in its own coordinate system attached along one of the three characteristic lines defined by the equations (4.30, 4.32, 4.34), respectively. Equations (4.34, 4.35) contain a very interesting analytical solution $s = const$. It means that an infinitesimal control volume moving with the fluid velocity along the line defined by Eq. (4.34) in the (τ, z) plane does not change a specific property defined as specific entropy. This change of state is called isentropic. Comparing with Eq. (4.5) it simply means that within this control volume the energy required for volume expansion per unit flow volume reduces the specific internal energy per unit time by the same amount.

Now let us perform the same analysis starting from the system containing the first principle in terms of specific internal energy. We use this time as dependent variables vector,

$$U = (p, w, e), \qquad (4.36)$$

and the system describing the flow

$$\frac{\partial p}{\partial \tau} + w\frac{\partial p}{\partial z} + \rho a^2 \frac{\partial w}{\partial z} = 0 \qquad (4.21)$$

$$\frac{\partial w}{\partial \tau} + w\frac{\partial w}{\partial z} + \frac{1}{\rho}\frac{\partial p}{\partial z} + g_z = 0 \qquad (4.2)$$

$$\frac{\partial e}{\partial \tau} + w\frac{\partial e}{\partial z} + \frac{p}{\rho}\frac{\partial w}{\partial z} = 0. \qquad (4.4a)$$

4.8 The equivalence of the canonical forms

If you are not familiar with the analysis of the type of a system of partial differential equations by first computing the eigen values, eigenvectors and canonical forms it is recommended first to read Section 11 before continuing here.

From the characteristic equation

$$\begin{vmatrix} w-\lambda & \rho a^2 & 0 \\ 1/\rho & w-\lambda & 0 \\ 0 & \dfrac{p}{\rho} & w-\lambda \end{vmatrix} = 0 \qquad (4.37)$$

we obtain the same eigen values, Eqs. (4.23, 4.24, 4.25), as in the previous case, which is not surprising for equivalent systems. Using the eigen vectors of the transposed characteristic matrix

$$\begin{pmatrix} w-\lambda & 1/\rho & 0 \\ \rho a^2 & w-\lambda & \dfrac{p}{\rho} \\ 0 & 0 & w-\lambda \end{pmatrix} \begin{pmatrix} h_1 \\ h_2 \\ h_3 \end{pmatrix} = 0, \qquad (4.38)$$

we can transform the system of partial differential equations again into a system of ordinary equations along the characteristic lines. Along the first two characteristic lines both canonical equations are identical with the system using the specific entropy as a dependent variable. Along the third characteristic line we have a canonical equation which is in terms of specific internal energy and pressure,

$$\frac{dz}{d\tau} = w, \qquad \frac{de}{d\tau} - \left(\frac{p}{\rho a}\right)^2 \frac{1}{p}\frac{dp}{d\tau} = 0. \qquad (4.39, 4.40)$$

For the general case it can be shown that the following relation holds

$$\left[T + \frac{p}{\rho^2}\left(\frac{\partial \rho}{\partial s}\right)_p\right] ds = de - \left(\frac{p}{\rho a}\right)^2 \frac{1}{p} dp. \qquad (4.41)$$

Applying this relation to the canonical equation (4.40) results exactly in Eq. (4.35). This demonstrates that also the canonical forms of the system using different sets as depending variables are mathematically equivalent.

4.9 Equivalence of the analytical solutions

That the analytical solutions of the equivalent systems of equations written in terms of different dependent variables are equivalent to each other is obvious.

4.10 Equivalence of the numerical solutions?

4.10.1 Explicit first order method of characteristics

Next let us apply the simple explicit first order method of characteristics for solving the system for a horizontal equidistant discretized pipe,

$$\Delta z_1 = (w+a)\Delta \tau, \qquad \frac{1}{\rho a}\left(p^{\tau+\Delta\tau} - p_1\right) + w^{\tau+\Delta\tau} - w_1 = 0, \qquad (4.42, 4.43)$$

$$\Delta z_2 = (w-a)\Delta \tau, \qquad -\frac{1}{\rho a}\left(p^{\tau+\Delta\tau} - p_2\right) + w^{\tau+\Delta\tau} - w_2 = 0, \qquad (4.44, 4.45)$$

$$\Delta z_3 = w\Delta \tau, \qquad s^{\tau+\Delta\tau} - s_3 = 0, \qquad (4.46, 4.47)$$

where indices i refers to values of the dependent variables at $(\tau, z - \lambda_i \Delta \tau)$. There are a variety of methods to compute U_i as a function of the neighboring values. One of the simplest is the linear interpolation

$$U_i = \frac{1}{2}\left[1 + \text{sign}(Cu_i)\right]\left[U_{k-1} + (U - U_{k-1})(1 - Cu_i)\right]$$

$$+ \frac{1}{2}\left[1 - \text{sign}(Cu_i)\right]\left[U - (U_{k+1} - U)Cu_i\right], \qquad (4.48)$$

where

$$Cu_i = \frac{\lambda_i \Delta \tau}{\Delta z}. \qquad (4.49)$$

The result is

$$w^{\tau+\Delta\tau} = \frac{1}{2}\left[-\frac{1}{\rho a}(p_2 - p_1) + w_1 + w_2\right] \qquad (4.50)$$

$$p^{\tau+\Delta\tau} = \frac{1}{2}\left[p_2 + p_1 - (w_2 - w_1)\rho a\right] \tag{4.51}$$

$$s^{\tau+\Delta\tau} = s_3. \tag{4.52}$$

Obviously, the pressure and the velocity in the point $(\tau+\Delta\tau, z)$ depend on the pressure and the velocities at points 1 and 2. The old time value of the specific entropy is simply shifted along the third characteristic from the point $(\tau, z - w\Delta\tau)$ to the point $(\tau+\Delta\tau, z)$. This result is derived without any assumption of the type of the fluid.

In case of closed ends of the pipe we have the following results. For the left end $w_L = 0$, $s_3 = s_L$, and therefore $s^{\tau+\Delta\tau} = s_L$, and from Eq. (4.45) $p^{\tau+\Delta\tau} = p_2 - w_2\rho a$. Similarly for the right end $w_R = 0$, $s_3 = s_R$, and therefore $s^{\tau+\Delta\tau} = s_R$, and from Eq. (4.43) $p^{\tau+\Delta\tau} = p_1 + w_1\rho a$. For open ends only the canonical equations whose characteristics are inside the pipe are used. For the missing equations boundary conditions have to be provided.

To get an impression about the physical meaning of Eq. (4.52) we rewrite it for a perfect gas

$$\frac{T^{\tau+\Delta\tau}}{T_3} = \left(\frac{p^{\tau+\Delta\tau}}{p_3}\right)^{R/c_p}. \tag{4.53}$$

This is usually used as a benchmark for any other numerical solution using first order explicit methods.

Applying again the explicit method of characteristics to the system in terms of the specific internal energy we obtain the same results for the pressure and velocities. The relation along the third characteristic line should be considered very carefully. If the differential form

$$de = \left(\frac{p}{\rho a}\right)^2 \frac{dp}{p} \tag{4.54}$$

is analytically integrated for a perfect gas it results exactly in Eq. (4.53). This is not the case if the differential form is written in finite difference form,

$$e^{\tau+\Delta\tau} = e_3 + \frac{p}{(\rho a)^2}\left(p^{\tau+\Delta\tau} - p_3\right), \tag{4.55}$$

because for the computation of the coefficient $\dfrac{p}{(\rho a)^2}$ an assumption has to be made of how it has to be computed. For this reason the result,

$$T^{\tau+\Delta\tau} - T_3 = \dfrac{T}{p}\dfrac{R}{c_p}\left(p^{\tau+\Delta\tau} - p_3\right), \tag{4.56}$$

does not fit to the exact benchmark solution. Over many time steps it may accumulate to considerable error.

4.10.2 The perfect gas shock tube: benchmark for numerical methods

What is a perfect gas? The experiment that demonstrates the properties of a fluid classified as a perfect fluid was performed in 1806 by *Gay-Lussac* and repeated with a more sophisticated apparatus in 1845 by *Joule*. In this remarkable experiment a closed space with rigid and thermally insulated walls is separated by a membrane into two parts. The left part contained gas with temperature T and pressure p. The right part was evacuated. The separating membrane was broken and the gas expands decreasing first temperature and pressure in the left side and increasing temperature and pressure in the right side. After enough time something remarkable happens. The temperature everywhere approaches the initial temperature. For the same mass and internal energy, that is for the same specific internal energy, the temperature remains constant in spite of the change of the specific volume. Mathematically this means

$$\left(\dfrac{\partial e}{\partial v}\right)_T = 0, \tag{4.57}$$

and therefore

$$e = e(T). \tag{4.58}$$

This reflects the condition which has to be satisfied by the caloric equation of state of the fluid to classify it as a perfect fluid. Real fluids do not satisfy this condition [4]. That is why after such experiments the temperature will not be the same as the initial temperature even if the experimental apparatus could provide perfect energy conservation inside the control volume.

The modern expression for this apparatus is a shock tube. Shock tubes are important experimental facilities for studying dynamic processes in fluids. In the numerical fluid dynamics they serve as a benchmark problem on which the accuracy of the numerical methods can be examined. Text book analytical solution for a perfect gas is available; see for instance [7]. Procedure to its use is given in Appendix A4.1 and comparison of this solution with some numerical solutions is given in Appendix A4.2.

Remember that there are two important error sources in the numerical solution: discretization errors and truncation errors. The discretitization errors are associated with the approximations used to represent derivatives at given points. Usually series expansion is used and the terms not taken into account are computed to estimate the error. To make the error plausible it can be expressed in changes of measurable variables. The truncation errors themselves are associated with inability of the digital computers to work with as much digits as necessary. In the course of the computation errors generated from both sources accumulate. *Under given circumstances the accumulated error can be so big that the obtained solutions are no longer consistent with the initial conditions.*

Let us perform such a test in order to quantify the error of the numerical solution of the system of partial equations. Consider a pipe with constant cross section A and length L. Discretize it equidistantly by using k_{max} points. k_{max} is an even number. Assign for time τ initial values for T and p as follows: $T = T^*$, $w = 0$ for $k = 1, k_{max}$, $p = p_L$ for $k = 1, k_{max}/2$ and $p = p_R$ for $k = k_{max}/2, k_{max}$. Having the temperature and pressure in each point all other variables required for the analysis can be computed as follows $\rho = p/(RT)$, $a = \sqrt{\kappa RT}$, where $\kappa = c_p/(c_p - R)$, $(\partial p/\partial s)_p = -p/(RTc_p)$, $s = c_p \ln(T/T_0) - R \ln(p/p_0)$, $e = c_v(T - T_0)$, $h = c_p(T - T_0)$. Select the time step $\Delta \tau$ satisfying the criterion

$$\max(Cu_i) < 1. \tag{4.59}$$

Compute all dependent variables in the new time plane $\tau + \Delta \tau$ taking into account that the velocities at the both ends are zero. If the entropy or the specific internal energy is one of the dependent variables compute the temperature as either using $T = T_0(p/p_0)^{R/c_p} \exp(s/c_p)$ or using $T = T_0 + e/c_v$, respectively. Then all properties can be computed as function of pressure and temperatures. Compare the differences between the initial total energy of the system at the beginning of the process to the actual energy and estimate the error made. Advance the computation up to obtaining the steady state solution and compare the temperature with the initial temperature.

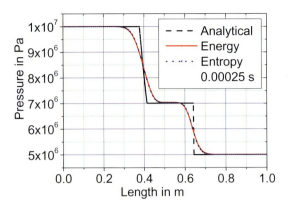

Fig. 4.1. Pressure as function of the distance from the left closed end of the pipe at $\tau = 250\,\mu s$. Comparison between the analytical and numerical solutions. The systems of PDE's contain the energy conservation principle in terms of specific internal energy or specific entropy. Both numerical solutions are indistinguishable

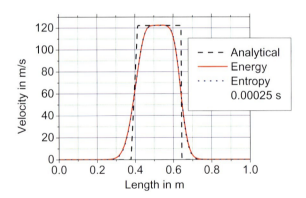

Fig. 4.2. Velocity as function of the distance from the left closed end of the pipe at $\tau = 250\,\mu s$. Comparison between the analytical and numerical solutions. The systems of PDE's contain the energy conservation principle in terms of specific internal energy or specific entropy. Both numerical solutions are indistinguishable

4.10 Equivalence of the numerical solutions? 193

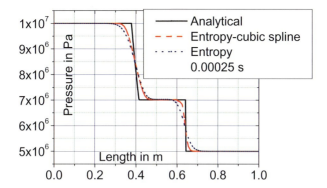

Fig. 4.3. Pressure as function of the distance from the left closed end of the pipe at $\tau = 250\mu s$. Comparison between the analytical and two numerical solutions obtained by using the explicit method of characteristics: one with piecewise linear interpolation, the second with cubic spline interpolation. The systems of PDE's contain the energy conservation principle in terms of specific internal entropy. As expected increasing the order of the spatial discretization increases the accuracy

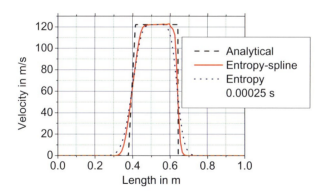

Fig. 4.4. Velocity as function of the distance from the left closed end of the pipe at $\tau = 250\mu s$. Comparison between the analytical and two numerical solutions obtained by using the explicit method of characteristics: one with piecewise linear interpolation, the second with cubic spline interpolation. The systems of PDE's contain the energy conservation principle in terms of specific internal entropy. As expected increasing the order of the spatial discretization increases the accuracy

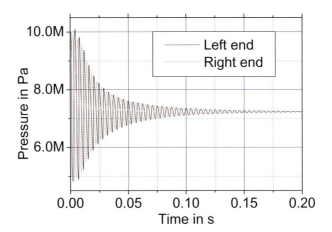

Fig. 4.5. Pressures at the both ends of the pipe as functions of time. The energy conservation is used in terms of specific internal energy. The result using the energy conservation in terms of specific entropy is almost indistinguishable

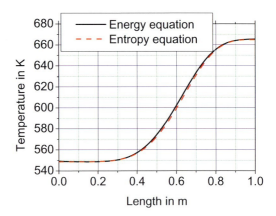

Fig. 4.6. Spatial distribution of the gas temperature at the 0.4th second. Comparison between the predictions using the energy conservation in terms of specific internal energy or entropy

Taking into account the equidistant discretization and the piecewise linear dependence of the variables between to adjacent points, the initial mass, the total energy and the total entropy in the pipe are

$$\frac{m_0}{\Delta z A} = \frac{1}{2} \sum_{k=1}^{k_{max}-1} (\rho_k + \rho_{k+1}), \tag{4.60}$$

$$\frac{E_0}{\Delta z A} = \frac{1}{2} \sum_{k=1}^{k_{max}-1} \left[\rho_k \left(e_k + \frac{1}{2} w_k^2 \right) + \rho_{k+1} \left(e_{k+1} + \frac{1}{2} w_{k+1}^2 \right) \right], \tag{4.61}$$

$$\frac{S_0}{\Delta z A} = \frac{1}{2} \sum_{k=1}^{k_{max}-1} (\rho_k s_k + \rho_{k+1} s_{k+1}), \tag{4.62}$$

respectively. After each time step we compute the mass, the total energy and the total entropy of the system. Thus, the relative mass and total energy errors are

$$\varepsilon_m = \frac{m}{m_0} - 1, \tag{4.63}$$

$$\varepsilon_E = \frac{E}{E_0} - 1, \tag{4.64}$$

respectively. Next we show a numerical computation for initial conditions of 100 and 50 bar respectively and initial temperature 600 K. The material properties used are $c_p = 1034.09$, $R = 287.04$, $T_0 = 1$ and $\rho_0 = 1$. The small computer code used here was written in FORTRAN 90 and runs using 32 or 16 bits floating point operations on real numbers. The computation was performed using a SGI Octane work station with a R10000 processor.

Figures 4.1 and 4.2 show comparisons between an analytical solution [13] for the problem defined at $\tau = 250 \mu s$. The time step used here was 0.1 μs. We see that both systems deliver indistinguishable numerical solutions. Both numerical solutions suffer from considerable numerical diffusion. Repeating the computation with single precision does not change the numerical diffusion. This is evidence that the numerical solution suffers from a considerable *truncation* error introduced by the lower order approximation of the PDE's with finite difference algebraic equations in time and space. Improvement is, as expected, achievable by increasing the order of the discretization as demonstrated in Figs. 4.3 and 4.4. The latter results are obtained by using instead of piecewise linear interpolation piece wise cubic spline interpolation.

Next we perform a computation for a much longer time. The time step was forced to satisfy the condition $Cu_i \equiv \lambda_i \Delta\tau / \Delta z = 0.1$. Figure 4.5 shows the pressure at the both ends of the pipe as a function of time. The computation was performed with a system containing the energy conservation in terms of specific energy. We see pressure waves are traveling through the pipe up to 0.2 s. The computation was stopped at 0.4th second when the "mechanical equilibrium" inside the pipe is almost reached. As it will be explained later the damping of the oscillations is an artificial product of the numerical method with a large truncation error acting as dissipative force. Now we repeat the same computation with a system containing the energy conservation in terms of specific entropy. The comparison of the two results shows that they are hardly distinguishable. The spatial distribution of the temperature at the 0.4th second is presented in Fig. 4.6. We found that the difference between the two results is very small.

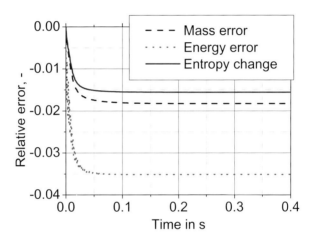

Fig. 4.7. Accumulated relative error as a function of time. The conservation of energy is used in terms of specific entropy

Figure 4.7 presents the accumulation of the numerical errors as a function of time. 3.5% energy error means that the averaged temperature is predicted with an error $\Delta T = \varepsilon_E (T^* - T_0) \approx 21K$. Again the accuracy of both numerical solutions is hardly distinguishable.

As noted at the beginning of our discussion, there are no dissipative diffusion terms in the solved equations. A prominent consequence of this model simplification is manifested by the temperature distribution after the "end" of the dynamic oscillations, Fig. 4.6. Actually, the nature will enforce spatial thermal equilibrium after some time due to molecular diffusion.

4.10 Equivalence of the numerical solutions? 197

Remarkable is the fact that the largest error originates in the initial phase of the oscillation process where the replacement of complicated non-linear spatial distribution with piecewise linear distribution is obviously not appropriate. Lower order numerical methods predict acceptably only the first few cycles but then the solution degrades destroyed by numerical diffusion. The inability of the lower order numerical methods to solve fluid mechanics equations for initial and boundary conditions defining strong gradients is not evidence that the equations are wrong. It is beyond question that high accuracy solution methods are necessary to satisfy future needs of the industry. Such a method is discussed in Section 12.15.3.

This computational analysis shows a behavior of the numerical solutions which is indeed expected. The solutions varies slightly only due to the additional error made by discretization of two terms in the energy conservation equation (4.4a) instead of one in the entropy equation (4.6).

Secondly, the viscosity of the fluid resists the relative movement of the fluid particles. To illustrate this effect let us introduce the viscous force in accordance with the *Stokes* hypothesis for a compressible fluid into the momentum equation [19]. For simplicity, we use the one-dimensional notation. The corresponding energy change per unit time and unit flow volume is introduced into the energy conservation equation. For comparison we also introduce a term $\dot{q}'''$ representing the heat introduction into the flow per unit time and unit flow volume. The modified systems is then

$$\frac{\partial \rho}{\partial \tau} + \frac{\partial}{\partial z}(\rho w) = 0 \qquad (4.65)$$

$$\frac{\partial}{\partial \tau}(\rho w) + \frac{\partial}{\partial z}(\rho w w) + \frac{\partial p}{\partial z} + \rho g_z = \frac{4}{3}\frac{\partial}{\partial z}\left(\rho v \frac{\partial w}{\partial z}\right) \qquad (4.66)$$

$$\frac{\partial}{\partial \tau}\left[\rho\left(e + \frac{1}{2}w^2\right)\right] + \frac{\partial}{\partial z}\left[\rho\left(e + p/\rho + \frac{1}{2}w^2\right)w\right] + \rho g_z w = \frac{4}{3}\frac{\partial}{\partial z}\left(\rho v w \frac{\partial w}{\partial z}\right) + \dot{q}''' \qquad (4.67)$$

Performing the same transformation as described in the previous chapter we obtain the corresponding equations

$$\frac{\partial p}{\partial \tau} + w\frac{\partial p}{\partial z} + \rho a^2 \frac{\partial w}{\partial z} = -\left(\frac{\partial \rho}{\partial s}\right)_s \frac{a^2}{\rho T}\left[\dot{q}''' + \frac{4}{3}\rho v \left(\frac{\partial w}{\partial z}\right)^2\right] \qquad (4.68)$$

$$\frac{\partial w}{\partial \tau} + w\frac{\partial w}{\partial z} + \frac{1}{\rho}\frac{\partial p}{\partial z} = -g_z + \frac{1}{\rho}\frac{4}{3}\frac{\partial}{\partial z}\left(\rho v \frac{\partial w}{\partial z}\right) \qquad (4.69)$$

$$\frac{\partial e}{\partial \tau}+w\frac{\partial e}{\partial z}+\frac{p}{\rho}\frac{\partial w}{\partial z}=\frac{1}{\rho}\left[\dot{q}'''+\frac{4}{3}\rho v\left(\frac{\partial w}{\partial z}\right)^2\right] \qquad (4.70)$$

or

$$\frac{\partial s}{\partial \tau}+w\frac{\partial s}{\partial z}=\frac{1}{\rho T}\left[\dot{q}'''+\frac{4}{3}\rho v\left(\frac{\partial w}{\partial z}\right)^2\right] \qquad (4.71)$$

Again the systems containing the energy conservation in terms of specific internal energy and those in terms of specific entropy are identical. We see that the viscous dissipation per unit flow volume $w\frac{4}{3}\frac{\partial}{\partial z}\left(\rho v \frac{\partial w}{\partial z}\right)$ cancels. This is the reversible component of the viscous dissipation of mechanical energy. The parts which remains $\frac{4}{3}\rho v\left(\frac{\partial w}{\partial z}\right)^2$ is called irreversible dissipation. Looking at Eq. (4.70) we immediately recognize that it acts in a way transferring mechanical energy in thermal energy. The molecular cinematic viscosity has typically very small values, e.g. 40/100000 m^2/s for air at atmospheric conditions. Numerical solutions with large truncation errors such as the solutions discussed above introduce much more numerical diffusion so that for such methods the effect of this term can not be actually seen. In nature this effect is also small except for singularities in the flow. Such a singularity is a shock front. In a shock front the dissociation processes can make this value effectively considerable higher [13].

The pseudo-canonical forms of the two systems are

$$\frac{dz}{d\tau}=w+a, \qquad (4.72)$$

$$\frac{1}{\rho a}\frac{dp}{d\tau}+\frac{dw}{d\tau}=-\frac{1}{\rho}\left(\frac{\partial p}{\partial s}\right)_p \frac{a}{\rho T}\left[\dot{q}'''+\frac{4}{3}\rho v\left(\frac{\partial w}{\partial z}\right)^2\right]-g_z+\frac{1}{\rho}\frac{4}{3}\frac{\partial}{\partial z}\left(\rho v \frac{\partial w}{\partial z}\right) \qquad (4.73)$$

$$\frac{dz}{d\tau}=w-a, \qquad (4.74)$$

$$-\frac{1}{\rho a}\frac{dp}{d\tau}+\frac{dw}{d\tau}=\frac{1}{\rho}\left(\frac{\partial p}{\partial s}\right)_p \frac{a}{\rho T}\left[\dot{q}'''+\frac{4}{3}\rho v\left(\frac{\partial w}{\partial z}\right)^2\right]-g_z+\frac{1}{\rho}\frac{4}{3}\frac{\partial}{\partial z}\left(\rho v \frac{\partial w}{\partial z}\right) \qquad (4.75)$$

$$\frac{dz}{d\tau} = w, \qquad \frac{ds}{d\tau} = \frac{1}{\rho T}\left[\dot{q}''' + \frac{4}{3}\rho v \left(\frac{\partial w}{\partial z}\right)^2\right]. \qquad (4.76, 4.77)$$

or

$$\frac{dz}{d\tau} = w, \qquad \frac{de}{d\tau} - \frac{p}{(\rho a)^2}\frac{dp}{d\tau} = \left[1 + \frac{p}{\rho T}\frac{1}{\rho}\left(\frac{\partial \rho}{\partial s}\right)_p\right]\frac{1}{\rho}\left[\dot{q}''' + \frac{4}{3}\rho v \left(\frac{\partial w}{\partial z}\right)^2\right].$$

$$(4.78, 4.79)$$

Pseudo-canonical because the dissipative terms are considered as a sources for the hyperbolic part.

4.11 Interpenetrating fluids

Mixtures are an example of interpenetrating fluids. For an *infinite time* gas species penetrate the space occupied by the total mixture in a way that in each point the same concentration is observed. This is known as a *Dalton*'s law. What does *infinite time* mean? The time scale for complete mixing depends on the size of the considered volume, of the specific properties of the constituents to each other and of course on the initial state of the system. The conservation equations for mass, momentum and energy can be written in the same way as for the single component by considering that the single component occupies all the volume, possesses locally the same temperature as the mixture temperature and exerts on walls partial pressure corresponding to each component so that the sum of the partial pressures is equal to the total pressure. It is common practice to introduce some averaged properties and to sum the component equations. The equations are first local volume averaged and then time averaged. For simplicity we do not consider any dissipative effects like diffusion, viscous effects etc. except fluid mixing due to convection and local mass sources. We give here the final results for one-dimensional flow in a thermally isolated non-deformable pipe with constant cross section without derivation.

$$\frac{\partial \rho}{\partial \tau} + \frac{\partial}{\partial z}(\rho w) = \mu \qquad (4.80)$$

$$\frac{\partial C_i}{\partial \tau} + w\frac{\partial C_i}{\partial z} = \frac{1}{\rho}(\mu_i - C_i \mu) \qquad i = 2, i_{max} \qquad (4.81)$$

$$\frac{\partial w}{\partial \tau} + w\frac{\partial w}{\partial z} + \frac{1}{\rho}\frac{\partial p}{\partial z} + g_z = 0, \qquad (4.82)$$

$$\rho\left(\frac{\partial h}{\partial \tau}+w\frac{\partial h}{\partial z}\right)-\left(\frac{\partial p}{\partial \tau}+w\frac{\partial p}{\partial z}\right)=\sum_{i=1}^{i_{max}}\mu_i\left(h_i^\sigma - h_i\right) \quad (4.83)$$

where

$$h = \sum_{i=1}^{i_{max}} C_i h_i . \quad (4.84)$$

The first equation reflects the mass conservation for the mixture. The second group of $i_{max} - 1$ equations reflects the mass conservation for each velocity field. Note that the sum of all mass concentrations is equal to one,

$$\sum_{i=1}^{i_{max}} C_i = 1, \quad (4.85)$$

or

$$C_1 = 1 - \sum_{i=2}^{i_{max}} C_i \quad (4.86)$$

Equation (4.83) is remarkable. It says that at constant pressure the injected mass sources contribute to the total enthalpy change with their deviation of the external component enthalpies from the corresponding component enthalpy inside the flow. More information about the rigorous derivation of the above set of equations for the general case of muti-phase multi-component flows is presented in the next Chapter (first published in [8]). Next we will write the energy conservation equation in terms of entropy. For this purpose define the mixture entropy in a similar way as the mixture enthalpy

$$s = \sum_{i=1}^{i_{max}} C_i s_i \quad (4.87)$$

Obviously

$$Tds = T\sum_{i=1}^{i_{max}} C_i ds_i + T\sum_{i=2}^{i_{max}} (s_i - s_1) dC_i . \quad (4.88)$$

Having in mind that

$$C_i = \frac{\rho_i}{\rho} \quad (4.89)$$

and replacing with the *Gibbs* equation (4.14) for each component $T\rho_i ds_i = \rho_i dh_i - dp_i$ or

$$TC_i ds_i = C_i dh_i - \frac{dp_i}{\rho} \qquad (4.90)$$

and rearranging results in the *Gibbs* mixture equation (compare with Eq. (12) in [6])

$$Tds = \sum_{i=1}^{i_{max}} C_i dh_i - \frac{dp}{\rho} + T\sum_{i=2}^{i_{max}}(s_i - s_1)dC_i = dh - \frac{dp}{\rho} - \sum_{i=2}^{i_{max}}\left[h_i - Ts_i - (h_1 - Ts_1)\right]dC_i. \qquad (4.91)$$

Now, Eq. (4.83) is easily transformed into the entropy equation

$$\rho T\left(\frac{\partial s}{\partial \tau} + w\frac{\partial s}{\partial z}\right) = \sum_{i=1}^{i_{max}} \mu_i \left(h_i^\sigma - h_i\right) + T\sum_{i=1}^{i_{max}}(\mu_i - \mu C_i)s_i \qquad (4.92)$$

using Eqns. (4.91) and (4.81). Again we see that the entropy equation is simpler then Eq. (4.83). The temperature equation can be derived from Eq. (4.92) by using the equation of state, Eq. (3.98),

$$\rho Tds = \rho c_p dT - \left(1 - \rho\left(\frac{\partial h}{\partial p}\right)_{T,all\_C's}\right)dp + \rho T\sum_{i=2}^{i_{max}}\left(\frac{\partial s}{\partial C_i}\right)_{p,T,all\_C's\_except\_C_i} dC_i \qquad (4.93)$$

derived by this author in [9]. Here

$$\left(\frac{\partial s}{\partial C_i}\right)_{p,T,all\_C's\_except\_C_i} = s_i - s_1 + \Delta s_i^{np} \qquad (4.94)$$

In the intermediate result

$$\rho c_p\left(\frac{\partial T}{\partial \tau} + w\frac{\partial T}{\partial z}\right) - \left(1 - \rho\left(\frac{\partial h}{\partial p}\right)_{T,all\_C's}\right)\left(\frac{\partial p}{\partial \tau} + w\frac{\partial p}{\partial z}\right)$$

$$+\rho T \sum_{i=2}^{i_{max}} \left(\frac{\partial s}{\partial C_i}\right)_{p,T,all\_C's\_except\_C_i} \left(\frac{\partial C_i}{\partial \tau} + w \frac{\partial C_i}{\partial z}\right) = \sum_{i=1}^{i_{max}} \mu_i \left(h_i^\sigma - h_i\right) + T \sum_{i=1}^{i_{max}} (\mu_i - \mu C_i) s_i \quad (4.95)$$

we replace the RHS of the concentration equations instead of the total variation of the concentrations. Keeping in mind that

$$\sum_{i=1}^{i_{max}} (\mu_i - \mu C_i) s_i - \sum_{i=2}^{i_{max}} (s_i - s_1)(\mu_i - C_i \mu) = 0 \quad (4.96)$$

we obtain a very informative form of the temperature equation

$$\rho c_p \left(\frac{\partial T}{\partial \tau} + w \frac{\partial T}{\partial z}\right) - \left[1 - \rho \left(\frac{\partial h}{\partial p}\right)_{T,all\_C's}\right] \left(\frac{\partial p}{\partial \tau} + w \frac{\partial p}{\partial z}\right)$$

$$= \sum_{i=1}^{i_{max}} \mu_i \left(h_i^\sigma - h_i\right) - T \sum_{i=2}^{i_{max}} \Delta s_i^{np} (\mu_i - C_i \mu). \quad (4.97)$$

The mixture mass conservation equation can be rearranged similarly as in the single component case by using the equation of state

$$d\rho = \frac{dp}{a^2} + \left(\frac{\partial \rho}{\partial s}\right)_{p,all\_C's} ds + \sum_{i=2}^{i_{max}} \left(\frac{\partial \rho}{\partial C_i}\right)_{p,s,all\_C's\_except\_C_i} dC_i \quad (4.98)$$

namely

$$\frac{1}{a^2}\left(\frac{\partial p}{\partial \tau} + w \frac{\partial p}{\partial z}\right) + \left(\frac{\partial \rho}{\partial s}\right)_{p,all\_C's} \left(\frac{\partial s}{\partial \tau} + w \frac{\partial s}{\partial z}\right) +$$

$$\sum_{i=2}^{i_{max}} \left(\frac{\partial \rho}{\partial C_i}\right)_{p,s,all\_C's\_except\_C_i} \left(\frac{\partial C_i}{\partial \tau} + w \frac{\partial C_i}{\partial z}\right) + \rho \frac{\partial w}{\partial z} = \mu \quad (4.99)$$

or

$$\frac{\partial p}{\partial \tau} + w \frac{\partial p}{\partial z} + \rho a^2 \frac{\partial w}{\partial z} = Dp \quad (4.100)$$

where

$$Dp = a^2 \left\{ \begin{array}{l} \left[\mu - \dfrac{1}{\rho T}\left(\dfrac{\partial p}{\partial s}\right)_{p,all\_C's} \right] \left[\sum_{i=1}^{i_{max}} \mu_i \left(h_i^\sigma - h_i \right) + T\sum_{i=1}^{i_{max}} (\mu_i - \mu C_i) s_i \right] \\ -\dfrac{1}{\rho} \sum_{i=2}^{i_{max}} \left(\dfrac{\partial p}{\partial C_i}\right)_{p,s,all\_C's\_except\_C_i} (\mu_i - C_i \mu) \end{array} \right\}. \quad (4.101)$$

The source term can be further simplified by using the following expressions for the derivatives

$$\left(\dfrac{\partial p}{\partial s}\right)_{p,all\_C's} = \left(\dfrac{\partial p}{\partial T}\right)_{p,all\_C's} \dfrac{T}{c_p}, \qquad (4.102)$$

$$\left(\dfrac{\partial p}{\partial C_i}\right)_{p,s,all\_C's\_except\_C_i} = \left(\dfrac{\partial p}{\partial C_i}\right)_{p,T,all\_C's\_except\_C_i}$$

$$-\left(\dfrac{\partial p}{\partial T}\right)_{p,all\_C's} \dfrac{T}{c_p} \left(\dfrac{\partial s}{\partial C_i}\right)_{p,T,all\_C's\_except\_C_i}, \qquad (4.103)$$

and Eq. (4.96). The result is

$$Dp = a^2 \left\{ \begin{array}{l} \left[\mu - \dfrac{1}{\rho c_p}\left(\dfrac{\partial p}{\partial T}\right)_{p,all\_C's} \right] \left[\sum_{i=1}^{i_{max}} \mu_i \left(h_i^\sigma - h_i \right) - T\sum_{i=2}^{i_{max}} \Delta s_i^{np} (\mu_i - C_i \mu) \right] \\ -\dfrac{1}{\rho} \sum_{i=2}^{i_{max}} \left(\dfrac{\partial p}{\partial C_i}\right)_{p,T,all\_C's\_except\_C_i} (\mu_i - C_i \mu) \end{array} \right\}. \quad (4.104)$$

For mixture of ideal gases the specific component enthalpies is not a function of pressure, therefore $\left(\dfrac{\partial h}{\partial p}\right)_{T,all\_C's} = 0$, and $\Delta s_i^{np} = 0$, $\left(\dfrac{\partial \rho}{\partial p}\right)_{T,all\_C's} = \dfrac{1}{RT}$,

$\left(\dfrac{\partial \rho}{\partial T}\right)_{p,all\_C's} = -\dfrac{\rho}{T}$, $\left(\dfrac{\partial \rho}{\partial C_i}\right)_{p,T,all\_C's\_except\_C_i} = -\dfrac{\rho}{R}(R_i - R_1)$. Consequently

$$\rho c_p \left(\dfrac{\partial T}{\partial \tau} + w\dfrac{\partial T}{\partial z} \right) - \left(\dfrac{\partial p}{\partial \tau} + w\dfrac{\partial p}{\partial z} \right) = \sum_{i=1}^{i_{max}} \mu_i \left(h_i^\sigma - h_i \right), \qquad (4.105)$$

and

$$Dp = a^2 \left[\mu + \frac{1}{c_p T} \sum_{i=1}^{i_{max}} \mu_i \left(h_i^\sigma - h_i \right) + \frac{1}{R} \sum_{i=2}^{i_{max}} \left(R_i - R_1 \right) \left(\mu_i - C_i \mu \right) \right] \quad (4.106)$$

The equation of state for a mixture of i_{max} ideal gases is $\rho = p/(RT)$, where

$$R = \sum_{i=1}^{i_{max}} C_i R_i, \quad a = \sqrt{\kappa RT}, \quad \text{where } \kappa = c_p/(c_p - R), \quad c_p = \sum_{i=1}^{i_{max}} C_i c_{pi}, \quad c_v = \sum_{i=1}^{i_{max}} C_i c_{vi},$$

$$p_i = Y_i p, \quad \text{where } Y_i = \frac{R_i}{R} C_i, \quad s = \sum_{i=1}^{i_{max}} C_i s_i = \sum_{i=1}^{i_{max}} C_i c_{pi} \ln(T/T_{0i}),$$

$$- \sum_{i=1}^{i_{max}} C_i R_i \ln(p_i/p_{0i}) \quad \text{where } s_i = c_{pi} \ln(T/T_{0i}) - R_i \ln(p_i/p_{0i}),$$

$$e = \sum_{i=1}^{i_{max}} C_i e_i = \sum_{i=1}^{i_{max}} C_i c_{vi} (T - T_{0i}), \quad \text{where } e_i = c_{vi} (T - T_{0i}),$$

$$h = \sum_{i=1}^{i_{max}} C_i h_i = \sum_{i=1}^{i_{max}} C_i c_{pi} (T - T_{0i}) \quad \text{where } h_i = c_{pi} (T - T_{0i}).$$

The caloric equation of state can be considerably simplified if one uses the same reference point for integration constants set to zero, that is, $T_{0i} = T_0$ and $p_{0i} = p_0$. The result is

$$s = \sum_{i=1}^{i_{max}} C_i s_i = c_p \ln(T/T_0) - \sum_{i=1}^{i_{max}} C_i R_i \ln(p_i/p_0)$$

$$= c_p \ln(T/T_0) - R \ln\left(\frac{p}{p_o}\right) - \sum_{i=1}^{i_{max}} C_i R_i \ln\left(\frac{C_i R_i}{R}\right),$$

$$T = T_0 \exp\left[\frac{s}{c_p} + \frac{R}{c_p} \ln\left(\frac{p}{p_o}\right) + \frac{1}{c_p} \sum_{i=1}^{i_{max}} C_i R_i \ln\left(\frac{C_i R_i}{R}\right) \right]$$

$$= T_0 \left(\frac{p}{p_o}\right)^{\frac{R}{c_p}} \exp\left(\frac{s}{c_p}\right) \prod_{i=1}^{i_{max}} \left(\frac{C_i R_i}{R}\right)^{\frac{C_i R_i}{c_p}},$$

$$e = c_v (T - T_0),$$

$$h = c_p (T - T_0).$$

For two components M and n we have

$$T = T_0 \left(\frac{p}{p_o}\right)^{\frac{R}{c_p}} \exp\left(\frac{s}{c_p}\right) \left(\frac{C_n R_n}{R}\right)^{\frac{C_n R_n}{c_p}} \left[\frac{(1-C_n)R_M}{R}\right]^{\frac{(1-C_n)R_M}{c_p}}.$$

4.12 Summary of different notations for interpenetrating fluids

For convenience we summarize the results of Section 4.11. The system of partial differential equations describing the multi-component flow is

$$\frac{\partial p}{\partial \tau} + w\frac{\partial p}{\partial z} + \rho a^2 \frac{\partial w}{\partial z} = Dp \qquad (4.107)$$

$$\frac{\partial C_i}{\partial \tau} + w\frac{\partial C_i}{\partial z} = \frac{1}{\rho}(\mu_i - C_i \mu) \qquad i = 2, i_{max} \qquad (4.108)$$

$$\frac{\partial w}{\partial \tau} + w\frac{\partial w}{\partial z} + \frac{1}{\rho}\frac{\partial p}{\partial z} + g_z = 0, \qquad (4.109)$$

$$\rho T \left(\frac{\partial s}{\partial \tau} + w\frac{\partial s}{\partial z}\right) = \sum_{i=1}^{i_{max}} \mu_i \left(h_i^\sigma - h_i\right) + T\sum_{i=1}^{i_{max}} (\mu_i - \mu C_i) s_i \qquad (4.110)$$

or

$$\rho\left(\frac{\partial h}{\partial \tau} + w\frac{\partial h}{\partial z}\right) - \left(\frac{\partial p}{\partial \tau} + w\frac{\partial p}{\partial z}\right) = \sum_{i=1}^{i_{max}} \mu_i \left(h_i^\sigma - h_i\right) \qquad (4.111)$$

or

$$\rho c_p \left(\frac{\partial T}{\partial \tau} + w\frac{\partial T}{\partial z}\right) - \left[1 - \rho\left(\frac{\partial h}{\partial p}\right)_{T, all\_C's}\right]\left(\frac{\partial p}{\partial \tau} + w\frac{\partial p}{\partial z}\right)$$

$$= \sum_{i=1}^{i_{max}} \mu_i \left(h_i^\sigma - h_i\right) - T\sum_{i=2}^{i_{max}} \Delta s_i^{np} (\mu_i - C_i\mu). \qquad (4.112)$$

As for the single-component flows, the energy conservation can be noted in terms of different variables. Equations (4.110), (4.111) and (4.112) which are completely equivalent are few of the possible examples. Preferring one of them for practical analysis is only a matter of convenience. Obviously the entropy concept offers again the simplest equation. Analysis of the type of the system is again greatly simplified by using the entropy equation. Thus, using the following vector of dependent variables

$$U = \left(p, w, s, C_{i=2,i_{max}}\right) \qquad (4.113)$$

the flow is completely described by the system of quasi linear partial differential equations

$$\frac{\partial p}{\partial \tau} + w\frac{\partial p}{\partial z} + \rho a^2 \frac{\partial w}{\partial z} = Dp \qquad (4.114)$$

$$\frac{\partial w}{\partial \tau} + w\frac{\partial w}{\partial z} + \frac{1}{\rho}\frac{\partial p}{\partial z} + g_z = 0, \qquad (4.115)$$

$$\frac{\partial s}{\partial \tau} + w\frac{\partial s}{\partial z} = Ds \qquad (4.116)$$

$$\frac{\partial C_i}{\partial \tau} + w\frac{\partial C_i}{\partial z} = DC_i \qquad i = 2, i_{max} \qquad (4.117)$$

where

$$Ds = \frac{1}{\rho T}\left[\sum_{i=1}^{i_{max}} \mu_i \left(h_i^\sigma - h_i\right) + T\sum_{i=1}^{i_{max}} \left(\mu_i - \mu C_i\right)s_i\right] \qquad (4.118)$$

$$DC_i = \frac{1}{\rho}\left(\mu_i - C_i \mu\right) \qquad (4.119)$$

The eigen values of this system are real and different from each other

$$\lambda_{1,2} = w \pm a, \quad \lambda_{3,i=2,i_{max}} = w. \qquad (4.120, 4.121, 4.122)$$

The eigen vectors of the transposed characteristic matrix are linearly independent. Therefore the quasi linear system of partial differential equations is hyperbolic and can be transformed into the very simple and useful canonical form

$$\frac{dz}{d\tau} = w + a, \quad \frac{1}{\rho a}\frac{dp}{d\tau} + \frac{dw}{d\tau} = \frac{1}{\rho a}Dp - g \qquad (4.123, 4.124)$$

$$\frac{dz}{d\tau} = w - a, \quad -\frac{1}{\rho a}\frac{dp}{d\tau} + \frac{dw}{d\tau} = -\frac{1}{\rho a}Dp - g \qquad (4.125, 4.126)$$

$$\frac{dz}{d\tau} = w, \qquad \frac{ds}{d\tau} = Ds, \qquad (4.127, 4.128)$$

$$\frac{dz}{d\tau} = w, \qquad \frac{dC_i}{d\tau} = DC_i. \qquad (4.129, 4.130)$$

We see the remarkable behavior that concentrations and entropy change along the characteristic line defined with the mixture velocity. For the here considered simplified case if there are no mass sources in the pipe the concentrations and the entropy do not change along the characteristic line for a given time step - convective transport only. In case of mass sources, the mixing process changes the concentration and increases the entropy along the characteristic line for a given time step. One should again not forget that at the beginning of Section 4.11 we have neglected all other dissipative effects which cause entropy increase in the real nature and that our final system is only an idealization of the nature.

Appendix 4.1 Analytical solution of the shock tube problem

Consider a pipe with a constant cross section filled with perfect gas. The left halve of the pipe is separated from the right with a membrane that separates high pressure region 4, from low pressure region 1 – see Fig- 4.1-1. The membrane is removed and a shock wave propagates from the left to the right and a refraction wave from right to the left. There is a text book solution for this case that is widely used as a benchmark for numerical methods. This solution is given below.

1) First the pressure ratio p_2/p_1 is found by solving the following equations by iterations

$$\frac{p_4}{p_1} = \frac{p_2}{p_1} \left\{ 1 - \frac{(\kappa_4 - 1)\frac{a_1}{a_4}\left(\frac{p_2}{p_1} - 1\right)}{\left[4\kappa_1^2 + 2\kappa_1(\kappa_1 + 1)\left(\frac{p_2}{p_1} - 1\right)\right]^{1/2}} \right\}^{-\frac{2\kappa_4}{\kappa_4 - 1}}. \qquad (1)$$

2) The shock Mach number is found by solving the following equation by iterations

$$\frac{p_4}{p_1} = \frac{\kappa_1 - 1}{\kappa_1 + 1}\left(\frac{2\kappa_1}{\kappa_1 - 1}Ma_s^2 - 1\right)\left[1 - \frac{\frac{\kappa_4 - 1}{\kappa_4 + 1}\frac{a_1}{a_4}\left(Ma_s^2 - 1\right)}{Ma_s}\right]^{-\frac{2\kappa_4}{\kappa_4 - 1}}. \quad (2)$$

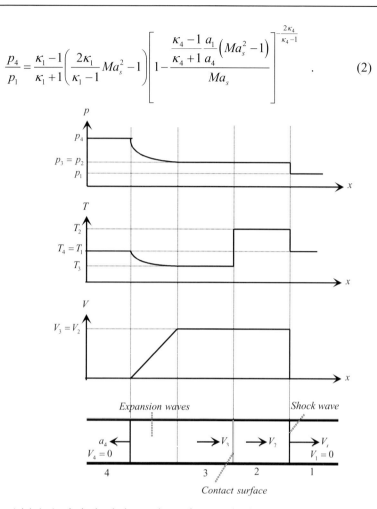

Fig. A4.1-1. Analytical solution to the perfect gas shock wave problem in a pipe with constant cross section

3) Knowing the pressure ratio p_2/p_1 all parameter behind the shock are computed as follows

$$\frac{T_2}{T_1} = \frac{p_2}{p_1}\frac{\frac{\kappa_1 + 1}{\kappa_1 - 1} + \frac{p_2}{p_1}}{1 + \frac{\kappa_1 + 1}{\kappa_1 - 1}\frac{p_2}{p_1}}, \quad (3)$$

$$\frac{p_2}{p_1} = \frac{1 + \frac{\kappa_1 + 1}{\kappa_1 - 1} \frac{p_2}{p_1}}{\frac{\kappa_1 + 1}{\kappa_1 - 1} + \frac{p_2}{p_1}}. \tag{4}$$

4) The shock velocity and the shock Mach number are computed alternatively by

$$V_s = a_1 \left[1 + \frac{\kappa_1 + 1}{2\kappa_1} \left(\frac{p_2}{p_1} - 1 \right) \right]^{1/2}, \tag{5}$$

$$Ma_s = V_s / a_1. \tag{6}$$

The velocity of the gas behind the shock is then

$$V_2 = \frac{a_1}{\kappa_1} \left(\frac{p_2}{p_1} - 1 \right) \left(\frac{\frac{2\kappa_1}{\kappa_1 + 1}}{\frac{\kappa_1 - 1}{\kappa_1 + 1} + \frac{p_2}{p_1}} \right)^{1/2}. \tag{7}$$

The parameter in the region 3 are computed as follows

$$\frac{p_3}{p_4} = \frac{p_3}{p_1} \frac{p_1}{p_4} = \frac{p_2}{p_1} \frac{p_1}{p_4}, \tag{8}$$

$$\frac{T_3}{T_4} = \left(\frac{p_3}{p_4} \right)^{\frac{\kappa_4 - 1}{\kappa_4}}, \tag{9}$$

$$\frac{\rho_3}{\rho_4} = \left(\frac{p_3}{p_4} \right)^{\frac{1}{\kappa_4}}, \tag{10}$$

$$\frac{a_4}{a_3} = \left(\frac{p_4}{p_3} \right)^{\frac{\kappa_4 - 1}{2\kappa_4}}, \tag{11}$$

$$Ma_3 = \frac{2}{\kappa_4 - 1} \left[\left(\frac{p_4}{p_3} \right)^{\frac{\kappa_4 - 1}{2\kappa_4}} - 1 \right], \tag{11}$$

$$V_3 = Ma_3 a_3. \tag{12}$$

The flow parameters in the refraction wave are functions of time and pace. The relation between the position and time along the backwards characteristic

$$\frac{dx}{d\tau} = V - a, \tag{13}$$

is

$$x = (V - a)\tau. \tag{14}$$

From

$$V + \frac{2a}{\kappa - 1} = const, \tag{15}$$

The following relations results

$$\frac{a}{a_4} = 1 - \frac{\kappa - 1}{2} \frac{V}{a_4}, \tag{16}$$

$$V = \frac{2}{\kappa + 1}\left(a_4 + \frac{x}{\tau}\right), \tag{17}$$

The other parameters in the refraction region are then

$$a = (\kappa RT)^{1/2}, \tag{18}$$

$$\frac{\rho}{\rho_4} = \left(1 - \frac{\kappa - 1}{2}\frac{V}{a_4}\right)^{\frac{2}{\kappa-1}}, \tag{19}$$

$$\frac{p}{p_4} = \left(1 - \frac{\kappa - 1}{2}\frac{V}{a_4}\right)^{\frac{2\kappa}{\kappa-1}}. \tag{20}$$

Appendix 4.2 Achievable accuracy of the donor-cell method for single-phase flows

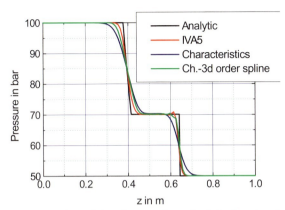

Fig. A4.2-1 Pressure as function of the distance from the left closed end of the pipe at $\tau = 250\,\mu s$. Comparison between the analytical and four numerical solutions obtained by using the explicit method of characteristics: piecewise linear interpolation using entropy equation, the second solution with cubic spline interpolation, and the third solution is the IVA5 solution

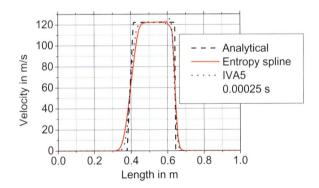

Fig. A4.2-2 Velocity as function of the distance from the left closed end of the pipe at $\tau = 250\,\mu s$. Comparison between the analytical and numerical solutions. The systems of PDE's contain energy conservation principle in terms of specific internal energy in the first case or specific entropy as used in IVA5 computer code in the second case

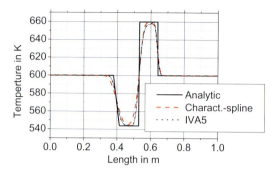

Fig. A4.2-3 Temperature as function of the distance from the left closed end of the pipe at $\tau = 250\mu s$. Comparison between analytical and two numerical solutions. The systems of PDE's contain the energy conservation principle in terms of specific internal energy in the first case or specific entropy as used in IVA5 computer code in the second case

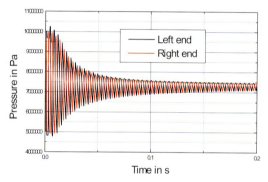

Fig. A4.2-4 Pressures at the both ends of the pipe as a functions of time (air). The energy conservation is used in terms of specific entropy in IVA5

For many years the first order donor-cell method has been widely used in the multi-phase fluid mechanics because of its simplicity and stability. It is interesting to know how this method compares with the other already discussed methods based on the shock tube problem. Figure A4.2-1 shows the pressure as a function of time computed with the computer code IVA5 which exploits the first order donor-cell method as discussed in [10, 12]. The initial conditions are the same as discussed before. The code uses equation of state for air as real gas. As we see from Figs. A4.2-1 through A4.2-3 the IVA5 donor-cell method is more accurate than the first order method of characteristics and less accurate then the method of characteristic using third order spatial interpolation. The pressure at both ends of the pipe as a function of time is presented in Fig. A4.2-4. Comparing Fig. A4.2-4 with Fig. 4.5 we see almost indistinguishable solutions. The total mass and energy

Appendix 4.2 Achievable accuracy of the donor-cell method for single-phase flows   213

conservation error is show in Fig. A4.2-5. Again we observe from Fig. A4.2-5 that the IVA5 numerical method is better than the first order method of characteristics.

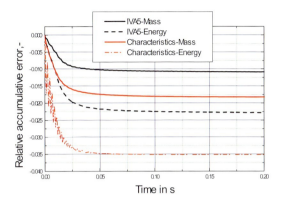

Fig. A4.2-5 Comparison between the overall mass and energy conservation error of the IVA5 numerical method and the first order method of characteristics

Now we perform a similar test with IVA5 replacing the air with a molten oxide mixture called corium at 3300 K initial temperature. The initial pressure distribution is the same as in the previous examples. The velocity of sound of the system is around 1500 m/s. For this test we developed a special set of equation of state and their derivatives which are strictly consistent to each other. The pressure at both ends as function of time is shown in Fig. A4.2-6 and the corresponding accumulative relative mass and energy conservation error is given in Fig. A4.2-7. We see the order of magnitude is 10^{-7} which is much better than for the case of the strongly compressible gas. The slight difference in steady state pressures is due to the geodetic pressure difference because the pipe is considered to be vertical.

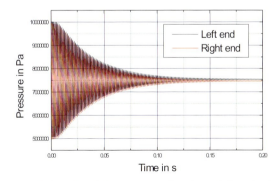

Fig. A4.2-6 Pressure at the ends of the pipe as a function of time (corium melt 3300 K). The energy conservation is used in terms of specific entropy in IVA5

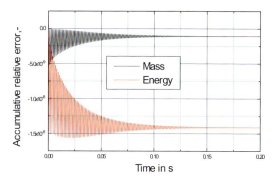

Fig. A4.2-7 Relative overall mass and energy conservation error as function of time (corium melt 3300 K). The energy conservation is used in terms of specific entropy in IVA5

For practical applications in large scale facilities spatial resolution of one centimeter is hardly achievable and therefore the expected errors for single phase-flows will be higher corresponding, among others, to the size of the cells.

For multi-phase flow in *Euler* representation there are additional sources of conservation errors, with one of them being associated with the limited minimum of the time steps used. An example for three-phase flow case is given in [11] where the accumulative relative mass error for two seconds transient was below 0.1% and the accumulative energy conservation error was below 1%.

References

1. Boltzman L (1909) Wissenschaftliche Abhandlungen. Bd. 1. Leipzig
2. Clausius R (1854) Ann. Phys., vol 125 p 390 (1868) Phil. Mag., vol 35 p 419 (1875) Die mechanische Wärmetheorie, Bonn, vol 1
3. D'Alembert (1743) Traitè de dynamique
4. Elsner N (1974) Grundlagen der Thechnischen Thermodynamik, Akademie Verlag, Berlin, p 86
5. Gibbs W (1878) Thermodynamics. Statistical mechanics
6. Gibbs JW (1992) Thermodynamische Studien, translated in German by W. Ostwald in 1892, Leipzig, Verlag von Wilchelm Engelmann, p 76
7. Hoffmann KA and Chang ST (2000) Computational fluid dynamics, vol 2, A Publication of Engineering Education System, Wichita, Kansass
8. Kolev NI, Kerntechnik, vol 59 no 4-5 pp 226-237, no.6 pp 249-258, vol 60 no 1 pp 1-39, vol 62 no 1 pp 67-70, Nuclear Science and Engineering, vol 108 pp 74-87
9. Kolev NI, Nuclear Science and Engineering, vol 108 pp 74-87
10. Kolev NI (1996) Three Fluid Modeling With Dynamic Fragmentation and Coalescence Fiction or Daily practice? 7th FARO Experts Group Meeting Ispra, October 15-16, 1996; Proceedings of OECD/CSNI Workshop on Transient thermal-hydraulic and

neutronic codes requirements, Annapolis, Md, U.S.A., 5th-8th November 1996; 4th World Conference on Experimental Heat Transfer, Fluid Mechanics and Thermodynamics, ExHFT 4, Brussels, June 2-6, 1997; ASME Fluids Engineering Conference & Exhibition, The Hyatt Regency Vancouver, Vancouver, British Columbia, CANADA June 22-26, 1997, Invited Paper; Proceedings of 1997 International Seminar on Vapor Explosions and Explosive Eruptions (AMIGO-IMI), May 22-24, Aoba Kinen Kaikan of Tohoku University, Sendai-City, Japan.
11. Kolev NI (1997) An pretest computation of FARO-FAT L27 experiment with IVA5 computer code, 8th FARO Expert Group Meeting, 9-10 december 1997, JRC-Ispra
12. Kolev NI (1999) Verification of IVA5 computer code for melt-water interaction analysis, Part 1: Single phase flow, Part 2: Two-phase flow, three-phase flow with cold and hot solid spheres, Part 3: Three-phase flow with dynamic fragmentation and coalescence, Part 4: Three-phase flow with dynamic fragmentation and coalescence – alumna experiments, CD Proceedings of the Ninth International Topical Meeting on Nuclear Reactor Thermal Hydraulics (NURETH-9), San Francisco, California, October 3-8, 1999, Log. Nr. 315.
13. Oertel H (1966) Stossrohre, Springer Verlag, Wien-New York
14. Oswatitsch K (1952) Gasdynamik, Wien, Springer
15. Newton I (1872) Matematische Prinzipien der Naturlehre, Berlin
16. Plank M (1964) Vorlesungen über Thermodynamic. 11. Aufl. S.91, Verlag Walter de Gruyter&Co., Berlin
17. Reynolds WC, Perkins HC (1977) Engineering Thermodynamics, Second Edition, McGraw-Hill, p 266
18. Riemann B (1858-1859) Über die Fortpflanzung ebener Luftwellen von endlicher Schwingungsweite, Abhandlungen der Königlichen Gesellschaft der Wissenschaften zu Göttingen, Band 8, S 43-65
19. Stokes GG (1845) On the theories of the internal friction of fluids in motion and the equilibrium and motion of elastic solids, Trans. Cambr. Phil. Soc., vol 8 pp 287-305

5 First and second laws of the thermodynamics

Local volume and time averaging is used to derive rigorous energy equations for multi-phase flows in heterogeneous porous media. The flow is conditionally divided into three velocity fields. Each of the fields consists of several chemical components. Using the conservation equations for mass and momentum and the Gibbs equation, entropy equations are rigorously derived. It is shown that the use of the specific entropy as one of the dependent variables results in the simplest method for describing and modeling such a complicated thermodynamic system. A working form of the final entropy equation is recommended for general use in multi-phase flow dynamics.

5.1 Introduction

This is a continuation of Chapters 1 and 2 which dealt with the basic equations used in the IVA computer code development. As in these Chapters, the heterogeneous porous media formulation proposed by *Sha, Chao* and *Soo* [31] is exploited, with some modifications, so as to derive the local volume and time average equations for the conservation of energy. The energy equations obtained in this way are then rearranged into the entropy equation using the mass and momentum equations, thereby reflecting the second law of thermodynamics.

As far as the author is aware, it was the first time in [23] that a formulation of the second law of thermodynamics has been presented for such a complicated thermodynamic system as the multi-phase flows consisting of three velocity fields in porous structure, with each of these consisting of several chemical components. The most interesting result of this work is the simplicity of the local volume and time-averaged entropy equation (5.125) finally obtained

$$\rho_l \left[\alpha_l \gamma_v \frac{\partial s_l}{\partial \tau} + \left(\alpha_l^e \mathbf{V}_l \gamma \cdot \nabla \right) s_l \right] - \frac{1}{T_l} \nabla \cdot \left(\alpha_l^e \lambda_l^* \gamma \nabla T_l \right)$$

$$-\nabla \cdot \left\{ \alpha_l^e \rho_l \gamma \left[\sum_{i=2}^{i_{\max}} \left(s_{il} - s_{1l} \right) D_{il}^* \nabla C_{il} \right] \right\} = \gamma_v \left[\frac{1}{T_l} DT_l^N + \sum_{i=1}^{i_{\max}} \mu_{il} \left(s_{il} - s_l \right) \right] \quad \text{for} \quad \alpha_l \geq 0,$$

where

$$DT_l^N = \alpha_l \rho_l \left(\delta_l P_{kl}^\tau + \varepsilon_l \right) + \dot{q}''' + \sum_{i=1}^{i_{max}} \mu_{iwl} \left(h_{iwl} - h_{il} \right) + \sum_{\substack{m=1 \\ m \neq l}}^{l_{max}} \left[\mu_{Mml} \left(h_{Ml}^\sigma - h_{Ml} \right) + \sum_{n=1}^{n_{max}} \mu_{nml} \left(h_{nm} - h_{nl} \right) \right]$$

$$+ \frac{1}{2} \left[\begin{array}{c} \mu_{wl} \left(\mathbf{V}_{wl} - \mathbf{V}_l \right)^2 - \mu_{lw} \left(\mathbf{V}_{lw} - \mathbf{V}_l \right)^2 + \sum_{m=1}^{3} \mu_{ml} \left(\mathbf{V}_m - \mathbf{V}_l \right)^2 \\ + \mu_{wl} \overline{\left(\mathbf{V}'_{wl} - \mathbf{V}'_l \right)^2} - \mu_{lw} \overline{\left(\mathbf{V}'_{lw} - \mathbf{V}'_l \right)^2} + \sum_{m=1}^{3} \mu_{ml} \overline{\left(\mathbf{V}'_m - \mathbf{V}'_l \right)^2} \end{array} \right].$$

This equation is more suitable for general use than the various forms of the energy equation eg. Eq. (5.109) written in terms of the specific internal energy,

$$\frac{\partial}{\partial \tau} \left(\alpha_l \rho_l e_l \gamma_v \right) + \nabla \cdot \left(\alpha_l^e \rho_l e_l \mathbf{V}_l \gamma \right) + p_l \left[\frac{\partial}{\partial \tau} \left(\alpha_l \gamma_v \right) + \nabla \cdot \left(\alpha_l^e \mathbf{V}_l \gamma \right) \right]$$

$$- \nabla \cdot \left(\alpha_l^e \rho_l D_l^e \gamma \cdot \nabla e_l \right) + \overline{p'_l \nabla \cdot \left(\alpha_l^e \mathbf{V}'_l \gamma \right)} = \gamma_v De_l^*,$$

where

$$De_l^* = \alpha_l \rho_l \left(\delta_l P_{kl} + \varepsilon_l \right) + \dot{q}''' + \frac{1}{2} \left[\begin{array}{c} \mu_{wl} \left(\mathbf{V}_{wl} - \mathbf{V}_l \right)^2 - \mu_{lw} \left(\mathbf{V}_{lw} - \mathbf{V}_l \right)^2 \\ + \sum_{m=1}^{3} \mu_{ml} \left(\mathbf{V}_m - \mathbf{V}_l \right)^2 \end{array} \right]$$

$$+ \frac{1}{2} \left[\mu_{wl} \overline{\left(\mathbf{V}'_{wl} - \mathbf{V}'_l \right)^2} - \mu_{lw} \overline{\left(\mathbf{V}'_{lw} - \mathbf{V}'_l \right)^2} + \sum_{m=1}^{3} \mu_{ml} \overline{\left(\mathbf{V}'_m - \mathbf{V}'_l \right)^2} \right]$$

$$+ \sum_{i=1}^{i_{max}} \left(\mu_{iwl} h_{iwl} - \mu_{ilw} h_{il} \right) + \sum_{m=1}^{3} \left[\mu_{Mml} h_{Ml}^\sigma + \sum_{n=1}^{n_{max}} \left(\mu_{nml} h_{nm} - \mu_{nlm} h_{nl} \right) \right]$$

or Eq. (5.176) written in terms of temperature

$$\rho_l c_{pl} \left[\alpha_l \gamma_v \frac{\partial T_l}{\partial \tau} + \left(\alpha_l^e \mathbf{V}_l \gamma \cdot \nabla \right) T_l \right] - \left[1 - \rho_l \left(\frac{\partial h_l}{\partial p} \right)_{T_l, all\_C's} \right] \left[\alpha_l \gamma_v \frac{\partial p}{\partial \tau} + \left(\alpha_l^e \mathbf{V}_l \gamma \cdot \nabla \right) p \right]$$

$$-\nabla \cdot \left(\alpha_l^e \lambda_l^* \gamma \nabla T\right) + T_l \sum_{i=2}^{i_{max}} \Delta s_{il}^{np} \nabla \cdot \left(\alpha_l^e \rho_l D_{il}^* \nabla C_{il}\right) = \gamma_v \left[DT_l^N - T_l \sum_{i=2}^{i_{max}} \Delta s_{il}^{np} \left(\mu_{il} - C_{il}\mu_l\right)\right],$$

or Eq. (5.115) written in terms of specific enthalpy

$$\frac{\partial}{\partial \tau}\left(\alpha_l \rho_l h_l \gamma_v\right) + \nabla \cdot \left(\alpha_l^e \rho_l \mathbf{V}_l h_l \gamma\right) - \left(\alpha_l \gamma_v \frac{\partial p_l}{\partial \tau} + \alpha_l^e \mathbf{V}_l \gamma \cdot \nabla p_l\right)$$

$$-\nabla \cdot \left\{\alpha_l^e \rho_l \left[c_{pl}\left(\delta_l \frac{\nu_l}{\mathrm{Pr}_l} + \frac{\nu_l^t}{\mathrm{Pr}_l^t}\right)\nabla T_l + \sum_{i=2}^{i_{max}}(h_{il} - h_{1l})D_{il}^* \nabla C_{il}\right]\gamma\right\} - \alpha_l^e \overline{\mathbf{V}_l' \gamma \cdot \nabla p_l'}$$

$$+ \delta_l\left(\alpha_l^e \gamma \nabla p_l\right) \cdot \sum_{i=1}^{i_{max}} D_{il}^l \nabla \ln C_{il} = \gamma_v DT_l^N + \gamma_v \sum_{\substack{m=1 \\ m \neq l}}^{3,w}\sum_{i=1}^{i_{max}}\left(\mu_{iml} - \mu_{ilm}\right)h_{il}.$$

Another important result is the so called volume conservation equation

$$\frac{\gamma_v}{\rho a^2}\frac{\partial p}{\partial \tau} + \sum_{l=1}^{l_{max}}\frac{\alpha_l}{\rho_l a_l^2}\left(\mathbf{V}_l \gamma \cdot \nabla\right) p + \nabla \cdot \sum_{l=1}^{l_{max}}\left(\alpha_l \mathbf{V}_l \gamma\right) = \sum_{l=1}^{l_{max}} D\alpha_l - \frac{\partial \gamma_v}{\partial \tau},$$

where

$$D\alpha_l = \frac{1}{\rho_l}\left\{\gamma_v \mu_l - \frac{1}{\rho_l}\left[\left(\frac{\partial \rho_l}{\partial s_l}\right)_{p,all\_C_{li}'s} Ds_l^N + \sum_{i=2}^{i_{max}}\left(\frac{\partial \rho_l}{\partial C_{li}}\right)_{p,s,all\_C_{li}'s\_except\_C_{l1}} DC_{il}^N\right]\right\}$$

$$DC_{il}^N = \nabla\left(\alpha_l \rho_l D_{il}^* \nabla C_{il}\right) + \gamma_v\left(\mu_{il} - C_{il}\mu_l\right),$$

$$Ds_l^N = \frac{1}{T_l}\nabla\cdot\left(\alpha_l^e \lambda_l^* \gamma \nabla T_l\right) + \nabla\cdot\left\{\alpha_l^e \rho_l \gamma\left[\sum_{i=1}^{i_{max}} s_{il} D_{il}^* \nabla C_{il}\right]\right\}$$

$$+ \gamma_v\left[\frac{1}{T_l}DT_l^N + \sum_{i=1}^{i_{max}}\mu_{il}\left(s_{il} - s_l\right)\right].$$

This equation can be used instead of the one of the field mass conservation equations.

This Chapter is an extended version of the work published in [23] and [24].

5.2 Instantaneous local volume average energy equations

The energy principle formulated for a volume occupied by the velocity field *l* only is as follows:

> The sum of the rates of energy added to the velocity field from the surroundings due to conduction and convection plus the rate of work done on the velocity field is equal to the rate of change in the energy of the velocity field as it flows through a volume occupied by this velocity field.

The instantaneous energy conservation equation written per unit field volume is thus

$$\sum_{i=1}^{i_{max}} \left\{ \frac{\partial}{\partial \tau} \left[\rho_{il} \left(e_{il}^\tau + \frac{1}{2} V_{il}^{\tau 2} \right) \right] + \nabla \cdot \left[\rho_{il} \left(e_{il}^\tau + p_{il}^\tau / \rho_{il} + \frac{1}{2} V_{il}^{\tau 2} \right) \mathbf{V}_{il}^\tau \right] \right\}$$

$$-\nabla \cdot \left(\mathbf{T}_l^\tau \cdot \mathbf{V}_l^\tau \right) - \nabla \cdot \left(\lambda_l^l \nabla T_l \right) + \rho_l \mathbf{g} \cdot \mathbf{V}_l^\tau = 0, \tag{5.1}$$

where $V_{il}^{\tau 2} = \mathbf{V}_{il}^\tau \cdot \mathbf{V}_{il}^\tau$ is a scalar. There is no doubt about the validity of this equation for velocities much less than the velocity of light. In contrast to the *Sha, Chao* and *Soo* derivation [31], a multi-component velocity field is considered here instead of a single component. This will allow for derivation of mixture properties for the field that are strictly consistent with the first law of thermodynamics.

The *specific enthalpy* of each component

$$h_{il}^\tau = e_{il}^\tau + p_{il}^\tau / \rho_{il} \tag{5.2}$$

naturally arises in the second differential term. Historically, the specific enthalpy was introduced as a very convenient variable to describe steady-state processes. We replace the specific internal energy also in the first differential term by using

$$e_{il}^\tau = h_{il}^\tau - p_{il}^\tau / \rho_{il} . \tag{5.3}$$

Performing local volume averaging on Eq. (5.1) as already described in Chapter 1, one obtains

$$\sum_{i=1}^{i_{max}} \left\{ \left\langle \frac{\partial}{\partial \tau} \left[\rho_{il} \left(h_{il}^\tau + \frac{1}{2} V_{il}^{\tau 2} \right) \right] \right\rangle + \left\langle \nabla \cdot \left[\rho_{il} \left(h_{il}^\tau + \frac{1}{2} V_{il}^{\tau 2} \right) \mathbf{V}_{il}^\tau \right] \right\rangle - \left\langle \frac{\partial p_{il}^\tau}{\partial \tau} \right\rangle \right\}$$

5.2 Instantaneous local volume average energy equations

$$-\left\langle \nabla \cdot \left(\mathbf{T}_l^\tau \cdot \mathbf{V}_l^\tau \right) \right\rangle + \nabla \cdot \left\langle \lambda_l^l \nabla T_l \right\rangle + \left\langle \rho_l \mathbf{g} \cdot \mathbf{V}_l^\tau \right\rangle = 0 . \tag{5.4}$$

The time derivative of the pressure is averaged using Eq. (1.32) (*Leibnitz* rule) as follows:

$$-\left\langle \frac{\partial p_{il}^\tau}{\partial \tau} \right\rangle = -\frac{\partial}{\partial \tau} \left(\gamma_v \alpha_{il} \left\langle p_{il}^\tau \right\rangle^{il} \right) + \frac{1}{Vol} \int_{F_{l\sigma}+F_{lw}} p_{il}^\tau \mathbf{V}_{il\sigma}^\tau \cdot \mathbf{n}_l dF . \tag{5.5}$$

As in Chapter 2, the pressure at the interface $F_{l\sigma}$ is expressed in the form of the sum of bulk averaged pressure, which is independent on the location of the interface, and of the deviation from the bulk pressure, which depends on the location of the surface:

$$p_{il}^\tau = \left\langle p_{il}^\tau \right\rangle^{il} + \Delta p_{il\sigma}^\tau \tag{5.6}$$

An analogous separation can be performed at the structure interface F_{lw}

$$p_{il}^\tau = \left\langle p_{il}^\tau \right\rangle^{il} + \Delta p_{ilw}^\tau . \tag{5.7}$$

Substituting Eqs. (5.6) and (5.7) into Eq. (5.5), and using Eq. (1.33), the following is obtained

$$\frac{1}{Vol} \int_{F_{l\sigma}+F_{lw}} p_{il}^\tau \mathbf{V}_{il\sigma}^\tau \cdot \mathbf{n}_l dF$$

$$= \left\langle p_{il}^\tau \right\rangle^{il} \frac{1}{Vol} \int_{F_{l\sigma}+F_{lw}} \mathbf{V}_{il\sigma}^\tau \cdot \mathbf{n}_l dF + \frac{1}{Vol} \int_{F_{l\sigma}} \Delta p_{il\sigma}^\tau \mathbf{V}_{il\sigma}^\tau \cdot \mathbf{n}_l dF + \frac{1}{Vol} \int_{F_{lw}} \Delta p_{ilw}^\tau \mathbf{V}_{ilw}^\tau \cdot \mathbf{n}_l dF$$

$$= \left\langle p_{il}^\tau \right\rangle^{il} \frac{\partial}{\partial \tau} (\alpha_{il} \gamma_v) + \frac{1}{Vol} \int_{F_{l\sigma}} \Delta p_{il\sigma}^\tau \mathbf{V}_{il\sigma}^\tau \cdot \mathbf{n}_l dF + \frac{1}{Vol} \int_{F_{lw}} \Delta p_{ilw}^\tau \mathbf{V}_{ilw}^\tau \cdot \mathbf{n}_l dF . \tag{5.8}$$

The first term is the power introduced into the velocity field due to a change of the field component volume per unit time and unit control volume (expansion or compression power). Thus, the final result for the averaging of the time derivative of the pressure is

$$-\left\langle \frac{\partial p_{il}^\tau}{\partial \tau} \right\rangle = -\frac{\partial}{\partial \tau} \left(\gamma_v \alpha_{il} \left\langle p_{il}^\tau \right\rangle^{il} \right) + \left\langle p_{il}^\tau \right\rangle^{il} \frac{\partial}{\partial \tau} (\alpha_{il} \gamma_v)$$

$$+\frac{1}{Vol}\int_{F_{l\sigma}}\Delta p_{il\sigma}^{\tau}\mathbf{V}_{il\sigma}^{\tau}\cdot\mathbf{n}_{l}dF+\frac{1}{Vol}\int_{F_{lw}}\Delta p_{ilw}^{\tau}\mathbf{V}_{ilw}^{\tau}\cdot\mathbf{n}_{l}dF$$

$$=-\alpha_{il}\gamma_{v}\frac{\partial}{\partial\tau}\left\langle p_{il}^{\tau}\right\rangle^{il}+\frac{1}{Vol}\int_{F_{l\sigma}}\Delta p_{il\sigma}^{\tau}\mathbf{V}_{il\sigma}^{\tau}\cdot\mathbf{n}_{l}dF+\frac{1}{Vol}\int_{F_{lw}}\Delta p_{ilw}^{\tau}\mathbf{V}_{ilw}^{\tau}\cdot\mathbf{n}_{l}dF. \quad (5.9)$$

Applying Eqs. (5.28), (5.25) and (1.32) to the other terms of Eq. (5.4), and taking into account Eq. (5.9), one obtains

$$\sum_{i=1}^{i_{\max}}\left\{\begin{array}{l}\dfrac{\partial}{\partial\tau}\left[\gamma_{v}\alpha_{il}\left\langle\rho_{il}\left(h_{il}^{\tau}+\dfrac{1}{2}V_{il}^{\tau2}\right)\right\rangle^{il}\right]\\ \\ +\nabla\cdot\left[\gamma\alpha_{il}^{e}\left\langle\rho_{il}\left(h_{il}^{\tau}+\dfrac{1}{2}V_{il}^{\tau2}\right)\mathbf{V}_{il}^{\tau}\right\rangle^{ile}\right]-\gamma_{v}\alpha_{il}\dfrac{\partial}{\partial\tau}\left\langle p_{il}^{\tau}\right\rangle^{il}\end{array}\right\}$$

$$-\nabla\cdot\left(\alpha_{l}^{e}\gamma\left\langle\mathbf{T}_{l}^{\tau}\cdot\mathbf{V}_{l}^{\tau}\right\rangle^{le}\right)-\nabla\cdot\left(\alpha_{l}^{e}\gamma\left\langle\lambda_{l}^{\prime}\nabla T_{l}\right\rangle^{le}\right)+\alpha_{l}\gamma_{v}\left\langle\rho_{l}\left(\mathbf{g}\cdot\mathbf{V}_{l}^{\tau}\right)\right\rangle^{l}$$

$$+\frac{1}{Vol}\int_{F_{l\sigma}+F_{lw}}\left[\sum_{i=1}^{i_{\max}}\rho_{il}\left(h_{il}^{\tau}+\frac{1}{2}V_{il}^{\tau2}\right)\left(\mathbf{V}_{il}^{\tau}-\mathbf{V}_{il\sigma}^{\tau}\right)\cdot\mathbf{n}_{l}dF\right]$$

$$+\frac{1}{Vol}\int_{F_{l\sigma}}\Delta p_{l\sigma}^{\tau}\mathbf{V}_{l\sigma}^{\tau}\cdot\mathbf{n}_{l}dF+\frac{1}{Vol}\int_{F_{lw}}\Delta p_{lw}^{\tau}\mathbf{V}_{lw}^{\tau}\cdot\mathbf{n}_{l}dF$$

$$-\frac{1}{Vol}\int_{F_{l\sigma}+F_{lw}}\mathbf{T}_{l}^{\tau}\cdot\mathbf{V}_{l}^{\tau}\cdot\mathbf{n}_{l}dF-\frac{1}{Vol}\int_{F_{l\sigma}+F_{lw}}\lambda_{l}^{\prime}\nabla T_{l}\cdot\mathbf{n}_{l}dF=0. \quad (5.10)$$

The δ_{l}-identifier for the dispersed field is now introduced into the heat conduction term, which thus becomes

$$\nabla\cdot\left(\alpha_{l}^{e}\gamma\delta_{l}\left\langle\lambda_{l}^{\prime}\nabla T_{l}\right\rangle^{le}\right). \quad (5.11)$$

The above simply means that there is no heat transfer through heat conduction for dispersed field, $\delta_{l}=0$, and that heat conduction is taken into account for a continuous field, $\delta_{l}=1$.

The heat source term

$$\frac{1}{Vol}\int_{F_{l\sigma}+F_{lw}} \lambda_l' \nabla T_l \cdot \mathbf{n}_l dF = \gamma_v \left(\dot{q}_{wl}''' + \sum_{\substack{m=1 \\ m \ne l}}^{3} \dot{q}_{m\sigma l}''' \right) = \gamma_v \dot{q}_l''' \qquad (5.12)$$

takes into account the sum of the heat power per unit control volume introduced into the velocity field through the interface, $\gamma_v \dot{q}_{m\sigma l}'''$, and through the wall, $\gamma_v \dot{q}_{wl}'''$.

The integral terms containing the velocity difference $\mathbf{V}_{il}^{\tau} - \mathbf{V}_{il\sigma}^{\tau}$ represent the energy transfer due to (a) evaporation, or (b) condensation, or (c) entrainment, or (d) deposition at the interface $F_{l\sigma}$, or (e) due to injection or (f) suction through the solid interface F_{lw}. It is very important to note that h_{il}^{τ} under the interface integral is taken *inside the field l* at the immediate interface neighborhood denoted with σ. The subscript M is introduced here to designate the non-inert component within the field. n used to designate the inert component. The total number of inert components in each field is $i_{max} - 1$.

Consequently

$$h_{Ml}^{\tau\sigma} = h''(T_{12}^{\sigma}) \quad \text{or} \quad h_{Ml}^{\tau\sigma} = h''(T_{13}^{\sigma}) \quad \text{for } l = 1, \qquad (5.13)$$

$$h_{M2}^{\tau\sigma} = h'(p) \quad \text{for } l = 2, \qquad (5.14)$$

$$h_{M3}^{\tau\sigma} = h'(p) \quad \text{for } l = 3, C_{M3} > 0, \qquad (5.15)$$

for all mass flows entering field l. Mass flows leaving field l possess the l-properties. One now assumes that $\mathbf{V}_{il}^{\tau} = \mathbf{V}_l^{\tau}$ at the l side of the interface, with these terms then split into non-negative components. It is necessary to distinguish between mass transfer due to change in the state of aggregate, and mass transfer such as entrainment, deposition, etc. resulting from mechanical macroscopic forces.

(a) The mass transfer term at the field-solid interface is decomposed as follows:

$$-\frac{1}{Vol}\int_{F_{lw}}\left[\sum_{i=1}^{i_{max}} \rho_{il}\left(h_{il}^{\tau} + \frac{1}{2}V_{il}^{\tau 2}\right)\left(\mathbf{V}_{il}^{\tau} - \mathbf{V}_{il\sigma}^{\tau}\right) \cdot \mathbf{n}_l dF\right]$$

$$= \gamma_v \left\{ \sum_{i=1}^{i_{max}}\left[\mu_{iwl}^{\tau}\langle h_{iw}^{\tau}\rangle^{iw} - \mu_{ilw}^{\tau}\langle h_{il}^{\tau}\rangle^{il}\right] + \mu_{wl}^{\tau}\frac{1}{2}\langle V_{wl}^{\tau 2}\rangle^{we} - \mu_{lw}^{\tau}\frac{1}{2}\langle V_{lw}^{\tau 2}\rangle^{le}\right\}; (5.16)$$

(b) The mass transfer term at the field interface is decomposed as follows

$$-\frac{1}{Vol}\int_{F_{l\sigma}}\left[\sum_{i=1}^{i_{max}}\rho_{il}\left(h_{il}^{\tau}+\frac{1}{2}V_{il}^{\tau 2}\right)\left(\mathbf{V}_{il}^{\tau}-\mathbf{V}_{il\sigma}^{\tau}\right)\cdot\mathbf{n}_{l}dF\right]$$

$$=\gamma_{v}\left\{\begin{array}{l}\sum_{m=1}^{3}\left\{\left(\mu_{Mml}^{\tau}h_{Ml}^{\tau\sigma}-\mu_{Mlm}^{\tau}\left\langle h_{Ml}^{\tau}\right\rangle^{Ml}\right)+\sum_{n=1}^{n_{max}}\left(\mu_{nml}^{\tau}\left\langle h_{nm}^{\tau}\right\rangle^{nm}-\mu_{nlm}^{\tau}\left\langle h_{nl}^{\tau}\right\rangle^{nl}\right)\right\}\\ +\mu_{ml}^{\tau}\frac{1}{2}\left\langle V_{m}^{\tau 2}\right\rangle^{me}-\mu_{lm}^{\tau}\frac{1}{2}\left\langle V_{l}^{\tau 2}\right\rangle^{le}\end{array}\right\};$$

(5.17)

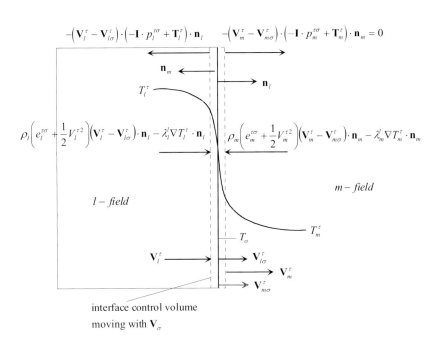

Fig. 5.1. Interfacial energy transfer

The sum of Eqs. (5.12), (5.16) and (5.17) gives

$$q_{l}^{*}=\dot{q}_{l}^{m}+\sum_{i=1}^{i_{max}}\left[\left(\mu_{iwl}^{\tau}\left\langle h_{iw}^{\tau}\right\rangle^{iw}-\mu_{ilw}^{\tau}\left\langle h_{il}^{\tau}\right\rangle^{il}\right)\right]$$

$$+\sum_{m=1}^{3}\left\{\mu_{Mml}^{\tau}h_{Ml}^{\tau\sigma}-\mu_{Mlm}^{\tau}\left\langle h_{Ml}^{\tau}\right\rangle^{Ml}\right\}+\sum_{n=1}^{n_{\max}}\left(\mu_{nml}^{\tau}\left\langle h_{nm}^{\tau}\right\rangle^{nm}-\mu_{nlm}^{\tau}\left\langle h_{nl}^{\tau}\right\rangle^{nl}\right)\right\}$$

$$+\mu_{wl}^{\tau}\frac{1}{2}\left\langle V_{wl}^{\tau 2}\right\rangle^{we}-\mu_{lw}^{\tau}\frac{1}{2}\left\langle V_{lw}^{\tau 2}\right\rangle^{le}+\frac{1}{2}\sum_{m=1}^{3}\left(\mu_{ml}^{\tau}\left\langle V_{m}^{\tau 2}\right\rangle^{me}-\mu_{lm}^{\tau}\left\langle V_{l}^{\tau 2}\right\rangle^{le}\right). \quad (5.18)$$

The *interfacial energy jump condition* is introduced at this point, with interface *lm* considered as *non-material*, see Fig. 5.1,

$$\rho_{l}\left(e_{l}^{\tau\sigma}+\frac{1}{2}V_{l}^{\tau 2}\right)\left(\mathbf{V}_{l}^{\tau}-\mathbf{V}_{l\sigma}^{\tau}\right)\cdot\mathbf{n}_{l}-\lambda_{l}^{l}\nabla T_{l}^{\tau}\cdot\mathbf{n}_{l}$$

$$+\rho_{m}\left(e_{m}^{\tau\sigma}+\frac{1}{2}V_{m}^{\tau 2}\right)\left(\mathbf{V}_{m}^{\tau}-\mathbf{V}_{m\sigma}^{\tau}\right)\cdot\mathbf{n}_{m}-\lambda_{m}^{l}\nabla T_{m}^{\tau}\cdot\mathbf{n}_{m}$$

$$-\left(\mathbf{V}_{l}^{\tau}-\mathbf{V}_{l\sigma}^{\tau}\right)\cdot\left(-\mathbf{I}\cdot p_{l}^{\tau\sigma}+\mathbf{T}_{l}^{\tau}\right)\cdot\mathbf{n}_{l}-\left(\mathbf{V}_{m}^{\tau}-\mathbf{V}_{m\sigma}^{\tau}\right)\cdot\left(-\mathbf{I}\cdot p_{m}^{\tau\sigma}+\mathbf{T}_{m}^{\tau}\right)\cdot\mathbf{n}_{m}=0 \quad (5.19)$$

or

$$\left[\begin{array}{l}\rho_{l}\left(h_{l}^{\tau\sigma}+\frac{1}{2}V_{l}^{\tau 2}\right)\left(\mathbf{V}_{l}^{\tau}-\mathbf{V}_{l\sigma}^{\tau}\right)-\rho_{m}\left(h_{m}^{\tau\sigma}+\frac{1}{2}V_{m}^{\tau 2}\right)\left(\mathbf{V}_{m}^{\tau}-\mathbf{V}_{m\sigma}^{\tau}\right)\\ -\lambda_{l}^{l}\nabla T_{l}^{\tau}+\lambda_{m}^{l}\nabla T_{m}^{\tau}-\left(\mathbf{V}_{l}^{\tau}-\mathbf{V}_{l\sigma}^{\tau}\right)\cdot\mathbf{T}_{l}^{\tau}+\left(\mathbf{V}_{m}^{\tau}-\mathbf{V}_{m\sigma}^{\tau}\right)\cdot\mathbf{T}_{m}^{\tau}\end{array}\right]\cdot\mathbf{n}_{l}=0. \quad (5.20)$$

Using Eqs. (1.42b) and (1.42c), Eq. (5.20) can be simplified as follows

$$\left[\begin{array}{l}\left[h_{l}^{\tau,m\sigma}-h_{m}^{\tau,l\sigma}+\frac{1}{2}\left(V_{l}^{\tau 2}-V_{m}^{\tau 2}\right)\right]\frac{\rho_{m}\rho_{l}}{\rho_{l}-\rho_{m}}\left(\mathbf{V}_{m}^{\tau}-\mathbf{V}_{l}^{\tau}\right)\\ -\lambda_{l}^{l}\nabla T_{l}^{\tau}+\lambda_{m}^{l}\nabla T_{m}^{\tau}+\frac{\rho_{l}\rho_{m}}{\rho_{l}-\rho_{m}}\left(\mathbf{V}_{m}^{\tau}-\mathbf{V}_{l}^{\tau}\right)\cdot\left(\frac{\mathbf{T}_{m}^{\tau}}{\rho_{m}}-\frac{\mathbf{T}_{l}^{\tau}}{\rho_{l}}\right)\end{array}\right]\cdot\mathbf{n}_{l}=0, \quad (5.21)$$

or

$$\left[h_{l}^{\tau,m\sigma}-h_{m}^{\tau,l\sigma}+\frac{1}{2}\left(V_{l}^{\tau 2}-V_{m}^{\tau 2}\right)+\frac{\mathbf{T}_{m}^{\tau}}{\rho_{m}}-\frac{\mathbf{T}_{l}^{\tau}}{\rho_{l}}\right](\rho w)_{lm}^{\tau}+\left(\lambda_{m}^{l}\nabla T_{m}^{\tau}-\lambda_{l}^{l}\nabla T_{l}^{\tau}\right)\cdot\mathbf{n}_{l}=0.$$

or

$$(\rho w)_{lm}^{\tau} = -\frac{\left(\lambda_m^l \nabla T_m^{\tau} - \lambda_l^l \nabla T_l^{\tau}\right) \cdot \mathbf{n}_l}{h_l^{\tau,m\sigma} - h_m^{\tau,l\sigma} + \frac{1}{2}\left(V_l^{\tau 2} - V_m^{\tau 2}\right) + \frac{\mathbf{T}_m^{\tau}}{\rho_m} - \frac{\mathbf{T}_l^{\tau}}{\rho_l}}.$$

We realize that if there is no mass transfer across the interfacial contact discontinuity heat conduction is the only mechanism transferring energy across. In the simple case of no heat conduction at both sides of the interface and zero stress tensors the energy jump condition simplifies to

$$h_l^{\tau\sigma} - h_m^{\tau\sigma} + \frac{1}{2}\left(V_l^{\tau 2} - V_m^{\tau 2}\right) = 0. \tag{5.22}$$

Integrating Eq. (5.20) over the interface inside the control volume and dividing by the control volume we obtain

$$\frac{1}{Vol} \int_{F_{l\sigma}+F_{lw}} \begin{bmatrix} \rho_l \left(h_l^{\tau\sigma} + \frac{1}{2}V_l^{\tau 2}\right)\left(\mathbf{V}_l^{\tau} - \mathbf{V}_{l\sigma}^{\tau}\right) - \rho_m \left(h_m^{\tau\sigma} + \frac{1}{2}V_m^{\tau 2}\right)\left(\mathbf{V}_m^{\tau} - \mathbf{V}_{m\sigma}^{\tau}\right) \\ -\lambda_l^l \nabla T_l^{\tau} + \lambda_m^l \nabla T_m^{\tau} - \left(\mathbf{V}_l^{\tau} - \mathbf{V}_{l\sigma}^{\tau}\right) \cdot \mathbf{T}_l^{\tau} + \left(\mathbf{V}_m^{\tau} - \mathbf{V}_{m\sigma}^{\tau}\right) \cdot \mathbf{T}_m^{\tau} \end{bmatrix} \cdot \mathbf{n}_l dF = 0.$$

(5.23)

In the event of evaporation or condensation this yields

$$\mu_{Ml}^{\tau}\left(h_{Ml}^{\tau\sigma} + \frac{1}{2}V_l^{\tau 2}\right) - \dot{q}_{m\sigma l}^{'''} - \mu_{Mm}^{\tau}\left(h_{Mm}^{\tau\sigma} + \frac{1}{2}V_m^{\tau 2}\right) - \dot{q}_{l\sigma m}^{'''}$$

$$+ \frac{1}{Vol} \int_{F_{l\sigma}+F_{lw}} \left[-\left(\mathbf{V}_l^{\tau} - \mathbf{V}_{l\sigma}^{\tau}\right) \cdot \mathbf{T}_l^{\tau} + \left(\mathbf{V}_m^{\tau} - \mathbf{V}_{m\sigma}^{\tau}\right) \cdot \mathbf{T}_m^{\tau}\right] \cdot \mathbf{n}_l dF = 0. \tag{5.24}$$

Postulating the weighted average

$$\left\langle \rho_{il} h_{il}^{\tau} \right\rangle^{il} / \left\langle \rho_{il} \right\rangle^{il} = \left\langle h_{il}^{\tau} \right\rangle^{il}, \tag{5.25}$$

$$\left\langle \rho_{il} V_{il}^{\tau 2} \right\rangle^{ile} / \left\langle \rho_{il} \right\rangle^{ile} = \left\langle V_{il}^{\tau 2} \right\rangle^{ile}, \tag{5.26}$$

$$\left\langle \rho_{il} V_{il}^{\tau 2} \mathbf{V}_{il}^{\tau} \right\rangle^{ile} / \left\langle \rho_{il} \right\rangle^{ile} = \left\langle V_{il}^{\tau 2} \right\rangle^{ile} \left\langle \mathbf{V}_{il}^{\tau} \right\rangle^{ile}, \tag{5.27}$$

one can then write the *conservative form* of the energy conservation equation in the following form

$$\sum_{i=1}^{i_{max}} \left\{ \begin{array}{l} \dfrac{\partial}{\partial \tau}\left[\gamma_v \alpha_{il} \langle \rho_{il} \rangle^{il} \left(\langle h_{il}^\tau \rangle^{il} + \dfrac{1}{2}\langle V_{il}^{\tau 2} \rangle^{ile} \right)\right] \\[2ex] +\nabla \cdot \left[\gamma \alpha_{il}^e \langle \rho_{il} \rangle^{il} \langle \mathbf{V}_{il}^\tau \rangle^{ile} \left(\langle h_{il}^\tau \rangle^{il} + \dfrac{1}{2}\langle V_{il}^{\tau 2} \rangle^{ile} \right) \right] - \gamma_v \alpha_{il} \dfrac{\partial}{\partial \tau}\langle p_{il}^\tau \rangle^{il} \end{array} \right\}$$

$$-\nabla \cdot \left(\alpha_l^e \gamma \langle \mathbf{T}_l^\tau \cdot \mathbf{V}_l^\tau \rangle^{le} \right) - \nabla \cdot \left(\alpha_l^e \gamma \delta_l \langle \lambda_l^l \nabla T_l \rangle^{le} \right) + \alpha_l \gamma_v \langle \rho_l (\mathbf{g} \cdot \mathbf{V}_l^\tau) \rangle^l$$

$$+\dfrac{1}{Vol}\int_{F_{l\sigma}} \Delta p_{l\sigma}^\tau \mathbf{V}_{l\sigma}^\tau \cdot \mathbf{n}_l dF + \dfrac{1}{Vol}\int_{F_{lw}} \Delta p_{lw}^\tau \mathbf{V}_{lw}^\tau \cdot \mathbf{n}_l dF - \dfrac{1}{Vol}\int_{F_{l\sigma}+F_{lw}} \mathbf{T}_l^\tau \cdot \mathbf{V}_l^\tau \cdot \mathbf{n}_l dF = \gamma_v q_l^* \; .$$

(5.28)

Using the chain rule, differentiation can now be performed on the first two terms. Comparing them with the mass conservation equations (1.38) one then obtains the *non-conservative form* of the energy equation:

$$\sum_{i=1}^{i_{max}} \langle \rho_{il} \rangle^{il} \left\{ \begin{array}{l} \alpha_{il}\gamma_v \dfrac{\partial}{\partial \tau}\left(\langle h_{il}^\tau \rangle^{il} + \dfrac{1}{2}\langle V_{il}^{\tau 2} \rangle^{ile} \right) \\[2ex] + \alpha_{il}^e \langle \mathbf{V}_{il}^\tau \rangle^{ile} \gamma \cdot \nabla \left(\langle h_{il}^\tau \rangle^{il} + \dfrac{1}{2}\langle V_{il}^{\tau 2} \rangle^{ile} \right) \end{array} \right\} - \gamma_v \alpha_{il} \dfrac{\partial}{\partial \tau}\langle p_{il}^\tau \rangle^{il}$$

$$-\nabla \cdot \left(\alpha_l^e \gamma \langle \mathbf{T}_l^\tau \cdot \mathbf{V}_l^\tau \rangle^{le} \right) - \nabla \cdot \left(\alpha_l^e \gamma \delta_l \langle \lambda_l^l \nabla T_l \rangle^{le} \right) + \alpha_l \gamma_v \langle \rho_l (\mathbf{g} \cdot \mathbf{V}_l^\tau) \rangle^l$$

$$+\dfrac{1}{Vol}\int_{F_{l\sigma}} \Delta p_{l\sigma}^\tau \mathbf{V}_{l\sigma}^\tau \cdot \mathbf{n}_l dF + \dfrac{1}{Vol}\int_{F_{lw}} \Delta p_{lw}^\tau \mathbf{V}_{lw}^\tau \cdot \mathbf{n}_l dF - \dfrac{1}{Vol}\int_{F_{l\sigma}+F_{lw}} \mathbf{T}_l^\tau \cdot \mathbf{V}_l^\tau \cdot \mathbf{n}_l dF = \gamma_v q_l^{*N}$$

(5.29)

where

$$q_l^{*N} = \dot{q}_l^{\prime\prime\prime} + \sum_{i=1}^{i_{max}} \mu_{iwl}^\tau \left(\langle h_{iw}^\tau \rangle^{iw} - \langle h_{il}^\tau \rangle^{il} \right)$$

$$+\sum_{m=1}^{3}\left[\mu_{Mml}^{\tau}\left(h_{Ml}^{\tau\sigma}-\left\langle h_{Ml}^{\tau}\right\rangle^{Ml}\right)+\sum_{n=1}^{n_{max}}\mu_{nml}^{\tau}\left(\left\langle h_{nm}^{\tau}\right\rangle^{nm}-\left\langle h_{nl}^{\tau}\right\rangle^{nl}\right)\right]$$

$$+\mu_{wl}^{\tau}\frac{1}{2}\left(\left\langle V_{wl}^{\tau 2}\right\rangle^{we}-\left\langle V_{l}^{\tau 2}\right\rangle^{le}\right)-\mu_{lw}^{\tau}\frac{1}{2}\left(\left\langle V_{lw}^{\tau 2}\right\rangle^{le}-\left\langle V_{l}^{\tau 2}\right\rangle^{le}\right)$$

$$+\frac{1}{2}\sum_{m=1}^{3}\mu_{ml}^{\tau}\left(\left\langle V_{m}^{\tau 2}\right\rangle^{me}-\left\langle V_{l}^{\tau 2}\right\rangle^{le}\right). \tag{5.30}$$

The subscript N stands here to remember that this RHS belongs to the non-conservative notation of the energy conservation equation.

5.3 *Dalton* and *Fick*'s laws, center of mass mixture velocity, caloric mixture properties

As mentioned in Chapter 1, for a gas mixture it follows from the *Dalton*'s law that $\alpha_{il}=\alpha_{l}$, whereas for mixtures consisting of liquid and macroscopic solid particles $\alpha_{il}\neq\alpha_{l}$. The instantaneous mass concentration of the component i in l is defined by Eq. (1.49),

$$C_{il}^{\tau}=\alpha_{il}\left\langle \rho_{il}\right\rangle^{l}/\left(\alpha_{l}\left\langle \rho_{l}\right\rangle^{l}\right). \tag{5.31}$$

The center of mass (c. m.) velocity is given by intrinsic surface-averaged field velocity $\left\langle \mathbf{V}_{l}^{\tau}\right\rangle^{le}$. Equation (1.50) can be rewritten as

$$\alpha_{l}\left\langle \rho_{l}\right\rangle^{l}\left\langle \mathbf{V}_{l}^{\tau}\right\rangle^{le}=\sum_{i=1}^{i_{max}}\alpha_{il}\left\langle \rho_{il}\right\rangle^{l}\left\langle \mathbf{V}_{il}^{\tau}\right\rangle^{le}=\alpha_{l}\left\langle \rho_{l}\right\rangle^{l}\sum_{i=1}^{i_{max}}C_{il}^{\tau}\left\langle \mathbf{V}_{il}^{\tau}\right\rangle^{le}. \tag{5.32}$$

Consequently

$$\left\langle \mathbf{V}_{l}^{\tau}\right\rangle^{le}=\sum_{i=1}^{i_{max}}C_{il}^{\tau}\left\langle \mathbf{V}_{il}^{\tau}\right\rangle^{le}. \tag{5.33}$$

As for the derivation of the mass conservation equation, it is convenient for description of transport of the microscopic component il in the velocity field l to replace the velocity component $\left\langle \mathbf{V}_{il}^{\tau}\right\rangle^{le}$ by the sum of the center of mass velocity for the particular field $\langle \mathbf{V}_l^\tau \rangle^{le}$ and the deviation from the c. m. velocity or the so called *diffusion velocity* of the inert component $\delta_i \langle \mathbf{V}_l^\tau \rangle^{le}$, this yielding the following

$$\langle \mathbf{V}_{il}^\tau \rangle^{le} = \langle \mathbf{V}_l^\tau \rangle^{le} + \left(\delta \langle \mathbf{V}_l^\tau \rangle^{le} \right)_i . \tag{5.34}$$

Fick [3] (1855) noticed that the mass flow rate of the inert component with respect to the total mass flow rate of the continuous mixture including the inert component is proportional to the gradient of the concentration of this inert component

$$\alpha_{il} \langle \rho_{il} \rangle^l \left(\delta \langle \mathbf{V}_l^\tau \rangle^{le} \right)_i = -\alpha_l^e \delta_l D_{il}^l \nabla \langle \rho_{il} \rangle^{il} = -\alpha_l \langle \rho_l \rangle^l \delta_l D_{il}^l \nabla \langle C_{il}^\tau \rangle^{il}, \tag{5.35}$$

or divided by the component density

$$\alpha_{il}^e \delta \mathbf{V}_{il}^\tau = -\alpha_l^e \delta_l D_{il}^l \frac{1}{\langle C_{il}^\tau \rangle^{il}} \nabla \langle C_{il}^\tau \rangle^{il} = -\alpha_l^e \delta_l D_{il}^l \nabla \ln \langle C_{il}^\tau \rangle^{il} . \tag{5.36}$$

The coefficient of proportionality, D_{il}^l, is known as the coefficient of molecular diffusion. The diffusion mass flow rate is directed from regions with higher concentration to regions with lower concentration, with this reflected by the minus sign in the assumption made by *Fick* (which has subsequently come to be known as *Fick*'s law), because many processes in nature and industrial equipment are successfully described mathematically by the above approach – so called diffusion processes. Here $\delta_l = 1$ for a continuous field and $\delta_l = 0$ for a disperse field. Molecular diffusion has microscopic character, as it is caused by molecular interactions. The special theoretical treatment and experimental experience of how to determine the molecular diffusion constant in multi-component mixtures is a science in its own right. This topic is beyond the scope of this chapter. In this context, it should merely be only noted that in line with the thermodynamics of irreversible processes, the thermal diffusivity and the diffusion coefficients influence each other. The interested reader can find useful information in *Reid* et al. [30].

One should keep in mind that there is no molecular net mass diffusively transported across a cross section perpendicular to the strongest concentration gradient. This is mathematically expressed as follows

$$\sum_{i=1}^{i_{max}} \left(D_{il}^l \nabla \langle C_{il}^\tau \rangle^{il} \right) = 0, \tag{5.37}$$

or

$$D_{1l}^l \nabla \langle C_{1l}^\tau \rangle^{il} = -\sum_{i=2}^{i_{max}} \left(D_{il}^l \nabla \langle C_{il}^\tau \rangle^{il} \right). \tag{5.38}$$

The caloric mixture properties naturally arise after summing all the energy conservation equations in conservative form. These are defined as follows

$$\langle \varphi_l^\tau \rangle^l = \frac{\sum_{i=1}^{i_{max}} \alpha_{il} \langle \rho_{il} \rangle^{il} \langle \varphi_{il}^\tau \rangle^{il}}{\alpha_l \langle \rho_l \rangle^l} = \sum_{i=1}^{i_{max}} \left(\langle C_{il} \rangle^{il} \langle \varphi_{il}^\tau \rangle^{il} \right), \tag{5.39}$$

where

$$\varphi = h, e, s \,. \tag{5.40}$$

5.4 Enthalpy equation

For practical applications it is extremely convenient to simplify the energy equation in a way that the mechanical energy terms disappear. The resulting equation is called the *enthalpy equation*. The enthalpy equation will now be derived.

The non-conservative form of the momentum equation (2.31) using Eq. (1.38) is

$$\langle \rho_l \rangle^l \left[\alpha_l \gamma_v \frac{\partial}{\partial \tau} \langle \mathbf{V}_l^\tau \rangle^{le} + \alpha_l^e \langle \mathbf{V}_l^\tau \rangle^{le} \gamma \nabla \cdot \langle \mathbf{V}_l^\tau \rangle^{le} \right] - \nabla \cdot \left(\alpha_l^e \gamma \langle \mathbf{T}_l^\tau \rangle^{le} \right) + \alpha_l^e \gamma \nabla \cdot \langle p_l^\tau \rangle^{le}$$

$$+ \frac{1}{Vol} \int_{F_{l\sigma}} \Delta p_{l\sigma}^\tau \cdot \mathbf{n}_{l\sigma} dF + \alpha_l \gamma_v \langle \rho_l \rangle^l \mathbf{g} + \frac{1}{Vol} \int_{F_{lw}} \Delta p_{lw}^\tau \cdot \mathbf{n}_{l\sigma} dF - \frac{1}{Vol} \int_{F_{l\sigma}+F_{lw}} \mathbf{T}_l^\tau \cdot \mathbf{n}_l dF$$

$$= \gamma_v \left\{ \mu_{wl}^\tau \left(\langle \mathbf{V}_{wl}^\tau \rangle^{we} - \langle \mathbf{V}_l^\tau \rangle^{le} \right) - \mu_{lw}^\tau \left(\langle \mathbf{V}_{lw}^\tau \rangle^{we} - \langle \mathbf{V}_l^\tau \rangle^{le} \right) + \sum_{m=1}^{3} \mu_{ml}^\tau \left(\langle \mathbf{V}_m^\tau \rangle^{me} - \langle \mathbf{V}_l^\tau \rangle^{le} \right) \right\}. \tag{5.41}$$

Here, Eqs. (5.6) and (5.7) are used, and the integral containing $\langle p_l^\tau \rangle^{le}$ is rearranged using Eq. (1.29). The scalar product of Eq. (5.41) is constructed with the field velocity. The result is a scalar expressing the mechanical energy balance. Subtracting this result from the energy equation and bearing in mind that

$$\frac{1}{2}\left(\left\langle V_m^{\tau 2}\right\rangle^{me}-\left\langle \mathbf{V}_l^{\tau 2}\right\rangle^{le}\right)-\left\langle \mathbf{V}_l^{\tau}\right\rangle^{le}\cdot\left(\left\langle \mathbf{V}_m^{\tau}\right\rangle^{me}-\left\langle \mathbf{V}_l^{\tau}\right\rangle^{le}\right)=\frac{1}{2}\left(\left\langle \mathbf{V}_m^{\tau}\right\rangle^{me}-\left\langle \mathbf{V}_l^{\tau}\right\rangle^{le}\right)^2,$$
(5.42)

and

$$\mu_{wl}^{\tau}\frac{1}{2}\left(\left\langle V_{wl}^{\tau 2}\right\rangle^{we}-\left\langle V_l^{\tau 2}\right\rangle^{le}\right)-\mu_{lw}^{\tau}\frac{1}{2}\left(\left\langle V_{lw}^{\tau 2}\right\rangle^{we}-\left\langle V_l^{\tau 2}\right\rangle^{le}\right)$$

$$-\left\langle \mathbf{V}_l^{\tau}\right\rangle^{le}\left[\mu_{wl}^{\tau}\left(\left\langle \mathbf{V}_{wl}^{\tau}\right\rangle^{we}-\left\langle \mathbf{V}_l^{\tau}\right\rangle^{le}\right)-\mu_{lw}^{\tau}\left(\left\langle \mathbf{V}_{lw}^{\tau}\right\rangle^{we}-\left\langle \mathbf{V}_l^{\tau}\right\rangle^{le}\right)\right]$$

$$=\mu_{wl}^{\tau}\frac{1}{2}\left(\left\langle \mathbf{V}_{wl}^{\tau}\right\rangle^{we}-\left\langle \mathbf{V}_l^{\tau}\right\rangle^{le}\right)^2-\mu_{lw}^{\tau}\frac{1}{2}\left(\left\langle \mathbf{V}_{lw}^{\tau}\right\rangle^{we}-\left\langle \mathbf{V}_l^{\tau}\right\rangle^{le}\right)^2,\qquad(5.43)$$

one then obtains the *non-conservative form of the enthalpy equation*:

$$\sum_{i=1}^{i_{max}}\left\{\begin{array}{l}\left\langle \rho_{il}\right\rangle^{il}\left[\alpha_{il}\gamma_v\frac{\partial}{\partial\tau}\left\langle h_{il}^{\tau}\right\rangle^{il}+\alpha_{il}^{e}\left\langle \mathbf{V}_{il}^{\tau}\right\rangle^{ile}\gamma\cdot\nabla\left\langle h_{il}^{\tau}\right\rangle^{il}\right]\\ -\left[\alpha_{il}\gamma_v\frac{\partial}{\partial\tau}\left\langle p_{il}^{\tau}\right\rangle^{il}+\alpha_{il}^{e}\left\langle \mathbf{V}_{il}^{\tau}\right\rangle^{ile}\gamma\cdot\nabla\left\langle p_{il}^{\tau}\right\rangle^{il}\right]\end{array}\right\}$$

$$-\nabla\cdot\left(\alpha_l^e\gamma\left\langle \lambda_l^I\nabla T_l\right\rangle^{le}\right)=\gamma_v\left(\alpha_l\rho_l P_{kl}^{\tau}+q_l^{\tau N}\right).\qquad(5.44)$$

where

$$q_l^{\tau N}=\dot{q}_l^{m}+\sum_{i=1}^{i_{max}}\mu_{iwl}^{\tau}\left(\left\langle h_{iw}^{\tau}\right\rangle^{iw}-\left\langle h_{il}^{\tau}\right\rangle^{il}\right)$$

$$+\sum_{m=1}^{3}\left[\mu_{Mml}^{\tau}\left(h_{Ml}^{\tau\sigma}-\left\langle h_{Ml}^{\tau}\right\rangle^{Ml}\right)+\sum_{n=1}^{n_{max}}\mu_{nml}^{\tau}\left(\left\langle h_{nm}^{\tau}\right\rangle^{nm}-\left\langle h_{nl}^{\tau}\right\rangle^{nl}\right)\right]$$

$$+\mu_{wl}^{\tau}\frac{1}{2}\left(\left\langle \mathbf{V}_{wl}^{\tau}\right\rangle^{we}-\left\langle \mathbf{V}_l^{\tau}\right\rangle^{le}\right)^2-\mu_{lw}^{\tau}\frac{1}{2}\left(\left\langle \mathbf{V}_{lw}^{\tau}\right\rangle^{le}-\left\langle \mathbf{V}_l^{\tau}\right\rangle^{le}\right)^2$$

$$+\frac{1}{2}\sum_{m=1}^{3}\mu_{ml}^{\tau}\left(\left\langle \mathbf{V}_m^{\tau}\right\rangle^{me}-\left\langle \mathbf{V}_l^{\tau}\right\rangle^{le}\right)^2.\qquad(5.45)$$

The corresponding conservative form is

$$\sum_{i=1}^{i_{max}} \left\{ \begin{array}{l} \dfrac{\partial}{\partial \tau} \left(\alpha_{il} \left\langle \rho_{il} \right\rangle^{il} \left\langle h_{il}^{\tau} \right\rangle^{il} \gamma_{v} \right) + \nabla \cdot \left(\alpha_{il}^{e} \left\langle \rho_{il} \right\rangle^{il} \left\langle \mathbf{V}_{il}^{\tau} \right\rangle^{ile} \left\langle h_{il}^{\tau} \right\rangle^{il} \gamma \right) \\ - \left[\alpha_{il} \gamma_{v} \dfrac{\partial}{\partial \tau} \left\langle p_{il}^{\tau} \right\rangle^{il} + \alpha_{il}^{e} \left\langle \mathbf{V}_{il}^{\tau} \right\rangle^{ile} \gamma \cdot \nabla \left\langle p_{il}^{\tau} \right\rangle^{il} \right] \end{array} \right\}$$

$$-\nabla \cdot \left(\alpha_{l}^{e} \gamma \delta_{l} \left\langle \lambda_{l}^{l} \nabla T_{l} \right\rangle^{le} \right) = \gamma_{v} \left(\alpha_{l} \rho_{l} P_{kl}^{\tau} + q_{l} \right), \qquad (5.46)$$

where

$$q_{l} = \dot{q}_{l}''' + \sum_{i=1}^{i_{max}} \left(\mu_{iwl}^{\tau} \left\langle h_{iw}^{\tau} \right\rangle^{iw} - \mu_{ilw}^{\tau} \left\langle h_{il}^{\tau} \right\rangle^{il} \right)$$

$$+ \sum_{m=1}^{3} \left[\left(\mu_{Mml}^{\tau} h_{Ml}^{\tau\sigma} - \mu_{Mlm}^{\tau} \left\langle h_{Ml}^{\tau} \right\rangle^{Ml} \right) + \sum_{n=1}^{n_{max}} \left(\mu_{nml}^{\tau} \left\langle h_{nm}^{\tau} \right\rangle^{nm} - \mu_{nlm}^{\tau} \left\langle h_{nl}^{\tau} \right\rangle^{nl} \right) \right]$$

$$+ \mu_{wl}^{\tau} \dfrac{1}{2} \left(\left\langle \mathbf{V}_{wl}^{\tau} \right\rangle^{we} - \left\langle \mathbf{V}_{l}^{\tau} \right\rangle^{le} \right)^{2} - \mu_{lw}^{\tau} \dfrac{1}{2} \left(\left\langle \mathbf{V}_{lw}^{\tau} \right\rangle^{le} - \left\langle \mathbf{V}_{l}^{\tau} \right\rangle^{le} \right)^{2}$$

$$+ \dfrac{1}{2} \sum_{m=1}^{3} \mu_{ml}^{\tau} \left(\left\langle \mathbf{V}_{m}^{\tau} \right\rangle^{me} - \left\langle \mathbf{V}_{l}^{\tau} \right\rangle^{le} \right)^{2} = q_{l}^{\tau N} + \sum_{i=1}^{i_{max}} \mu_{il}^{\tau} \left\langle h_{il} \right\rangle^{il}. \qquad (5.47)$$

Here

$$\gamma_{v} \alpha_{l} \rho_{l} P_{kl}^{\tau} = \nabla \cdot \left(\alpha_{l}^{e} \gamma \left\langle \mathbf{T}_{l}^{\tau} \cdot \mathbf{V}_{l}^{\tau} \right\rangle^{le} \right) - \left\langle \mathbf{V}_{l}^{\tau} \right\rangle^{le} \cdot \nabla \cdot \left(\alpha_{l}^{e} \gamma \left\langle \mathbf{T}_{l}^{\tau} \right\rangle^{le} \right) + E_{l}^{*\tau}$$

$$= \alpha_{l}^{e} \gamma \cdot \left\langle \mathbf{T}_{l}^{\tau} : \nabla \cdot \mathbf{V}_{l}^{\tau} \right\rangle^{le} + E_{l}^{*\tau} \qquad (5.48)$$

is the *irreversible bulk viscous dissipation*. The last term,

$$E_{l}^{*\tau} = \dfrac{1}{Vol} \int_{F_{l\sigma}} \Delta p_{l\sigma}^{\tau} \left(\left\langle \mathbf{V}_{l}^{\tau} \right\rangle^{le} - \mathbf{V}_{l\sigma}^{\tau} \right) \cdot \mathbf{n}_{l} dF + \dfrac{1}{Vol} \int_{F_{lw}} \Delta p_{lw}^{\tau} \left(\left\langle \mathbf{V}_{l}^{\tau} \right\rangle^{le} - \mathbf{V}_{lw}^{\tau} \right) \cdot \mathbf{n}_{l} dF$$

$$- \dfrac{1}{Vol} \int_{F_{l\sigma}+F_{lw}} \mathbf{T}_{l}^{\tau} \cdot \left(\left\langle \mathbf{V}_{l}^{\tau} \right\rangle^{le} - \mathbf{V}_{l}^{\tau} \right) \cdot \mathbf{n}_{l} dF = \dfrac{1}{Vol} \int_{F_{l\sigma}} \dfrac{\Delta p_{l\sigma}^{\tau}}{\rho_{l}} \rho_{l} \left(\mathbf{V}_{l}^{\tau} - \mathbf{V}_{l\sigma}^{\tau} \right) \cdot \mathbf{n}_{l} dF$$

5.4 Enthalpy equation

$$+\frac{1}{Vol}\int_{F_{lw}}\frac{\Delta p_{lw}^{\tau}}{\rho_l}\rho_l\left(\mathbf{V}_l^{\tau}-\mathbf{V}_{lw}^{\tau}\right)\cdot\mathbf{n}_l dF - \frac{1}{Vol}\int_{F_{l\sigma}}\left(-\Delta p_{l\sigma}^{\tau}+\mathbf{T}_l^{\tau}\right)\cdot\left(\langle\mathbf{V}_l^{\tau}\rangle^{le}-\mathbf{V}_l^{\tau}\right)\cdot\mathbf{n}_l dF$$

$$-\frac{1}{Vol}\int_{F_{lw}}\left(-\Delta p_{lw}^{\tau}+\mathbf{T}_l^{\tau}\right)\cdot\left(\langle\mathbf{V}_l^{\tau}\rangle^{le}-\mathbf{V}_l^{\tau}\right)\cdot\mathbf{n}_l dF \qquad (5.49)$$

is the irreversible power dissipation caused by the interface mass transfer between two regions with different velocities. A good approximation of the last two terms is obtained if one assumes

$$\mathbf{V}_l^{\tau}\approx\langle\mathbf{V}_m^{\tau}\rangle^{me} \quad \text{at} \quad F_{l\sigma} \qquad (5.50)$$

and

$$\mathbf{V}_l^{\tau}=0 \quad \text{at} \quad F_{lw} \qquad (5.51)$$

(non-slip condition), namely

$$-\frac{1}{Vol}\int_{F_{l\sigma}}\left(-\Delta p_{l\sigma}^{\tau}+\mathbf{T}_l^{\tau}\right)\cdot\left(\langle\mathbf{V}_l^{\tau}\rangle^{le}-\mathbf{V}_l^{\tau}\right)\cdot\mathbf{n}_l dF$$

$$\approx-\left(\langle\mathbf{V}_l^{\tau}\rangle^{le}-\langle\mathbf{V}_m^{\tau}\rangle^{me}\right)\frac{1}{Vol}\int_{F_{l\sigma}}\left(-\Delta p_{l\sigma}^{\tau}+\mathbf{T}_l^{\tau}\right)\cdot\mathbf{n}_l dF \qquad (5.52)$$

and

$$-\frac{1}{Vol}\int_{F_{lw}}\left(-\Delta p_{lw}^{\tau}+\mathbf{T}_l^{\tau}\right)\cdot\left(\langle\mathbf{V}_l^{\tau}\rangle^{le}-\mathbf{V}_l^{\tau}\right)\cdot\mathbf{n}_l dF \approx -\langle\mathbf{V}_l^{\tau}\rangle^{le}\frac{1}{Vol}\int_{F_{lw}}\left(-\Delta p_{lw}^{\tau}+\mathbf{T}_l^{\tau}\right)\cdot\mathbf{n}_l dF . \qquad (5.53)$$

The order of magnitude of the first two terms in Eq. (5.49) is

$$\frac{1}{Vol}\int_{F_{l\sigma}}\Delta p_{l\sigma}^{\tau}\left(\langle\mathbf{V}_l^{\tau}\rangle^{le}-\mathbf{V}_{l\sigma}^{\tau}\right)\cdot\mathbf{n}_l dF + \frac{1}{Vol}\int_{F_{lw}}\Delta p_{lw}^{\tau}\left(\langle\mathbf{V}_l^{\tau}\rangle^{le}-\mathbf{V}_{lw}^{\tau}\right)\cdot\mathbf{n}_l dF$$

$$\approx\sum_{m=1}^{3,w}\frac{\overline{\Delta p_{lm\sigma}^{\tau}}}{\rho_l}(\mu_{ml}-\mu_{lm}), \qquad (5.54)$$

where $\dfrac{\Delta p^\tau_{lm\sigma}}{\rho_l}$ is an averaged pressure difference between the bulk pressure and the boundary layer pressure inside the velocity field l.

Performing the summation in Eq. (5.46), using the *Dalton*'s law, substituting the *Fick*'s laws in the thus obtained equation, and applying the definitions given in Chapter 3 yields

$$\frac{\partial}{\partial \tau}\left(\alpha_l \langle \rho_l \rangle^l \langle h_l^\tau \rangle^l \gamma_v\right) + \nabla \cdot \left(\alpha_l^e \langle \rho_l \rangle^l \langle \mathbf{V}_l^\tau \rangle^{le} \langle h_l^\tau \rangle^l \gamma\right)$$

$$-\left[\alpha_l \gamma_v \frac{\partial}{\partial \tau} \langle p_l^\tau \rangle^l + \alpha_l^e \langle \mathbf{V}_l^\tau \rangle^{le} \gamma \cdot \nabla \langle p_l^\tau \rangle^l \right]$$

$$-\nabla \cdot \left\{ \alpha_l^e \gamma \delta_l \left[\langle \lambda_l^l \nabla T_l \rangle^{le} + \langle \rho_l \rangle^l \sum_{i=1}^{i_{max}} \left(\langle h_{il}^\tau \rangle^{il} D_{il}^l \nabla \langle C_{il}^\tau \rangle^{il} \right) \right] \right\}$$

$$+\delta_l \left(\alpha_l^e \gamma \nabla \langle p_l^\tau \rangle^l \right) \cdot \sum_{i=1}^{i_{max}} D_{il}^l \nabla \ln \langle C_{il}^\tau \rangle^{il} = \gamma_v \left(\alpha_l \rho_l P_{kl}^\tau + q_l \right). \qquad (5.55)$$

The non-conservative form of the equation (5.55) is readily obtained by differentiating the first two terms and comparing them with the field mass conservation equation. The result is

$$\boxed{\begin{aligned}
& \alpha_l \langle \rho_l \rangle^l \gamma_v \frac{\partial \langle h_l^\tau \rangle^l}{\partial \tau} + \alpha_l^e \langle \rho_l \rangle^l \langle \mathbf{V}_l^\tau \rangle^{le} \gamma \nabla \langle h_l^\tau \rangle^l \\
& -\left[\alpha_l \gamma_v \frac{\partial}{\partial \tau} \langle p_l^\tau \rangle^l + \alpha_l^e \langle \mathbf{V}_l^\tau \rangle^{le} \gamma \cdot \nabla \langle p_l^\tau \rangle^l \right] \\
& -\nabla \cdot \left\{ \alpha_l^e \gamma \delta_l \left[\langle \lambda_l^l \nabla T_l \rangle^{le} + \sum_{i=1}^{i_{max}} \left(\langle \rho_{il} \rangle^{il} \langle h_{il}^\tau \rangle^{il} D_{il}^l \nabla \ln \langle C_{il}^\tau \rangle^{il} \right) \right] \right\} \\
& +\delta_l \left(\alpha_l^e \gamma \nabla \langle p_l^\tau \rangle^l \right) \cdot \sum_{i=1}^{i_{max}} D_{il}^l \nabla \ln \langle C_{il}^\tau \rangle^{il} = \gamma_v \left(\alpha_l \rho_l P_{kl}^\tau + q_l - \mu_l^\tau \langle h_l^\tau \rangle^l \right). \quad (5.56)
\end{aligned}}$$

Remember that the sum of all *C*'s inside the velocity field is equal to one so that one of the concentrations depends on all the others.

5.5 Internal energy equation

Engineers sometimes have in their personal library approximations for the state variables and transport properties in terms of the specific internal energy. For practical use in this case, Eq. (5.46) can be rewritten in terms of the specific internal energy:

$$\sum_{i=1}^{i_{max}} \left\{ \begin{array}{c} \dfrac{\partial}{\partial \tau}\left(\alpha_{il}\langle \rho_{il}\rangle^{il}\langle e_{il}^{\tau}\rangle^{il}\gamma_v\right) + \nabla \cdot \left(\alpha_{il}^{e}\langle \rho_{il}\rangle^{il}\langle \mathbf{V}_{il}^{\tau}\rangle^{ile}\langle e_{il}^{\tau}\rangle^{il}\gamma\right) \\ -\langle p_{il}^{\tau}\rangle^{il}\left[\dfrac{\partial}{\partial \tau}(\alpha_{il}\gamma_v) + \nabla\cdot\left(\alpha_{il}^{e}\langle \mathbf{V}_{il}^{\tau}\rangle^{ile}\gamma\right)\right] \end{array} \right\}$$

$$-\nabla\cdot\left(\alpha_l^e\gamma\delta_l\langle \lambda_l^l\nabla T_l\rangle^{le}\right) = \gamma_v\left(\alpha_l\rho_l P_{kl}^{\tau} + q_l\right), \qquad (5.57)$$

or using the definitions introduced in Chapter 3

$$\dfrac{\partial}{\partial \tau}\left(\alpha_l\langle \rho_l\rangle^l\langle e_l^{\tau}\rangle^l\gamma_v\right) + \nabla\cdot\left(\alpha_l^e\langle \rho_l\rangle^l\langle \mathbf{V}_l^{\tau}\rangle^{le}\langle e_l^{\tau}\rangle^l\gamma\right)$$

$$-\langle p_l^{\tau}\rangle^l\left[\dfrac{\partial}{\partial \tau}(\alpha_l\gamma_v) + \nabla\cdot\left(\alpha_l^e\langle \mathbf{V}_l^{\tau}\rangle^{le}\gamma\right)\right]$$

$$-\nabla\cdot\left\{\alpha_l^e\gamma\delta_l\left[\langle \lambda_l^l\nabla T_l\rangle^{le} + \langle \rho_l\rangle^l\sum_{i=1}^{i_{max}}\left(\langle e_i^{\tau}\rangle^{il} D_{il}^l\nabla\langle C_{il}^{\tau}\rangle^{il}\right)\right]\right\}$$

$$+\delta_l\sum_{i=1}^{i_{max}}\left[\langle p_{il}^{\tau}\rangle^{il}\nabla\left(\alpha_l^e D_{il}^l\gamma\nabla\ln\langle C_{il}^{\tau}\rangle^{il}\right)\right] = \gamma_v\left(\alpha_l\rho_l P_{kl}^{\tau} + q_l\right). \qquad (5.58)$$

5.6 Entropy equation

At this point, use will be made of *Gibbs* equation (1878)

$$Tds = de + pdv. \qquad (5.59)$$

The thermodynamic definitions of temperature (absolute temperature - *Kelvin* 1848) and pressure are used here. This is a differential equation of state which is extremely important in the thermodynamic theory of compressible substances. It relates the difference in entropy between any two infinitesimally separated states to the infinitesimal differences in internal energy and volume between those states. Note that the *Gibbs* equation is valid only if s is a *smooth* function of e and v, i.e. where the differentials de and dv are uniquely defined (smooth equation of state). Using the definition of enthalpy (a mixture of thermodynamic and mechanical properties)

$$h = e + pv \qquad (5.60)$$

the *Gibbs* equation can be written as

$$T \rho ds = \rho dh - dp . \qquad (5.61)$$

Equation (5.59) and consequently Eq. (5.61) represent a *Legendre* transformation, which transforms independent variables to first order derivatives. After substituting for $\langle \rho_{il} \rangle^{il} d \langle h_{il}^{\tau} \rangle^{il} - d \langle p_{il}^{\tau} \rangle^{il}$ in Eq. (5.44) with the *Gibbs* definitions of the *specific entropy* of the corresponding components,

$$\langle \rho_{il} \rangle^{il} d \langle h_{il}^{\tau} \rangle^{il} - d \langle p_{il}^{\tau} \rangle^{il} = T_l \langle \rho_{il} \rangle^{il} d \langle s_{il}^{\tau} \rangle^{il} , \qquad (5.61a)$$

one obtains

$$T_l \sum_{i=1}^{i_{max}} \left\{ \langle \rho_{il} \rangle^{il} \left[\alpha_{il} \gamma_v \frac{\partial}{\partial \tau} \langle s_{il}^{\tau} \rangle^{il} + \alpha_{il}^e \langle \mathbf{V}_{il}^{\tau} \rangle^{ile} \gamma \cdot \nabla \langle s_{il}^{\tau} \rangle^{il} \right] \right\}$$

$$- \nabla \cdot \left(\alpha_l^e \gamma \langle \lambda_l^t \nabla T_l \rangle^{le} \right) = \alpha_l^e \gamma \cdot \langle \mathbf{T}_l^{\tau} : \nabla \cdot \mathbf{V}_l^{\tau} \rangle^{le} + E_l^{*\tau} + \gamma_v q_l^N . \qquad (5.62)$$

The following rearrangements are performed to give the conservative form of Eq. (5.62). Each of the mass conservation equations is multiplied by the corresponding component-specific entropy and field temperature. All the mass conservation equations are then added to Eq. (5.62). The differential terms are lumped together using the reverse chain rule of differentiation. The resulting *conservative form* is:

$$T_l \left\{ \frac{\partial}{\partial \tau} \left[\gamma_v \sum_{i=1}^{i_{max}} \left(\alpha_{il} \langle \rho_{il} \rangle^{il} \langle s_{il}^{\tau} \rangle^{il} \right) \right] + \nabla \cdot \left[\gamma \sum_{i=1}^{i_{max}} \left(\alpha_{il}^e \langle \rho_{il} \rangle^{il} \langle \mathbf{V}_{il}^{\tau} \rangle^{ile} \langle s_{il}^{\tau} \rangle^{il} \right) \right] \right\}$$

$$-\nabla \cdot \left(\alpha_l^e \gamma \delta_l \left\langle \lambda_l^l \nabla T_l \right\rangle^{le} \right) = \gamma_v \left(\alpha_l \rho_l P_{kl}^\tau + q_l^N + T_l \sum_{\substack{m=1 \\ m \neq l}}^{3,w} \sum_{i=1}^{i_{max}} \left(\mu_{iml}^\tau - \mu_{ilm}^\tau \right) \left\langle s_{il}^\tau \right\rangle^{il} \right). \quad (5.63)$$

After replacing the instantaneous values of the component velocities with the sums of the c. m. field velocities plus the deviations from the c. m. velocities, Eq. (5.34), and applying *Fick's* law, Eq. (5.36) and keeping in mind that there is no molecular net mass diffusively transported across a cross section perpendicular to the strongest concentration gradient, Eqs. (5.37, 5.38), defining the *specific mixture entropy of the velocity field*, $\left\langle s_l^\tau \right\rangle^l$, by Eq. (5.39), introducing the *Prandtl number*

$$\text{Pr}_l^l = \rho_l c_{pl} v_l^l / \lambda_l^l, \quad (5.64)$$

and dividing by T_l, the following form for the local instantaneous entropy equations is obtained:

$$\frac{\partial}{\partial \tau} \left(\alpha_l \left\langle \rho_l \right\rangle^l \left\langle s_l^\tau \right\rangle^l \gamma_v \right) + \nabla \cdot \left(\alpha_l^e \left\langle \rho_l \right\rangle^l \left\langle s_l^\tau \right\rangle^l \left\langle \mathbf{V}_l^\tau \right\rangle^{le} \gamma \right)$$

$$-\nabla \cdot \left[\alpha_l^e \delta_l \left\langle \rho_l \right\rangle^l \gamma \left(\sum_{i=2}^{i_{max}} \left(\left\langle s_{il}^\tau \right\rangle^{il} - \left\langle s_{1l}^\tau \right\rangle^{1l} \right) D_{il}^l \nabla \left\langle C_{il}^\tau \right\rangle^{il} \right) \right]$$

$$-\frac{1}{T_l} \nabla \cdot \left(\alpha_l^e \gamma \delta_l \left\langle \rho_l c_{pl} \frac{v_l^l}{\text{Pr}_l^l} \nabla T_l \right\rangle^{le} \right)$$

$$= \gamma_v \left[\left(\alpha_l \rho_l P_{kl}^\tau + q_l^N \right) / T_l + \sum_{\substack{m=1 \\ m \neq l}}^{3,w} \sum_{i=1}^{i_{max}} \left(\mu_{iml}^\tau - \mu_{ilm}^\tau \right) \left\langle s_{il}^\tau \right\rangle^{il} \right]. \quad (5.65)$$

We discussed in Chapter 4 a variety of notations of the energy conservation principle for single-phase flow. All of them are more complicated than the entropy notation. As for the single-phase flow comparing the multi-phase entropy equation (5.65) with the energy conservation equations in terms of specific enthalpy and specific internal energy (5.56, 5.58), it is evident that the computational effort required to approximate

1 time derivative,

238 5 First and second laws of the thermodynamics

1 divergence term,
2 diffusion terms, and
1 tensor product

is much less than the computational effort required to discretize

2 time derivatives,
2 divergence terms,
2 diffusion terms, and
1 tensor product.

Bearing in mind that each divergence term contains at least 3 differential terms, it is obvious that *for computational analysis the use of the entropy equation instead of energy equation in any of its variants is the most cost effective way for the numerical modeling of flows.*

> CONCLUSION: The use of the specific entropies as components of the dependent variable vector, *the entropy concept, gives the simplest form for the mathematical description of the flow.* The use of any other state variables instead of the entropies makes the description more complicated.

5.7 Local volume- and time-averaged entropy equation

The procedure we use to obtain a time-averaged entropy equation will be used also for deriving the time-averaged enthalpy, or specific internal energy equations. Here we introduce the time averaging rules. The instantaneous surface-averaged velocity of the field l, $\left\langle \mathbf{V}_l^\tau \right\rangle^{le}$, can be expressed as the sum of the surface-averaged velocity which is subsequently time averaged,

$$V_l := \overline{\left\langle \mathbf{V}_l^\tau \right\rangle^{le}} \tag{5.66}$$

and a fluctuation component V_l',

$$\left\langle \mathbf{V}_l^\tau \right\rangle^{le} = V_l + V_l' \tag{5.67}$$

as proposed by *Reynolds*. The same is performed for the pressure

$$\left\langle p_l^\tau \right\rangle^l = p_l + p_l', \tag{5.68}$$

and for any specific property

$$\left\langle \varphi_{il}^{\tau} \right\rangle^{l} = \varphi_{il} + \varphi_{il}', \tag{5.69}$$

where $\varphi = s, h, e, T, C$. Here the fluctuations of densities, volume fractions are neglected as they are only of small magnitude relative to the velocity and entropy fluctuations.

The time averaging rules we use are:

$$\overline{\alpha_l \left\langle \rho_l \right\rangle^l \varphi_{il}} = \alpha_l \left\langle \rho_l \right\rangle^l \varphi_{il}, \tag{5.70}$$

$$\overline{\alpha_l \left\langle \rho_l \right\rangle^l \varphi_{il}'} = 0, \tag{5.71}$$

$$\overline{\alpha_l^e \left\langle \rho_l \right\rangle^l \varphi_{il} \mathbf{V}_l} = \alpha_l^e \left\langle \rho_l \right\rangle^l \varphi_{il} \mathbf{V}_l, \tag{5.72}$$

$$\overline{\alpha_l^e \left\langle \rho_l \right\rangle^l \varphi_{il} \mathbf{V}_l'} = 0, \tag{5.73}$$

$$\overline{\alpha_l^e \left\langle \rho_l \right\rangle^l \varphi_{il}' \mathbf{V}_l} = 0, \tag{5.74}$$

$$\overline{\varphi_{il}' D_{il}^l \nabla C_{il}} = 0, \tag{5.75}$$

$$\overline{\varphi_{il} D_{il}^l \nabla C_{il}'} = 0, \tag{5.76}$$

$$\overline{\varphi_{il} D_{il}^l \nabla C_{il}} = \varphi_{il} D_{il}^l \nabla C_{il}, \tag{5.77}$$

$$\overline{\left(\mu_{iml}^{\tau} - \mu_{ilm}^{\tau} \right) \left\langle \varphi_{il}^{\tau} \right\rangle^{il}} = \left(\mu_{iml} - \mu_{ilm} \right) \varphi_{il} \tag{5.78}$$

$$\overline{\mu_{il}^{\tau}} = \mu_{il}, \tag{5.79}$$

$$\overline{\mu_{ml}^{\tau} \left(\mathbf{V}_m' + \mathbf{V}_m' - \mathbf{V}_l - \mathbf{V}_l' \right)^2} = \mu_{ml} \overline{\left(\mathbf{V}_m - \mathbf{V}_l \right)^2} + \mu_{ml} \overline{\left(\mathbf{V}_m' - \mathbf{V}_l' \right)^2}, \tag{5.80}$$

$$\overline{\mathbf{T}_l' : \nabla \cdot \mathbf{V}_l} = 0, \tag{5.81}$$

$$\overline{\mathbf{T}_l : \nabla \cdot \mathbf{V}_l'} = 0, \tag{5.82}$$

$$\overline{\mathbf{T}_l : \nabla \cdot \mathbf{V}_l} = \overline{\mathbf{T}_l} : \nabla \cdot \overline{\mathbf{V}_l}, \tag{5.83}$$

$$\overline{\mu_{iwl}^\tau \left(\langle h_{iw}^\tau \rangle^{iw} - \langle h_{il}^\tau \rangle^{il} \right)} = \mu_{iwl} \left(h_{iw} - h_{il} \right) \tag{5.84}$$

$$\overline{\mu_{Mml}^\tau \left(h_{Ml}^{\tau\sigma} - \langle h_{Ml}^\tau \rangle^{Ml} \right)} = \mu_{Mml} \left(h_{Ml}^\sigma - h_{Ml} \right) \tag{5.85}$$

$$\overline{\mu_{nml}^\tau \left(\langle h_{nm}^\tau \rangle^{nm} - \langle h_{nl}^\tau \rangle^{nl} \right)} = \mu_{nml} \left(h_{nm} - h_{nl} \right) \tag{5.86}$$

$$\overline{q_l^{\tau N}} = q_l^N \tag{5.87}$$

$$q_l^N = \dot{q}_l''' + \sum_{i=1}^{i_{max}} \mu_{iwl} \left(h_{iw} - h_{il} \right) + \sum_{m=1}^{3} \left[\mu_{Mml} \left(h_{Ml}^\sigma - h_{Ml} \right) + \sum_{n=1}^{n_{max}} \mu_{nml} \left(h_{nm} - h_{nl} \right) \right]$$

$$+ \frac{1}{2} \left[\mu_{wl} \left(\mathbf{V}_{wl} - \mathbf{V}_l \right)^2 - \mu_{lw} \left(\mathbf{V}_{lw} - \mathbf{V}_l \right)^2 + \sum_{m=1}^{3} \mu_{ml} \left(\mathbf{V}_m - \mathbf{V}_l \right)^2 \right]$$

$$+ \frac{1}{2} \left[\mu_{wl} \overline{\left(\mathbf{V}_{wl}' - \mathbf{V}_l' \right)^2} - \mu_{lw} \overline{\left(\mathbf{V}_{lw}' - \mathbf{V}_l' \right)^2} + \sum_{m=1}^{3} \mu_{ml} \overline{\left(\mathbf{V}_m' - \mathbf{V}_l' \right)^2} \right] \tag{5.88}$$

Substituting in Eq. (5.65) the above mentioned variables as consisting of mean values and fluctuations and performing time averaging following the above rules we obtain

$$\frac{\partial}{\partial \tau} \left(\alpha_l \rho_l s_l \gamma_v \right) + \nabla \cdot \left(\alpha_l^e \rho_l s_l \mathbf{V}_l \gamma \right)$$

$$- \frac{1}{T_l} \nabla \cdot \left(\alpha_l^e \gamma \delta_l \rho_l c_{pl} \frac{v_l'}{\Pr_l'} \nabla T_l \right) - \nabla \cdot \left[\alpha_l^e \delta_l \rho_l \gamma \left(\sum_{i=1}^{i_{max}} s_{il} D_{il}^l \nabla C_{il} \right) \right]$$

$$+ \nabla \cdot \left[\alpha_l^e \rho_l \left(\overline{s_l' \mathbf{V}_l'} - \sum_{i=1}^{i_{max}} \overline{s_{il}' D_{il}^l \nabla C_{il}'} \right) \gamma \right] = \gamma_v \frac{1}{T_l} DT_l^N + \gamma_v \sum_{\substack{m=1 \\ m \neq l}}^{3,w} \sum_{i=1}^{i_{max}} \left(\mu_{iml} - \mu_{ilm} \right) s_{il}.$$

$$\tag{5.89}$$

Here

5.7 Local volume- and time-averaged entropy equation

$$DT_l^N = \alpha_l \rho_l \left(\delta_l P_{kl}^\tau + \varepsilon_l\right) + \dot{q}''' + \sum_{i=1}^{i_{max}} \mu_{iwl}\left(h_{iwl} - h_{il}\right) + \sum_{\substack{m=1 \\ m\neq l}}^{l_{max}} \left[\begin{array}{c} \mu_{Mml}\left(h_{Ml}^\sigma - h_{Ml}\right) \\ + \sum_{n=1}^{n_{max}} \mu_{nml}\left(h_{nm} - h_{nl}\right) \end{array}\right]$$

$$+ \frac{1}{2}\left[\begin{array}{c} \mu_{wl}\left(\mathbf{V}_{wl} - \mathbf{V}_l\right)^2 - \mu_{lw}\left(\mathbf{V}_{lw} - \mathbf{V}_l\right)^2 + \sum_{m=1}^{3}\mu_{ml}\left(\mathbf{V}_m - \mathbf{V}_l\right)^2 \\ \overline{+\mu_{wl}\left(\mathbf{V}'_{wl} - \mathbf{V}'_l\right)^2} - \overline{\mu_{lw}\left(\mathbf{V}'_{lw} - \mathbf{V}'_l\right)^2} + \sum_{m=1}^{3}\overline{\mu_{ml}\left(\mathbf{V}'_m - \mathbf{V}'_l\right)^2} \end{array}\right], \quad (5.90)$$

$$\gamma_v \alpha_l \rho_l \delta_l P_{kl} = \alpha_l^e \gamma \cdot \left(\mathbf{T}_l : \nabla \cdot \mathbf{V}_l\right) + E_l^* \qquad (5.91)$$

is the *irreversible dissipated power* caused by the viscous forces *due to deformation of the mean values of the velocities* in the space, and

$$\gamma_v \alpha_l \rho_l \varepsilon_l = \alpha_l^e \gamma \cdot \left(\mathbf{T}'_l : \nabla \cdot \mathbf{V}'_l\right) + E_l'^* \qquad (5.92)$$

is the *irreversibly dissipated power* in the viscous fluid due to *turbulent pulsations*.

It is evident that the mass transfer between the velocity fields and between the fields and external sources causes additional entropy transport as a result of the different pulsation characteristics of the donor and receiver fields.

For simplicity's sake, averaging signs are omitted except in the turbulent diffusion term, which will be discussed next. The diffusion can also have *macroscopic* character, being caused by the macroscopic strokes between eddies with dimensions considerably larger than the molecular dimensions: turbulent diffusion. In a mixture at rest, *molecular* strokes represent the only mechanism driving diffusion. In real flows both mechanisms are observed. The higher the velocity of the flow, the higher the effect of turbulent diffusion.

To permit practical application of the entropy equation, it is necessary to define more accurately the term

$$\nabla \cdot \left[\alpha_l^e \rho_l \left(\overline{s'_l \mathbf{V}'}_l - \sum_{i=1}^{i_{max}} \overline{s'_{il} D_{il}^l \nabla C'_{il}}\right)\gamma\right].$$

A possible assumption for this case is that the mechanism of entropy transport caused by fluctuations is a diffusion-like mechanism, which means that

$$\nabla \cdot \left[\alpha_l^e \rho_l \left(\overline{s_l' \mathbf{V}_l'} - \sum_{i=1}^{i_{max}} \overline{s_{il}' D_{il}^l \nabla C_{il}'} \right) \gamma \right]$$

$$= -\frac{1}{T_l} \nabla \cdot \left[\alpha_l^e \rho_l \left(c_{pl} \frac{v_l^t}{\Pr_l^t} \nabla T_l \right) \gamma \right] - \nabla \cdot \left[\alpha_l^e \rho_l \left(\sum_{i=1}^{i_{max}} s_{il} \frac{v_l^t}{Sc_l^t} \nabla C_{il} \right) \gamma \right] \quad (5.93)$$

where

$$\Pr_l^t = \rho_l c_{pl} v_l^t / \lambda_l^t \quad (5.94)$$

is the turbulent *Prandtl* number and λ_l^t is the *turbulent coefficient of thermal conductivity* or *eddy conductivity*. Note that the thermal diffusion

$$\lambda_l = \rho_l c_{pl} v_l^t / \Pr_l^t \quad (5.95)$$

is a thermodynamic property of the continuum *l*, and

$$\lambda_l^t = \rho_l c_{pl} v_l^t / \Pr_l^t \quad (5.96)$$

is a mechanical property of the flowing field *l*. In channels,

$$\Pr_l^t \approx 0.7...0.9, \quad (5.97)$$

whereas for flow in jets, i.e., in *free turbulence*, the value is closer to 0.5 -see *Bird et al.* [1] (1960),

$$\Pr_l^t \approx 0.25...2.5 \quad (5.98)$$

for water, air and steam - see *Hammond* [5] (1985)]. The turbulent *Schmidt* number is defined as

$$Sc_l^t = v_l^t / D_l^t. \quad (5.99)$$

Here again the turbulent diffusion coefficient

$$D_l^t = v_l^t / Sc_l^t \quad (5.100)$$

is a mechanical property of the flowing field l. As a result, the final form of the entropy equation is

$$\frac{\partial}{\partial \tau}(\alpha_l \rho_l s_l \gamma_v) + \nabla \cdot (\alpha_l^e \rho_l s_l \mathbf{V}_l \gamma) - \frac{1}{T_l} \nabla \cdot \left[\alpha_l^e \rho_l c_{pl} \left(\delta_l \frac{v_l^l}{\Pr_l^l} + \frac{v_l^t}{\Pr_l^t} \right) \gamma \nabla T_l \right]$$

$$- \nabla \cdot \left\{ \alpha_l^e \rho_l \gamma \left[\sum_{i=1}^{i_{max}} s_{il} \left(\delta_l \frac{v_l^l}{Sc_l^l} + \frac{v_l^t}{Sc_l^t} \right) \nabla C_{il} \right] \right\} = \gamma_v \frac{1}{T_l} DT_l^N + \sum_{\substack{m=1 \\ m \ne l}}^{3,w} \sum_{i=1}^{i_{max}} (\mu_{iml} - \mu_{ilm}) s_{il} ,$$

(5.101a)

with this called the *conservative form*.

Keeping in mind that there is no net mass diffusively transported across a cross section perpendicular to the strongest concentration gradient

$$\sum_{i=1}^{i_{max}} (D_{il}^* \nabla C_{il}) = 0 , \qquad (5.102)$$

results in

$$D_{1l}^* \nabla C_{1l} = -\sum_{i=2}^{i_{max}} (D_{il}^* \nabla C_{il}) . \qquad (5.103)$$

where the effective conductivity, and the diffusivity are

$$\lambda_l^* = \delta_l \lambda_l + \rho_l c_{pl} \frac{v_l^t}{\Pr_l^t} = \rho_l c_{pl} \left(\delta_l \frac{v_l^l}{\Pr_l^l} + \frac{v_l^t}{\Pr_l^t} \right), \qquad (5.104)$$

$$D_{il}^* = \delta_l D_{il} + \frac{v_l^t}{Sc_l^t} = \delta_l \frac{v_l^l}{Sc_l^l} + \frac{v_l^t}{Sc_l^t} , \qquad (5.105)$$

we obtain finally

$$\boxed{\frac{\partial}{\partial \tau}(\alpha_l \rho_l s_l \gamma_v) + \nabla \cdot \left\{ \alpha_l^e \rho_l \gamma \left[s_l \mathbf{V}_l - \sum_{i=2}^{i_{max}} (s_{il} - s_{1l}) D_{il}^* \nabla C_{il} \right] \right\}}$$

$$\boxed{-\frac{1}{T_l}\nabla\cdot\left(\alpha_l^e \lambda_l^* \nabla T\right) = \gamma_v \left[\frac{1}{T_l}DT_l^N + \sum_{\substack{m=1\\m\ne l}}^{3,w}\sum_{i=1}^{i_{max}}\left(\mu_{iml}-\mu_{ilm}\right)s_{il}\right]}. \quad (5.101b)$$

For a mixture consisting of several inert components and one no-inert component it is advisable to select subscript $i = 1 = M$ for the non-inert component and consider the mixture of the inert components as an ideal gas. In this case we have

$$... = \nabla\cdot\left[\alpha_l^e \rho_l \gamma \left(s_{nl}-s_{Ml}\right) D_{nl}^* \nabla C_{nl}\right]... \quad (5.101c)$$

This method is used in IVA6 computer code [25] developed by this author.

5.8 Local volume- and time-averaged internal energy equation

Following the same procedure as for the entropy equation we perform time averaging of Eq. (5.58). The result is

$$\frac{\partial}{\partial \tau}\left(\alpha_l \rho_l e_l \gamma_v\right) + \nabla\cdot\left(\alpha_l^e \rho_l e_l \mathbf{V}_l \gamma\right) + p_l\left[\frac{\partial}{\partial \tau}\left(\alpha_l \gamma_v\right) + \nabla\cdot\left(\alpha_l^e \mathbf{V}_l \gamma\right)\right]$$

$$-\nabla\cdot\left[\alpha_l^e \rho_l \left(\delta_l c_{pl} \frac{v_l^t}{\Pr_l^t}\nabla T_l + \overline{e_l' \mathbf{V}_l'}\right)\gamma\right]$$

$$-\nabla\cdot\left[\alpha_l^e \rho_l \left(\delta_l \sum_{i=1}^{i_{max}} e_{il} D_{il}' \nabla C_{il} + \sum_{i=1}^{i_{max}} \overline{e_{il}' D_{il}' \nabla C_{il}'}\right)\gamma\right] + \overline{p_l' \nabla\cdot\left(\alpha_l^e \mathbf{V}_l' \gamma\right)} = \gamma_v De_l^*$$

$$(5.106)$$

where

$$\gamma_v De_l^* = \gamma_v \alpha_l \rho_l \left(\delta_l P_{kl} + \varepsilon_l\right)$$

$$+\gamma_v\left\{\dot{q}'''+\frac{1}{2}\left[\mu_{wl}\left(\mathbf{V}_{wl}-\mathbf{V}_l\right)^2 - \mu_{lw}\left(\mathbf{V}_{lw}-\mathbf{V}_l\right)^2 + \sum_{\substack{m=1\\m\ne l}}^{l_{max}}\mu_{ml}\left(\mathbf{V}_m-\mathbf{V}_l\right)^2\right]\right\}$$

$$+\gamma_v \frac{1}{2} \left[\overline{\mu_{wl}(\mathbf{V}'_{wl} - \mathbf{V}'_l)^2} - \overline{\mu_{lw}(\mathbf{V}'_{lw} - \mathbf{V}'_l)^2} + \sum_{\substack{m=1 \\ m \neq l}}^{l_{max}} \overline{\mu_{ml}(\mathbf{V}'_m - \mathbf{V}'_l)^2} \right]$$

$$+\gamma_v \sum_{i=1}^{i_{max}} \left(\mu_{iwl} h_{iwl} - \mu_{ilw} h_{il} \right) + \gamma_v \sum_{\substack{m=1 \\ m \neq l}}^{l_{max}} \left[\mu_{Mml} h_{Ml}^\sigma + \sum_{n=1}^{n_{max}} \left(\mu_{nml} h_{nm} - \mu_{nlm} h_{nl} \right) \right] \quad (5.107)$$

or introducing

$$\nabla \cdot \left[\alpha_l^e \rho_l \left(\overline{e'_l \mathbf{V}'_l} - \delta_l \sum_{i=1}^{i_{max}} \overline{e'_{il} D'_{il} \nabla C'_{il}} \right) \gamma \right]$$

$$= -\nabla \cdot \left[\alpha_l^e \rho_l \left(c_{pl} \frac{v_l^t}{\Pr_l^t} \nabla T_l + \sum_{i=1}^{i_{max}} e_{il} \frac{v_l^t}{\text{Sc}_l^t} \nabla C_{il} \right) \gamma \right] \quad (5.108)$$

we obtain finally

$$\boxed{\begin{aligned} &\frac{\partial}{\partial \tau}(\alpha_l \rho_l e_l \gamma_v) + \nabla \cdot (\alpha_l^e \rho_l e_l \mathbf{V}_l \gamma) + p_l \left[\frac{\partial}{\partial \tau}(\alpha_l \gamma_v) + \nabla \cdot (\alpha_l^e \mathbf{V}_l \gamma) \right] \\ &-\nabla \cdot \left\{ \alpha_l^e \rho_l \gamma \left[c_{pl} \left(\delta_l \frac{v_l^l}{\Pr_l^l} + \frac{v_l^t}{\Pr_l^t} \right) \nabla T_l + \sum_{i=2}^{i_{max}} (e_{il} - e_{1l}) D_{il}^* \nabla C_{il} \right] \right\} \\ &+ \overline{p'_l \nabla \cdot (\alpha_l^e \mathbf{V}'_l \gamma)} = \gamma_v D e_l^* . \end{aligned}} \quad (5.109)$$

The expression defining the diffusion coefficient for the specific internal mixture energy is then

$$\nabla \cdot (\alpha_l^e \rho_l D_l^e \gamma \cdot \nabla e_l)$$

$$= \nabla \cdot \left\{ \alpha_l^e \rho_l \gamma \left[c_{pl} \left(\delta_l \frac{v_l^l}{\Pr_l^l} + \frac{v_l^t}{\Pr_l^t} \right) \nabla T_l + \sum_{i=2}^{i_{max}} (e_{il} - e_{1l}) D_{il}^* \nabla C_{il} \right] \right\} \quad (5.110a)$$

which for ideal gas mixtures where the specific internal energy is not a function of pressure reduces to

$$\nabla \cdot \left(\alpha_l^e \rho_l D_l^e \gamma \cdot \nabla e_l \right) = \nabla \cdot \left\{ \alpha_l^e \rho_l \left[\frac{\lambda_l^*}{\rho_l c_{pl}} + \sum_{i=1}^{i_{max}} e_{il} D_{il}^* / \left(\frac{\partial e_l}{\partial C_{il}} \right)_{p,T_l} \right] \gamma \nabla e_l \right\} \quad (5.110b)$$

where

$$D_l^e = \frac{\lambda_l^*}{\rho_l c_{pl}} + \sum_{i=1}^{i_{max}} e_{il} D_{il}^* / \left(\frac{\partial e_l}{\partial C_{il}} \right)_{p,T_l} \quad (5.111)$$

This equation can be used to derive an alternative form for computation of the entropy diffusion coefficient

$$\nabla \cdot \left(\alpha_l^e \rho_l D_l^s \gamma \cdot \nabla s_l \right) = \left[\nabla \cdot \left(\alpha_l^e \rho_l D_l^e \gamma \cdot \nabla e_l \right) - \overline{p_l' \nabla \cdot \left(\alpha_l^e \mathbf{V}_l' \gamma \right)} \right] / T_l . \quad (5.112)$$

5.9 Local volume- and time-averaged specific enthalpy equation

Following the same procedure as for the entropy equation we perform time averaging of Eq. (5.55). The result is

$$\frac{\partial}{\partial \tau} \left(\alpha_l \rho_l h_l \gamma_v \right) + \nabla \cdot \left(\alpha_l^e \rho_l \mathbf{V}_l h_l \gamma \right) - \left(\alpha_l \gamma_v \frac{\partial p_l}{\partial \tau} + \alpha_l^e \mathbf{V}_l \gamma \cdot \nabla p_l \right)$$

$$- \nabla \cdot \left\{ \alpha_l^e \gamma \left[\delta_l \lambda_l \nabla T_l - \rho_l \overline{\mathbf{V}_l' h_l'} + \delta_l \rho_l \sum_{i=1}^{i_{max}} \left(h_{il} D_{il} \nabla C_{il} \right) + \rho_l \sum_{i=1}^{i_{max}} \left(h_{il}' D_{il} \nabla C_{il}' \right) \right] \right\}$$

$$- \alpha_l^e \overline{\mathbf{V}_l' \gamma \cdot \nabla p_l'} + \delta_l \left(\alpha_l^e \gamma \nabla p_l \right) \cdot \sum_{i=1}^{i_{max}} D_{il}^l \nabla \ln C_{il}$$

$$= \gamma_v DT_l^N + \gamma_v \sum_{m=1, m \neq l}^{3,w} \sum_{i=1}^{i_{max}} \left(\mu_{iml} - \mu_{ilm} \right) h_{il} \quad (5.113)$$

Introducing

$$\nabla \cdot \left\{ \alpha_l^e \rho_l \left[\overline{\mathbf{V}_l' h_l'} - \sum_{i=1}^{i_{max}} \left(h_{il}' D_{il} \nabla C_{il}' \right) \right] \gamma \right\}$$

$$= -\nabla \cdot \left[\alpha_l^e \rho_l \left(\delta_l c_{pl} \frac{v_l^t}{\Pr_l^t} \nabla T_l + \sum_{i=1}^{i_{max}} h_{il} \frac{v_l^t}{Sc_l^t} \nabla C_{il} \right) \gamma \right] \quad (5.114)$$

we obtain

$$\frac{\partial}{\partial \tau}(\alpha_l \rho_l h_l \gamma_v) + \nabla \cdot (\alpha_l^e \rho_l \mathbf{V}_l h_l \gamma) - \left(\alpha_{l\gamma_v} \frac{\partial p_l}{\partial \tau} + \alpha_l^e \mathbf{V}_l \gamma \cdot \nabla p_l \right)$$

$$-\nabla \cdot \left\{ \alpha_l^e \rho_l \left[c_{pl} \left(\delta_l \frac{v_l}{\Pr_l} + \frac{v_l^t}{\Pr_l^t} \right) \nabla T_l + \sum_{i=2}^{i_{max}} (h_{il} - h_{1l}) D_{il}^* \nabla C_{il} \right] \gamma \right\} - \alpha_l^e \overline{\mathbf{V}_l' \gamma \cdot \nabla p_l'}$$

$$+ \delta_l (\alpha_l^e \gamma \nabla p_l) \cdot \sum_{i=1}^{i_{max}} D_{il}^l \nabla \ln C_{il} = \gamma_v DT_l^N + \gamma_v \sum_{\substack{m=1 \\ m \neq l}}^{3,w} \sum_{i=1}^{i_{max}} (\mu_{iml} - \mu_{ilm}) h_{il}. \quad (5.115)$$

The non-conservative form is then

$$\alpha_l \rho_l \gamma_v \frac{\partial h_l}{\partial \tau} + \alpha_l^e \rho_l (\mathbf{V}_l \gamma \cdot \nabla) h_l - \left(\alpha_{l\gamma_v} \frac{\partial p_l}{\partial \tau} + \alpha_l^e \mathbf{V}_l \gamma \cdot \nabla p_l \right)$$

$$-\nabla \cdot \left\{ \alpha_l^e \rho_l \left[c_{pl} \left(\delta_l \frac{v_l}{\Pr_l} + \frac{v_l^t}{\Pr_l^t} \right) \nabla T_l + \sum_{i=2}^{i_{max}} (h_{il} - h_{1l}) D_{il}^* \nabla C_{il} \right] \gamma \right\} - \alpha_l^e \overline{\mathbf{V}_l' \gamma \cdot \nabla p_l'}$$

$$+ \delta_l (\alpha_l^e \gamma \nabla p_l) \cdot \sum_{i=1}^{i_{max}} D_{il}^l \nabla \ln C_{il} = \gamma_v DT_l^N. \quad (5.115a)$$

The definition equation for an enthalpy diffusion coefficient is then

$$\nabla \cdot (\alpha_l^e \rho_l D_l^h \gamma \nabla h_l) = \nabla \cdot \left\{ \alpha_l^e \rho_l \left[\frac{\lambda_l^*}{\rho_l} \nabla T_l + \sum_{i=2}^{i_{max}} (h_{il} - h_{1l}) D_{il}^* \nabla C_{il} \right] \gamma \right\}. \quad (5.116)$$

For mixtures of two ideal gases n and M we have

$$\nabla \cdot (\alpha_l^e \rho_l D_l^h \gamma \cdot \nabla h_l) = \nabla \cdot \left\{ \alpha_l^e \rho_l \left[\frac{\lambda_l^*}{\rho_l} \nabla T_l + (h_{nl} - h_{Ml}) D_{nl}^* \nabla C_{nl} \right] \gamma \right\} \quad (5.117a)$$

and an equation of state in differential form

$$\nabla h_l = c_{pl}\nabla T_l + (h_{nl} - h_{Ml})\nabla C_{nl},$$

which finally gives

$$\nabla \cdot (\alpha_l^e \rho_l D_l^h \gamma \nabla h_l) = \nabla \cdot \left\{\alpha_l^e \gamma \left[\lambda_l^* (1 - Le_{il})\nabla T + \rho_l D_{nl}^* \nabla h_l\right]\right\} \quad (5.117b)$$

where

$$Le_{il} = \frac{\rho_l c_{pl} D_{nl}^*}{\lambda_l^*}$$

is the *Lewis-Semenov* number. For a lot of binary gas mixtures $Le_{il} \approx 1$ and

$$\nabla \cdot (\alpha_l^e \rho_l D_l^h \gamma \nabla h_l) = \nabla \cdot (\alpha_l^e \rho_l D_{nl}^* \gamma \nabla h_l), \quad (5.117c)$$

which results in

$$D_l^h = D_{nl}^*,$$

[4] p. 267. This relation is widely used for simulation of diffusion in binary gas mixtures. If both components possess equal specific capacities at constant pressures, and therefore $h_{nl} \approx h_{Ml}$, we have

$$\nabla \cdot (\alpha_l^e \rho_l D_l^h \gamma \cdot \nabla h_l) = \nabla \cdot (\alpha_l^e \lambda_l^* \gamma \nabla T_l). \quad (5.117d)$$

This condition is fulfilled for many liquid mixtures and solutions. In this case the diffusion enthalpy transport looks like the *Fick*'s law and is totally controlled by the temperature gradient.

5.10 Non-conservative and semi-conservative forms of the entropy equation

The *non-conservative* form of the entropy equation is obtained by differentiating the first and second terms of the conservative form appropriately and comparing with the mass conservation equation. The result is

$$\rho_l \left[\alpha_l \gamma_v \frac{\partial s_l}{\partial \tau} + (\alpha_l^e \mathbf{V}_l \gamma \cdot \nabla) s_l\right]$$

$$-\frac{1}{T_l}\nabla\cdot\left(\alpha_l^e\lambda_l^*\nabla T\right)-\nabla\cdot\left\{\alpha_l^e\rho_l\gamma\left[\sum_{i=2}^{i_{max}}(s_{il}-s_{1l})D_{il}^*\nabla C_{il}\right]\right\}$$

$$=\gamma_v\left(\frac{1}{T_l}DT_l^N+\sum_{\substack{m=1\\m\neq l}}^{l_{max},w}\sum_{i=1}^{i_{max}}(\mu_{iml}-\mu_{ilm})s_{il}-\mu_l s_l\right)\equiv\gamma_v Ds_l^{N\,\prime}. \qquad (5.118)$$

The superscript N stands to remember that the RHS is for the non-conservative form of the energy conservation equation written in entropy form. *Note* that in addition to the enthalpy source terms divided by the field temperature, new terms arise. In fact, these denote the difference between the sum of the mass source per unit time and mixture volume multiplied by the corresponding specific component entropies and the product of the specific field entropy and the field mass source density.

$$\gamma_v\left[\sum_{\substack{m=1\\m\neq l}}^{l_{max},w}\sum_{i=1}^{i_{max}}(\mu_{iml}-\mu_{ilm})s_{il}-s_l\sum_{\substack{m=1\\m\neq l}}^{l_{max},w}\sum_{i=1}^{i_{max}}(\mu_{iml}-\mu_{ilm})\right]. \qquad (5.119)$$

Splitting the mass source term into *sources* and *sinks*

$$\mu_l=\sum_{\substack{m=1\\m\neq l}}^{l_{max},w}\sum_{i=1}^{i_{max}}(\mu_{iml}-\mu_{ilm})=\mu_l^+-\mu_l^- \qquad (5.120)$$

where

$$\mu_l^+=\sum_{\substack{m=1\\m\neq l}}^{l_{max},w}\sum_{i=1}^{i_{max}}\mu_{iml}\geq 0, \qquad (5.121)$$

$$\mu_l^-=\sum_{\substack{m=1\\m\neq l}}^{l_{max},w}\sum_{i=1}^{i_{max}}\mu_{ilm}\geq 0, \qquad (5.122)$$

$$\mu_l^+ s_l\geq 0, \qquad (5.123)$$

$$\mu_l^- s_l\geq 0, \qquad (5.124)$$

one then obtains the *final semi-conservative* form for the entropy equation

$$\boxed{\rho_l \left[\alpha_l \gamma_v \frac{\partial s_l}{\partial \tau} + \left(\alpha_l^e \mathbf{V}_l \gamma \cdot \nabla \right) s_l \right] - \frac{1}{T_l} \nabla \cdot \left(\alpha_l^e \lambda_l^* \nabla T \right)}$$

$$\boxed{- \nabla \cdot \left\{ \alpha_l^e \rho_l \gamma \left[\sum_{i=2}^{i_{\max}} (s_{il} - s_{1l}) D_{il}^* \nabla C_{il} \right] \right\} + \gamma_v \mu_l^+ s_l = \gamma_v D s_l}, \quad (5.125)$$

in which

$$D s_l = \frac{1}{T_l} D T_l^N + \sum_{\substack{m=1 \\ m \neq l}}^{3,w} \sum_{i=1}^{i_{\max}} (\mu_{iml} - \mu_{ilm}) s_{il} - \mu_l s_l + \mu_l^+ s_l$$

$$= \frac{1}{T_l} D T_l^N + \sum_{\substack{m=1 \\ m \neq l}}^{3,w} \sum_{i=1}^{i_{\max}} (\mu_{iml} - \mu_{ilm}) s_{il} + \mu_l^- s_l$$

$$= \frac{1}{T_l} D T_l^N + \gamma_v \left\{ \sum_{\substack{m=1 \\ m \neq l}}^{3,w} \left[\mu_{lm} s_l + \sum_{i=1}^{i_{\max}} (\mu_{iml} - \mu_{ilm}) s_{il} \right] \right\} \quad (5.126)$$

The three forms of the entropy equation, conservative (5.110b), non-conservative (5.118), and semi-conservative (5.125) are mathematically identical.

> The introduction of the semi-conservative form is perfectly suited to numerical integration because it ensures proper initialization of the value for the entropy in a computational cell in which a previously non-existent field is just in the course of origination.

5.11 Comments on the source terms in the mixture entropy equation

The following terms in the semi-conservative entropy equation (5.125),

$$D s_l = \ldots + \gamma_v \left\{ \sum_{\substack{m=1 \\ m \neq l}}^{3,w} \left[\mu_{lm} s_l + \sum_{i=1}^{i_{\max}} (\mu_{iml} - \mu_{ilm}) s_{il} \right] \right\}$$

5.11 Comments on the source terms in the mixture entropy equation

$$+\frac{\gamma_v}{T_l}\left\{\sum_{i=1}^{i_{max}}\mu_{iwl}\left(h_{iwl}-h_{il}\right)+\sum_{m=1}^{3}\left[\mu_{Mml}\left(h_{Ml}^{\sigma}-h_{Ml}\right)+\sum_{n=1}^{n_{max}}\mu_{nml}\left(h_{nm}-h_{nl}\right)\right]\right\}, \quad (5.127)$$

are discussed in some detail in this section in order to facilitate their practical application. For the wall source terms, that is, for injection from the wall into the flow and suction from the flow into the wall one can write

$$\mu_{iwl} = C_{iwl}\mu_{wl}, \quad (5.128)$$

$$\mu_{ilw} = C_{il}\mu_{lw}, \quad (5.129)$$

and therefore

$$Ds_l = \ldots + \gamma_v \mu_{wl} \sum_{i=1}^{i_{max}} C_{iwl}\left[s_{il}+\left(h_{iwl}-h_{il}\right)/T_l\right] + \gamma_v \sum_{\substack{m=1\\m\neq l}}^{l_{max}}\left\{\mu_{lm}s_l+\sum_{i=1}^{i_{max}}\left(\mu_{iml}-\mu_{ilm}\right)s_{il}\right\}$$

$$+\frac{\gamma_v}{T_l}\sum_{m=1}^{3}\left[\mu_{Mml}\left(h_{Ml}^{\sigma}-h_{Ml}\right)+\sum_{n=1}^{n_{max}}\mu_{nml}\left(h_{nm}-h_{nl}\right)\right]. \quad (5.130)$$

Similar relationships as Eqs. (5.128) and (5.129) are valid for entrainment from field 2 to 3 and deposition from field 3 to 2, that is for $lm = 23, 32$. Note that, say, for steam condensation from multi-component gas mixtures, $lm = 12, 13$, and for evaporation, $lm = 21, 31$, the following then applies: $\mu_{ilm} \neq \mu_{lm}C_{il}$. For practical applications it is convenient to group all inert and non-inert components in pseudo-two-component mixtures inside the field l designated simply with n and M, respectively.

The most important mass transfer mechanisms in the three velocity fields flow will now be taken into account in accordance with the following assumptions:

a) No solution of gas into field 2 and/or 3;

b) No gas dissolution from field 2 and/or 3;

c) Evaporation from field 2 and/or 3 into the gas field 1 is allowed;

d) Condensation of the component M from the gas field onto the interface of fields 2 and/or 3 is allowed;

e) Injection from the wall into all of the fields is allowed;

f) Suction from all of the fields through the wall is allowed.

The mass transfer terms are written in the following form:

$$\mu_{nlm} = \begin{vmatrix} 0 & 0 & 0 & \mu_{n1w} = C_{n1}\mu_{1w} \\ 0 & 0 & \mu_{n23} = C_{n2}\mu_{23} & \mu_{n2w} = C_{n2}\mu_{2w} \\ 0 & \mu_{n32} = C_{n3}\mu_{32} & 0 & \mu_{n3w} = C_{n3}\mu_{3w} \\ \mu_{nw1} = C_{nw1}\mu_{w1} & \mu_{nw2} = C_{nw2}\mu_{w2} & \mu_{nw3} = C_{nw3}\mu_{w3} & 0 \end{vmatrix}, \quad (5.131)$$

$$\mu_{Mlm} = \begin{vmatrix} 0 & \mu_{M12} = \mu_{12} & \mu_{M13} = \mu_{13} & \mu_{M1w} = C_{M1}\mu_{1w} \\ \mu_{M21} = \mu_{21} & 0 & \mu_{M23} = C_{M2}\mu_{23} & \mu_{M2w} = C_{M2}\mu_{2w} \\ \mu_{M31} = \mu_{31} & \mu_{M32} = C_{M3}\mu_{32} & 0 & \mu_{M3w} = C_{M3}\mu_{5w} \\ \mu_{Mw1} = C_{Mw1}\mu_{w1} & \mu_{Mw2} = C_{Mw2}\mu_{w2} & \mu_{Mw3} = C_{Mw3}\mu_{w3} & 0 \end{vmatrix}. \quad (5.132)$$

Except for the diagonal elements, the zeros in Eq. (5.132) result from assumptions a) and b). Assumptions c) and d) give rise to the terms

$$\mu_{Mlm} = \begin{vmatrix} 0 & \mu_{M12} = \mu_{12} & \mu_{M13} = \mu_{13} & \dots \\ \mu_{M21} = \mu_{21} & 0 & \dots & \dots \\ \mu_{M31} = \mu_{31} & \dots & 0 & \dots \\ \dots & \dots & \dots & 0 \end{vmatrix} \quad (5.133)$$

in the above equation.

Figures 5.2 and 5.3 illustrate the l-field side interface properties for the three fields under consideration.

5.11 Comments on the source terms in the mixture entropy equation

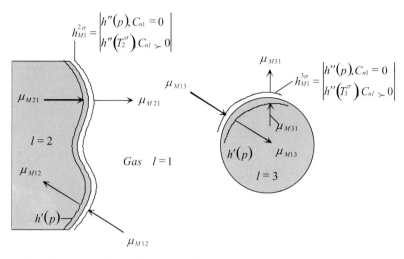

Fig. 5.2. Evaporation and condensation mass transfer. Interface properties

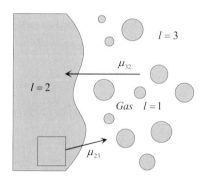

Fig. 5.3. Entrainment and deposition mass transfer

Taking into account Eqs. (5.131) and (5.132), the specific form for the source terms is:

$$Ds_1 = \dot{q}_1''' / T_1$$

$$+ \gamma_v \mu_{w1} \left\{ (1 - C_{nw1}) \left[s_{M1} + (h_{Mw1} - h_{M1})/T_1 \right] + C_{nw1} \left[s_{n1} + (h_{nw1} - h_{n1})/T_1 \right] \right\}$$

$$+ \gamma_v \begin{bmatrix} \mu_{21} \left[s_{M1} + (h_{M1}^{2\sigma} - h_{M1})/T_1 \right] - \mu_{12} C_{n1} (s_{M1} - s_{n1}) \\ \mu_{31} \left[s_{M1} + (h_{M1}^{3\sigma} - h_{M1})/T_1 \right] - \mu_{13} C_{n1} (s_{M1} - s_{n1}) \end{bmatrix} + \alpha_1 \rho_1 (\delta_1 P_{k1}^\tau + \varepsilon_1)/T_1$$

$$+\frac{1}{2}\left[\mu_{w1}\left(\mathbf{V}_{w1}-\mathbf{V}_1\right)^2 - \mu_{1w}\left(\mathbf{V}_{1w}-\mathbf{V}_1\right)^2 + \sum_{m=2,3}\mu_{m1}\left(\mathbf{V}_m-\mathbf{V}_1\right)^2\right]/T_1$$

$$+\frac{1}{2}\left[\mu_{w1}\overline{\left(\mathbf{V}'_{w1}-\mathbf{V}'_1\right)^2} - \mu_{1w}\overline{\left(\mathbf{V}'_{1w}-\mathbf{V}'_1\right)^2} + \sum_{m=2,3}\mu_{m1}\overline{\left(\mathbf{V}'_m-\mathbf{V}'_1\right)^2}\right]/T_1, \quad (5.134)$$

$$Ds_2 = \dot{q}_2'''/T_2$$

$$+\gamma_v \mu_{w2}\left\{(1-C_{nw2})\left[s_{M2}+\left(h_{Mw2}-h_{M2}\right)/T_2\right] + C_{nw2}\left[s_{n2}+\left(h_{nw2}-h_{n2}\right)/T_2\right]\right\}$$

$$+\gamma_v \mu_{32}\left\{(1-C_{n3})\left[s_{M2}+\left(h_{M3}-h_{M2}\right)/T_2\right] + C_{n3}\left[s_{n2}+\left(h_{n3}-h_{n2}\right)/T_2\right]\right\}$$

$$+\gamma_v \left\{\mu_{12}\left[s_{M2}+\left(h^\sigma_{M2}-h_{M2}\right)/T_2\right] - \mu_{21}C_{n2}\left(s_{M2}-s_{n2}\right)\right\} + \alpha_2\rho_2\left(\delta_2 P^\tau_{k2}+\varepsilon_2\right)/T_2$$

$$+\frac{1}{2}\left[\mu_{w2}\left(\mathbf{V}_{w2}-\mathbf{V}_2\right)^2 - \mu_{2w}\left(\mathbf{V}_{2w}-\mathbf{V}_2\right)^2 + \sum_{m=1,3}\mu_{m2}\left(\mathbf{V}_m-\mathbf{V}_2\right)^2\right]/T_2$$

$$+\frac{1}{2}\left[\mu_{w2}\overline{\left(\mathbf{V}'_{w2}-\mathbf{V}'_2\right)^2} - \mu_{2w}\overline{\left(\mathbf{V}'_{2w}-\mathbf{V}'_2\right)^2} + \sum_{m=1,3}\mu_{m2}\overline{\left(\mathbf{V}'_m-\mathbf{V}'_2\right)^2}\right]/T_2, \quad (5.135)$$

$$Ds_3 = \dot{q}_3'''/T_3$$

$$+\gamma_v \mu_{w3}\left\{(1-C_{nw3})\left[s_{M3}+\left(h_{Mw3}-h_{M3}\right)/T_3\right] + C_{nw3}\left[s_{n3}+\left(h_{nw3}-h_{n3}\right)/T_3\right]\right\}$$

$$+\gamma_v \mu_{23}\left\{(1-C_{n2})\left[s_{M3}+\left(h_{M2}-h_{M3}\right)/T_3\right] + C_{n2}\left[s_{n3}+\left(h_{n2}-h_{n3}\right)/T_3\right]\right\}$$

$$+\gamma_v \left\{\mu_{13}\left[s_{M3}+\left(h^\sigma_{M3}-h_{M3}\right)/T_3\right] - \mu_{31}C_{n3}\left(s_{M3}-s_{n3}\right)\right\} + \alpha_3\rho_3\left(\delta_3 P^\tau_{k3}+\varepsilon_3\right)/T_3$$

$$+\frac{1}{2}\left[\mu_{w3}\left(\mathbf{V}_{w3}-\mathbf{V}_3\right)^2 - \mu_{3w}\left(\mathbf{V}_{3w}-\mathbf{V}_3\right)^2 + \sum_{m=1,2}\mu_{m3}\left(\mathbf{V}_m-\mathbf{V}_3\right)^2\right]/T_3$$

5.11 Comments on the source terms in the mixture entropy equation

$$+\frac{1}{2}\left[\overline{\mu_{w3}\left(\mathbf{V}'_{w3}-\mathbf{V}'_3\right)^2}-\overline{\mu_{3w}\left(\mathbf{V}'_{3w}-\mathbf{V}'_3\right)^2}+\sum_{m=1,2}\overline{\mu_{m3}\left(\mathbf{V}'_m-\mathbf{V}'_3\right)^2}\right]/T_3. \quad (5.136)$$

The above equations can be simplified by taking into account

$$s_{iwl} = s_{il} + \left(h_{iwl} - h_{il}\right)/T_l, \quad (5.137)$$

$$s_{wl} = \sum_{i=1}^{i_{max}} C_{iwl} s_{iwl}, \quad (5.138)$$

$$s_{il} = s_{im} + \left(h_{il} - h_{im}\right)/T_m, \quad (5.139)$$

$$s_l = \sum_{i=1}^{i_{max}} C_{il} s_{il}, \quad (5.140)$$

$$s_{Ml}^{\sigma} = s_{Ml} + \left(h_{Ml}^{\sigma} - h_{Ml}\right)/T_l, \quad (5.141)$$

for $l = 2,3$ and $m = 3,2$ respectively.

$$Ds_1 = \ldots + \gamma_v \mu_{w1}\left\{(1-C_{nw1})\left[s_{M1}+\left(h_{Mw1}-h_{M1}\right)/T_1\right]+C_{nw1}\left[s_{n1}+\left(h_{nw1}-h_{n1}\right)/T_1\right]\right\}$$

$$+\gamma_v\begin{bmatrix}\left(\mu_{21}+\mu_{31}\right)s_{M1}+\left[\mu_{21}\left(h_{M1}^{2\sigma}-h_{M1}\right)+\mu_{31}\left(h_{M1}^{3\sigma}-h_{M1}\right)\right]/T_1\\-\left(\mu_{12}+\mu_{13}\right)C_{n1}\left(s_{M1}-s_{n1}\right)\end{bmatrix}+\ldots, \quad (5.142)$$

$$Ds_2 = \ldots + \gamma_v\left[\mu_{w2}s_{w2}+\mu_{32}s_3+\mu_{12}s'(p)\right]+\ldots, \quad (5.143)$$

$$Ds_3 = \ldots + \gamma_v\left[\mu_{w3}s_{w3}+\mu_{23}s_2+\mu_{13}s'(p)\right]+\ldots. \quad (5.144)$$

Note that the mass transfer terms can give rise to entropy change even if mass leaves the field with entropy at the interface not equal to the intrinsic average field entropy. There is an additional source of entropy change if the non-inert component leaves the field with entropy at the interface not equal to the intrinsic average field entropy as a result of evaporation or condensation.

For evaporation only within the closed control volume one obtains

$$(\mu_{12}+\mu_{13})s_1 = \mu_{21}\left[s_{M1}+\left(h''(T_2^\sigma)-h_{M1}\right)/T_1\right]+\mu_{31}\left[s_{M1}+\left(h''(T_3^\sigma)-h_{M1}\right)/T_1\right] \tag{5.145}$$

or

$$s_1 = \left\{\mu_{21}\left[s_{M1}+\left(h''(T_2^\sigma)-h_{M1}\right)/T_1\right]+\mu_{31}\left[s_{M1}+\left(h''(T_3^\sigma)-h_{M1}\right)/T_1\right]\right\}/(\mu_{12}+\mu_{13}). \tag{5.146}$$

For condensation and deposition the following is obtained for the specific entropy of the second velocity field

$$s_2 = \left[\mu_{32}s_3+\mu_{12}s'(p)\right]/(\mu_{12}+\mu_{32}). \tag{5.147}$$

For condensation and entrainment the following is obtained for the specific entropy of the third velocity field

$$s_3 = \left[\mu_{23}s_2+\mu_{13}s'(p)\right]/(\mu_{13}+\mu_{23}). \tag{5.148}$$

A frequently used simplification of the energy jump condition for computing the resulting evaporation or condensation is

$$\dot{q}'''_{m\sigma l}+\dot{q}'''_{l\sigma m}+\mu_{Ml}^\tau\left(h_{Mm}^\sigma-h_{Ml}^\sigma\right)=0, \tag{5.149}$$

or if $-\left(\dot{q}'''_{m\sigma l}+\dot{q}'''_{l\sigma m}\right)/\left(h_{Mm}^\sigma-h_{Ml}^\sigma\right) > 0$, then

$$\mu_{Mml} = -\left(\dot{q}'''_{m\sigma l}+\dot{q}'''_{l\sigma m}\right)/\left(h_{Mm}^\sigma-h_{Ml}^\sigma\right) \quad \text{and} \quad \mu_{Mlm}=0, \tag{5.150}$$

else

$$\mu_{Mlm} = \left(\dot{q}'''_{m\sigma l}+\dot{q}'''_{l\sigma m}\right)/\left(h_{Ml}^\sigma-h_{Mm}^\sigma\right) \quad \text{and} \quad \mu_{Mml}=0. \tag{5.151}$$

5.12 Viscous dissipation

One can next compute the *irreversible power dissipation caused by the viscous forces due to deformation as a result of the mean velocities in the space*:

5.12 Viscous dissipation

$$\gamma_v \alpha_l \rho_l P_{kl} = \alpha_l^e \gamma \cdot (\mathbf{T}_l : \nabla \cdot \mathbf{V}_l) = \nabla \cdot (\alpha_l^e \mathbf{T}_l \cdot \mathbf{V}_l \gamma) - \mathbf{V}_l \cdot \nabla \cdot (\alpha_l^e \mathbf{T}_l \gamma)$$

$$= \alpha_l^e \left\{ \begin{array}{l} \gamma_r \left(\tau_{l,rr} \dfrac{\partial u_l}{\partial r} + \tau_{l,r\theta} \dfrac{\partial v_l}{\partial r} + \tau_{l,rz} \dfrac{\partial w_l}{\partial r} \right) \\[2ex] \gamma_\theta \left[\tau_{l,\theta r} \dfrac{1}{r^\kappa} \left(\dfrac{\partial u_l}{\partial \theta} - \kappa v_l \right) + \tau_{l,\theta\theta} \dfrac{1}{r^\kappa} \left(\dfrac{\partial v_l}{\partial \theta} + \kappa u_l \right) + \tau_{l,\theta z} \dfrac{1}{r^\kappa} \dfrac{\partial w_l}{\partial \theta} \right] \\[2ex] \gamma_z \left(\tau_{l,zr} \dfrac{\partial u_l}{\partial z} + \tau_{l,z\theta} \dfrac{\partial v_l}{\partial z} + \tau_{l,zz} \dfrac{\partial w_l}{\partial z} \right) \end{array} \right\}. \quad (5.152)$$

Since for a symmetric stress tensor

$$\tau_{l,r\theta} = \tau_{l,\theta r}, \quad \tau_{l,\theta z} = \tau_{l,z\theta}, \quad \tau_{l,zr} = \tau_{l,rz}$$

one has

$$\gamma_v \alpha_l \rho_l P_{kl} = \alpha_l^e \left\{ \begin{array}{l} \gamma_r \tau_{l,rr} \dfrac{\partial u_l}{\partial r} + \gamma_\theta \tau_{l,\theta\theta} \dfrac{1}{r^\kappa} \left(\dfrac{\partial v_l}{\partial \theta} + \kappa u_l \right) + \gamma_z \tau_{l,zz} \dfrac{\partial w_l}{\partial z} \\[2ex] + \tau_{l,r\theta} \left[\gamma_r \dfrac{\partial v_l}{\partial r} + \gamma_\theta \dfrac{1}{r^\kappa} \left(\dfrac{\partial u_l}{\partial \theta} - \kappa v_l \right) \right] + \tau_{l,rz} \left[\gamma_r \dfrac{\partial w_l}{\partial r} + \gamma_z \dfrac{\partial u_l}{\partial z} \right] \\[2ex] + \tau_{l,z\theta} \left[\gamma_\theta \dfrac{1}{r^\kappa} \dfrac{\partial w_l}{\partial \theta} + \gamma_z \dfrac{\partial v_l}{\partial z} \right] \end{array} \right\}$$

(5.153)

Substituting for the stress tensor components using the *Helmholtz* and *Stokes* hypothesis, the following final expression is then obtained

$$\gamma_v \dfrac{\alpha_l}{\alpha_l^e} \dfrac{P_{kl}}{v_l} = 2 \left\{ \gamma_r \left(\dfrac{\partial u_l}{\partial r} \right)^2 + \gamma_\theta \left[\dfrac{1}{r^\kappa} \left(\dfrac{\partial v_l}{\partial \theta} + \kappa u_l \right) \right]^2 + \gamma_z \left(\dfrac{\partial w_l}{\partial z} \right)^2 \right\}$$

$$+ \left[\dfrac{\partial v_l}{\partial r} + \dfrac{1}{r^\kappa} \left(\dfrac{\partial u_l}{\partial \theta} - \kappa v_l \right) \right] \left[\gamma_r \dfrac{\partial v_l}{\partial r} + \gamma_\theta \dfrac{1}{r^\kappa} \left(\dfrac{\partial u_l}{\partial \theta} - \kappa v_l \right) \right]$$

$$+ \left[\dfrac{\partial w_l}{\partial r} + \dfrac{\partial u_l}{\partial z} \right] \left[\gamma_r \dfrac{\partial w_l}{\partial r} + \gamma_z \dfrac{\partial u_l}{\partial z} \right]$$

$$+ \left[\frac{1}{r^\kappa}\frac{\partial w_l}{\partial \theta} + \frac{\partial v_l}{\partial z}\right]\left[\gamma_\theta \frac{1}{r^\kappa}\frac{\partial w_l}{\partial \theta} + \gamma_z \frac{\partial v_l}{\partial z}\right]$$

$$-\frac{2}{3}\left[\frac{\partial u_l}{\partial r} + \frac{1}{r^\kappa}\left(\frac{\partial v_l}{\partial \theta} + \kappa u_l\right) + \frac{\partial w_l}{\partial z}\right]\left[\gamma_r \frac{\partial u_l}{\partial r} + \gamma_\theta \frac{1}{r^\kappa}\left(\frac{\partial v_l}{\partial \theta} + \kappa u_l\right) + \gamma_z \frac{\partial w_l}{\partial z}\right].$$

(5.154)

$P_{kl} \geq 0$ for an existing continuous velocity field ($\alpha_l > 0$ and $\delta_l = 1$). $P_{kl} = 0$ for a non-existent velocity field, or for an existent but discrete field ($\delta_l = 0$). It is evident that, as for single-phase flow, the *direct viscous dissipation*, P_{kl}, is (a) a non-negative quadratic form, b) its mathematical description does not depend on the rotation of the coordinate system [see *Zierep* [34] (1983)], and (c) this contains no derivatives of the viscosity. For single-phase flow, $\alpha_l = 1$, in free three-dimensional space, $\gamma = 1$, the above equation then reducing to the form obtained for the first time by *Rayleigh*.

The indirect power dissipated irreversibly due to *turbulent pulsations* in the viscous fluid is

$$\gamma_v \alpha_l \rho_l \varepsilon_l = \alpha_l^e \gamma \cdot \left(\mathbf{T}_l' : \nabla \cdot \mathbf{V}_l'\right),$$

(5.155)

where

$$\frac{\varepsilon_l}{\nu_l} = 2\left\{\gamma_r \overline{\left(\frac{\partial u_l'}{\partial r}\right)^2} + \gamma_\theta \overline{\left[\frac{1}{r^\kappa}\left(\frac{\partial v_l'}{\partial \theta} + \kappa u_l'\right)\right]^2} + \gamma_z \overline{\left(\frac{\partial w_l'}{\partial z}\right)^2}\right\}$$

$$+ \overline{\left[\frac{\partial v_l'}{\partial r} + \frac{1}{r^\kappa}\left(\frac{\partial u_l'}{\partial \theta} - \kappa v_l'\right)\right]\left[\gamma_r \frac{\partial v_l'}{\partial r} + \gamma_\theta \frac{1}{r^\kappa}\left(\frac{\partial u_l'}{\partial \theta} - \kappa v_l'\right)\right]}$$

$$+ \overline{\left[\frac{\partial w_l'}{\partial r} + \frac{\partial u_l'}{\partial z}\right]\left[\gamma_r \frac{\partial w_l'}{\partial r} + \gamma_z \frac{\partial u_l'}{\partial z}\right]} + \overline{\left[\frac{1}{r^\kappa}\frac{\partial w_l'}{\partial \theta} + \frac{\partial v_l'}{\partial z}\right]\left[\gamma_\theta \frac{1}{r^\kappa}\frac{\partial w_l'}{\partial \theta} + \gamma_z \frac{\partial v_l'}{\partial z}\right]}$$

$$-\frac{2}{3}\overline{\left[\frac{\partial u_l'}{\partial r} + \frac{1}{r^\kappa}\left(\frac{\partial v_l'}{\partial \theta} + \kappa u_l'\right) + \frac{\partial w_l'}{\partial z}\right]\left[\gamma_r \frac{\partial u_l'}{\partial r} + \gamma_\theta \frac{1}{r^\kappa}\left(\frac{\partial v_l'}{\partial \theta} + \kappa u_l'\right) + \gamma_z \frac{\partial w_l'}{\partial z}\right]}.$$

(5.156)

All terms in the right-hand side of this equation are time averages. In fact, this is the definition equation for the viscous dissipation rate, ε_l of the turbulent kinetic energy k_l. Here, it is once again evident that ε_l is (a) a non-negative quadratic form, $\varepsilon_l \geq 0$, (b) its mathematical description does not depend on the rotation of the coordinate system, and (c) it contains no derivatives of the viscosity.

For the case of steady-state single-phase, $\alpha_l = 1$, incompressible ($\nabla \mathbf{V}_l = 0$), isentropic, $P_{kl} + \varepsilon_l = 0$, flow with equal velocity gradients in all directions, *Taylor* [32] (1935) noticed that $\varepsilon_l \approx \frac{1}{2} 15 \nu_l \left(\frac{\partial w_l}{\partial z} \right)^2$. In this case the turbulence does not depend on the spatial direction (isotropic turbulence).

For axially symmetric flow in a pipe the Eqs. (5.154) and (5.155) reduce to

$$-\alpha_l^e \rho_l P_{kl} = \frac{1}{Vol} \int_{Vol} \alpha_l^e \left(\mathbf{T}_l : \nabla \mathbf{V}_l \right) dVol = \frac{4}{\pi D_h^2 \Delta z} \int_0^{\Delta z} \int_0^{R_h} \alpha_l^e \left(\mathbf{T}_l : \nabla \mathbf{V}_l \right) 2\pi r \, dr \, dz$$

$$= \int_0^1 \int_0^1 \alpha_l^e \left(\mathbf{T}_l : \nabla \mathbf{V}_l \right) d\left(r / R_h \right)^2 d\left(z / \Delta z \right) \tag{5.157}$$

$$-\alpha_l^e \rho_l \varepsilon_l = \frac{1}{Vol} \int_{Vol} \alpha_l^e \left(\mathbf{T}_l' : \nabla \mathbf{V}_l' \right) dVol = \int_0^1 \int_0^1 \alpha_l^e \left(\mathbf{T}_l' : \nabla \mathbf{V}_l' \right) d\left(r / R_h \right)^2 d\left(z / \Delta z \right). \tag{5.158}$$

For a single-phase flow Eq. (5.157) reduces to

$$\rho_l P_{kl} = \int_0^1 \left(\frac{dw}{dr} \right)^2 d\left(r / R_h \right)^2. \tag{5.159}$$

As an illustration of how to estimate the above integral, laminar flow in a pipe in line with the *Hagen-Poiseuille* law will now be considered, with the velocity distribution

$$w(r) = -\frac{dp}{dz} \frac{1}{4\eta} \left(R_h^2 - r^2 \right), \tag{5.160}$$

$$\rho P_k = \left(\frac{dp}{dz} \right)^2 \frac{D_h^2}{32\eta}. \tag{5.161}$$

Bearing in mind that

$$\frac{dp}{dz} = 32\eta \overline{w} / D_h^2, \qquad (5.162)$$

one then obtains

$$\rho P_k = \overline{w} \frac{-dp}{dz}, \qquad (5.163)$$

i.e., the *frictional pressure loss per unit mixture volume multiplied by the averaged flow velocity gives the irreversible part of the dissipated energy due to friction per unit time and unit flow volume*. There are a number of industrial processes where this component is important, e.g., heating of liquid in a circuit due to heat dissipation from pumping, and gas flow in very long pipes.

The modeling of turbulence in multi-phase flows is in its initial stage. This constitutes an exciting challenge for theoreticians and experimental scientists.

5.13 Temperature equation

The purpose of this section is to rewrite the entropy equation in terms of the field temperature and system pressure. It will be shown that the temperature and pressure changes do not depend on the absolute values of the specific component entropics, and therefore on the selection of the reference temperature and pressure to define the zero specific entropies. Important differences between mixtures of perfect and non-perfect gases will also be demonstrated.

In Chapter 3 the relationship between the field temperature, T_l, and the field properties (s_l, C_{il}, p), Eq. (3.106), is found to be

$$c_{pl} \frac{dT_l}{T_l} - \overline{R}_l \frac{dp_l}{p_l} = ds_l - \sum_{i=2}^{i_{\max}} \left(\frac{\partial s_l}{\partial C_{il}} \right)_{p, T_l, all\_C's\_except\_C_{il}} dC_{il}, \qquad (5.164)$$

where

$$\left(\frac{\partial s}{\partial C_i} \right)_{p, T, all\_C's\_except\_C_i} = s_{il} - s_{1l} + \Delta s_{il}^{np}. \qquad (5.165)$$

One of the mass concentrations, arbitrarily numbered with subscript 1, C_{1l}, depends on all others and is computed as all others are known,

$$C_{1l} = 1 - \sum_{i=2}^{i_{max}} C_{il} . \qquad (5.166)$$

Equation (3.86) consists of two parts. For the case of a mixture consisting of ideal gases the second part is equal to zero,

$$\Delta s_{il}^{np} = 0 , \qquad (5.167)$$

this also demonstrating the meaning of the subscript np, which stands for non-perfect fluid.

The non-conservative form of the entropy equation, Eq. (5.118), is

$$\rho_l \left[\alpha_l \gamma_v \frac{\partial s_l}{\partial \tau} + \left(\alpha_l^e \mathbf{V}_l \gamma \cdot \nabla \right) s_l \right] - \frac{1}{T_l} \nabla \cdot \left(\alpha_l^e \lambda_l^* \gamma \nabla T_l \right) - \nabla \cdot \left\{ \alpha_l^e \rho_l \gamma \left[\sum_{i=1}^{i_{max}} s_{il} D_{il}^* \nabla C_{il} \right] \right\}$$

$$= \gamma_v \left[\frac{1}{T_l} DT_l^N + \sum_{i=1}^{i_{max}} \mu_{il} \left(s_{il} - s_l \right) \right] , \qquad (5.168)$$

The non-conservative form of the mass conservation for each component inside the velocity field, Eq. (1.95), is

$$\alpha_l \rho_l \left(\gamma_v \frac{\partial C_{il}}{\partial \tau} + \mathbf{V}_l \gamma \nabla C_{il} \right) - \nabla \left(\alpha_l \rho_l D_{il}^* \gamma \nabla C_{il} \right) = \gamma_v \left(\mu_{il} - C_{il} \mu_l \right) . \qquad (5.169)$$

The non-conservative form of the entropy equation in terms of temperature and pressure is obtained by multiplying the $i_{max} - 1$ mass conservation equations (5.169) by $s_{il} - s_{1l} + \Delta s_{il}^{np}$ and subtracting them from the equation (5.168). The result is simplified by using

$$\sum_{i=1}^{i_{max}} \mu_{il} \left(s_{il} - s_l \right) - \sum_{i=2}^{i_{max}} \left(s_{il} - s_{1l} + \Delta s_{il}^{np} \right) \left(\mu_{il} - C_{il} \mu_l \right) = -\sum_{i=2}^{i_{max}} \Delta s_{il}^{np} \left(\mu_{il} - C_{il} \mu_l \right) ,$$

$$(5.170)$$

namely

$$\rho_l \left[\alpha_l \gamma_v \frac{\partial s_l}{\partial \tau} + \left(\alpha_l^e \mathbf{V}_l \gamma \cdot \nabla \right) s_l \right] - \rho_l \sum_{i=2}^{i_{max}} \left(s_{il} - s_{1l} + \Delta s_{il}^{np} \right) \left(\alpha_l \gamma_v \frac{\partial C_{il}}{\partial \tau} + \alpha_l^e \mathbf{V}_l \gamma \nabla C_{il} \right)$$

$$-\frac{1}{T_l}\nabla\cdot\left[\alpha_l^e \lambda_l^* \gamma \nabla T_l\right] - \nabla\cdot\left[\alpha_l^e \gamma \left(\sum_{i=1}^{i_{max}} s_{il} \rho_l D_{il}^* \nabla C_{il}\right)\right]$$

$$+\sum_{i=2}^{i_{max}} (s_{il} - s_{1l} + \Delta s_{il}^{np}) \nabla\cdot(\alpha_l^e \rho_l D_{il}^* \nabla C_{il}) = \gamma_v \left[\frac{1}{T_l} DT_l^N - \sum_{i=2}^{i_{max}} \Delta s_{il}^{np} (\mu_{il} - C_{il}\mu_l)\right].$$

(5.171)

Further simplification is obtained by using

$$\rho_l \left[\alpha_l \gamma_v \frac{\partial s_l}{\partial \tau} + (\alpha_l^e \mathbf{V}_l \gamma \cdot \nabla) s_l\right]$$

$$-\rho_l \sum_{i=2}^{i_{max}} \left(\frac{\partial s_l}{\partial C_{il}}\right)_{p, T_l, all\_C's\_except\_C_{il}} \left(\alpha_l \gamma_v \frac{\partial C_{il}}{\partial \tau} + \alpha_l^e \mathbf{V}_l \gamma \nabla C_{il}\right)$$

$$= \left\{\rho_l c_{pl}\left[\alpha_l \gamma_v \frac{\partial T_l}{\partial \tau} + (\alpha_l^e \mathbf{V}_l \gamma \cdot \nabla) T_l\right] - \frac{\rho_l \overline{R}_l T_l}{p}\left[\alpha_l \gamma_v \frac{\partial p}{\partial \tau} + (\alpha_l^e \mathbf{V}_l \gamma \cdot \nabla) p\right]\right\} / T_l,$$

(5.172)

where Eq. (3.121) is

$$\frac{\rho_l \overline{R}_l T_l}{p} = \left[1 - \rho_l \left(\frac{\partial h_l}{\partial p}\right)_{T_l, all\_C's}\right].$$

(5.173)

The last two terms of the left-hand side are simplified by using Eq. (5.103)

$$\nabla\cdot\left[\alpha_l^e \gamma \left(\sum_{i=1}^{i_{max}} s_{il} \rho_l D_{il}^* \nabla C_{il}\right)\right] - \sum_{i=2}^{i_{max}} (s_{il} - s_{1l} + \Delta s_{il}^{np}) \nabla\cdot(\alpha_l^e \rho_l D_{il}^* \nabla C_{il})$$

$$= \nabla\cdot\left[\alpha_l^e \gamma \left(\sum_{i=2}^{i_{max}} (s_{il} - s_{1l}) \rho_l D_{il}^* \nabla C_{il}\right)\right] - \sum_{i=2}^{i_{max}} (s_{il} - s_{1l} + \Delta s_{il}^{np}) \nabla\cdot(\alpha_l^e \rho_l D_{il}^* \nabla C_{il})$$

$$= \alpha_l^e \rho_l \gamma \sum_{i=2}^{i_{max}} (D_{il}^* \nabla C_{il}) \nabla (s_{il} - s_{1l}) - \sum_{i=2}^{i_{max}} \Delta s_{il}^{np} \nabla\cdot(\alpha_l^e \rho_l D_{il}^* \nabla C_{il}) \quad (5.174)$$

Neglecting the second order terms

5.13 Temperature equation

$$\alpha_l^e \rho_l \gamma \sum_{i=2}^{i_{max}} \left(D_{il}^* \nabla C_{il} \right) \nabla \left(s_{il} - s_{1l} \right), \tag{5.175}$$

we obtain the form below that is very convenient for practical applications

$$\boxed{\begin{aligned} & \rho_l c_{pl} \left[\alpha_l \gamma_v \frac{\partial T_l}{\partial \tau} + \left(\alpha_l^e \mathbf{V}_l \gamma \cdot \nabla \right) T_l \right] \\ & - \nabla \cdot \left(\alpha_l^e \lambda_l^* \gamma \nabla T \right) - \left[1 - \rho_l \left(\frac{\partial h_l}{\partial p} \right)_{T_l, all\_C's} \right] \left[\alpha_l \gamma_v \frac{\partial p}{\partial \tau} + \left(\alpha_l^e \mathbf{V}_l \gamma \cdot \nabla \right) p \right] \\ & + T_l \sum_{i=2}^{i_{max}} \Delta s_{il}^{np} \nabla \left(\alpha_l^e \rho_l D_{il}^* \nabla C_{il} \right) = \gamma_v \left[DT_l^N - T_l \sum_{i=2}^{i_{max}} \Delta s_{il}^{np} \left(\mu_{il} - C_{il} \mu_l \right) \right]. \end{aligned}} \tag{5.176}$$

This is the entropy equation rewritten in terms of temperature and pressure. For perfect gas mixtures, the following is obtained

$$\rho_l c_{pl} \left[\alpha_l \gamma_v \frac{\partial T_l}{\partial \tau} + \left(\alpha_l^e \mathbf{V}_l \gamma \cdot \nabla \right) T_l \right] - \nabla \cdot \left(\alpha_l^e \lambda_l^* \gamma \nabla T \right)$$

$$- \left[\alpha_l \gamma_v \frac{\partial p}{\partial \tau} + \left(\alpha_l^e \mathbf{V}_l \gamma \cdot \nabla \right) p \right] = \gamma_v DT_l^N. \tag{5.177}$$

Conclusions:

a) The temperature change caused by the injection into the velocity field depends on the difference between the specific enthalpy of the donor and the specific enthalpy of the velocity field. It is important to note that this is an *enthalpy difference* but not, say, differences in the specific internal energies, entropies, etc.

b) Equation (5.176) does not contain any specific entropies. As a result, the temperature change described by Eq. (5.176) does not depend on the reference temperature and pressure for computation of the specific entropies for the specific components.

c) Consider diffusion in a gas mixture consisting of non-perfect gases in adiabatic closed space. The initial state is characterized by spatially uniform temperature. At nearly constant system pressure Eq. (5.176) reduces to

$$\gamma_v \rho_l c_{pl} \frac{\partial T_l}{\partial \tau} \approx -T_l \sum_{i=2}^{i_{\max}} \Delta s_{il}^{np} \nabla \left(\rho_l D_{il}^* \nabla C_{il} \right),$$

which demonstrates that a temporal temperature change takes place at the locations of considerable diffusion. Bad numerical resolution can amplify this effect, leading to considerable local error in the numerical analysis.

d) Injection of non-perfect gas components into a volume initially filled with perfect gas can also give rise to behavior different from that of injection of a perfect gas component, due to the differences in the temperature and pressure equations describing both cases. This will be demonstrated in a simple case in the next section.

5.14 Second law of the thermodynamics

The entropy equation reflects very interesting physical phenomena. It is evident that velocity gradients in continuum of viscous fluid cause energy dissipation, $P_{kl} \geq 0$, which may generate turbulent kinetic energy. The turbulent kinetic energy increases the turbulent viscosity according to the *Prandtl-Kolmogorov* law, and helps to reduce the velocity gradients. The irreversible dissipation of kinetic energy caused by the turbulent pulsation increases the specific internal energy of the continuum field, $\varepsilon_l \geq 0$. This dissipation decreases the specific turbulent kinetic energy directly.

If the equation is applied to a single velocity field in a closed system without interaction with external mass, momentum or energy sources, the change in the specific entropy of the system will be non-negative, as the sum of the dissipation terms, $P_{kl} + \varepsilon_l$, is non-negative. This expresses the *second law of thermodynamics*. The second law tells us in what direction a process will develop in nature for closed and isolated systems.

> The process will proceed in a direction such that the entropy of the system always increases, or at best stays the same, $P_{kl} + \varepsilon_l = 0$, - entropy principle.

This information is not contained in the first law of thermodynamics. It results only after combining the three conservation principles (mass, momentum and energy) and introducing a *Legendre* transformation in the form of a *Gibbs* equation. In a way, it is a general expression of these conservation principles.

The entropy equation is not only very informative, but as already mentioned, it is very convenient for numerical integration because of its *simplicity* compared to the primitive form of the energy principle. This is the reason why the *specific entropies* of the velocity fields together with the concentrations of the inert component C_{nl} were chosen for use as *components of the dependent variables* vector as already mentioned in Ref. [18]. This unique combination of the dependent variables *simply minimizes the computational effort* associated with numerical integration and therefore makes the computer code *faster* and the analysis *cheaper*. It also makes the computer code architecture simple and allows the inclusion of more physical phenomena within a general flow model.

I call the flow modeling concept which makes use of the specific entropies of the velocity fields as components of the dependent variables vector the *entropy concept*.

5.15 Mixture volume conservation equation

Any numerical method in fluid mechanics should provide correct coupling between pressure changes and velocity changes. There are different ways of achieving this. One of these is faster than the others. Probably the fastest of these methods uses a specially derived equation, referred to as here the *mixture volume conservation equation*, MVCE. The MVCE is a linear combination of the mass conservation equations. The purpose of this section is to derive the analytical form of the MVCE, to discuss the physical meaning of each separate term and finally to show the reasons that make this equation so appropriate for use in constructing numerical schemes for complicated multi-phase flows.

The MVCE was obtained as follows:

(a) The mass conservation equations (1.62) were differentiated using the chain rule.

(b) Each equation was divided by the associated density. The resulting equations are dimensionless, m^3/m^3, and reflect the volume change balance for each velocity field per unit time and per unit mixture volume.

$$\gamma_v \frac{\partial \alpha_l}{\partial \tau} + \alpha_l \frac{\partial \gamma_v}{\partial \tau} + \frac{1}{\rho_l}\left[\alpha_l \gamma_v \frac{\partial \rho_l}{\partial \tau} + (\alpha_l \mathbf{V}_l \gamma \cdot \nabla)\rho_l\right] + \nabla \cdot (\alpha_l \mathbf{V}_l \gamma) = \gamma_v \frac{\mu_l}{\rho_l} \quad (5.178)$$

(c) The volume conservation equations obtained in this way were added and the fact that

$$\sum_{l=1}^{3} \alpha_l = 1 \qquad (5.179)$$

and

$$\sum_{l=1}^{3} d\alpha_l = 0 \qquad (5.180)$$

was used to cancel the sum of the time derivatives of the volume concentrations.

$$\sum_{l=1}^{l_{max}} \frac{1}{\rho_l}\left[\alpha_l \gamma_v \frac{\partial \rho_l}{\partial \tau} + (\alpha_l \mathbf{V}_l \gamma \cdot \nabla)\rho_l\right] + \nabla \cdot \sum_{l=1}^{l_{max}}(\alpha_l \mathbf{V}_l \gamma) = \gamma_v \sum_{l=1}^{l_{max}} \frac{\mu_l}{\rho_l} - \frac{\partial \gamma_v}{\partial \tau} \qquad (5.181)$$

(d) The density derivatives were substituted using the differential form of the equation of state for each velocity field (3.137)

$$d\rho = \frac{dp}{a^2} + \left(\frac{\partial \rho}{\partial s}\right)_{p,all\_C's} ds + \sum_{i=2}^{i_{max}}\left(\frac{\partial \rho}{\partial C_i}\right)_{p,s,all\_C's\_except\_C_i} dC_i . \qquad (5.182)$$

The result is

$$\gamma_v \left(\sum_{l=1}^{l_{max}} \frac{\alpha_l}{\rho_l a_l^2}\right)\frac{\partial p}{\partial \tau} + \sum_{l=1}^{l_{max}} \frac{\alpha_l}{\rho_l a_l^2}(\mathbf{V}_l \gamma \cdot \nabla)p + \nabla \cdot \sum_{l=1}^{l_{max}}(\alpha_l \mathbf{V}_l \gamma)$$

$$= \sum_{l=1}^{l_{max}} \frac{1}{\rho_l}\left\{\gamma_v \mu_l - \frac{1}{\rho_l}\left[\begin{array}{l}\left(\frac{\partial \rho_l}{\partial s_l}\right)_{p,all\_C_l's} \rho_l\left[\alpha_l \gamma_v \frac{\partial s_l}{\partial \tau} + (\alpha_l \mathbf{V}_l \gamma \cdot \nabla)s_l\right] \\ + \sum_{i=2}^{i_{max}}\left(\frac{\partial \rho_l}{\partial C_{li}}\right)_{p,s,all\_C's\_except\_C_i} \rho_l\left[\alpha_l \gamma_v \frac{\partial C_{li}}{\partial \tau} + (\alpha_l \mathbf{V}_l \gamma \cdot \nabla)C_{li}\right]\end{array}\right]\right\}$$

$$- \frac{\partial \gamma_v}{\partial \tau} \qquad (5.183)$$

(e) The concentration equations (1.96) in Chapter 1 and the non-conservative form of the entropy equation (5.118) in this Chapter are compared with the LHS of Eq. (5.183). Making the substitution

5.15 Mixture volume conservation equation

$$\rho_l \left[\alpha_l \gamma_v \frac{\partial C_{il}}{\partial \tau} + \left(\alpha_l^e \mathbf{V}_l \gamma . \nabla \right) C_{il} \right] = DC_{il}^N , \qquad (5.184)$$

$$\rho_l \left[\alpha_l \gamma_v \frac{\partial s_l}{\partial \tau} + \left(\alpha_l^e \mathbf{V}_l \gamma . \nabla \right) s_l \right] = Ds_l^N , \qquad (5.185)$$

where

$$\overline{DC_{il}^N} = \nabla \left(\alpha_l \rho_l D_{il}^* \nabla C_{il} \right) + \gamma_v \left(\mu_{il} - C_{il} \mu_l \right), \qquad (5.186)$$

$$\overline{Ds_l^N} = \frac{1}{T_l} \nabla \cdot \left(\alpha_l^e \lambda_l^* \nabla T_l \right) + \nabla \cdot \left\{ \alpha_l^e \rho_l \gamma \left[\sum_{i=1}^{i_{max}} s_{il} D_{il}^* \nabla C_{il} \right] \right\}$$

$$+ \gamma_v \left[\frac{1}{T_l} DT_l^N + \sum_{i=1}^{i_{max}} \mu_{il} \left(s_{il} - s_l \right) \right], \qquad (5.187)$$

we obtain the final form for the MVCE

$$\boxed{\frac{\gamma_v}{\rho a^2} \frac{\partial p}{\partial \tau} + \sum_{l=1}^{l_{max}} \frac{\alpha_l}{\rho_l a_l^2} \left(\mathbf{V}_l \gamma \cdot \nabla \right) p + \nabla \cdot \sum_{l=1}^{l_{max}} \left(\alpha_l \mathbf{V}_l \gamma \right) = \sum_{l=1}^{l_{max}} D\alpha_l - \frac{\partial \gamma_v}{\partial \tau} , \qquad (5.188)}$$

or in scalar form for practical application in Cartesian and cylindrical coordinates

$$\frac{\gamma_v}{\rho a^2} \frac{\partial p}{\partial \tau} + \sum_{l=1}^{3} \frac{\alpha_l}{\rho_l a_l^2} \left(u_l \gamma_r \frac{\partial p}{\partial r} + v_l \gamma_\theta \frac{\partial p}{\partial \theta} + w_l \gamma_z \frac{\partial p}{\partial z} \right)$$

$$+ \frac{\partial}{\partial r} \left(\alpha_l u_l \gamma_r \right) + \frac{1}{r^\kappa} \frac{\partial}{\partial \theta} \left(r^\kappa \alpha_l v_l \gamma_\theta \right) + \frac{\partial}{\partial z} \left(\alpha_l w_l \gamma_z \right) = \sum_{l=1}^{3} D\alpha_l - \frac{\partial \gamma_v}{\partial \tau} \qquad (5.189)$$

where

$$D\alpha_l = \frac{1}{\rho_l} \left\{ \gamma_v \mu_l - \frac{1}{\rho_l} \left[\left(\frac{\partial \rho_l}{\partial s_l} \right)_{p, all\_C_{li}'s} \overline{Ds_l^N} + \sum_{i=2}^{i_{max}} \left(\frac{\partial \rho_l}{\partial C_{li}} \right)_{p, s, all\_C_{li}'s\_except\_C_{l1}} \overline{DC_{il}^N} \right] \right\}$$

$$(5.190)$$

Replacing the derivatives of the mixture density with the Eqs. (110) and (111) obtained in [22], one obtains

$$D\alpha_l = \gamma_v \frac{\mu_l}{\rho_l} - \frac{1}{\rho_l^2} \left[\left(\frac{\partial \rho_l}{\partial T_l}\right)_{p,all\_C's} \frac{T_l}{c_{pl}} \left[\overline{Ds_l^N} - \sum_{i=2}^{i_{max}} \left(\frac{\partial s_l}{\partial C_{il}}\right)_{p,T_l,all\_C's\_except\_C_{il}} \overline{DC_{il}^N} \right] \right.$$
$$\left. + \sum_{i=2}^{i_{max}} \left(\frac{\partial \rho_l}{\partial C_{il}}\right)_{p,T_l,all\_C's\_except\_C_{il}} \overline{DC_{il}^N} \right].$$

(5.191)

Further simplification is obtained by using Eqs. (5.186) and (5.187), subsequently the Chain Rules and neglecting the second order term

$$\alpha_l \rho_l \gamma T_l \sum_{i=2}^{i_{max}} \nabla \cdot \left(D_{il}^* \nabla C_{il}\right) \nabla \cdot \left(s_{il} - s_{1l}\right).$$ (5.192)

The result is

$$T_l \left[\overline{Ds_l^N} - \sum_{i=2}^{i_{max}} \left(\frac{\partial s_l}{\partial C_{il}}\right)_{p,T_l,all\_C's\_except\_C_{il}} \overline{DC_{il}^N} \right]$$

$$\approx \nabla \cdot \left(\alpha_l^e \lambda_l^* \nabla T\right) + \gamma_v \overline{DT_l^N} - T_l \sum_{i=2}^{i_{max}} \Delta s_{il}^{np} \left[\nabla \cdot \left(\alpha_l^e \rho_l D_{il}^* \nabla C_{il}\right) + \gamma_v \left(\mu_{il} - C_{il}\mu_l\right)\right],$$

(5.193)

For perfect gas mixtures the following alone is obtained

$$T_l \left[\overline{Ds_l^N} - \sum_{i=2}^{i_{max}} \left(\frac{\partial s_l}{\partial C_{il}}\right)_{p,T_l,all\_C's\_except\_C_{il}} \overline{DC_{il}^N} \right] \approx \nabla \cdot \left(\alpha_l^e \lambda_l^* \nabla T\right) + \gamma_v \overline{DT_l^N}$$

(5.194)

a is the *sonic velocity* in a homogeneous multi-phase mixture and is defined as follows

$$\frac{1}{\rho a^2} = \sum_{l=1}^{3} \frac{\alpha_l}{\rho_l a_l^2} = \frac{1}{p} \sum_{l=1}^{3} \frac{\alpha_l}{\kappa_l} = \frac{1}{\kappa p},$$ (5.195)

and

$$\rho = \sum_{l=1}^{3} \alpha_l \rho_l \tag{5.196}$$

is the mixture density. Equation (5.188) was already derived in [9] in 1986 and published in [10] in 1987 p.100.

The MVCE equation can be directly discretized and incorporated into the numerical scheme. Another possibility is to follow the same scheme as for deriving the MVCE analytically but starting with already discretized mass conservation equations. The coupling finally obtained is then *strictly consistent* with the discretized form of the mass conservation equations.

Alternative forms of the MVCE can also be used, e.g.

$$\frac{\gamma_v}{\kappa p} \frac{\partial p}{\partial \tau} + \sum_{l=1}^{3} \frac{\alpha_l}{\kappa_l p} \mathbf{V}_l \gamma . \nabla p + \nabla . \left(\sum_{l=1}^{3} \alpha_l \mathbf{V}_l \gamma \right) = \sum_{l=1}^{3} D\alpha_l - \frac{\partial \gamma_v}{\partial \tau}, \tag{5.197}$$

or certain integrated forms

$$\frac{\gamma_v}{\kappa} \frac{\partial}{\partial \tau} \ln p + \sum_{l=1}^{3} \frac{\alpha_l}{\kappa_l} \mathbf{V}_l \gamma . \nabla \ln p + \nabla . \left(\sum_{l=1}^{3} \alpha_l \mathbf{V}_l \gamma \right) = \sum_{l=1}^{3} D\alpha_l - \frac{\partial \gamma_v}{\partial \tau}, \tag{5.198}$$

where

$$\kappa = \rho a^2 / p, \tag{5.199}$$

$$\kappa_l = \rho_l a_l^2 / p \tag{5.200}$$

are the mixture isentropic exponent and the isentropic exponent of each particular velocity field, respectively.

For completeness of the theory the MVCE equation valid in the case of steady-state non-compressible flow will also be given

$$\nabla . \left(\sum_{l=1}^{3} \alpha_l \mathbf{V}_l \gamma \right) = \nabla . (\mathbf{J}\gamma) = \sum_{l=1}^{3} D\alpha_l, \tag{5.201}$$

where

$$\mathbf{J} = \sum_{l=1}^{3} \alpha_l \mathbf{V}_l = \sum_{l=1}^{3} \mathbf{j}_l \tag{5.202}$$

is the *volumetric mixture flow rate* in $m^3/(m^2s)$. If diffusion is neglected and no mass exchange takes place between the velocity fields or between the flow and external sources, this gives

$$\nabla \cdot (\mathbf{J}\gamma) = 0 \tag{5.203}$$

or

$$\mathbf{J}\gamma = const. \tag{5.204}$$

The MVCE has the remarkable feature that it couples the temporal pressure change through the compressibility $1/(\rho a^2)$ with

a) the convective specific volume change $\nabla \cdot \left(\sum_{l=1}^{3} \alpha_l \mathbf{V}_l \gamma \right)$ for the control volume;

b) the change in the specific volume of the mixture associated with the net specific volume change for the mixture due to the mass sources $\gamma_v \sum_{l=1}^{3} \mu_l / \rho_l$;

c) the density change due to the spatial pressure, entropy and concentration changes for the associated velocity field.

Another specific property of Eq. (5.188) should be emphasized. The second term on the left-hand side represents the dimensionless changes in the density. This fact allows one to use *up-wind* discretization even for the pressure terms (*donor-cell* concept), because one in fact discretizes the dimensionless density change within the interval $\Delta r, \Delta \theta, \Delta z$.

The above features of this equation make it very suitable for coupling with the momentum equations in order to derive an equation for the mixtures which is similar to the *Poisson* equation for single-phase flow.

For the case of negligible diffusion the right-hand side of the MVCE contains no differential terms:

$$D\alpha_l = \frac{\gamma_v}{\rho_l} \left\{ \mu_l - \left[\frac{\partial \rho_l}{\partial s_l} (Ds_l - \mu_l^+ s_l) + \sum_{i=1}^{i_{max}} \frac{\partial \rho_l}{\partial C_{li}} (\mu_{li} - \mu_l^+ C_{li}) \right] / \rho_l \right\}. \tag{5.205}$$

This means that during numerical integration the influence of the changes of entropies and concentrations on the pressure field can be taken into account in a single step only (without outer iterations). The computer code architecture is thus extremely simplified, speeding up the numerical integration and therefore making it cheaper. This is an important feature of the entropy concept presented here and used in the IVA computer code development compared to the concepts of other computer codes, e.g. TRAC [28, 29, 6, 33] and AFDM [2].

The right-hand side of the pressure equation, Eq. (5.188), does not contain any specific entropies. As a result, the pressure change described by Eq. (5.188) is not dependent on the reference temperature and pressure for computation of the specific entropies for the specific components.

Comparing Eqs. (5.193) and (5.194) it is evident that if systems in which at least one of the gas components deviates from the perfect gas behavior are approximated as consisting of perfect gasses, an entropy imbalance of about

$$-\sum_{i=2}^{i_{\max}} \Delta s_{il}^{np} \left[\nabla \left(\alpha_l^e \rho_l D_{il}^* \nabla C_{il} \right) + \gamma_v \left(\mu_{il} - C_{il} \mu_l \right) \right] \tag{5.206}$$

is introduced. This is a very surprising result. I call this result the *non-perfect gas paradox*.

5.16 Linearized form of the source term for the temperature equation

The source terms with fluctuation components neglected

$$DT_l^N = \alpha_l \rho_l \left(\delta_l P_{kl}^\tau + \varepsilon_l \right) + \dot{q}''' + \sum_{i=1}^{i_{\max}} \mu_{iwl} \left(h_{iwl} - h_{il} \right)$$

$$+ \sum_{m=1}^{3} \left[\mu_{Mml} \left(h_{Ml}^\sigma - h_{Ml} \right) + \sum_{n=1}^{n_{\max}} \mu_{nml} \left(h_{nm} - h_{nl} \right) \right] \tag{5.207}$$

will be next rewritten for each velocity field in a form that allows the use of implicit integration schemes. The conditions governing heat and mass transfer can change during the time step, thereby influencing the averaged rate for the transported quantity. This feedback effect during a single time step may be crucial in the case of a number of applications associated with strong heat and mass transfer processes. The first work known to this author formalizing source terms is those by *Solbrig* et al. [35] published in 1977. *Solbrig* et al. considered mass and energy equations in a closed system for two single component phases. The meta-stable state of the phases was not allowed in their work. Instead the pressure depending mass sources are defined adjusting the state of each phase to the saturation with the changing pressure. But even the meta-stable state is the driving force for spontaneous evaporation and condensation, and therefore has to be taken into account as we do in our analysis.

The following definitions and assumptions are used here:

$$c_{pw2} = \sum_{i=1}^{i_{max}} C_{iw2} c_{piw2}, \tag{5.208}$$

$$c_{pw3} = \sum_{i=1}^{i_{max}} C_{iw3} c_{piw3}, \tag{5.209}$$

$$c_{p2} = \sum_{i=1}^{i_{max}} C_{i2} c_{pi2}, \tag{5.210}$$

$$c_{p3} = \sum_{i=1}^{i_{max}} C_{i3} c_{pi3}, \tag{5.211}$$

$$h_{M1}^{2\sigma} - h_{M1} \approx c_{pM1}\left(T_{M1}^{2\sigma} - T_1\right) \text{ and } p_{M1}^{2\sigma} \approx p_{M1}, \tag{5.212}$$

$$h_{M1}^{3\sigma} - h_{M1} \approx c_{pM1}\left(T_{M1}^{3\sigma} - T_1\right) \text{ and } p_{M1}^{3\sigma} \approx p_{M1}, \tag{5.213}$$

$$h_{M2}^{1\sigma} - h_{M2} \approx c_{pM2}\left(T_2^{1\sigma} - T_2\right), \tag{5.214}$$

$$h_{M3}^{1\sigma} - h_{M3} \approx c_{pM3}\left(T_3^{1\sigma} - T_3\right), \tag{5.215}$$

$$C_{M2}\left(h_{M2} - h_{M3}\right) + \sum_{n=1}^{n_{max}} C_{n2}\left(h_{n2} - h_{n3}\right) \approx c_{p2}\left(T_2 - T_3\right), \tag{5.216}$$

$$C_{M3}\left(h_{M3} - h_{M2}\right) + \sum_{n=1}^{n_{max}} C_{n3}\left(h_{n3} - h_{n2}\right) \approx c_{p3}\left(T_3 - T_2\right), \tag{5.217}$$

$$\sum_{i=1}^{i_{max}} C_{i2}\left(h_{iw2} - h_{i2}\right) \approx c_{pw2}\left(T_w - T_2\right), \tag{5.218}$$

5.16 Linearized form of the source term for the temperature equation

$$\sum_{i=1}^{i_{max}} C_{i3}\left(h_{iw3} - h_{i3}\right) \approx c_{pw3}\left(T_w - T_3\right). \qquad (5.219)$$

It is assumed that the properties of the mass leaving the velocity field are equal to the properties of this donor field. Here I introduce the product of the effective heat transfer coefficient and the interfacial area density and designate this by $\chi_l^{m\sigma}$. The subscript l designates the location inside the velocity field l and the superscript $m\sigma$ designates the location at the interface σ dividing field m from field l. Superscripts are only used if the interfacial heat transfer is associated with mass transfer. If there is heat transfer only, the linearized interaction coefficient is assigned the subscript ml only, this indicating the interface at which the heat transfer takes place. These rules are valid also for the wall. Actually, the wall is treated as an additional field. The result is

$$DT_1^N = \alpha_1\rho_1\left(\delta_1 P_{k1}^\tau + \varepsilon_1\right) + \chi_1^{w\sigma}\left(T_1^{w\sigma} - T_1\right) + \left(\mu_{21}^w + \mu_{31}^w\right)c_{pM1}\left(T_{M1}^{w\sigma} - T_1\right)$$

$$+ \chi_{w1}\left(T_w - T_1\right) + \chi_{21}\left(T_2 - T_1\right) + \chi_{31}\left(T_3 - T_1\right) + \chi_1^{2\sigma}\left(T_1^{2\sigma} - T_1\right) + \chi_1^{3\sigma}\left(T_1^{3\sigma} - T_1\right)$$

$$+ \sum_{i=1}^{i_{max}}\mu_{iw1}\left(h_{iw1} - h_{i1}\right) + c_{pM1}\left[\mu_{21}\left(T_1^{2\sigma} - T_1\right) + \mu_{31}\left(T_1^{3\sigma} - T_1\right)\right], \qquad (5.220)$$

$$DT_2^N = \alpha_2\rho_2\left(\delta_2 P_{k2}^\tau + \varepsilon_2\right) + \chi_2^{w\sigma}\left(T_2^{w\sigma} - T_2\right) + \mu_{12}^w c_{pM2}\left(T_2^{w\sigma} - T_2\right)$$

$$+ \left(\chi_{w2} + \mu_{w2}c_{pw2}\right)\left(T_w - T_2\right) + \left(\chi_{32} + \mu_{32}c_{p3}\right)\left(T_3 - T_2\right) - \chi_{21}\left(T_2 - T_1\right)$$

$$+ \chi_2^{1\sigma}\left(T_2^{1\sigma} - T_2\right) + \mu_{12}c_{pM2}\left(T_2^{1\sigma} - T_2\right), \qquad (5.221)$$

$$DT_3^N = \alpha_3\rho_3\left(\delta_3 P_{k3}^\tau + \varepsilon_3\right) + \chi_3^{w\sigma}\left(T_3^{w\sigma} - T_3\right) + \mu_{13}^w c_{pM3}\left(T_3^{w\sigma} - T_3\right)$$

$$+ \left(\chi_{w3} + \mu_{w3}c_{pw3}\right)\left(T_w - T_3\right) - \left(\chi_{32} + \mu_{23}c_{p2}\right)\left(T_3 - T_2\right) - \chi_{31}\left(T_3 - T_1\right)$$

$$+ \chi_3^{1\sigma}\left(T_3^{1\sigma} - T_3\right) + \mu_{13}c_{pM3}\left(T_3^{1\sigma} - T_3\right). \qquad (5.222)$$

The energy jump condition at the interfaces yields the following for the *condensation* sources

$$\mu_{12} = \psi_{12} \frac{\chi_2^{1\sigma}\left(T_2^{1\sigma}-T_2\right)+\chi_1^{2\sigma}\left(T_1^{2\sigma}-T_1\right)}{h_{M1}^{2\sigma}-h_{M2}^{1\sigma}} \geq 0, \qquad (5.223)$$

$$\mu_{13} = \psi_{13} \frac{\chi_3^{1\sigma}\left(T_3^{1\sigma}-T_3\right)+\chi_1^{3\sigma}\left(T_1^{3\sigma}-T_1\right)}{h_{M1}^{3\sigma}-h_{M3}^{1\sigma}} \geq 0, \qquad (5.224)$$

for $C_{M1} > 0$. Here the integer switches

$$\psi_{12} = \frac{1}{2}\left\{1+\text{sign}\left[\chi_2^{1\sigma}\left(T_2^{1\sigma}-T_2\right)+\chi_1^{2\sigma}\left(T_1^{2\sigma}-T_1\right)\right]\right\}, \qquad (5.225)$$

$$\psi_{13} = \frac{1}{2}\left\{1+\text{sign}\left[\chi_3^{1\sigma}\left(T_3^{1\sigma}-T_3\right)+\chi_1^{3\sigma}\left(T_1^{3\sigma}-T_1\right)\right]\right\} \qquad (5.226)$$

guarantee that the mass sources are non-negative. For $C_{M1} = 0$ we have $\mu_{12} = 0$, and $\mu_{13} = 0$ because there is nothing to condense. The energy jump condition at the interfaces yields the following for the *evaporation* sources:

$$\mu_{21} = -\left(1-\psi_{12}\right)\frac{\chi_2^{1\sigma}\left(T_2^{1\sigma}-T_2\right)+\chi_1^{2\sigma}\left(T_1^{2\sigma}-T_1\right)}{h_{M1}^{2\sigma}-h_{M2}^{1\sigma}} \geq 0, \text{ for } C_{M2} > 0, \quad (5.227)$$

$$\mu_{31} = -\left(1-\psi_{13}\right)\frac{\chi_3^{1\sigma}\left(T_3^{1\sigma}-T_3\right)+\chi_1^{3\sigma}\left(T_1^{3\sigma}-T_1\right)}{h_{M1}^{2\sigma}-h_{M3}^{1\sigma}} \geq 0, \text{ for } C_{M3} > 0 \quad (5.228)$$

For $C_{M2} = 0$ we have $\mu_{21} = 0$ because there is nothing to evaporate. Similarly for $C_{M3} = 0$ we have $\mu_{31} = 0$.

The mass transfer at heated or cooled walls is defined similarly:

Condensation:

$$\mu_{12}^w = \psi_{12}^w f_{cond\to film}\frac{\chi_w^{1\sigma}\left(T_w^{1\sigma}-T_w\right)+\chi_1^{w\sigma}\left(T_1^{w\sigma}-T_1\right)}{h_{M1}^{2\sigma}-h_{M2}^{1\sigma}} \geq 0, \qquad (5.229)$$

$$\mu_{13}^w = \psi_{13}^w \left(1-f_{cond\to film}\right)\frac{\chi_w^{1\sigma}\left(T_w^{1\sigma}-T_w\right)+\chi_1^{w\sigma}\left(T_1^{w\sigma}-T_1\right)}{h_{M1}^{3\sigma}-h_{M3}^{1\sigma}} \geq 0. \qquad (5.230)$$

5.16 Linearized form of the source term for the temperature equation

Here we introduce the factor $f_{cond \to film}$ which determines how much of the condensed vapor is going into the film. For $f_{cond \to film} = 1$ we have film condensation at the wall. For $f_{cond \to film} = 0$ we have droplet condensation at the wall. The decision which of the mechanisms is active depends on the wetability of the surface and its orientation in space.

Evaporation:

$$\mu_{21}^w = -\psi_{21}^w \frac{\chi_w^{2\sigma}\left(T_w^{2\sigma} - T_w\right) + \chi_2^{w\sigma}\left(T_2^{w\sigma} - T_2\right)}{h_{M1}^{2\sigma} - h_{M2}^{1\sigma}}$$

$$= -\psi_{21}^w \frac{\chi_w^{2\sigma}\left[T'(p) - T_w\right] + \chi_2^{w\sigma}\left[T'(p) - T_2\right]}{h''(p) - h'(p)} \geq 0, \text{ for } C_{M2} > 0, \quad (5.231)$$

$$\mu_{31}^w = -\psi_{31}^w \frac{\chi_w^{3\sigma}\left(T_w^{3\sigma} - T_w\right) + \chi_3^{w\sigma}\left(T_3^{w\sigma} - T_3\right)}{h_{M1}^{3\sigma} - h_{M3}^{1\sigma}}$$

$$= -\psi_{31}^w \frac{\chi_w^{3\sigma}\left[T'(p) - T_w\right] + \chi_3^{w\sigma}\left[T'(p) - T_3\right]}{h''(p) - h'(p)} \geq 0, \text{ for } C_{M3} > 0. \quad (5.232)$$

In deriving Eqs. (5.231) and (5.232) the following assumptions are made

$$T_w^{2\sigma}, T_2^{w\sigma}, T_w^{3\sigma}, T_3^{w\sigma} = T'(p),$$

$$h_{M1}^{2\sigma}, h_{M1}^{3\sigma} = h''(p),$$

and

$$h_{M2}^{1\sigma}, h_{M3}^{1\sigma} = h'(p).$$

Here again the integer switches

$$\psi_{21}^w = \frac{1}{2}\left\{1 - sign\left(\chi_w^{2\sigma}\left[T'(p) - T_w\right] + \chi_2^{w\sigma}\left[T'(p) - T_2\right]\right)\right\}, \quad (5.233)$$

$$\psi_{31}^w = \frac{1}{2}\left\{1 - sign\left(\chi_w^{3\sigma}\left[T'(p) - T_w\right] + \chi_3^{w\sigma}\left[T'(p) - T_3\right]\right)\right\}, \quad (5.234)$$

$$\psi_{12}^{w} = \frac{1}{2}\left\{1 + sign\left[\chi_{w}^{1\sigma}\left(T_{w}^{1\sigma} - T_{w}\right) + \chi_{1}^{w\sigma}\left(T_{1}^{w\sigma} - T_{1}\right)\right]\right\}, \tag{5.235}$$

and conditions for film condensation otherwise $\psi_{12}^{w} = 0$,

$$\psi_{13}^{w} = \frac{1}{2}\left\{1 + sign\left[\chi_{w}^{1\sigma}\left(T_{w}^{1\sigma} - T_{w}\right) + \chi_{1}^{w\sigma}\left(T_{1}^{w\sigma} - T_{1}\right)\right]\right\} \tag{5.236}$$

and conditions for droplet condensation otherwise $\psi_{13}^{w} = 0$, guarantee that the mass sources are non-negative.

Substituting for the evaporation and condensation mass sources one obtains

$$DT_{1}^{N} = \alpha_{1}\rho_{1}\left(\delta_{1}P_{k1}^{\tau} + \varepsilon_{1}\right) + \chi_{1}^{w\sigma}\left(T_{1}^{w\sigma} - T_{1}\right)$$

$$-\begin{pmatrix}\psi_{21}^{w}\left\{\chi_{w}^{2\sigma}\left[T'(p) - T_{w}\right] + \chi_{2}^{w\sigma}\left[T'(p) - T_{2}\right]\right\} \\ +\psi_{31}^{w}\left\{\chi_{w}^{3\sigma}\left[T'(p) - T_{w}\right] + \chi_{3}^{w\sigma}\left[T'(p) - T_{3}\right]\right\}\end{pmatrix}\frac{c_{pM1}}{h''(p) - h'(p)}\left(T_{1}^{w\sigma} - T_{1}\right)$$

$$+\chi_{w1}\left(T_{w} - T_{1}\right) + \chi_{21}\left(T_{2} - T_{1}\right) + \chi_{31}\left(T_{3} - T_{1}\right) + \sum_{i=1}^{i_{\max}}\mu_{iw1}\left(h_{iw1} - h_{i1}\right)$$

$$+\chi_{1}^{2\sigma}\left(T_{1}^{2\sigma} - T_{1}\right) - \left(1 - \psi_{12}^{w}\right)\left[\chi_{2}^{1\sigma}\left(T_{2}^{1\sigma} - T_{2}\right) + \chi_{1}^{2\sigma}\left(T_{1}^{2\sigma} - T_{1}\right)\right]\frac{c_{pM1}\left(T_{1}^{2\sigma} - T_{1}\right)}{h_{M1}^{2\sigma} - h_{M2}^{1\sigma}}$$

$$+\chi_{1}^{3\sigma}\left(T_{1}^{3\sigma} - T_{1}\right) - \left(1 - \psi_{13}^{w}\right)\left[\chi_{3}^{1\sigma}\left(T_{3}^{1\sigma} - T_{3}\right) + \chi_{1}^{3\sigma}\left(T_{1}^{3\sigma} - T_{1}\right)\right]\frac{c_{pM1}\left(T_{1}^{3\sigma} - T_{1}\right)}{h_{M1}^{3\sigma} - h_{M3}^{1\sigma}}, \tag{5.237}$$

$$DT_{2}^{N} = \alpha_{2}\rho_{2}\left(\delta_{2}P_{k2}^{\tau} + \varepsilon_{2}\right) + \chi_{2}^{w\sigma}\left(T_{2}^{w\sigma} - T_{2}\right)$$

$$+\psi_{12}^{w}f_{cond \to film}\left[\chi_{w}^{1\sigma}\left(T_{w}^{1\sigma} - T_{w}\right) + \chi_{1}^{w\sigma}\left(T_{1}^{w\sigma} - T_{1}\right)\right]\frac{c_{pM2}}{h_{M1}^{2\sigma} - h_{M2}^{1\sigma}}\left(T_{2}^{w\sigma} - T_{2}\right)$$

$$+\left(\chi_{w2} + \mu_{w2}c_{pw2}\right)\left(T_{w} - T_{2}\right) + \left(\chi_{32} + \mu_{32}c_{p3}\right)\left(T_{3} - T_{2}\right) - \chi_{21}\left(T_{2} - T_{1}\right)$$

$$+\chi_2^{1\sigma}\left(T_2^{1\sigma}-T_2\right)+\psi_{12}\left[\chi_2^{1\sigma}\left(T_2^{1\sigma}-T_2\right)+\chi_1^{2\sigma}\left(T_1^{2\sigma}-T_1\right)\right]\frac{c_{pM2}\left(T_2^{1\sigma}-T_2\right)}{h_{M1}^{2\sigma}-h_{M2}^{1\sigma}},$$

(5.238)

$$DT_3^N = \alpha_3\rho_3\left(\delta_3 P_{k3}^{\tau}+\varepsilon_3\right)+\chi_3^{w\sigma}\left(T_3^{w\sigma}-T_3\right)$$

$$+\psi_{13}^w\left(1-f_{cond\to film}\right)\left[\chi_w^{1\sigma}\left(T_w^{1\sigma}-T_w\right)+\chi_1^{w\sigma}\left(T_1^{w\sigma}-T_1\right)\right]\frac{c_{pM3}}{h_{M1}^{3\sigma}-h_{M3}^{1\sigma}}\left(T_3^{w\sigma}-T_3\right)$$

$$+\left(\chi_{w3}+\mu_{w3}c_{pw3}\right)\left(T_w-T_3\right)-\left(\chi_{32}+\mu_{23}c_{p2}\right)\left(T_3-T_2\right)-\chi_{31}\left(T_3-T_1\right)$$

$$+\chi_3^{1\sigma}\left(T_3^{1\sigma}-T_3\right)+\psi_{13}\left[\chi_3^{1\sigma}\left(T_3^{1\sigma}-T_3\right)+\chi_1^{3\sigma}\left(T_1^{3\sigma}-T_1\right)\right]\frac{c_{pM3}\left(T_3^{1\sigma}-T_3\right)}{h_{M1}^{3\sigma}-h_{M3}^{1\sigma}}.$$

(5.239)

The above expressions can be rewritten in the compact form

$$DT_l^N = c_l^T - \sum_{k=1}^{3} a_{lk}^{*T} T_k \qquad (5.240)$$

where

$$c_1^T = \alpha_1\rho_1\left(\delta_1 P_{k1}^{\tau}+\varepsilon_1\right)+\sum_{i=1}^{i_{max}}\mu_{iw1}\left(h_{iw1}-h_{i1}\right)+\chi_{w1}T_w$$

$$+\chi_1^{w\sigma}T_1^{w\sigma} - \frac{\left[\psi_{21}^w\left(\chi_w^{2\sigma}+\chi_2^{w\sigma}\right)+\psi_{31}^w\left(\chi_w^{3\sigma}+\chi_3^{w\sigma}\right)\right]T'(p)}{-\left(\psi_{21}^w\chi_w^{2\sigma}+\psi_{31}^w\chi_w^{3\sigma}\right)T_w}\frac{c_{pM1}}{h''(p)-h'(p)}T_1^{w\sigma}$$

$$+\left[\chi_1^{2\sigma}-\frac{(1-\psi_{12})c_{pM1}}{h_{M1}^{2\sigma}-h_{M2}^{1\sigma}}\left(\chi_2^{1\sigma}T_2^{1\sigma}+\chi_1^{2\sigma}T_1^{2\sigma}\right)\right]T_1^{2\sigma}$$

$$+\left[\chi_1^{3\sigma}-\frac{(1-\psi_{13})c_{pM1}}{h_{M1}^{3\sigma}-h_{M3}^{1\sigma}}\left(\chi_3^{1\sigma}T_3^{1\sigma}+\chi_1^{3\sigma}T_1^{3\sigma}\right)\right]T_1^{3\sigma}, \qquad (5.241)$$

$$a_{11}^{*T} = +\chi_1^{w\sigma} + \chi_{21} + \chi_{31} + \chi_1^{2\sigma} + \chi_1^{3\sigma} + \chi_{w1}$$

$$-\left(\begin{array}{l}\psi_{21}^w\left\{\chi_w^{2\sigma}\left[T'(p)-T_w\right]+\chi_2^{w\sigma}\left[T'(p)-T_2\right]\right\} \\ +\psi_{31}^w\left\{\chi_w^{3\sigma}\left[T'(p)-T_w\right]+\chi_3^{w\sigma}\left[T'(p)-T_3\right]\right\}\end{array}\right)\frac{c_{pM1}}{h''(p)-h'(p)}$$

$$-\frac{\chi_2^{1\sigma}(1-\psi_{12})c_{pM1}}{h_{M1}^{2\sigma}-h_{M2}^{1\sigma}}\left(T_2^{1\sigma}-T_2\right)+\frac{\chi_1^{2\sigma}(1-\psi_{12})c_{pM1}}{h_{M1}^{2\sigma}-h_{M2}^{1\sigma}}\left(T_1-2T_1^{2\sigma}\right)$$

$$-\frac{\chi_3^{1\sigma}(1-\psi_{13})c_{pM1}}{h_{M1}^{3\sigma}-h_{M3}^{1\sigma}}\left(T_3^{1\sigma}-T_3\right)+\frac{\chi_1^{3\sigma}(1-\psi_{13})c_{pM1}}{h_{M1}^{3\sigma}-h_{M3}^{1\sigma}}\left(T_1-2T_1^{3\sigma}\right), \quad (5.242)$$

$$a_{12}^{*T} = -\chi_{21} - \frac{\chi_2^{1\sigma}(1-\psi_{12})c_{pM1}}{h_{M1}^{2\sigma}-h_{M2}^{1\sigma}}T_1^{2\sigma} - \frac{\psi_{21}^w\chi_2^{w\sigma}c_{pM1}}{h''(p)-h'(p)}T_1^{w\sigma}, \quad (5.243)$$

$$a_{13}^{*T} = -\chi_{31} - \frac{\chi_3^{1\sigma}(1-\psi_{13})c_{pM1}}{h_{M1}^{3\sigma}-h_{M3}^{1\sigma}}T_1^{3\sigma} - \frac{\psi_{31}^w\chi_3^{w\sigma}c_{pM1}}{h''(p)-h'(p)}T_1^{w\sigma}, \quad (5.244)$$

$$c_2^T = \alpha_2\rho_2\left(\delta_2 P_{k2}^\tau + \varepsilon_2\right) + \chi_2^{w\sigma}T_2^{w\sigma} + \chi_2^{1\sigma}T_2^{1\sigma} + \left(\chi_{w2}+\mu_{w2}c_{pw2}\right)T_w$$

$$+\left\{\begin{array}{l}\psi_{12}^w f_{cond\to film}\left[\chi_w^{1\sigma}\left(T_w^{1\sigma}-T_w\right)+\chi_1^{w\sigma}T_1^{w\sigma}\right]T_2^{w\sigma} \\ +\psi_{12}\left[\chi_2^{1\sigma}T_2^{1\sigma}+\chi_1^{2\sigma}T_1^{2\sigma}\right]T_2^{1\sigma}\end{array}\right\}\frac{c_{pM2}}{h_{M1}^{2\sigma}-h_{M2}^{1\sigma}}, \quad (5.245)$$

$$a_{21}^{*T} = \left(\psi_{12}^w f_{cond\to film}\chi_1^{w\sigma}T_2^{w\sigma}+\psi_{12}\chi_1^{2\sigma}T_2^{1\sigma}\right)\frac{c_{pM2}}{h_{M1}^{2\sigma}-h_{M2}^{1\sigma}}-\chi_{21}, \quad (5.246)$$

$$a_{22}^{*T} = \chi_2^{w\sigma} + \chi_2^{1\sigma} + \chi_{w2} + \mu_{w2}c_{pw2} + \chi_{32} + \mu_{32}c_{p3} + \chi_{21}$$

$$+\begin{Bmatrix} \psi_{12}^w f_{cond \to film} \left[\chi_w^{1\sigma} \left(T_w^{1\sigma} - T_w \right) + \chi_1^{w\sigma} \left(T_1^{w\sigma} - T_1 \right) \right] \\ + \psi_{12} \left[\chi_2^{1\sigma} \left(2T_2^{1\sigma} - T_2 \right) + \chi_1^{2\sigma} \left(T_1^{2\sigma} - T_1 \right) \right] \end{Bmatrix} \frac{c_{pM2}}{h_{M1}^{2\sigma} - h_{M2}^{1\sigma}}, \quad (5.247)$$

$$a_{23}^{*T} = -\chi_{32} - \mu_{32} c_{p3}, \quad (5.248)$$

$$c_3^T = \alpha_3 \rho_3 \left(\delta_3 P_{k3}^r + \varepsilon_3 \right) + \chi_3^{w\sigma} T_3^{w\sigma} + \chi_3^{1\sigma} T_3^{1\sigma} + \left(\chi_{w3} + \mu_{w3} c_{pw3} \right) T_w$$

$$+\begin{Bmatrix} \psi_{13}^w \left(1 - f_{cond \to film} \right) \left[\chi_w^{1\sigma} \left(T_w^{1\sigma} - T_w \right) + \chi_1^{w\sigma} T_1^{w\sigma} \right] T_3^{w\sigma} \\ + \psi_{13} \left(\chi_3^{1\sigma} T_3^{1\sigma} + \chi_1^{3\sigma} T_1^{3\sigma} \right) T_3^{1\sigma} \end{Bmatrix} \frac{c_{pM3}}{h_{M1}^{3\sigma} - h_{M3}^{1\sigma}}, \quad (5.249)$$

$$a_{31}^{*T} = \left[\psi_{13}^w \left(1 - f_{cond \to film} \right) \chi_1^{w\sigma} T_3^{w\sigma} + \psi_{13} \chi_1^{3\sigma} T_3^{1\sigma} \right] \frac{c_{pM3}}{h_{M1}^{3\sigma} - h_{M3}^{1\sigma}} - \chi_{31}, \quad (5.250)$$

$$a_{32}^{*T} = -\chi_{32} - \mu_{23} c_{p2}, \quad (5.251)$$

$$a_{33}^{*T} = \chi_3^{w\sigma} + \chi_3^{1\sigma} + \chi_{w3} + \mu_{w3} c_{pw3} + \chi_{32} + \mu_{23} c_{p2} + \chi_{31}$$

$$+\begin{Bmatrix} \psi_{13}^w \left(1 - f_{cond \to film} \right) \left[\chi_w^{1\sigma} \left(T_w^{1\sigma} - T_w \right) + \chi_1^{w\sigma} \left(T_1^{w\sigma} - T_1 \right) \right] \\ + \psi_{13} \left[\chi_3^{1\sigma} \left(2T_3^{1\sigma} - T_3 \right) + \chi_1^{3\sigma} \left(T_1^{3\sigma} - T_1 \right) \right] \end{Bmatrix} \frac{c_{pM3}}{h_{M1}^{3\sigma} - h_{M3}^{1\sigma}}, \quad (5.252)$$

5.17 Interface conditions

Steam only: For the case of steam only in the gas field, i.e. $C_{M1} = 1$, one has

$$T_3^{1\sigma} = T_2^{1\sigma} = T_1^{2\sigma} = T_1^{3\sigma} = T'(p) \approx T'(p_a) + \frac{dT'}{dp}(p - p_a)$$

280   5 First and second laws of the thermodynamics

$$\approx T'(p_a) + \frac{dT'}{dp}\frac{dp}{d\tau}\Delta\tau, \tag{5.253}$$

$$h_{M3}^{1\sigma} = h_{M2}^{1\sigma} = h'(p), \tag{5.254}$$

$$h_{M1}^{2\sigma} = h_{M1}^{3\sigma} = h''(p). \tag{5.255}$$

Here p_a is the reference pressure for the previous time step. Note that the pressure p can change during the time interval considered, which can reduce or even stop the condensation and promote evaporation, for example.

Flashing: For the case of spontaneous evaporation of the water, e.g. from the second velocity field,

$$T_2 > \left(1 + \frac{\chi_1^{2\sigma}}{\chi_2^{1\sigma}}\right)T'(p) - \frac{\chi_1^{2\sigma}}{\chi_2^{1\sigma}}T_1 \tag{5.256}$$

one has

$$T_2^{1\sigma} = T_1^{2\sigma} = T'(p) \approx T'(p_a) + \frac{dT'}{dp}(p - p_a), \tag{5.257}$$

$$h_{M1}^{2\sigma} = h''(p). \tag{5.258}$$

Similarly, for the case of spontaneous evaporation of the water from the third velocity field,

$$T_3 > \left(1 + \frac{\chi_1^{3\sigma}}{\chi_3^{1\sigma}}\right)T'(p) - \frac{\chi_1^{3\sigma}}{\chi_3^{1\sigma}}T_1 \tag{5.259}$$

one has

$$T_3^{1\sigma} = T_1^{3\sigma} = T'(p) \approx T'(p_a) + \frac{dT'}{dp}(p - p_a), \tag{5.260}$$

$$h_{M1}^{3\sigma} = h''(p). \tag{5.261}$$

Boiling at the wall: If boiling conditions are in force at the wall one has

$$T_{M1}^{w\sigma} = T'(p) \approx T'(p_a) + \frac{dT'}{dp}(p - p_a). \qquad (5.262)$$

Non-condensibles present in the gas field: This case is much more complicated than the single component case, especially for diffusion-controlled interfacial mass transfer.

The liquid side interface temperature during condensation processes is the saturation temperature at the local partial steam pressure

$$T_3^{1\sigma} = T_2^{1\sigma} = T'(p_{M1}) = T'(p_{M1a}) + \frac{dT'}{dp}(p_{M1} - p_{M1a}). \qquad (5.263)$$

5.18 Lumped parameter volumes

If in a practical application a good mixing in any time within a control volume can be considered as a good approximation the energy conservation equations in any of their variants simplify considerable. All convection and diffusion terms disappear. What remains are the time derivatives and the source terms. We summarize the results of this section for this particular case in order to make easy practical applications. Thus, we have for the entropy, specific internal energy, the temperature and the specific enthalpy equations the following result:

$$\rho_l \alpha_l \frac{ds_l}{d\tau} = \frac{1}{T_l} DT_l^N + \sum_{i=1}^{i_{\max}} \mu_{il}(s_{il} - s_l) \quad \text{for } \alpha_l \geq 0, \qquad (5.264)$$

$$\frac{d}{d\tau}(\alpha_l \rho_l e_l \gamma_v) + p_l \frac{d}{d\tau}(\alpha_l \gamma_v) = \gamma_v De_l^*, \qquad (5.265)$$

$$\rho_l c_{pl} \alpha_l \frac{dT_l}{d\tau} - \left[1 - \rho_l \left(\frac{\partial h_l}{\partial p}\right)_{T_l, all\_C's}\right] \alpha_l \frac{dp}{d\tau}$$

$$= DT_l^N - T_l \sum_{i=2}^{i_{\max}} \Delta s_{il}^{np}(\mu_{il} - C_{il}\mu_l), \qquad (5.266)$$

$$\frac{d}{d\tau}(\alpha_l \rho_l h_l \gamma_v) - \alpha_l \gamma_v \frac{dp_l}{d\tau} = \gamma_v DT_l^N + \gamma_v \sum_{\substack{m=1 \\ m \neq l}}^{3,w} \sum_{i=1}^{i_{\max}}(\mu_{iml} - \mu_{ilm})h_{il}. \qquad (5.267)$$

The volume conservation equation divided by the volumetric porosity for this case reduces to

$$\frac{1}{\rho a^2}\frac{dp}{d\tau} = \sum_{l=1}^{l_{max}} D\alpha_l - \frac{d}{d\tau}\ln\gamma_v, \qquad (5.268)$$

where

$$D\alpha_l = \frac{\mu_l}{\rho_l} - \frac{1}{\rho_L^2}\left[\left(\frac{\partial \rho_l}{\partial s_l}\right)_{p,all\_C'_{li}s}\left[\frac{1}{T_l}DT_l^N + \sum_{i=1}^{i_{max}}\mu_{il}\left(s_{il}-s_l\right)\right] + \sum_{i=2}^{i_{max}}\left(\frac{\partial \rho_l}{\partial C_{li}}\right)_{p,s,all\_C'_{li}s\_except\_C_{l1}}\left(\mu_{il}-C_{il}\mu_l\right)\right]. \qquad (5.269)$$

5.19 Final remarks

It should be emphasized that in the temperature equation, just as in the energy equation, in the enthalpy equation and in the internal energy equations, the specific enthalpy occurs in the mass transfer terms. In the entropy equations the interfacial mass source terms are associated with specific entropies.

The most important result of this chapter is the rigorous derivation of the entropy equation, which reflects the second law of thermodynamics for multi-phase systems consisting of several chemical components which are conditionally divided into three velocity fields. It was shown that the use of specific entropy, introduced by *Clausius* and recognized by *Gibbs* as a very important quantity more than one hundred years ago, provides the simplest method for modern description and modeling of complicated flow systems. The experience gained by the author of this work in the development of the various versions of the IVA computer code, [7-27], which are based on the entropy concept allows for the recommendation of the local volume and time average entropy equation for general use in multi-phase flow dynamics.

References

1. Bird BR, Stewart WE, Lightfoot E N (1960) Transport Phenomena, John Wiley & Sons, New York, Chichester, Brisbane, Toronto, Singapore

2. Bohl WR et al (1988) Multiphase flow in the advanced fluid dynamics model, ANS Proc. 1988 Nat. Heat Transfer Conf., HTC-vol 3 July 24-27, Houston, Texas, pp 61-70
3. Fick A (1855) Über Diffusion. Ann. der Physik, vol 94 p 59
4. Grigorieva VA, Zorina VM, eds. (1988) Handbook of thermal engineering, thermal engineering experiment, in Russian, 2d Edition, Moskva, Atomisdat, vol 2
5. Hammond GP (Nov. 1985) Turbulent Prandtl number within a near-wall flow, AIAA Journal, vol 23 no 11 pp 1668-1669
6. Kelly JM, Kohrt R J (1983) COBRA-TF: Flow blockage heat transfer program. Proc. Eleventh Water Reactor Safety Research Information Meeting (Oct. 24-28 1983) Gaithersbury - Maryland, NUREG/CP-0048, vol 1 pp 209-232
7. Kolev NI (March 1985) Transiente Drephasen Dreikomponenten Stroemung, Teil 1: Formulierung des Differentialgleichungssystems, KfK 3910
8. Kolev NI (1986) Transient three-dimensional three-phase three-component nonequilibrium flow in porous bodies described by three-velocity fields, Kernenergie, vol 29 no 10 pp 383-392
9. Kolev NI (1986) Transiente Dreiphasen Dreikomponenten Stroemung, Part 3: 3D-Dreifluid-Diffusionsmodell, KfK 4080
10. Kolev NI (August 1987) A Three Field-Diffusion Model of Three-Phase, Three-Component Flow for the Transient 3D-Computer Code IVA2/01, Nuclear Technology, vol 78 pp 95-131
11. Kolev NI (Sept. 1991) A three-field model of transient 3D multi-phase, three-component flow for the computer code IVA3, Part 1: Theoretical basics: Conservation and state equations, numerics. KfK 4948 Kernforschungszentrum Karlsruhe
12. Kolev NI (1991) IVA3: A transient 3D three-phase, three-component flow analyzer, Proc. of the Int. Top. Meeting on Safety of Thermal Reactors, Portland, Oregon, July 21-25, 1991, pp 171-180. The same paper was presented at the 7th Meeting of the IAHR Working Group on Advanced Nuclear Reactor Thermal-Hydraulics, Kernforschungszentrum Karlsruhe, August 27 to 29, 1991
13. Kolev NI (1993) Berechnung der Fluiddynamischen Vorgänge bei einem Sperrwasser-kühlerrohrbruch, Projekt KKW Emsland, Siemens KWU Report R232/93/0002
14. Kolev NI (1993) IVA3-NW A three phase flow network analyzer. Input description, Siemens KWU Report R232/93/E0041
15. Kolev NI (1993) IVA3-NW components: Relief valves, pumps, heat structures, Siemens KWU Report R232/93/E0050.
16. Kolev NI (1993) The code IVA3 for modelling of transient three-phase flows in complicated 3D geometry, Kerntechnik, vol 58 no 3 pp 147-156
17. Kolev NI (1993) IVA3 NW: Computer code for modelling of transient three phase flow in complicated 3D geometry connected with industrial networks, Proc. of the Sixth Int. Top. Meeting on Nuclear Reactor Thermal Hydraulics, Oct. 5-8, 1993, Grenoble, France
18. Kolev NI (1994) IVA4: Modelling of mass conservation in multi-phase multi-component flows in heterogeneous porous media, Siemens KWU Report NA-M/94/E029, July 5, 1994 also in Kerntechnik, vol 59 no 4-5 pp 226-237
19. Kolev NI (1994) IVA4: Modelling of momentum conservation in multi-phase flows in heterogeneous porous media, Siemens KWU Report NA-M/94/E030, July 5, 1994, also in Kerntechnik, vol 59 no 6 pp 249-258
20. Kolev NI (1994) The influence of the mutual bubble interaction on the bubble departure diameter, Experimental Thermal and Fluid Science, vol 8 pp 167-174

21. Kolev NI (1996) Three fluid modeling with dynamic fragmentation and coalescence fiction or daily practice? 7th FARO Experts Group Meeting Ispra, October 15-16
22. Kolev NI (1991) Derivatives for the equation of state of multi-component mixtures for universal multi-component flow models, Nuclear Science and Engineering: vol 108 pp 74-87
23. Kolev NI (1995), The code IVA4: Second law of thermodynamics for multi-phase multi-component flows in heterogeneous media, Kerntechnik, vol 60 no 1, pp 1-39
24. Kolev NI (1997) Comments on the entropy concept, Kerntechnik, vol 62 no 1 pp 67-70
25. Kolev NI (1997) Three fluid modeling with dynamic fragmentation and coalescence fiction or daily practice? Proceedings of OECD/CSNI Workshop on Transient thermal-hydraulic and neutronic codes requirements, Annapoliss, Md, U.S.A., 5th-8th November 1996; 4th World Conference on Experimental Heat Transfer, Fluid Mechanics and Thermodynamics, ExHFT 4, Brussels, June 2-6, 1997; ASME Fluids Engineering Conference & Exhibition, The Hyatt Regency Vancouver, Vancouver, British Columbia, CANADA June 22-26, 1997, Invited Paper; Proceedings of 1997 International Seminar on Vapor Explosions and Explosive Eruptions (AMIGO-IMI), May 22-24, Aoba Kinen Kaikan of Tohoku University, Sendai-City, Japan
26. Kolev NI (1998) On the variety of notation of the energy conservation principle for single phase flow, Kerntechnik, vol 63 no 3 pp 145-156
27. Kolev NI (1999) Verification of IVA5 computer code for melt-water interaction analysis, Part 1: Single phase flow, Part 2: Two-phase flow, three-phase flow with cold and hot solid spheres, Part 3: Three-phase flow with dynamic fragmentation and coalescence, Part 4: Three-phase flow with dynamic fragmentation and coalescence – alumna experiments, CD Proceedings of the Ninth International Topical Meeting on Nuclear Reactor Thermal Hydraulics (NURETH-9), San Francisco, California, October 3-8, 1999, Log. Nr. 315.
28. Liles DR et al (June 1978) TRAC-P1: An advanced best estimate computer program for PWR LOCA analysis. I. Methods, Models, User Information and Programming Details, NUREG/CR-0063, LA-7279-MS vol 1
29. Liles DR et al (April 1981) TRAC-FD2 An advanced best-estimate computer program for pressurized water ractor loss-of-coolant accident analysis. NUREG/CR-2054, LA-8709 MS
30. Reid RC, Prausnitz JM, Sherwood TK (1982) The Properties of Gases and Liquids, Third Edition, McGraw-Hill Book Company, New York
31. Sha T, Chao BT, Soo SL (1984) Porous media formulation for multi phase flow with heat transfer, Nuclear Engineering and Design, vol 82 pp 93-106
32. Taylor GI (1935) Proc. Roy. Soc. A, vol 151 p 429
33. Thurgood MJ et al (1983) COBRA/TRAC - A thermal hydraulic code for transient analysis of nuclear reactor vessels and primary coolant systems. NUREG/CR-346, vol 1-5
34. Zierep J (1983) Einige moderne Aspekte der Stroemungsmechanik, Zeitschrift fuer Flugwissenschaften und Weltraumforschung, vol 7 no 6 pp 357-361
35. Solbrig CW, Hocever CH and Huges ED (14-17 August 1977) A model for a heterogeneous two-phase unequal temperature fluid, 17th National heat transfer conference, Salt Lake Sity, Utah, pp 139-151

6 Some simple applications of the mass and energy conservation

This Chapter contains some simple cases illustrating the use of the mass and the energy conservation for describing the thermodynamic state of multi-component single-phase systems. The results are useful themselves or for use as a benchmarks for testing the performance of computer codes.

6.1 Infinite heat exchange without interfacial mass transfer

Consider the simple case of a three-field mixture having constant component mass concentration without wall interaction, without any mass transfer, and with instantaneous heat exchange that equalizes the field temperatures at any moment. Equation (5.176) reduces to

$$\alpha_l \rho_l c_{pl} \frac{dT}{T} - \alpha_l \rho_l \overline{R}_l \frac{dp}{p} = \frac{\dot{q}_l'''}{T}. \tag{6.1}$$

After dividing by the mixture density

$$\rho = \sum_{l=1}^{3} \alpha_l \rho_l, \tag{6.2}$$

and summing all three equations one obtains

$$\frac{dT}{T} = \frac{\overline{R}}{c_p} \frac{dp}{p} = \frac{n-1}{n} \frac{dp}{p}, \tag{6.3}$$

where the mixture specific heat is

$$c_p = \sum_{l=1}^{3} \frac{\alpha_l \rho_l}{\rho} c_{pl}, \tag{6.4}$$

and the effective gas constant is

$$R = \sum_{l=1}^{3} \frac{\alpha_l \rho_l}{\rho} \overline{R}_l \, . \tag{6.5}$$

Note that for the second and the third field the pseudo-gas constant is negligibly small. The polytrophic exponent is

$$n = \frac{c_p}{c_p - R} \, . \tag{6.6}$$

Integration of Eq. (6.3) yields

$$\frac{T}{T_0} = \left(\frac{p}{p_0}\right)^{\frac{n-1}{n}} \, . \tag{6.7}$$

This change of state is associated with considerable entropy change for the gas phase

$$ds_1 = -c_{p1}\left(\frac{1}{n} - \frac{1}{\kappa_1}\right)\frac{dp}{p}, \tag{6.8}$$

and for the other fields

$$ds_l = c_{pl}\frac{n-1}{n}\frac{dp}{p}, \quad \text{for } l = 2,3. \tag{6.9}$$

There are a number of cases where such simplification forms a useful approximation instead of modeling all details of the heat transfer for inert velocity fields. This result was obtained for two velocity field by *Tangren* et al. [28] in 1949. In expanding steady state flow the steady state energy conservation for each field gives

$$\frac{ds_1}{dp}\frac{dp}{dz} = \frac{1}{X_1 G T_1}(\dot{q}_{21}''' + \dot{q}_{31}'''), \tag{6.10}$$

$$\frac{ds_2}{dp}\frac{dp}{dz} = -\frac{1}{X_2 G T_2}\dot{q}_{21}''', \tag{6.11}$$

$$\frac{ds_3}{dp}\frac{dp}{dz} = -\frac{1}{X_3 GT_3}\dot{q}_{31}'''. \tag{6.12}$$

Therefore the transferred thermal power per unit mixture volume from the liquid to the gas necessary to equalize the temperatures is a function of the pressure gradient

$$\dot{q}_{21}''' + \dot{q}_{31}''' = -c_{p1}\left(\frac{1}{n} - \frac{1}{\kappa_1}\right)\frac{X_1 GT_1}{p}\frac{dp}{dz}, \tag{6.13}$$

$$\dot{q}_{21}''' = -X_2 GT_2 \frac{ds_2}{dp}\frac{dp}{dz} = -X_2 GT_2 c_{p2}\frac{n-1}{n}\frac{1}{p}\frac{dp}{dz}, \tag{6.14}$$

$$\dot{q}_{31}''' = -X_3 GT_3 \frac{ds_3}{dp}\frac{dp}{dz} = -X_3 GTc_{p3}\frac{n-1}{n}\frac{1}{p}\frac{dp}{dz}. \tag{6.15}$$

Useful relations are

$$\frac{d\rho_1}{dp} = \frac{\rho_1}{np}, \tag{6.16}$$

the integrated form

$$\frac{\rho_1}{\rho_{10}} = \left(\frac{p}{p_0}\right)^{\frac{1}{n}} \tag{6.17}$$

and

$$\frac{d\rho_1}{ds_1} = -\frac{\rho_1}{nc_{p1}\left(\frac{1}{n} - \frac{1}{\kappa_1}\right)}. \tag{6.18}$$

6.2 Discharge of gas from a volume

Consider a volume V with initial pressure and temperature p_0 and T_0, respectively. The discharged mass per unit time and volume is

$$\mu_w = \rho_{out} w \frac{F}{V}. \tag{6.19}$$

Here F is the flow cross section. The transient behavior of the pressure and temperature is then described by

$$\frac{1}{a^2}\frac{dp}{d\tau} = -\rho_{out} w \frac{F}{V} \qquad (6.20)$$

$$\rho c_p \frac{dT}{d\tau} - \frac{dp}{d\tau} = 0 \quad \text{i.e.} \quad p = p_0 \left(\frac{T}{T_0}\right)^{\frac{\kappa}{\kappa-1}}. \qquad (6.21)$$

Note that for perfect gases the last equation results in the isentropic relation. The discharge velocity and the corresponding densities are

$$w = \sqrt{\kappa RT}\sqrt{\frac{2}{\kappa+1}}, \qquad (6.22)$$

$$\rho_{out} = \rho\left(\frac{2}{\kappa+1}\right)^{\frac{1}{\kappa-1}}, \qquad (6.23)$$

for $\varepsilon \geq \varepsilon^*$, and

$$w^2 = 2\frac{p}{\rho}\frac{\kappa}{\kappa-1}\left(1-\varepsilon^{\frac{\kappa-1}{\kappa}}\right) = \kappa RT\frac{2}{\kappa-1}\left(1-\varepsilon^{\frac{\kappa-1}{\kappa}}\right), \qquad (6.24)$$

$$\rho_{out} = \rho\left[1-\frac{1}{2}(\kappa-1)\frac{w^2}{\kappa RT}\right]^{\frac{1}{\kappa-1}} = \rho\varepsilon^{\frac{1}{\kappa}} \qquad (6.25)$$

for $\varepsilon < \varepsilon^*$, which means for critical and sub-critical flow, respectively. This result was obtained by *de Saint Venant* and *Wantzel* in 1838. Here

$$\varepsilon = \frac{p_{out}}{p} \qquad (6.26)$$

is the pressure ratio and

$$\varepsilon^* = \left(\frac{2}{\kappa+1}\right)^{\frac{\kappa}{\kappa-1}} \qquad (6.27)$$

is the critical pressure ratio – see *Oswatitsch* [25] (1952), *Landau* and *Lifshitz* [19], p. 319. For the case of critical outflow we have

$$\frac{dp}{d\tau} = -\frac{F}{V}\kappa p \left(\frac{2}{\kappa+1}\right)^{\frac{1}{2}\frac{\kappa+1}{\kappa-1}} \sqrt{\kappa RT}, \qquad (6.28)$$

$$\rho c_p \frac{dT}{d\tau} - \frac{dp}{d\tau} = 0. \qquad (6.29)$$

Replacing the pressure derivative in the second equation we obtain

$$\rho c_p \frac{dT}{d\tau} = -\frac{F}{V}\kappa p \left(\frac{2}{\kappa+1}\right)^{\frac{1}{2}\frac{\kappa+1}{\kappa-1}} \sqrt{\kappa RT}, \qquad (6.30)$$

or

$$\frac{1}{T^{3/2}}\frac{dT}{d\tau} = -\frac{F}{V}(\kappa-1)\left(\frac{2}{\kappa+1}\right)^{\frac{1}{2}\frac{\kappa+1}{\kappa-1}} \sqrt{\kappa R}, \qquad (6.31)$$

or after integrating

$$T = \left[\frac{1}{T_0^{1/2}} + \frac{1}{2}\frac{F}{V}(\kappa-1)\left(\frac{2}{\kappa+1}\right)^{\frac{1}{2}\frac{\kappa+1}{\kappa-1}} \sqrt{\kappa R}\ \tau\right]^{-2}. \qquad (6.32)$$

The pressure as a function of time is computed from the isentropic equation as mentioned before.

For the sub-critical case we have

$$\frac{1}{p}\frac{dp}{d\tau} = -\kappa\sqrt{\kappa RT}\left(\frac{2}{\kappa-1}\right)^{1/2} \varepsilon^{\frac{1}{\kappa}}\left(1-\varepsilon^{\frac{\kappa-1}{\kappa}}\right)^{1/2}\frac{F}{V}. \qquad (6.33)$$

Replacing the temperature using the isentropic relation we obtain

$$\frac{dp}{d\tau} = -\kappa p_0 \sqrt{\kappa RT_0}\left(\frac{p_{out}}{p_0}\right)^{\frac{3\kappa-1}{2\kappa}}\left(\frac{2}{\kappa-1}\right)^{1/2} \varepsilon^{\frac{1-3\kappa}{2\kappa}}\left(1-\varepsilon^{\frac{\kappa-1}{\kappa}}\right)^{1/2}\frac{F}{V}, \qquad (6.34)$$

or

$$\frac{d\left(1-\varepsilon^{\frac{\kappa-1}{\kappa}}\right)}{\varepsilon\left(1-\varepsilon^{\frac{\kappa-1}{\kappa}}\right)^{1/2}} = \frac{\kappa^2}{\kappa-1}\left(\frac{2}{\kappa-1}\right)^{1/2} p_0\sqrt{\kappa RT_0}\left(\frac{p_{out}}{p_0}\right)^{\frac{\kappa-1}{2\kappa}} \frac{F}{V} d\tau \ . \qquad (6.35)$$

Here the subscript 0 means an arbitrary state, e.g. the initial state. The equation has to be integrated numerically.

6.3 Injection of inert gas in a closed volume initially filled with inert gas

For the case of inert gas injection with constant mass source density μ_{nw} and constant temperature T_w in a closed volume initially filled with inert gas at temperature and pressure T_0 and p_0, respectively, we have

$$\frac{1}{a^2}\frac{dp}{d\tau} = \mu_{nw} + \frac{1}{Tc_p}\left[\mu_{nw}\left(h_{nw} - h_n\right)\right] \qquad (6.36)$$

$$\rho c_p \frac{dT}{d\tau} - \frac{dp}{d\tau} = \mu_{nw}\left(h_{nw} - h_n\right). \qquad (6.37)$$

Rearranging

$$\frac{dp}{d\tau} = \mu_{nw} a^2 \frac{T_w}{T} \quad \text{or} \quad \frac{dp}{d\tau} = \mu_{nw} \kappa RT_w \quad \text{or} \quad p = p_0 + \mu_{nw}\kappa RT_w\tau \ , \qquad (6.38)$$

$$\rho c_p \frac{dT}{d\tau} - \frac{dp}{d\tau} = \mu_{nw} c_p \left(T_w - T\right) \quad \text{or} \quad \frac{dT}{\kappa T_w T - T^2} = \frac{d\tau}{\frac{p_0}{\mu_{nw}R} + \kappa T_w \tau} \ , \qquad (6.39)$$

and integrating we obtain

$$p = p_0 + \mu_{nw}\kappa RT_w\tau \qquad (6.40)$$

$$T = T_0 \frac{1 + \kappa \frac{\mu_{nw}RT_w}{p_0}\tau}{1 + \frac{\mu_{nw}RT_0}{p_0}\tau} \ , \qquad \frac{T - T_0}{T_0} = \frac{\kappa T_w - T_0}{\frac{p_0}{\mu_{nw}R\tau} + T_0} \ . \qquad (6.41)$$

We see that in this case the pressure increases linearly with time and the temperature tends to an asymptotic for infinite time

$$T(\tau \to \infty) = kT_w. \tag{6.42}$$

6.4 Heat input in a gas in a closed volume

For the case of no chemical reaction but introduction of heat only into the gas mixture we have

$$\frac{dp}{d\tau} = \frac{a^2}{Tc_p} \dot{q}''', \tag{6.43}$$

$$\rho c_p \frac{dT}{d\tau} - \frac{dp}{d\tau} = \dot{q}'''. \tag{6.44}$$

Bearing in mind that

$$a^2 = \kappa RT, \tag{6.45}$$

$$\frac{R}{c_p} = \frac{\kappa - 1}{\kappa}, \tag{6.46}$$

and after rearrangement we obtain

$$\frac{dp}{d\tau} = (\kappa - 1)\dot{q}''', \tag{6.47}$$

$$\frac{d}{d\tau} \ln T = (\kappa - 1)\frac{\dot{q}'''}{p}. \tag{6.48}$$

We recognize that

$$\frac{d}{d\tau} \ln T = \frac{d}{d\tau} \ln p \tag{6.49}$$

results in a polytrophic change of state with polytrophic exponent equal to one,

$$T = \frac{T_0}{p_0} p. \tag{6.50}$$

Integrating over the time $\Delta\tau$ for initial conditions p_0 and T_0 and assuming that the isentropic exponent does not depend on temperature and pressure we finally obtain

$$p = p_0 + (\kappa - 1) \int_0^{\Delta\tau} \dot{q}''' d\tau, \qquad (6.51)$$

$$T = \frac{T_0}{p_0} p = T_0 \left[1 + \frac{\kappa - 1}{p_0} \int_0^{\Delta\tau} \dot{q}''' d\tau \right]. \qquad (6.52)$$

If the heat source term is not a time function we have

$$p = p_0 + (\kappa - 1)\dot{q}'''\Delta\tau, \qquad (6.53)$$

$$T = \frac{T_0}{p_0} p = T_0 \left[1 + \frac{\kappa - 1}{p_0} \dot{q}''' \Delta\tau \right]. \qquad (6.54)$$

6.5 Steam injection in a steam-air mixture

Consider a closed volume filled with air, $C_n = 1$, at initial temperature $T = 273.15 + 20K$ and pressure $p = 1 \times 10^5 Pa$. The volume is adiabatic, $\dot{q}''' = 0$. Inject into this volume steam at a temperature $T_w = 273.15 + 100K$ and intensity $\mu = \mu_M - \mu^+ - const$. There is no mass generation for the inert component (air), that is $\mu_{n1} = 0$. Compute the change in the concentration, pressure and temperature with time. The air concentration in the volume is governed by

$$\frac{dC_n}{d\tau} = -C_n \mu / \rho, \qquad (6.55)$$

or integrating within a time step $\Delta\tau$,

$$C_{n,\tau+\Delta\tau} = C_n / \exp\left(\int_\tau^{\tau+\Delta\tau} \frac{\mu}{\rho} d\tau \right). \qquad (6.56)$$

For a closed adiabatic volume occupied by a single-phase mixture, one has the simple form of the entropy equation in terms of temperature and pressure

$$\rho c_p \frac{dT}{d\tau} - \left[1 - \rho \left(\frac{\partial h}{\partial p}\right)_{T, all\_C's}\right] \frac{dp}{d\tau} = \mu \left[c_{pM}(T_w - T) + C_n T \Delta s_n^{np}\right]. \quad (6.57)$$

Taking into account that

$$\Delta s_n^{np} = -(1-C_n)\left(\frac{\partial h_M}{\partial p_M}\right)_T \rho R_n / \left\{1 - C_n\left[1 - R_n T\left(\frac{\partial \rho_M}{\partial p_M}\right)_T\right]\right\} \geq 0 \quad (6.58)$$

one obtains

$$\rho c_p \frac{dT}{d\tau} - \left[1 - \rho \left(\frac{\partial h}{\partial p}\right)_{T, all\_C's}\right] \frac{dp}{d\tau}$$

$$= \mu \left[c_{pM}(T_w - T) - C_n(1-C_n)\rho \frac{R_n T \left(\frac{\partial h_M}{\partial p_M}\right)_T}{1 - C_n\left[1 - R_n T\left(\frac{\partial \rho_M}{\partial p_M}\right)_T\right]}\right]. \quad (6.59)$$

For steam

$$\left(\frac{\partial h_M}{\partial p_M}\right)_T < 0, \quad (6.60)$$

$$R_n T \left(\frac{\partial \rho_M}{\partial p_M}\right)_T \approx \frac{R_n}{R_M} = \frac{287.22}{461.631} < 1, \quad (6.61)$$

and therefore

$$\frac{R_n T \left(\frac{\partial h_M}{\partial p_M}\right)_T}{1 - C_n\left[1 - R_n T \left(\frac{\partial \rho_M}{\partial p_M}\right)_T\right]} < 0. \quad (6.62)$$

Consider the case of a open volume in which in addition to the inlet flow there is an outlet flow that guarantees nearly constant pressure. We assume an ideal, immediate intermixing of the introduced steam with the air in the volume. For this particular case and ideal gas mixtures one has

$$\rho c_p \frac{dT}{d\tau} = \mu c_{pM}(T_w - T). \tag{6.63}$$

It is evident that for $T_w > T$ the volume temperature should increase monotonically until reaching T_w. For steam considered as a real gas one obtains

$$\rho c_p \frac{dT}{d\tau} = \mu \left[c_{pM}(T_w - T) - C_n(1 - C_n)\rho \frac{R_n T \left(\frac{\partial h_M}{\partial p_M}\right)_T}{1 - C_n\left[1 - R_n T\left(\frac{\partial \rho_M}{\partial p_M}\right)_T\right]} \right]. \tag{6.64}$$

On the basis of Eq. (6.62), the temperature will increase somewhat more strongly and monotonically if the mixture is considered to be a non-perfect gas mixture. For single component flow the non-perfect gas term is equal to zero because of the multiplier $C_n(1 - C_n)$.

For a closed volume, the pressure is described by

$$\frac{1}{a^2}\frac{\partial p}{\partial \tau} = \mu \left\{ \begin{array}{c} \left[1 + C_n \frac{1}{\rho}\left(\frac{\partial \rho}{\partial C_n}\right)_{p,T,all\_C's\_except\_C_i}\right] \\ -\frac{1}{\rho}\left(\frac{\partial \rho}{\partial T}\right)_{p,all\_C's} \frac{1}{c_p}\left[c_{pM}(T_w - T) + T\Delta s_n^{np}C_n\right] \end{array} \right\}. \tag{6.65}$$

After substituting the derivatives (A-1), (A-2) and (A-3) from Chapter 3 one obtains

$$\frac{1}{\mu a^2}\frac{\partial p}{\partial \tau} = 1 + \left(\frac{\partial \rho}{\partial p}\right)_{T,all\_C's}\left\{ C_n\left[\frac{1}{(\partial \rho_M/\partial p_M)_T} - R_n T\right]\right.$$

$$\left. + \frac{1}{\rho c_p}\left[\rho_n R_n - \frac{\left(\frac{\partial \rho_M}{\partial T}\right)_{p_M}}{\left(\frac{\partial \rho_M}{\partial p_{M1}}\right)_T}\right]\left[\begin{array}{c}c_{pM}(T_w - T)\\+T\Delta s_n^{np}C_n\end{array}\right]\right\}. \tag{6.66}$$

For a mixture of perfect gases this gives

$$\frac{1}{a_1^2}\frac{\partial p}{\partial \tau} = \mu_1 \left(1 + C_{n1}\frac{R_{M1} - R_{n1}}{R_1} + \frac{c_{pM1}}{c_{p1}}\frac{T_{w1} - T_1}{T_1} \right). \tag{6.67}$$

In a similar manner to the temperature changes, the pressure change will be somewhat more pronounced in a non-perfect gas than in a perfect gas mixture.

For practical application the perfect gas mixture approximation can be used:

$$\frac{d}{d\tau}\ln C_n = -\frac{\mu_{Mw} RT}{p}, \tag{6.68}$$

$$\frac{dp}{d\tau} = \mu_{Mw} c_{pM} \left[\left(1 - \frac{\kappa}{\kappa_M}\right) T + (\kappa - 1) T_w \right], \tag{6.69}$$

$$\frac{dT}{d\tau} = \frac{\mu_{Mw} R}{p}\frac{c_{pM}}{c_p}\left(\kappa T_w T - \frac{\kappa}{\kappa_M} T^2 \right) \quad \text{or} \quad \frac{d}{d\tau}\ln\frac{T}{\kappa T_w - \frac{\kappa}{\kappa_M}T} = \frac{\mu_{Mw} R}{p}\frac{c_{pM}}{c_p}\kappa T_w, \tag{6.70}$$

$$R = C_n R_n + (1 - C_n) R_M, \tag{6.71}$$

$$c_p = C_n c_{pn} + (1 - C_n) c_{pM}, \tag{6.72}$$

$$\kappa = \frac{c_p}{c_p - R}, \tag{6.73}$$

$$\kappa_M = \frac{c_{pM}}{c_{pM} - R_M}. \tag{6.74}$$

A simple numerical method can be easily constructed to integrate this system of three non-linear ordinary differential equations.

6.6 Chemical reaction in a gas mixture in a closed volume

For the case of a closed control volume filled with a multi-component gas mixture we have

$$\rho c_p \frac{dT}{d\tau} - \frac{dp}{d\tau} = DT^N, \tag{6.75}$$

$$\frac{1}{\rho a^2} \frac{dp}{d\tau} = D, \tag{6.76}$$

where for a mixture consisting of perfect gases

$$DT^N = \dot{q}''' + \sum_{i=1}^{i_{\max}} \mu_i \left(h_i^* - h_i \right), \tag{6.77}$$

$$\rho D = \frac{1}{c_p T} DT^N + \frac{1}{R} \sum_{i=2}^{i_{\max}} \mu_i \left(R_i - R_1 \right)$$

$$= \frac{1}{c_p T} \left[\dot{q}''' + \sum_{i=1}^{i_{\max}} \mu_i \left(h_i^* - h_i \right) \right] + \frac{1}{R} \sum_{i=2}^{i_{\max}} \mu_i \left(R_i - R_1 \right). \tag{6.78}$$

For $\mu_i < 0$, $h_i^* = h_i$. For the case $\mu_i > 0$ the assumption that the origination enthalpy is equal to the component specific enthalpy at some specified reference temperature and pressure,

$$h_i^* = h_{i,ref} \left(T_{ref}, p_{ref} \right), \tag{6.79}$$

simplifies the computation but should be considered in computing a proper energy source due to chemical reaction. The final form of the temperature and the pressure change equations is

$$\frac{dT}{d\tau} = \frac{1}{\rho c_p} \left(1 + \frac{a^2}{c_p T} \right) \left[\dot{q}''' + \sum_{i=1}^{i_{\max}} \mu_i \left(h_i^* - h_i \right) \right] + \frac{a^2}{\rho c_p R} \sum_{i=2}^{i_{\max}} \mu_i \left(R_i - R_1 \right), \tag{6.80}$$

$$\frac{dp}{d\tau} = \frac{a^2}{c_p T} \left[\dot{q}''' + \sum_{i=1}^{i_{\max}} \mu_i \left(h_i^* - h_i \right) \right] + \frac{a^2}{R} \sum_{i=2}^{i_{\max}} \mu_i \left(R_i - R_1 \right). \tag{6.81}$$

Bearing in mind that

$$a^2 = \kappa RT, \tag{6.82}$$

$$\frac{R}{c_p} = \frac{\kappa - 1}{\kappa}, \qquad (6.83)$$

$$\frac{d}{d\tau} \ln T = (\kappa - 1) \frac{1}{p} \left[\dot{q}''' + \sum_{i=1}^{i_{max}} \mu_i \left(h_i^* - h_i \right) + T \sum_{i=2}^{i_{max}} \mu_i \left(R_i - R_1 \right) \right]$$

$$= (\kappa - 1) \frac{1}{p} \left[\dot{q}''' + \sum_{i=1}^{i_{max}} \mu_i \left(h_i^* - h_i + TR_i \right) \right], \qquad (6.84)$$

$$\frac{dp}{d\tau} = (\kappa - 1) \left[\dot{q}''' + \sum_{i=1}^{i_{max}} \mu_i \left(h_i^* - h_i \right) \right] + \kappa T \sum_{i=2}^{i_{max}} \mu_i \left(R_i - R_1 \right). \qquad (6.85)$$

These are remarkable equations. From Eq. (6.84) we realize that in the case of a chemical reaction in order to have a constant temperature the following condition has to hold

$$\dot{q}''' = -\sum_{i=1}^{i_{max}} \mu_i \left(h_i^* - h_i + TR_i \right). \qquad (6.86)$$

6.7 Hydrogen combustion in an inert atmosphere

Mixtures of hydrogen and oxygen react at atmospheric conditions over millions of years to produce water [16]. At an elevated temperature of 673.15K this process happens within 80 *days* and at 773.15K within 2h [8]. A real combustion is possible if some specific thermodynamic conditions are satisfied as will be discussed in a moment.

6.7.1 Simple introduction to combustion kinetics

A single chemical combustion reaction containing N initial components and final products can be mathematically described by the so called stoichiometric equation

$$\sum_{i=1}^{N} v_i' M_i \rightarrow \sum_{i=1}^{N} v_i'' M_i, \qquad (6.87)$$

see in *Kuo* [17]. Here M_i is the chemical identification symbol of the i-th species before and after the reaction, v_i' is the stoichiometric coefficient of the *initial* substances, and v_i'' is the stoichiometric coefficient of the *final* substances. N is the total number of the components participating in the reaction. The *order of the chemical reaction* is defined by

$$n = \sum_{i=1}^{N} v_i' . \tag{6.88}$$

The reaction velocity as found by *Beketov* in 1865, and by *Guldberg* and *Baare* in 1867, has the general form

$$\frac{dY_i^*}{d\tau} = (v_i'' - v_i') k(T) \prod_{i=1}^{N} Y_i^{*v'} , \tag{6.89}$$

where $k(T)$ is the velocity coefficient of the reaction and

$$Y_{i,1}^* = \frac{\rho_1 C_{i,1}}{M_{i,1}} \tag{6.90}$$

is the molar density in $kg\text{-}mole/m^3$. The dependence of the reaction velocity on the temperature is described by the *Arrhenius law*,

$$k(T) = A \exp\left[-E_a / (RT)\right], \tag{6.91}$$

see in *Bartlmä* [1], where $A = 5 \times 10^6 [m^3/(kg\text{-}mole)]^2/s$, $R = 8\ 314\ J/(kg\text{-}mole\ K)$ is the general gas constant, and $E = 78 \times 10^6\ J/(kg\text{-}mole)$ is the activation energy. Instead of complete modeling of the complicated H_2-O_2 kinetics, as is usually done in rocket propulsion systems design, there is a simplified possibility to describe the global chemical combustion that is frequently used in the literature:

$$2H_2 + O_2 \rightarrow 2H_2O . \tag{6.92}$$

Therefore for a closed control volume we have

$$\mu_{H_2,1} = \alpha_1 M_{H_2,1} \frac{dY_{H_2,1}^*}{d\tau} = -\alpha_1 M_{H_2,1} 2k(T) Y_{H_2,1}^{*2} Y_{O_2,1}^* , \tag{6.93}$$

$$\mu_{O_2,1} = \alpha_1 M_{O_2,1} \frac{dY_{O_2,1}^*}{d\tau} = -\alpha_1 M_{O_2,1} k(T) Y_{H_2,1}^{*2} Y_{O_2,1}^* . \tag{6.94}$$

One can easily see in the kinetic model the expected relations: a) the mass of hydrogen and oxygen is consumed to produce water-steam

$$\mu_{H_2,1} + \mu_{O_2,1} + \mu_{H_2O,1} = 0,\qquad(6.95)$$

and b) the ratio between the reacted mass source terms is related to the mole masses as follows

$$\frac{\mu_{O_2,1}}{\mu_{H_2,1}} = \frac{1}{2}\frac{M_{O_2}}{M_{H_2}} = 8,\qquad(6.96)$$

$$\frac{\mu_{H_2O,1}}{\mu_{H_2,1}} = -\left(1+\frac{1}{2}\frac{M_{O_2}}{M_{H_2}}\right) = -9.\qquad(6.97)$$

Therefore, once the mass source term of the hydrogen mass conservation equation is computed all other mass sources are easily derived from Eqs. (6.96) and (6.97).

Note that usually the *fuel equivalence ratio*

$$\Phi = \frac{C_{H_2}}{1-C_{H_2}}\frac{1-C_{st,H_2}}{C_{st,H_2}} = \frac{Y_{H_2}}{1-Y_{H_2}}\frac{1-Y_{st,H_2}}{Y_{st,H_2}},\qquad(6.98)$$

with a stoichiometric mass concentration C_{st,H_2}, is used to classify the premixed combustion process into three groups,

rich combustion: $\quad\Phi > 1,\quad$ (6.99)
stoichiometric combustion: $\quad\Phi = 1,\quad$ (6.100)
lean combustion: $\quad\Phi < 1.\quad$ (6.101)

In the limiting case of complete consumption of the fuel or of the oxidizer over the time step $\Delta\tau$ we have

$$\mu_{H_2,1} = -\min\left(\frac{\rho_1 C_{H_2,1}}{\Delta\tau},\frac{\rho_1 C_{O_2,1}/8}{\Delta\tau}\right) = -\frac{\rho_1}{\Delta\tau}\min\left(C_{H_2,1},C_{O_2,1}/8\right).\qquad(6.102)$$

The first term in the brackets is taken for either lean or stoichiometric mixtures, and the second for rich mixtures.

6.7.2 Ignition temperature and ignition concentration limits

In accordance with *Isserlin* [13] (1987), if the mole-fraction of H_2 in the air is between 4.1 and 74.2%, or if the mole-fraction of H_2 in the hydrogen-oxygen mixtures is between 4 and 94%, and if the gas temperature increases above $T_{1,ign}$ = 783.15 - 863.15K, ignition is possible. *Bröckerhoff* et al. [4] reviewed 10 sources and summarized that if the mole-fraction of H_2 in the air is between 4.0 and 75% the burning gas temperature is 803.15 to 953.15 K with most of the authors reporting temperatures between 833.15 and 857.15 K. The ignition temperature is a function of the local species concentrations and of the pressure of the mixture. *Belles* [2] reported in 1958 a method for prediction of the ignition temperature based on equalizing the rate of the chain-branching reaction to the half of the rate of the chain-breaking reaction resulting in a transcendental equation

$$\frac{3.11 T_{1,ign} \exp(-8550/T_{1,ign})}{Y_X p_{in\_atm}} = 1, \qquad (6.103)$$

where the effective mole fraction of the third bodies for the formation of HO_2 radicals is

$$Y_X = Y_{H_2} + 0.35 Y_{O_2} + 0.43 Y_{N_2} + 0.20 Y_{Ar} + 1.47 Y_{CO_2}. \qquad (6.104)$$

In a similar way *Maas* and *Wanatz* [22] proposed in 1988 a method for explicit computation of the ignition temperature defined as a temperature dividing the slow reaction from the rapid reaction region. *Moser* [24] p. 38 approximated in 1997 this method by the following equation

$$T_{1,ign} = 1.03015 \times 10^3 + 1.08375 \times 10^2 \ln \{ 8.47444 \times 10^{-2}$$

$$+ p_{bar} \zeta \left[1 + p_{bar} \zeta \left(7.40849 \times 10^{-2} + p_{bar} \zeta 1.45577 \times 10^{-3} \right) \right] \}, \qquad (6.105)$$

where

$$\zeta = Y_H + 0.3 Y_{O_2} + 0.5 Y_{N_2} + 6.5 Y_{H_2O}, \qquad (6.106)$$

is the stokes efficiency and $p_{bar} = p/10^5$. The approximation is reported to be valid in the region of p_{bar} = 1 to 40 bar, T_1 = 700 to 2500 K for a fuel equivalence ratio

$$\Phi = 0.2 \text{ to } 0.4, \qquad (6.107)$$

with a stoichiometric mass concentration

$$C_{st,H_2} = 0.02818 \qquad (6.108)$$

in air.

As already mentioned the ignition may start at a much lower temperature but the velocity of the reaction is very low. That is why the term ignition temperature is associated with a single-step reaction because the steps absorbing energy necessary to create mutually interacting radicals are absorbed in a single reaction which simply releases the energy. The term ignition temperature is not necessary if one uses appropriate multi-step reactions and in addition resolves the spatial variation of the controlling variables properly. In this case, as in nature, if the local energy dissipation is higher than the locally released energy the reaction cannot propagate.

6.7.3 Detonability concentration limits

For the history of the detonation analysis see [18, 32, 5, 14, 8, 33]. *Belles* [2] analyzed how strong a shock wave must be in order to obtain parameters satisfying Eq. (6.86). Then he computed the enthalpy of the gas behind the shock for different hydrogen concentrations and compared it with the available enthalpy increase due to combustion.

Table 6.1. Predicted and observed limits of detonability of hydrogen mixtures at $300K$ and 1 atm

System	Lean limit, molar concentration H_2 in %	Rich limit, molar concentration H_2 in %
H_2-O_2	16.3 (15*)	92.3 (90*)
H_2-Air	15.8 (18.3*)	59.7 (58.9*)

* Experimentally observed detonability limits by [23].

As a result he obtained the detonability limits given in Table 6.1.

6.7.4 The heat release due to combustion

The *heat* release per 1 kg of the final reaction product (in this case steam) is usually called the *enthalpy* of formation of steam. One should pay attention to the exact definition of the measured values. *Dorsey* [9] provided a review of all the measurements up to 1940 and selected the value

$$\Delta h_{H_2O}\left[T_{H_2O,ref}=(273.1+15)K, p_{ref}=1.01325\ bar\right]=13\ 425\ 000\ J/kg. \quad (6.109)$$

defined as follow:

> Enthalpy of formation of steam = the heat produced after reaction of 1/9 *kg* hydrogen with 8/9 *kg* oxygen at the condition $T_{H_2O,ref} = 288.1K$ and $p_{H_2O,ref} = 1.101325 bar$ and then removed from the produced $1kg$ water-steam in order to obtain the same initial temperature and pressure.

Note that the enthalpy of formation of water is about six times larger then the latent heat of evaporation. The enthalpy of formation of steam is almost not a function of the pressure but is a slight function of the temperature due to the differences of the specific capacities at constant pressure of the stoichiometric mixture of the initial products and of the resulting steam.

$$h_{H_2O} = \Delta h_{H_2O,ref} - \int_{T_{H_2O,ref}}^{T} \left[c_{p,H_2O} - c_{p,(H_2+O_2)_{stoichiometric}} \right] dT' \approx$$

$$\approx \left[240.227 + 3.93108\bar{T} + 6.9003\bar{T}^2 - 3.094968\bar{T}^3 \right] 10^6 / 18.0154$$

$$= \left(13.334\ 536\ 01 + 0.118206\bar{T} + 0.383022\bar{T}^2 - 0.1717957\bar{T}^3 \right) 10^6 \ J/kg, \quad (6.110)$$

where $\bar{T} = T/1000$ [9]. The difference of the measurements of the enthalpy of formation and of the specific heat at constant pressure gives an agreement of the computed values within 10% deviation from different authors. In *Zwerev* and *Smirnov* [34] p. 169 the value $\Delta h_{H_2O} = 13\ 275\ 363\ J/kg$ is used, which gives an adiabatic temperature of the burned gases of about 2773 K. In [19] p. 655 the value $\Delta h_{H_2O} = 13\ 333\ kJ/kg$ is reported. *Lewis* and *von Elbe* [21] p. 696 reported in 1987 the value of 13 281 kJ/kg. *Cox* [7] recommended in 1989 the following value $\Delta h_{H_2O}(298.15K, 1atm) = 13\ 431 \pm 0.017\%\ kJ/kg$. A summary is given in Table 3. Further information on this topic can be found in [34, 7, 21].

Table 6.2. Enthalpy of formation of steam

Author	Δh_{H_2O} in MJ/kg	$T_{burned\ gases}$ in K
Dorsey [9], 1940	13.425	
Zwerev and *Smirnov* [34] p. 169, 1987	13.275	2 773*
Landau and *Lifshitz* [19] p. 655, 1987	13.333	
Lewis and *von Elbe* [21] p. 696, 1987	13.281	
Cox [7], 1989	13.435 ±0.017%	

*computed

6.7.5 Equilibrium dissociation

Identification of the chemical system. Consider the idealized single-step reaction

$$2H_2 + O_2 + m_1 H_2O \rightarrow (2+m_1)H_2O, \tag{6.111}$$

followed by dissociation. Six compounds are usually identified in the chemical literature as important with their chemical identification symbols

$$Symb_{j=1,6} = (H_2O, H_2, O_2, HO, H, O) \tag{6.112}$$

consisting of two different chemical elements (O, H). Therefore $6 - 2 = 4$ chemical reactions are required to describe the system. Water dissociates in two ways,

$$2H_2 + O_2 = 2H_2O, \tag{6.113}$$

$$2OH + H_2 = 2H_2O. \tag{6.114}$$

Hydrogen and oxygen dissociate as follows

$$2H = H_2, \tag{6.115}$$

$$2O = O_2. \tag{6.116}$$

The four expressions can be rewritten formally as

$$2 \times H_2O - 2 \times H_2 - 1 \times O_2 + 0 \times HO + 0 \times H + 0 \times O = 0 \tag{6.117}$$

$$2 \times H_2O - 1 \times H_2 + 0 \times O_2 - 2 \times HO + 0 \times H + 0 \times O = 0 \tag{6.118}$$

$$0 \times H_2O + 1 \times H_2 + 0 \times O_2 + 0 \times HO - 2 \times H + 0 \times O = 0 \tag{6.119}$$

$$0 \times H_2O + 0 \times H_2 + 1 \times O_2 + 0 \times HO + 0 \times H - 2 \times O = 0, \tag{6.120}$$

or

$$\sum_{j=1}^{6} n_{ij} Symb_j = 0, \quad \text{for} \quad i = 1, 4. \tag{6.121}$$

Here n_{ij} is the stoichiometric coefficient, < 0 for reactants and > 0 for products. The temperature T and the total pressure p of the mixture of compounds are given. We look for a solution of six molar concentrations

$$\mathbf{Y}^T = (Y_{H_2O}, Y_{H_2}, Y_{O_2}, Y_{OH}, Y_H, Y_O), \qquad (6.122)$$

for which the system is in a chemical equilibrium. We know from *Dalton*'s law that the system pressure is the sum of the partial pressures

$$p = \sum_{j=1}^{6} p_j, \quad Y_i = \frac{p_j}{p}, \quad \sum_{j=1}^{6} Y_j = 1 \qquad (6.123)$$

and that the total number of hydrogen atoms is twice that of oxygen in non-dissociated as well in dissociated water

$$\frac{2Y_1 + 2Y_2 + Y_4 + Y_5}{Y_1 + 2Y_3 + Y_4 + Y_6} = 1. \qquad (6.124)$$

We need four additional equations to close the system. The four equation are the enforced chemical equilibrium for the considered four reactions for the prescribed temperature T and pressure p. The chemical equilibrium condition is a complicated relation between partial pressure ratios and the system pressure and temperature. For a mixture of perfect fluids the resulting system consists of implicit transcendental equations. It is solvable by iteration. Now we consider this task in more detail.

Chemical equilibrium conditions. The so called *pressure equilibrium factors* are defined as follows:

$$K_p = \prod_{i=1}^{i_{max}} p_i^{n_i} = p^{\sum_{i=1}^{i_{max}} n_i} \prod_{i=1}^{i_{max}} Y_i^{n_i}. \qquad (6.125)$$

For a reference pressure equal to $1 bar$ and for pressures in dimensions of *bar* we have

$$K_{p1} = \frac{1}{p} \frac{Y_{H_2O}^2}{Y_{H_2}^2 Y_{O_2}} \qquad (6.126)$$

$$K_{p2} = \frac{1}{p} \frac{Y_{H_2O}^2}{Y_{H_2} Y_{HO}^2}, \qquad (6.127)$$

$$K_{p3} = \frac{1}{p}\frac{Y_{H_2}}{Y_H^2}, \qquad (6.128)$$

$$K_{p4} = \frac{1}{p}\frac{Y_{O_2}}{Y_O^2}. \qquad (6.129)$$

Each of the pressure equilibrium factors can be computed for the reference pressure and temperature $p_0 = 1bar$ and $T_0 = 0K$ using the chemical equilibrium condition

$$\ln K_{pi} = -\frac{\Delta H_{ch,0,i}}{TR} + \sum_{j=1}^{j_{max}} n_{ij}\frac{s_{j0}M_j}{R} = -\frac{\Delta H_{ch,0,i}}{TR} + \sum_{j=1}^{j_{max}} n_{ij}\,\overline{s}_{0j}/R_j\,. \qquad (3.130)$$

Here the R is the universal gas constant, R_j is the gas constant of the j-th component,

$$\Delta H_{ch,0,i} = \sum_{j=1}^{j_{max}} h_{0j} n_{ij} M_j \qquad (6.131)$$

and

$$\overline{s}_{0j} = s_{0j}/R_j \qquad (6.132)$$

are the dimensionless reference entropies of the species j. Equation (6.130) is applied for each of the four selected chemical reactions resulting in

$$\ln K_{p1} = -\left(\Delta H_{ch,0,1}/R\right)/T - 2\overline{s}_{01} + 2\overline{s}_{02} + \overline{s}_{03}\,, \qquad (6.133)$$

$$\ln K_{p2} = -\left(\Delta H_{ch,0,2}/R\right)/T - 2\overline{s}_{01} + \overline{s}_{02} + 2\overline{s}_{04}\,, \qquad (6.134)$$

$$\ln K_{p1} = -\left(\Delta H_{ch,0,3}/R\right)/T - \overline{s}_{02} + 2\overline{s}_{05}\,, \qquad (6.135)$$

$$\ln K_{p4} = -\left(\Delta H_{ch,0,4}/R\right)/T - \overline{s}_{03} + 2\overline{s}_{06}\,. \qquad (6.136)$$

Thermophysical data required. Thermophysical data are available in the chemical literature e.g. [6, 26]. Examples for molar masses and the enthalpies of formation for a reference pressure of $1bar$ for each of the compounds are given in Table 6.3.

Table 6.3. The molar mass and the enthalpy of formation for components of water vapor dissociation at pressure $p = 0.1MPa$, by *Chase* et al. [6] (1985)

j	Symbol	M_j	$\overline{\Delta h_j}$, $T = 0K$	Δh_j $T = 0K$	$\overline{\Delta h_j}$, $T = 298.15K$	Δh_j $T = 298.15K$
		kg/kg-mole	MJ/kg-mole	MJ/kg	MJ/kg-mole	MJ/kg
1	H_2O	18.01528	−238.921	13.262	−241.826	13.424
2	H_2	2.01588	0	0	0	0
3	O_2	31.9988	0	0	0	0
4	HO	17.00734	38.39	2.257	38.987	2.292
5	H	1.00794	216.035	214.333	217.999	216.28
6	O	15.9994	246.79	15.425	249.173	15.574

Using the data in the table we obtain

$$\Delta H_{ch,0,1} / R = 57.471546K, \tag{6.137}$$

$$\Delta H_{ch,0,2} / R = 66.706116K, \tag{6.138}$$

$$\Delta H_{ch,0,3} / R = 51.966405K, \tag{6.139}$$

$$\Delta H_{ch,0,4} / R = 59.364404K. \tag{6.140}$$

Ihara [11, 12] provided the following approximation for the dimensionless entropies at the reference pressure and temperature for each species.

$$\overline{s}_{01} = \overline{s}_{0,H_2O} = -\ln\left[1 - \exp(-5262/T)\right] - \ln\left[1 - \exp(-2294/T)\right]$$

$$-\ln\left[1 - \exp(-5404/T)\right] + 4\ln(T) - 4.1164 - \ln(p/10^5), \tag{6.141}$$

$$\overline{s}_{02} = \overline{s}_{0,H_2} = \ln\left[\frac{T}{682.6} + \frac{1}{24} + \frac{0.711}{T} + \frac{104}{T^2} + \left(\frac{T}{227.53} + 0.875 + \frac{87.457}{T} + \frac{6136.3}{T^2}\right)\exp\left(-\frac{171}{T}\right)\right]$$

$$+ \ln\left\{\left[1-\exp\left(-\frac{6338}{T}\right)\right]^{-1} + \frac{360.65}{T}\exp\left(-\frac{6338}{T}\right)\left[1-\exp\left(-\frac{6338}{T}\right)\right]^{-3}\right\}$$

$$+2.5\ln(T) - 2.6133 - \ln(p/10^5), \tag{6.142}$$

$$\overline{s}_{03} = \overline{s}_{0,O_2} = -\ln\left[1-\exp(-2239/T)\right]$$

$$+\ln\left[3+2\exp(-11340/T)+\exp(-18878/T)\right]$$

$$+3.5\ln(T)+0.114-\ln(p/10^5), \qquad (6.143)$$

$$\overline{s}_{04} = \overline{s}_{0,OH} = \ln(T/26.638+1/3+1.776/T)-\ln\left[1-\exp(-5136/T)\right]$$

$$+\ln\left[1+\exp(-201/T)\right]+2.5\ln(T)+1.2787-\ln(p/10^5), \qquad (6.144)$$

$$\overline{s}_{0,H} = 2.5\ln(T)-2.9604-\ln(p/10^5), \qquad (6.145)$$

$$\overline{s}_{0,O} = \ln\left[5+3\exp(-228/T)+\exp(-326/T)+5\exp(-22830/T)\right]$$

$$+2.5\ln(T)+0.4939-\ln(p/10^5). \qquad (6.146)$$

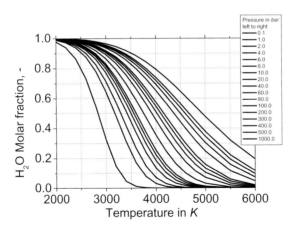

Fig. 6.1. Molar fraction of dissociated steam as a function of temperature with pressure as a parameter

Figure 6.1 demonstrates the result of the solution of the system of equations (6.123), (6.124), (6.126) through (6.129) for different temperatures and pressures with the algorithm developed by *Vasic* [30]. For comparison see the method reported by *Kessel* et al. [15] and the tables provided by *Vargaftik* [30]. We immediately realize how important it is to take into account the dissociation physics in analyzing combustion processes leading to temperatures higher then $1600K$.

Thus initially we have a gas consisting of

$$C_{1,H_2} + C_{1,O_2} + C_{1,H_2O} = 1, \qquad (6.147)$$

and finally a mixture of gases and radicals

$$C_{2,H_2} + C_{2,O_2} + C_{2,H_2O} + C_{2,H} + C_{2,O} + C_{2,OH} = 1. \qquad (6.148)$$

Note that

$$C_{1,H_2} + C_{1,O_2} - C_{2,H_2} - C_{2,O_2} - C_{2,H} - C_{2,O} - C_{2,OH} = C_{2,H_2O} - C_{1,H_2O} \qquad (6.149)$$

reflects in fact the net generation of stable steam and therefore

$$h_{formation} = (\Delta h_{ref} - c_{p2} T_{ref})(C_{2,H_2O} - C_{1,H_2O}). \qquad (6.150)$$

6.7.6 Source terms of the energy conservation of the gas phase

Bearing in mind the information already introduced the expression in the brackets in Eq. (6.84) can then be rearranged as follows:

$$\dot{q}''' + \sum_{i=1}^{i_{max}} \mu_i (h_i^* - h_i) + T \sum_{i=2}^{i_{max}} \mu_i (R_i - R_1)$$

$$= \dot{q}'''_{ref} + \mu_{H_2O}(h_{H_2O,ref} - h_{H_2O}) + T\left[\mu_{H_2}(R_{H_2} - R_{H_2O}) + \mu_{O_2}(R_{O_2} - R_{H_2O})\right]$$

$$= \dot{q}'''_{ref} + \mu_{H_2O}(h_{H_2O,ref} - h_{H_2O}) + T(\mu_{H_2} R_{H_2} + \mu_{O_2} R_{O_2} + \mu_{H_2O} R_{H_2O}). \qquad (6.151)$$

Note that for perfect gases $\mu RT = \mu pv$ which results in a very interesting interpretation of the term

$$T(\mu_{H_2} R_{H_2} + \mu_{O_2} R_{O_2} + \mu_{H_2O} R_{H_2O}) \equiv \mu pv, \qquad (6.152)$$

which is in fact the μpv work associated with the disappearance and appearance of components with respect to zero reference pressure.

6.7 Hydrogen combustion in an inert atmosphere

Expressing on the right hand side all mass sources as a function of the hydrogen mass source only results in

$$= \dot{q}'''_{ref} + \mu_{H_2}\left[-\left(1+\frac{1}{2}\frac{M_{O_2}}{M_{H_2}}\right)\left(h_{H_2O,ref} - h_{H_2O}\right)\right.$$

$$\left.+T\left(R_{H_2} + \frac{1}{2}\frac{M_{O_2}}{M_{H_2}}R_{O_2} - \left(1+\frac{1}{2}\frac{M_{O_2}}{M_{H_2}}\right)R_{H_2O}\right)\right]$$

$$= \dot{q}'''_{ref} - \mu_{H_2}\left[9\left(h_{H_2O,ref} - h_{H_2O}\right) - T\left(R_{H_2} + 8R_{O_2} - 9R_{H_2O}\right)\right]. \quad (6.152)$$

Keeping in mind that

$$h_{H_2O,ref} - h_{H_2O} = c_{p,H_2O}\left(T_{H_2O,ref} - T\right), \quad (6.153)$$

and

$$\dot{q}'''_{ref} = \mu_{H_2O}\Delta h_{H_2O} = -\left(9\mu_{H_2}\right)\Delta h_{H_2O}, \quad (6.154)$$

we finally obtain

$$= \dot{q}'''_{ref} - \mu_{H_2}\left[9c_{p,H_2O}\left(T_{H_2O,ref} - T\right) - T\left(R_{H_2} + 8R_{O_2} - 9R_{H_2O}\right)\right]$$

$$= \dot{q}'''_{ref} + \mu_{H_2}\left[T\left(R_{H_2} + 8R_{O_2} - 9R_{H_2O} + 9c_{p,H_2O}\right) - 9c_{p,H_2O}T_{H_2O,ref}\right]$$

$$= -\mu_{H_2}\left[9\left(\Delta h_{H_2O} + c_{p,H_2O}T_{H_2O,ref}\right) - T\left(R_{H_2} + 8R_{O_2} - 9R_{H_2O} + 9c_{p,H_2O}\right)\right]$$

$$= -\mu_{H_2}\left(1.19478267 \times 10^8 - 10515.3 \times T\right). \quad (6.155)$$

Consequently the term $\mu_{H_2} 10515.3 \times T < 0$ obviously has a cooling effect. At $T = 3000K$ the cooling effect amounts to about 26% of the origination enthalpy.

6.7.7 Temperature and pressure changes in a closed control volume; adiabatic temperature of the burned gases

Thus, the temperature and pressure changes of the burning mixture in a closed volume are governed by the equations

$$\frac{1}{aT-cT^2}\frac{dT}{d\tau} = -\mu_{H_2}(\kappa-1)\frac{1}{p}, \qquad (6.156)$$

$$\frac{dp}{d\tau} = -\mu_{H_2}(\kappa-1)(a-bT), \qquad (6.157)$$

where $a = 1.2 \times 10^8$, $c = 10515.3$, $b = 8466 + \dfrac{\kappa}{\kappa-1}2049.3$. The hydrogen and oxygen concentrations change obeying the mass conservation

$$\frac{dC_{H_2}}{d\tau} = -\mu_{H_2}/\rho, \qquad (6.158)$$

$$\frac{dC_{O_2}}{d\tau} = -8\mu_{H_2}/\rho, \qquad (6.159)$$

where

$$\rho = \rho_0 = const. \qquad (6.160)$$

Therefore

$$\int_0^{\Delta\tau} \mu_{H_2} d\tau = -\rho_0 \Delta C_{H_2}. \qquad (6.161)$$

Given the initial conditions, the system can be integrated numerically until the concentrations of the fuel and the oxidizer are below given limits for which no more oxidation is possible. The duration of the process gives the inherent time scale of the burning process as presented in Fig. 6.2. The higher the initial pressure the faster is the burning. This information is important for selecting of an appropriate integration time step for large scale computational analysis.

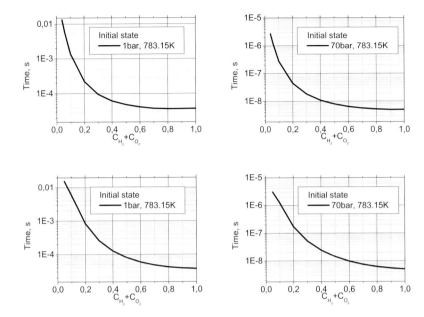

Fig. 6.2. Time elapsed from ignition to complete burning as a function of the initial mass concentration of the stoichiometric mixture fuel + oxidizer. Initial temperature 783.15K. Inert component is nitrogen. a) 1*bar*; b) 70*bar*; Inert component is steam: c) 1*bar*; d) 70*bar*

The hydrogen mass source term can be eliminated from Eqs. (6.156) and (6.157) and the equation obtained written in a compact form

$$\frac{a - bT}{aT - cT^2} dT = d\ln p \quad \text{or} \quad d\ln \frac{Tb^{1/c}}{(a - cT)^{\frac{c-1}{c}}} = d\ln p \tag{6.162}$$

can be integrated analytically assuming that $\kappa \approx const$. The result

$$\frac{T}{T_0} \frac{(a - cT_0)^{\frac{c-1}{c}}}{(a - cT)^{\frac{c-1}{c}}} = \frac{p}{p_0} \tag{6.163}$$

can be further simplified by noting that $c \gg 1$

$$\frac{T}{T_0} = \frac{p}{p_0} \bigg/ \left(1 + \frac{c}{a} T_0 \frac{p - p_0}{p_0}\right). \tag{6.164}$$

Bearing in mind that $1 \gg \dfrac{c}{a} T_0 \dfrac{p - p_0}{p_0}$ the solution converges to those of heat input only in a closed volume obtained previously. Using Eq. (6.163) and substituting for the pressure into Eq. (6.156) we obtain

$$\dfrac{1}{(a - cT)^m} \dfrac{dT}{d\tau} = -\mu_{H_2} \dfrac{T_0 (\kappa - 1)}{p_0 (a - cT_0)^{\frac{c-1}{c}}}, \qquad (6.165)$$

where $m = 2 - 1/c \approx 2$. After integrating with respect to time and solving with respect to the temperature we obtain

$$T = \dfrac{1}{c} \left\{ a - \left[\dfrac{1}{(a - cT_0)^{m-1}} - \dfrac{(\kappa - 1) c (m - 1)}{(a - cT_0)^{\frac{c-1}{c}}} \dfrac{T_0}{p_0} \int_0^{\Delta\tau} \mu_{H_2} d\tau \right]^{-\frac{1}{m-1}} \right\}$$

$$\approx \dfrac{1}{c} \left\{ a - \dfrac{a - cT_0}{1 - (\kappa - 1) c \dfrac{T_0}{p_0} \int_0^{\Delta\tau} \mu_{H_2} d\tau} \right\}. \qquad (6.166)$$

For the limiting case of complete consumption of the fuel or of the oxidizer over the time step $\Delta\tau$ we have

$$T = \dfrac{1}{c} \left\{ a - \left[\dfrac{1}{(a - cT_0)^{m-1}} + \dfrac{(\kappa - 1) c (m - 1)}{(a - cT_0)^{\frac{c-1}{c}}} \dfrac{T_0 p_0}{p_0} \min\left(\Delta C_{H_2}, \Delta C_{O_2}/8 \right) \right]^{-\frac{1}{m-1}} \right\}$$

$$\approx a/c - \dfrac{a/c - T_0}{1 + (\kappa - 1) c \dfrac{T_0 p_0}{p_0} \min\left(\Delta C_{H_2}, \Delta C_{O_2}/8 \right)}. \qquad (6.167)$$

This result is valid for the case of no steam dissociation, i.e. for a burned products (steam) temperature below 1600K. For higher temperatures κ and R are no longer constant and the integration has to be performed numerically. Note that for temperature higher then 1600K the specific capacity at constant pressure changes dramatically due to the thermal dissociation and the "effective isentropic exponent" tends to unity which clearly reduces the right hand side of Eqs. (6.156) and (6.157).

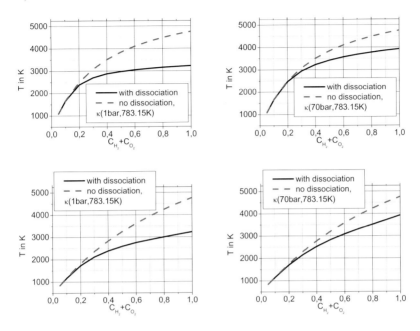

Fig. 6.3. Temperature after burning as a function of the initial mass concentration of the stoichiometric mixture fuel + oxidizer. Initial temperature 783.15K. Initial pressure: inert component is nitrogen. a) 1*bar*; b) 70*bar*; inert component is steam: c) 1*bar*; d) 70*bar*

Figures 6.3 and 6.4 demonstrate the error in the final temperature and pressure made if the dissociation is not taken into account. For low pressure the difference in the final temperatures may be higher then 1500K, and for high initial pressure the difference in the final pressures may be greater then 450bar.

An easy method of indication of combustible gases in pipelines intended to work with saturated steam is to measure their temperature. The accumulation of combustible gases is manifested in a temperature reduction. An example is given in Fig. 6.5 for steam lines at 70*bar* nominal pressure.

In a coordinate system moving with the burning front velocity $w_{1,b}$, the temperature before the front is governed by heat conduction mainly, $-w_{1,b} dT_1/dz - \lambda_1 d^2T_1/dz = 0$, with the solution $T_1 = T_{10} \exp(-w_{1,b}z/\lambda_1)$, where T_{10} if the flame temperature at the burning front, *Landau* and *Lifschitz* [19] p. 645.

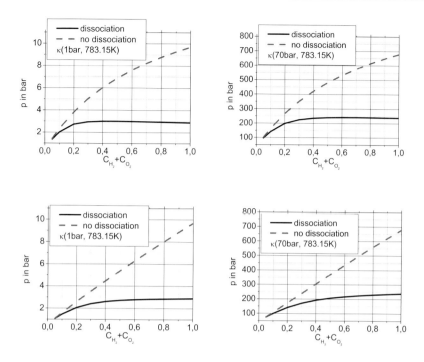

Fig. 6.4. Pressure after burning as a function of the initial mass concentration of the stoichiometric mixture fuel + oxidizer. Initial temperature 783.15K. Initial pressure: inert component is nitrogen. a) 1*bar*; b) 70*bar*; inert component is steam: c) 1*bar*; d) 70*bar*

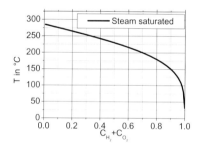

Fig. 6.5. Initial temperature − saturated temperature under the given partial steam pressure as a function of the mass concentration of a stoichiometric hydrogen-oxygen mixture with initial presence of steam. Initial total pressure 70*bar*

Here are some velocities characterizing the interaction process for stoichiometric mixture: The propagation of a laminar combustion front is reported to be about 1 to 30 *m/s* see [19] p. 639 and *Schmidt* [12], respectively. The sound velocity in stagnant mixture at $783.15K$ is about $664 m/s$. The detonation velocity of 2819 *m/s* was measured by *Lewis* and *Friauf* [10] in 1930 for initial mixture temperature of $291K$ (see also *Oswatitsch* [11] p.59). The authors computed a temperature of the burning products of about $3583K$. From these data we see that the propagation velocity of a combustion front is very small compared to the sound velocity and to the detonation velocity.

References

1. Bartlmä F (1975) Gasdynamik der Verbrennung, Springer, Berlin Heidelberg New York
2. Belles FE (28 August – 3 September, 1958) Detonability and chemical kinetics: Prediction of the limits of detonability of hydrogen, Seventh Symposium (international) on combustion, At London and Oxford, pp 745-751
3. Bertelot M and Vieille P (1881) Compt. Rend. Acad. Sci. Paris, vol 93 p. 18
4. Bröckerhoff P, Kugeler K, Reinecke A-E, Tragsdorf IM (4.4.2002) Untersuchungen zur weiteren Verbesserung der Methoden zur sicherheitstechnischen Bewertung der katalytischen Rekombinatoren von Wasserstoff in Sicherheitsbehältern von Kernkraftwerken bei schweren Störfällen, Institut für Sicherheitsforschung und Reaktortechnik (IRS-2), Forschungszentrum Jülich
5. Chapman DL (1899) Philos. Mag., vol 47 no 5 p 90
6. Chase MW ed (1998) NIST-JANAF Thermochemical Tables, 4th ed Part I, II, American Institute of Physics and American Chemical Society, Woodbury, New York
7. Cox RA (ed) (1989) Kinetics and mechanisms of elementary chemical processes of importance in combustion: Research within the frame of the CEC non-nuclear energy R&D program; Final report of the period 1.4.1986 to 30.6.1989, Engineering Science Division, Harwell Laboratory, Didcot/UK
8. Crussard L (1907) Bull. De la Soc. De l'industrie Minérale St.-Etienne, vol 6 pp 1-109
9. Dorsey NE (November 1951) Properties of ordinary water-substance, Reihold, 1940, Second Printing
10. Gaydon A and Wolfhard H (1979) Flames, their structure, radiation, and temperature. Chapman and Hall, London
11. Ihara S (1977) Approximation for the thermodynamic properties of high-temperature dissociated water vapour, Bulletin of the Electrotechnical Laboratory, Tokyo, Japan vol 41 no 4 pp 259-280
12. Ihara S (1979) Direct thermal decomposition of water, Solar-Hydrogen Energy Systems, Ch 4, ed Ohta T, Pergamon, Oxford, pp 58-79
13. Isserlin AS (1987) Osnovy zzhiganija gazovogo topliva, Nedra, Leningrad
14. Jouguet E (1905) J. Mathématique, p 347; (1906) p 6; (1917) Mécanique des Explosifs, Doin O, Paris

15. Kesselman PM, Blank JuI and Mogilevskij (1968) Thermodynamical properties of thermally dissociated water steam for temperatures 1600-6000K and pressures 0.1-1000bar, High temperature physics, vol 6 no 4, in Russian
16. Kolarov NCh (1970) Inorganic Chemistry, Sofia, Technika, in Bulgarian
17. Kuo KK (1986) Principles of Combustion, Wiley-Intersience Publication
18. Laffitte PF (1938) Flames of high-speed detonation, Science of Petrolium, Oxford University Press, London, pp 2995-3003
19. Landau LD, Lifshitz EM (1987) Course of theoretical physics, vol 6, Fluid Mechanics, Second Edition, Pergamon, Oxford
20. Lewis B and Friauf JB (1930) Explosives in detonating gas mixtures. 1. Calculation of rates of explosions in mixtures of hydrogen and oxigen and the influence of rare gases. J. Amer. Chem. Soc. LII pp 3905-3929
21. Lewis B and von Elbe G (1987) Combustion, flames and explosion of gases, Academic Press, Harcourt Brace Jovanovich, Orlando
22. Maas U and Wanatz J (1988) Ignition process in hydrogen-oxygen mixtures, Comb. Flame vol 74 pp 53-69
23. Mallard E and Le Chatelier HL (1881) Compt. Rend. Acad. Sci. Paris, vol 93 p. 145
24. Moser V (6.2.1997) Simulation der Explosion magerer Wasserstoff-Luft-Gemische in großskaligen Geometrien, PhD, Achener Beiträge zum Kraftahr- und Maschinenwesen, Band 11
25. Oswatitsch K (1952) Gasdynamik, Springer-Verlag, Vienna
26. Robert JK, Rupley and Miller JA (April 1987) The CHEMKIN thermodynamic data base, SAND-87-8215, DE87 009358
27. Schmidt E (1945) Thermodynamik, 3. Aufl., Springer-Verlag, Vienna, p. 272
28. Tangren RF, Dodge CH, Seifert HS (1949) Compressibility effects in two-phase flow, Journal of Applied Physics, vol 20 p 736
29. Taylor GI (1935) Proc. Roy. Soc. London Ser. A, vol 151 p 429
30. Vargaftik NB (1983) Handbook of physical properties of liquids and gases: pure substances and mixtures, The English translation of the second edition, Hemisphere Publishing Corporation
31. Vasic AZ (1993) High temperature properties and heat transfer phenomena for steam at temperatures up to 5000K, MS Thesis, Ottawa-Carleton Institute for Mechanical and Aeronautical Engineering, Ottawa, Ontario
32. Wendlandt R (1924) Z. für phys. Chemie, vol. 110, pp. 637
33. Zeldovich JB (1940) To the theory of detonation propagation in gas systems, Journal of experimental and theoretical physics, vol 10 no 5 pp 542-568
34. Zwerev IN, Smirnov NN (1987) Gasodinamika gorenija, Izdatelstvo Moskovskogo universiteta

7 Exergy of multi-phase multi-component systems

7.1 Introduction

Fluids at pressures and temperatures higher then the environment pressures and temperatures may perform technical work at the costs of their internal energy. Experience shows that not all available internal energy of fluids may be consumed for performing technical work but only part of it. The industrial revolution initiated with the invention of the steam machine started also the discussion on how much of the internal fluid energy may be transferred under given circumstances in technical work. The result of this discussion is well presented in the references and text books [1, 3, 4, 10, 11, 12, 14]. We will shortly demonstrate the main ideas on a single-phase multi-component open system in which the spatial intermixing at any time is assumed to be perfect.

We will emphasize the different definitions of the exergy used in the Anglo-Saxon thermodynamic literature and in the German literature. Thereafter, we will discuss different limiting cases. On the example of the heat pump we will demonstrate a practical application of the exergy.

Finally, we consider the exergy of multi-fluid mixtures for which each of the fluids consists of many chemical components.

7.2 The pseudo-exergy equation for single-fluid systems

The energy conservation equation written in a specific enthalpy form is

$$\rho \frac{dh}{d\tau} - \frac{dp}{d\tau} = DT^N, \qquad (7.1)$$

where

$$DT^N = \rho\left(P_k + \varepsilon\right) + \dot{q}''' + \sum_{i=1}^{i_{max}} \mu_{i,in}\left(h_{i,in} - h_i\right) - \sum_{i=1}^{i_{max}} \mu_{i,out}\left(h_{i,out} - h_i\right)$$

$$+ \frac{1}{2}\left[\mu_{in}\left(\mathbf{V}_{in} - \mathbf{V}\right)^2 - \mu_{out}\left(\mathbf{V}_{out} - \mathbf{V}\right)^2\right] \tag{7.2}$$

are different components of the energy input into the flow. P_k and ε are irreversible components of the energy dissipation per unit volume of the flow for whatever reason. The thermal energy (heat) introduced per unit time into unit volume of the flow is $\dot{q}'''$. The specific energy of the system is influenced by the difference between the in-flowing and the system enthalpy of the chemical components $\sum_{i=1}^{i_{max}} \mu_{i,in}\left(h_{i,in} - h_i\right)$ as well as by the difference between the out-flowing and the system enthalpy of the chemical components $\sum_{i=1}^{i_{max}} \mu_{i,out}\left(h_{i,out} - h_i\right)$. Sources and sinks of fluid mass in the system with a velocity different from the system fluid velocity also give rise to the internal energy change, $\frac{1}{2}\left[\mu_{in}\left(\mathbf{V}_{in} - \mathbf{V}\right)^2 - \mu_{out}\left(\mathbf{V}_{out} - \mathbf{V}\right)^2\right]$. $\mu_{in} \geq 0$ and $\mu_{out} \geq 0$ are the inflow and outflow mass per unit volume and unit time, $\mu_i = \mu_{in} - \mu_{out}$. The complete equivalent to this equation in its entropy form is

$$\rho \frac{ds}{d\tau} = \frac{1}{T}DT^N + \sum_{i=1}^{i_{max}} \mu_i\left(s_i - s\right) \tag{7.3}$$

For the derivation see Chapter 5 or [6-8].

In order to come to the exergy definition *Elsner* [3], *Reynolds* and *Perkins* [11] used the energy conservation equation for an infinite time step *without* irreversible entropy sources and combined it with the entropy conservation equation with irreversible entropy sources. Then they simply eliminate the thermal energy from both equations. In this way the exergy of the thermal energy is not defined, which is a disadvantage of their approach, the reason why we will not follow this approach but only use their exergy definition. Note that in our quantitative statement we take into account in the energy conservation equation the same irreversible terms as in the entropy equations. We emphasize that the two equations are mathematically equivalent (*Legendre* transformed by using the *Gibbs* definition of specific entropy).

Now we multiply the entropy equation by the environment temperature, subtract it from Eq. (7.1), multiply the resulting equation by the time differential and divide by the fluid density. The result is called here the *pseudo-exergy equation*

$$dh - vdp - T_\infty ds = v\left[\left(1 - \frac{T_\infty}{T}\right)DT^N + T_\infty \sum_{i=1}^{i_{max}} \mu_i (s_i - s)\right] d\tau. \quad (7.4)$$

Note the appearance in the right-hand site of the so called efficiency coefficient $1 - T_\infty/T$ called *Carnot* coefficient – compare with *Carnot* [2] in 1824.

As already mentioned, there are different definitions of the exergy used in the Anglo-Saxon thermodynamic literature and in the German literature. Next we discuss the differences.

7.3 The fundamental exergy equation

7.3.1 The exergy definition in accordance with *Reynolds* and *Perkins*

Assuming constant environmental conditions, $p_\infty = const$ and $T_\infty = const$, Reynolds and Perkins [11] defined the specific quantity

$$e_x^{pdv} := e + p_\infty v - T_\infty s, \quad (7.5)$$

as a specific internal *pdv* exergy. *pdv* here is not an exponent but a superscript, whose meaning will be clear in a moment. Note that the exergy is a combination of state variables. Consequently the exergy itself is a state variable. The exergy is a remarkable quantity. As noted by the authors, it is obviously a function of the environmental temperature and pressure besides the two other state variables selected as thermodynamically independent,

$$e_x^{pdv} = e_x^{pdv}(e, v, T_\infty, p_\infty). \quad (7.6)$$

Note the similarity and the difference to the *Gibbs* function $e + pv - Ts$, which does not depend on the environmental parameter. The differential form of the exergy is

$$de_x^{pdv} = \left(\frac{\partial e_x^{pdv}}{\partial e}\right)_v de + \left(\frac{\partial e_x^{pdv}}{\partial v}\right)_e dv = \left(1 - \frac{T_\infty}{T}\right)de + \left(p_\infty - \frac{T_\infty}{T}p\right)dv. \quad (7.7)$$

The last form is obtained having in mind the thermodynamic definition of temperature and pressure, see Eqs. (9) and (20) in [11],

$$\left(\frac{\partial e_x^{pdv}}{\partial e}\right)_v = 1 - T_\infty \left(\frac{\partial s}{\partial e}\right)_v = 1 - \frac{T_\infty}{T}, \qquad (7.8)$$

and

$$\left(\frac{\partial e_x^{pdv}}{\partial v}\right)_e = p_\infty - T_\infty \left(\frac{\partial s}{\partial v}\right)_e = p_\infty - \frac{T_\infty}{T} p, \qquad (7.9)$$

see in [11]. Note that the partial derivative of the exergy with respect to the specific internal energy at constant specific volume, Eq. (7.8) gives exactly the *Carnot* coefficient. The minimum of this function is defined by simultaneously equating the partial derivatives to zero,

$$\left(\frac{\partial e_x^{pdv}}{\partial e}\right)_v = 1 - \frac{T_\infty}{T} = 0, \qquad (7.10)$$

and

$$\left(\frac{\partial e_x^{pdv}}{\partial v}\right)_e = p_\infty - \frac{T_\infty}{T} p = 0. \qquad (7.11)$$

This leads to the conclusion that the *pdv* exergy has an extremum and this extremum is at $T = T_\infty$ and $p = p_\infty$, compare with *Reynolds* and *Perkins* [11]. Because the exergy is decreasing function with decreasing pressure and temperature we conclude that this extreme is a minimum.

7.3.2 The exergy definition in accordance with *Gouy* (l'énergie utilisable, 1889)

Assuming constant environmental conditions, $p_\infty = const$ and $T_\infty = const$, *Gouy* [4] in 1889 Eq. 2, p.506 defined the specific *vdp* exergy as

$$e_x^{vdp} := h - T_\infty s = e + pv - T_\infty s = e + p_\infty v - T_\infty s + (p - p_\infty)v = e_x^{pdv} + (p - p_\infty)v, \qquad (7.12)$$

which differs from the *Reynolds* and *Perkins* [11] definition

$$e_x^{pdv} := e + p_\infty v - T_\infty s, \qquad (7.13)$$

by the residual

$$e_x^{vdp} - e_x^{pdv} = (p - p_\infty)v. \tag{7.14}$$

Note again that *vdp* here is not an exponent but a superscript, whose meaning will be clear in a moment. The name $\varepsilon\chi \ \varepsilon\rho\gamma o\nu$ = available work, was given by *Rant* [10]. Note again the similarity and the difference to the *Gibbs* function $e + pv - Ts$, which does not depend on the environmental parameter. Of course for relaxed fluid at environmental conditions $p = p_\infty$, $e_x^{vdp} = e_x^{pdv}$, and both definitions possess a minimum. The differential form of the exergy definition is then

$$e_x^{vdp} = e_x^{pdv} + d\left[(p-p_\infty)v\right] = \left[1 - \frac{T_\infty}{T} + v\left(\frac{\partial p}{\partial e}\right)_v\right]de + \left[1 - \frac{T_\infty}{T} + \frac{v}{p}\left(\frac{\partial p}{\partial v}\right)_e\right]pdv. \tag{7.15}$$

The difference

$$e - e_x^{vdp} = p_\infty v - T_\infty s, \tag{7.16}$$

is called *anergy* - *Ruppel*. The anergy can not be transferred into mechanical work.

Using the two above introduced definitions of the exergy the pseudo-exergy equation (7.4) can be transferred into two alternative forms of the exergy equation.

7.3.3 The exergy definition appropriate for estimation of the volume change work

Using the differential form of the definition equation of the specific enthalpy

$$h := e + pv , \qquad (7.17)$$

and replacing the specific enthalpy differential we obtain from Eq. (7.4)

$$de + pdv - T_\infty ds = v\left[\left(1 - \frac{T_\infty}{T}\right)DT^N + T_\infty \sum_{i=1}^{i_{max}} \mu_i (s_i - s)\right]d\tau . \qquad (7.18)$$

Replacing with

$$p = p + p_\infty - p_\infty , \qquad (7.19)$$

and rearranging we obtain

$$de + p_\infty dv - T_\infty ds + (p - p_\infty)dv = v\left[\left(1 - \frac{T_\infty}{T}\right)DT^N + T_\infty \sum_{i=1}^{i_{max}} \mu_i (s_i - s)\right]d\tau . \qquad (7.20)$$

The assumption $p_\infty = const$ and $T_\infty = const$ allows us to write

$$de + p_\infty dv - T_\infty ds = d(e + p_\infty v - T_\infty s) = de_r \qquad (7.21)$$

where the specific quantity e_x^{pdv}, defined by Eq. (7.5), arises. With this definition of the specific exergy we obtain an abbreviated notation of the integral form of Eq. (7.20)

$$\int_1^2 (p - p_\infty)dv = e_{x,1}^{pdv} - e_{x,2}^{pdv} + \int_{\tau_1}^{\tau_2} v\left[\left(1 - \frac{T_\infty}{T}\right)DT^N + T_\infty \sum_{i=1}^{i_{max}} \mu_i (s_i - s)\right]d\tau . \quad (7.22)$$

The expression

$$-w_t^{pdv} = \int_1^2 (p - p_\infty)dv > 0 , \qquad (7.23)$$

defines the specific work which is required to change the volume of the fluid and is called *useful work* (Nutzarbeit), see *Baer* [1], p. 50. In the literature this work is called also *expansion* or *compression work* depending on the sign. For $p > p_\infty$

and $dv > 0$ the expression defines the *specific expansion work per unit mass of the fluid*. Technical expansion work taken away from the fluid is considered negative per definition. The understanding that Eq. (7.21) is a total differential was clearly expressed first by *Gouy* in 1889.

Thus the available technical work is defined exactly by the *fundamental exergy equation* in integral form

$$-w_{t,12}^{pdv} = e_{x,1}^{pdv} - e_{x,2}^{pdv} + \int_{\tau_1}^{\tau_2} v\left[\left(1 - \frac{T_\infty}{T}\right)DT^N + T_\infty \sum_{i=1}^{i_{max}} \mu_i (s_i - s)\right] d\tau . \qquad (7.24)$$

7.3.4 The exergy definition appropriate for estimation of the technical work

In the sense of the *Gouy* exergy definition, Eq. (7.4) reads

$$-vdp = -de_x^{vdp} + v\left[\left(1 - \frac{T_\infty}{T}\right)DT^N + T_\infty \sum_{i=1}^{i_{max}} \mu_i (s_i - s)\right] d\tau , \qquad (7.25)$$

or in the integral form

$$-w_{t,12}^{vdp} = e_{x,1}^{vdp} - e_{x,2}^{vdp} + \int_1^2 v\left[\left(1 - \frac{T_\infty}{T}\right)DT^N + T_\infty \sum_{i=1}^{i_{max}} \mu_i (s_i - s)\right] d\tau , \qquad (7.26)$$

with

$$w_{t,12}^{vdp} = \int_1^2 vdp , \qquad (7.27)$$

interpreted as a *specific technical work*. Now the meaning of the superscript *vdp* is clear.

7.4 Some interesting consequences of the fundamental exergy equation

Several limiting cases can be derived from Eq. (7.26). Some of them are given as examples.

A very interesting consequence of the two different forms (7.24) and (7.26) of the fundamental exergy equation in integral form is obtained for closed cycle processes in which the end and the initial fluid parameters are equal and therefore the specific exergy difference between the states 1 and 2 is zero. For this case we obtain

$$\int_1^2 (p - p_\infty) dv = \int_1^2 v dp. \qquad (7.28)$$

This conclusion is manifested in the classical example for computing the technical work of a piston air compressor, see e.g. *Stephan* and *Mayinger* [12] p. 107.

Let as discuss some important cases for the engineering applications:

Case 1: For a *closed* and *adiabatic* system without internal irreversible energy dissipation, without external thermal energy supply or removal, the available technical work is exactly equal to the differences of the initial and the final exergy of the fluid

$$-w_{t,12}^{pdv} = e_{x,1}^{pdv} - e_{x,2}^{pdv}. \qquad (7.29a)$$

$$-w_{t,12}^{vdp} = e_{x,1}^{vdp} - e_{x,2}^{vdp}. \qquad (7.29b)$$

The above expression is valid also for closed systems with irreversible energy dissipation and heat exchange with the environment if it happens at environmental temperature.

Case 2: If the system is closed, and there is no irreversible energy dissipation, but there is thermal energy exchange with the environment under temperatures different from the environmental temperatures, the available technical work is

$$-w_{t,12}^{pdv} = e_{x,1}^{pdv} - e_{x,2}^{pdv} + \int_{\tau_1}^{\tau_2} \left(1 - \frac{T_\infty}{T}\right) v\dot{q}''' d\tau. \qquad (7.30a)$$

$$-w_{t,12}^{pdv} = e_{x,1}^{pdv} - e_{x,2}^{pdv} + \int_{\tau_1}^{\tau_2} \left(1 - \frac{T_\infty}{T}\right) v\dot{q}''' d\tau. \qquad (7.30b)$$

Case 3: If the real system is closed, and there is irreversible energy dissipation, and there is thermal energy exchange with the environment under temperatures different from the environmental temperatures, the available volume change work and the available technical work are

7.5 Judging the efficiency of a heat pump as an example of application of the exergy

$$-w_{t,12,irev}^{pdv} = e_{x,1}^{pdv} - e_{x,2}^{pdv} + \int_{\tau_1}^{\tau_2}\left(1-\frac{T_\infty}{T}\right)v\dot{q}'''d\tau + \int_{\tau_1}^{\tau_2}\left(1-\frac{T_\infty}{T}\right)(P_k+\varepsilon)d\tau, \quad (7.31a)$$

$$-w_{t,12,irev}^{vdp} = e_{x,1}^{vdp} - e_{x,2}^{vdp} + \int_{\tau_1}^{\tau_2}\left(1-\frac{T_\infty}{T}\right)v\dot{q}'''d\tau + \int_{\tau_1}^{\tau_2}\left(1-\frac{T_\infty}{T}\right)(P_k+\varepsilon)d\tau, \quad (7.31b)$$

which leads to

$$w_{t,12,rev}^{pdv} > w_{t,12}^{pdv}, \quad (7.32a)$$

$$w_{t,12,rev}^{pdv} > w_{t,12}^{pdv}. \quad (7.32b)$$

Case 4: If there is a cyclic process with the final state equal to the initial state and thermal energy supply only, only part of the thermal energy can be transferred in the technical work

$$-w_{t,12,rev}^{pdv} = -w_{t,12,rev}^{vdp} = \int_{\tau_1}^{\tau_2}\left(1-\frac{T_\infty}{T}\right)v\dot{q}'''d\tau, \quad (7.33)$$

compare with *Carnot* [2] 1824.

Case 5: If there is a process resulting in irreversible energy dissipation only, the available technical work is only a part of this dissipation

$$-w_{t,12,irev}^{pdv} = -w_{t,12,irev}^{vdp} = \int_{\tau_1}^{\tau_2}\left(1-\frac{T_\infty}{T}\right)vP_k\,d\tau. \quad (7.34)$$

7.5 Judging the efficiency of a heat pump as an example of application of the exergy

Let as consider a heat pump (*Thomson* [13], 1852) working with a cyclic steady state process for which at the same spatial point in the system at any time $e_{x,1}^{vdp} = e_{x,2}^{vdp}$. The circuit of the system is closed and therefore there are no mass sinks or sources. Equation (7.26) therefore reads

$$-w_{t,12}^{vdp} = \int_1^2\left(1-\frac{T_\infty}{T}\right)v\dot{q}'''d\tau + \int_1^2\left(1-\frac{T_\infty}{T_{dissipation}}\right)v\dot{q}'''_{dissipation}d\tau. \quad (7.35)$$

Thermal power $\dot{q}'''_{in}$ is introduced in the system at constant temperature T_{in}. Thermal power $\dot{q}'''_{out}$ is removed by the system at T_{out}. Somewhere under way from the lower to the higher level there are also unwished heat losses, $\dot{q}'''_{dissipation}$ happening at some $T_{dissipation}$. In this case the exergy equation reads

$$-w^{vdp}_{t,12} = \left(1 - \frac{T_\infty}{T_{in}}\right)\int_1^2 v\dot{q}'''_{in}d\tau - \left(1 - \frac{T_\infty}{T_{out}}\right)\int_1^2 v\dot{q}'''_{out}d\tau - \left(1 - \frac{T_\infty}{T_{dissipation}}\right)\int_1^2 v\dot{q}'''_{dissipation}d\tau .$$
(7.36)

We immediately realize that the mechanic work, which has to be performed on the fluid by the compressor, depends on the temperature levels at which energy is supplied from the environment and provided to the house. In the idealized case of introducing heat at environmental temperature and negligible energy dissipation the obtained heat from unit fluid mass $\int_1^2 v\dot{q}'''_{out}d\tau$ is proportional to the compressor work $w^{vdp}_{t,12}$ but not equal to it. The ratio

$$cop := \frac{\int_1^2 v\dot{q}'''_{out}d\tau}{w^{vdp}_{t,12}} = \frac{T_{out}}{T_{out} - T_\infty} > 1 \qquad (7.37)$$

is called *coefficient of performance*. The smaller the temperature difference which has to be overcome the larger is the amount of the transferred thermal energy per unit technical work. Usually the heat is introduced into the heat pump condenser at a temperature higher then the environmental temperature for instance if heat from another process is available. In this case the exergy of the heat introduced into the fluid,

$$\left(1 - \frac{T_\infty}{T_{in}}\right)\int_1^2 v\dot{q}'''_{in}d\tau ,$$

is a measure of the technical work which can be gained by an ideal *Carnot* cycle, so not utilizing this exergy is considered as a loss. This leads some authors to introduce the so called exergetic coefficient of performance

$$ex\_cop := \frac{\left(1 - \frac{T_\infty}{T_{out}}\right)\int_1^2 v\dot{q}'''_{out}d\tau}{w^{vdp}_{t,12} + \left(1 - \frac{T_\infty}{T_{in}}\right)\int_1^2 v\dot{q}'''_{in}d\tau} = 1 - \frac{\left(1 - \frac{T_\infty}{T_{dissipation}}\right)\int_1^2 v\dot{q}'''_{dissipation}d\tau}{w^{vdp}_{t,12} + \left(1 - \frac{T_\infty}{T_{in}}\right)\int_1^2 v\dot{q}'''_{in}d\tau} < 1 .$$
(7.38)

This coefficient reflects the self-understanding that the heat $\int_1^2 v\dot{q}_{out}^{'''} d\tau$ has an exergetic equivalence (value) of $\left(1 - \dfrac{T_\infty}{T_{out}}\right)\int_1^2 v\dot{q}_{out}^{'''} d\tau$, which can be used as useful technical work. Thus, the exergetic coefficient of performance is a ratio of exergies. Only if there are no heat losses in the system is the exergetic coefficient of performance equal to one. If the introduction of the thermal energy happens under the environmental temperature we have

$$ex\_cop := \left(1 - \dfrac{T_\infty}{T_{out}}\right) cop. \qquad (7.39)$$

7.6 Three-fluid multi-component systems

Now we consider a system consisting of three different fluids dessignated with subscript l. Each of the fluids takes only $\alpha_l \geq 0$ part of the total system volume. Each of the fluids consists of several chemical components, designated with subscript i. Evaporation and condensation are allowed as well as mechanical transfer between the fluids having the same aggregate state as a constitutive fluid component. Details of the definitions of such a system are given in Chapter 1 or in [5]. Again as in the energy conservation equation for each fluid written in a specific enthalpy form we have

$$\alpha_l \left(\rho_l \dfrac{dh_l}{d\tau} - \dfrac{dp_l}{d\tau} \right) = DT_l^N, \qquad (7.40)$$

where

$$DT_l^N = \alpha_l \rho_l \left(\delta_l P_{kl} + \varepsilon_l \right)$$

$$+ \dot{q}_l''' + \sum_{i=1}^{i_{max}} \mu_{il,in}\left(h_{il,in} - h_{il}\right) + \sum_{i=1}^{i_{max}} \mu_{il,out}\left(h_{il,out} - h_{il}\right)$$

$$+ \sum_{\substack{m=1 \\ m \neq l}}^{l_{max}} \left[\mu_{Mml}\left(h_{Ml}^\sigma - h_{Ml}\right) + \sum_{n=1}^{n_{max}} \mu_{nml}\left(h_{nm} - h_{nl}\right) \right]$$

$$+ \dfrac{1}{2}\left[\mu_{l,in}\left(\mathbf{V}_{l,in} - \mathbf{V}_l\right)^2 - \mu_{l,out}\left(\mathbf{V}_{l,out} - \mathbf{V}_l\right)^2 + \sum_{m=1}^{3} \mu_{ml}\left(\mathbf{V}_m - \mathbf{V}_l\right)^2 \right]. \qquad (7.41)$$

are different components of the energy input into the flow. P_{kl} and ε_l are irreversible components of the energy dissipation per unit volume of the fluid mixture for whatever reason. The thermal energy introduced into the fluid per unit time into unit volume of the fluid mixture is $\dot{q}_l'''$. The specific energy of the fluid is influenced by the difference between the in-flowing and the fluid enthalpy of the chemical components $\sum_{i=1}^{i_{max}} \mu_{il,in} \left(h_{il,in} - h_{il} \right)$ as well as by the difference between the out-flowing and the system enthalpy of the chemical components $\sum_{i=1}^{i_{max}} \mu_{il,out} \left(h_{il,out} - h_{il} \right)$. Injecting and removing of mass of the system with a velocity different to the system velocity also gives rise to the internal energy change, $\frac{1}{2}\left[\mu_{l,in} \left(\mathbf{V}_{l,in} - \mathbf{V}_l \right)^2 - \mu_{l,out} \left(\mathbf{V}_{l,out} - \mathbf{V}_l \right)^2 + \sum_{m=1}^{3} \mu_{ml} \left(\mathbf{V}_m - \mathbf{V}_l \right)^2 \right]$. $\mu_{l,in}$ and $\mu_{l,out}$ are the inflow and outflow mass per unit volume of the fluid mixture and unit time. μ_{ml} is the mass transferred from fluid m to fluid l per unit volume of the mixture and unit time. The complete equivalent to this equation in its entropy form is

$$\rho_l \alpha_l \frac{ds_l}{d\tau} = \frac{1}{T_l} DT_l^N + \sum_{\substack{m=1 \\ m \neq l}}^{l_{max},w} \sum_{i=1}^{i_{max}} \left(\mu_{iml} - \mu_{ilm} \right) s_{il} - \mu_l s_l . \tag{7.42}$$

For the derivation see Chapter 5 or [6-8].

Now we multiply the entropy equation by the environment temperature, subtract it from the energy conservation equation, multiply the resulting equation by the time differential and divide by the fluid density and fluid volume fraction. The result for $\alpha_l \geq 0$ is

$$de_l + p_l dv_l - T_\infty ds_l = \frac{1}{\alpha_l \rho_l} \left[\left(1 - \frac{T_\infty}{T_l} \right) DT_l^N + T_\infty \sum_{\substack{m=1 \\ m \neq l}}^{l_{max},w} \sum_{i=1}^{i_{max}} \left(\mu_{iml} - \mu_{ilm} \right) s_{il} - \mu_l s_l \right] d\tau . \tag{7.43}$$

Defining the fluid exergy as

$$e_{x,l}^{pdv} := e_l + p_{l\infty} v_l - T_\infty s_l , \tag{7.44}$$

and the useful volume change work

$$-w_{l,l,12}^{pdv} = \int_1^2 \left(p_l - p_\infty \right) dv_l , \tag{7.45}$$

and after integration we obtain

$$-w_{t,l,12}^{pdv} = e_{x,l,1}^{pdv} - e_{x,l,2}^{pdv} + \int_{\tau_1}^{\tau_2} \frac{1}{\alpha_l \rho_l} \left[\begin{array}{c} \left(1 - \dfrac{T_\infty}{T_l}\right) DT_l^N \\ + T_\infty \sum_{\substack{m=1 \\ m \neq l}}^{l_{max}, w} \sum_{i=1}^{i_{max}} (\mu_{iml} - \mu_{ilm}) s_{il} - \mu_l s_l \end{array} \right] d\tau, \quad (7.46)$$

the fundamental exergy equation for each fluid inside the fluid mixture. The essential difference to the single phase formulation is that mass and thermal energy transport inside the mixture is also taken into account. This equation can be reduced to all simple cases discussed before by setting the corresponding simplifying assumptions.

In practical analysis not the available technical work from the each fluid but that from the mixture is of interests. To compute it we sum all the exergy equations and introduce the instantaneous local mass fraction of each fluid

$$x_l := \frac{\alpha_l \rho_l}{\rho}, \quad (7.47)$$

where

$$\rho = \frac{1}{v} := \sum_{l=1}^{l_{max}} \alpha_l \rho_l, \quad (7.48)$$

is the mixture density. The result obtained after the integration

$$-w_{t,12}^{pdv} = -\int_1^{2} \sum_{l=1}^{l_{max}} x_l de_{x,l}^{pdv} + \int_{\tau_1}^{\tau_2} v \sum_{l=1}^{l_{max}} \left\{ \begin{array}{c} \left(1 - \dfrac{T_\infty}{T_l}\right) DT_l^N \\ + T_\infty \sum_{\substack{m=1 \\ m \neq l}}^{l_{max}, w} \sum_{i=1}^{i_{max}} (\mu_{iml} - \mu_{ilm}) s_{il} - \mu_l s_l \end{array} \right\} d\tau, \quad (7.49)$$

is the specific volume change work of the mixture.

$$-w_{t,12}^{pdv} = \int_1^{2} \sum_{l=1}^{l_{max}} x_l (p_l - p_\infty) dv_l, \quad (7.50)$$

or for the case of assumed equal fluid pressures (single pressure model) we have

$$-w_{t,12}^{pdv} = \int_1^2 (p - p_\infty) \sum_{l=1}^{l_{max}} x_l dv_l . \qquad (7.51)$$

In a similar way the alternative exergy equation is

$$-w_{t,12}^{vdp} = -\int_1^2 \sum_{l=1}^{l_{max}} x_l de_{x,l}^{vdp} + \int_{\tau_1}^{\tau_2} v \sum_{l=1}^{l_{max}} \left\{ \begin{array}{c} \left(1 - \dfrac{T_\infty}{T_l}\right) DT_l^N \\ \\ +T_\infty \left[\displaystyle\sum_{\substack{m=1 \\ m \neq l}}^{l_{max}} \sum_{i=1}^{w_{i_{max}}} (\mu_{iml} - \mu_{ilm}) s_{il} - \mu_l s_l \right] \end{array} \right\} d\tau, \qquad (7.52)$$

where the specific technical work of the mixture is

$$-w_{t,12}^{vdp} = \int_1^2 \sum_{l=1}^{l_{max}} x_l v_l dp_l . \qquad (7.53)$$

For the case of assumed equal fluid pressures (single pressure model) we have

$$-w_{t,12}^{vdp} = \int_1^2 \left(\sum_{l=1}^{l_{max}} x_l v_l \right) dp = \int_1^2 v dp . \qquad (7.54)$$

Note that due to evaporation, condensation, or other mass transfer processes the mass concentration between the two selected states may change, which make the estimation of the integrals more complicated. For constant mass concentrations the exergy equations simplify.

7.7 Practical relevance

The general exergy equations for mixtures provide a simple way for computing the available volume change work or the technical work for a variety of processes. The multi-fluid exergy equations allow us to estimate the changes of the available technical work of the system due to processes like evaporation or condensation in non-equilibrium mixtures, which seems to be not known up to now. It is a generalization of the exergy principal to multi-phase multi-component systems. For the limiting case of a single fluid the equations reduce to the known relationships of classical thermodynamics. To the author's knowledge, Eqs. (46), (49) and (52) are derived for the first time in [9].

References

1. Baer HD (1996) Thermodynamik, Springer
2. Carnot NLS (1824) Réflexions sur la puissance motrice du feu et sur les machines propres à développer cette puissance, See also in "Oswalds Klassiker der exakten Wissenschaften", W. Engelmann, Leipzig.
3. Elsner N (1974) Grundlagen der Technischen Thermodynamik, Akademie Verlag, Berlin
4. Gouy M (Novembre 1889) Sur L'énrgie Utilisable, Journal de physique teoretique et appliquee, II Ser. 8 pp 501-518
5. Kolev NI (1994) The code IVA4: Modeling of mass conservation in multi-phase multi componet flows in heterogeneous porous media, Kerntechnik, vol 59 no 4-5 pp 226-237
6. Kolev NI (1995) The code IVA4: Second law of thermodynamics for multi phase flows in heterogeneous porous media, Kerntechnik, vol 6 no 1 pp 1-39
7. Kolev NI (1997) Comments on the entropy concept, Kerntechnik, vol 62 no 1 pp 67-70
8. Kolev NI (1998) On the variety of notation of the energy conservation principle for single phase flow, Kerntechnik, vol 63 no 3 pp 145-156
9. Kolev NI (2001) Exergie von Mehrphasen-Mehrkomponenten-Systemen, Seminar „Verfahrenstechnik/ Thermodynamik" Institut für Energie- und Verfahrenstechnik Universität Paderborn, reported at 12.02.2001
10. Rant Z (1956-1964) Exergie, ein neues Wort für „technische Arbeitsfähigkeit", Forsch.-Ing. Wes., vol 22 (1956) S 36-37; Die Thermodynamik von Heizprozessen. Strojniki vertnik vol 8 (1962) S 1-2 (Slowenisch). Die Heiztechnik und der zweite Hauptsatz der Thermodynamik, Gaswärme, vol 12 (1962) S 297-304. Siehe auch in: Exergie und Anergie. Wissenschaftliche Zeitschrift der TU Dresden, vol 13 (1964) S 1145 – 1149.
11. Reynolds WC, Perkins H (1977) Engineering Thermodynamics, McGraw-Hill Book Company, New York, Second Edition
12. Stephan K, Mayinger F (1998) Techn. Thermodynamik, Bd.1, Springer, 15. Auflage
13. Thomson W (1852), On the economy of the heating and cooling of buildings by means of current of air. Proc. Phil. Soc. (Glasgow) 3 pp 268-272
14. Zwicker A (1976) Wärmepumpen, in Taschenbuch Maschinenbau, Bd.2, VEB Verlag Technik, Berlin, pp 858-864

8 One-dimensional three-fluid flows

We call "one dimensional" the flow in pipes or in a pipe networks. We understand here a flow with cross section averaged flow characteristics with special wall boundary layer treatment, like pressure loss modeling, heat transfer modeling etc. The flow axis is of course arbitrarily oriented in space. Therefore this class of flows is one dimensional along a curvilinear pipe axis. A *network* consists of pipes and knots. An example is illustrated in Fig. 8.1.

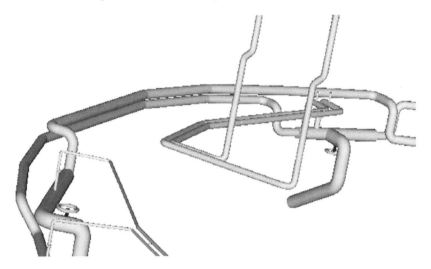

Fig. 8.1. Pipe network

8.1 Summary of the local volume- and time-averaged conservation equations

In order to facilitate the practical application of the conservation equations derived in the previous chapters we give here a summary of the equations simplified for the case of flow in the axial direction only. The mass conservation equation (1.62) derived in Chapter 1 simplifies as follows

$$\frac{\partial}{\partial \tau}(\alpha_l \rho_l \gamma_v) + \frac{\partial}{\partial z}(\alpha_l \rho_l w_l \gamma_z) = \gamma_v \sum_{m=1}^{l_{\max},w}(\mu_{ml} - \mu_{lm}). \tag{8.1}$$

The mass conservation equation for each species inside the velocity field (1.90) derived in Chapter 1 simplifies as follows

$$\frac{\partial}{\partial \tau}(\alpha_l \rho_l C_{il} \gamma_v) + \frac{\partial}{\partial z}\left[\alpha_l \rho_l \left(w_l C_{il} - D_{il}^* \frac{\partial C_{il}}{\partial z}\right) \gamma_z\right] = \gamma_v \sum_{m=1}^{l_{\max},w}(\mu_{iml} - \mu_{ilm})$$

for $\alpha_l \geq 0$, $i = 1, i_{\max}$ \hfill (8.2)

or alternatively Eq. (1.96)

$$\alpha_l \rho_l \left(\gamma_v \frac{\partial C_{il}}{\partial \tau} + w_l \gamma_z \frac{\partial C_{il}}{\partial z}\right) - \frac{\partial}{\partial z}\left[\alpha_l \rho_l \left(D_{il}^* \frac{\partial C_{il}}{\partial z}\right) \gamma_z\right] = \gamma_v (\mu_{il} - C_{il} \mu_l). \tag{8.3}$$

The particle number density equation (1.109) derived in Chapter 1 simplifies as follows

$$\frac{\partial}{\partial \tau}(n_l \gamma_v) + \frac{\partial}{\partial z}\left[\left(w_l n_l - \frac{v_l^t}{Sc_l^t} \frac{\partial n_l}{\partial z}\right) \gamma_z\right] = \gamma_v (\dot{n}_{l,kin} - \dot{n}_{l,coal} + \dot{n}_{l,sp}) \quad \text{for } \alpha_l \geq 0. \tag{8.4}$$

The momentum equation (2.233) derived in Chapter 2 simplifies as follows

$$\alpha_l \rho_l \gamma_v \frac{\partial w_l}{\partial \tau} + \alpha_l^e \rho_l \gamma_z \frac{1}{2} \frac{\partial w_l^2}{\partial z} - \frac{\partial}{\partial z}\left(\alpha_l^e \rho_l v_l^* \gamma_z \frac{\partial w_l}{\partial z}\right)$$

$$+ \alpha_l^e \gamma_z \frac{\partial p_l}{\partial z} - \Delta p_m^{w\sigma*} \frac{\partial \gamma_z}{\partial z} + f_{l\alpha} + \gamma_v \alpha_l \rho_l g \cos\varphi$$

$$- \gamma_v \left\{ \sum_{\substack{m=1 \\ m \neq l}}^{3} \left[\bar{c}_{ml}^{vm} \left(\frac{\partial \Delta w_{ml}}{\partial \tau} + w_l \frac{\partial \Delta w_{ml}}{\partial z} \right) \right] - \rho_l c_{lw}^{vm} \frac{\partial w_l}{\partial \tau} \right\}$$

8.1 Summary of the local volume- and time-averaged conservation equations

$$= \gamma_v \left\{ \begin{array}{c} \sum_{\substack{m=1 \\ m \neq l}}^{3} \left[\overline{c}_{ml}^{d} \left| \Delta w_{ml} \right| \cdot \Delta w_{ml} + \mu_{ml} \left(w_m - w_l \right) \right] + \overline{c}_{lw}^{d} \left| w_l \right| w_l \\ + \mu_{wl} \left(w_{wl} - w_l \right) - \mu_{lw} \left(w_{lw} - w_l \right) \end{array} \right\}. \qquad (8.5)$$

Here φ is the polar angle between the flow direction and the upwards directed vertical as shown in Fig. 8.2

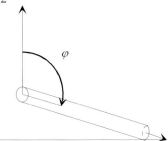

Fig. 8.2. Definition of the polar angle

The entropy equation (5.125) derived in Chapter 5 simplifies as follows

$$\rho_l \left(\alpha_l \gamma_v \frac{\partial s_l}{\partial \tau} + \alpha_l^e w_l \gamma_z \frac{\partial s_l}{\partial z} \right) - \frac{1}{T_l} \frac{\partial}{\partial z} \left[\left(\alpha_l^e \lambda_l^* \frac{\partial T_l}{\partial z} \right) \gamma_z \right]$$

$$- \frac{\partial}{\partial z} \left\{ \alpha_l^e \rho_l \left[\sum_{i=2}^{i_{\max}} \left(s_{il} - s_{1l} \right) D_{il}^* \frac{\partial C_{il}}{\partial z} \right] \gamma_z \right\}$$

$$= \gamma_v \left[\frac{1}{T_l} DT_l^N + \sum_{i=1}^{i_{\max}} \mu_{il} \left(s_{il} - s_l \right) \right] \text{ for } \alpha_l \geq 0, \qquad (8.6)$$

Alternatively, the temperature equation (5.176) derived in Chapter 5 can be used instead of the entropy equation for some applications

$$\rho_l c_{pl} \left(\alpha_l \gamma_v \frac{\partial T_l}{\partial \tau} + \alpha_l^e w_l \gamma_z \frac{\partial T_l}{\partial z} \right) - \left[1 - \rho_l \left(\frac{\partial h_l}{\partial p} \right)_{T_l, all\_C's} \right] \left(\alpha_l \gamma_v \frac{\partial p}{\partial \tau} + \alpha_l^e w_l \gamma_z \frac{\partial p}{\partial z} \right)$$

$$-\frac{\partial}{\partial z}\left[\left(\alpha_l^e \lambda_l^* \frac{\partial T_l}{\partial z}\right)\gamma_z\right] + T_l \sum_{i=2}^{i_{\max}} \Delta s_{il}^{np} \frac{\partial}{\partial z}\left[\left(\alpha_l^e \rho_l D_{il}^*\frac{\partial C_{il}}{\partial z}\right)\gamma_z\right]$$

$$= \gamma_v\left[DT_l^N - T_l \sum_{i=2}^{i_{\max}} \Delta s_{il}^{np}\left(\mu_{il} - C_{il}\mu_l\right)\right]. \tag{8.7}$$

The volume conservation equation (5.188) derived in Chapter 5 simplifies for one-dimensional flow as follows

$$\frac{\gamma_v}{\rho a^2}\frac{\partial p}{\partial \tau} + \gamma_z\left(\sum_{l=1}^{l_{\max}}\frac{\alpha_l w_l}{\rho_l a_l^2}\right)\frac{\partial p}{\partial z} + \frac{\partial}{\partial z}\left(\sum_{l=1}^{l_{\max}}\alpha_l w_l \gamma_z\right) = \sum_{l=1}^{l_{\max}}D\alpha_l - \frac{\partial \gamma_v}{\partial \tau}, \tag{8.8}$$

where

$$D\alpha_l = \frac{1}{\rho_l}\left\{\gamma_v \mu_l - \frac{1}{\rho_l}\left[\frac{\partial \rho_l}{\partial s_l}\overline{Ds_l^N} + \sum_{i=2}^{i_{\max}}\frac{\partial \rho_l}{\partial C_{il}}\overline{DC_{il}^N}\right]\right\} \tag{8.9}$$

and

$$\overline{DC_{il}^N} = \frac{\partial}{\partial z}\left[\alpha_l^e \rho_l\left(D_{il}^*\frac{\partial C_{il}}{\partial z}\right)\gamma_z\right] + \gamma_v\left(\mu_{il} - C_{il}\mu_l\right), \tag{8.10}$$

$$\overline{Ds_l^N} = \frac{1}{T_l}\frac{\partial}{\partial z}\left[\left(\alpha_l^e \lambda_l^*\frac{\partial T_l}{\partial z}\right)\gamma_z\right] + \frac{\partial}{\partial z}\left\{\alpha_l^e \rho_l\left[\sum_{i=2}^{i_{\max}}(s_{il} - s_{1l})D_{il}^*\frac{\partial C_{il}}{\partial z}\right]\gamma_z\right\}$$

$$+\gamma_v\left[\frac{1}{T_l}DT_l^N + \sum_{i=1}^{i_{\max}}\mu_{il}(s_{il} - s_l)\right]. \tag{8.11}$$

Remember that a is the *sonic velocity* in a homogeneous multi-phase mixture defined as

$$\frac{1}{\rho a^2} = \sum_{l=1}^{3}\frac{\alpha_l}{\rho_l a_l^2} = \frac{1}{p}\sum_{l=1}^{3}\frac{\alpha_l}{\kappa_l} = \frac{1}{\kappa p}, \tag{8.12}$$

and

$$\rho = \sum_{l=1}^{3}\alpha_l \rho_l \tag{8.13}$$

is the mixture density.

8.2 Treatment of the field pressure gradient forces

8.2.1 Dispersed flows

The pressure terms in the momentum equation are now written for a continuous and a disperse phase

$$f_{d\alpha} = \left(p_d - p_c - \delta_d \sigma_{dc} K_d - \Delta p_c^{d\sigma*}\right) \frac{\partial}{\partial z}\left(\alpha_d^e \gamma_z\right), \qquad (8.14)$$

and

$$f_{c\alpha} = \Delta p_c^{d\sigma*} \frac{\partial}{\partial z}\left(\alpha_d^e \gamma_z\right). \qquad (8.15)$$

where

$$\Delta p_c^{d\sigma*} = -0.37 c_{cd}^d \rho_c \left(w_c - w_d\right)^2 \qquad (8.16)$$

is Eq. (2.178) in Chapter 2. An approximation is frequently used in the literature in the following form

$$\ldots + \alpha_c \gamma_z \frac{\partial p}{\partial z} - \Delta p_c^{w\sigma*} \frac{\partial \gamma_z}{\partial z} + \gamma_v \alpha_c \rho_c g \cos\varphi \ldots, \qquad (8.17)$$

$$\ldots + \alpha_d \gamma_z \frac{\partial p}{\partial z} - 0.37 c_{cd}^d \rho_c \left|w_c - w_d\right|\left(w_c - w_d\right)\frac{\partial}{\partial z}\left(\alpha_d \gamma_z\right) + \gamma_v \alpha_d \rho_d g \cos\varphi \ldots . \qquad (8.18)$$

8.2.2 Stratified flow

In stratified flow the gravitation is the dominant force. It is manifested in different pressure distribution over the cross section which gives rise to differences in the averaged system pressure and the averaged field pressures. In Chapter 2 the Eqs. (2.103) and (2.104)

$$\ldots + \gamma_z \left[\alpha_1 \frac{\partial p}{\partial z} - \alpha_1 \alpha_2 \left(\rho_1 - \rho_2\right) g H \frac{\partial \alpha_1}{\partial z}\right] - \Delta p_1^{w\sigma} \frac{\partial \gamma_z}{\partial z} \ldots \qquad (8.19)$$

$$\ldots + \gamma_z \left[\alpha_2 \frac{\partial p}{\partial z} + \alpha_1 \alpha_2 \left(\rho_1 - \rho_2\right) g H \frac{\partial \alpha_1}{\partial z}\right] - \Delta p_2^{w\sigma} \frac{\partial \gamma_z}{\partial z} \ldots \qquad (8.20)$$

are derived for rectangular channels and Eqs. (2.148) and (2.149)

$$\ldots + \gamma_z \left[\alpha_1 \frac{\partial p}{\partial z} + \alpha_1 \alpha_2 (\rho_2 - \rho_1) g \frac{\pi D_h}{4 \sin \theta} \frac{\partial \alpha_1}{\partial z} \right] - \Delta p_1^{w\sigma} \frac{\partial \gamma_z}{\partial z} \ldots \quad (8.21)$$

$$\ldots + \gamma_z \left[\alpha_2 \frac{\partial p}{\partial z} - \alpha_1 \alpha_2 (\rho_2 - \rho_1) g \frac{\pi D_h}{4 \sin \theta} \frac{\partial \alpha_1}{\partial z} \right] - \Delta p_2^{w\sigma} \frac{\partial \gamma_z}{\partial z} \ldots \quad (8.22)$$

for pipes. Note that the term $\ldots \gamma_v \alpha_l \rho_l g \cos \varphi \ldots$ does not appear any more in these equations.

8.3 Pipe deformation due to temporal pressure change in the flow

Pipes with elastic walls change the propagation velocity of pressure pulses because of the energy dissipation for mechanical deformation. The effect can be taken into account into the pressure equation (8.8) as follows

$$\gamma_v \left(\frac{1}{\rho a^2} + \frac{1}{\gamma_v} \frac{d\gamma_v}{dp} \right) \frac{\partial p}{\partial \tau} + \gamma_z \left[\left(\sum_{l=1}^{3} \frac{\alpha_l w_l}{\rho_l a_l^2} \right) + \frac{1}{\gamma_z} \frac{d\gamma_z}{dp} \right] \frac{\partial p}{\partial z}$$

$$+ \frac{\partial}{\partial z} \left(\sum_{l=1}^{3} \alpha_l w_l \gamma_z \right) = \sum_{l=1}^{l_{\max}} D\alpha_l . \quad (8.23)$$

With the terms $d\gamma_v / dp$ and $d\gamma_z / dp$ we take into account the elasticity of the pipe. The method for computation of the two terms is given below.

For closed pipe with uniform pressure the deformation caused by the pressure can be computed using text book solution

$$dz / \Delta z = (\sigma_z - \mu^* \sigma_\theta - \mu^* \sigma_r) / E, \quad (8.24)$$

$$dr_i / r_i = (\sigma_\theta - \mu^* \sigma_z - \mu^* \sigma_r) / E, \quad (8.25)$$

$$d\delta / \delta = (\sigma_r - \mu^* \sigma_\theta - \mu^* \sigma_z) / E, \quad (8.26)$$

where the subscripts i and a stand for inner and outer wall radius, respectively. Here E is the *elasticity modulus* e.g. for steel $E=1 \times 10^{11}$ Nm^2, μ^* is the contraction number, e.g. 0.3. The axial, the radial and the angular stresses are functions of the pressure change of the fluid

$$\sigma_r = dp / \left[(r_a/r_i)^2 - 1 \right], \qquad (8.27)$$

$$\sigma_r = -dp, \qquad (8.28)$$

$$\sigma_\theta = dp \left(r_a^2 + r_i^2 \right) / \left(r_a^2 - r_i^2 \right). \qquad (8.29)$$

Here r_i and r_a are the inner and the outer radius.

$$\delta = r_a - r_i \qquad (8.30)$$

is the wall thickness. The change of the flow volume due to elastic pipe deformation is therefore

$$\frac{1}{V_{pipe}} \frac{dV_{pipe}}{dp} = \frac{1}{\gamma_v} \frac{d\gamma_v}{dp} = \frac{1}{\pi r_i^2 \Delta z} \frac{d}{dp}\left(\pi r_i^2 \Delta z \right) = \frac{2}{r_i} \frac{dr_i}{dp} + \frac{1}{\Delta z} \frac{d\Delta z}{dp} = \frac{\psi}{E}, \qquad (8.31)$$

where

$$\psi = \left[2(1+\mu^*) r_a^2 + (3 - 6\mu^*) r_i^2 \right] / \left(r_a^2 - r_i^2 \right) \qquad (8.32)$$

is frequently called *geometry factor*. For long pipes the local distribution of pressure makes it difficult to compute the axial component. Often, only the radial component is used

$$\psi = 2\left\{ \left[r_a^2 + (1-\mu^*) r_i^2 \right] / \left(r_a^2 - r_i^2 \right) + \mu^* \right\}. \qquad (8.33)$$

From the geometry factor we see that the smaller the pipe wall thickness the stronger the effect of the pipe elasticity on the pressure wave propagation.

8.4 Some simple cases

Concentration and specific entropy propagation in incompressible flow with prescribed velocity: In many practical applications where the convection is the governing phenomenon the diffusion terms in the concentration and the entropy equations can be neglected. The resulting equations

$$\alpha_1 \rho_1 \left(\gamma_v \frac{\partial C_{il}}{\partial \tau} + w_1 \gamma_z \frac{\partial C_{il}}{\partial z} \right) = \gamma_v \left(\mu_{il} - \mu_1 C_{il} \right), \qquad (8.34)$$

$$\alpha_l \rho_l \left(\gamma_v \frac{\partial s_l}{\partial \tau} + w_l \gamma_z \frac{\partial s_l}{\partial z} \right) = \gamma_v \left[\frac{1}{T_l} DT_l^N + \sum_{i=1}^{i_{max}} \mu_{il} \left(s_{il} - s_l \right) \right], \qquad (8.35)$$

can be easily analytically integrated. The eigenvalues of the characteristic matrix of the above system are $\lambda_l = w_l \gamma_z / \gamma_v$, where $l = 1,2,3$ for each velocity field and equation, respectively. This means that changes of the concentrations and of the specific entropies travel with the corresponding field velocity multiplied by γ_z / γ_v. For pipes with constant cross section the multiplier is unity. If one changes the coordinate system for each equation with an rectangular system having one of the axes tangential to the curve defined with inclination $dz/d\tau = \lambda_l$ in the time - space plane, called characteristic curves, the equations take a very simple form

$$\frac{dC_{il}}{d\tau} = \left(\mu_{il} - \mu_l C_{il} \right) / \left(\alpha_l \rho_l \right), \qquad (8.36)$$

$$\frac{ds_l}{d\tau} = \left(\frac{1}{T_l} DT_l^N + \sum_{i=1}^{i_{max}} \mu_{il} s_{il} - \mu_l s_l \right) / \left(\alpha_l \rho_l \right), \qquad (8.37)$$

called characteristic form (*Rieman*). In case of no sources, that is the right-hand side of the equations is equal to zero,

$$\frac{dC_{il}}{d\tau} = 0, \qquad (8.38)$$

$$\frac{ds_l}{d\tau} = 0, \qquad (8.39)$$

the concentrations and the specific entropies are constant

$$C_{il} = const, \qquad (8.40)$$

$$s_l = const, \qquad (8.41)$$

along the characteristic line. The asymptotic solutions are

$$C_{il\infty} = \mu_{il} / \mu_l, \qquad (8.42)$$

$$s_{l\infty} = \frac{1}{\mu_l} \left(\frac{1}{T_l} DT_l^N + \sum_{i=1}^{i_{max}} \mu_{il} s_{il} \right). \qquad (8.43)$$

Rewriting the system in the following form

$$\frac{dC_{il}}{d\tau} = (C_{il\infty} - C_{il})\mu_l / (\alpha_l \rho_l), \qquad (8.44)$$

$$\frac{ds_l}{d\tau} = (s_{l\infty} - s_l)\mu_l / (\alpha_l \rho_l), \qquad (8.45)$$

and assuming that during a time step $\Delta\tau$ the expressions $C_{il\infty} = const$, $s_{l\infty} = const$, the equations can be analytically integrated with the result

$$C_{il,\tau+\Delta\tau} = C_{il\infty} - (C_{il\infty} - C_{il})/\exp\left[\int_\tau^{\tau+\Delta\tau} \mu_l/(\alpha_l \rho_l)\,d\tau\right], \qquad (8.46)$$

$$s_{l,\tau+\Delta\tau} = s_{l\infty} - (s_{l\infty} - s_l)/\exp\left[\int_\tau^{\tau+\Delta\tau} \mu_l/(\alpha_l \rho_l)\,d\tau\right]. \qquad (8.47)$$

Note once more that the so obtained analytical solutions are valid along the characteristic curve.

Steady state flow: In steady state the time derivatives are equal to zero. I split the resulting system in two groups of equations. For the first group

$$\frac{dC_{il}}{dz} = \gamma_v (C_{il\infty} - C_{il})\mu_l / (\alpha_l \rho_l w_l \gamma_z), \qquad (8.48)$$

$$\frac{ds_l}{dz} = \gamma_v (s_{l\infty} - s_l)\mu_l / (\alpha_l \rho_l w_l \gamma_z), \qquad (8.49)$$

assuming that along Δz the expressions $C_{il\infty} = const$, $s_{l\infty} = const$ hold, the equations can be analytically integrated with the result

$$C_{il,z+\Delta z} = C_{il\infty} - (C_{il\infty} - C_{il})/\exp\left[\int_z^{z+\Delta z} \gamma_v \mu_l/(\alpha_l \rho_l w_l \gamma_z)\,dz\right], \qquad (8.50)$$

$$s_{l,z+\Delta z} = s_{l\infty} - (s_{l\infty} - s_l)/\exp\left[\int_z^{z+\Delta z} \gamma_v \mu_l/(\alpha_l \rho_l w_l \gamma_z)\,dz\right]. \qquad (8.51)$$

The second group of equations is discussed below. The mass conservation equation

$$\frac{d}{dz}\left(\alpha_l \rho_l w_l \gamma_z\right) = \gamma_v \mu_l ,\qquad(8.52)$$

is expanded and the densities are replaced by their equals from the series expansion of the state equations. The result is

$$\frac{d\alpha_l}{dz} + \frac{\alpha_l}{w_l}\frac{dw_l}{dz} + \frac{\alpha_l}{\rho_l a_l^2}\frac{dp}{dz} + \alpha_l \frac{1}{\gamma_z}\frac{d\gamma_z}{dz} = D\alpha_l .\qquad(8.53)$$

Using the concentration and the entropy equations we obtain

$$D\alpha_l = \frac{1}{\rho_l w_l}\frac{\gamma_v}{\gamma_z}\mu_l - \frac{\alpha_l}{\rho_l}\left[\frac{\partial \rho_l}{\partial s_l}\frac{ds_l}{dz} + \sum_{i=2}^{i_{\max}}\frac{\partial \rho_l}{\partial C_{il}}\frac{dC_{il}}{dz}\right]$$

$$= \frac{1}{\rho_l w_l}\frac{\gamma_v}{\gamma_z}\left\{\mu_l - \frac{1}{\rho_l}\left[\frac{\partial \rho_l}{\partial s_l}(s_{l\infty} - s_l)\mu_l + \mu_l \sum_{i=2}^{i_{\max}}\frac{\partial \rho_l}{\partial C_{il}}(C_{il\infty} - C_{il})\right]\right\}.\qquad(8.54)$$

Having in mind that

$$\sum_{l=1}^{3}\frac{d\alpha_l}{dz} = 0,\qquad(8.55)$$

the sum of the volume fraction equations become the so called pressure equation

$$\sum_{l=1}^{3}\frac{\alpha_l}{w_l}\frac{dw_l}{dz} + \frac{1}{\rho a^2}\frac{dp}{dz} = \sum_{l=1}^{3} D\alpha_l - \frac{1}{\gamma_z}\frac{d\gamma_z}{dz}.\qquad(8.56)$$

Next we look for expression for the velocity gradients obtained from the momentum equations with the purpose to substitute them into the pressure equation and solve the so obtained equation with respect to the pressure gradient. The momentum equation is

$$\left(\alpha_l^e \rho_l + \frac{\gamma_v}{\gamma_z}\sum_{\substack{m=1\\m\neq l}}^{3}\bar{c}_{ml}^{vm}\right)\frac{dw_l}{dz} - \frac{\gamma_v}{\gamma_z}\sum_{\substack{m=1\\m\neq l}}^{3}\bar{c}_{ml}^{vm}\frac{dw_m}{dz} + \frac{\alpha_l}{w_l}\frac{dp}{dz} = \frac{Z_l}{w_l}\qquad(8.57)$$

where

$$Z_l = \frac{\gamma_v}{\gamma_z} \left\{ \begin{array}{l} -\alpha_l \rho_l g + \sum_{\substack{m=1 \\ m \neq l}}^{3} \left[\overline{c}_{ml}^d \left| \Delta w_{ml} \right| \Delta w_{ml} + \mu_{ml} \left(w_m - w_l \right) \right] - \overline{c}_{lw}^d \left| w_l \right| w_l \\ \\ + \mu_{wl} \left(w_{wl} - w_l \right) - \mu_{lw} \left(w_{lw} - w_l \right) \end{array} \right\}, \qquad (8.58)$$

contains no differential terms. The system of algebraic equations containing the velocity gradients is therefore

$$\mathbf{A} \left[\frac{dw}{dz} \right] = \mathbf{B} - \mathbf{C} \frac{dp}{dz}. \qquad (8.59)$$

The components of the algebraic vectors **C** and **B** are $c_l = \alpha_l / w_l$ and $b_l = Z_l / w_l$, respectively. The diagonal and the non-diagonal elements of the matrix **A** are

$$a_{ll} = \alpha_l^e \rho_l + \frac{\gamma_v}{\gamma_z} \sum_{\substack{m=1 \\ m \neq l}}^{3} \overline{c}_{ml}^{vm}, \qquad (8.60)$$

$$a_{lm} = a_{ml} = -\frac{\gamma_v}{\gamma_z} \sum_{\substack{m=1 \\ m \neq l}}^{3} \overline{c}_{ml}^{vm}, \qquad (8.61)$$

respectively. Having in mind that $\overline{c}_{ml}^{vm} = \overline{c}_{lm}^{vm}$, we realize that the coefficient matrix **A** of the velocity derivatives is symmetric. If the virtual mass coefficients are set to zero there is only one velocity gradient in each momentum equation

$$\frac{1}{2} \frac{dw_l^2}{dz} + \frac{1}{\rho_l} \frac{dp}{dz} = \frac{Z_l}{\alpha_l \rho_l}. \qquad (8.62)$$

The algebraic system (8.59) can be solved with respect to the velocity gradients to provide

$$\det |\mathbf{A}| = a_{11} a_{22} a_{33} + 2 a_{12} a_{23} a_{13} - a_{13}^2 a_{22} - a_{23}^2 a_{11} - a_{12}^2 a_{33} \neq 0. \qquad (8.63)$$

Note that if one field does not exists the coupling coefficients with the other fields are by definition zeros. In such case the rank of the matrix is reduced by one. Solving with respect to the velocity gradients gives

$$\frac{dw_l}{dz} = w_l^* - R_l \frac{dp}{dz} \tag{8.64}$$

where

$$w_l^* = \left[\sum_{m=1}^{3}(Z_l/w_l)\overline{a}_{lm}\right]/\det|\mathbf{A}|, \tag{8.65}$$

$$R_l = \left[\sum_{m=1}^{3}(\alpha_l/w_l)\overline{a}_{lm}\right]/\det|\mathbf{A}|, \tag{8.66}$$

$$\overline{a}_{11} = a_{22}a_{33} - a_{23}^2, \; \overline{a}_{22} = a_{11}a_{33} - a_{13}^2, \; \overline{a}_{33} = a_{11}a_{22} - a_{12}^2, \tag{8.67-8.69}$$

$$\overline{a}_{12} = \overline{a}_{21} = a_{32}a_{13} - a_{12}a_{33}, \; \overline{a}_{13} = \overline{a}_{31} = a_{12}a_{23} - a_{22}a_{13}, \; \overline{a}_{23} = \overline{a}_{32} = a_{21}a_{13} - a_{23}a_{11}.$$
$$\tag{8.70-8.72}$$

Replacing the so obtained velocity gradients into the pressure equation we obtain

$$\frac{dp}{dz} = -\left\{\sum_{l=1}^{3}D\alpha_l - \frac{1}{\gamma_z}\frac{d\gamma_z}{dz} - \sum_{l=1}^{3}\frac{\alpha_l}{w_l}w_l^*\right\}/\left(1-Ma^2\right)\sum_{l=1}^{3}\frac{\alpha_l}{w_l}R_l. \tag{8.73}$$

Here

$$Ma^2 = \frac{1}{\rho a^2}/\sum_{l=1}^{3}\frac{\alpha_l}{w_l}R_l \tag{8.74}$$

is the definition of a dimensionless number corresponding to the *Mach* number for single-phase flows. Obviously if this number tends to unity the pressure gradient tends to minus infinity. If this happens at some position in the channel no pressure disturbances coming from flow downwards can influence the flow. This state is called *critical flow*. It plays an important role in the technology. The critical flow is expected to happen in the smallest cross section of the pipe where the velocities have local maximums and the pressure local minimum, or at the end of pipes with constant cross section.

Let us summarize the resulting system in a form very convenient for numerical integration.

$$\frac{dp}{dz} = -\left\{\sum_{l=1}^{3}D\alpha_l - \frac{1}{\gamma_z}\frac{d\gamma_z}{dz} - \sum_{l=1}^{3}\frac{\alpha_l}{w_l}w_l^*\right\}/\left(1-Ma^2\right)\sum_{l=1}^{3}\frac{\alpha_l}{w_l}R_l, \tag{8.75}$$

8.4 Some simple cases

$$\frac{dw_l}{dz} = w_l^* - R_l \frac{dp}{dz}, \qquad (8.76)$$

or alternatively

$$\frac{1}{2}\frac{dw_l^2}{dz} = \frac{1}{\det|A|}\left[\sum_{m=1}^{3}\left(Z_l - \alpha_l \frac{dp}{dz}\right)\bar{a}_{lm}\right] = \frac{1}{\det|A|}\left[\sum_{m=1}^{3}Z_l\bar{a}_{lm} - \left(\sum_{m=1}^{3}\alpha_l\bar{a}_{lm}\right)\frac{dp}{dz}\right] \qquad (8.77)$$

and

$$\frac{d\alpha_l}{dz} = D\alpha_l - \frac{\alpha_l}{w_l}\frac{dw_l}{dz} - \frac{\alpha_l}{\rho_l a_l^2}\frac{dp}{dz} - \alpha_l \frac{1}{\gamma_z}\frac{d\gamma_z}{dz}, \qquad (8.78)$$

$$\frac{dC_{il}}{dz} = \gamma_v \left(C_{il\infty} - C_{il}\right)\mu_l / \left(\alpha_l \rho_l w_l \gamma_z\right), \qquad (8.79)$$

$$\frac{ds_l}{dz} = \gamma_v \left(s_{l\infty} - s_l\right)\mu_l / \left(\alpha_l \rho_l w_l \gamma_z\right), \qquad (8.80)$$

$$\frac{dn_l}{dz} = \frac{\gamma_v}{w_l \gamma_z}\left(\dot{n}_{l,kin} - \dot{n}_{l,coal} + \dot{n}_{l,sp}\right) - \frac{n_l}{w_l}\frac{dw_l}{dz} - \frac{n_l}{\gamma_z}\frac{d\gamma_z}{dz}. \qquad (8.81)$$

Nozzles frozen flow: For steady state with neglected virtual mass forces the momentum equation takes the simple form (8.64). Replacing the density in the gas momentum equation with the expression for the isentropic state of change we obtain

$$\frac{1}{2}\frac{dw_1^2}{dz} + \frac{p_0}{\rho_{10}}\frac{\kappa_1}{\kappa_1 - 1}\frac{d}{dz}\varepsilon^{\frac{\kappa_1-1}{\kappa_1}} = \frac{Z_1}{\alpha_1 \rho_1}, \qquad (8.82)$$

where

$$\varepsilon = \frac{p}{p_0} \qquad (8.83)$$

is the pressure ratio. Here we designate with 0 some reference state, e.g. the state at the pipe inlet. For the second and the third velocity field Eq. (8.62) remains unchanged. Integrating between two points we obtain

$$\frac{1}{2}\left(w_1^2 - w_{10}^2\right) = \int_{z_0}^{z}\frac{Z_1}{\alpha_1 \rho_1}dz - \frac{p_0}{\rho_{10}}\frac{\kappa_1}{\kappa_1 - 1}\left(\varepsilon^{\frac{\kappa_1-1}{\kappa_1}} - 1\right), \qquad (8.84)$$

$$\frac{1}{2}\left(w_l^2 - w_{l0}^2\right) = \int_{z_0}^{z} \frac{Z_l}{\alpha_l \rho_l} dz - \frac{p_0}{\rho_{l0}}(\varepsilon - 1), \text{ for } l = 2,3. \tag{8.85}$$

For short pipes, nozzles and orifices with negligible interfacial drag we obtain

$$w_1^2 = w_{10}^2 + 2\frac{p_0}{\rho_{10}} \frac{\kappa_1}{\kappa_1 - 1}\left(1 - \varepsilon^{\frac{\kappa_1 - 1}{\kappa_1}}\right), \tag{8.86}$$

$$w_l^2 = w_{l0}^2 + 2\frac{p_0}{\rho_{l0}}(1 - \varepsilon), \text{ for } l = 2,3. \tag{8.87}$$

The last two equations were obtained in 1949 by *Tangren* et al. [25]. In the literature concerning critical two-phase flow the velocity ratio $S_1 = w_1 / w_2$ is frequently called *slip ratio* or *slip*. For two velocity fields the slip ratio at the outlet of the nozzle

$$S_1 = \left\{ \frac{w_{10}^2 + 2\frac{p_0}{\rho_{10}} \frac{\kappa_1}{\kappa_1 - 1}\left(1 - \varepsilon^{\frac{\kappa_1 - 1}{\kappa_1}}\right)}{w_{20}^2 + 2\frac{p_0}{\rho_{20}}(1 - \varepsilon)} \right\}^{1/2} \tag{8.88}$$

depends also on the inlet history. For discharge from a vessel with stagnant mixture or flushing inlet flow we have

$$S_1 = \left\{ \frac{\rho_{20}}{\rho_{10}} \frac{\kappa_1}{\kappa_1 - 1}\left(1 - \varepsilon^{\frac{\kappa_1 - 1}{\kappa_1}}\right) / (1 - \varepsilon) \right\}^{1/2}. \tag{8.89}$$

In the reality the interfacial drag will reduce considerable this value especially in the limiting case of diminishing velocity field.

The sum of Eqs. (8.86) and (8.87) multiplied by the corresponding densities and volume fractions at the nozzle outlet gives the interesting expression

$$\frac{1}{2}\sum_{l=1}^{3}\alpha_l \rho_l w_l^2 = \frac{1}{2}\sum_{l=1}^{3}\alpha_l \rho_l w_{l0}^2 + p_0\left[\alpha_1 \varepsilon^{1/\kappa_1} \frac{\kappa_1}{\kappa_1 - 1}\left(1 - \varepsilon^{\frac{\kappa_1 - 1}{\kappa_1}}\right) + (1 - \alpha_1)(1 - \varepsilon)\right]. \tag{8.90}$$

Equation (8.90) allows us to compute the mass flow rate by known slip and pressure ratios. Note that in accordance with the slip definition from the next section Eq. (8.90) can be rewritten as follows

$$\frac{1}{2}v_S f_0 G^2 = \frac{1}{2}v_{S0} f_{00} G_0^2 + p_0 \left[\alpha_1 \varepsilon^{1/\kappa_1} \frac{\kappa_1}{\kappa_1 - 1} \left(1 - \varepsilon^{\frac{\kappa_1 - 1}{\kappa_1}}\right) + (1 - \alpha_1)(1 - \varepsilon) \right].$$

Nozzle flow with instantaneous heat exchange without mass exchange: For this case

$$\frac{1}{2}\sum_{l=1}^{3} \alpha_l \rho_l w_l^2 = \frac{1}{2}\sum_{l=1}^{3} \alpha_l \rho_l w_{l0}^2 + p_0 \left[\alpha_1 \varepsilon^{1/n} \frac{n}{n-1}\left(1 - \varepsilon^{\frac{n-1}{n}}\right) + (1 - \alpha_1)(1 - \varepsilon) \right] \tag{8.91}$$

we have the polytrophic state of change with the polytrophic exponent

$$n = \frac{c_p}{c_p - R}, \tag{8.92}$$

and mixture specific heat defined as

$$c_p = \sum_{l=1}^{3} \frac{\alpha_l \rho_l}{\rho} c_{pl}, \tag{8.93}$$

and the effective gas constant

$$R = \sum_{l=1}^{3} \frac{\alpha_l \rho_l}{\rho} \overline{R}_l. \tag{8.94}$$

Note that for the second and the third field the pseudo gas constant is negligibly small.

8.5 Slip model – transient flow

Historically for the mathematical description of the mechanical interfacial interaction the so called slip models are used among others. In this technique field velocity ratios are modeled with empirical correlation replacing the complete description of the mechanical interaction by means of separated momentum equations. Only the mixture momentum equation is used instead. Thus only the mechanical behavior of the mixture as a whole is considered properly. The adjustment of the slip ratio as a function of the local flow parameter is assumed to be instantaneous that is – inertialess. The disadvantages of the slip model are obvious. Nevertheless, the slip model provides for some application a reasonable simplicity and therefore will be described here. We described one of the families of the slip models in which the *slip velocity ratio* is defined as the field velocity divided by the center of mass mixture velocity.

$$S_l = w_l / w, \qquad (8.95)$$

where the center of mass velocity

$$w = G / \rho, \qquad (8.96)$$

is defined as the mixture mass flow rate

$$G = \sum_{l=1}^{3} \alpha_l \rho_l w_l, \qquad (8.97)$$

divided by the mixture density

$$\rho = \sum_{l=1}^{3} \alpha_l \rho_l. \qquad (8.98)$$

Instead of using local volume fractions the local mass flow concentrations defined as follows

$$X_l = \alpha_l \rho_l w_l / G, \qquad (8.99)$$

will be used as very convenient. Obviously

$$\sum_{l=1}^{3} X_l = 1, \qquad (8.100)$$

per definition. Some useful consequences of the introduction of the slip ratios and the mass flow concentrations are

$$f_0 = \sum_{l=1}^{3} \frac{X_l}{S_l}, \qquad (8.101)$$

$$f_1 = \rho v_S = \sum_{l=1}^{3} X_l S_l, \qquad (8.102)$$

$$\rho w \equiv G, \qquad (8.103)$$

$$v_S = \sum_{l=1}^{3} X_l S_l v_l ,\qquad(8.104)$$

$$v_l = \left(\sum_{l=1}^{3} \alpha_l \rho_l w_l^2 \right)/G^2 = v_S f_0 ,\qquad(8.105)$$

$$\alpha_l = \frac{X_l v_l S_l}{v_S} ,\qquad(8.106)$$

$$w_l = v_S G / S_l .\qquad(8.107)$$

With these definitions the conservation equations for field mass, mixture mass and mixture momentum

$$\frac{\partial}{\partial \tau}(\alpha_l \rho_l \gamma_v) + \frac{\partial}{\partial z}(\alpha_l \rho_l w_l \gamma_z) = \gamma_v \mu_l ,\qquad(8.108)$$

$$\frac{\partial}{\partial \tau}\left(\gamma_v \sum_{l=1}^{3} \alpha_l \rho_l \right) + \frac{\partial}{\partial z}\left(\gamma_z \sum_{l=1}^{3} \alpha_l \rho_l w_l \right) = \gamma_v \mu_l ,\qquad(8.109)$$

$$\frac{\partial}{\partial \tau}\left(\gamma_v \sum_{l=1}^{3} \alpha_l \rho_l w_l \right) + \frac{\partial}{\partial z}\left(\gamma_z \sum_{l=1}^{3} \alpha_l^e \rho_l w_l^2 \right) + \gamma_z \frac{\partial p}{\partial z} + \gamma_v g \cos\varphi \sum_{l=1}^{3} \alpha_l \rho_l$$

$$= \gamma_v \sum_{l=1}^{3} \left[\overline{c}_{lw}^d |w_l| w_l + \mu_{wl}(w_{wl} - w_l) - \mu_{lw}(w_{lw} - w_l)\right],\qquad(8.110)$$

can be rewritten as follows

$$\frac{\partial}{\partial \tau}\left(\frac{X_l S_l}{v_S}\gamma_v\right) + \frac{\partial}{\partial z}(X_l G \gamma_z) = \gamma_v \mu_l ,\qquad(8.111)$$

$$\frac{\partial}{\partial \tau}\left(\frac{f_1}{v_S}\gamma_v\right) + \frac{\partial}{\partial z}(G\gamma_z) = \gamma_v \mu ,\qquad(8.112)$$

$$\frac{\partial}{\partial \tau}(G\gamma_v) + \frac{\partial}{\partial z}(v_l G^2 \gamma_z) + \gamma_z \frac{\partial p}{\partial z} = \gamma_v (f_{fr} + f_\mu - \rho g \cos\varphi).\qquad(8.113)$$

Here f_{fr} is the frictional pressure drop gradient of the mixture depending on the local parameter. f_μ is the force due to injection of suction. We assign the subscript m to those particular fields whose mass flow concentration is dependent and defined by the other ones through Eq.(8.100). The dependent variable vector is

$$\mathbf{U}^T = \left[G, p, X_{l,l \neq m}, s_l, C_{il} \right].\tag{8.114}$$

The conservative equations can be rewritten in the non-conservative form by using the chain rule. At this place the assumption of quasi-constant slip ratio is introduced. With this assumption we obtain

$$dv_l = -\frac{dp}{G^{*2}} - f_0 \sum_{l=1}^{3} \frac{X_l S_l}{\rho_l^2} \left(\frac{\partial \rho_l}{\partial s_l} ds_l + \sum_{\substack{l=1 \\ l \neq m}}^{3} \frac{\partial \rho_l}{\partial C_{il}} dC_{il} \right)$$

$$+ \sum_{\substack{l=1 \\ l \neq m}}^{3} \left[f_0 \left(v_l S_l - v_m S_m \right) + v_S \left(\frac{1}{S_l} - \frac{1}{S_m} \right) \right] dX_l,\tag{8.115}$$

$$d\rho = \frac{\rho^2}{f_1} \frac{1}{f_0 G^{*2}} \left[\begin{array}{l} dp + f_0 G^{*2} \sum_{l=1}^{3} \frac{X_l S_l}{\rho_l^2} \left(\frac{\partial \rho_l}{\partial s_l} ds_l + \sum_{i=2}^{i_{\max}} \frac{\partial \rho_l}{\partial C_{il}} dC_{il} \right) \\ + f_0 G^{*2} \sum_{\substack{l=1 \\ l \neq m}}^{3} \left(\frac{S_l - S_m}{\rho} - v_l S_l + v_m S_m \right) dX_l \end{array} \right],\tag{8.116}$$

where the local critical mass flow rate is

$$\frac{1}{G_l^{*2}} = f_0 \sum_{l=1}^{3} \frac{X_l S_l}{G_l^{*2}} = f_0 \left(\frac{X_1 S_1}{\kappa_1 \rho_1 p} + \frac{X_2 S_2}{G_2^{*2}} + \frac{X_3 S_3}{G_3^{*2}} \right).\tag{8.117}$$

Replacing dv_l in the mixture mass equation we obtain Eq. (8.118). Equation (8.119) is obtained after differentiating the field mass conservation equations, adding them, and taking into account Eq. (8.100). For the limiting case of three velocity fields this equations reduces to the equation obtained by *Kolev* in [11] (1985). The third group of $l_{\max} - 1$ equations (8.120) are obtained from the mixture mass conservation equation by using the above derived differential relationships, and after some rearrangements. For two velocity fields these equations reduce to the equation obtained in [12, 13] (1986). The last two equations are obtained easily from the entropy and momentum equations neglecting the diffusion terms. The

8.5 Slip model – transient flow

equations below are obtained assuming also that $\gamma_v = \gamma_z \neq f(\tau)$, and setting $\gamma^* = \dfrac{1}{\gamma}\dfrac{\partial \gamma}{\partial z}$.

$$\frac{\partial G}{\partial \tau} + 2Gv_l \frac{\partial G}{\partial z} + \left(1 - \frac{G^2}{G^{*2}}\right)\frac{\partial p}{\partial z} + G^2 \sum_{\substack{l=1 \\ l \neq m}}^{3}\left[f_0\left(v_l S_l - v_m S_m\right) + v_S\left(\frac{1}{S_l} - \frac{1}{S_m}\right)\right]\frac{\partial X_l}{\partial z}$$

$$= \gamma_v\left(f_{fr} + f_\mu - \rho g \cos\varphi\right) - G^2\left[v_l \gamma^* - f_0 \sum_{l=1}^{3}\frac{X_l S_l}{\rho_l^2}\left(\frac{\partial \rho_l}{\partial s_l}\frac{\partial s_l}{\partial z} + \sum_{i=2}^{i_{max}}\frac{\partial \rho_l}{\partial C_{il}}\frac{\partial C_{il}}{\partial z}\right)\right], \quad (8.118)$$

$$\frac{\partial p}{\partial \tau} + f_0 G^{*2}\left[\frac{f_1}{\rho^2}\frac{\partial G}{\partial z} + \sum_{\substack{l=1 \\ l \neq m}}^{3}\left(\frac{S_l - S_m}{\rho} - v_l S_l + v_m S_m\right)\frac{\partial X_l}{\partial \tau}\right]$$

$$= -f_0 G^{*2}\left[\frac{f_1}{\rho^2}\left(G\gamma^* - \mu\right) + \sum_{l=1}^{3}\frac{X_l S_l}{\rho_l^2}\left(\frac{\partial \rho_l}{\partial s_l}\frac{\partial s_l}{\partial \tau} + \sum_{i=2}^{i_{max}}\frac{\partial \rho_l}{\partial C_{il}}\frac{\partial C_{il}}{\partial \tau}\right)\right], \quad (8.119)$$

$$\frac{\partial}{\partial \tau}\left(\frac{X_l S_l}{v_S}\gamma_v\right) + \frac{\partial}{\partial z}\left(X_l G \gamma_z\right) = \gamma_v \mu_l, \quad (8.120)$$

$$\frac{1}{v_S}\left(\sum_{k \neq l} X_k S_k\right)\frac{\partial}{\partial \tau}(X_l S_l) - X_l S_l \frac{1}{v_S}\sum_{k \neq l}\frac{\partial}{\partial \tau}(X_k S_k)$$

$$+ f_1 G \frac{\partial X_l}{\partial z} + (f_1 - S_l) X_l \frac{1}{\gamma}\frac{\partial}{\partial z}(G\gamma) = f_1 \mu_l - X_l S_l \mu, \quad (8.121)$$

$l \neq m$

$$\frac{\partial s_l}{\partial \tau} + \frac{G v_S}{S_l}\frac{\partial s_l}{\partial z} = \frac{v_S}{X_l S_l}\left[\frac{1}{T_l}DT_l^N + \sum_{i=1}^{i_{max}}\mu_{il}\left(s_{il} - s_l\right)\right], \quad l = 1, l_{max}, \quad (8.122)$$

$$\frac{\partial C_{il}}{\partial \tau} + \frac{G v_S}{S_l}\frac{\partial C_{il}}{\partial z} = \frac{v_S}{X_l S_l}\left(\mu_{il} - \mu_l C_{il}\right), \quad l = 1, l_{max}, i = 1, i_{max}. \quad (8.123)$$

This non-conservative form

$$A\frac{\partial \mathbf{U}}{\partial \tau} + B\frac{\partial \mathbf{U}}{\partial z} = C \qquad (8.124)$$

of the system is very suitable for numerical integration. The entropy and concentration equations can be integrated in each time step separately. With the thus-obtained values for the new time plane, the terms in the first two equations, containing entropies and concentrations can be calculated explicitly, and the resulting system can be integrated with one of the known integration methods (see *Kolev* [12] (1986) p.157), as it was made for a three velocity fields by *Kolev* [11] (1985).

8.6 Slip model – steady state. Critical mass flow rate

The local critical mass flow rate plays an important role for the modeling of the three-phase flow in technological equipment, because of the fact that the local mass flow rate in a confined geometry can not exceed the local critical mass flow rate, which is in many cases the limitation of the productivity of the particular apparatus. The local critical mass flow rate can be used as a boundary condition for the analysis of mixture discharge from three-dimensional pressurized vessels. The purpose of this section is to derive from the steady state part of the system (8.118-8.123) an expression defining the local critical mass flow rate.

We obtain the definition of the local critical mass flow rate from the steady state part of the system, solved as an algebraic equation with respect to the space derivatives. The resulting form of the equivalent system of ordinary differential equations is Eqs. (8.125) to (8.129). As we see the critical condition is reached if the mass flow rate tends to the local critical mass flow rate. This leads to an infinite negative pressure gradient either in the smallest cross section of the channel, or at the end of the duct with constant cross section. In case of three velocity fields the expression defining the local critical mass flow rate reduces to the one obtained by *Kolev* [11] (1985). For three velocity fields without inert components it is reduced to the one derived by *Kolev* [9] (1977). In case of three velocity fields without inert components, incompressible liquid, and a perfect gas the general expression reduces to one obtained by *Nigmatulin* and *Ivandeev* [18] (1977). In case of homogeneous two-phase flow the general expression reduces to the one obtained by *Wood* (1937) - see in [26].

$$\frac{dp}{dz} = -\left\{ -f_{fr} - f_{\mu} + 2Gv_l\mu + \rho\mathbf{g} + G\sum_{\substack{l=1\\l\neq m}}^{3}\left[\frac{f_0\left(v_l S_l - v_m S_m\right)}{+v_s\left(\frac{1}{S_l} - \frac{1}{S_m}\right)}\right](\mu_l - X_l\mu)\right\}$$

$$-G^2 v_1 \gamma^* - f_0 G \sum_{l=1}^{3} \frac{S_l}{\rho_l^2} \left(\frac{\partial \rho_l}{\partial s_l} \left[\frac{1}{T_l} DT_l^N + \sum_{i=1}^{i_{\max}} \mu_{il} (s_{il} - s_l) \right] + \sum_{i=1}^{i_{\max}-1} \frac{\partial \rho_l}{\partial C_{il}} (\mu_{il} - \mu_l C_{il}) \right) \Bigg/ \left(1 - \frac{G^2}{G^{*2}}\right) \quad (8.125)$$

$$\frac{dG}{dz} = \mu - G\gamma^*, \quad (8.126)$$

$$\frac{dX_l}{dz} = (\mu_l - X_l \mu)/G, \; l \neq m, \quad (8.127)$$

$$\frac{ds_l}{dz} = \frac{1}{\gamma X_l G} Ds_l^N = \frac{1}{X_l G} \left[\frac{1}{T_l} DT_l^N + \sum_{i=1}^{i_{\max}} \mu_{il} (s_{il} - s_l) \right], \; l=1, l_{\max}, \quad (8.128)$$

$$\frac{dC_{il}}{dz} = \frac{1}{\gamma X_l G} DC_{il}^N = \frac{1}{X_l G} (\mu_{il} - \mu_l C_{il}), \; l=1,l_{\max}, \; i=1,i_{\max}. \quad (8.129)$$

Single component flashing nozzle flow without external sources: For a nozzle flow we have no friction forces. The gravitation is negligible. The system simplifies to the following form:

$$\frac{dp}{dz} = -\frac{G}{1 - \frac{G^2}{G^{*2}}} \left\{ -f_0 \sum_{l=1}^{3} \frac{S_l}{\rho_l^2} \frac{\partial \rho_l}{\partial s_l} \frac{1}{T_l} DT_l^N + \sum_{\substack{l=1 \\ l\neq m}}^{3} \left[f_0 (v_l S_l - v_m S_m) + v_S \left(\frac{1}{S_l} - \frac{1}{S_m} \right) \right] \mu_l \right\}$$

$$(8.130)$$

$$G = const, \quad (8.131)$$

$$\frac{dX_l}{dz} = \frac{\mu_l}{G}, \; l \neq m, \quad (8.132)$$

$$\frac{ds_l}{dz} = \frac{1}{X_l G T_l} DT_l^N, \; l=1, l_{\max}. \quad (8.133)$$

The knowledge of the particular interfacial heat, mass and momentum transfer is the prerequisite for integrating the above system. In the 1960s several authors proposed different assumptions replacing the mechanistic description of the interfacial transfer mechanism, like homogeneity for the interfacial momentum description, thermodynamic equilibrium or limited non-equilibrium for the description of the heat and mass transfer etc. Probably the most successful approach was those proposed by *Henry* and *Fauske* [5] in 1969. We will implement the main ideas of their model applying them consequently to three-field flow. We first compute the heat fluxes required to keep the thermal equilibrium between the fields. Then we will assume that only a part of this heat transfer happens in the real nozzle flow. Then we will compute the evaporation required to keep the liquid fields saturated and assume that only part of this evaporation happens in the real discharge after the nozzle. We will define under this assumption the critical mass flow rate in the throat of the nozzle and then use the simplified integrated *Euler* equations (8.90) to compute the critical pressure ratio. Then knowing the pressure ratio we compute the critical mass flow rate.

Modeling of the interfacial heat transfer: The transferred heat between the velocity fields will be a b_1-part of those required to equalize instantaneously their temperatures defined by Eqs. (6.13) and (6.15):

$$\dot{q}_{21}''' + \dot{q}_{31}''' = -b_1 c_{p1} \left(\frac{1}{n} - \frac{1}{\kappa_1} \right) \frac{X_1 GT_1}{p} \frac{dp}{dz}, \tag{8.134}$$

$$\dot{q}_{21}''' = -b_1 X_2 GT_2 c_{p2} \frac{n-1}{n} \frac{1}{p} \frac{dp}{dz}, \tag{8.135}$$

$$\dot{q}_{31}''' = -b_1 X_3 GT c_{p3} \frac{n-1}{n} \frac{1}{p} \frac{dp}{dz}. \tag{8.136}$$

The heat removed by the liquids due to the evaporation is connected through the interfacial energy jump condition with the evaporated masses per unit time and unit mixture volume as follows

$$\dot{q}_2'''^{\sigma 1} = -\mu_{21} (h'' - h'), \tag{8.137}$$

$$\dot{q}_3'''^{\sigma 1} = -\mu_{31} (h'' - h'). \tag{8.138}$$

Thus, the sources in the entropy equations (8.133) are

$$DT_1^N = \dot{q}_{21}''' + \dot{q}_{21}''' + (\mu_{21} + \mu_{31})(h'' - h_1), \tag{8.139}$$

$$DT_2^N = -\dot{q}_{21}''' + \dot{q}_2'''^{\sigma 1} - \mu_{21}(h' - h_2) = -\dot{q}_{21}''' - \mu_{21}(h'' - h_2), \tag{8.140}$$

$$DT_3^N = -\dot{q}_{31}''' + \dot{q}_3'''^{\sigma 1} - \mu_{31}(h' - h_2) = -\dot{q}_{31}''' - \mu_{31}(h'' - h_3). \tag{8.141}$$

8.6 Slip model – steady state. Critical mass flow rate

Modeling of the interfacial mass transfer: The evaporation required to keep the fields 2 and 3 saturated is computed from the entropy equations

$$\mu_{21} = -X_2 G \frac{T_2}{h''-h_2} \left(\frac{ds'}{dp} - b_1 \frac{c_{p2}}{p} \frac{n-1}{n} \right) \frac{dp}{dz}, \tag{8.142}$$

$$\mu_{31} = -X_3 G \frac{T_3}{h''-h_3} \left(\frac{ds'}{dp} - b_1 \frac{c_{p3}}{p} \frac{n-1}{n} \right) \frac{dp}{dz}. \tag{8.143}$$

The real evaporation may be a b_2-part of this amount

$$\mu_{21} = -b_2 X_2 G \frac{T_2}{h''-h_2} \left(\frac{ds'}{dp} - b_1 \frac{c_{p2}}{p} \frac{n-1}{n} \right) \frac{dp}{dz}, \tag{8.144}$$

$$\mu_{31} = -b_2 X_3 G \frac{T_3}{h''-h_3} \left(\frac{ds'}{dp} - b_1 \frac{c_{p3}}{p} \frac{n-1}{n} \right) \frac{dp}{dz}, \tag{8.145}$$

and therefore

$$\mu_{21} + \mu_{31} = -b_2 G \left[\frac{X_2 T_2}{h''-h_2} \left(\frac{ds'}{dp} - b_1 \frac{c_{p2}}{p} \frac{n-1}{n} \right) + \frac{X_3 T_3}{h''-h_3} \left(\frac{ds'}{dp} - b_1 \frac{c_{p3}}{p} \frac{n-1}{n} \right) \right] \frac{dp}{dz}. \tag{8.146}$$

The change of the gas mass concentration is then

$$dX_1 = -b_2 \left[\left(\frac{X_2 T_2}{h''-h_2} + \frac{X_3 T_3}{h''-h_3} \right) \frac{ds'}{dp} - b_1 \left(\frac{X_2 T_2 c_{p2}}{h''-h_2} + \frac{X_3 T_3 c_{p3}}{h''-h_3} \right) \frac{1}{p} \frac{n-1}{n} \right] dp. \tag{8.146}$$

The critical mass flow rate: The momentum equation is then rewritten as follows

$$\frac{dp}{dz} = -\frac{G}{1-\frac{G^2}{G^{*2}}} \left\{ \begin{array}{l} -f_0 \left(\frac{S_1}{\rho_1^2} \frac{\partial \rho_1}{\partial s_1} \frac{1}{T_1} DT_1^N + \frac{S_2}{\rho_2^2} \frac{\partial \rho_2}{\partial s_2} \frac{1}{T_2} DT_2^N + \frac{S_3}{\rho_3^2} \frac{\partial \rho_3}{\partial s_3} \frac{1}{T_3} DT_3^N \right) \\ + \left[f_0 (v_1 S_1 - v_2 S_2) + v_S \left(\frac{1}{S_1} - \frac{1}{S_2} \right) \right] (\mu_{21} + \mu_{31}) \\ - \left[f_0 (v_3 S_3 - v_2 S_2) + v_S \left(\frac{1}{S_3} - \frac{1}{S_2} \right) \right] \mu_{31} \end{array} \right. \tag{8.147}$$

Substituting the sources we obtain

$$\frac{1}{G^2} = \frac{1}{G^{*2}} + f_0 \frac{b_1}{p} \left[\begin{array}{c} -\frac{S_1}{\rho_1^2} \frac{\partial \rho_1}{\partial s_1} X_1 c_{p1} \left(\frac{1}{n} - \frac{1}{\kappa_1} \right) \\ + \left(\frac{S_2}{\rho_2^2} \frac{\partial \rho_2}{\partial s_2} X_2 c_{p2} + \frac{S_3}{\rho_3^2} \frac{\partial \rho_3}{\partial s_3} X_3 c_{p3} \right) \frac{n-1}{n} \end{array} \right]$$

$$+ b_2 \left[\begin{array}{c} \left[f_0 (v_1 S_1 - v_2 S_2) + v_S \left(\frac{1}{S_1} - \frac{1}{S_2} \right) \right] \left[\frac{X_2 T_2}{h'' - h_2} \left(\frac{ds'}{dp} - b_1 \frac{c_{p2}}{p} \frac{n-1}{n} \right) \right. \\ \left. + \frac{X_3 T_3}{h'' - h_3} \left(\frac{ds'}{dp} - b_1 \frac{c_{p3}}{p} \frac{n-1}{n} \right) \right] \\ - f_0 (h'' - h_1) \frac{S_1}{\rho_1^2} \frac{\partial \rho_1}{\partial s_1} \frac{1}{T_1} \end{array} \right]$$

$$+ b_2 X_2 \frac{S_2}{\rho_2^2} \frac{\partial \rho_2}{\partial s_2} f_0 (h'' - h_2) \left(\frac{ds'}{dp} - b_1 \frac{c_{p2}}{p} \frac{n-1}{n} \right)$$

$$- b_2 X_3 \left[\begin{array}{c} \left[f_0 (v_3 S_3 - v_2 S_2) + v_S \left(\frac{1}{S_3} - \frac{1}{S_2} \right) \right] \\ + \frac{S_3}{\rho_3^2} \frac{\partial \rho_3}{\partial s_3} \frac{1}{T_3} f_0 (h'' - h_3) \end{array} \right] \frac{T_3}{h'' - h_3} \left(\frac{ds'}{dp} - b_1 \frac{c_{p3}}{p} \frac{n-1}{n} \right). \quad (8.148)$$

It is a very good approximation to neglect the change of the liquid density with the change of the liquid entropy. In addition the relative deviation of the gas temperature with respect to the saturation is very small. Taking into account this simplification the final result is therefore

$$\frac{1}{f_0 G^2} = \frac{X_1 S_1}{\rho_1 p} \left(\frac{1}{\kappa_1} + \frac{b_1}{n} \right) + \frac{X_2 S_2}{G_2^{*2}} + \frac{X_3 S_3}{G_3^{*2}}$$

$$+ b_2 \left\{ \begin{array}{c} \left[v_1 S_1 - v_2 S_2 + \frac{v_S}{f_0} \left(\frac{1}{S_1} - \frac{1}{S_2} \right) \right] \frac{X_2 T_2}{h'' - h_2} \left(\frac{ds'}{dp} - b_1 \frac{c_{p2}}{p} \frac{n-1}{n} \right) \\ + \left[v_1 S_1 - v_3 S_3 + \frac{v_S}{f_0} \left(\frac{1}{S_1} - \frac{1}{S_3} \right) \right] \frac{X_3 T_3}{h'' - h_3} \left(\frac{ds'}{dp} - b_1 \frac{c_{p3}}{p} \frac{n-1}{n} \right) \end{array} \right\}. \quad (8.149)$$

Thus at the critical cross section the critical mass flow rate is defined by the above equation. If the b_2 coefficient is set to zero the model results in the so called frozen model. If the b_2 coefficient is set to unity the model is close to the so called thermal equilibrium model. For gas mass concentrations different from zero the second and the third terms are much smaller then the first one. For homogeneous flow we have

$$\frac{1}{G^2} = \frac{X_1}{\rho_1 p}\left(\frac{1}{\kappa_1} + \frac{b_1}{n}\right)$$

$$+ b_2 \left[(v_1 - v_2)\frac{X_2 T_2}{h''-h_2}\left(\frac{ds'}{dp} - b_1 \frac{c_{p2}}{p}\frac{n-1}{n}\right) + (v_1 - v_3)\frac{X_3 T_3}{h''-h_3}\left(\frac{ds'}{dp} - b_1 \frac{c_{p3}}{p}\frac{n-1}{n}\right) \right]. \quad (8.150)$$

Comparing this result with the derivation by *Henry* and *Fauske* 1956

$$\frac{1}{G^2} = \frac{X_1}{n\rho_1 p} + (v_1 - v_2)\left[\frac{(1-X_1)b_2}{s''-s'}\frac{ds'}{dp} - X_1 \frac{c_{p1}}{p(s_1-s_2)}\left(\frac{1}{n} - \frac{1}{\kappa_1}\right)\right], \quad (8.151)$$

we see a slight difference.

The gas density changes rapidly in the process in accordance with $\rho_1 = \rho_{10}\varepsilon^{\frac{1}{n}}$. Replacing in Eq. (8.149) and introducing the pressure ratios we obtain

$$\frac{1}{f_0 G^2} = \frac{X_1 S_1}{\rho_{10} p_0 \varepsilon^{\frac{n+1}{n}}}\left(\frac{1}{\kappa_1} + \frac{b_1}{n}\right) + \frac{X_2 S_2}{G_2^{*2}} + \frac{X_3 S_3}{G_3^{*2}}$$

$$+ b_2 \left\{ \begin{array}{l} \left[\dfrac{v_{10}S_1}{\varepsilon^{\frac{1}{n}}} - v_2 S_2 + \dfrac{v_S}{f_0}\left(\dfrac{1}{S_1} - \dfrac{1}{S_2}\right)\right]\dfrac{X_2 T_2}{h''-h_2}\left(\dfrac{ds'}{dp} - b_1 \dfrac{c_{p2}}{p_0 \varepsilon}\dfrac{n-1}{n}\right) \\ \\ + \left[\dfrac{v_{10}S_1}{\varepsilon^{\frac{1}{n}}} - v_3 S_3 + \dfrac{v_S}{f_0}\left(\dfrac{1}{S_1} - \dfrac{1}{S_3}\right)\right]\dfrac{X_3 T_3}{h''-h_3}\left(\dfrac{ds'}{dp} - b_1 \dfrac{c_{p3}}{p_0 \varepsilon}\dfrac{n-1}{n}\right) \end{array} \right\}, \quad (8.152)$$

where

$$v_S = X_1 S_1 \frac{v_{10}}{\varepsilon^{\frac{1}{n}}} + X_2 S_2 v_2 + X_3 S_3 v_3, \quad (8.153)$$

$$X_1 = X_{10} - b_2 \left[\left(\frac{X_2 T_2}{h'' - h_2} + \frac{X_3 T_3}{h'' - h_3} \right) \frac{ds'}{dp} - b_1 \left(\frac{X_2 T_2 c_{p2}}{h'' - h_2} + \frac{X_3 T_3 c_{p3}}{h'' - h_3} \right) \frac{1}{p} \frac{n-1}{n} \right] p_0 (\varepsilon - 1), \tag{8.154}$$

$$X_2 = X_{20} + b_2 X_2 G \frac{T_2}{h'' - h_2} \left(\frac{ds'}{dp} - b_1 \frac{c_{p2}}{p} \frac{n-1}{n} \right) p_0 (\varepsilon - 1), \tag{8.155}$$

and

$$X_3 = 1 - X_1 - X_2. \tag{8.156}$$

Equation (8.152) together with Eq. (8.90)

$$\frac{1}{2} v_S f_0 G^2 = p_0 \left[\alpha_1 \varepsilon^{1/\kappa_1} \frac{\kappa_1}{\kappa_1 - 1} \left(1 - \varepsilon^{\frac{\kappa_1 - 1}{\kappa_1}} \right) + (1 - \alpha_1)(1 - \varepsilon) \right] \tag{8.157}$$

define the critical pressure ratio. Both equations have to be solved by iterations with respect to G and ε. For estimation of the slip ratios, a reasonable assumption is $S_2 = S_3 = 1$ and S_1 computed using

$$S_1 = 1 + \frac{X_1}{X_{1,\max}} \left(\frac{1 - X_1}{1 - X_{1,\max}} \right)^{\frac{1 - X_1}{X_{1,\max}}} (S_{1,\max} - 1), \tag{8.158}$$

$$S_{1,\max} = \left(\frac{\rho_2}{\rho_1} \right)^{0.1}, \tag{8.159}$$

$$X_{1,\max} = 0.1. \tag{8.160}$$

Henry and *Fauske* proposed to use

$$b_2 = \frac{X_{10,e}}{0.14} \quad \text{for} \quad X_{10,e} < 0.14 \tag{8.161}$$

and $b_2 = 1$ otherwise, where

$$X_{1,e} = \frac{\sum_{l=1}^{3} X_{l0} s_{l0} - s_0'}{s_0'' - s_0'}. \tag{8.162}$$

As already mentioned the pressure gradient at the critical flow location is minus infinite. Such gradients are very difficult to be resolved numerically and special attention have to be paid to the discretization. Non-equidistant discretization with sizes getting smaller in the proximity of the critical cross section is the right answer of this challenge.

8.7 Forces acting on the pipes due to the flow – theoretical basics

In many applications the analysis of thermo-hydraulic processes in a networks or 3D facilities is done with the intention to estimate the thrusts acting on the pipes and facilities. These thrusts are used to design the mechanical supports of the constructions.

The purpose of this chapter is to describe the algorithm needed for computation of the pipe thrusts in the network operating with multi-phase flows.

In the following we assume

(a) that the effect of gravitational forces is negligible,

and postulate that

(b) the positive flow force direction is the direction opposite to the positive flow velocity direction.

The solid structure experiences forces from the continuum flow wetting its internal surface and from the ambient fluid. There are normal and shear forces caused by the flow and acting on the internal wall surface and normal forces due the action of the ambient pressure on the external wall surface.

We designate with $\mathbf{f}_{wp}$ the normal pressure force acting on the structure. This force consists of two components due to internal and ambient pressure, respectively

$$\mathbf{f}_{wp} = \iint_{A_{wall-ambient}} p_\infty \mathbf{n}_w da + \iint_{A_{wall-flow}} p \mathbf{n}_f da = \iint_{A_{wall-ambient}} p_\infty \mathbf{n}_w da - \iint_{A_{wall-flow}} p \mathbf{n}_w da \,.$$

(8.163)

Here the subscript w stands for wall, f for flow, p is the fluid pressure inside the pipe acting on the infinitesimal surface da, p_∞ is the ambient pressure acting outside of the pipe on the same surface. $\mathbf{n}_w$ is the unit vector normal to the internal wall surface pointing into the flow - see Fig. 8.3.

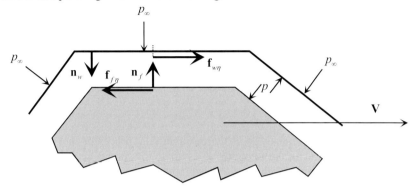

Fig. 8.3. Definitions of the force directions signs

The flow friction shear force, $\mathbf{f}_{f\eta}$, resists the flow and is positive per definition. The wall experiences a force, $\mathbf{f}_{w\eta}$, with magnitude equal to the magnitude of the flow friction force but with the opposite direction,

$$\mathbf{f}_{f\eta} + \mathbf{f}_{w\eta} = 0. \tag{8.164}$$

The total force acting on the wall is therefore

$$\mathbf{f}_{wall} = \mathbf{f}_{wp} + \mathbf{f}_{w\eta} = \mathbf{f}_{wp} - \mathbf{f}_{f\eta}. \tag{8.165}$$

Direct computation of the fluid friction force: The fluid friction force can be computed directly by integrating the friction pressure loss through the total flow volume

$$\mathbf{f}_{f\eta} = -\iiint_{flow\ volume} (\nabla p)_{friction}\, dVol \approx -\iiint_{flow\ volume} \left(\frac{dp}{dz}\right)_{friction} dVol. \tag{8.166}$$

Indirect computation of the fluid friction force: The fluid friction force can be computed indirectly, using the momentum balance on the fluid in the control volume,

$$\frac{\partial}{\partial \tau} \iiint_{flow\ volume} \sum_{l=1}^{l_{max}} \alpha_l \rho_l \mathbf{V}_l \gamma_v dVol - \iint_{A_{flow}} p\, \mathbf{n}_f\, da_f$$

8.7 Forces acting on the pipes due to the flow – theoretical basics

$$+ \iint\limits_{A_{open}} \sum_{l=1}^{l_{max}} \left(\alpha_l \rho_l \mathbf{V}_l \cdot \mathbf{n}_f \right) \mathbf{V}_l \gamma_n \, da + \mathbf{f}_\eta = 0 . \qquad (8.167)$$

i.e.

$$-\mathbf{f}_\eta = \frac{\partial}{\partial \tau} \iiint\limits_{flow\ volume} \sum_{l=1}^{l_{max}} \alpha_l \rho_l \mathbf{V}_l \gamma_v dVol$$

$$- \iint\limits_{A_{flow}} p\, \mathbf{n}_f da_f + \iint\limits_{A_{open}} \sum_{l=1}^{l_{max}} \left(\alpha_l \rho_l \mathbf{V}_l \cdot \mathbf{n}_f \right) \mathbf{V}_l \gamma_n \, da . \qquad (8.168)$$

Unlike the momentum flux

$$\iint\limits_{A_{open}} \sum_{l=1}^{l_{max}} \left(\alpha_l \rho_l \mathbf{V}_l \cdot \mathbf{n}_f \right) \mathbf{V}_l \gamma_n da$$

that acts only at the open flow areas and vanishes on the wet surface the friction tension acts at the wet walls and vanishes at the open areas. We split the area surrounding our flow control volume into a part contacting the wall and a open part crossed by the flow

$$A_{flow} = A_{open} + A_{wall\ flow} , \qquad (8.169)$$

and therefore

$$\iint\limits_{A_{flow}} p\, \mathbf{n}_f da = \iint\limits_{A_{open}} p\, \mathbf{n}_f da + \iint\limits_{A_{wall\ flow}} p\, \mathbf{n}_f da . \qquad (8.170)$$

The total wall force is therefore

$$\mathbf{f}_{wall} = \mathbf{f}_{wp} - \mathbf{f}_{f\eta} = \iint\limits_{A_{wall\ ambient}} p_\infty \mathbf{n}_w da - \iint\limits_{A_{wall\ flow}} p\mathbf{n}_w da - \iint\limits_{A_{open}} p\mathbf{n}_f da - \iint\limits_{A_{wall\ flow}} p\mathbf{n}_f da$$

$$+ \frac{\partial}{\partial \tau} \iiint\limits_{flow\ volume} \sum_{l=1}^{l_{max}} \alpha_l \rho_l \mathbf{V}_l \gamma_v dVol + \iint\limits_{A_{open}} \sum_{l=1}^{l_{max}} \left(\alpha_l \rho_l \mathbf{V}_l \cdot \mathbf{n}_f \right) \mathbf{V}_l \gamma_n \, da . \qquad (8.171)$$

Having in mind that the unit vectors of the wall surface and the flow surface are opposite at the wet wall

$$\mathbf{n}_w = -\mathbf{n}_f \qquad (8.172)$$

we recognize that in all surfaces where the pressure acts simultaneously on the wall and at the flow $A_{wall\ flow}$ the pressure force cancels and therefore

$$\mathbf{f}_{wall} = -\iint_{A_{wall\ ambient}} p_\infty \mathbf{n}_f da - \iint_{A_{open}} p\mathbf{n}_f da$$

$$+ \frac{\partial}{\partial \tau} \iiint_{flow\ volume} \sum_{l=1}^{l_{max}} \alpha_l \rho_l \mathbf{V}_l \gamma_v dVol + \iint_{A_{open}} \sum_{l=1}^{l_{max}} \left(\alpha_l \rho_l \mathbf{V}_l \cdot \mathbf{n}_f \right) \mathbf{V}_l \gamma_n da \, . \qquad (8.173)$$

Therefore the wall force consists of three components

(i) pressure force

$$- \iint_{A_{wall\ ambient}} p_\infty \mathbf{n}_f da - \iint_{A_{open}} p\mathbf{n}_f da \, , \qquad (8.174)$$

(ii) "wave" force

$$\frac{\partial}{\partial \tau} \iiint_{flow\ volume} \sum_{l=1}^{l_{max}} \alpha_l \rho_l \mathbf{V}_l \gamma_v dVol \, , \qquad (8.175)$$

and

(iii) reaction thrust

$$\iint_{A_{open}} \sum_{l=1}^{l_{max}} \alpha_l \rho_l V_l^2 \gamma_n \mathbf{n}_f da \, . \qquad (8.176)$$

The estimation of wall forces from the primary Eqs. (8.163-8.166) appears to be much simple than the final expression Eq. (8.173). Often, however, pressure distribution on the wet wall is not known well enough to obtain reasonable results. That is way Eq. (8.173) is successfully used in practice, see e.g. [1], [16]. As illustration of the application of the final equation (8.173) we consider a few practical cases.

Forces on a pipe segment with two bends: The pipe segment as given in Fig. 8.4 consists of one straight part and two elbows. The respective angles of the elbows are θ_{k1}, and θ_{k2}. The velocity in the segment is directed from left to the right. Compute the projection of all wall forces on the main axis of the straight part.

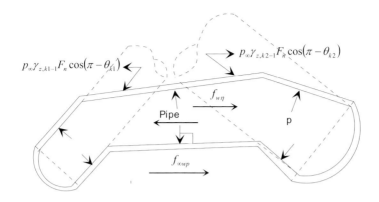

Fig. 8.4. Forces acting at the flow control volume

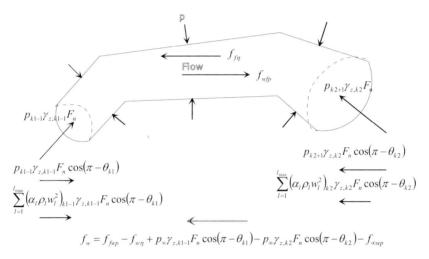

Fig. 8.5. Pressure forces acting on the flow control volume

The pipe segment may be divided into several control volumes starting with *k1* and ending with *k2*. Each control volume has a length Δz_k and volume $\gamma_v \Delta z_k F_n$, where γ_v is the part of the control volume $\Delta z_k F_n$ occupied by flow. Here F_n is some constant cross section. The left surface passed by the flow at the left-hand side control volume *k1*, is $\gamma_{z,k1-1} F_n$, and right hand side surface passed by the flow of the right-hand side control volume *k2* is $\gamma_{z,k2} F_n$. The normal velocity at the entrance flow cross section of each velocity field is $w_{l,k1-1}$, and at the outlet velocity at the outlet flow cross section of each velocity field is $w_{l,k2}$, respectively. The flow pressure before the cell *k1* inside the pipe at the left is p_{k1-1}, and the pressure after the cell *k2* at the right is p_{k2+1}. The ambient pressure is p_∞.

The flow entering the control volume accelerates the flow with a force component in the z direction

$$-\cos(\pi - \theta_{k1})\left(\sum_{l=1}^{l_{max}} \dot{m}_l w_l\right)_1 = -\gamma_{z,k1-1} F_n \cos(\pi - \theta_{k1})\left(\sum_{l=1}^{l_{max}} \alpha_l \rho_l w_l^2\right)_{k1-1}, \quad (8.177)$$

which is directed from the left to the right and is parallel to the flow axis, see Fig. 8.4 and 8.5. The flow leaving the control volume resists the control flow volume with the force component in the z direction

$$\cos(\pi - \theta_{k2})\left(\sum_{l=1}^{l_{max}} \dot{m}_l w_l\right)_2 = \gamma_{z,k2} F_n \cos(\pi - \theta_{k2})\left(\sum_{l=1}^{l_{max}} \alpha_l \rho_l w_l^2\right)_{k2} \quad (8.178)$$

which is directed from the right to the left and is parallel to the flow axis. The pressure forces acting on the flow control volumes, see Fig. 8.5, are computed as follows. The projection of the pressure force acting at the inlet cross section on the axis,

$$-p_{k1-1}\gamma_{z,k1-1} F_n \cos(\pi - \theta_{k1}), \quad (8.179)$$

is directed from left to the right. The projection of the pressure force acting at the left side of the flow control volume on the axis,

$$p_{k2+1}\gamma_{z,k2} F_n \cos(\pi - \theta_{k2}), \quad (8.180)$$

is directed from right to the left. In addition at the projection of the pressure force acting on the flow volume from the side of the wet wall surface on the flow axis is $\mathbf{f}_{wfp}$, and is directed from the left to the right. The opposite force $\mathbf{f}_{fwp}$ acts on the structure at the wet side. Both forces cancel after summation. Thus, the friction force resisting the flow is

$$-f_\eta = F_n \frac{\partial}{\partial \tau} \sum_{k=k1}^{k2}\left(\gamma_v \sum_{l=1}^{l_{max}} \alpha_l \rho_l w_l \Delta z\right)_k$$

$$-\left(p_{k1-1} + \left(\sum_{l=1}^{l_{max}} \alpha_l \rho_l w_l^2\right)_{k1-1}\right)\gamma_{z,k1-1} F_n \cos(\pi - \theta_{k1})$$

$$+\left(p_{k2+1} + \left(\sum_{l=1}^{l_{max}} \alpha_l \rho_l w_l^2\right)_{k2}\right)\gamma_{z,k2} F_n \cos(\pi - \theta_{k2}). \quad (8.181)$$

8.7 Forces acting on the pipes due to the flow – theoretical basics

Now we consider as a control volume the structure only, see Fig. 8.6.

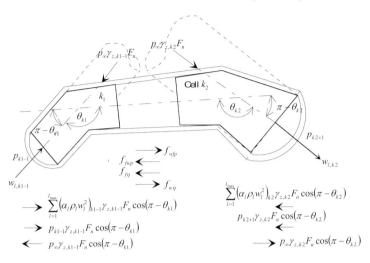

Fig. 8.6. Total force acting on the pipe in axial direction

The projection of the ambient pressure forces acting at the left cross section $\gamma_{z,k1-1} F_n$ on the axis is

$$p_\infty \gamma_{z,k1-1} F_n \cos(\pi - \theta_{k1}), \qquad (8.182)$$

and is directed from the right to the left. The projection of the ambient pressure forces acting at the right cross section $\gamma_{z,k2} F_n$ on the axis is

$$-p_\infty \gamma_{z,k2} F_n \cos(\pi - \theta_{k2}), \qquad (8.183)$$

and is directed from the left to the right. In addition there is a pressure force acting from the fluid side at the wet wall. Its projection on the axis is $\mathbf{f}_{fwp}$. The resulting pressure force acting on the structure is

$$\mathbf{f}_{wp} = p_\infty \gamma_{z,k1-1} F_n \cos(\pi - \theta_{k1}) - p_\infty \gamma_{z,k2} F_n \cos(\pi - \theta_{k2}) + \mathbf{f}_{fwp}. \qquad (8.184)$$

Now we have all components we need to compute the resulting force acting on the structure

$$\mathbf{f}_w = \mathbf{f}_{wp} + \mathbf{f}_{w\eta} = \mathbf{f}_{wp} - \mathbf{f}_{f\eta} = F_n \frac{\partial}{\partial \tau} \sum_{k=k1}^{k2} \left(\gamma_v \sum_{l=1}^{l_{max}} \alpha_l \rho_l w_l \Delta z \right)_k$$

$$-\left[p_{k1-1}-p_{\infty}+\left(\sum_{l=1}^{l_{max}}\alpha_{l}\rho_{l}w_{l}^{2}\right)_{k1-1}\right]\gamma_{z,k1-1}F_{n}\cos(\pi-\theta_{k1})$$

$$+\left[p_{k2+1}-p_{\infty}+\left(\sum_{l=1}^{l_{max}}\alpha_{l}\rho_{l}w_{l}^{2}\right)_{k2}\right]\gamma_{z,k2}F_{n}\cos(\pi-\theta_{k2}). \quad (8.185)$$

The wave force during a time step is computed using the old time level values, designated with a, and the new time level values - without indices - as follows

$$F_{n}\frac{\partial}{\partial\tau}\sum_{k=k1}^{k2}\left(\gamma_{v}\sum_{l=1}^{l_{max}}\alpha_{l}\rho_{l}w_{l}\Delta z\right)_{k}=\frac{1}{\Delta\tau}F_{n}\frac{\partial}{\partial\tau}\sum_{k=k1}^{k2}\left(\gamma_{v}\sum_{l=1}^{l_{max}}\left[\alpha_{l}\rho_{l}w_{l}-(\alpha_{l}\rho_{l}w_{l})_{a}\right]\Delta z\right)_{k}.$$

$$(8.186)$$

For the simple case of a discharging flow from a pipe with dead end, $\gamma_{z,k1-1}=0$, and constant cross section F_{n}, $\gamma_{v,k}=1$, we obtain

$$\mathbf{f}_{w}=\mathbf{f}_{wp}-\mathbf{f}_{f\eta}=\frac{1}{\Delta\tau}F_{n}\frac{\partial}{\partial\tau}\sum_{k=k1}^{k2}\left\{\gamma_{v}\sum_{l=1}^{l_{max}}\left[\alpha_{l}\rho_{l}w_{l}-(\alpha_{l}\rho_{l}w_{l})_{a}\right]\Delta z\right\}_{k}$$

$$+\left[p_{k2+1}-p_{\infty}+\left(\sum_{l=1}^{l_{max}}\alpha_{l}\rho_{l}w_{l}^{2}\right)_{k2}\right]\gamma_{z,k2}F_{n}. \quad (8.187)$$

For the same case as that previously discussed but with critical discharge the critical pressure, p_c, should replace p_{k2+1} in the above equation - see the discussion by *Yano* et al. [27] (1982).

$$\mathbf{f}_{w}=\mathbf{f}_{wp}-\mathbf{f}_{f\eta}=\frac{1}{\Delta\tau}F_{n}\frac{\partial}{\partial\tau}\sum_{k=k1}^{k2}\left\{\gamma_{v}\sum_{l=1}^{l_{max}}\left[\alpha_{l}\rho_{l}w_{l}-(\alpha_{l}\rho_{l}w_{l})_{a}\right]\Delta z\right\}_{k}$$

$$+\left[p_{c}-p_{\infty}+\left(\sum_{l=1}^{l_{max}}\alpha_{l}\rho_{l}w_{l}^{2}\right)_{k2}\right]\gamma_{z,k2}F_{n}. \quad (8.188)$$

8.8 Relief valves

8.8.1 Introduction

Relief valves are designed to keep the system pressure under a prescribed value, and therefore are important safety components of industrial networks. Operating valves excites considerable forces in pipe networks. If the design of the support structures is based on the *steady state analysis* only the structures may be destroyed by *transient forces*. That is why proper design of dynamic valve behavior is a necessary step towards modeling industrial pipe networks. Usually the second order ordinary differential equation describing the piston motion is solved by finite difference methods, e.g. [14] (1988) with time step limited by the nondumped oscillation time constant of the valve. In this Section we demonstrate a piecewise analytical solution which removes this limitation and increases the stability of the numerical solution for the flow itself.

8.8.2 Valve characteristics, model formulation

The commonly used components of a relief valve are: valve housing, inlet, outlet, piston, rod assembly, spring, bellows, and valve adjusting ring assembly, see Fig. 8.7. *Back pressing* valves belong to the same category and can also be described by the models considered in this section. The main difference between relief and back pressing valves is their mode of operation. Relief valves are usually closed during regular operation and open if the inlet pressure exceeds a prescribed value, whereas back pressing valves are open under regular conditions and closed in case of a sudden pressure loss. Also the damping mechanisms may be different.

The dynamic behavior of a valve is uniquely defined at any moment τ if we know the piston position z and the dependence of the smallest flow cross section as a function of the piston position $A^*(z)$ - see the Nomenclature at the end of Section 8. The inlet-outlet flow paths are usually absorbed as a part of the pipe model. Knowing $A^*(z)$ the dimensionless surface permeability at the contraction cross section used in the network analysis,

$$\gamma_z = A^*(z)/A_{norm} \qquad (8.189)$$

368   8 One-dimensional three-fluid flows

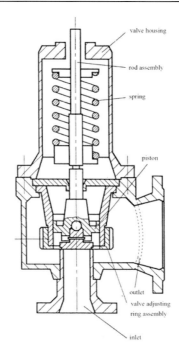

Fig. 8.7. Valve components

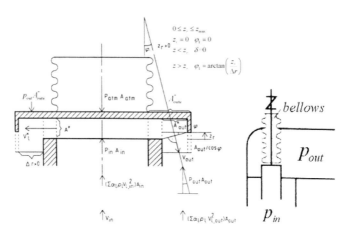

Fig. 8.8. Valve characteristics

can be computed, where A_{norm} is some normalizing cross section in m^2, e.g. the maximum cross section in the pipe where the valve is installed. The quantitative characteristics needed to describe the dynamic behavior of the valve given in Fig. 8.8 are given in the Nomenclature. Valves without adjusting ring assembly,

$A_{out} = 0$, do not develop additional reaction force. Valves provided with adjusting ring assembly with lower end being below the closing surface of the valve in closed position,

$$z_r \leq 0, \qquad (8.190)$$

$$\varphi = 0, \qquad (8.191)$$

develop the largest additional flow reaction force at the valve piston and consequently have the lowest closing pressure for the particulate valve geometry. If the position of the lower end of the valve adjusting ring assembly is above the closing surface of the valve during the valve operation

$$z_r > 0, \qquad (8.192)$$

there are two possible regimes for the flow to develop additional reaction force at the piston. The first regime is defined if the piston position is below the lower end of the valve adjusting ring assembly

$$z \leq z_r. \qquad (8.193)$$

In this case the adjusting ring assembly has no influence on the flow behavior. We describe this by setting a multiplier

$$\varphi = \pi/2, \qquad (8.194)$$

for the additional flow reaction force. In the second regime the piston position is above the lower end of the valve adjusting ring assembly

$$z > z_r. \qquad (8.195)$$

In this case the angle between the velocity vectors of the inlet and outlet flows is

$$\varphi = \arctan(z_r / \Delta r). \qquad (8.196)$$

In this case the additional flow reaction force at the piston is between zero and the maximum possible value.

Before writing down the momentum equation for the moving assembly of the valve let us summarize the forces acting on the moving valve mechanism:

Inertia force $m_v \dfrac{d^2 z}{d\tau^2}$; $\qquad (8.197)$

Friction force $C_{fr}\dfrac{dz}{d\tau}$; (8.198)

Spring force $C_{sp}(z+z_0)$; (8.199)

Gravitation force $m_v g$; (8.200)

Pressure difference forces for open valve

$$p_{in}(\tau)A_{in} + p_{out}(\tau)A_{out} - p_{outu}(\tau)A_{outu} - p_{atm}(\tau)A_{atm}, \qquad (8.201)$$

or for

$$p_{out}(\tau) \approx p_{outu}(\tau), \qquad (8.202)$$

$$p_{in}(\tau)A_{in} + p_{out}(\tau)(A_{out} - A_{outu}) - p_{atm}(\tau)A_{atm}. \qquad (8.203)$$

The fluid momentum flux force for open valve is

$$f_{fl} = A_{in}\sum_{l=1}^{l_{max}}\alpha_l\rho_l V_{l,in}^2 + A_{out}\sum_{l=1}^{l_{max}}\alpha_l\rho_l V_{l,out}^2 \approx \left(\sum_{l=1}^{l_{max}}\alpha_l^*\rho_l^* V_l^*\right)A^{*2}\left(\frac{1}{A_{in}} + \frac{1}{A_{out}}\cos^2\varphi\right)$$

$$\approx \left(\sum_{l=1}^{l_{max}}\alpha_l^*\rho_l^* V_l^*\right)A^{*2}\left(\frac{1}{A_{in}} + \frac{1}{A_{out}}\frac{\Delta r^2}{z_r^2 + \Delta r^2}\right) \qquad (8.204)$$

for $A_{out} > 0$, and

$$f_{fl} \approx \left(\sum_{l=1}^{l_{max}}\alpha_l^*\rho_l^* V_l^*\right)A^{*2} / A_{in} \qquad (8.205)$$

for $A_{out} = 0$. Here $A_{out}\sum_{l=1}^{l_{max}}\alpha_l\rho_l V_{l,out}^2$ is the additional flow reaction force at the piston. The field velocities V_l^* at the *vena contracta*, A^*, for open valve are computed with the fluid dynamic model of the flow in the network. Assuming

$$(\alpha_l\rho_l)_{in} \approx (\alpha_l\rho_l)_{out} = (\alpha_l\rho_l)^* \qquad (8.206)$$

the inlet and outlet velocities can be approximately estimated as follows

$$V_{1,in} = V_l^* A^* / A_{in}, \tag{8.207}$$

$$V_{1,out} = V_l^* \left(A^* / A_{out} \right) \cos\varphi \quad \text{for} \quad A_{out} > 0. \tag{8.208}$$

The momentum principle applied to the valve moving mechanism gives

$$m_v \frac{d^2 z}{d\tau^2} + C_{fr} \frac{dz}{d\tau} + C_{sp}(z + z_0) + m_v g -$$

$$\left[p_{in}(\tau) A_{in} + p_{out}(\tau)(A_{out} - A_{outu}) - p_{atm}(\tau) A_{atm} \right] + f_{fl} = 0. \tag{8.209}$$

Having in mind that for steady state the following force equilibrium is valid at the moment the valve is just starting to open,

$$m_v g + C_{sp} z_0 = p_{on}(\tau) A_{in} + p_{out}(\tau)(A_{out} - A_{outu}) - p_{atm}(\tau) A_{atm} \tag{8.210}$$

we replace

$$m_v g + C_{sp} z_0 - p_{out}(\tau)(A_{out} - A_{outu}) + p_{atm}(\tau) A_{atm} = p_{on}(\tau) A_{in} \tag{8.211}$$

and obtain a simpler dynamic force balance

$$m_v \frac{d^2 z}{d\tau^2} + C_{fr} \frac{dz}{d\tau} + C_{sp} z = \left[p_{in}(\tau) - p_{on}(\tau) \right] A_{in} + f_{fl} = 0. \tag{8.212}$$

In case $A_{out} \approx A_{outu}$, $p_{on}(\tau) \approx p_{on}$ is no longer a function of time. In this case instead of the input information z_0, A_{out}, A_{outu}, A_{atm} only p_{on} is necessary for the description of the valve behavior.

Next we look for a piecewise analytical solution of this equation assuming that within short time interval

$$\Delta\tau, \tag{8.213}$$

the input pressure is constant

$$p_{in}(\tau) \approx const. \tag{8.214}$$

8.8.3 Analytical solution

As already mentioned the differential equation describing the piston motion is solved usually by finite difference methods – see [21]. This is associated with a time step controlled by the criterion

$$\Delta \tau \approx 0.01 \Delta \tau_v, \qquad (8.215)$$

where $\Delta \tau_v$ is the oscillation time constant. In order to avoid this limitation a piecewise quasi analytical method is applied in this work. Next we describe the theoretical basics for this solution – for more information see [4, 23, 17]. First we use the text book solution of the homogeneous equation

$$m_v \frac{d^2 z}{d\tau^2} + C_{fr} \frac{dz}{d\tau} + C_{sp} z = 0. \qquad (8.216)$$

The eigenvalues $\lambda_{1,2}$ are the roots of the equation

$$m_v \lambda^2 + C_{fr} \lambda + C_{sp} = 0, \qquad (8.217)$$

$$\lambda_{1,2} = -\frac{1}{2} \frac{C_{fr}}{m_v} \pm \sqrt{\left(\frac{1}{2} \frac{C_{fr}}{m_v}\right)^2 - \frac{C_{sp}}{m_v}} = -\frac{1}{\Delta \tau_d} \pm \sqrt{-1} \frac{2\pi}{\Delta \tau_v}. \qquad (8.218)$$

a) For the *harmonic oscillations* case,

$$\frac{C_{sp}}{m_v} > \left(\frac{1}{2} \frac{C_{fr}}{m_v}\right)^2, \qquad (8.219)$$

the eigenvalues are complex:

$$\lambda_{1,2} = -\frac{1}{\Delta \tau_d} \pm i \frac{2\pi}{\Delta \tau_v}. \qquad (8.220)$$

Here the time constants of the process are

(i) the *dumping time constant*

$$\Delta \tau_d = 2 m_v / C_{fr}, \qquad (8.221)$$

and

(ii) the *oscillation period*

$$\Delta \tau_v = \frac{2\pi}{\sqrt{\dfrac{C_{sp}}{m_v} - \dfrac{1}{\Delta \tau_d^2}}} \,. \tag{8.222}$$

The oscillation frequency is then

$$\omega = 2\pi / \Delta \tau_v \,. \tag{8.223}$$

The text book parametric solution of the homogeneous equation is therefore

$$z(\tau) = e^{-\tau/\Delta \tau_d} \left[C_1 \cos(\omega \tau) + C_2 \sin(\omega \tau) \right] \tag{8.224}$$

and the general parametric solution of the non-homogeneous equation

$$z(\tau) = e^{-\tau/\Delta \tau_d} \left[C_1 \cos(\omega \tau) + C_2 \sin(\omega \tau) \right] + z_{max}^{**}/2 \,, \tag{8.225}$$

where

$$z_{max}^{**} = \frac{2}{C_{sp}} \left\{ \left[p_{in}(\tau) - p_{on}(\tau) \right] A_{in} + f_{fl} \right\} . \tag{8.226}$$

b) For the *aperiodic* case of

$$\frac{C_{sp}}{m_v} < \left(\frac{1}{2} \frac{C_{fr}}{m_v} \right)^2 \tag{8.227}$$

$$\Delta \tau_v = \frac{2\pi}{\sqrt{\dfrac{1}{\Delta \tau_d^2} - \dfrac{C_{sp}}{m_v}}} \,. \tag{8.228}$$

the eigenvalues are real

$$\lambda_{1,2} = -\frac{1}{\Delta \tau_d} \pm \frac{2\pi}{\Delta \tau_v} \tag{8.229}$$

and

$$\lambda_1 \neq \lambda_2 . \tag{8.230}$$

In this case the text book solution is

$$z(\tau) = C_1 e^{\lambda_1 \tau} + C_2 e^{\lambda_2 \tau} + z_{max}^{**}/2 . \tag{8.231}$$

c) For the *asymptotic* case

$$\frac{C_{sp}}{m_v} = \left(\frac{1}{2} \frac{C_{fr}}{m_v} \right)^2 \tag{8.232}$$

the eigenvalues are equal to each other

$$\lambda_{1,2} = \lambda = -\frac{1}{\Delta \tau_d} . \tag{8.233}$$

In this case the text book solution is

$$z(\tau) = \frac{z_{max}^{**}}{2} + C_1 e^{\lambda \tau} + C_2 \tau e^{\lambda \tau} . \tag{8.234}$$

8.8.4 Fitting the piecewise solution on two known position – time points

a) *Harmonic oscillations case*:

Knowing two arbitrary (τ, z) points

$$\tau = \tau_1, \qquad z = z_1, \tag{8.235}$$

and

$$\tau = \tau_2, \qquad z = z_2, \tag{8.236}$$

we compute the constants C_1 and C_2. Replacing the denominator of the so obtained solution with its equivalent

$$\cos(\omega \tau_1)\sin(\omega \tau_2) - \cos(\omega \tau_2)\sin(\omega \tau_1) = \sin\left[\omega(\tau_2 - \tau_1)\right], \tag{8.237}$$

and substituting

$$z_1^* = \left(z_1 - z_{max}^{**}/2\right)e^{\tau_1/\Delta\tau_d}, \tag{8.238}$$

and

$$z_2^* = \left(z_2 - z_{max}^{**}/2\right)e^{\tau_2/\Delta\tau_d}, \tag{8.239}$$

we obtain finally

$$C_1 = \frac{z_1^* \sin(\omega\tau_2) - z_2^* \sin(\omega\tau_1)}{\sin\left[\omega(\tau_2 - \tau_1)\right]}, \tag{8.240}$$

$$C_2 = \frac{z_2^* \cos(\omega\tau_1) - z_1^* \cos(\omega\tau_2)}{\sin\left[\omega(\tau_2 - \tau_1)\right]}, \tag{8.241}$$

or after inserting into Eq. (8.225) and some rearrangements we obtain

$$z(\tau) = \frac{z_{max}^{**}}{2} + \frac{z_1^{**} \sin\left[\omega(\tau_2 - \tau)\right] - z_2^{**} \sin\left[\omega(\tau_1 - \tau)\right]}{\sin\left[\omega(\tau_2 - \tau_1)\right]}, \tag{8.242}$$

where

$$z_1^{**} = \left(z_1 - z_{max}^{**}/2\right)e^{(\tau_1 - \tau)/\Delta\tau_d}, \tag{8.243}$$

and

$$z_2^{**} = \left(z_2 - z_{max}^{**}/2\right)e^{(\tau_2 - \tau)/\Delta\tau_d}. \tag{8.244}$$

b) *Aperiodic case*

Again knowing the solutions in two previous points we can estimate the constants. Inserting into Eq. (8.231) and rearranging we obtain finally

$$z(\tau) = \frac{z_{max}^{**}}{2} + \frac{z_1^{**} \sinh\left[\omega(\tau_2 - \tau)\right] - z_2^{**} \sinh\left[\omega(\tau_1 - \tau)\right]}{\sinh\left[\omega(\tau_2 - \tau_1)\right]}. \tag{8.245}$$

Note the formal similarity to Eq. (8.242). The sine functions are here replaced by hyperbolic sines.

c) *Asymptotic case*

Again knowing the solutions in two previous points we can estimate the constants. Inserting into Eq. (8.234) and rearranging we obtain finally

$$z(\tau) = \frac{z^{**}_{max}}{2} + \frac{z^{**}_1(\tau_2 - \tau) - z^{**}_2(\tau_1 - \tau)}{\tau_2 - \tau_1}. \tag{8.246}$$

Note that Eqs. (8.242) and (8.245) reduces to the above equation for $\Delta \tau_v \to \infty$.

8.8.5 Fitting the piecewise solution on known velocity and position for a given time

For a description of reflection of the piston after the impact with the upper or lower lift limitation structures it is more appropriate to use a solution fitted to a single time point position and velocity $\tau = \tau_1, z = z_1, \frac{dz}{d\tau} = w_{v,1}$.

a) For the *harmonic oscillations* case we have for the integration constants

$$z(\tau) = e^{-\tau/\Delta\tau_d}\left[C_1 \cos(\omega\tau) + C_2 \sin(\omega\tau)\right] + z^{**}_{max}/2, \tag{8.247}$$

$$w_v = -\frac{1}{\Delta\tau_d}(z - z^{**}_{max}/2) + \omega e^{-\tau/\Delta\tau_d}\left[-C_1 \sin(\omega\tau) + C_2 \cos(\omega\tau)\right], \tag{8.248}$$

$$C_1 = e^{\tau_1/\Delta\tau_d}\left\{(z_1 - z^{**}_{max}/2)\left[\cos(\omega\tau_1) - \frac{\sin(\omega\tau_1)}{\omega\Delta\tau_d}\right] - \frac{w_{v,1}}{\omega}\sin(\omega\tau_1)\right\}, \tag{8.249}$$

$$C_2 = e^{\tau_1/\Delta\tau_d}\left\{(z_1 - z^{**}_{max}/2)\left[\sin(\omega\tau_1) + \frac{\cos(\omega\tau_1)}{\omega\Delta\tau_d}\right] + \frac{w_{v,1}}{\omega}\cos(\omega\tau_1)\right\}. \tag{8.250}$$

b) For the *aperiodic* case

$$z = C_1 e^{\lambda_1 \tau} + C_2 e^{\lambda_2 \tau} + z^{**}_{max}/2, \tag{8.251}$$

$$w_v = C_1 \lambda_1 e^{\lambda_1 \tau} + C_2 \lambda_2 e^{\lambda_2 \tau}, \tag{8.252}$$

$$C_1 = \frac{-\left(z_1 - z_{max}^{**}/2\right) \lambda_2 e^{\lambda_2 \tau_1} + w_{v,1} e^{\lambda_2 \tau_1}}{\lambda_1 e^{(\lambda_1 + \lambda_2)\tau_1} - \lambda_2 e^{(\lambda_1 + \lambda_2)\tau_1}}, \tag{8.253}$$

$$C_2 = \frac{\left(z_1 - z_{max}^{**}/2\right) \lambda_1 e^{\lambda_1 \tau_1} - w_{v,1} e^{\lambda_1 \tau_1}}{\lambda_1 e^{(\lambda_1 + \lambda_2)\tau_1} - \lambda_2 e^{(\lambda_1 + \lambda_2)\tau_1}}. \tag{8.254}$$

c) For the *asymptotic* case

$$z(\tau) = \frac{z_{max}^{**}}{2} + C_1 e^{\lambda \tau} + C_2 \tau e^{\lambda \tau}, \tag{8.255}$$

$$w_v = C_1 \lambda e^{\lambda \tau} + C_2 \left(1 + \lambda \tau\right) e^{\lambda \tau}, \tag{8.256}$$

$$C_1 = \left[\left(1 + \lambda \tau_1\right)\left(z_1 - z_{max}^{**}/2\right) - w_v \tau_1\right] e^{-\lambda \tau_1}, \tag{8.257}$$

$$C_2 = \left[w_v - \left(z_1 - z_{max}^{**}/2\right) \lambda\right] e^{-\lambda \tau_1}. \tag{8.258}$$

8.8.6 Idealized valve characteristics

In order to understand some characterizing features of the valve dynamic we consider next some interesting simple cases.

a) No friction force

If there is no friction force,

$$C_{fr} = 0 \tag{8.259}$$

$$\Delta \tau_d \to \infty, \tag{8.260}$$

the solution is

$$z(\tau) = \frac{z_{max}^{**}}{2} + C_1 \cos(\omega \tau) + C_2 \sin(\omega \tau), \tag{8.261}$$

which means that the oscillations are not damped.

Some important features of the dynamic behavior of the valve can be studied on the non-damped solution.

b) Opening just starts, idealized opening time:

Let us consider the opening process at the very beginning. For this case we have

$$\tau = 0, \qquad (8.262)$$

$$z = 0, \qquad (8.263)$$

and

$$dz/d\tau = 0, \qquad (8.264)$$

and consequently

$$C_{fr} \, dz/d\tau = 0, \qquad (8.265)$$

the friction force is small due to the small averaged velocity gradient at the moment of the opening. Further it can be assumed that

$$f_{fl} = 0, \qquad (8.266)$$

i. e. there is no flow reaction force.

$$z(\tau) = \frac{z^*_{max}}{2}\left[1 - \cos(\omega\tau)\right] \geq 0, \qquad (8.267)$$

where

$$z^*_{max} = \frac{2A_{in}}{C_{sp}}\left[p_{in}(\tau) - p_{on}(\tau)\right] \qquad (8.268)$$

is the maximum of the piston position which can be reached if there where no hardware limitation of the piston motion. Obviously z^*_{max} depends an $p_{in}(\tau)$. If

$$z_{max} < z^*_{max} \qquad (8.269)$$

we have stable opening process. The time needed for opening the valve is

$$\Delta\tau_{op} = \omega \arccos\left(1 - 2\frac{z_{max}}{z^*_{max}}\right), \qquad (8.270)$$

where

$$-1 < 1 - 2\frac{z_{max}}{z^*_{max}} \le 1 \quad \text{or} \quad 0 < z_{max} < z^*_{max}. \qquad (8.271)$$

If

$$z_{max} \ge z^*_{max} \qquad (8.272)$$

the non-damping harmonic oscillation of the piston position within

$$0 < z < z^*_{max} \qquad (8.273)$$

is expected. This operation regime of the valve is called *piston fluttering*.

c) Closing just starts, idealized closing time:

Now let as consider the case where the valve was completely open and starts to close, i.e.

$$\tau = 0, \qquad (8.274)$$

$$z = z_{max}, \qquad (8.275)$$

$$dz/d\tau = 0, \qquad (8.276)$$

and consequently

$$C_{fr}\, dz/d\tau = 0, \qquad (8.277)$$

the solution is

$$z(\tau) = \frac{z^*_{max}}{2}\left[1 - \cos(\omega\tau)\right] + z_{max}\cos(\omega\tau) < z_{max}. \qquad (8.278)$$

The condition $z < z_{max}$ leads to

$$z^{**}_{max} < 2z_{max}, \qquad (8.279)$$

which is the necessary condition that the open valve starts to close. Obviously

$$\left|z_{\max}^{**}\right| > \left|z_{\max}^{*}\right| \quad \text{for the some} \quad \left|p_{in}(\tau) - p_{on}\right| \tag{8.280}$$

due to the existence of the flow reaction thrust

$$f_{fl} > 0. \tag{8.281}$$

Consequently the closing time

$$\Delta \tau_{cl} = \omega \arccos\left(1 - 2\frac{z_{\max}}{z_{\max}^{**}}\right)^{-1} \tag{8.282}$$

is larger than the closing time for the same value of $\left|p_{in}(\tau) - p_{on}\right|$. For $z_{\max}^{**} = 0$, $\Delta\tau_{cl} = \Delta\tau_v/4$. For $z_{\max}^{**} > 0$, $\Delta\tau_{cl} > \Delta\tau_v/4$ and for $z_{\max}^{**} < 0$, $\Delta\tau_{cl} < \Delta\tau_v/4$. We see that in the case of open valve even for pressure difference $\left|p_{in}(\tau) - p_{on}\right| \approx 0$ the valve remain open at least $\Delta\tau_v/4$ due to the action of the flow reaction force.

8.8.7 Recommendations for the application of the model in system computer codes

Finally the following algorithm can be recommended for modeling the dynamic valve behavior. The algorithm consists of logical conditions needed to identified the valve regime and the appropriate solution of the valve dynamic equation.

1. The valve is closed and remains closed if

$$(z = 0 \text{ and } p_{in} \leq p_{on}). \tag{8.283}$$

2. The valve starts to open if

$$(z = 0 \text{ and } p_{in} > p_{on}). \tag{8.284}$$

For

$$\tau = \tau + \Delta\tau, \tag{8.285}$$

the piston is lifted to

$$z(\tau) = \frac{z_{\max}^{*}}{2}\left[1 - \cos(\omega\Delta\tau)\right]. \tag{8.286}$$

In this case if $z < \varepsilon$ set $z = \varepsilon$, where

$$\varepsilon \to 0, \tag{8.287}$$

is the "computer zero". This is very important for the practical application of the method for all possible time steps.

3. The valve was completely open and remains completely open if

$$z = z_{max} \quad \text{and} \quad z_{max}^{**} \geq 2z_{max}. \tag{8.288}$$

4. The valve was completely open and starts to close

$$z = z_{max} \quad \text{and} \quad z_{max}^{**} < 2z_{max}. \tag{8.289}$$

In this case after $\Delta\tau$ the plug position is removed to the position

$$z(\tau) = \frac{z_{max}^*}{2}\left[1 - \cos(\omega\Delta\tau)\right] + z_{max}\cos(\omega\Delta\tau) < z_{max} \tag{8.290}$$

In this case if $z > z_{max} - \varepsilon$ set $z = z_{max} - \varepsilon$. This is very important for the practical application of the method for all possible time steps.

5. The valve is in operation if

$$\varepsilon < z = z_{max} - \varepsilon. \tag{8.291}$$

In this case the preceding information for two points is used

$$\tau = \tau_1, \qquad z = z_1, \tag{8.292}$$

and

$$\tau = \tau_2, \qquad z = z_2, \tag{8.293}$$

to compute the constants of the piecewise analytical solution C_1 and C_2. The plug position at τ for $\dfrac{C_{sp}}{m_v} > \left(\dfrac{1}{2}\dfrac{C_{fr}}{m_v}\right)^2$ is

$$z(\tau) = \frac{z_{max}^{**}}{2} + \frac{z_1^{**}\sin\left[\omega(\tau_2 - \tau)\right] - z_2^{**}\sin\left[\omega(\tau_1 - \tau)\right]}{\sin\left[\omega(\tau_2 - \tau_1)\right]} \tag{8.294}$$

for $\dfrac{C_{sp}}{m_v} < \left(\dfrac{1}{2}\dfrac{C_{fr}}{m_v}\right)^2$,

$$z(\tau) = \dfrac{z_{max}^{**}}{2} + \dfrac{z_1^{**}\sinh\left[\omega(\tau_2-\tau)\right]-z_2^{**}\sinh\left[\omega(\tau_1-\tau)\right]}{\sinh\left[\omega(\tau_2-\tau_1)\right]} \qquad (8.295)$$

and for $\dfrac{C_{sp}}{m_v} = \left(\dfrac{1}{2}\dfrac{C_{fr}}{m_v}\right)^2$,

$$z(\tau) = \dfrac{z_{max}^{**}}{2} + \dfrac{z_1^{**}(\tau_2-\tau)-z_2^{**}(\tau_1-\tau)}{\tau_2-\tau_1}, \qquad (8.296)$$

where if

$$z < 0, \qquad (8.297)$$

z is set to zero

$$z = 0, \qquad (8.298)$$

and the valve is considered as completely closed, and if

$$z > z_{max}, \qquad (8.299)$$

z is set to z_{max},

$$z = z_{max}, \qquad (8.300)$$

and the valve is considered completely open. Note that if $\varphi < 0.00001$ $\sin\varphi \approx \varphi$ and $\sinh\varphi \approx \varphi$.

Thus the mathematical description of the plug motion is completed. Having the prescribed dependence of the flow surface at the vena contracta as a function of the plug position

$$A^* = A^*(z) \qquad (8.301)$$

we can estimate in any moment the surface available for the flow and consequently the flow reaction force f_{fl}.

8.8.8 Some illustrations of the valve performance model

This chapter is devoted to an example of the analytical work required to be done before implementing a single-component model into system computer codes. The system computer codes are very complex. The behavior of any single pipe network component has to be examined very carefully outside the code before coupling the model with the system code. The valve model is a typical example. It introduce such a non-linear interaction with the flows in pipe systems that without knowing exactly the valve model behavior it is very difficult to make error diagnostics during the development. Here we present some examples documented in [14] and [22]. The tests have been performed by *Iris Roloff* [22].

We will have a look on the valve response to various forms of the driving force function $F(\tau)$. The fluid momentum flux force F_{fl} is set to zero, therefore the driving force is directly proportional to the inlet pressure. For this purpose simply the inlet pressure p_{in} is provided as a time function, which will be either a harmonic oscillation or a step function.

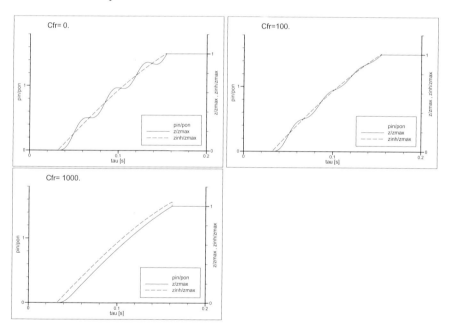

Fig. 8.9. Valve response to a slowly varying driving function with frequency $f_v = 1\ Hz$

1) Figure 8.9 shows the *response to a slowly varying driving function* with several values of the friction coefficient C_{fr}. The valve was closed at the beginning. When the inlet pressure exceeds the onset pressure p_{on}, the valve starts to open as mentioned in the previous chapter.

If there is no friction, i. e. $C_{fr} = 0$, the valve piston oscillates around the equilibrium position $z_{inh}(\tau)$, which follows the driving function, until the upper boundary at z_{max} is reached. No reflection is taken into account here, so the valve stays open.

If the friction coefficient is small with respect to the effect of the spring, the oscillations around the equilibrium position are damped.

If the friction coefficient is large, the valve piston follows the equilibrium position without oscillating, but with a delay.

2) Figure 8.10 shows the *response to a harmonic oscillation function*. The frequency of the driving function is of the same order of magnitude as the eigenfrequency of the free oscillating valve, which is in this case $f_v = 28\ Hz$. The valve is closed at the beginning. It starts to open, when the onset pressure p_{on} is exceeded by the inlet pressure p_{in}. Reflection is not taken into account here. Therefore the valve stays open, until the inlet pressure drops according to case 4 discussed in the previous Section.

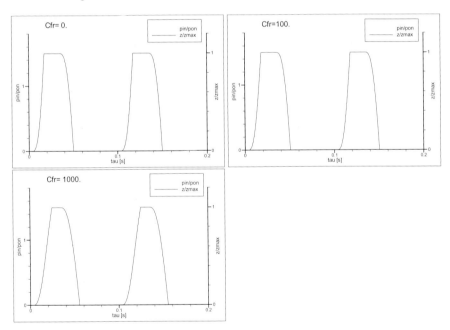

Fig. 8.10. Valve response to a harmonic oscillation with frequency $v = 10\ Hz$

8.8 Relief valves 385

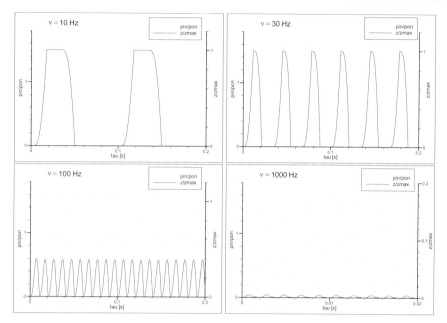

Fig. 8.11. Dependence on the frequency ν of the driving function

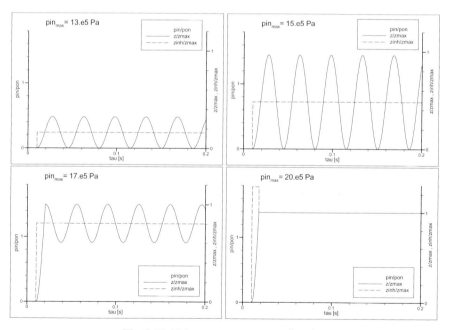

Fig. 8.12. Valve response to a step function

386 8 One-dimensional three-fluid flows

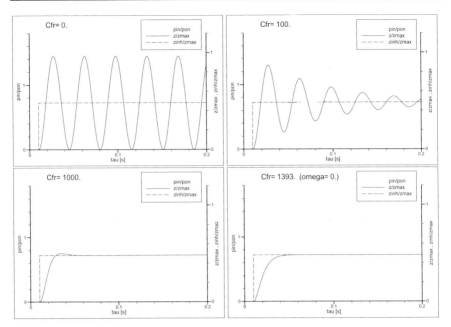

Fig. 8.13. Influence of the friction coefficient, oscillating and aperiodic case

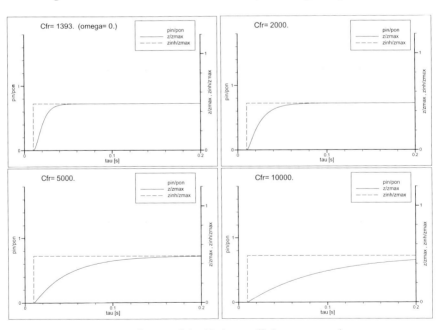

Fig. 8.14. Influence of the friction coefficient, asymptotic case

When the lower boundary at $z = 0$ is reached, the valve remains closed, until the inlet pressure again exceeds the onset pressure. Also in this case a delay due to the friction can be clearly seen.

3) Figure 8.11 shows the dependence of the valve response on the frequency of the driving function. No friction is taken into account this time. Reflection is also not considered here.

If the frequency v of the driving function is much larger than the eigenfrequency of the free oscillating valve, which in this example is $f_v = 28\ Hz$, then the amplitude of the valve oscillation drops according to the ratio of the two frequencies.

Figure 8.12 shows the valve response to a step function. The height of the step is varied. The previously closed valve starts oscillating around the new equilibrium position, which corresponds to the height of the step, as long as the value of the inlet pressure does not satisfy the conditions of case 3 discussed in the previous Section to hold up the valve in a completely open stage.

4) Figure 8.13 and 8.14 show the effect of the friction coefficient C_{fr} and the transition from the oscillation case to the aperiodic solution case and the asymptotic solution case. If the friction coefficient is small, the valve does some damped oscillations around the new equilibrium position. When the friction coefficient takes a value that counteracts the spring effect, the new equilibrium position is reached within one period and the oscillations are stopped. As the friction coefficient increases, the valve needs more and more time to reach the new equilibrium position.

5) Figure 8.15 shows the effect of the reflection coefficients. The reflection coefficients for both boundaries are varied simultaneously. As an example the response to a large step satisfying the conditions of case 3 discussed in the previous chapter to hold up the valve in a completely open stage is considered.

With no reflection at the upper boundary, i. e. $r_u = 0$, the piston sticks to the boundary once it is reached.

A reflection coefficient of $r_u = 1$ means full reflection. In this case the valve piston retains all its velocity, but the direction of the movement is inverted. Therefore the piston goes back exactly in reverse manner.

With a reflection coefficient value of $r_u = 0.5$, half of the velocity amount is preserved. That means that every reversion goes only half the way back.

388 8 One-dimensional three-fluid flows

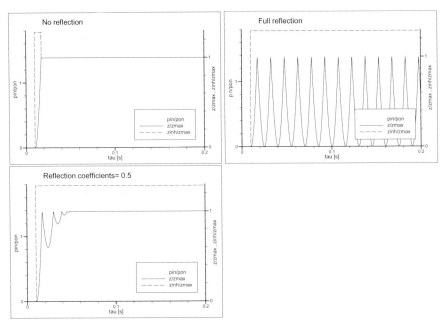

Fig. 8.15. Influence of the reflection coefficients

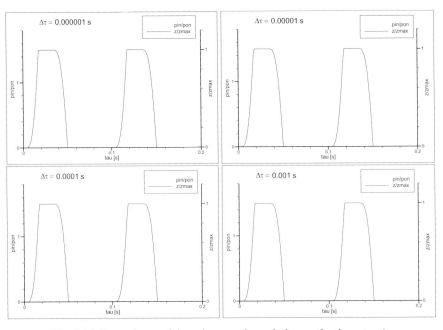

Fig. 8.16. Dependence of the valve equation solution on the time step $\Delta\tau$

6) Figure 8.16 shows the dependence of the solution on the time step $\Delta \tau$. As can be seen clearly, the solution method gives good results also for very large time steps compared to a finite difference method approach.

8.8.9 Nomenclature for Section 8.8

Latin

$A^*(z)$	flow cross section or critical flow cross section as a function of the piston position, m^2. This function is defined by the particular valve geometry.
A^*	vena contracta area i.e. the valve ring area m^2
A_{norm}	some normalizing cross section, m^2
A_{in}	> 0, valve piston face area exposed to the inlet flow stream, m^2
A_{out}	≥ 0, valve piston face ring area exposed to the outlet pressure by open and closed valve, m^2
A_{outu}	back valve piston area outside the bellow, m^2
A_{atm}	back valve piston area inside the bellow, m^2
f_{fl}	fluid force acting on the piston for open valve, N
f_v	eigen frequency of the piston, $1/s$
C_{sp}	spring constant, N/m
C_{ft}	valve damping coefficient, Ns/m
m_v	mass of the valve mechanism that is in motion (i.e. the valve piston, rod assembly combined with the spring and bellows), kg
$p_{in}(\tau)$	valve inlet pressure, Pa
$p_{out}(\tau)$	valve outlet pressure, Pa
p_{on}	pressure at which previously closed valve starts to open, Pa
p_{atm}	atmospheric back pressure inside the bellow, Pa
p_{outu}	valve back pressure outside the bellow, Pa
r_u	reflection coefficient, *dimensionless*
w_v	velocity of the valve piston, m/s
z	piston position (i.e., z coordinate, $0 \leq z \leq z_{max}$), m
	$= 0$, piston position for closed valve, m
	$= z_{max}$, piston position for completely open valve, valve lift, m
z_0	spring pressing distance for the normal valve operation, m
z_r	position of the lower end of the valve adjusting ring assembly, $0 \leq z_r \leq z_{max}$, m

Greek

α_l	volume fraction of the field l, *dimensionless*
γ_z	$= A^*(z)/A_{norm}$, surface permeability at the contraction cross section used in the network analysis, *dimensionless*
Δr	width of the expansion ring area formed by the inner surface of the valve adjusting ring assembly and the outer surface of the inlet, m
$\Delta \tau_{cl}$	closing time, s
$\Delta \tau_v$	oscillation period, s
$\Delta \tau_d$	damping time constant, s
$\lambda_{1,2}$	eigenvalues, $1/s$
ρ_l	density of field l, kg/m^3
φ	$\leq \pi/2$, angle of the flow path trajectory between valve inlet and valve outlet, *rad*
τ	time, s
ω	oscillation frequency, $2\pi/s$

Subscripts

1	field 1, gas
2	field 2, liquid
3	field 3, droplets
l	field l
in	valve inlet
out	valve outlet
atm	at atmospheric pressure
norm	scaling
outu	outside the bellow
sp	spring
fr	friction
on	previously closed valve starts to open
r	ring

Superscripts

*	at the smallest flow cross section called in Latin vena contracta

8.9 Pump model

In the previous section we considered the relief valves as an important component of pipe networks. Another important component is the pump. In this Section we describe the dynamic behavior of the *centrifugal pumps* by simple model. The model is appropriate for use in computer codes simulating system behavior of complex pipe networks e.g. [15].

The derived pump model is based on the following simplifying assumptions:

(a) Transient flow processes into the pump impeller can be represented by a sequence of steady state processes because the time needed by the flow particle to pass the pump is very short. The smaller the pumps dimension the more correct is this assumption and vice versa.

(b) The flow through the pump is incompressible.

8.9.1 Variables defining the pump behavior

The variables required to describe mathematically the pump behavior are defined below.

Geometry:

D_{p1} inner pump impeller diameter, m
D_{p2} outer pump impeller diameter, m
β_1 angle between the relative flow velocity V_1^r and the impeller angular velocity $V_{\theta 1}$, rad
β_2 angle between the relative flow velocity V_2^r and the impeller angular velocity $V_{\theta 2}$, rad
I_{pr} moment of inertia of the pump rotor, kgm^2

Velocities:

V_r radial (meridian) velocity, m/s
V^r relative flow velocity for an observer rotating with the impeller, m/s
V_1^r $= \dot{V} / A_1$, relative flow velocity for an observer rotating with the impeller at the entrance cross section of the impeller, m/s
V_2^r $= \dot{V} / A_2$, relative flow velocity for an observer rotating with the impeller at the exit cross section of the impeller, m/s
V_θ angular impeller velocity, m/s

$V_{\theta i}$ $= \omega D_{pi}/2 = \pi D_{pi} n$, impeller angular velocity at the position defined with diameter D_{pi}, m/s

Fluid characteristics:

ρ fluid density, kg/m³

V_m $= \left(\sum_{l=1}^{l_{max}} \alpha_l \rho_l V_l \right) \Big/ \sum_{l=1}^{l_{max}} \alpha_l \rho_l$, fixture velocity, m/s

A flow cross section at which the mixture velocity, m²

Pump characteristics:

g = 9.81, acceleration due to gravity, m/s²

$\dot{V}$ volumetric through flow, m³/s. The volumetric flow is positive if it is in the same direction as the positive velocity in the control volume i.e. if it has the direction of the increasing cell indices.

$\dot{V}_R$ rated volumetric through flow characterizing the normal pump operation, m³/s

$\dot{V}^*$ $= \dot{V}/\dot{V}_R$, volumetric flow ratio, *dimensionless*

Δp_{pump} pump pressure difference i.e. pressure at the pump outlet minus pressure at the pump inlet (used in the momentum balance), Pa. The pressure difference is positive if it would accelerate the flow in the positive velocity direction.

H $= \Delta p_{pump}/(\rho g)$, total head rise of the pump (defined by empirical homologous pump performance model as a function of the volumetric through flow $\dot{V}$), m

H_R rated total head rise of the pump characterizing the normal pump operation, m

H^* $= H/H_R$, head ratio, *dimensionless*

M_{fl} torque acting at the pump rotor and caused by the flow in the pump (defined by empirical homologous pump performance model as a function of the volumetric through flow $\dot{V}$), Nm. Negative, if it tends to decelerate the pump.

$M_{fl,R}$ rated pump torque characterizing the normal pump operation, Nm
M^* $= M_{fl}/M_{fl,R}$, torque ratio, *dimensionless*
M_{fr} friction torque, Nm
M_m motor torque, Nm
ω pump angular speed (rotational speed defined by a pump drive model). A pump operating in a normal pump regime has positive angular velocity, rad/s

8.9 Pump model

ω_R rated pump angular speed characterizing the normal pump operation, rad/s
ω^* ω/ω_R, angular velocity ratio, *dimensionless*
n rotation per second, 1/s
n_R rated rotation per second characterizing the normal pump operation, 1/s
n^* $= n/n_R$, rotation ratio, *dimensionless*
P_{pump} $= M\omega = \eta_p \Delta p_{pump} A V_m = \eta_p \Delta p_{pump} \dot{V}$, pump power introduced into the flow (used into the energy balance), W
η_p efficiency, *dimensionless*

The rated values $\dot{V}_R$, ω_R, H_R and $M_{fl,R}$ are required input data for the pump model.

Dimensionless similarity criteria:

φ $= \dfrac{\dot{V}}{V_\theta \pi D_{pi}^2 / 4} = \dfrac{8\dot{V}}{\pi \omega D_{pi}^3} = \dfrac{4\dot{V}}{\pi^2 n D_{pi}^3}$, specific capacity, *Keller* 1934, *dimensionless*

ψ $= \dfrac{\Delta p_{pump}}{\rho V_{\theta i}^2 / 2} = \dfrac{\rho g H}{\rho V_{\theta i}^2 / 2} = \dfrac{gH}{V_{\theta i}^2 / 2} = \dfrac{8gH}{(\omega D_{pi})^2} = \dfrac{2gH}{(\pi n D_{pi})^2}$, specific head, *L. Prandtl* 1912, *dimensionless*. $\Delta p_{pump}/\rho$ can be considered as a specific work done on the fluid and $V_{\theta i}^2/2$ the specific kinetic energy at the outer impeller diameter.

λ $= \varphi \psi / \eta_{eff}$, power number, *dimensionless*
η_{eff} ratio of the power inserted into the flow to the pump motor power, *dimensionless*
    $M/(\rho \omega^2 D_{p2}^5)$ specific torque, *dimensionless*

In order to obtain bounded values in the homologous pump curve the following modified similarity numbers are used too:

$gH/\dot{V}^2$ modified specific head, *dimensionless*
$M/(\rho \dot{V}^2 D_{p2}^3)$ modified specific torque, *dimensionless*
n_s specific speed, 1/s

The *specific speed* is the speed necessary for a geometrically similar model pump to provide a head of $H = 1$ m and a volume flow rate of $\dot{V} = 1$ m³/s. Using the similarity relationships and the steady state operation point (H_R, $\dot{V}_R$ and n_R) the specific speed is computed with Eq. 8.5 in [10] p. 251 as follows

$$n_s = n_R \left(\frac{\dot{V}_R}{1m^3 s} \right)^{1/2} \Big/ \left(\frac{H_R}{1m} \right)^{3/4}. \tag{8.302}$$

The dimension of the specific speed is the same as the dimension of the rated speed n_R. One should be careful with the definition of the specific speed in older sources because at least three metric systems were widely used in the past.

8.9.2 Theoretical basics

The text book theoretical basics, see *Pohlenz* [20], p. 53, are briefly given here in order to understand the world wide accepted pump modeling technique. Consider a centrifugal pump with impeller rotating with n rotations per second that is with angular velocity

$$\omega = 2\pi n, \tag{8.303}$$

inner radius r_1 and outer radius r_2. The angular velocity at the inner impeller radius is

$$V_{\theta 1} = \omega r_1 \tag{8.304}$$

and at the outer impeller radius

$$V_{\theta 2} = \omega r_2. \tag{8.305}$$

The relative flow velocity for an observer rotating with the impeller is V^r. The relative flow velocity is equal to the volumetric flow divided by the cross section normal to V^r. For incompressible flow we have

$$V_1^r = \dot{V} / A_1, \tag{8.306}$$

$$V_2^r = \dot{V} / A_2. \tag{8.307}$$

The vector sum of the relative flow velocity and the angular impeller velocity gives the absolute flow velocity $\mathbf{V}^a$.

$$\mathbf{V}^a = \mathbf{V}^r + \mathbf{V}_\theta^a. \tag{8.306}$$

8.9 Pump model

We designate the angular component of this velocity magnitude with V_θ^a.

The mass flow in *kg/s* entering the pump impeller is

$$\dot{m} = \rho \dot{V}. \tag{8.307}$$

The angle between the relative flow velocity V_1^r and the impeller angular velocity $V_{\theta 1}$ is β_1 and that the angle between the relative flow velocity V_2^r and the impeller angular velocity $V_{\theta 2}$ is β_2. Therefore

$$V_{\theta 1}^a = V_{\theta 1} - V_1^r \cos \beta_1, \tag{8.308}$$

$$V_{\theta 2}^a = V_{\theta 2} - V_2^r \cos \beta_2. \tag{8.309}$$

The flow is stationary and incompressible. Thus the force acting on the flow in the angular direction at the inlet is

$$F_{\theta 1} = \dot{m} V_{\theta 1}^a \tag{8.310}$$

and at the impeller outlet

$$F_{\theta 2} = -\dot{m} V_{\theta 2}^a. \tag{8.311}$$

The corresponding torques are $F_{\theta 1} r_1$ and $F_{\theta 2} r_2$, respectively. The resulting torque acting on the flow inside the impeller is

$$M_{fl} = F_{\theta 2} r_2 + F_{\theta 1} r_1 = \dot{m} \left(V_{\theta 2}^a r_2 - V_{\theta 1}^a r_1 \right) \tag{8.312}$$

The power inserted into the flow is therefore $M_{fl} \omega$ or

$$\Delta p_{pump} \dot{V} = M_{fl} \omega = \omega \dot{m} \left(V_{\theta 2}^a r_2 - V_{\theta 1}^a r_1 \right) = \rho \dot{V} \left(V_{\theta 2}^a \omega r_2 - V_{\theta 1}^a \omega r_1 \right), \tag{8.313}$$

or after canceling the volumetric flux

$$\boxed{\Delta p_{pump} = \rho \left(V_{\theta 2}^a \omega r_2 - V_{\theta 1}^a \omega r_1 \right) = \rho \left(V_{\theta 2}^a V_{\theta 2} - V_{\theta 1}^a V_{\theta 1} \right).} \tag{8.314}$$

This is the *main pump equation* describing the pressure rise as a function of the impeller speed and geometry. Replacing Eqs. (8.308) and (8.309) in the main pump equation we obtain

$$\Delta p_{pump} = \rho \left[\left(V_{\theta 2} - V_2^r \cos \beta_2 \right) V_{\theta 2} - \left(V_{\theta 1} - V_1^r \cos \beta_1 \right) V_{\theta 1} \right]$$

$$= \rho \left(V_{\theta 2}^2 - V_{\theta 1}^2 + V_1^r V_{\theta 1} \cos \beta_1 - V_2^r V_{\theta 2} \cos \beta_2 \right). \qquad (8.315)$$

or after using the mass conservation equations (8.306) and (8.307) one obtains

$$\Delta p_{pump} = \rho \left[\left(r_2^2 - r_1^2 \right) \omega^2 + \left(\frac{r_1}{A_1} \cos \beta_1 - \frac{r_2}{A_2} \cos \beta_2 \right) \omega \dot{V} \right]$$

$$= \rho \left[\left(r_2^2 - r_1^2 \right) 4\pi^2 n^2 + \left(\frac{r_1}{A_1} \cos \beta_1 - \frac{r_2}{A_2} \cos \beta_2 \right) 2\pi n \dot{V} \right]. \qquad (8.316)$$

If we consider additionally that the flow exerts some pressure loss due to friction with the pump structures in the impeller region we finally obtain

$$\Delta p_{pump} = \rho \left[\left(r_2^2 - r_1^2 \right) \omega^2 + \left(\frac{r_1}{A_1} \cos \beta_1 - \frac{r_2}{A_2} \cos \beta_2 \right) \omega \dot{V} \right] - \xi \frac{1}{2} \frac{\rho}{A_m} \dot{V}^2$$

$$= \rho \left[\left(r_2^2 - r_1^2 \right) 4\pi^2 n^2 + \left(\frac{r_1}{A_1} \cos \beta_1 - \frac{r_2}{A_2} \cos \beta_2 \right) 2\pi n \dot{V} \right] - \xi \frac{1}{2} \frac{\rho}{A_m} \dot{V}^2. \qquad (8.317)$$

where ξ is some friction loss coefficient and A_m some averaged flow cross section. The pressure rise of the pump is usually expressed for liquids with the head H

$$\Delta p_{pump} = \rho g H, \qquad (8.318)$$

where g is the gravitational acceleration. Usually the main pump equation is written in the form

$$\boxed{H = An^2 + Bn\dot{V} - C\dot{V}^2,} \qquad (8.319)$$

where

$$A = (r_2^2 - r_1^2) 4\pi^2 / g ,\qquad(8.320)$$

$$B = \left(\frac{r_1}{A_1}\cos\beta_1 - \frac{r_2}{A_2}\cos\beta_2\right) 2\pi / g ,\qquad(8.321)$$

$$C = \xi \frac{1}{2}\frac{P}{A_m} / g .\qquad(8.322)$$

The coefficients A, B and C are not dependent on the fluid density and can be computed approximately from the particular pump geometry. The above equation contains the most important effect in the first quadrant i.e. $\dot{V} > 0$ and $H > 0$. In fact such analysis does not take into account several effects. That is why in the practical applications the dependence $H = H(n,\dot{V})$ is obtained experimentally for each particular pump family by the pump manufacturer. In the reality the ratio of the power inserted into the flow to the impeller power called efficiency is less than one

$$\eta_p = \rho g H \dot{V} / (M_{fl}\omega) .\qquad(8.323)$$

The power

$$W_{diss} = (1-\eta_p)(M_{fl}\omega)\qquad(8.324)$$

is transferred irreversibly from mechanical energy into internal energy of the flow per unit second due to internal friction, turbulization etc.

Obviously H is a function of n and $\dot{V}$. It is easy to show that introduction of a simple variable transformation reduces the three-dimensional relationship to a two-dimensional one. Dividing either by $\dot{V}^2$ or by n^2 one obtains

$$\boxed{H/\dot{V}^2 = A(n/\dot{V})^2 + Bn/\dot{V} - C = f(n/\dot{V})}\qquad(8.325)$$

or

$$\boxed{H/n^2 = A + B\dot{V}/n - C(\dot{V}/n)^2 = f(\dot{V}/n) .}\qquad(8.326)$$

This is a remarkable quality of the main pump equation. In spite of the fact that this dependence is only an approximate one, the practical measurements

$$f(H, \dot{V}, n) = 0 \tag{8.327}$$

are surprisingly reduced to a single curve named the *similarity* or *homologous* curve

$$f_1(H/\dot{V}^2, n/\dot{V}) = 0 \tag{8.328}$$

or

$$f_2(H/n^2, \dot{V}/n) = 0. \tag{8.329}$$

Both above discussed forms are used in the literature in order to avoid division by zero in the four quadrant if n or $\dot{V}$ tends to zero. This fact was recognized several decades ago. It was *Prandtl* in 1912 who introduced the so called specific head,

$$\psi = \frac{\Delta p_{pump}}{\rho V_{\theta 2}^2 / 2} = \frac{\rho g H}{\rho V_{\theta 2}^2 / 2} = \frac{gH}{V_{\theta 2}^2 / 2} = \frac{8gH}{(\omega D_{p2})^2} = \frac{2gH}{(\pi n D_{p2})^2}. \tag{8.330}$$

Here $\Delta p_{pump} / \rho$ can be considered as a specific work done on the fluid and $V_{\theta 2}^2 / 2$ the specific kinetic energy at the outer impeller diameter. It was *Keller* in 1934 who introduced the so called specific capacity of the pump defined as follows

$$\varphi = \frac{\dot{V}}{V_{\theta,2} \pi D_{p2}^2 / 4} = \frac{8\dot{V}}{\pi \omega D_{p2}^3} = \frac{4\dot{V}}{\pi^2 n D_{p2}^3}. \tag{8.331}$$

Like the measured dependence $f(H, \dot{V}, n) = 0$ the presentation in the form

$$f_1(\varphi, \psi) = 0. \tag{8.332}$$

reduces with remarkable accuracy the data to a single curve. The very important practical significance of the above relationship is that keeping φ constant in two neighboring steady state points of operation by

$$\dot{V}_1 / n_1 = \dot{V}_2 / n_2 \tag{8.333}$$

ψ is also constant, which means

$$H_1 / n_1^2 = H_2 / n_2^2. \tag{8.334}$$

Recently the main pump relationship was further modified by introducing dimensionless quantities namely quantities divided by the corresponding quantities characterizing the normal pump operation called some times rated quantities, H_R, n_R, $\dot{V}_R$,

$$H^* = H / H_R, \tag{8.335}$$

$$n^* = n / n_R, \tag{8.336}$$

$$\dot{V}^* = \dot{V} / \dot{V}_R, \tag{8.337}$$

$$H^* / \dot{V}^{*2} = A^* \left(n^* / \dot{V}^* \right)^2 + B^* n^* / \dot{V}^* - C^* = f\left(n^* / \dot{V}^* \right), \tag{8.338}$$

or

$$H^* / n^{*2} = A^* + B^* \dot{V}^* / n^* - C^* \left(\dot{V}^* / n^* \right)^2 = f\left(\dot{V}^* / n^* \right), \tag{8.339}$$

where

$$A^* = A n_R^2 / H_R, \tag{8.340}$$

$$B^* = B n_R \dot{V}_R / H_R \tag{8.341}$$

and

$$C^* = C \dot{V}_R^2 / H_R. \tag{8.342}$$

Table 1 gives a possible way how the homologous curve can be stored for all four quadrants avoiding division with zero as already exercised in RELAP 5/Mod 2 computer code, see *Ransom* et al. [21] 1988, p. 27, 179. Instead of the rotation ratio in Table 8.1 the angular velocity ratio is used. Note that the rotation ratio is equal to the angular velocity ratio

$$n^* = n / n_R = \omega / \omega_R = \omega^*. \tag{8.343}$$

Table 8.1. Pump homologous curve definitions

Regime mode	Regime	ω^*	$\dot{V}^*$	$\dfrac{\dot{V}^*}{\omega^*}$	Head		Key	Torque		Key
1	Normal pump, or energy dissipating pump, or reverse turbine	>0	≥ 0	≤ 1	$\dfrac{\dot{V}^*}{\omega^*}$	$\dfrac{H^*}{\omega^{*2}}$	HON	$\dfrac{M^*}{\omega^{*2}}$		TON
2		>0	≥ 0	>1	$\dfrac{\omega^*}{\dot{V}^*}$	$\dfrac{H^*}{\dot{V}^{*2}}$	HVN	$\dfrac{M^*}{\dot{V}^{*2}}$		TVN
3	Energy dissipation only	>0	<0	≥ -1	$\dfrac{\dot{V}^*}{\omega^*}$	$\dfrac{H^*}{\omega^{*2}}$	HOD	$\dfrac{M^*}{\omega^{*2}}$		TOD
4		>0	<0	<-1	$\dfrac{\omega^*}{\dot{V}^*}$	$\dfrac{\omega^{*2}}{H^*}$	HVD	$\dfrac{M^*}{\dot{V}^{*2}}$		TVD
5	Normal turbine, or energy dissipating turbine	≤ 0	≤ 0	≤ 1	$\dfrac{\dot{V}^*}{\omega^*}$	$\dfrac{H^*}{\omega^{*2}}$	HOT	$\dfrac{M^*}{\omega^{*2}}$		TOT
6		≤ 0	≤ 0	>1	$\dfrac{\omega^*}{\dot{V}^*}$	$\dfrac{H^*}{\dot{V}^{*2}}$	HVT	$\dfrac{M^*}{\dot{V}^{*2}}$		TVT
7	Reverse pump or energy dissipating reverse pump	≤ 0	>0	≥ -1	$\dfrac{\dot{V}^*}{\omega^*}$	$\dfrac{H^*}{\omega^{*2}}$	HOR	$\dfrac{M^*}{\omega^{*2}}$		TOR
8		≤ 0	>0	<-1	$\dfrac{\omega^*}{\dot{V}^*}$	$\dfrac{H^*}{\dot{V}^{*2}}$	HVR	$\dfrac{M^*}{\dot{V}^{*2}}$		TVR

ω^* = rotational velocity ratio, $\dot{V}^*$ = volumetric flow ratio, H^* = head ratio, and M^* = torque ratio. The key indicates which of the homologous parameters in each octant. The first sign is either H or T and means either head or torque, the second O or V indicates division either by ω^{*2} or $\dot{V}^{*2}$, and the third N, D, T and R indicates the 1, 2, 3 and 4 quadrant, respectively.

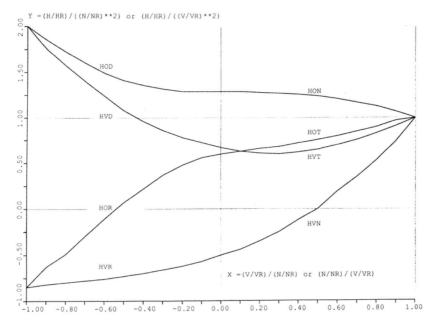

Fig. 8.17. Typical pump homologous head curves

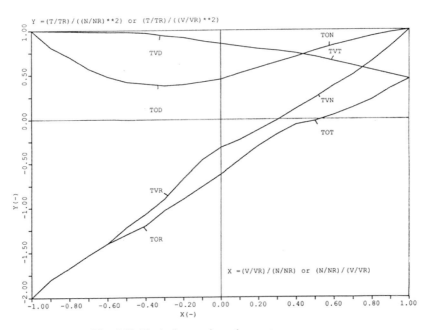

Fig. 8.18. Typical pump homologous torque curves

402   8 One-dimensional three-fluid flows

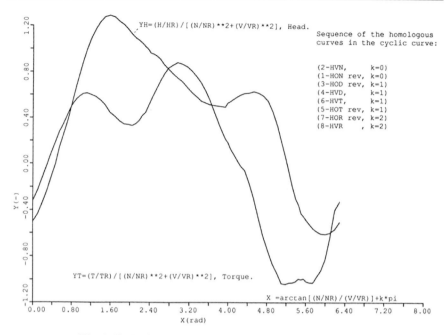

Fig. 8.19. Cyclic head and torque curves for single phase flow

Similarly the hydraulic torque acting on the pump rotor can be presented.

One of the possibilities to provide the pump data for computer codes is to read the 16 curves as discussed above as input - see Figs. 8.17 and 8.18. This is getting popular through the using of the RELAP computer code.

But there is more efficient way to use the pump characteristics: It is recommended then to store the data in the form of the so called *Suter* diagram, which is much simpler to use in a computer code - see Fig. 8.19. The idea how to construct a *Suter* diagram is described in the next section.

8.9.3 *Suter* diagram

In order to avoid the division of the homologous curve into eight segments the proposal made by *Suter* [24] in 1966 and modified by *Lang*, see in *Kastner* et al. [6, 7] (1983), is very useful. Instead of using either $H*/n*^2$ or $H*/\dot{V}*^2$ it is convenient to use only one function quantity

$$H*/\left(n*^2 + \dot{V}*^2\right),\tag{8.344}$$

which can be easily computed from the known eight homologous curves as follows

$$H*/\left(n*^2 + \dot{V}*^2\right) = \frac{H*}{n*^2}\left/\left[1+\left(\dot{V}*/n*\right)^2\right]\right. = \frac{H*}{\dot{V}*^2}\left/\left[1+\left(n*/\dot{V}*\right)^2\right]\right., \quad (8.345)$$

or directly deduced from the experimental observation. The region splitting for the argument can also be avoided if one uses as argument

$$X = \arctan\left(n*/\dot{V}*\right) + k\pi, \quad (8.346)$$

where for the first $n - \dot{V}$ quadrant $k = 0$, for the second and third quadrants $k = 1$, and for the fourth quadrant $k = 2$. The eight head curves presented in Figs. 8.17 and 8.18 reduce to only one curve, as shown in Fig. 8.19. In similar way the eight torque curves can be reduced to a single curve. Thus finally we can use the *Suter* diagram in the form

$$H*/\left(n*^2 + \dot{V}*^2\right) = f\left[\arctan\left(n*/\dot{V}*\right) + k\pi\right], \quad (8.347)$$

and

$$M_{fl}^*/\left(n*^2 + \dot{V}*^2\right) = f\left[\arctan\left(n*/\dot{V}*\right) + k\pi\right]. \quad (8.348)$$

Examples of such pump characteristics were obtained by *Hollander* in 1953 - see *Donsky* [2] 1961 for three types of pumps in wide range of the so called specific speed defined as follows

$$n_s = n_R \left(\frac{\dot{V}_R}{1m^3 s}\right)^{1/2} \left/ \left(\frac{H_R}{1m}\right)^{3/4}\right., \quad (8.349)$$

n_s = 34.8 rotations per minute for a *centrifugal* pump, 147.17 for a *diagonal* pump and 261.42 for an *axial* pump. *Fox* gives in [3] 1977 the numerical values needed to draw the *Suter* diagram in the first three quadrants of the *Hollander* characteristics (see pp. 135 - 142 in the Russian translation). We supplement these data with the data for the fourth quadrant obtained from the original diagrams as given in [2] and use in the computer code series IVA the resulting curves as given in Table 8.2 and plotted in Figs. 8.20 to 8.22 as one of the possible options to generate *Suter* diagrams knowing only the pump data for the point having optimum pump efficiency.

Table 8.2. *Suter* diagram for three different specific speeds n_s

X	n_s =0.581		2.452		4.357	
	YH	YT	YH	YT	YH	YT
0.000	-0.549	-0.433	-1.644	-1.417	-0.952	-0.574
0.168	-0.408	-0.155	-1.098	-0.904	-0.874	-0.602
0.318	-0.198	0.009	-0.623	-0.424	-0.686	-0.542
0.464	-0.032	0.160	-0.350	-0.088	-0.399	-0.312
0.588	0.158	0.297	0.035	0.200	-0.076	0.021
0.695	0.332	0.415	0.308	0.397	0.219	0.303
0.785	0.500	0.500	0.500	0.500	0.500	0.500
0.876	0.650	0.555	0.626	0.579	0.803	0.619
0.983	0.817	0.596	0.776	0.651	1.088	0.741
1.107	0.984	0.616	0.968	0.728	1.409	0.904
1.249	1.143	0.594	1.197	0.882	1.817	1.214
1.406	1.254	0.526	1.479	1.147	2.268	1.626
1.571	1.290	0.440	1.960	1.481	2.729	1.960
1.736	1.274	0.370	2.103	1.538	3.183	2.310
1.893	1.214	0.342	2.187	1.548	3.474	2.647
2.034	1.145	0.345	2.265	1.623	3.576	2.934
2.159	1.080	0.367	2.359	1.711	3.508	3.031
2.266	1.020	0.437	2.474	1.907	3.451	2.945
2.356	0.994	0.520	2.637	2.079	3.272	2.756
2.447	0.958	0.604	2.802	2.356	2.853	2.547
2.554	0.897	0.691	2.900	2.541	2.484	2.182
2.678	0.865	0.783	2.976	2.722	2.161	1.801
2.820	0.812	0.857	2.890	2.749	1.823	1.442
2.976	0.767	0.884	2.624	2.496	1.440	1.030
3.142	0.691	0.859	2.170	2.103	1.082	0.669
3.307	0.623	0.787	1.555	1.525	0.787	0.417
3.463	0.569	0.686	0.992	1.036	0.704	0.415
3.605	0.529	0.552	0.616	0.664	0.616	0.504
3.730	0.504	0.428	0.415	0.387	0.462	0.372
3.836	0.503	0.319	0.279	0.183	0.260	0.106
3.972	0.506	0.187	0.150	-0.100	0.065	-0.075
4.018	0.520	0.141	0.112	-0.171	-0.166	-0.325
4.124	0.548	0.069	0.042	-0.318	-0.416	-0.582

4.249	0.584	-0.024	-0.096	-0.503	-0.687	-0.880
4.391	0.621	-0.144	-0.252	-0.711	-1.026	-1.171
4.547	0.642	-0.360	-0.448	-1.061	-1.508	-1.538
4.712	0.630	-0.671	-0.671	-1.501	-2.190	-2.329
4.791	0.616	-0.868	-0.786	-1.577	-2.427	-2.427
4.869	0.514	-1.035	-0.800	-1.681	-2.625	-2.625
4.948	0.367	-1.182	-0.859	-1.729	-2.785	-2.785
5.027	0.202	-1.321	-0.937	-1.745	-2.928	-2.928
5.105	0.068	-1.450	-1.008	-1.777	-3.071	-3.071
5.184	-0.120	-1.571	-1.137	-1.892	-3.192	-3.192
5.262	-0.239	-1.618	-1.367	-2.008	-3.113	-3.113
5.341	-0.325	-1.646	-1.582	-2.185	-3.034	-3.034
5.419	-0.389	-1.663	-1.751	-2.355	-2.955	-2.955
5.498	-0.470	-1.652	-1.888	-2.451	-2.876	-2.876
5.576	-0.534	-1.641	-2.026	-2.547	-2.716	-2.742
5.655	-0.586	-1.563	-2.163	-2.644	-2.436	-2.525
5.733	-0.625	-1.475	-2.258	-2.742	-2.155	-2.309
5.812	-0.653	-1.374	-2.341	-2.765	-1.875	-2.093
5.890	-0.672	-1.245	-2.286	-2.750	-1.595	-1.850
5.969	-0.682	-1.112	-2.232	-2.708	-1.519	-1.588
6.048	-0.671	-0.957	-2.130	-2.583	-1.448	-1.326
6.126	-0.652	-0.774	-2.021	-2.328	-1.262	-1.053
6.205	-0.616	-0.546	-1.840	-1.889	-1.092	-0.788
6.283	-0.549	-0.433	-1.644	-1.417	-0.952	-0.574

There are situations in the engineering practice where transients have to be analyzed for networks with old pumps for which limited data are available. One usually knows the optimum working point and sometimes the first quadrant characteristics. But the data for all four quadrants are required for transient analysis. In this case one can generate *approximate* four quadrant characteristics by interpolation between the *Hollander's* data. The way used to generate the *Suter* diagrams is the following: First the specific speed for the particular pump under consideration is computed. If

$$n_s < 0.581 \; rps \qquad (8.350)$$

406   8 One-dimensional three-fluid flows

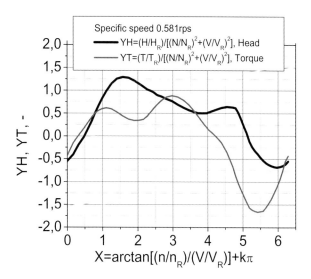

Fig. 8.20. Cyclic head and torque curves for single phase flow centrifugal pump. Measurements made by Hollander 1953

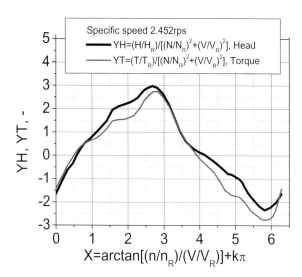

Fig. 8.21. Cyclic head and torque curves for single phase flow diagonal pump. Measurements made by Hollander 1953

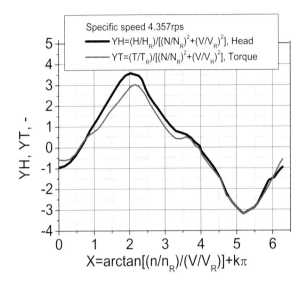

Fig. 8.22. Cyclic head and torque curves for single phase flow axial pump. Measurements made by Hollander 1953

the *Suter* diagram for $n_s = 0.581$ *rps* is used. If

$$n_s > 4.357\ rps \tag{8.351}$$

the *Suter* diagram for $n_s = 4.357$ *rps* is used. Otherwise the interpolation between two neighboring characteristics is used for 0.581 and 2.452, or 2.452 and 4.357, respectively, depending on the particular specific velocity. The code prints comprehensive data for the first quadrant in order to check the so obtained pump characteristics - see Figs. 8.23 and 8.24.

As already mentioned, usually the first quadrant information is available. One should compare the so obtained characteristics with the known ones for the first quadrant and if the agreement is satisfactory the computation can continue. If the characteristics disagree the actual data should be obtained from the pump manufacturer. In the case that no data except first quadrant data are available the *Sute* diagram can be separately computed from the available first quadrant data and the residual part can be drawn parallel to some known characteristics for a pump with similar geometry.

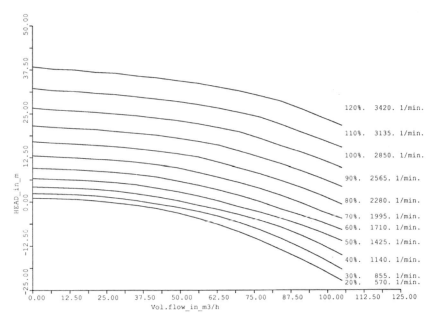

Fig. 8.23. Head as a function of the volumetric flow. Parameter – rotational frequency

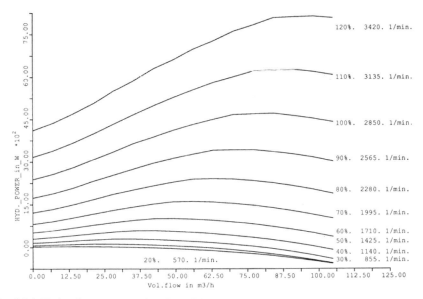

Fig. 8.24. Hydraulic power as a function of the volumetric flow. Parameter – rotational frequency

Characteristics of 1:4 and 1:5 models of large scale mixed flow and axial pumps are reported by *Kastner* et al. in [6, 7] 1983 and *Kennedy* et al.[8] 1982 where the specific speed mass $n_s \approx 130$ *rpm*.

Large scale pump characteristics for two pumps are used by *Ransom* [21] 1988.

Characteristics of centrifugal pumps with different constructions are available in *Pohlenz* [19] (1977).

Data in a form of the *Suter* diagram can be approximated by a quadratic polynomial of the main pump equation. In order to compute the coefficients A^*, B^*, and C^* we need three characteristic points, e.g. the point with the maximum efficiency,

$$X = 1, YH = 1, \qquad (8.352)$$

the point with the zero volumetric flux

$$X = 0, YH = YH_0, \qquad (8.353)$$

and some third point

$$X = X_1, YH = YH_1. \qquad (8.354)$$

Substituting into the main pump equation we obtain

$$1 = A^* + B^* - C^*, \qquad (8.355)$$

$$C^* = -YH_0, \qquad (8.356)$$

$$YH_1 = A^* X_1^2 + B^* X_1 + C^*. \qquad (8.357)$$

Solving with respect to the unknown constants we obtain

$$A^* = [(1 - YH_0) X_1 - YH_1 + YH_0]/(X_1 - X_1^2), \qquad (8.358)$$

$$B^* = [-(1 - YH_0) X_1^2 + YH_1 - YH_0]/(X_1 - X_1^2), \qquad (8.359)$$

$$C^* = -YH_0. \qquad (8.360)$$

An approximate model of the two-phase performance can be easily derived having in mind that the term C is in fact the frictional pressure loss coefficient for one-phase flow. The method of *Lockhart* and *Martinelli* can be applied here in order to compute also the frictional pressure drop exerted by the flow passing the pump.

Thus a liquid only two-phase flow multiplier Φ_{fo}^2 should be computed for volumetric flow $\dot{V}$ considered to consist only of liquid and with some averaged characteristic hydraulic diameter of the impeller. Thereafter the coefficient C^* is corrected as follows

$$C^* = -YH_0 \Phi_{fo}^2 . \tag{8.361}$$

An immediate result of this approximation is that, if the pump is entered by a two-phase mixture instead of water only, the pressure head decreases due to the dramatic increase of the frictional pressure drop. Additional effects are the flow non-homogeneity and the disturbance of the flow velocity triangles in the impeller.

8.9.4 Computational procedure

Input for each time τ:

$\dot{V}_R$, ω_R, H_R and $M_{fl,R}$ a priori defined.

$\dot{V} = \sum_{l=1}^{l_{max}} \alpha_l w_l$ provided from the hydraulic modeling in the pipe network.

ω either prescribed or computed from the pump drive model.

Compute

$\dot{V}* = \dot{V} / \dot{V}_R$,

$\omega* = \omega / \omega_R$.

From the homologous curve compute by interpolation

$H* = H*(V*, \omega*)$,

$M* = M*(V*, \omega*)$,

Correct $H*$ in case of multi-phase flow.

Compute

$\Delta p_{pump} = \rho g H * H_R$,

$M_{fl} = M * M_{fl,R}$.

Use further

$$\Delta p_{pump}$$

in the momentum equations and the energy dissipation

$$\left| M_{fl} \omega - \Delta p_{pump} \dot{V} \right|$$

in the energy balance.

8.9.5 Centrifugal pump drive model

The angular momentum balance at the pump rotor reads

$$I_{pr} \frac{d\omega}{d\tau} + M_{fl} + M_{fr} - M_m = 0. \qquad (8.362)$$

Here the motor torque is provided by the motor manufacturer as a function of the angular speed (e.g. for induction motors at constant voltage)

$$M_R = M_R(\omega). \qquad (8.363)$$

M_{fl} is the hydraulic torque computed from the homologous curve as a function of pump angular speed and flow rate. The frictional torque is a prescribed function of the angular velocity

$$M_{fr} = M_{fr}(\omega), \qquad (8.364)$$

where for $M_{fr} \geq 0 > 0$ for $\omega \geq 0$ and $M_{fr} \leq 0$ for $\omega \leq 0$ (if reverse rotation is allowed at all).

8.9.6 Extension of the theory to multi-phase flow

The theoretical basics for approximate description of the pump performance for multi-phase flow are briefly given here. Consider a centrifugal pump with impeller rotating with n rotations per second or with angular velocity

$$\omega = 2\pi n, \qquad (8.365)$$

inner radius r_1 and outer radius r_2. The angular velocity at the inner impeller radius is

$$V_{\theta 1} = \omega r_1 \tag{8.366}$$

and at the outer impeller radius

$$V_{\theta 2} = \omega r_2. \tag{8.367}$$

The corresponding velocity field components are at the inner impeller radius

$$V_{\theta 1,l} = \omega r_1 \tag{8.368}$$

and at the outer impeller radius

$$V_{\theta 2,l} = \omega r_2. \tag{8.369}$$

The relative flow field velocity for an observer rotating with the impeller is V_l^r. The vector sum of the relative flow field velocity and the angular impeller velocity gives the absolute flow velocity V_l^a. We designate the angular component of this velocity with $V_{\theta,l}^a$. The mass flow in kg/s of each velocity field entering the pump impeller is $\dot{m}_l$. The flow is stationary and incompressible. Thus the force acting on the flow in the angular direction at the inlet is

$$F_{\theta 1,l} = \dot{m}_l V_{\theta 1,l}^a \tag{8.370}$$

and at the impeller outlet

$$F_{\theta 2,l} = -\dot{m}_l V_{\theta 2,l}^a \tag{8.371}$$

The corresponding torques are $F_{\theta 1,l} r_1$ and $F_{\theta 2,l} r_2$. The resulting torque acting on the flow inside the impeller is

$$M_{fl} = \sum_{l=1}^{l_{max}} \left(F_{\theta 2,l} r_2 + F_{\theta 1,l} r_1 \right) = \sum_{l=1}^{l_{max}} \dot{m}_l \left(V_{\theta 2,l}^a r_2 - V_{\theta 1,l}^a r_1 \right). \tag{8.372}$$

The power inserted into the flow is therefore $M_{fl} \omega$ or

$$\Delta p_{pump} \dot{V} = M_{fl} \omega = \omega \sum_{l=1}^{l_{max}} \dot{m}_l \left(V_{\theta 2,l}^a r_2 - V_{\theta 1,l}^a r_1 \right). \tag{8.373}$$

Having in mind that

$$\dot{m} = \rho \dot{V} \tag{8.374}$$

one obtains

$$\Delta p_{pump} = \omega \sum_{l=1}^{l_{max}} \frac{\dot{m}_l}{\dot{V}} \left(V_{\theta 2,l}^a r_2 - V_{\theta 1,l}^a r_1 \right) = \sum_{l=1}^{l_{max}} \frac{\dot{m}_l}{\dot{V}} \left(V_{\theta 2,l}^a V_{\theta 2,l} - V_{\theta 1,l}^a V_{\theta 1,l} \right). \tag{8.375}$$

This is the *main pump equation* describing the pressure rise as a function of the impeller speed and geometry. Having in mind that the angle between the relative flow velocity $V_{1,l}^r$ and the impeller angular velocity $V_{\theta 1,l}$ is β_1 and that the angle between the relative flow velocity $V_{2,l}^r$ and the impeller angular velocity $V_{\theta 2,l}$ is β_2 the projections of the absolute flow velocities on the angular impeller velocities are

$$V_{\theta 1,l}^a = V_{\theta 1,l} - V_{1,l}^r \cos \beta_1, \tag{8.376}$$

$$V_{\theta 2,l}^a = V_{\theta 2,l} - V_{2,l}^r \cos \beta_2. \tag{8.377}$$

Substituting into the main pump equation we obtain

$$\Delta p_{pump} = \sum_{l=1}^{l_{max}} \frac{\dot{m}_l}{\dot{V}} \left[\left(V_{\theta 2,l} - V_{2,l}^r \cos \beta_2 \right) V_{\theta 2,l} - \left(V_{\theta 1,l} - V_{1,l}^r \cos \beta_1 \right) V_{\theta 1,l} \right]$$

$$= \sum_{l=1}^{l_{max}} \frac{\dot{m}_l}{\dot{V}} \left(V_{\theta 2,l}^2 - V_{\theta 1,l}^2 + V_{\theta 1,l} V_{1,l}^r \cos \beta_1 - V_{\theta 2,l} V_{2,l}^r \cos \beta_2 \right). \tag{8.378}$$

Having in mind that the relative flow velocity is equal to the volumetric flow divided by the cross section normal to V^r i.e.

$$V_{1,l}^r = \dot{V}_l / (\alpha_{1,l} A_1), \tag{8.379}$$

$$V_{2,l}^r = \dot{V}_l / (\alpha_{2,l} A_2), \tag{8.380}$$

one obtains

414   8 One-dimensional three-fluid flows

$$\Delta p_{pump} = \sum_{l=1}^{l_{max}} \frac{\dot{m}_l}{\dot{V}} \left[\left(r_2^2 - r_1^2\right)\omega^2 + \left(\frac{r_1 \cos\beta_1}{\alpha_{1,l} A_1} - \frac{r_2 \cos\beta_2}{\alpha_{2,l} A_2}\right)\omega \dot{V}_l \right]$$

$$= \sum_{l=1}^{l_{max}} \frac{\dot{m}_l}{\dot{V}} \left[\left(r_2^2 - r_1^2\right) 4\pi^2 n^2 + \left(\frac{r_1 \cos\beta_1}{\alpha_{1,l} A_1} - \frac{r_2 \cos\beta_2}{\alpha_{2,l} A_2}\right) 2\pi n \dot{V}_l \right]. \tag{8.381}$$

If we consider additionally that the flow exerts some pressure loss due to friction with the pump structures in the impeller region we finally obtain

$$\Delta p_{pump} = \sum_{l=1}^{l_{max}} \frac{\dot{m}_l}{\dot{V}} \left[\left(r_2^2 - r_1^2\right)\omega^2 + \left(\frac{r_1 \cos\beta_1}{\alpha_{1,l} A_1} - \frac{r_2 \cos\beta_2}{\alpha_{2,l} A_2}\right)\omega \dot{V}_l \right] - \xi \frac{1}{2} \frac{\rho}{A_m} \dot{V}^2 \Phi_{20}^2$$

$$= \sum_{l=1}^{l_{max}} \frac{\dot{m}_l}{\dot{V}} \left[\left(r_2^2 - r_1^2\right) 4\pi^2 n^2 + \left(\frac{r_1 \cos\beta_1}{\alpha_{1,l} A_1} - \frac{r_2 \cos\beta_2}{\alpha_{2,l} A_2}\right) 2\pi n \dot{V}_l \right] - \xi \frac{1}{2} \frac{\rho}{A_m} \dot{V}^2 \Phi_{20}^2, \tag{8.382}$$

where ξ is some friction loss coefficient computed as with the multiphase mass flow rate and liquid properties, Φ_{20}^2 is the *Lochart* and *Martinelli* two-phase multiplier and A_m some averaged flow cross section. The pressure rise of the pump is usually expressed with the head H

$$\Delta p_{pump} = \rho_h g H, \tag{8.383}$$

where g is the gravitational acceleration and

$$\rho_h = \sum_{l=1}^{l_{max}} \alpha_l \rho_l \tag{8.384}$$

is the homogeneous mixture density. Usually the main pump equation is written in the form

$$\rho_h g H = \left(r_2^2 - r_1^2\right) 4\pi^2 \left[\left(\sum_{l=1}^{l_{max}} \dot{m}_l\right) \middle/ \dot{V}\right] n^2$$

$$+ 2\pi n \dot{V} \sum_{l=1}^{l_{max}} \left[\frac{\dot{m}_l}{\dot{V}} \left(\frac{r_1 \cos\beta_1}{\alpha_{1,l} A_1} - \frac{r_2 \cos\beta_2}{\alpha_{2,l} A_2}\right) \frac{\dot{V}_l}{\dot{V}}\right] - \xi \frac{1}{2} \frac{\rho}{A_m} \dot{V}^2 \Phi_{20}^2 \tag{8.385}$$

or

$$H^+ = A^+ n^2 + B^+ n \dot{V} - C^+ \dot{V}^2, \tag{8.386}$$

where

$$A^+ = \left(r_2^2 - r_1^2\right) \frac{4\pi^2}{\rho_h g} \left[\left(\sum_{l=1}^{l_{max}} \dot{m}_l\right) \bigg/ \dot{V}\right] = A \left(\sum_{l=1}^{l_{max}} \dot{m}_l\right) \bigg/ \left(\rho_h \dot{V}\right) \tag{8.387}$$

$$B^+ = \frac{2\pi}{\rho_h g} \sum_{l=1}^{l_{max}} \left[\frac{\dot{m}_l}{\dot{V}} \left(\frac{r_1 \cos\beta_1}{\alpha_{1,l} A_1} - \frac{r_2 \cos\beta_2}{\alpha_{2,l} A_2}\right) \frac{\dot{V}_l}{\dot{V}}\right]$$

$$= B \frac{\sum_{l=1}^{l_{max}} \left[\frac{\dot{m}_l}{\dot{V}} \left(\frac{r_1 \cos\beta_1}{\alpha_{1,l} A_1} - \frac{r_2 \cos\beta_2}{\alpha_{2,l} A_2}\right) \frac{\dot{V}_l}{\dot{V}}\right]}{\rho_h \left(\frac{r_1 \cos\beta_1}{A_1} - \frac{r_2 \cos\beta_2}{A_2}\right)} \tag{8.388}$$

$$C^+ = \xi \frac{1}{2} \frac{\rho}{\rho_h g A_m} \Phi_{20}^2 = C \frac{\rho}{\rho_h} \Phi_{20}^2 \approx C \Phi_{20}^2. \tag{8.389}$$

Obviously for single phase flow

$$\frac{1}{\rho_h \dot{V}} \sum_{l=1}^{l_{max}} \dot{m}_l = 1 \tag{8.390}$$

$$\frac{\sum_{l=1}^{l_{max}} \left[\frac{\dot{m}_l}{\dot{V}} \left(\frac{r_1 \cos\beta_1}{\alpha_{1,l} A_1} - \frac{r_2 \cos\beta_2}{\alpha_{2,l} A_2}\right) \frac{\dot{V}_l}{\dot{V}}\right]}{\rho_h \left(\frac{r_1 \cos\beta_1}{A_1} - \frac{r_2 \cos\beta_2}{A_2}\right)} = 1 \tag{8.391}$$

$$\Phi_{20}^2 = 1 \tag{8.392}$$

and therefore

$$A^+ = A, \tag{8.393}$$

$$B^+ = B, \quad (8.394)$$

$$C^+ = C. \quad (8.395)$$

References

1. Abedin S, Takeuchi K, Zoung MY (1986) A method of computing hydraulic reaction force due to a fluid jet at steamline break, Nuclear Science and Engineering, vol 92 pp 162-169
2. Donsky B (1961) Complete pump characteristics and the effects of specific speeds an hydraulic transients, J. Basic Eng., December, pp 685-699
3. Fox JA (1977) Hydraulic analysis of unsteady flow in pipe networks, Macmillan Press Ltd.
4. Greiner W (1984) Theoretische Physik, Band 1: Mechanik I, Verlag Harri Deutsch
5. Henry RE, Fauske HK (1969) The two-phase critical flow of one-component mixtures in nozzles orifices, and short tubes, Journal of Heat Transfer, vol 2 pp 47-56
6. Kastner W, Riedle K, Seeberger G (1983) Experimentelle Untersuchungen über das Verhalten von Hauptkühlmittelpumpen bei Kühlmittelverluststörfäallen, Brennstoff-Wärme-Kraft 35, Heft 6
7. Kastner W, Seeberger G (Feb. 1983) Pump behavior and its impact an a loss-of-coolant accident in pressurized water reactor, Nuclear Technology, vol 60 pp 268 277
8. Kennedy WG et al. (1982) Two-phase flow behavior of axial pumps, International Meeting on Thermal Nuclear Reactor Safety, August 29 - September 2, Americana Congress Hotel Chicago, Illinois, USA
9. Kolev NI (1977) Zweiphasen - Zweikomponentenströmung (Luft- Wasserdampf- Wasser) zwischen den Sicherheitsräumen der KKW mit wassergekühlten Reaktoren bei Kühlmittelverlusthavarie, Dissertation TU – Dresden
10. Kolev NI (1982) To the modeling of transient non equilibrium, non homogeneous systems, Proc. of the seminar "Thermal Physics 82 (Thermal Safety of VVER Type Nuclear Reactors)" held in Karlovy Vary, May 1982, Czechoslovakia, vol 2 pp 129 - 147 (in Russian)
11. Kolev NI (August 1985) Transiente Dreiphasen Dreikomponenten Strömung, Teil 2: Eindimensionales Schlupfmodell Vergleich Theorie-Experiment, KfK 3926, Kernforschungs-zentrum Karlsruhe
12. Kolev NI (1986) Transiente Zweiphasenströmung. Springer
13. Kolev NI (1986) Transient three phase three component nonequilibrium nonhomogeneous flow, Nuclear Engineering and Design, vol 91 pp 373-390
14. Kolev NI (1993) IVA3-NW Components: relief valves, pumps, heat structures, Siemens Work Report No. KWU R232/93/e0050
15. Kolev NI (Oct. 5-8, 1993) IVA3 NW: Computer code for modeling of transient three phase flow in complicated 3D geometry connected with industrial networks, Proc. of the Sixth Int. Top. Meeting on Nuclear Reactor Thermal Hydraulics, Grenoble, France
16. Lahey RT, Moody FJ (1977) The thermal - hydraulics of a boiling water nuclear reactor, American Nuclear Society, 2th printing 1979
17. Magnus K (1986) Schwingungen, BG Teubner

18. Nigmatulin B, Ivandeev AI (1977) Investigation of the hydrodynamic boiling crisis in two phase flow, High Temperature Thermal Physics, vol 15 no 1 pp 129 - 136
19. Pohlenz W (1977) Pumpen für Flüssigkeiten, VEB Verlag Technik, Berlin
20. Polenz W (1975) Grundlagen für Pumpen, VEB Verlag Technik Berlin
21. Ransom VH et al., (1988) RELAP5/MOD2 Code Manual Vol 1: Code structure, system models and solution methods, NUREG/CR-4312 EGG-2396, rev 1 pp 209-216.
22. Roloff-Bock I, Kolev NI (1.7.1998) IVA5 Computer Code: Relief and back pressing valve model, KWU Work Report No. NA-T/1998/E058, Project: IVA5-Development
23. Sass F, Bouché C, Leitner A (Hrsg.) (1969) Dubbels Taschenbuch für den Maschinenbau, Springer
24. Suter P (1966) Darstellung der Pumpencharakteristik für Druckstoßrechnungen, Sulzer Technical Review, pp 45-48
25. Tangren RF, Dodge CH, Seifert HS (July 1949) Compressibility effects in two-phase flow, Journal of Applied Physics, vol 20 no 7 pp 673-645
26. Wallis GB (1969) One-dimensional two-phase Flow. New York: McGraw Hill
27. Yano T, Miyazaki N, Isozaki T (1982) Transient analysis of blow down thrust force under PWR LOCA, Nuclear Engineering and Design, vol 75 pp 157-168

9 Detonation waves caused by chemical reactions or by melt-coolant interactions

9.1 Introduction

Analyzing a fascinating physical phenomenon such as the melt-water detonation in this chapter we will give an interesting application of the theory of multi-phase flows – namely the analysis of the detonation wave propagation during the interaction of molten materials with liquids such as that of iron with water.

Melt-coolant interaction analysis plays an important role in risk estimation for facilities in which high-temperature molten material comes or may come into contact with low-temperature liquid coolant. Contacts of iron with water or uranium dioxide with water are some examples of this phenomenon. One of the intriguing discussions in this field is motivated by the observation that molten alumina produces *severe interactions* when dropped into water whereas molten uranium dioxide at a similar temperature produces only *moderate interactions*. Use of the solution of the *Fourier* equation for contact of melt and coolant during a prescribed time as presented in [12] does not yield any explanation for these differences. It was shown in [12] that the ratio of the discharged to the available thermal energy for UO_2 at an initial temperature $T_3 = 3000K$ in contact over $\Delta\tau_{23} \approx 0.001s$ with water at temperature $T_2 = 30°C$ and atmospheric pressure is $\eta_{th} = 0.000133/D_3$. For the same conditions for Al_2O_3 this ratio is $\eta_{th} = 0.000129/D_3$. Here D_3 is the particle size. It is evident that the thermal properties for transient heat conduction by these fluid pairs do not differ to such a large extent that may explain the differences in the interaction. Thus the question still remains why these materials behave so differently. We will demonstrate in this chapter that the reason is *mainly the differences in their caloric equations of state* – this is the first target of this chapter.

The next interesting problem is associated with the question of how to judge the explosivity of melt-water mixtures in engineering premixing computations used for upper bound estimates of the possible detonation energy release. In [13] the criterion that mixtures of melt and water can only experience severe detonations if one of the liquids continuously surrounds the other was introduced. The roots of this idea go back to the works by *Henry* and *Fauske* [7, 8] and are discussed in detail by *Park* and *Corradini* [19]. In the same work the so called "stochiometric" mixtures were introduced which are defined as mixtures in which there is an optimum amount of water available locally to permit the melt to transfer all its thermal energy by evaporation. The question whether dispersed systems such as melt droplets flowing together with water droplets may also be of some risk is still unresolved. We will answer in this work two questions associated with this issue. *First, is it possible to have detonation waves in such disperse-disperse systems and second, if so, what are the expected pressure wave magnitudes and can they be considered as dangerous?*

A further interesting aspect results from the detonation theory of melt-water interaction proposed by *Board* et al. [2]. Just as in detonation of reacting gases, *Board* et al. considered the fine fragmentation behind a shock front and the subsequent steam production and expansion as the driving force for transferring shock waves in self-sustaining detonation waves. The ideas propagated by *Board* have been discussed worldwide. In particular, his remark that not all the melt behind the front may necessarily participate in the energy transfer process was widely accepted and introduced in the detonation analysis – see for instance the work by *Shamoun* and *Corradini* [24]. *Board's* remark that not all the coolant behind the front necessarily has to participate in the energy transfer process was controversially discussed. For instance, some authors thought that assuming that all the melt and coolant participate in the interaction is conservative, e.g. [6], and this may serve as an upper bound estimate. This was criticized by *Yuen* and *Theofanous* [29]. Based on their own experiments, these authors introduced the idea of not only a limited amount of melt but also of a limited amount of coolant participating in the interaction – the so-called micro-interaction concept. *Yuen* and *Theofanous* came to an important conclusion for practical applications: *mixtures judged as non-explosive assuming that all the water is participating in the debris quenching may be explosive if only part of the water is participating in the debris quenching.* Of course, this problem naturally disappears if numerical computations are used that feature a very fine discretization grid for resolving regions of near and far interactions numerically. In simulations with a characteristic grid size much larger than the near interaction length this problem is acute and a quasi-micro-interaction approach is thus inevitable. One of the targets in the analysis presented here is to discriminate for the three important materials pairs (UO_2-water, Al_2O_3-water and *Fe*-water) between coolant entrainment leading to detonations and entrainment leading to quenching, thereby preventing detonations.

The assumption of a homogeneous mixture was relaxed by *Berthoud* [1] and *Scott* and *Berthoud* [23] in a steady-state analysis and resulted in interesting estimates, one of them being the "thickness of the discontinuity front" which depends on the initial particle size. Nevertheless, all applied computations use a detonation theory based on homogeneity, inclusive of those of *Scott* and *Berthoud*.

9.2 Single-phase theory

9.2.1 Continuum sound waves (*Laplace*)

Consider a small pressure pulse propagating through a continuum. The pressure disturbance causes velocity disturbance. We are interested in the velocity of propagation of this disturbance. Consider a control olume with infinitesimal thickness located around a plane that is perpendicular to the flow velocity. The control volume moves with the plane. The steady-state conservation equations for mass, momentum (neglecting body and viscous forces) , energy (neglecting all dissipative effects such as work done by viscous stress and heat transfer due to thermal conduction and diffusion) are

$$d(\rho w) = 0, \qquad (9.1)$$

$$d(\rho w^2) + dp = 0, \qquad (9.2)$$

$$d\left[\rho w\left(h + \frac{1}{2}w^2\right)\right] = 0. \qquad (9.3)$$

Now remember that all flow parameters are continuous in space. Using the mass conservation equation, the momentum and energy conservation equations reduce to

$$\rho d\left(\frac{1}{2}w^2\right) + dp = 0 \quad \text{or} \quad -w^2 d\rho + dp = 0 \qquad (9.4)$$

$$d\left(h + \frac{1}{2}w^2\right) = 0 \qquad (9.5)$$

If we exclude the differential of the kinetic energy, this yields $\rho dh - dp = 0$ which after application of the *Gibbs* equation is simply

$$ds = 0.\tag{9.6}$$

Using the equation of state in the form

$$d\rho = \left(\frac{\partial \rho}{\partial s}\right)_p ds + \frac{dp}{a^2},\tag{9.7}$$

where

$$\frac{1}{a^2} = \left(\frac{\partial \rho}{\partial p}\right)_s\tag{9.8}$$

the momentum equation becomes

$$\left[1 - \left(\frac{w}{a}\right)^2\right] dp = 0.\tag{9.9}$$

For a pressure difference that is non-zero, this is satisfied only if the velocity of the flow with respect to the plane is equal to a value defined solely by the local fluid properties,

$$w = a,$$

and called the *sound velocity*. This remarkable result was first obtained by *Laplace* [18] in 1816 and is generally accepted today. It simply says that:

1) Small pressure-velocity disturbances called *weak waves* or *acoustic waves* propagate with a velocity which is a function only of the local continuum parameters.

2) Across small pressure-velocity disturbances (neglecting dissipative effects) the continuity of the flow variables implies entropy conservation across any infinitesimal distance.

9.2.2 Discontinuum shock waves (*Rankine-Hugoniot*)

Now consider a plane normal to the flow velocity separated into two continuous regions featuring discontinuity at that plane. We designate the medium ahead of wave motion and the medium behind the wave with indices 1 and 2, respectively. Following *Landau* and *Lifshitz* [17] we call the side of the plane facing the non-disturbed medium the down-flow side and the side facing the disturbed medium the up-flow side. Note that the major difference from the case with continuous flow parameters is that the *differentiation across the plane is not possible*. We will see that velocities before and after the plane are related through relationships that differ from the isentropic wave disturbance as discussed above. The conservation equations governing this process are similar to Eqs. (9.1), (9.2) and (9.3) but written as a primitive balance applied on the control volume:

$$\Delta(\rho w) = 0, \quad \text{or} \quad \rho w = const \quad \text{or} \quad w_1 = \rho w / \rho_1 \quad \text{or} \quad w_2 = \rho w / \rho_2, \quad (9.10)$$

$$\Delta(\rho w^2) + \Delta p = 0, \quad \text{or} \quad \rho_2 w_2^2 - \rho_1 w_1^2 + p_2 - p_1 = 0, \quad (9.11)$$

$$\Delta\left[\rho w \left(h + \frac{1}{2} w^2\right)\right] = 0, \quad \text{or} \quad \rho_2 w_2 \left(h_2 + \frac{1}{2} w_2^2\right) - \rho_1 w_1 \left(h_1 + \frac{1}{2} w_1^2\right) = 0. \quad (9.12)$$

Thus, the parameters in front of and behind the discontinuity plane are related to each other by Eqs. (9.10) through (9.12). Using mass conservation, momentum and energy conservation can be rewritten as

$$(\rho w)^2 = \frac{p_2 - p_1}{\frac{1}{\rho_1} - \frac{1}{\rho_2}}, \quad (9.13)$$

$$h_2 - h_1 + \frac{1}{2}(\rho w)^2 \left(\frac{1}{\rho_2^2} - \frac{1}{\rho_1^2}\right) = 0. \quad (9.14)$$

By inserting the mass flow rate from the momentum equation into the energy equation we obtain

$$h_2 - h_1 - \frac{1}{2}(p_2 - p_1)\left(\frac{1}{\rho_2} + \frac{1}{\rho_1}\right) = 0. \quad (9.15)$$

Using the equation of state

$$h_2 = f(p_2, \rho_2) \quad (9.16)$$

(e.g. $h_2 = c_{p2} T_2 = \frac{\kappa_2}{\kappa_2 - 1} \frac{p_2}{\rho_2}$ for perfect gases) the state behind the front is a unique function of the state in the undisturbed region. Equations (9.15) and (9.16) have to be solved with respect to p_2 by iterations. With the pressure difference known, the mass flow rate is then readily computed using Eq. (9.13). The velocities are then computed from the mass conservation equation. A very convenient formula for the velocity difference is

$$w_1 - w_2 = -\sqrt{(p_2 - p_1)\left(\frac{1}{\rho_1} - \frac{1}{\rho_2}\right)}, \qquad (9.17)$$

which is the *shock wave velocity in a laboratory frame*. Equation (9.15) is frequently written in terms of specific internal energies

$$e_2 - e_1 + \frac{p_2 + p_1}{2}\left(\frac{1}{\rho_2} - \frac{1}{\rho_1}\right) = 0. \qquad (9.18)$$

A similar procedure can be applied by using the equation of state in the form

$$e_2 = f(p_2, \rho_2). \qquad (9.19)$$

This remarkable result was first obtained by *Rankine* [20] in 1870 and independently by *Hugoniot* [9] in 1887 and is widely accepted today. It says that *there is a velocity discontinuity in the shock plane*. The velocities in front of and behind the shock plane relative to the shock plane are a unique function of state of the undisturbed region 1.

The pressure and the density behind the shock front are higher than those in front of the plane. There is an entropy increase across the shock front. This conclusion was reached by *Rayleigh* [21] and *Taylor* [26] in 1910. It is the only known case where flow processes in a non-viscous and non-dissipating medium take place irreversibly, that is with entropy increase, see the discussion by *Landau* and *Lifshitz* [17] p. 443. The velocity in front of the shock with respect to the front is greater than the local sound velocity, $w_1 > a_1$, and the velocity behind the shock with respect to the front is less than the local sound velocity, $w_2 < a_2$. Whatever happens before the shock cannot influence the shock propagation. The events behind the shock influence the shock form and propagation.

In the laboratory frame of reference we have

$$w_1 = -w_{cs},$$

$$w_2 = -(w_{cs} - w_2^a).$$

Here w_{cs} is the velocity of our moving frame of reference and w_2^a is the absolute velocity behind the shock. Therefore

$$w_1 - w_2 = -w_2^a.$$

In the case of shock wave propagation in an explosive mixture, the compression may increase the temperature behind the shock to values that are higher than the ignition temperature. In the case of an exothermic chemical reaction, it is possible for the released energy behind the front to cause additional drive, resulting in a rise in the *velocity behind the wave with respect to the front* to the local velocity of sound, $w_2 = a_2$. This process is usually visualized in the literature as a piston moving with a velocity $-(w_1 - a_2)$ pushing the reaction front ahead and leaving behind a refraction wave. In this case, neither the disturbance behind the shock nor the disturbance ahead of the shock can have any influence on the shock wave propagation. The propagation process becomes self-sustaining.

A shock wave fulfilling the condition

$$w_2 = a_2 \qquad (9.20)$$

is called a *detonation wave, Chapman-Jouguet* [3, 11].

This condition is called the *Chapman-Jouguet* condition – see *Chapman* [3] who found in 1899 that the velocity behind the front with respect to the front in reality take the minimum of the possible detonation states, and *Jouguet* [11] who found in 1905 that the velocity behind the front with respect to the front is equal to the sound velocity of the burned products. *Crussard* [4] showed in 1907 that both statements are identical.

It is important to note that the equations of state of the mixture ahead of the front and behind the front are different.

The energy conservation equation also has to take into account the reaction energy release. An additional term then occurs, $\Delta h_{reaction}$,

$$h_2 - h_1 + \frac{1}{2}w_2^2 - \frac{1}{2}w_1^2 - \Delta h_{reaction} = 0$$

or

$$h_2 - h_1 - \Delta h_{reaction} - \frac{1}{2}(p_2 - p_1)\left(\frac{1}{\rho_2} + \frac{1}{\rho_1}\right) = 0. \qquad (9.21)$$

The formation enthalpy $\Delta h_{reaction}$ in J per kg of the final mixture at zero absolute temperature is related to the formation enthalpy Δh_{ref} at the reference conditions $\left(p_{ref}, T_{ref}\right)$ by the relation

$$\Delta h_{reaction} = \left(\Delta h_{ref} - c_{p2} T_{ref}\right) \Delta C_{st\_mix}. \qquad (9.22)$$

Here the multiplier is first demonstrated for the case of an initially stoichiometric gas mixture containing hydrogen and oxygen without radicals in equilibrium

$$\Delta C_{st\_mix} = C_{1,H_2} + C_{1,O_2} - C_{2,H_2} - C_{2,O_2}.$$

The multiplier converts the formation enthalpy related to $1kg$ of hydrogen-oxygen mixture to that related to $1kg$ gas mixture. If for instance other gas components are in the mixture that do not participate into the reaction the heating effect of the reaction on the final products is smaller. If the resulting temperature after the detonation wave is so low that the dissociation can be neglected we have $C_{2,H_2} = 0$, $C_{2,O_2} = 0$. The general form of the multiplier reads

$$\Delta C_{st\_mix} = \Delta C_{H_2} + \Delta C_{O_2} = \left(n_{H_2} M_{H_2} + n_{O_2} M_{O_2}\right) \min\left(\frac{\Delta C_{H_2}}{n_{H_2} M_{H_2}}, \frac{\Delta C_{O_2}}{n_{O_2} M_{O_2}}\right)$$

$$= 9 \min\left(\Delta C_{H_2}, \Delta C_{O_2}/8\right),$$

where $n_{H_2} = 2$, $n_{O_2} = 1$ are the stoichiometric coefficients equal to the number of kg-moles of each species that participates in the reaction and $M_{H_2} = 2kg$, $M_{O_2} = 32kg$ are the kg-mole masses.

The *Rankine-Hugoniot* curve constructed on the basis of the equation of state before the shock is called the *shock adiabatic*. The *Rankine-Hugoniot* curve constructed on the basis of the equation of state after the shock is called the *detonation adiabatic*. The thermodynamic properties in front of and behind the detonation discontinuity therefore lie on different thermodynamic curves.

Detonation waves have a destructive potential. It is interesting to know how much energy can be released in this case and transferred into technical work. Using Eq. (7.26) we obtain

$$-w_{t,21}^{vdp} = h_2 - h_1 + T_\infty (s_1 - s_2). \tag{9.23}$$

9.2.3 The *Landau* and *Liftshitz* analytical solution for detonation in perfect gases

Landau and *Liftshitz* obtained in [17] an analytical detonation solution for perfect gas mixtures. The main idea of the procedure is to express the state of the gas behind the detonation front as a function of the parameters before the detonation front. Starting with the definition equation for having detonation

$$w_2 = a_2 = \sqrt{\kappa_2 p_2 / \rho_2}, \tag{9.24}$$

and therefore $(\rho w)^2 = \kappa_2 p_2 \rho_2$ or

$$\rho_2 = (\rho w)^2 / (\kappa_2 p_2), \tag{9.25}$$

the authors solved the system consisting of the last equation and the momentum Eq. (9.13) with respect to the pressure and density of the burned products. The result is

$$p_2 = \frac{p_1 + (\rho w)^2 \frac{1}{\rho_1}}{\kappa_2 + 1} = \frac{p_1 + \rho_1 w_1^2}{\kappa_2 + 1}. \tag{9.26}$$

$$\rho_2 = \frac{(\rho w)^2}{\kappa_2 p_2} = \frac{\kappa_2 + 1}{\kappa_2} \frac{(\rho w)^2}{p_1 + (\rho w)^2 \frac{1}{\rho_1}} = \frac{\kappa_2 + 1}{\kappa_2} \frac{\rho_1^2 w_1^2}{p_1 + \rho_1 w_1^2}, \tag{9.27}$$

Using the burned gas density, the burned gas velocity relative to the moving front is

$$w_2 = \frac{\rho_1 w_1}{\rho_2} = \frac{\kappa_2}{\kappa_2 + 1} \frac{p_1 + \rho_1 w_1^2}{\rho_1 w_1}. \tag{9.28}$$

The enthalpies of the perfect gas mixtures are computed with the zero point selected at zero temperature as follows

$$h_1 = c_{p1}T_1 = \frac{\kappa_1}{\kappa_1 - 1}\frac{p_1}{\rho_1}, \tag{9.29}$$

$$h_2 = c_{p2}T_2 = \frac{\kappa_2}{\kappa_2 - 1}\frac{p_2}{\rho_2} = \frac{\kappa_2^2}{(\kappa_2^2 - 1)(\kappa_2 + 1)}\frac{(p_1 + \rho_1 w_1^2)^2}{\rho_1^2 w_1^2}. \tag{9.30}$$

Here $T = \frac{p}{\rho R}$, $\frac{c_p}{R} = \frac{\kappa}{\kappa - 1}$, Eqs. (9.26) and (9.27) are used. Therefore

$$h_2 + \frac{1}{2}w_2^2 = \frac{1}{2}\frac{\kappa_2^2}{\kappa_2^2 - 1}\frac{(p_1 + \rho_1 w_1^2)^2}{\rho_1^2 w_1^2}. \tag{2.31}$$

Replacing all the terms in the energy conservation equation

$$h_2 - h_1 + \frac{1}{2}w_2^2 - \frac{1}{2}w_1^2 - \Delta h_{reaction} = 0, \tag{2.32}$$

one obtains

$$\frac{1}{2}\frac{\kappa_2^2}{\kappa_2^2 - 1}\frac{(p_1 + w_1^2 \rho_1)^2}{w_1^2 \rho_1^2} - \frac{\kappa_1}{\kappa_1 - 1}\frac{p_1}{\rho_1} - \frac{1}{2}w_1^2 - \Delta h_{reaction} = 0, \tag{2.33}$$

or

$$w_1^4 - 2\left[\left(\kappa_2^2 - 1\right)\Delta h_{reaction} + \frac{\kappa_2^2 - \kappa_1}{\kappa_1 - 1}\frac{p_1}{\rho_1}\right]w_1^2 + \left(\kappa_2 \frac{p_1}{\rho_1}\right)^2 = 0. \tag{2.34}$$

This is a bi-quadratic equation of the type

$$x^4 - 2px^2 + q = 0, \tag{2.35}$$

having the largest positive solution

$$x = \sqrt{p + \sqrt{p^2 - q}} = \sqrt{(p + \sqrt{q})/2} + \sqrt{(p - \sqrt{q})/2}. \tag{3.36}$$

For a strong detonation waves the equation simplifies to

$$w_1 = \sqrt{2(\kappa_2^2 - 1)\Delta h_{reaction}}. \tag{3.37}$$

Knowing the velocity of the unburned products with respect to the detonation front, the pressure and the density behind the front are computed using Eqs. (2.26) and (2.27). The velocity of the burned products with respect to the front is then computed using Eq. (2.24). Note the important limitation of this theory

$$\kappa_1 = const, \qquad (3.38)$$

$$\kappa_2 = const. \qquad (3.39)$$

A systematic summary of data for the specific capacity was obtained in 1987 by *Robert* et al. [22] for 165 gases and radicals. Some of these data are based on measurements up to $5000K$ and others are provided by quantum chemistry computations. These data help us to compute properly the specific capacity at constant pressure for mixtures of gases or radicals and therefore to compute properly the isentropic exponents of the burned products. This approach results in more realistic temperatures.

> For large values of $\Delta h_{reaction}$, the computed temperatures, even with an appropriate isentropic exponent of the burned products, are unrealistically high. At high temperature the products obtained start to dissociate absorbing energy. This process limits the temperature increase itself.

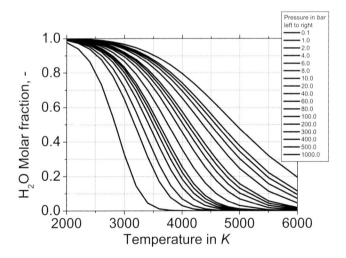

Fig. 9.1. Molar fraction of dissociated steam as a function of temperature with pressure as a parameter

This was found first by *Zeldovich* in 1940 [30] p. 544 who proposed using the specific heat as a function of temperature taking into account the dissociation.

Figure 9.1 shows the molar fraction of the dissociated steam as a function of the local temperature for different pressures. We compute this concentration as discussed in Chapter 3. Other authors have introduced a coefficient of completion of the reaction less then unity depending on the temperature of the burned gases.

We give here two examples of detonation CJ pressures, CJ temperatures and CJ velocities in Figs. 9.2 through 9.4 (index 2 is replaced with CJ if the *Chapman-Jouguet* condition is fulfilled). We consider two cases one with nitrogen as an inert component and the second with steam. We present the results as a function of the concentration of the stoichiometric mixture of hydrogen and oxygen.

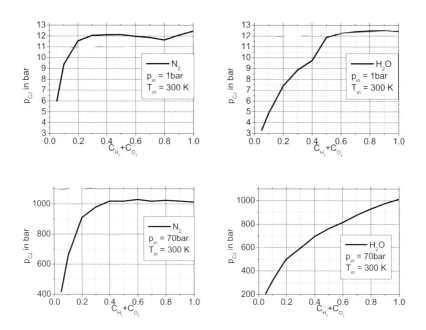

Fig. 9.2. CJ pressure for stoichiometric hydrogen-oxygen mixtures with nitrogen or steam as an inert component

For the case of steam as an inert component we use the equation of state for dissociated steam. We see the effect of dissociation. It is manifested in the reduction of the effective isentropic exponent of the products close to unity. This makes the increases temperature and pressure with increasing energy release considerable smaller than if computed without consideration of the dissociation.

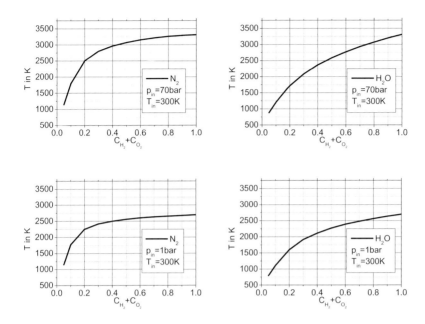

Fig. 9.3. CJ temperature for stoichiometric hydrogen-oxygen mixtures with nitrogen or steam as an inert component

9.2.4 Numerical solution for detonation in closed pipes

Combustion in pipes with closed ends usually starts generating a shock wave which travels forwards and backwards between the front and the dead end. The reflected waves interact with the shock so that the picture of reacting detonation conditions is much more complex than that described by the idealized steady state theory. To demonstrate this we analyze a detonation process in a pipe $0.1m$ in diameter and $11.65m$ in length. The left half of the pipe is filled with steam, and the right hand side with an explosive mixture of stoichiometric hydrogen-oxygen and steam as an inert species. The initial pressure is selected to be 70 *bar*. The initial temperature is 400 *K*. The solution is obtained with single step combustion kinetic postulating an ignition temperature of 783 *K*. We use a donor-cell discretization for the convective terms and second order discretization for the diffusion terms with IVA_5M computer code. The axial discretization has 1165 1-*cm* cells. The results are given below.

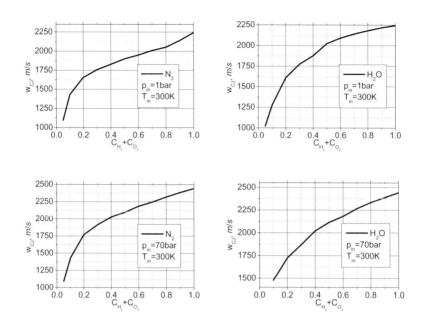

Fig. 9.4. CJ velocity for stoichiometric hydrogen-oxygen mixtures with nitrogen or steam as an inert component

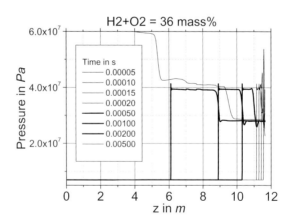

Fig. 9.5. Pressure as a function of the distance for different times after the ignition

Figure 9.5 gives a typical development of the shock wave into a detonation wave after ignition at one end. The initial transient process, followed by the semi-steady-state formation of the detonation front, is clearly recognizable. The extension of the CJ pressure region followed by a refraction wave is also clearly visible. There is the effect of reflection of the wave at the moment it reaches the separation point and penetrates the inert gases without fuel continuing to propagate and without the pushing effect of the combustion behind the shock front. The reflected shock wave is clearly visible from Fig. 9.6. The destructive potential of such waves for pipes not designed to sustain such pressures is obvious.

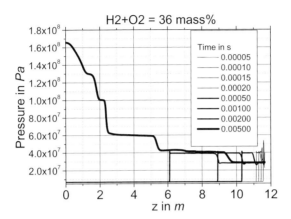

Fig. 9.6. Pressure as a function of distance for different times after the ignition

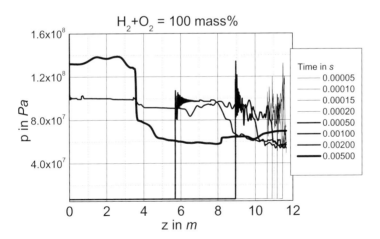

Fig. 9.7. Pressure as a function of distance for different times after the ignition

Increasing the concentration of the stoichiometric explosive mixture leads to an increase in the CJ pressure. The velocity of sound in the burning products is then higher. Therefore, the frequency of the wave reflections between the front and the dead end is much higher, as demonstrated in Fig. 9.7.

9.3 Multi-phase flow

9.3.1 Continuum sound waves

a) No mechanical phase coupling (Wallis). Let us now examine the similar geometrical case of sound wave propagation in multi-phase flows. The local volume-averaged equations give:

$$d(\alpha_l \rho_l w_l) = 0, \qquad (9.40)$$

$$d(\alpha_l \rho_l w_l^2) + \alpha_l dp = 0 \quad \text{or} \quad \rho_l d\left(\frac{1}{2} w_l^2\right) + dp = 0, \qquad (9.41)$$

$$d\left[\alpha_l \rho_l w_l \left(h_l + \frac{1}{2} w_l^2\right)\right] = 0 \quad \text{or} \quad d\left(h_l + \frac{1}{2} w_l^2\right) = 0 \quad \text{or}$$

$$\rho_l dh_l - dp = 0 \quad \text{or} \quad ds_l = 0, \qquad (9.42)$$

$$d(\alpha_l \rho_l w_l C_{il}) = 0 \quad \text{or} \quad dC_{il} = 0 \quad \text{for} \quad \alpha_l \geq 0, \qquad (9.43)$$

$$d(w_l n_l) = 0 \quad \text{for} \quad \alpha_l \geq 0. \qquad (9.44)$$

Equation (9.40) says that mass is conserved. Equation (9.41) says that force equals the time rate of change of momentum (neglecting body and viscous sources). Equation (9.42) says that energy is conserved (neglecting all dissipative effects such as work done by viscous stress, and heat transfer due to thermal conduction and diffusion). Beyond *Wallis* [27], we also take into account Eqs. (9.43) and (9.44) here. Equation (9.43) says that the species mass is conserved, neglecting diffusion inside the shock front. Equation (9.44) says that there is no fragmentation across the shock. Differentiating the mass conservation equations using the chain rule, dividing by $\rho_l w_l$, summing the resulting equations and allowing for the fact that the sum of the volume fraction is constant we then obtain

$$\sum_{l=1}^{l_{max}} \frac{\alpha_l}{\rho_l} d\rho_l + \sum_{l=1}^{l_{max}} \frac{\alpha_l}{w_l} dw_l = 0, \qquad (9.45)$$

Using the equation of state

$$d\rho_l = \frac{dp}{a_l^2} + \left(\frac{\partial \rho_l}{\partial s_l}\right)_{p,\text{all C's}} ds_l + \sum_{i_l=1,}^{i_{l,\max}} \left(\frac{\partial \rho_l}{\partial C_{il}}\right)_{s_l,p} dC_{il}, \qquad (9.46)$$

where

$$\frac{1}{a_l^2} = \left(\frac{\partial \rho_l}{\partial p}\right)_{s_l,\text{all C's}} \qquad (9.47)$$

and taking into account the last form of Eqs. (9.42) and (9.43) we obtain

$$dp \left(\sum_{l=1}^{l_{\max}} \frac{\alpha_l}{\rho_l a_l^2}\right) + \sum_{l=1}^{l_{\max}} \frac{\alpha_l}{w_l} dw_l = 0. \qquad (9.48)$$

Excluding the velocity differential by using the momentum equation we obtain

$$\left(\sum_{l=1}^{l_{\max}} \frac{\alpha_l}{\rho_l a_l^2}\right) \left(1 - \frac{\sum_{l=1}^{l_{\max}} \frac{\alpha_l}{\rho_l w_l^2}}{\sum_{l=1}^{l_{\max}} \frac{\alpha_l}{\rho_l a_l^2}}\right) dp = 0. \qquad (9.49)$$

This equation says that for pressure perturbations that are not equal to zero the following condition must be satisfied:

$$\sum_{l=1}^{l_{\max}} \frac{\alpha_l}{\rho_l w_l^2} = \sum_{l=1}^{l_{\max}} \frac{\alpha_l}{\rho_l a_l^2}. \qquad (9.50)$$

This result was obtained by *Wallis* [27] p.142 in 1969 and is applicable to stratified flows *without strong coupling* between the phases and *no interfacial heat and mass transfer*. Equation (9.50) does not give any information on the magnitude of the phase velocities.

b) *Homogeneous multiphase flow (Wood).* Note that the sound velocity in a multi-phase mixture is flow-pattern dependent and may be influenced by the amplitude and frequency. Let us recall the case for homogeneous multi-phase flow. In this case the sum of the momentum equations, the so called mixture momentum equation, is used instead of the separated momentum equations.

$$d\left(\rho_h w_h^2\right) + dp = 0 \quad \text{or} \quad dw_h = -\frac{1}{\rho_h w_h} dp, \qquad (9.51)$$

where

$$\rho_h = \sum_{l=1}^{l_{max}} \alpha_l \rho_l \qquad (9.52)$$

is the mixture density. Equation (9.48) simplifies in this case to

$$dp \left(\sum_{l=1}^{l_{max}} \frac{\alpha_l}{\rho_l a_l^2} \right) + \frac{1}{w_h} dw_h = 0. \qquad (9.53)$$

By inserting the velocity differential from the mixture momentum equation, we obtain

$$\left[\left(\sum_{l=1}^{l_{max}} \frac{\alpha_l}{\rho_l a_l^2} \right) - \frac{1}{\rho_h w_h^2} \right] dp = 0. \qquad (9.54)$$

For disturbances of small pressure amplitude we have the velocity of sound

$$w_h = \frac{1}{\sqrt{\rho \sum_{l=1}^{l_{max}} \frac{\alpha_l}{\rho_l a_l^2}}}. \qquad (9.55)$$

This is the so called homogeneous mixture sound velocity. For two-phase flow this equation was first derived by *Wood* [28] in 1930. It is widely accepted to be valid for bubbly flow, see the discussion by *Wallis* [27] in 1969.

9.3.2 Discontinuum shock waves

We assume that the thickness of the shock front is small and that no fragmentation, interfacial heat and mass transfer essentially take place. The control volume balance is therefore

$$\Delta(\alpha_l \rho_l w_l) = 0, \qquad (9.56)$$

$$\Delta(\alpha_l \rho_l w_l^2) + \alpha_l \Delta p = 0, \qquad (9.57)$$

$$\Delta \left[\alpha_l \rho_l w_l \left(h_l + \frac{1}{2} w_l^2 \right) \right] = 0, \qquad (9.58)$$

$$\Delta\left(\alpha_l \rho_l w_l C_{il}\right) = 0 \quad \text{for} \quad \alpha_l \geq 0, \tag{9.59}$$

$$\Delta\left(w_l n_l\right) = 0 \quad \text{for} \quad \alpha_l \geq 0. \tag{9.60}$$

Homogeneous multiphase flow. The problem is considerably simplified by the assumption that all the velocities are equal,

$$w_l = w_h. \tag{9.61}$$

The mass conservation for a single velocity field is then

$$\alpha_{l,2}\rho_{l,2}w_{h,2} - \alpha_{l,1}\rho_{l,1}w_{h,1} = 0 \quad \text{or} \quad \frac{\alpha_{l,2}\rho_{l,2}}{\rho_{h,2}} - \frac{\alpha_{l,1}\rho_{l,1}}{\rho_{h,1}} = 0 \quad \text{or}$$

$$x_{l,2} - x_{l,1} = 0, \tag{9.62}$$

which says the field mass concentration remains constant across the shock front. This is a remarkable equation. It indicates that such calculations can be readily performed if one uses the following definition for the mass concentrations:

$$x_l = \frac{\alpha_l \rho_l}{\rho_h}. \tag{9.63}$$

The homogeneous density can thus be expressed as

$$\rho_h = \frac{1}{\sum_{l=1}^{l_{\max}} \frac{x_l}{\rho_l}}. \tag{9.64}$$

The mixture mass conservation results in

$$\rho_{h,1} w_{h,1} = \rho_{h,2} w_{h,2} = \rho_h w_h. \tag{9.65}$$

Using the mixture mass conservation, the mixture momentum equation can then be transformed into

$$\rho_{h,2} w_{h,2}^2 - \rho_{h,1} w_{h,1}^2 + p_2 - p_1 = 0 \quad \text{or} \quad \left(\rho_h w_h\right)^2 = \frac{p_2 - p_1}{\dfrac{1}{\rho_{h,1}} - \dfrac{1}{\rho_{h,2}}}. \tag{9.66}$$

The energy conservation for each velocity field is then

$$\alpha_{l,2}\rho_{l,2}w_{h,2}\left(h_{l,2}+\frac{1}{2}w_{h,2}^2\right)-\alpha_{l,1}\rho_{l,1}w_{h,1}\left(h_{l,1}+\frac{1}{2}w_{h,1}^2\right)=0. \tag{9.67}$$

After applying the field- and the mixture-mass conservation equation and using the last form of the momentum equation, the single field energy conservation that results is as follows:

$$h_{l,2}-h_{l,1}-\frac{1}{2}(p_2-p_1)\left(\frac{1}{\rho_{h,1}}+\frac{1}{\rho_{h,2}}\right)=0. \tag{9.68}$$

The assumption that the multi-phase flow consists of steam and water, which are in thermodynamic equilibrium, reduces Eq. (9.66) to those derived by *Fischer* [5] in 1967.

From Eq. (7.52) we obtain the capability of the wave to perform technical work

$$-w_{t,21}^{vdp}=\sum_{l=1}^{l_{\max}}x_l\left[h_{l,2}-h_{l,1}+T_\infty\left(s_{l,1}-s_{l,2}\right)\right]. \tag{9.69}$$

9.3.3 Melt-coolant interaction detonations

In the melt-coolant interaction, molten material is premixed with liquid that is in thermal non-equilibrium before the shock wave (Fig. 9.8).

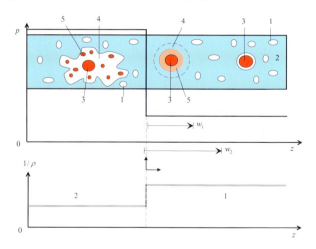

Fig. 9.8. Definitions of the velocity field in the five-field concept

The melt is in the state of film boiling. The incoming shock wave causes instabilities and fragments the f_3 part of the melt mass. The microscopic particles generated release their thermal energy by direct contact with the f_2 part of the surrounding coolant mass. This interaction is a short distance interaction. $1-f_2$ of the coolant mass outside the short distance remains unaffected by this interaction. The physical problem can be thus specified using the following quantities:

$$p_1, T_{1,1}, T_{2,1}, T_{3,1}, x_1, x_3, \qquad (9.50)$$

where the mass concentrations in the undisturbed mixture are

vapor $\quad x_1,$ $\hfill (9.71)$

liquid $\quad x_2 = 1 - x_1 - x_3,$ $\hfill (9.72)$

melt $\quad x_3,$ $\hfill (9.73)$

and the mass concentrations of the newly born fields behind the front are

short distance liquid $\quad x_4 = f_2 x_2 = f_2(1 - x_1 - x_3),$ $\hfill (9.74)$

micro particles $\quad x_5 = f_3 x_3.$ $\hfill (9.75)$

Here f_2 indicates the fraction of x_2 that is within the short distance liquid. f_3 is the fraction of the melt that is fine fragmented. These fractions are controlled by complex interactions not considered in detail in this study. We use the two fractions as parameters in our analysis.

One very useful simplification for such analyses is proposed by *Shamoun* and *Corradini* [25]. The authors proposed that the mixture of the entrained melt intermixed with the entrained liquid reaches thermal equilibrium so that both components have a common pressure p_2 and a common temperature $(T_m)_2$. In accordance with this we seek for the following unknown variables:

$$p_2, T_{1,2}, T_{2,2}, T_{3,2}, T_m, \qquad (9.76)$$

which satisfy the *Chapman-Jouguet* condition behind the detonation front. *Shamoun* and *Corradini* [25] used the assumption that the non-entrained melt and water do not change their initial temperatures and densities, and that the vapor phase experiences an isentropic change of state. These assumptions are in contradiction with the discontinuity of the properties across the shock. We relax these assumptions by using instead the corresponding conservation equations.

The densities and enthalpies required for this computation are functions of the pressure and temperature in front of and behind the shock. The following results are obtained for the vapor

$$T_{1,1},\ T_{1,2},\ (\rho_1)_1 = \rho_1(p_1,T_{1,1}),\ (\rho_1)_2 = \rho_1(p_2,T_{1,2}),\ (h_1)_1 = h_1(p_1,T_{1,1}),$$

$$(h_1)_2 = h_1(p_2,T_{1,2}), \qquad (9.77)$$

for the liquid

$$T_{2,1},\ T_{2,2},\ (\rho_2)_1 = \rho_2(p_1,T_{2,1}),\ (\rho_2)_2 = \rho_2(p_2,T_{2,2}),\ (h_2)_1 = h_2(p_1,T_{2,1}),$$

$$(h_2)_2 = h_1(p_2,T_{2,2}), \qquad (9.78)$$

for the melt

$$T_{3,1},\ T_{3,2},\ (\rho_3)_1 = \rho_3(p_1,T_{3,1}),\ (\rho_3)_2 = \rho_3(p_2,T_{3,2}),\ (h_3)_1 = h_3(p_1,T_{3,1}),$$

$$(h_3)_2 = h_3(p_2,T_{3,2}), \qquad (9.79)$$

for the short distance entrained liquid

$$T_{2,1},\ T_{1,2},\ (\rho_4)_1 = \rho_2(p_1,T_{2,1}),\ (\rho_4)_2 = \rho_1(p_2,T_m),\ (h_4)_1 = h_2(p_1,T_{1,2}),$$

$$(h_4)_2 = h_1(p_2,T_m), \qquad (9.80)$$

and for the micro particles

$$T_{3,1},\ T_{1,2},\ (\rho_5)_1 = \rho_3(p_1,T_{3,1}),\ (\rho_5)_2 = \rho_3(p_2,T_m),\ (h_5)_1 = h_3(p_1,T_{3,1}),$$

$$(h_5)_2 = h_3(p_2,T_m). \qquad (9.81)$$

We assume that the pressure wave first causes the micro-fragmentation and acceleration of the fine particles into the short distance liquid. The shock discontinuity then occurs. In this case the mass conservation equation (9.62) holds. The homogeneous mixture densities are then

$$\rho_{h,1} = \cfrac{1}{\cfrac{x_1}{\rho_1(p_1,T_{1,1})} + \cfrac{1-x_1-x_3}{\rho_2(p_1,T_{2,1})} + \cfrac{x_3}{\rho_3(p_1,T_{3,1})}} \qquad (9.82)$$

$$\rho_{h,2} = \left[\cfrac{x_1}{\rho_1(p_2,T_{1,2})} + \cfrac{(1-f_2)(1-x_1-x_3)}{\rho_2(p_2,T_{2,2})} + \cfrac{(1-f_3)x_3}{\rho_3(p_2,T_{3,2})} + \cfrac{f_2(1-x_1-x_3)}{\rho_1(p_2,T_m)} + \cfrac{f_3 x_3}{\rho_3(p_2,T_m)} \right]^{-1}. \qquad (9.83)$$

Then the detonation adiabatic are for each phase as follows

1) primary steam

$$(h_1)_2 = h_1(p_1,T_{1,1}) + \Delta h, \qquad (9.84)$$

where

$$\Delta h = \frac{1}{2}(p_2 - p_1)\left(\frac{1}{\rho_{h,1}} + \frac{1}{\rho_{h,2}}\right), \qquad (9.85)$$

$$h_1(p_2,T_{1,2}) = (h_1)_2 \quad \text{or} \quad T_{1,2} = T_1\left[p_2,(h_1)_2\right], \qquad (9.86)$$

$$(\rho_1)_2 = \rho_1(p_2,T_{1,2}). \qquad (9.88)$$

2) non-entrained liquid

$$(h_2)_2 = h_2(p_1,T_{2,1}) + \Delta h, \qquad (9.88)$$

$$h_2(p_2,T_{2,2}) = (h_2)_2, \qquad (9.89)$$

$$(\rho_2)_2 = \rho_2(p_2,T_{2,2}). \qquad (9.90)$$

3) non-entrained melt

$$(h_3)_2 = h_3(p_1, T_{3,1}) + \Delta h, \tag{9.91}$$

$$h_3(p_2, T_{3,2}) = (h_3)_2, \tag{9.92}$$

$$(\rho_3)_2 = \rho_3(p_2, T_{3,2}). \tag{9.93}$$

4) entrained liquid

$$(h_4)_2 = h_2(p_1, T_{2,1}) + \Delta h, \tag{9.94}$$

$$h_1(p_2, T_m) = (h_4)_2, \tag{9.95}$$

$$(\rho_4)_2 = \rho_1(p_2, T_m). \tag{9.96}$$

5) entrained melt

$$(h_5)_2 = h_3(p_1, T_{3,1}) + \Delta h, \tag{9.97}$$

$$h_3(p_2, T_m) = (h_5)_2, \tag{9.98}$$

$$(\rho_5)_2 = \rho_3(p_2, T_m). \tag{9.99}$$

The definition of the homogeneous velocity of sound is rewritten in terms of the mass concentrations to give

$$\frac{1}{(\rho w_h)^2} = \sum_{l=1}^{l_{max}} \frac{x_l}{(\rho_l a_l)^2}. \tag{9.100}$$

The *Chapman-Jouguet* condition is thus defined as

$$\frac{1}{(\rho_h w_h)_2^2} = \frac{1}{(\rho_h w_h)^2} = \frac{\dfrac{1}{\rho_{h,1}} - \dfrac{1}{\rho_{h,2}}}{p_2 - p_1} \tag{9.101}$$

or

$$\sum_{l=1}^{l_{\max}} \frac{x_l}{\left(\rho_l a_l\right)_2^2} = \frac{\dfrac{1}{\rho_{h,1}} - \dfrac{1}{\rho_{h,2}}}{p_2 - p_1}, \qquad (9.102)$$

or

$$\frac{x_1}{\left[\rho_1(p_2,T_{1,2})a_1(p_2,T_{1,2})\right]^2} + \frac{(1-f_2)(1-x_1-x_3)}{\left[\rho_2(p_2,T_{2,2})a_2(p_2,T_{2,2})\right]^2}$$

$$+ \frac{(1-f_3)x_3}{\left[\rho_3(p_2,T_{3,2})a_3(p_2,T_{3,2})\right]^2} + \frac{f_2(1-x_1-x_3)}{\left[\rho_1(p_2,T_m)a_1(p_2,T_m)\right]^2}$$

$$+ \frac{f_3 x_3}{\left[\rho_3(p_2,T_m)a_3(p_2,T_m)\right]^2} = \frac{\dfrac{1}{\rho_{h,1}} - \dfrac{1}{\rho_{h,2}}}{p_2 - p_1}. \qquad (9.103)$$

The temperature of the mixture consisting of entrained melt and coolant has to satisfy the sum of the detonation equations

$$f_2(1-x_1-x_3)(h_4)_2 + f_3 x_3 (h_5)_2 \equiv f_2(1-x_1-x_3)h_1(p_2,T_m) + f_3 x_3 h_3(p_2,T_m)$$

$$= \Delta h_{45,2}, \qquad (9.104)$$

where

$$\Delta h_{45,2} = f_2(1-x_1-x_3)h_2(p_1,T_{2,1}) + f_3 x_3 h_3(p_1,T_{3,1}) + \left[f_2(1-x_1-x_3) + f_3 x_3\right]\Delta h. \qquad (9.105)$$

The temperature inversion can be performed using the *Newton* iteration method

$$T_m^{n+1} = T_m^n - \frac{f_2(1-x_1-x_3)h_1(p_2,T_m^n) + f_3 x_3 h_3(p_2,T_m^n) - \Delta h_{45,2}}{f_2(1-x_1-x_3)c_{p,1}(p_2,T_m^n) + f_3 x_3 c_{p,3}(p_2,T_m^n)}. \qquad (9.106)$$

The superscripts n and $n + 1$ denote the old and the new iteration values, respectively. Similarly for the vapor, non-entrained liquid and non-entrained melt we have

$$T_{1,2}^{n+1} = T_{1,2}^n - \frac{h_1(p_2, T_{1,2}^n) - h_1(p_1, T_{1,1}) - \Delta h}{c_{p,1}(p_1, T_{1,1})}, \quad (9.107)$$

$$T_{2,2}^{n+1} = T_{2,2}^n - \frac{h_2(p_2, T_{2,2}^n) - h_2(p_1, T_{2,1}) - \Delta h}{c_{p,2}(p_1, T_{2,1})}, \quad (9.108)$$

$$T_{3,2}^{n+1} = T_{3,2}^n - \frac{h_3(p_2, T_{3,2}^n) - h_3(p_1, T_{3,1}) - \Delta h}{c_{p,3}(p_1, T_{3,1})}. \quad (9.109)$$

9.3.4 Similarity to and differences from the *Yuen* and *Theofanous* formalism

The exact mathematical formalism used by *Yuen* and *Theofanous* is not reported in [29]. It seems that we use very much the same primitive conservation equations as these authors used. They then constructed *numerically* the *Rankine-Hugoniot* shock and detonation adiabatic by using the primitive conservation equations and looked for the tangent to the detonation adiabatic corresponding to the initial state, the so called *Chapman-Jouguet* point.

Instead, we use the formalism *analytically* transformed to the detonation adiabatic for each of the five fields together with the condition that the speed of propagation behind the shock is equal to the homogeneous velocity of sound that corresponds to the local conditions behind the shock. The computer code written by *Huang Hu Lin* to solve this system of nonlinear transcendental equations allows rapid performance of a variety of computations. In addition, the initial vapor mass is compressed by the shock adiabatic and the resulting vapor temperature behind the shock is different from that of the mixture of entrained melt and entrained liquid. Results of such computations will be shown in the next section.

9.3.5 Numerical solution method

In fact for the five unknowns, $p_2, T_{1,2}, T_{2,2}, T_{3,2}, T_m$, we have a system of five nonlinear transcendental equations (9.103) and (9.106-9.109). The system was coupled with the material properties library of the IVA6 computer code [14]. This library contains properties for water and water vapor in sub- and supercritical states, properties of different inert gases, properties for ten materials, which can be either solid or two-phase or liquid with iron, uranium dioxide and alumina among them. The equations are solved by iteration as follows:

a) Assume initial values at the very beginning of the iterations as follows:

$p_2 = p_1 + \Delta p$, where

$$\Delta p > 0, \ T_{1,2} = T_{1,1} \left(\frac{p_2}{p_1} \right)^{\frac{\kappa_1 - 1}{\kappa_1}}, \ T_{2,2} = T_{2,1}, \ T_{3,2} = T_{3,1}, \ T_m = T_{1,1}.$$

b) Compute the densities behind the front by using the state equations.

c) Compute the pressure jump across the shock from the multi-phase *Chapman-Jouguet* condition.

d) Compute Δh using Eq. (9.85).

e) Compute the temperatures using Eqs. (9.106) to (9.109).

f) Repeat steps c) through d) until the change in the pressure jump from iteration to iteration becomes smaller than the prescribed.

Conclusions. The following conclusions can be drawn from Sections 9.2 and 9.3. The maximum of the pressure jump for multiphase detonations in melt-water systems can be described by using the assumption of homogeneous flows and relying on the *Rankine-Hugoniot* relations for multiple concentration fields having the same speed. The resulting transcendental system of algebraic equations has to be solved by iteration. Several assumptions with respect of the part of the fragmented particles and of the part of the entrained coolant in the interacting zone are required to close the system. Parametric study can provide the upper limit of the produced detonation shock waves which can be used in estimation of the explosive potential of mixtures of different material pairs.

Such a study will be presented in the next section. In Section 9.4 we present a parametric detonation study for the systems of UO_2-water, Al_2O_3-water and Fe-water. We will draw the conclusion that at the same initial temperatures, mass concentrations, entrainment factors and local volume fractions, the Al_2O_3-water system produces the highest detonation pressures, and that the dispersed systems consisting of melted particles and water droplets can create detonation pressures less than $200 bar$.

9.4 Detonation waves in water mixed with different molten materials

In Sections 9.2 and 9.3, see also [15], we recalled the basics of the detonation theory for single- and multi-phase flows. We ended up with a model for detonation analysis for melt-coolant interaction systems. In this section, see also [16], we will apply the detonation theory of melt-water interaction to gain information required for an upper bound estimate of the energetic interactions for the nuclear industry. We will consider melt initially mixed with water that is in saturation at atmospheric pressure. In all examples the initial temperature of the melt is considered to be 3000 K. We will study the influence of the following parameters on the detonation pressure of self-sustaining waves: initial void fraction, initial mass concentration of the melt, melt entrainment ratio, water entrainment ratio, different melt material, etc.

In particular, we will answer the following questions:

a) How can we judge the risk potential of dispersed-dispersed systems for in-vessel and ex-vessel analysis?

b) Are there any differences between the detonation behavior of the following three pairs of materials: UO_2 water, Al_2O_2 water and Fe water that result from the detonation theory?

c) Do detonation solutions exist for lean mixtures with a limited amount of water entrained?

d) What is the maximum water entrainment ratio that allows detonation solutions for the above mentioned three material pairs and what are the associated maximum pressures for these solutions?

Note that postulating a given set of initial conditions that theoretically lead to detonations does not necessarily mean that such an event can really take place. The present analysis provides only an upper bound of solution sets.

9.4.1 UO$_2$ water system

Consider a mixture consisting of uranium dioxide and water. We will represent the detonation pressure behind a self-sustaining shock as a function of the void volume fraction rather than as a function of the mixture specific volume as is usually done in the literature. This is due to our aim of deriving conclusions for systems for which the local void fraction controls the flow pattern. Figure 9.9 shows the pressure behind the self-sustained detonation wave as a function of vapor volume fraction for different mass concentrations of UO_2. The C-J pressures sharply increase with decreasing vapor volume fraction. Increasing the melt mass concentration x_3 up to about 30% leads to an increase in the C-J pressure.

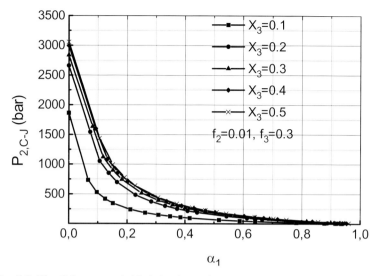

Fig. 9.9. The C-J pressure behind the detonation wave as a function of the volume fraction of vapor for large melt mass concentrations of UO_2 (rich mixtures)

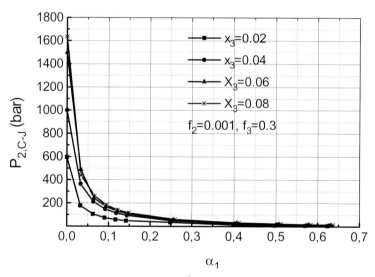

Fig. 9.10. The C-J pressure behind the detonation wave as a function of the volume fraction of vapor for small melt mass concentrations of UO_2 (lean mixtures)

448   9 Detonation waves caused by chemical reactions or by melt-coolant interactions

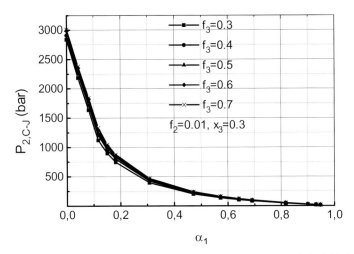

Fig. 9.11. The C-J pressure behind the detonation wave as a function of the initial volume fraction of vapor. Parameter: melt entrained fraction f_3 of UO_2

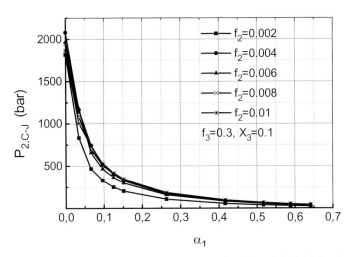

Fig. 9.12. The C-J pressure behind the shock wave as a function of the initial volume fraction of vapor. Parameter: liquid entrained fraction f_2 for UO_2

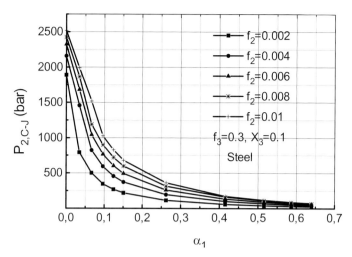

Fig. 9.13. The steel C-J pressure behind the shock wave as a function of the initial volume fraction of vapor. Parameter: liquid entrained fraction f_2

Figure 9.10 demonstrates the behavior of the lean mixtures – mixtures having small amounts of melt. One sees that even a very small amount of entrained water may cause considerable detonation pressures at low void fractions. This result confirms the warning expressed by *Yuen* and *Theofanous* to consider lean mixtures as potentially explosive. We also see that lean systems are very sensitive to void fraction. For void fractions larger then 10% the lean mixture can in fact detonate but the detonation pressure has a low risk potential. Homogeneous low void fraction and low water fraction mixtures are in fact possible only in the case of fast transient water entrapment or water injection into the melt. Such systems possess very high detonation pressures. That is the reason why such systems, despite the fact that their existence is only locally possible, are real triggers in nature. Systems not having depletion paths may realize such states locally. Large-scale existence of such mixtures is practically impossible in open systems having depletion paths for the participating materials. More discussion of this point is provided in [7, 8].

Figures 9.11 and 9.12 show the pressure as a function of the volume fraction for different entrained melt and liquid fractions, respectively. In Fig. 9.11 the entrained coolant controls the pressure and the amount of entrained melt does not influence the process for fixed coolant entrainment.

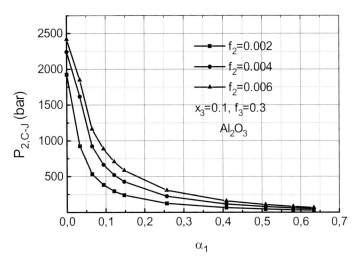

Fig. 9.14. The alumina C-J pressure behind the shock wave as a function of the initial volume fraction of vapor. Parameter: liquid entrained fraction f_2

Figure 9.12 shows that at fixed melt entrainment an increase of the coolant entrainment up to a given maximum value may increase the detonation pressure. Above this value we have predominant by fine debris quenching, and detonation solutions can no longer be obtained.

From this analysis we learn the following for *dispersed* systems: detonation solutions exist theoretically for mixtures consisting of *dispersed* melt and *dispersed* water. There is, however, no mechanism that would explain the degree of melt fragmentation and water entrainment required for these solutions. As a result, such mixtures cannot be considered to have explosive risk potential. Even assuming a hypothetical, as yet unknown, mechanism that may lead to the required fragmentation, the resulting detonation pressures are of no concern for the so called in-vessel analysis in nuclear reactor safety.

Comparison of the detonation behavior of different material pairs. Here, we keep the parameter the same as for Fig. 9.5 but change only the melt material. The computational results for steel are presented in Fig. 9.6 and for alumina in Fig. 9.7. Comparing Figs. 9.5, 9.6 and 9.7 we come to a very interesting conclusion: at the same initial conditions the steel and the alumina possess the potential to create stronger detonations than the uranium dioxide. This is a fact known from experiments e.g. KROTOS [13]. This means that it is solely the differences in the caloric thermal state properties (an example is given in Appendix 9.1 where the specific capacities at constant pressure for urania and zirconia as functions of the temperature are plotted) for the different materials that make any difference to the achievable detonation pressure under similar conditions.

9.4.2 Efficiencies

We will now compare for the three different pairs the fraction of the energy contained in the melt that is necessary to transport the mixture across the boundary of the control volume moving with the frame of reference,

$$\eta_{th} = \frac{p_2 v_2 - p_1 v_1}{e_{3,1}},$$

and the fraction of the thermal energy discharge that is transformed into this work

$$\eta = \frac{p_2 v_2 - p_1 v_1}{e_{3,1} - f_3 e_5 - (1 - f_3) e_{3,2}}.$$

From Figs. 9.15a, 9.16a and 9.17a we see that for all of the material pairs considered here lean mixtures with a melt mass fraction of 10%, and 30% melt entrainment cannot transform more than 20% of their thermal energy into mechanical flow transport energy. Increasing the void in the system reduces this figure dramatically.

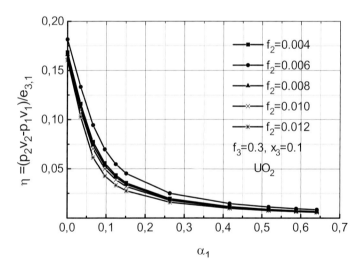

a) $\eta_{th} = \dfrac{p_2 v_2 - p_1 v_1}{e_{3,1}}$

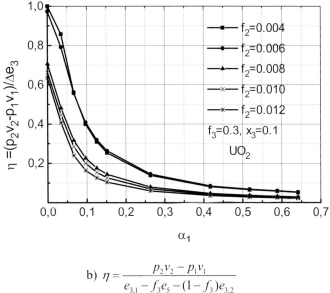

b) $\eta = \dfrac{p_2 v_2 - p_1 v_1}{e_{3,1} - f_3 e_5 - (1-f_3) e_{3,2}}$

Fig. 9.15. The two nominal efficiencies of uranium dioxide behind the shock wave as a function of the initial volume fraction of vapor. Parameter: liquid entrained fraction f_2

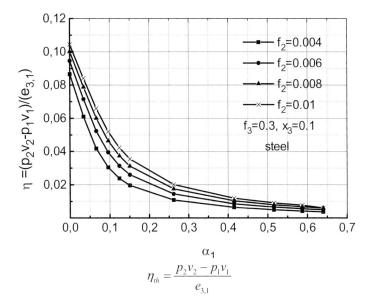

$\eta_{th} = \dfrac{p_2 v_2 - p_1 v_1}{e_{3,1}}$

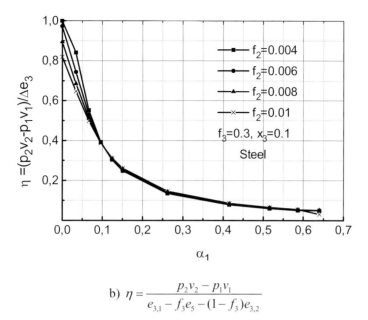

$$b)\ \eta = \frac{p_2 v_2 - p_1 v_1}{e_{3,1} - f_3 e_5 - (1 - f_3) e_{3,2}}$$

Fig. 9.16. The two nominal efficiencies of steel behind the shock wave as a function of the initial volume fraction of vapor. Parameter: liquid entrained fraction f_2

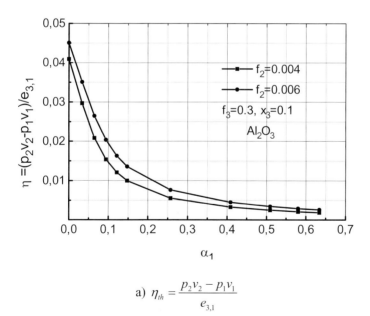

$$a)\ \eta_{th} = \frac{p_2 v_2 - p_1 v_1}{e_{3,1}}$$

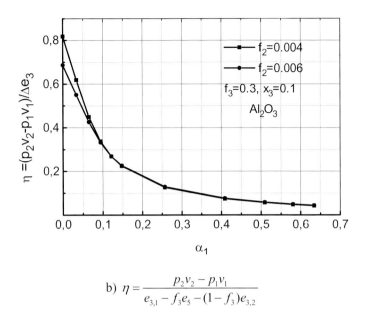

b) $\eta = \dfrac{p_2 v_2 - p_1 v_1}{e_{3,1} - f_3 e_5 - (1 - f_3) e_{3,2}}$

Fig. 9.17. The two nominal efficiencies of steel behind the shock wave as a function of the initial volume fraction of vapor. Parameter: liquid entrained fraction f_2

From Figs. 9.15b, 9.16b and 9.17b we learn that increasing the entrained coolant may cause quenching and reduce the transformation of all the discharged thermal energy into mechanical flow transport energy. Surprisingly, this effect is not so manifested for the alumina-water system.

9.4.3 The maximum coolant entrainment ratio

As already mentioned, even theoretically large coolant entrainment ratios do not result in detonation solutions. In this chapter we compare the three pairs by presenting the maximum of the achievable coolant entrainment ratio that allows self-sustaining detonation waves as a function of the void fractions and the corresponding detonation pressures – Figs. 9.18, 9.19 and 9.20. The jumps in the entrainment ratios in Figs. 9.18 and 9.19 are associated with the latent heat of solidification. The larger water entrainment ratios give higher detonation pressures because the volume of the produced vapor is larger.

Surprisingly, the alumina-water system once again allows the highest water entrainment rates and produces higher detonation pressures at very low void fractions. This is due to the fact that the specific heat of uranium dioxide is less than that of alumina.

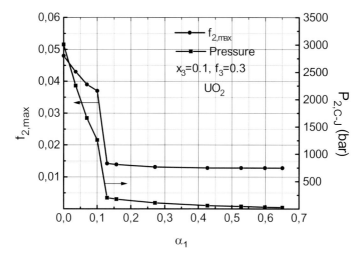

Fig. 9.18. The maximum water mass entrained fraction and C-J pressure behind the shock wave for uranium dioxide as a function of the initial volume fraction of vapor

Assuming small coolant entrainment leads to large superheating of the evaporated entrained coolant surrounding the melt and therefore a temperature difference between this coolant and the melt limits locally the energy transferred from the melt to the coolant. This means there is also in this sense a natural limit of the energy transfer.

9.5 Conclusions

In this chapter we applied the detonation theory in order to learn features of melt-water interaction required for an upper bound estimate of the energetic interactions for the nuclear industry. We found the following results:

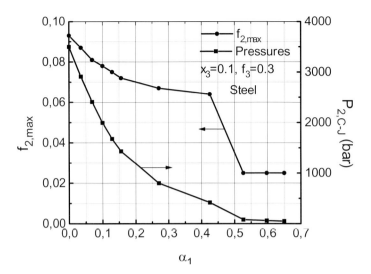

Fig. 9.19. The maximum water mass entrained fraction and C-J pressure behind the shock wave for steel as a function of the initial volume fraction of vapor

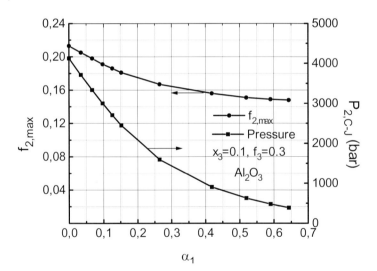

Fig. 9.20. The maximum water mass entrained fraction and C-J pressure behind the shock wave for alumina as a function of the initial volume fraction of vapor

1) Mixtures consisting of dispersed melt and dispersed water have detonation solutions but there is no mechanism to explain the degree of melt fragmentation and of water entrainment into brought contact with the melt debris. Therefore, such mixtures cannot be considered to have explosive risk poten-

tial. Even assuming a hypothetical mechanism that may lead to the required fragmentation, the resulting detonation pressures are of no concern for in-vessel melt-water interaction risk analysis.

2) Even a very small amount of entrained water may cause considerable detonation pressures at low void fractions. This result confirms the warning expressed by *Yuen* and *Theofanous* to consider lean mixtures as potentially explosive. We also see that lean systems are very sensitive to void fraction. For void fractions higher then 10% the lean mixture can in fact detonate but the detonation pressure has low risk potential.

3) Homogeneous low void fraction and low water fraction mixtures are in fact possible only in the case of fast transient water entrapment or water injection into the melt. Such systems possess very high detonation pressures. That is the reason why such systems, despite the fact that their existence is only locally possible, are real triggers in nature. Large-scale existence of such mixtures is practically impossible in open systems having depletion paths for the participating materials. Systems not having depletion paths may realize such states locally.

4) For each entrained melt ratio there is a maximum of entrained coolant ratio that controls the maximum of the detonation pressure.

5) For the same initial conditions the detonation pressures for Al_2O_3-water and *Fe*-water systems are definitively larger than that for UO_2-water systems. The capability for heat discharge during postulated melt-water contact for these systems does not explain the differences in their detonation behavior. As a result, only the differences in the caloric equation of state explain the differences in the detonation behavior. In addition, at the same initial condition, UO_2-water systems can entrain a lower mass fraction of liquid than the other analyzed material pairs. This system reaches the thermal equilibrium state at a lower mixture temperature than the other analyzed material pairs.

6) The efficiency of transformation of thermal energy into mechanical flow transport energy depends strongly a) on the vapor fraction and b) on the entrained water. The higher the void fraction the lower the efficiency of transformation of thermal into mechanical flow transport energy. At levels below some characteristic water entrainment rates for low void fraction mixtures lower efficiency results due to the local non-availability of an adequate amount of water to evaporate. The theoretical maximum of the highest total efficiency of transformation of the all-internal melt energy into mechanical flow transport energy is lower than 20%. This very high figure is achievable in a small-scale laboratory experiment with very low initial void fraction, mainly in cases of entrapment or very fast melt injection into water. That is why we consider this value as a local maximum for trigger efficiencies which

cannot be transferred to large scale systems with a variety of depletion paths such as those of in-vessel melt-water interaction.

9.6 Practical significance

Judging the explosivity of melt-water and vapor mixtures is important for nuclear safety. There is a consensus among researchers that mixtures in which one of the liquids is continuous are explosive. But up to this study [15, 16] it was not clear how to judge the risk of mixtures consisting of melt droplets, coolant droplets and gas. The practical outcome of this study is the demonstration that detonation pressure jumps in disperse systems are lower than 200 *bar* without considering the question whether they can be realized in nature or not. This knowledge classified such mixtures as not dangerous at all for in-vessel melt-water interactions.

Another unexpected outcome of this study is the different theoretical behavior of different melt-water pairs for the same initial temperatures, gas volume fractions and mass concentrations of the melt and water, resulting only from the difference of their caloric equations of state.

Appendix 9.1 Specific heat capacity at constant pressure for urania and alumina

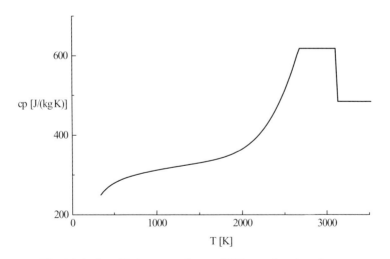

Fig. A9-1. Specific heat capacity c_p of UO_2 as a function of temperature

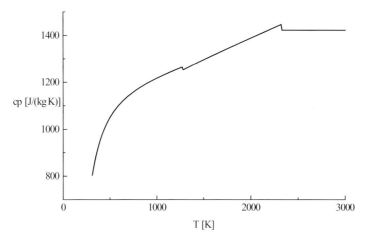

Fig. A9-2 Specific heat capacity c_p of Al_2O_3 as a function of temperature

References

1. Berthoud G (March 1999) Heat transfer modeling during a thermal detonation, CEA/Grenoble Report no SMTH/LM2/99-37
2. Board SJ, Hall RW, Hall RS (1975) Detonation of fuel coolant explosions, Nature 254, March 27, pp 319-321
3. Chapman DL (1899) Philos. Mag., vol 47 no 5 p 90
4. Crussard L (1907) Bull. De la Soc. De l'industrie Minérale St.-Etienne, vol 6 pp 1-109
5. Fischer M (1967) Zur Dynamik der Wellenausbreitung in den Zweiphasenströmungen unter Berücksichtigung von Verdichtungsstössen, Dissertation, TH Karlsruhe
6. Frost DL, Lee JHS, Ciccarelli (1991) The use of Hugoniot analysis for the propagation of vapor explosion waves, Shock Waves, Springer, Berlin Heidelberg New York, pp 99-110
7. Henry RE, Fauske HK (August 1981) Core melt progression and the attainment of a permanently coolable state, in Proc. of the ANS Topical Meeting on Reactor Safety Aspects of Fuel Behavior, San Valley, Idaho. American Nuclear Society
8. Henry RE, Fauske HK (1981) Required initial conditions for energetic steam explosions, J. Heat Transfer, vol 19 pp 99-107
9. Hugoniot PH (1887) Mémoire sur la propagation du mouvement dans les corps et spécialement dans les gazes parfaits, Journal de l'École Polytechnique
10. Huhtiniemi I, Magalon D, Hohmann H (1997) Results of recent KROTOS FCI tests: alumna vs. corium melts, OECD/CSNI Specialist Meeting on Fuel Coolant Interactions, JAERI-Tokai Research Establishment, Japan 19-21 May
11. Jouguet E (1905) J. Mathématique, p 347; (1906) p 6; (1917) Mécanique des Explosifs, Doin O, Paris

12. Kolev NI (1999) In-vessel melt-water interaction caused by core support plate failure under molten pool, Part 1: Choice of the solution method, Proceedings of the ninth International Topical Meeting on Nuclear Reactor Thermal Hydraulics (NURETH-9), San Francisco, California, 3-8 October, 1999, Log. No. 316_1
13. Kolev NI (1999) In-vessel melt-water interaction caused by core support plate failure under molten pool, Part 2: Analysis, Proceedings of the ninth International Topical Meeting on Nuclear Reactor Thermal Hydraulics (NURETH-9), San Francisco, California, 3-8 October, 1999, Log. No. 316_2
14. Kolev N I (1999) Verification of IVA5 computer code for melt-water interaction analysis, Part 1: Single phase flow, Part 2: Two-phase flow, three-phase flow with cold and hot solid spheres, Part 3: Three-phase flow with dynamic fragmentation and coalescence, Part 4: Three-phase flow with dynamic fragmentation and coalescence – alumna experiments, Proc. of the Ninth International Topical Meeting on Nuclear Reactor Thermal Hydraulics (NURETH-9), San Francisco, California, 3-8 October
15. Kolev NI (2000) Detonation waves in melt-coolant interaction, Part 1: Theory, EU Nr. INV-MFCI(99)-D038, Kcrntcchnik, vol 65 no 5-6 pp 254-260
16. Kolev NI and Hulin H (2001) Detonation waves in melt-coolant interaction, Part.2: Aplied analysis, MFCI Project, 6th progress meeting, CEA, Grenoble, 23-24 June 1999. EU Nr. INV-MFCI(99)-D038. Kerntechnik, vol 66 no 1-2 pp 21-25
17. Landau L, Lifshitz EM (1953) Hydrodynamics, Nauka i izkustwo, Sofia 1978 (in Bulgarian), translated from Russian: Theoretical physics: Continuum mechanics and hydrodynamics, Technikoistorizeskoy literatury, Moscu
18. Laplace PSM (1816) Sur la vitesse du son dans l'air at dan l'eau, Annales de Chimie et de Physique
19. Park GC, Corradini ML (July 1991) Estimates of limits for fuel-coolant mixing, in AIChE Proc. of the National Heat Transfer Conference, Minneapolis
20. Rankine WJM (1870) On the thermodynamic theory of waves of finite longitudinal disturbances, Philosophical Transactions of the Royal Society
21. Rayleigh L (Sept. 15 1910) Aerial plane waves of finite amplitude, Proc. of the Royal Society
22. Robert JK, Rupley and Miller JA (April 1987) The CHEMKIN thermodynamic data base, SAND-87-8215, DE87 009358
23. Scott EF, Berthoud GJ (Dec.10-15 1978) Multi-phase thermal detonation, Topics in two-phase heat transfer and flow, pp 11-16, ASME Winter annual meeting, San Francisco
24. Shamoun BI, Corradini ML (1995) Analysis of supercritical vapor explosions using thermal detonation wave theory, Proceedings of the seventh International Topical Meeting on Nuclear Reactor Thermal Hydraulics (NURETH-7) pp 1634-1652
25. Shamoun BI, Corradini ML (July 1996) Analytical study of subcritical vapor explosions using thermal detonation wave theory, Nuclear Technology, vol 115, pp 35-45
26. Taylor G I (Oct. 1910) The condition necessary for discontinuous motion in gases, Proc. of the Royal Society
27. Wallis GB (1969) One-dimensional two-phase flow, McGraw-Hill, New York
28. Wood B (1930) Textbook of sound, Macmillan, New York, p 327
29. Yuen WW, Theofanous TG (19th-21st May 1997) On the existence of multi-phase thermal detonation, Proceedings of OECD/CSNI Specialists Meeting on Fuel-Coolant Interactions (FCI), JAERI-Tokai Research Establishment, Japan

30. Zeldovich JB (1940) To the theory of detonation propagation in gas systems, Journal of experimental and theoretical physics, vol 10 no 5 pp 542-568

10 Conservation equations in general curvilinear coordinate systems

10.1 Introduction

In 1974 *Vivand* [14] and *Vinokur* [13] published their remarkable works on conservation equations of gas dynamics in curvilinear coordinate systems. Since that time there have been many publications on different aspects of this topic. Several providers of computational fluid dynamics tools use the developed strategy for single-phase flows in attempting to extend the algorithm for multi-phase flows. The usually used approach is to write all partial differential equation in convection-diffusion form and to use the already existing transformations and numerical algorithms for single-phase flow. What remains outside the convection-diffusion terms is pooled as a source into the right hand side. The problems with this approach are two: (a) the remaining terms can contain substantial physics represented in differential terms and (b) the realized coupling between the equations is weak. The latter is manifested if one tries to use such codes for processes with strong feedback of the interfacial heat and mass transfer processes on the pressure, e.g. steam explosion, spontaneous flashing etc.

In this Chapter, the conservation equations rigorously derived in Chapters 1,2,5 for multi-phase flows based on local time and volume average [4-9] are rewritten in general curvilinear coordinate in order to facilitate their numerical integration for arbitrary geometry. In addition the so called conservation of total volume equation is also transformed. The latter is extremely useful for creating strongly coupled numerical solution algorithms. Note that the derivation in Chapter 1,2,5, [4-9], is performed for flows in heterogeneous porous structures. Each of the velocity fields was considered to consist of several chemical components. The concept of dynamic fragmentation and coalescence is used. The derivation given here was first published in [10].

For the consideration presented below we assume that the flow happens in the physical space described by the Cartesian, left oriented coordinate systems (x,y,z) (Euclidean space) – Fig. 10.1. In this space the new curvilinear coordinate system is introduced having the coordinates ξ, η, ζ. Another notation simultaneously used is x_i $(i=1,2,3)$: x_1, x_2, x_3 and ξ^i $(i=1,2,3)$: ξ^1, ξ^2, ξ^3. The new coordinate system is called the transformed coordinate system. The curves defining the transformed system are smooth (at least one times differentiable). The transformation is invertible. The Jacobian, the metrics and inverted metrics tensors, the covariant and the contravariant vectors exist. Figure 10.1 shows the physical coordinate system, the transformed coordinate system and a control volume in the transformed coordinates system build by coordinate surfaces. The figure shows also the definition of the contravariant vectors perpendicular to the coordinate surfaces. In addition, the curvilinear coordinates system moves with velocity $\mathbf{V}_{cs}$.

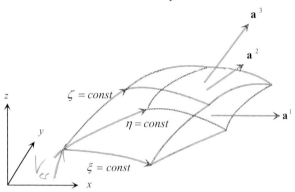

Fig. 10.1. The unit normal vectors to the coordinate surfaces form the contravariant base vectors

For better understanding of the material in this Chapter we present in Appendix 1 a brief introduction to vector analysis and in Appendix 2 some basics of the coordinate transformation theory. I strongly recommend the reader to go over these Appendixes before continuing reading this Chapter.

10.2 Field mass conservation equations

The local volume- and time-averaged mass conservation equation derived in Chapter 1, Eq. (1.62) [4] is

$$\frac{\partial}{\partial \tau}(\alpha_l \rho_l \gamma_v) + \nabla \cdot (\alpha_l \rho_l \mathbf{V}_l \gamma) = \gamma_v \mu_l \ . \qquad (10.1)$$

First we transform the time derivative having in mind that

$$\left(\frac{\partial \varphi}{\partial \tau}\right)_{x,y,z} = \left(\frac{\partial \varphi}{\partial \tau}\right)_{\xi,\eta,\zeta} - \mathbf{V}_{cs} \cdot \nabla \varphi \ . \qquad (10.2)$$

The result is

$$\frac{\partial}{\partial \tau}(\alpha_l \rho_l \gamma_v) + \nabla \cdot (\alpha_l \rho_l \mathbf{V}_l \gamma) - \mathbf{V}_{cs} \cdot \nabla(\alpha_l \rho_l \gamma) = \gamma_v \mu_l \ . \qquad (10.3)$$

Here the time derivative $(\partial / \partial \tau)_{\xi,\eta,\zeta}$ is understood to be at a fixed point in the transformed region and the subscripts are omitted for simplicity of the notation. Then having in mind that the divergence of a vector and the gradient of a scalar in the transformed region are

$$\nabla \cdot \mathbf{F} = \frac{1}{\sqrt{g}} \left[\frac{\partial}{\partial \xi}\left(\sqrt{g}\, \mathbf{a}^1 \cdot \mathbf{F}\right) + \frac{\partial}{\partial \eta}\left(\sqrt{g}\, \mathbf{a}^2 \cdot \mathbf{F}\right) + \frac{\partial}{\partial \zeta}\left(\sqrt{g}\, \mathbf{a}^3 \cdot \mathbf{F}\right) \right], \qquad (10.4)$$

$$\nabla \varphi = \frac{1}{\sqrt{g}} \left[\frac{\partial}{\partial \xi}\left(\sqrt{g}\, \mathbf{a}^1 \varphi\right) + \frac{\partial}{\partial \eta}\left(\sqrt{g}\, \mathbf{a}^2 \varphi\right) + \frac{\partial}{\partial \zeta}\left(\sqrt{g}\, \mathbf{a}^3 \varphi\right) \right], \qquad (10.5)$$

we obtain finally

$$\frac{\partial}{\partial \tau}(\alpha_l \rho_l \gamma_v)$$

$$+ \frac{1}{\sqrt{g}} \left[\frac{\partial}{\partial \xi}\left(\alpha_l \rho_l \gamma_\xi \sqrt{g}\, \mathbf{a}^1 \cdot \mathbf{V}_l\right) + \frac{\partial}{\partial \eta}\left(\alpha_l \rho_l \gamma_\eta \sqrt{g}\, \mathbf{a}^2 \cdot \mathbf{V}_l\right) + \frac{\partial}{\partial \zeta}\left(\alpha_l \rho_l \gamma_\zeta \sqrt{g}\, \mathbf{a}^3 \cdot \mathbf{V}_l\right) \right]$$

$$- \mathbf{V}_{cs} \cdot \frac{1}{\sqrt{g}} \left[\frac{\partial}{\partial \xi}\left(\alpha_l \rho_l \gamma_\xi \sqrt{g}\, \mathbf{a}^1\right) + \frac{\partial}{\partial \eta}\left(\alpha_l \rho_l \gamma_\eta \sqrt{g}\, \mathbf{a}^2\right) + \frac{\partial}{\partial \zeta}\left(\alpha_l \rho_l \gamma_\zeta \sqrt{g}\, \mathbf{a}^3\right) \right] = \gamma_v \mu_l \ .$$

$$(10.6)$$

The term containing the velocity of the transformed coordinate system can be rewritten as follows

$$\mathbf{V}_{cs} \cdot \frac{1}{\sqrt{g}} \left[\frac{\partial}{\partial \xi} \left(\alpha_1 \rho_1 \gamma_\xi \sqrt{g}\ \mathbf{a}^1 \right) + \frac{\partial}{\partial \eta} \left(\alpha_1 \rho_1 \gamma_\eta \sqrt{g}\ \mathbf{a}^2 \right) + \frac{\partial}{\partial \zeta} \left(\alpha_1 \rho_1 \gamma_\zeta \sqrt{g}\ \mathbf{a}^3 \right) \right]$$

$$= \frac{1}{\sqrt{g}} \left[\begin{array}{l} \dfrac{\partial}{\partial \xi} \left(\alpha_1 \rho_1 \gamma_\xi \sqrt{g}\ \mathbf{a}^1 \cdot \mathbf{V}_{cs} \right) + \dfrac{\partial}{\partial \eta} \left(\alpha_1 \rho_1 \gamma_\eta \sqrt{g}\ \mathbf{a}^2 \cdot \mathbf{V}_{cs} \right) \\ + \dfrac{\partial}{\partial \zeta} \left(\alpha_1 \rho_1 \gamma_\zeta \sqrt{g}\ \mathbf{a}^3 \cdot \mathbf{V}_{cs} \right) \end{array} \right]$$

$$- \alpha_1 \rho_1 \gamma_v \left(\mathbf{a}^1 \cdot \frac{\partial \mathbf{V}_{cs}}{\partial \xi} + \mathbf{a}^2 \cdot \frac{\partial \mathbf{V}_{cs}}{\partial \eta} + \mathbf{a}^3 \cdot \frac{\partial \mathbf{V}_{cs}}{\partial \zeta} \right). \tag{10.7}$$

Having in mind that

$$\frac{1}{\sqrt{g}} \frac{d\sqrt{g}}{dt} = \mathbf{a}^1 \cdot \frac{\partial \mathbf{V}_{cs}}{\partial \xi} + \mathbf{a}^2 \cdot \frac{\partial \mathbf{V}_{cs}}{\partial \eta} + \mathbf{a}^3 \cdot \frac{\partial \mathbf{V}_{cs}}{\partial \zeta},$$

we obtain after multiplying by $\sqrt{g}$ and using the reverse chain rule,

$$\sqrt{g} \frac{\partial}{\partial \tau} (\alpha_1 \rho_1 \gamma_v) + \alpha_1 \rho_1 \gamma_v \frac{\partial \sqrt{g}}{\partial \tau} = \frac{\partial}{\partial \tau} \left(\alpha_1 \rho_1 \gamma_v \sqrt{g} \right), \tag{10.8}$$

the following final conservative form of the mass conservation equation for multi-phase flows

$$\frac{\partial}{\partial \tau} \left(\alpha_1 \rho_1 \gamma_v \sqrt{g} \right) + \frac{\partial}{\partial \xi} \left[\alpha_1 \rho_1 \gamma_\xi \sqrt{g}\ \mathbf{a}^1 \cdot (\mathbf{V}_l - \mathbf{V}_{cs}) \right] + \frac{\partial}{\partial \eta} \left[\alpha_1 \rho_1 \gamma_\eta \sqrt{g}\ \mathbf{a}^2 \cdot (\mathbf{V}_l - \mathbf{V}_{cs}) \right]$$

$$+ \frac{\partial}{\partial \zeta} \left[\alpha_1 \rho_1 \gamma_\zeta \sqrt{g}\ \mathbf{a}^3 \cdot (\mathbf{V}_l - \mathbf{V}_{cs}) \right] = \gamma_v \sqrt{g}\ \mu_l. \tag{10.9}$$

We see that the scalar

$$\alpha_1 \rho_1 \sqrt{g}$$

with the flux

$$\alpha_1 \rho_1 \sqrt{g}\ \mathbf{a}^i \cdot (\mathbf{V}_l - \mathbf{V}_{cs})$$

10.3 Mass conservation equations for components inside the field – conservative form

is subject to conservation in the transformed coordinate system. The velocity vector components

$$\bar{V}_l^i = \mathbf{a}^i \cdot (\mathbf{V}_l - \mathbf{V}_{cs})$$

will be called contravariant field relative velocity vector components. It is perpendicular to the coordinate surface defined by $\xi^i = const$. Thus the mass conservation equation is simply

$$\frac{\partial}{\partial \tau}\left(\alpha_l \rho_l \gamma_v \sqrt{g}\right) + \frac{\partial}{\partial \xi}\left(\alpha_l \rho_l \gamma_\xi \sqrt{g}\,\bar{V}_l^1\right) + \frac{\partial}{\partial \eta}\left(\alpha_l \rho_l \gamma_\eta \sqrt{g}\,\bar{V}_l^2\right) + \frac{\partial}{\partial \zeta}\left(\alpha_l \rho_l \gamma_\zeta \sqrt{g}\,\bar{V}_l^3\right)$$

$$= \gamma_v \sqrt{g}\,\mu_l . \qquad (10.10)$$

Remember again that the time derivative is understood to be at a fixed point in the transformed region. Note that setting $\alpha_l \rho_l \gamma_v = \alpha_l \rho_l \gamma = const$, $\mathbf{V}_l = const$ and $\mu_l = 0$, and using the *first fundamental metric identity* [11]

$$\frac{\partial}{\partial \xi}\left(\sqrt{g}\,\mathbf{a}^1\right) + \frac{\partial}{\partial \eta}\left(\sqrt{g}\,\mathbf{a}^2\right) + \frac{\partial}{\partial \zeta}\left(\sqrt{g}\,\mathbf{a}^3\right) = 0, \qquad (10.11)$$

results in

$$\frac{\partial \sqrt{g}}{\partial \tau} - \frac{\partial}{\partial \xi}\left(\sqrt{g}\,\mathbf{a}^1 \cdot \mathbf{V}_{cs}\right) - \frac{\partial}{\partial \eta}\left(\sqrt{g}\,\mathbf{a}^2 \cdot \mathbf{V}_{cs}\right) - \frac{\partial}{\partial \zeta}\left(\sqrt{g}\,\mathbf{a}^3 \cdot \mathbf{V}_{cs}\right) = 0, \quad (10.12)$$

which is the *second fundamental metric identity* Thompson et al. [12], p.159. As pointed out by *Thompson* et al. [12] this identity should be used to numerically determine updated values of the *Jacobian*, $\sqrt{g}$, instead of updating it directly from the new values of the Cartesian coordinates. In the later case spurious source terms will appear [12].

10.3 Mass conservation equations for components inside the field – conservative form

The conservative form of the local volume- and time-averaged field mass conservation equations for each component inside a velocity field as derived in [4] is

$$\frac{\partial}{\partial \tau}\left(\alpha_l \rho_l C_{il} \gamma_v\right) + \nabla \cdot \left[\alpha_l \rho_l \gamma \left(\mathbf{V}_l C_{il} - D_{il}^* \nabla C_{il}\right)\right] = \gamma_v \mu_{il}. \tag{10.13}$$

The time derivative and the convection term are transformed as in the previous section. For the transformation of the diffusion term we use the *Diffusion Laplacian* in the non-conservative form

$$\sqrt{g}\, \nabla \cdot (\lambda \nabla \varphi)$$

$$= \frac{\partial}{\partial \xi}\left[\sqrt{g}\, \mathbf{a}^1 \cdot \lambda \left(\mathbf{a}^1 \frac{\partial \varphi}{\partial \xi} + \mathbf{a}^2 \frac{\partial \varphi}{\partial \eta} + \mathbf{a}^3 \frac{\partial \varphi}{\partial \zeta}\right)\right]$$

$$+ \frac{\partial}{\partial \eta}\left[\sqrt{g}\, \mathbf{a}^2 \cdot \lambda \left(\mathbf{a}^1 \frac{\partial \varphi}{\partial \xi} + \mathbf{a}^2 \frac{\partial \varphi}{\partial \eta} + \mathbf{a}^3 \frac{\partial \varphi}{\partial \zeta}\right)\right]$$

$$+ \frac{\partial}{\partial \zeta}\left[\sqrt{g}\, \mathbf{a}^3 \cdot \lambda \left(\mathbf{a}^1 \frac{\partial \varphi}{\partial \xi} + \mathbf{a}^2 \frac{\partial \varphi}{\partial \eta} + \mathbf{a}^3 \frac{\partial \varphi}{\partial \zeta}\right)\right], \tag{10.14}$$

where $\varphi = \alpha_l \rho_l \gamma D_{il}^*$ is a scalar valued function of the local flow parameters. With this we obtain the final form which is additionally multiplied by the Jacobian

$$\frac{\partial}{\partial \tau}\left(\alpha_l \rho_l C_{il} \sqrt{g}\, \gamma_v\right)$$

$$+ \frac{\partial}{\partial \xi}\left\{\alpha_l \rho_l \sqrt{g}\, \gamma_\xi \left[C_{il} \overline{V}_l^1 - D_{il}^* \left(\mathbf{a}^1 \cdot \mathbf{a}^1 \frac{\partial C_{il}}{\partial \xi} + \mathbf{a}^1 \cdot \mathbf{a}^2 \frac{\partial C_{il}}{\partial \eta} + \mathbf{a}^1 \cdot \mathbf{a}^3 \frac{\partial C_{il}}{\partial \zeta}\right)\right]\right\}$$

$$+ \frac{\partial}{\partial \eta}\left\{\alpha_l \rho_l \sqrt{g}\, \gamma_\eta \left[C_{il} \overline{V}_l^2 - D_{il}^* \left(\mathbf{a}^2 \cdot \mathbf{a}^1 \frac{\partial C_{il}}{\partial \xi} + \mathbf{a}^2 \cdot \mathbf{a}^2 \frac{\partial C_{il}}{\partial \eta} + \mathbf{a}^2 \cdot \mathbf{a}^3 \frac{\partial C_{il}}{\partial \zeta}\right)\right]\right\}$$

$$+ \frac{\partial}{\partial \zeta}\left\{\alpha_l \rho_l \sqrt{g}\, \gamma_\zeta \left[C_{il} \overline{V}_l^3 - D_{il}^* \left(\mathbf{a}^3 \cdot \mathbf{a}^1 \frac{\partial C_{il}}{\partial \xi} + \mathbf{a}^3 \cdot \mathbf{a}^2 \frac{\partial C_{il}}{\partial \eta} + \mathbf{a}^3 \cdot \mathbf{a}^3 \frac{\partial C_{il}}{\partial \zeta}\right)\right]\right\}$$

$$= \gamma_v \sqrt{g}\, \mu_{il}. \tag{10.15}$$

We see that the scalar

10.3 Mass conservation equations for components inside the field – conservative form

$$\alpha_l \rho_l C_{il} \sqrt{g}$$

with the convective flux

$$\alpha_l \rho_l C_{il} \sqrt{g}\, \mathbf{a}^i \cdot (\mathbf{V}_l - \mathbf{V}_{cs})$$

and a diffusion flux is subject to conservation in the transformed coordinate system. Using the elements of the inverted metric tensor which is symmetric per definition the notation simplifies to

$$\frac{\partial}{\partial \tau}\left(\alpha_l \rho_l C_{il} \sqrt{g}\, \gamma_v\right)$$

$$+ \frac{\partial}{\partial \xi}\left\{\alpha_l \rho_l \sqrt{g}\, \gamma_\xi \left[C_{il}\bar{V}_l^1 - D_{il}^*\left(g^{11}\frac{\partial C_{il}}{\partial \xi} + g^{12}\frac{\partial C_{il}}{\partial \eta} + g^{13}\frac{\partial C_{il}}{\partial \zeta}\right)\right]\right\}$$

$$+ \frac{\partial}{\partial \eta}\left\{\alpha_l \rho_l \sqrt{g}\, \gamma_\eta \left[C_{il}\bar{V}_l^2 - D_{il}^*\left(g^{21}\frac{\partial C_{il}}{\partial \xi} + g^{22}\frac{\partial C_{il}}{\partial \eta} + g^{23}\frac{\partial C_{il}}{\partial \zeta}\right)\right]\right\}$$

$$+ \frac{\partial}{\partial \zeta}\left\{\alpha_l \rho_l \sqrt{g}\, \gamma_\zeta \left[C_{il}\bar{V}_l^3 - D_{il}^*\left(g^{31}\frac{\partial C_{il}}{\partial \xi} + g^{32}\frac{\partial C_{il}}{\partial \eta} + g^{33}\frac{\partial C_{il}}{\partial \zeta}\right)\right]\right\} = \gamma_v \sqrt{g}\, \mu_{il}.$$

(10.16)

Thus the *isotropic* convection-diffusion problem in the physical space turns out to be a *anisotropic* in the transformed space. One immediately recognizes the advantage of the orthogonal coordinate systems for which only the diagonal elements of the inverse matrices are different from zero

$$\frac{\partial}{\partial \tau}\left(\alpha_l \rho_l C_{il} \sqrt{g}\, \gamma_v\right)$$

$$+ \frac{\partial}{\partial \xi}\left[\alpha_l \rho_l \sqrt{g}\, \gamma_\xi \left(C_{il}\bar{V}_l^1 - g^{11}D_{il}^*\frac{\partial C_{il}}{\partial \xi}\right)\right]$$

$$+ \frac{\partial}{\partial \eta}\left[\alpha_l \rho_l \sqrt{g}\, \gamma_\eta \left(C_{il}\bar{V}_l^2 - g^{22}D_{il}^*\frac{\partial C_{il}}{\partial \eta}\right)\right]$$

$$+ \frac{\partial}{\partial \zeta}\left[\alpha_l \rho_l \sqrt{g}\, \gamma_\zeta \left(C_{il}\bar{V}_l^3 - g^{33} D_{il}^* \frac{\partial C_{il}}{\partial \zeta}\right)\right] = \gamma_v \sqrt{g}\, \mu_{il}. \qquad (10.17)$$

10.4 Field mass conservation equations for components inside the field – non-conservative form

The non-conservative form is obtained by differentiating the time derivative and the convection term and comparing with the mass conservation equation. The result is

$$\alpha_l \rho_l \sqrt{g}\left(\gamma_v \frac{\partial C_{il}}{\partial \tau} + \gamma_\xi \bar{V}_l^1 \frac{\partial C_{il}}{\partial \xi} + \gamma_\eta \bar{V}_l^2 \frac{\partial C_{il}}{\partial \eta} + \gamma_\zeta \bar{V}_l^3 \frac{\partial C_{il}}{\partial \zeta}\right)$$

$$- \frac{\partial}{\partial \xi}\left[\alpha_l \rho_l \sqrt{g}\, \gamma_\xi D_{il}^* \left(g^{11} \frac{\partial C_{il}}{\partial \xi} + g^{12} \frac{\partial C_{il}}{\partial \eta} + g^{13} \frac{\partial C_{il}}{\partial \zeta}\right)\right]$$

$$- \frac{\partial}{\partial \eta}\left[\alpha_l \rho_l \sqrt{g}\, \gamma_\eta D_{il}^* \left(g^{21} \frac{\partial C_{il}}{\partial \xi} + g^{22} \frac{\partial C_{il}}{\partial \eta} + g^{23} \frac{\partial C_{il}}{\partial \zeta}\right)\right]$$

$$- \frac{\partial}{\partial \zeta}\left[\alpha_l \rho_l \sqrt{g}\, \gamma_\zeta D_{il}^* \left(g^{31} \frac{\partial C_{il}}{\partial \xi} + g^{32} \frac{\partial C_{il}}{\partial \eta} + g^{33} \frac{\partial C_{il}}{\partial \zeta}\right)\right] = \gamma_v \sqrt{g}\, (\mu_{il} - \mu_l C_{il}).$$

$$(10.18)$$

10.5. Particles number conservation equations for each velocity field

The local volume- and time-averaged particle number density conservation equation for each velocity field derived in [4] is

$$\frac{\partial}{\partial \tau}(n_l \gamma_v) + \nabla \cdot \left[\left(\mathbf{V}_l n_l - \frac{V_l^t}{Sc^t}\nabla n_l\right)\gamma\right] = \gamma_v \left(\dot{n}_{l,kin} - \dot{n}_{l,coal} + \dot{n}_{l,sp}\right). \qquad (10.19)$$

By analogy to the component mass conservation equation derived in the previous section we have

$$\frac{\partial}{\partial \tau}\left(n_l \sqrt{g}\, \gamma_v\right)$$

$$+\frac{\partial}{\partial \xi}\left\{\sqrt{g}\, \gamma_\xi \left[n_l \bar{V}_l^1 - \frac{v_l^t}{Sc^t}\left(g^{11}\frac{\partial n_l}{\partial \xi} + g^{12}\frac{\partial n_l}{\partial \eta} + g^{13}\frac{\partial n_l}{\partial \zeta}\right)\right]\right\}$$

$$+\frac{\partial}{\partial \eta}\left\{\sqrt{g}\, \gamma_\eta \left[n_l \bar{V}_l^2 - \frac{v_l^t}{Sc^t}\left(g^{21}\frac{\partial n_l}{\partial \xi} + g^{22}\frac{\partial n_l}{\partial \eta} + g^{23}\frac{\partial n_l}{\partial \zeta}\right)\right]\right\}$$

$$+\frac{\partial}{\partial \zeta}\left\{\sqrt{g}\, \gamma_\zeta \left[n_l \bar{V}_l^3 - \frac{v_l^t}{Sc^t}\left(g^{31}\frac{\partial n_l}{\partial \xi} + g^{32}\frac{\partial n_l}{\partial \eta} + g^{33}\frac{\partial n_l}{\partial \zeta}\right)\right]\right\}$$

$$= \gamma_v \sqrt{g}\, \left(\dot{n}_{l,kin} - \dot{n}_{l,coal} + \dot{n}_{l,sp}\right). \qquad (10.20)$$

10.6 Field entropy conservation equations – conservative form

The conservative form of the local volume- and time-averaged field entropy conservation equation Eq. (5.101b) as derived in Chapter 5 [4] is

$$\frac{\partial}{\partial \tau}\left(\alpha_l \rho_l s_l \gamma_v\right) + \nabla \cdot \left[\alpha_l^e \rho_l \gamma \left(s_l \mathbf{V}_l - \sum_{i=2}^{i_{max}} (s_{il} - s_{1l}) D_{il}^* \nabla C_{il}\right)\right] - \frac{1}{T_l}\nabla \cdot \left(\alpha_l^e \lambda_l^* \gamma \nabla T_l\right)$$

$$= \gamma_v \left[\frac{1}{T_l}DT_l^N + \sum_{i=1}^{i_{max}} \mu_{il}(s_{il} - s_l) + \sum_{\substack{m=1 \\ m \neq l}}^{3,w} \sum_{i=1}^{i_{max}}(\mu_{iml} - \mu_{ilm})s_{il}\right] \qquad (10.21)$$

for $\alpha_l \geq 0$. The entropy conservation equation in the transformed space is therefore

$$\frac{\partial}{\partial \tau}\left(\alpha_l \rho_l s_l \sqrt{g}\, \gamma_v\right)$$

$$+\frac{\partial}{\partial \xi}\left\{\alpha_l \rho_l \sqrt{g}\, \gamma_\xi \left\{s_l \bar{V}_l^1 - \sum_{i=2}^{i_{max}}\left[(s_{il} - s_{1l})D_{il}^*\left(g^{11}\frac{\partial C_{il}}{\partial \xi} + g^{12}\frac{\partial C_{il}}{\partial \eta} + g^{13}\frac{\partial C_{il}}{\partial \zeta}\right)\right]\right\}\right\}$$

$$+\frac{\partial}{\partial \eta}\left\{\alpha_l \rho_l \sqrt{g}\, \gamma_\eta \left\{ s_l \bar{V}_l^2 - \sum_{i=2}^{i_{max}} \left[(s_{il} - s_{1l}) D_{il}^* \left(g^{21} \frac{\partial C_{il}}{\partial \xi} + g^{22} \frac{\partial C_{il}}{\partial \eta} + g^{23} \frac{\partial C_{il}}{\partial \zeta} \right) \right] \right\} \right\}$$

$$+\frac{\partial}{\partial \zeta}\left\{\alpha_l \rho_l \sqrt{g}\, \gamma_\zeta \left\{ s_l \bar{V}_l^3 - \sum_{i=2}^{i_{max}} \left[(s_{il} - s_{1l}) D_{il}^* \left(g^{31} \frac{\partial C_{il}}{\partial \xi} + g^{32} \frac{\partial C_{il}}{\partial \eta} + g^{33} \frac{\partial C_{il}}{\partial \zeta} \right) \right] \right\} \right\}$$

$$-\frac{1}{T_l}\frac{\partial}{\partial \xi}\left[\alpha_l^e \lambda_l^* \sqrt{g}\, \gamma_\xi \left(g^{11} \frac{\partial T_l}{\partial \xi} + g^{12} \frac{\partial T_l}{\partial \eta} + g^{13} \frac{\partial T_l}{\partial \zeta} \right) \right]$$

$$-\frac{1}{T_l}\frac{\partial}{\partial \eta}\left[\alpha_l^e \lambda_l^* \sqrt{g}\, \gamma_\eta \left(g^{21} \frac{\partial T_l}{\partial \xi} + g^{22} \frac{\partial T_l}{\partial \eta} + g^{23} \frac{\partial T_l}{\partial \zeta} \right) \right]$$

$$-\frac{1}{T_l}\frac{\partial}{\partial \zeta}\left[\alpha_l^e \lambda_l^* \sqrt{g}\, \gamma_\zeta \left(g^{31} \frac{\partial T_l}{\partial \xi} + g^{32} \frac{\partial T_l}{\partial \eta} + g^{33} \frac{\partial T_l}{\partial \zeta} \right) \right]$$

$$= \gamma_v \sqrt{g} \left[\frac{1}{T_l} DT_l^N + \sum_{i=1}^{i_{max}} \mu_{il}(s_{il} - s_l) + \sum_{\substack{m=1 \\ m \neq l}}^{3,w} \sum_{i=1}^{i_{max}} (\mu_{iml} - \mu_{ilm}) s_{il} \right]. \qquad (10.22)$$

10.7 Field entropy conservation equations – non-conservative form

The non-conservative form is obtained by differentiating the time derivative and the convection term and comparing with the mass conservation equation. The result is

$$\alpha_l \rho_l \sqrt{g} \left(\gamma_v \frac{\partial s_l}{\partial \tau} + \gamma_\xi \bar{V}_l^1 \frac{\partial s_l}{\partial \xi} + \gamma_\eta \bar{V}_l^2 \frac{\partial s_l}{\partial \eta} + \gamma_\zeta \bar{V}_l^3 \frac{\partial s_l}{\partial \zeta} \right)$$

$$-\frac{\partial}{\partial \xi}\left\{\alpha_l \rho_l \sqrt{g}\, \gamma_\xi \sum_{i=2}^{i_{max}} \left[(s_{il} - s_{1l}) D_{il}^* \left(g^{11} \frac{\partial C_{il}}{\partial \xi} + g^{12} \frac{\partial C_{il}}{\partial \eta} + g^{13} \frac{\partial C_{il}}{\partial \zeta} \right) \right] \right\}$$

$$-\frac{\partial}{\partial \eta}\left\{\alpha_l \rho_l \sqrt{g}\, \gamma_\eta \sum_{i=2}^{i_{max}} \left[(s_{il} - s_{1l}) D_{il}^* \left(g^{21} \frac{\partial C_{il}}{\partial \xi} + g^{22} \frac{\partial C_{il}}{\partial \eta} + g^{23} \frac{\partial C_{il}}{\partial \zeta} \right) \right] \right\}$$

$$-\frac{\partial}{\partial \zeta}\left\{\alpha_l \rho_l \sqrt{g}\, \gamma_\zeta \sum_{i=2}^{i_{max}}\left[(s_{il}-s_{1l})D_{il}^*\left(g^{31}\frac{\partial C_{il}}{\partial \xi}+g^{32}\frac{\partial C_{il}}{\partial \eta}+g^{33}\frac{\partial C_{il}}{\partial \zeta}\right)\right]\right\}$$

$$-\frac{1}{T_l}\frac{\partial}{\partial \xi}\left[\alpha_l^e \lambda_l^* \sqrt{g}\, \gamma_\xi \left(g^{11}\frac{\partial T_l}{\partial \xi}+g^{12}\frac{\partial T_l}{\partial \eta}+g^{13}\frac{\partial T_l}{\partial \zeta}\right)\right]$$

$$-\frac{1}{T_l}\frac{\partial}{\partial \eta}\left[\alpha_l^e \lambda_l^* \sqrt{g}\, \gamma_\eta \left(g^{21}\frac{\partial T_l}{\partial \xi}+g^{22}\frac{\partial T_l}{\partial \eta}+g^{23}\frac{\partial T_l}{\partial \zeta}\right)\right]$$

$$-\frac{1}{T_l}\frac{\partial}{\partial \zeta}\left[\alpha_l^e \lambda_l^* \sqrt{g}\, \gamma_\zeta \left(g^{31}\frac{\partial T_l}{\partial \xi}+g^{32}\frac{\partial T_l}{\partial \eta}+g^{33}\frac{\partial T_l}{\partial \zeta}\right)\right]$$

$$= \gamma_v \sqrt{g}\left[\frac{1}{T_l}DT_l^N + \sum_{i=1}^{i_{max}}\mu_{il}(s_{il}-s_l)\right]. \quad (10.23)$$

10.8 Irreversible power dissipation caused by the viscous forces

The *irreversible power dissipation caused by the viscous forces due to deformation as a result of the mean velocities in the space* is an important term in the energy conservation. Substituting for the stress tensor components using the *Helmholtz* and *Stokes* hypothesis, the following expression was obtained in Chapter 5 Eq. (5.154), [4]

$$\gamma_v \frac{\alpha_l}{\alpha_l^e}\frac{P_{kl}}{V_l} = 2\left[\gamma_x\left(\frac{\partial u_l}{\partial x}\right)^2 + \gamma_y\left(\frac{\partial v_l}{\partial y}\right)^2 + \gamma_z\left(\frac{\partial w_l}{\partial z}\right)^2\right]$$

$$+\left(\frac{\partial v_l}{\partial x}+\frac{\partial u_l}{\partial y}\right)\left(\gamma_x\frac{\partial v_l}{\partial x}+\gamma_y\frac{\partial u_l}{\partial y}\right)+\left(\frac{\partial w_l}{\partial x}+\frac{\partial u_l}{\partial z}\right)\left(\gamma_x\frac{\partial w_l}{\partial x}+\gamma_z\frac{\partial u_l}{\partial z}\right)$$

$$+\left(\frac{\partial w_l}{\partial y}+\frac{\partial v_l}{\partial z}\right)\left(\gamma_y\frac{\partial w_l}{\partial y}+\gamma_z\frac{\partial v_l}{\partial z}\right)$$

$$-\frac{2}{3}\left(\frac{\partial u_I}{\partial x}+\frac{\partial v_I}{\partial y}+\frac{\partial w_I}{\partial z}\right)\left(\gamma_x\frac{\partial u_I}{\partial r}+\gamma_y\frac{\partial v_I}{\partial y}+\gamma_z\frac{\partial w_I}{\partial z}\right) \quad (10.24)$$

The transformed form is then

$$\gamma_v\frac{\alpha_I}{\alpha_I^e}\frac{P_{kl}}{v_I}=2\begin{bmatrix}\gamma_\xi\left(a^{11}\frac{\partial u_I}{\partial\xi}+a^{21}\frac{\partial u_I}{\partial\eta}+a^{31}\frac{\partial u_I}{\partial\zeta}\right)^2\\[6pt]+\gamma_\eta\left(a^{12}\frac{\partial v_I}{\partial\xi}+a^{22}\frac{\partial v_I}{\partial\eta}+a^{32}\frac{\partial v_I}{\partial\zeta}\right)^2\\[6pt]+\gamma_\zeta\left(a^{13}\frac{\partial w_I}{\partial\xi}+a^{23}\frac{\partial w_I}{\partial\eta}+a^{33}\frac{\partial w_I}{\partial\zeta}\right)^2\end{bmatrix}$$

$$+\begin{pmatrix}a^{11}\frac{\partial v_I}{\partial\xi}+a^{21}\frac{\partial v_I}{\partial\eta}+a^{31}\frac{\partial v_I}{\partial\zeta}\\[4pt]+a^{12}\frac{\partial u_I}{\partial\xi}+a^{22}\frac{\partial u_I}{\partial\eta}+a^{32}\frac{\partial u_I}{\partial\zeta}\end{pmatrix}\begin{pmatrix}\gamma_\xi\left(a^{11}\frac{\partial v_I}{\partial\xi}+a^{21}\frac{\partial v_I}{\partial\eta}+a^{31}\frac{\partial v_I}{\partial\zeta}\right)\\[4pt]+\gamma_\eta\left(a^{12}\frac{\partial u_I}{\partial\xi}+a^{22}\frac{\partial u_I}{\partial\eta}+a^{32}\frac{\partial u_I}{\partial\zeta}\right)\end{pmatrix}$$

$$+\begin{pmatrix}a^{11}\frac{\partial w_I}{\partial\xi}+a^{21}\frac{\partial w_I}{\partial\eta}+a^{31}\frac{\partial w_I}{\partial\zeta}\\[4pt]+a^{13}\frac{\partial u_I}{\partial\xi}+a^{23}\frac{\partial u_I}{\partial\eta}+a^{33}\frac{\partial u_I}{\partial\zeta}\end{pmatrix}\begin{pmatrix}\gamma_\xi\left(a^{11}\frac{\partial w_I}{\partial\xi}+a^{21}\frac{\partial w_I}{\partial\eta}+a^{31}\frac{\partial w_I}{\partial\zeta}\right)\\[4pt]+\gamma_\zeta\left(a^{13}\frac{\partial u_I}{\partial\xi}+a^{23}\frac{\partial u_I}{\partial\eta}+a^{33}\frac{\partial u_I}{\partial\zeta}\right)\end{pmatrix}$$

$$+\begin{pmatrix}a^{12}\frac{\partial w_I}{\partial\xi}+a^{22}\frac{\partial w_I}{\partial\eta}+a^{32}\frac{\partial w_I}{\partial\zeta}\\[4pt]+a^{13}\frac{\partial v_I}{\partial\xi}+a^{23}\frac{\partial v_I}{\partial\eta}+a^{33}\frac{\partial v_I}{\partial\zeta}\end{pmatrix}\begin{pmatrix}\gamma_\eta\left(a^{12}\frac{\partial w_I}{\partial\xi}+a^{22}\frac{\partial w_I}{\partial\eta}+a^{32}\frac{\partial w_I}{\partial\zeta}\right)\\[4pt]+\gamma_\zeta\left(a^{13}\frac{\partial v_I}{\partial\xi}+a^{23}\frac{\partial v_I}{\partial\eta}+a^{33}\frac{\partial v_I}{\partial\zeta}\right)\end{pmatrix}$$

$$\begin{pmatrix} a^{11}\dfrac{\partial u_l}{\partial \xi}+a^{21}\dfrac{\partial u_l}{\partial \eta}+a^{31}\dfrac{\partial u_l}{\partial \zeta} \\ +a^{12}\dfrac{\partial v_l}{\partial \xi}+a^{22}\dfrac{\partial v_l}{\partial \eta}+a^{32}\dfrac{\partial v_l}{\partial \zeta} \\ +a^{13}\dfrac{\partial w_l}{\partial \xi}+a^{23}\dfrac{\partial w_l}{\partial \eta}+a^{33}\dfrac{\partial w_l}{\partial \zeta} \end{pmatrix} \begin{pmatrix} \gamma_\xi \left(a^{11}\dfrac{\partial u_l}{\partial \xi}+a^{21}\dfrac{\partial u_l}{\partial \eta}+a^{31}\dfrac{\partial u_l}{\partial \zeta}\right) \\ +\gamma_\eta \left(a^{12}\dfrac{\partial v_l}{\partial \xi}+a^{22}\dfrac{\partial v_l}{\partial \eta}+a^{32}\dfrac{\partial v_l}{\partial \zeta}\right) \\ +\gamma_\zeta \left(a^{13}\dfrac{\partial w_l}{\partial \xi}+a^{23}\dfrac{\partial w_l}{\partial \eta}+a^{33}\dfrac{\partial w_l}{\partial \zeta}\right) \end{pmatrix} \cdot$$

$$-\dfrac{2}{3} \quad (10.25)$$

10.9 The non-conservative entropy equation in terms of temperature and pressure

The so called non-conservative entropy equation in terms of temperature and pressure was already derived in Chapter 5, Eq. (5.176) [7] in 1997:

$$\rho_l c_{pl}\left[\alpha_l \gamma_v \dfrac{\partial T_l}{\partial \tau}+\left(\alpha_l^e \mathbf{V}_l \gamma \cdot \nabla \right) T_l \right] - \left[1 - \rho_l \left(\dfrac{\partial h_l}{\partial p} \right)_{T_l, all\_C's} \right] \left[\alpha_l \gamma_v \dfrac{\partial p}{\partial \tau}+\left(\alpha_l^e \mathbf{V}_l \gamma \cdot \nabla \right) p \right]$$

$$-\nabla \cdot \left(\alpha_l^e \lambda_l^* \gamma \nabla T \right) + T_l \sum_{i=2}^{i_{max}} \Delta s_{il}^{np} \nabla \left(\alpha_l^e \rho_l D_{il}^* \gamma \nabla C_{il} \right) = \gamma_v \left[DT_l^N - T_l \sum_{i=2}^{i_{max}} \Delta s_{il}^{np} \left(\mu_{il} - C_{il}\mu_l \right) \right].$$

$$(10.26)$$

For the derivation of its counterpart in transformed coordinates we follow the same procedure as described in Chapter 5 for the derivation of Eq. (5.176) or in [7]. In Chapter 3 also in [3] the differential relationship, Eq. (3.106), between the field temperature, T_l, and the field properties (s_l, C_{il}, p) is found to be

$$c_{pl}\dfrac{dT_l}{T_l}-\bar{R}_l\dfrac{dp_l}{p_l}=ds_l-\sum_{i=2}^{i_{max}}\left(\dfrac{\partial s_l}{\partial C_{il}}\right)_{p,T_l,all\_C's\_except\_C_{il}}dC_{il}, \quad (10.27)$$

where

$$\left(\dfrac{\partial s}{\partial C_i}\right)_{p,T,all\_C's\_except\_C_i}=s_{il}-s_{1l}+\Delta s_{il}^{np}, \quad (10.28)$$

$$\frac{\rho_l \bar{R}_l T_l}{p} = \left[1 - \rho_l \left(\frac{\partial h_l}{\partial p}\right)_{T_l, all\_C's}\right]. \qquad (10.29)$$

One of the mass concentrations, arbitrarily numbered with subscript 1, C_{1l}, depends on all others and is computed as all others are known,

$$C_{1l} = 1 - \sum_{i=2}^{i_{max}} C_{il}. \qquad (10.30)$$

Equation (10.28) consists of two parts. For the case of a mixture consisting of ideal fluids the second part is equal to zero,

$$\Delta s_{il}^{np} = 0, \qquad (10.31)$$

this also demonstrating the meaning of the subscript np, which stands for non-perfect fluid. The non-conservative form of the entropy equation in terms of temperature and pressure is obtained by multiplying the $i_{max} - 1$ mass conservation equations (10.18) by $s_{il} - s_{1l} + \Delta s_{il}^{np}$ and subtracting them from Eq. (10.23). The result is simplified by using

$$\sum_{i=1}^{i_{max}} \mu_{il}(s_{il} - s_l) - \sum_{i=2}^{i_{max}} \left(s_{il} - s_{1l} + \Delta s_{il}^{np}\right)\left(\mu_{il} - C_{il}\mu_l\right) = -\sum_{i=2}^{i_{max}} \Delta s_{il}^{np}\left(\mu_{il} - C_{il}\mu_l\right), \qquad (10.32)$$

neglecting the second order terms, and making the same assumption about the difference in the diffusion coefficients as made in [3],

$$\sqrt{g}\, \alpha_l \rho_l c_{pl} \left(\gamma_v \frac{\partial T_l}{\partial \tau} + \gamma_\xi \bar{V}_l^1 \frac{\partial T_l}{\partial \xi} + \gamma_\eta \bar{V}_l^2 \frac{\partial T_l}{\partial \eta} + \gamma_\zeta \bar{V}_l^3 \frac{\partial T_l}{\partial \zeta} \right)$$

$$-\alpha_l \sqrt{g} \left[1 - \rho_l \left(\frac{\partial h_l}{\partial p}\right)_{T_l, all\_C's}\right] \left(\gamma_v \frac{\partial p_l}{\partial \tau} + \gamma_\xi \bar{V}_l^1 \frac{\partial p_l}{\partial \xi} + \gamma_\eta \bar{V}_l^2 \frac{\partial p_l}{\partial \eta} + \gamma_\zeta \bar{V}_l^3 \frac{\partial p_l}{\partial \zeta} \right)$$

$$+T_l \sum_{i=2}^{i_{max}} \left\{ \Delta s_{il}^{np} \left\{ \begin{array}{c} \dfrac{\partial}{\partial \xi}\left[\alpha_l \rho_l \sqrt{g}\, \gamma_\xi D_{il}^*\left(g^{11}\dfrac{\partial C_{il}}{\partial \xi}+g^{12}\dfrac{\partial C_{il}}{\partial \eta}+g^{13}\dfrac{\partial C_{il}}{\partial \zeta}\right)\right] \\[6pt] +\dfrac{\partial}{\partial \eta}\left[\alpha_l \rho_l \sqrt{g}\, \gamma_\eta D_{il}^*\left(g^{21}\dfrac{\partial C_{il}}{\partial \xi}+g^{22}\dfrac{\partial C_{il}}{\partial \eta}+g^{23}\dfrac{\partial C_{il}}{\partial \zeta}\right)\right] \\[6pt] +\dfrac{\partial}{\partial \zeta}\left[\alpha_l \rho_l \sqrt{g}\, \gamma_\zeta D_{il}^*\left(g^{31}\dfrac{\partial C_{il}}{\partial \xi}+g^{32}\dfrac{\partial C_{il}}{\partial \eta}+g^{33}\dfrac{\partial C_{il}}{\partial \zeta}\right)\right] \end{array} \right\} \right.$$

$$-\dfrac{\partial}{\partial \xi}\left[\alpha_l \lambda_l^* \sqrt{g}\, \gamma_\xi \left(g^{11}\dfrac{\partial T_l}{\partial \xi}+g^{12}\dfrac{\partial T_l}{\partial \eta}+g^{13}\dfrac{\partial T_l}{\partial \zeta}\right)\right]$$

$$-\dfrac{\partial}{\partial \eta}\left[\alpha_l \lambda_l^* \sqrt{g}\, \gamma_\eta \left(g^{21}\dfrac{\partial T_l}{\partial \xi}+g^{22}\dfrac{\partial T_l}{\partial \eta}+g^{23}\dfrac{\partial T_l}{\partial \zeta}\right)\right]$$

$$-\dfrac{\partial}{\partial \zeta}\left[\alpha_l \lambda_l^* \sqrt{g}\, \gamma_\zeta \left(g^{31}\dfrac{\partial T_l}{\partial \xi}+g^{32}\dfrac{\partial T_l}{\partial \eta}+g^{33}\dfrac{\partial T_l}{\partial \zeta}\right)\right]$$

$$= \gamma_v \sqrt{g}\left[DT_l^N - T_l \sum_{i=2}^{i_{max}} \Delta s_{il}^{np}\left(\mu_{il}-C_{il}\mu_l\right)\right]. \qquad (10.33)$$

Note that the prescribing of the velocity at the boundary as a boundary condition is associated with pressure decoupling across the interface.

10.10 The volume conservation equation

The so called volume conservation equation (5.188) was already derived in [1] in 1986 and published in [2] in 1987:

$$\frac{\gamma_v}{\rho a^2}\frac{\partial p}{\partial \tau}+\sum_{l=1}^{l_{max}}\frac{\alpha_l}{\rho_l a_l^2}(\mathbf{V}_l \gamma \cdot \nabla)p + \nabla \cdot \sum_{l=1}^{l_{max}}(\alpha_l \mathbf{V}_l \gamma) = \sum_{l=1}^{l_{max}} D\alpha_l - \frac{\partial \gamma_v}{\partial \tau},$$

where

$$D\alpha_l = \frac{1}{\rho_l}\left\{\gamma_v\mu_l - \frac{1}{\rho_l}\left[\left(\frac{\partial\rho_l}{\partial s_l}\right)_{p,all\_C'_{li}s} Ds_l^N + \sum_{i=2}^{i_{max}}\left(\frac{\partial\rho_l}{\partial C_{li}}\right)_{p,s,all\_C'_{li}s\_except\_C_{l1}} DC_{il}^N\right]\right\},$$

(10.34)

a is the *sonic velocity* in a "homogeneous" multi-phase mixture

$$\frac{1}{\rho a^2} = \sum_{l=1}^{3}\frac{\alpha_l}{\rho_l a_l^2} = \frac{1}{p}\sum_{l=1}^{3}\frac{\alpha_l}{\kappa_l} = \frac{1}{\kappa p},$$

(10.35)

and

$$\rho = \sum_{l=1}^{3}\alpha_l\rho_l$$

(10.36)

is the mixture density. It is very useful for designing numerical solution methods. We derive this equation for the transformed system following the same procedure as in Chapter 5 [1,2]. This means we start with the mass conservation for each velocity field, use the chain rule, divide by the field density and add the resulting field equations. The result is

$$\sqrt{g}\sum_{l=1}^{l_{max}}\frac{\alpha_l}{\rho_l}\left(\gamma_v\frac{\partial\rho_l}{\partial\tau} + \gamma_\xi\bar{V}_l^1\frac{\partial\rho_l}{\partial\xi} + \gamma_\eta\bar{V}_l^2\frac{\partial\rho_l}{\partial\eta} + \gamma_\zeta\bar{V}_l^3\frac{\partial\rho_l}{\partial\zeta}\right)$$

$$+\sum_{l=1}^{l_{max}}\left[\frac{\partial}{\partial\xi}\left(\alpha_l\gamma_\xi\sqrt{g}\,\bar{V}_l^1\right) + \frac{\partial}{\partial\eta}\left(\alpha_l\gamma_\eta\sqrt{g}\,\bar{V}_l^2\right) + \frac{\partial}{\partial\zeta}\left(\alpha_l\gamma_\zeta\sqrt{g}\,\bar{V}_l^3\right)\right]$$

$$=\gamma_v\sqrt{g}\sum_{l=1}^{l_{max}}\frac{\mu_l}{\rho_l} - \frac{\partial}{\partial\tau}\left(\gamma_v\sqrt{g}\right).$$

(10.37)

The density derivatives were substituted using the differential form of the equation of state for each velocity field, Eq. (3.157) [3]

$$d\rho = \frac{dp}{a^2} + \left(\frac{\partial\rho}{\partial s}\right)_{p,all\_C's} ds + \sum_{i=2}^{i_{max}}\left(\frac{\partial\rho}{\partial C_i}\right)_{p,s,all\_C's\_except\_C_i} dC_i.$$

Using the following substitutions

$$\alpha_l\rho_l\left(\gamma_\xi\bar{V}_l^1\frac{\partial C_{il}}{\partial\xi} + \gamma_\eta\bar{V}_l^2\frac{\partial C_{il}}{\partial\eta} + \gamma_\zeta\bar{V}_l^3\frac{\partial C_{il}}{\partial\zeta}\right) = DC_{il}^N$$

(10.38)

$$\alpha_l \rho_l \left(\gamma_\xi \overline{V}_l^1 \frac{\partial s_l}{\partial \xi} + \gamma_\eta \overline{V}_l^2 \frac{\partial s_l}{\partial \eta} + \gamma_\zeta \overline{V}_l^3 \frac{\partial s_l}{\partial \zeta} \right) = Ds_l^N \qquad (10.39)$$

we finally obtain

$$\gamma_v \sqrt{g} \frac{1}{\rho a^2} \frac{\partial p}{\partial \tau} + \sqrt{g} \sum_{l=1}^{l_{max}} \frac{\alpha_l}{\rho_l a_l^2} \left(\gamma_\xi \overline{V}_l^1 \frac{\partial p}{\partial \xi} + \gamma_\eta \overline{V}_l^2 \frac{\partial p}{\partial \eta} + \gamma_\zeta \overline{V}_l^3 \frac{\partial p}{\partial \zeta} \right)$$

$$+ \sum_{l=1}^{l_{max}} \left[\frac{\partial}{\partial \xi} \left(\alpha_l \gamma_\xi \sqrt{g} \, \overline{V}_l^1 \right) + \frac{\partial}{\partial \eta} \left(\alpha_l \gamma_\eta \sqrt{g} \, \overline{V}_l^2 \right) + \frac{\partial}{\partial \zeta} \left(\alpha_l \gamma_\zeta \sqrt{g} \, \overline{V}_l^3 \right) \right]$$

$$= \sqrt{g} \sum_{l=1}^{l_{max}} \frac{1}{\rho_l} \left[\gamma_v \mu_l - \frac{1}{\rho_l} \left(\frac{\partial \rho_l}{\partial s_l} Ds_l^N + \sum_{i=2}^{i_{max}} \frac{\partial \rho_l}{\partial C_{il}} DC_{il}^N \right) \right] - \frac{\partial}{\partial \tau} \left(\gamma_v \sqrt{g} \right). \quad (10.40)$$

Note that this equation is not more complicated then its counterpart in Cartesian coordinates and can be used instead of one of the mass conservation equations. The volume conservation equation can be directly discretized and incorporated into the numerical scheme. Another possibility is to follow the same scheme as for deriving it analytically but starting with already discretized mass conservation equations. The coupling finally obtained is then *strictly consistent* with the discretized form of the mass conservation equations.

10.11 The momentum equations

The conservative form of the local volume- and time-averaged field momentum conservation equations (2.232) as derived in Chapter 2 [4] is

$$\frac{\partial}{\partial \tau} \left(\alpha_l \rho_l \mathbf{V}_l \gamma_v \right) + \nabla \cdot \left(\alpha_l^e \rho_l \gamma \left\{ \mathbf{V}_l \mathbf{V}_l - v_l^* \left[\nabla \mathbf{V}_l + (\nabla \mathbf{V}_l)^T - \frac{2}{3} (\nabla \cdot \mathbf{V}_l) \mathbf{I} \right] \right\} \right)$$

$$+ \alpha_l^e \gamma \nabla p + \alpha_l \rho_l \mathbf{g} \gamma_v$$

$$- \gamma_v \sum_{\substack{m=1 \\ m \neq l}}^{3,w} \left\{ \overline{c}_{ml}^d \left| \Delta \mathbf{V}_{ml} \right| \cdot \Delta \mathbf{V}_{ml} + \overline{c}_{ml}^{vm} \left[\frac{\partial}{\partial \tau} \Delta \mathbf{V}_{ml} + (\mathbf{V}_l \cdot \nabla) \Delta \mathbf{V}_{ml} \right] + \overline{c}_{ml}^L (\mathbf{V}_l - \mathbf{V}_m) \times (\nabla \times \mathbf{V}_m) \right\}$$

$$= \gamma_v \sum_{m=1}^{3,w} \left(\mu_{ml} \mathbf{V}_m - \mu_{lm} \mathbf{V}_l \right). \tag{10.41}$$

For the transformation it is more convenient to write the components in each Cartesian direction:

x – direction:

$$\frac{\partial}{\partial \tau}\left(\alpha_l \rho_l u_l \gamma_v\right) + \nabla \cdot \left(\alpha_l^e \rho_l \gamma \left\{ \mathbf{V}_l u_l - v_l^* \left[\nabla u_l + \frac{\partial \mathbf{V}_l}{\partial x} - \frac{2}{3}(\nabla \cdot \mathbf{V}_l) \mathbf{i} \right] \right\} \right)$$

$$+ \alpha_l^e \gamma_x \frac{\partial p}{\partial x} + \alpha_l \rho_l g_x \gamma_v$$

$$- \gamma_v \sum_{\substack{m=1 \\ m \neq l}}^{3,w} \left\{ \begin{array}{l} \overline{c}_{ml}^d \left| \Delta \mathbf{V}_{ml} \right| \Delta u_{ml} + \overline{c}_{ml}^{vm} \left[\frac{\partial \Delta u_{ml}}{\partial \tau} + (\mathbf{V}_l \cdot \nabla) \Delta u_{ml} \right] \\ + \overline{c}_{ml}^L \left[(v_l - v_m) \left(\frac{\partial v_m}{\partial x} - \frac{\partial u_m}{\partial y} \right) + (w_l - w_m) \left(\frac{\partial w_m}{\partial x} - \frac{\partial u_m}{\partial z} \right) \right] \end{array} \right\}$$

$$= \gamma_v \sum_{m=1}^{3,w} \left(\mu_{ml} u_m - \mu_{lm} u_l \right). \tag{10.42}$$

y – direction:

$$\frac{\partial}{\partial \tau}\left(\alpha_l \rho_l v_l \gamma_v\right) + \nabla \cdot \left(\alpha_l^e \rho_l \gamma \left\{ \mathbf{V}_l v_l - v_l^* \left[\nabla v_l + \frac{\partial \mathbf{V}_l}{\partial y} - \frac{2}{3}(\nabla \cdot \mathbf{V}_l) \mathbf{j} \right] \right\} \right)$$

$$+ \alpha_l^e \gamma_y \frac{\partial p}{\partial y} + \alpha_l \rho_l g_y \gamma_v$$

$$- \gamma_v \sum_{\substack{m=1 \\ m \neq l}}^{l_{max},w} \left\{ \begin{array}{l} \overline{c}_{ml}^d \left| \Delta \mathbf{V}_{ml} \right| \Delta v_{ml} + \overline{c}_{ml}^{vm} \left[\frac{\partial \Delta v_{ml}}{\partial \tau} + (\mathbf{V}_l \cdot \nabla) \Delta v_{ml} \right] \\ - \overline{c}_{ml}^L \left[(u_l - u_m) \left(\frac{\partial v_m}{\partial x} - \frac{\partial u_m}{\partial y} \right) - (w_l - w_m) \left(\frac{\partial w_m}{\partial y} - \frac{\partial v_m}{\partial z} \right) \right] \end{array} \right\}$$

$$= \gamma_v \sum_{m=1}^{3,w}\left(\mu_{ml}v_m - \mu_{lm}v_l\right). \tag{10.43}$$

z – *direction*:

$$\frac{\partial}{\partial \tau}\left(\alpha_l \rho_l w_l \gamma_v\right) + \nabla \cdot \left(\alpha_l^e \rho_l \gamma \left\{ \mathbf{V}_l w_l - v_l^*\left[\nabla w_l + \frac{\partial \mathbf{V}_l}{\partial z} - \frac{2}{3}(\nabla \cdot \mathbf{V}_l)\mathbf{k}\right]\right\}\right)$$

$$+\alpha_l^e \gamma_z \frac{\partial p}{\partial z} + \alpha_l \rho_l g_z \gamma_v$$

$$-\gamma_v \sum_{\substack{m=1 \\ m \neq l}}^{l_{\max},w} \left\{ \begin{array}{l} \overline{c}_{ml}^d |\Delta \mathbf{V}_{ml}|\Delta w_{ml} + \overline{c}_{ml}^{vm}\left[\dfrac{\partial \Delta w_{ml}}{\partial \tau} + (\mathbf{V}_l \cdot \nabla)\Delta w_{ml}\right] \\ \\ -\overline{c}_{ml}^L\left[(u_l - u_m)\left(\dfrac{\partial w_m}{\partial x} - \dfrac{\partial u_m}{\partial z}\right) + (v_l - v_m)\left(\dfrac{\partial w_m}{\partial y} - \dfrac{\partial v_m}{\partial z}\right)\right] \end{array} \right\}$$

$$= \gamma_v \sum_{m=1}^{3,w}\left(\mu_{ml}w_m - \mu_{lm}w_l\right). \tag{10.44}$$

We see that the velocity components can be treated as scalars, which are subject to advection and diffusion as any other scalar variable describing the flow.

First we transfer the bulk viscosity term

$$\nabla \cdot \left(\alpha_l^e \rho_l \gamma v_l^* \frac{2}{3}(\nabla \cdot \mathbf{V}_l)\mathbf{i}\right),$$

because it appears in similar form in all the momentum equations. Using Eqs. (10.4) and (10.11) we find that the divergence of the field velocity is the number

$$\nabla \cdot \mathbf{V}_l = \mathbf{a}^1 \cdot \frac{\partial \mathbf{V}_l}{\partial \xi} + \mathbf{a}^2 \cdot \frac{\partial \mathbf{V}_l}{\partial \eta} + \mathbf{a}^3 \cdot \frac{\partial \mathbf{V}_l}{\partial \zeta}. \tag{10.45}$$

Using Eq. (10.4) we then find that the divergence of the vector is

$$\sqrt{g}\,\nabla\left[\alpha_l^e \rho_l \gamma v_l^* \frac{2}{3}(\nabla \cdot \mathbf{V}_l)\mathbf{i}\right]$$

$$= \frac{\partial}{\partial \xi}\left[\sqrt{g}\,\mathbf{a}^1 \cdot i\gamma_\xi \alpha_l^e \rho_l v_l^* \frac{2}{3}(\nabla \cdot \mathbf{V}_l)\right] + \frac{\partial}{\partial \eta}\left[\sqrt{g}\,\mathbf{a}^1 \cdot i\gamma_\eta \alpha_l^e \rho_l v_l^* \frac{2}{3}(\nabla \cdot \mathbf{V}_l)\right]$$

$$+ \frac{\partial}{\partial \zeta}\left[\sqrt{g}\,\mathbf{a}^1 \cdot i\gamma_\zeta \alpha_l^e \rho_l v_l^* \frac{2}{3}(\nabla \cdot \mathbf{V}_l)\right] \tag{10.46}$$

The term

$$\nabla \cdot \left(\alpha_l^e \rho_l \gamma v_l^* \frac{\partial \mathbf{V}_l}{\partial x}\right)$$

presents the divergence of a vector. Again using Eq. (10.4) we obtain

$$\sqrt{g}\,\nabla \cdot \left(\alpha_l^e \rho_l \gamma v_l^* \frac{\partial \mathbf{V}_l}{\partial x}\right)$$

$$= \frac{\partial}{\partial \xi}\left(\sqrt{g}\,\alpha_l^e \rho_l \gamma_\xi v_l^* \mathbf{a}^1 \cdot \frac{\partial \mathbf{V}_l}{\partial x}\right) + \frac{\partial}{\partial \eta}\left(\sqrt{g}\,\alpha_l^e \rho_l \gamma_\eta v_l^* \mathbf{a}^2 \cdot \frac{\partial \mathbf{V}_l}{\partial x}\right) + \frac{\partial}{\partial \zeta}\left(\sqrt{g}\,\alpha_l^e \rho_l \gamma_\zeta v_l^* \mathbf{a}^3 \cdot \frac{\partial \mathbf{V}_l}{\partial x}\right). \tag{10.47}$$

The component notation is

$$\mathbf{a}^1 \cdot \frac{\partial \mathbf{V}}{\partial x} = a^{11}\left(\frac{\partial u}{\partial \xi}a^{11} + \frac{\partial u}{\partial \eta}a^{21} + \frac{\partial u}{\partial \zeta}a^{31}\right) + a^{12}\left(\frac{\partial v}{\partial \xi}a^{11} + \frac{\partial v}{\partial \eta}a^{21} + \frac{\partial v}{\partial \zeta}a^{31}\right)$$

$$+ a^{13}\left(\frac{\partial w}{\partial \xi}a^{11} + \frac{\partial w}{\partial \eta}a^{21} + \frac{\partial w}{\partial \zeta}a^{31}\right) \tag{10.48}$$

$$\mathbf{a}^2 \cdot \frac{\partial \mathbf{V}}{\partial y} = a^{21}\left(\frac{\partial u}{\partial \xi}a^{12} + \frac{\partial u}{\partial \eta}a^{22} + \frac{\partial u}{\partial \zeta}a^{32}\right) + a^{22}\left(\frac{\partial v}{\partial \xi}a^{12} + \frac{\partial v}{\partial \eta}a^{22} + \frac{\partial v}{\partial \zeta}a^{32}\right)$$

$$+ a^{23}\left(\frac{\partial w}{\partial \xi}a^{12} + \frac{\partial w}{\partial \eta}a^{22} + \frac{\partial w}{\partial \zeta}a^{32}\right) \tag{10.49}$$

$$\mathbf{a}^3 \cdot \frac{\partial \mathbf{V}}{\partial z} = a^{31}\left(\frac{\partial u}{\partial \xi}a^{13} + \frac{\partial u}{\partial \eta}a^{23} + \frac{\partial u}{\partial \zeta}a^{33}\right) + a^{32}\left(\frac{\partial v}{\partial \xi}a^{13} + \frac{\partial v}{\partial \eta}a^{23} + \frac{\partial v}{\partial \zeta}a^{33}\right)$$

$$+ a^{33}\left(\frac{\partial w}{\partial \xi}a^{13} + \frac{\partial w}{\partial \eta}a^{23} + \frac{\partial w}{\partial \zeta}a^{33}\right) \tag{10.50}$$

with the second superscript indicating the Cartesian component of the contravariant vectors.

The interfacial virtual mass term is transformed as follows

$$\left(\frac{\partial \varphi}{\partial \tau}\right)_{x,y,z} + \mathbf{V}_I \cdot \nabla \varphi = \left(\frac{\partial \varphi}{\partial \tau}\right)_{\xi,\eta,\zeta} + V_I^1 \frac{\partial \varphi}{\partial \xi} + V_I^2 \frac{\partial \varphi}{\partial \eta} + V_I^3 \frac{\partial \varphi}{\partial \zeta},$$ (10.51)

after using the Eqs. (10.2), (10.5) and (10.11). In a similar way the virtual mass term for the wall-field force is transformed. Note that in the moving coordinate system we have

$$\Delta \mathbf{V}_{wl} = \mathbf{V}_{cs} - \mathbf{V}_I.$$ (10.52)

The conservative form of the transformed equations is given below. Note that the equations are still Cartesian components of the vector momentum equation. The component equations are multiplied by $\sqrt{g}$.

x – direction:

$$\frac{\partial}{\partial \tau}\left(\alpha_1 \rho_1 u_I \sqrt{g}\, \gamma_v\right)$$

$$+ \frac{\partial}{\partial \xi}\left\{\alpha_1^e \rho_1 \sqrt{g}\, \gamma_\xi \left[u_I \overline{V}_I^1 - v_I^* \left(\begin{array}{c} g^{11} \dfrac{\partial u_I}{\partial \xi} + g^{12} \dfrac{\partial u_I}{\partial \eta} + g^{13} \dfrac{\partial u_I}{\partial \zeta} \\ + \mathbf{a}^1 \cdot \dfrac{\partial \mathbf{V}_I}{\partial x} - \dfrac{2}{3} a^{11} \left(\nabla \cdot \mathbf{V}_I\right) \end{array}\right)\right]\right\}$$

$$+ \frac{\partial}{\partial \eta}\left\{\alpha_1^e \rho_1 \sqrt{g}\, \gamma_\eta \left[u_I \overline{V}_I^2 - v_I^* \left(\begin{array}{c} g^{21} \dfrac{\partial u_I}{\partial \xi} + g^{22} \dfrac{\partial u_I}{\partial \eta} + g^{23} \dfrac{\partial u_I}{\partial \zeta} \\ + \mathbf{a}^2 \cdot \dfrac{\partial \mathbf{V}_I}{\partial x} - \dfrac{2}{3} a^{21} \left(\nabla \cdot \mathbf{V}_I\right) \end{array}\right)\right]\right\}$$

$$+\frac{\partial}{\partial \zeta}\left\{\alpha_l^e \rho_l \sqrt{g}\, \gamma_\zeta \left[u_l \overline{V}_l^3 - v_l^* \begin{pmatrix} g^{31}\dfrac{\partial u_l}{\partial \xi} + g^{32}\dfrac{\partial u_l}{\partial \eta} + g^{33}\dfrac{\partial u_l}{\partial \zeta} \\ +\mathbf{a}^3 \cdot \dfrac{\partial \mathbf{V}_l}{\partial x} - \dfrac{2}{3}a^{31}(\nabla \cdot \mathbf{V}_l) \end{pmatrix}\right]\right\}$$

$$+\sqrt{g}\,\alpha_l^e \left(a^{11}\gamma_\xi \frac{\partial p}{\partial \xi} + a^{21}\gamma_\eta \frac{\partial p}{\partial \eta} + a^{31}\gamma_\zeta \frac{\partial p}{\partial \zeta}\right)$$

$$-\gamma_v \sqrt{g} \sum_{\substack{m=1 \\ m \neq l}}^{3,w} \left\{ \overline{c}_{ml}^{vm} \left(\frac{\partial \Delta u_{ml}}{\partial \tau} + V_l^1 \frac{\partial \Delta u_{ml}}{\partial \xi} + V_l^2 \frac{\partial u_{ml}}{\partial \eta} + V_l^3 \frac{\partial \Delta u_{ml}}{\partial \zeta} \right) + \overline{c}_{ml}^d \left|\Delta \mathbf{V}_{ml}\right| \Delta u_{ml} \right\}$$

$$-\gamma_v \sqrt{g} \sum_{\substack{m=1 \\ m \neq l}}^{3,w} \left\{ \overline{c}_{ml}^L \begin{bmatrix} (v_l - v_m)\begin{pmatrix} a^{11}\dfrac{\partial v_m}{\partial \xi} + a^{21}\dfrac{\partial v_m}{\partial \eta} + a^{31}\dfrac{\partial v_m}{\partial \zeta} \\ -a^{12}\dfrac{\partial u_m}{\partial \xi} - a^{22}\dfrac{\partial u_m}{\partial \eta} - a^{32}\dfrac{\partial u_m}{\partial \zeta} \end{pmatrix} \\ +(w_l - w_m)\begin{pmatrix} a^{11}\dfrac{\partial w_m}{\partial \xi} + a^{21}\dfrac{\partial w_m}{\partial \eta} + a^{31}\dfrac{\partial w_m}{\partial \zeta} \\ -a^{13}\dfrac{\partial u_m}{\partial \xi} - a^{23}\dfrac{\partial u_m}{\partial \eta} - a^{33}\dfrac{\partial u_m}{\partial \zeta} \end{pmatrix} \end{bmatrix} \right\}$$

$$= \gamma_v \sqrt{g}\left[-\alpha_l \rho_l g_x + \sum_{m=1}^{3,w}\left(\mu_{ml} u_m - \mu_{lm} u_l\right) \right], \tag{10.53}$$

10.11 The momentum equations

y – direction:

$$\frac{\partial}{\partial \tau}\left(\alpha_l \rho_l v_l \sqrt{g}\, \gamma_v\right)$$

$$+\frac{\partial}{\partial \xi}\left\{\alpha_l^e \rho_l \sqrt{g}\, \gamma_\xi \left[v_l \overline{V}_l^1 - v_l^* \left(\begin{array}{c} g^{11}\dfrac{\partial v_l}{\partial \xi} + g^{12}\dfrac{\partial v_l}{\partial \eta} + g^{13}\dfrac{\partial v_l}{\partial \zeta} \\ \\ +\mathbf{a}^1\cdot\dfrac{\partial \mathbf{V}_l}{\partial y} - \dfrac{2}{3}a^{12}\left(\nabla\cdot\mathbf{V}_l\right) \end{array}\right)\right]\right\}$$

$$+\frac{\partial}{\partial \eta}\left\{\alpha_l^e \rho_l \sqrt{g}\, \gamma_\eta \left[v_l \overline{V}_l^2 - v_l^* \left(\begin{array}{c} g^{21}\dfrac{\partial v_l}{\partial \xi} + g^{22}\dfrac{\partial v_l}{\partial \eta} + g^{23}\dfrac{\partial v_l}{\partial \zeta} \\ \\ +\mathbf{a}^2\cdot\dfrac{\partial \mathbf{V}_l}{\partial y} - \dfrac{2}{3}a^{22}\left(\nabla\cdot\mathbf{V}_l\right) \end{array}\right)\right]\right\}$$

$$+\frac{\partial}{\partial \zeta}\left\{\alpha_l^e \rho_l \sqrt{g}\, \gamma_\zeta \left[v_l \overline{V}_l^3 - v_l^* \left(\begin{array}{c} g^{31}\dfrac{\partial v_l}{\partial \xi} + g^{32}\dfrac{\partial v_l}{\partial \eta} + g^{33}\dfrac{\partial v_l}{\partial \zeta} \\ \\ +\mathbf{a}^3\cdot\dfrac{\partial \mathbf{V}_l}{\partial y} - \dfrac{2}{3}a^{32}\left(\nabla\cdot\mathbf{V}_l\right) \end{array}\right)\right]\right\}$$

$$+\sqrt{g}\,\alpha_l^e\left(a^{12}\gamma_\xi \frac{\partial p}{\partial \xi} + a^{22}\gamma_\eta \frac{\partial p}{\partial \eta} + a^{32}\gamma_\zeta \frac{\partial p}{\partial \zeta}\right)$$

$$-\gamma_v \sqrt{g}\sum_{\substack{m=1 \\ m\neq l}}^{3,w}\left\{\overline{c}_{ml}^{vm}\left(\frac{\partial \Delta v_{ml}}{\partial \tau} + V_l^1 \frac{\partial \Delta v_{ml}}{\partial \xi} + V_l^2 \frac{\partial \Delta v_{ml}}{\partial \eta} + V_l^3 \frac{\partial \Delta v_{ml}}{\partial \zeta}\right) + \overline{c}_{ml}^d \left|\Delta \mathbf{V}_{ml}\right| \Delta v_{ml}\right\}$$

$$-\gamma_v\sqrt{g}\sum_{\substack{m=1\\m\neq l}}^{3,w}\left\{\bar{c}_{ml}^L\left[\begin{array}{l}-(u_l-u_m)\left(a^{11}\dfrac{\partial v_m}{\partial\xi}+a^{21}\dfrac{\partial v_m}{\partial\eta}+a^{31}\dfrac{\partial v_m}{\partial\zeta}\right.\\ \left.-a^{12}\dfrac{\partial u_m}{\partial\xi}-a^{22}\dfrac{\partial u_m}{\partial\eta}-a^{32}\dfrac{\partial u_m}{\partial\zeta}\right)\\ +(w_l-w_m)\left(a^{12}\dfrac{\partial w_m}{\partial\xi}+a^{22}\dfrac{\partial w_m}{\partial\eta}+a^{32}\dfrac{\partial w_m}{\partial\zeta}\right.\\ \left.-a^{13}\dfrac{\partial v_m}{\partial\xi}-a^{23}\dfrac{\partial v_m}{\partial\eta}-a^{33}\dfrac{\partial v_m}{\partial\zeta}\right)\end{array}\right]\right\}$$

$$=\gamma_v\sqrt{g}\left[-\alpha_l\rho_l g_y+\sum_{m=1}^{3,w}\left(\mu_{ml}v_m-\mu_{lm}v_l\right)\right],\qquad(10.54)$$

$z-direction$:

$$\dfrac{\partial}{\partial\tau}\left(\alpha_l\rho_l w_l\sqrt{g}\,\gamma_v\right)$$

$$+\dfrac{\partial}{\partial\xi}\left\{\alpha_l^e\rho_l\sqrt{g}\,\gamma_\xi\left[w_l\overline{V}_l^1-v_l^*\begin{array}{l}\left(g^{11}\dfrac{\partial w_l}{\partial\xi}+g^{12}\dfrac{\partial w_l}{\partial\eta}+g^{13}\dfrac{\partial w_l}{\partial\zeta}\right.\\ \left.+\mathbf{a}^1\cdot\dfrac{\partial\mathbf{V}_l}{\partial z}-\dfrac{2}{3}a^{13}(\nabla\cdot\mathbf{V}_l)\right)\end{array}\right]\right\}$$

$$+\dfrac{\partial}{\partial\eta}\left\{\alpha_l^e\rho_l\sqrt{g}\,\gamma_\eta\left[w_l\overline{V}_l^2-v_l^*\begin{array}{l}\left(g^{21}\dfrac{\partial w_l}{\partial\xi}+g^{22}\dfrac{\partial w_l}{\partial\eta}+g^{23}\dfrac{\partial w_l}{\partial\zeta}\right.\\ \left.+\mathbf{a}^2\cdot\dfrac{\partial\mathbf{V}_l}{\partial z}-\dfrac{2}{3}a^{23}(\nabla\cdot\mathbf{V}_l)\right)\end{array}\right]\right\}$$

$$
\begin{aligned}
&+\frac{\partial}{\partial \zeta}\left\{\alpha_l^e \rho_l \sqrt{g}\, \gamma_\zeta \left[w_l \overline{V}_l^3 - v_l^* \left(\begin{array}{l} g^{31}\dfrac{\partial w_l}{\partial \xi} + g^{32}\dfrac{\partial w_l}{\partial \eta} + g^{33}\dfrac{\partial w_l}{\partial \zeta} \\[4pt] +\mathbf{a}^3 \cdot \dfrac{\partial \mathbf{V}_l}{\partial y} - \dfrac{2}{3}a^{33}(\nabla \cdot \mathbf{V}_l) \end{array} \right) \right] \right\} \\[6pt]
&+\sqrt{g}\,\alpha_l^e \left(a^{13}\gamma_\xi \frac{\partial p}{\partial \xi} + a^{23}\gamma_\eta \frac{\partial p}{\partial \eta} + a^{33}\gamma_\zeta \frac{\partial p}{\partial \zeta} \right) \\[6pt]
&-\gamma_v \sqrt{g}\, \sum_{\substack{m=1 \\ m\neq l}}^{3,w} \left\{ \overline{c}_{ml}^{vm}\left(\frac{\partial \Delta w_{ml}}{\partial \tau} + V_l^1 \frac{\partial \Delta w_{ml}}{\partial \xi} + V_l^2 \frac{\partial \Delta w_{ml}}{\partial \eta} + V_l^3 \frac{\partial \Delta w_{ml}}{\partial \zeta}\right) + \overline{c}_{ml}^d |\Delta \mathbf{V}_{ml}|\Delta w_{ml} \right\} \\[6pt]
&-\gamma_v \sqrt{g}\, \sum_{\substack{m=1 \\ m\neq l}}^{3,w} \left\{ \overline{c}_{ml}^L \left[\begin{array}{l} -(u_l - u_m)\left(a^{11}\dfrac{\partial w_m}{\partial \xi} + a^{21}\dfrac{\partial w_m}{\partial \eta} + a^{31}\dfrac{\partial w_m}{\partial \zeta} \right. \\[3pt] \left. \qquad -a^{13}\dfrac{\partial u_m}{\partial \xi} - a^{23}\dfrac{\partial u_m}{\partial \eta} - a^{33}\dfrac{\partial u_m}{\partial \zeta} \right) \\[6pt] -(v_l - v_m)\left(a^{12}\dfrac{\partial w_m}{\partial \xi} + a^{22}\dfrac{\partial w_m}{\partial \eta} + a^{32}\dfrac{\partial w_m}{\partial \zeta} \right. \\[3pt] \left. \qquad -a^{13}\dfrac{\partial v_m}{\partial \xi} - a^{23}\dfrac{\partial v_m}{\partial \eta} - a^{33}\dfrac{\partial v_m}{\partial \zeta} \right) \end{array} \right] \right\} \\[6pt]
&= \gamma_v \sqrt{g}\, \left[-\alpha_l \rho_l g_z + \sum_{m=1}^{3,w}(\mu_{ml}w_m - \mu_{lm}w_l) \right]. \tag{10.55}
\end{aligned}
$$

10.12 The flux concept, conservative and semi-conservative forms

The purpose of this section is to introduce the so called *flux concept*. Within the flux concept the integration over a control volume gives simple balance expressions which are very convenient for constructing of numerical algorithms.

10.12.1 Mass conservation equation

The conservative form of the mass conservation equation for each species i inside the velocity field l is

$$\frac{\partial}{\partial \tau}\left(\alpha_l \rho_l C_{il} \sqrt{g}\, \gamma_v\right) + \frac{\partial}{\partial \xi}\left(\gamma_\xi \sqrt{g}\, \mathbf{a}^1 \cdot \mathbf{G}_{il}^C\right) + \frac{\partial}{\partial \eta}\left(\gamma_\eta \sqrt{g}\, \mathbf{a}^2 \cdot \mathbf{G}_{il}^C\right)$$

$$+ \frac{\partial}{\partial \zeta}\left(\gamma_\zeta \sqrt{g}\, \mathbf{a}^3 \cdot \mathbf{G}_{il}^C\right) = \gamma_v \sqrt{g}\, \mu_{il}, \qquad (10.56)$$

where the species mass flow rate vector defined as follows

$$\mathbf{G}_{il}^C = \alpha_l \rho_l \left(\mathbf{V}_l - \mathbf{V}_{cs}\right) C_{il} - \alpha_l \rho_l D_{il}^* \left(\mathbf{a}^1 \frac{\partial C_{il}}{\partial \xi} + \mathbf{a}^2 \frac{\partial C_{il}}{\partial \eta} + \mathbf{a}^3 \frac{\partial C_{il}}{\partial \zeta}\right) = \mathbf{G}_l + \mathbf{F}_{il}^C,$$

$$(10.57)$$

consists of a convective

$$\mathbf{G}_l = \alpha_l \rho_l \left(\mathbf{V}_l - \mathbf{V}_{cs}\right) \qquad (10.58)$$

and of a diffusion

$$\mathbf{F}_{il}^C = -\alpha_l \rho_l D_{il}^* \left(\mathbf{a}^1 \frac{\partial C_{il}}{\partial \xi} + \mathbf{a}^2 \frac{\partial C_{il}}{\partial \eta} + \mathbf{a}^3 \frac{\partial C_{il}}{\partial \zeta}\right) \qquad (10.59)$$

component. The minus sign reflects the observation that the positive diffusion mass flow rate happens towards the decreasing concentrations. Note that for $C_{il} = 1$ we have $\mathbf{G}_{il}^C = \mathbf{G}_l$. The corresponding mass conservation equation for each velocity field is

$$\frac{\partial}{\partial \tau}\left(\alpha_l \rho_l \sqrt{g}\, \gamma_v\right) + \frac{\partial}{\partial \xi}\left(\gamma_\xi \sqrt{g}\, \mathbf{a}^1 \cdot \mathbf{G}_l\right) + \frac{\partial}{\partial \eta}\left(\gamma_\eta \sqrt{g}\, \mathbf{a}^2 \cdot \mathbf{G}_l\right)$$

$$+ \frac{\partial}{\partial \zeta}\left(\gamma_\zeta \sqrt{g}\, \mathbf{a}^3 \cdot \mathbf{G}_l\right) = \gamma_v \sqrt{g}\, \mu_{il}. \qquad (10.60)$$

We multiply Eq. (10.60) by the concentration and subtract the resulting equation from Eq. (10.56). Then the field mass source term is split in two non-negative parts $\mu_l = \mu_l^+ + \mu_l^-$. The result is the so called semi-conservative form of the species mass conservation equation

$$\sqrt{g}\left[\alpha_l \rho_l \gamma_v \frac{\partial C_{il}}{\partial \tau}+\left(\gamma_\xi \mathbf{a}^1 \frac{\partial C_{il}}{\partial \xi}+\gamma_\eta \mathbf{a}^2 \frac{\partial C_{il}}{\partial \eta}+\gamma_\zeta \mathbf{a}^3 \frac{\partial C_{il}}{\partial \zeta}\right)\cdot \mathbf{G}_l\right]$$

$$+\frac{\partial}{\partial \xi}\left(\gamma_\xi \sqrt{g}\, \mathbf{a}^1 \cdot \mathbf{F}_{il}^C\right)+\frac{\partial}{\partial \eta}\left(\gamma_\eta \sqrt{g}\, \mathbf{a}^2 \cdot \mathbf{F}_{il}^C\right)+\frac{\partial}{\partial \zeta}\left(\gamma_\zeta \sqrt{g}\, \mathbf{a}^3 \cdot \mathbf{F}_{il}^C\right)$$

$$+\gamma_v \sqrt{g}\, \mu_l^+ C_{il} = \gamma_v \sqrt{g}\left(\mu_{il} - C_{il}\mu_l^-\right). \tag{10.61}$$

We will keep in mind the procedure used to derive this equation which is simpler than the initial equation (10.56) because in the convection part it does not contain the derivatives of the contravariant vectors. By designing numerical methods for Eq. (10.61) we will follow the same procedure but applied to the discretized couple of equations (10.56) and (10.60).

10.12.2 Entropy equation

The flux notation of the entropy equation is

$$\sqrt{g}\left[\alpha_l \rho_l \gamma_v \frac{\partial s_l}{\partial \tau}+\left(\gamma_\xi \mathbf{a}^1 \frac{\partial s_l}{\partial \xi}+\gamma_\eta \mathbf{a}^2 \frac{\partial s_l}{\partial \eta}+\gamma_\zeta \mathbf{a}^3 \frac{\partial s_l}{\partial \zeta}\right)\cdot \mathbf{G}_l\right]$$

$$+\frac{1}{T}\left[\frac{\partial}{\partial \xi}\left(\gamma_\xi \sqrt{g}\, \mathbf{a}^1 \cdot \mathbf{F}_l^T\right)+\frac{\partial}{\partial \eta}\left(\gamma_\eta \sqrt{g}\, \mathbf{a}^2 \cdot \mathbf{F}_l^T\right)+\frac{\partial}{\partial \zeta}\left(\gamma_\zeta \sqrt{g}\, \mathbf{a}^3 \cdot \mathbf{F}_l^T\right)\right]$$

$$+\frac{\partial}{\partial \xi}\left(\gamma_\xi \sqrt{g}\, \mathbf{a}^1 \cdot \mathbf{F}_l^{\Delta sC}\right)+\frac{\partial}{\partial \eta}\left(\gamma_\eta \sqrt{g}\, \mathbf{a}^2 \cdot \mathbf{F}_l^{\Delta sC}\right)+\frac{\partial}{\partial \zeta}\left(\gamma_\zeta \sqrt{g}\, \mathbf{a}^3 \cdot \mathbf{F}_l^{\Delta sC}\right)$$

$$+\gamma_v \sqrt{g}\, \mu_l^+ s_l = \gamma_v \sqrt{g}\, Ds_l. \tag{10.62}$$

We see that two diffusion fluxes additionally appear, the heat flux

$$\mathbf{F}_l^T = -\alpha_l \lambda_l^*\left(\mathbf{a}^1 \frac{\partial T_l}{\partial \xi}+\mathbf{a}^2 \frac{\partial T_l}{\partial \eta}+\mathbf{a}^3 \frac{\partial T_l}{\partial \zeta}\right), \tag{10.63}$$

and the entropy flux due to diffusion of species with different thermal properties

$$\mathbf{F}_l^{\Delta sC} = \sum_{i=2}^{i_{max}}(s_{il} - s_{1l})\mathbf{F}_{il}^C. \tag{10.64}$$

10.12.3 Temperature equation

The flux notation of the temperature equation is

$$\sqrt{g}\, c_{pl} \left[\alpha_l \rho_l \gamma_v \frac{\partial T_l}{\partial \tau} + \left(\gamma_\xi \mathbf{a}^1 \frac{\partial T_l}{\partial \xi} + \gamma_\eta \mathbf{a}^2 \frac{\partial T_l}{\partial \eta} + \gamma_\zeta \mathbf{a}^3 \frac{\partial T_l}{\partial \zeta} \right) \cdot \mathbf{G}_l \right]$$

$$-\alpha_l \sqrt{g} \left[1 - \rho_l \left(\frac{\partial h_l}{\partial p} \right)_{T_l, all\_C's} \right] \left[\gamma_v \frac{\partial p_l}{\partial \tau} + \left(\gamma_\xi \mathbf{a}^1 \frac{\partial p_l}{\partial \xi} + \gamma_\eta \mathbf{a}^2 \frac{\partial p_l}{\partial \eta} + \gamma_\zeta \mathbf{a}^3 \frac{\partial p_l}{\partial \zeta} \right) \cdot \mathbf{G}_l \frac{1}{\rho_l} \right]$$

$$+ \frac{\partial}{\partial \xi} \left(\gamma_\xi \sqrt{g}\, \mathbf{a}^1 \cdot \mathbf{F}_l^T \right) + \frac{\partial}{\partial \eta} \left(\gamma_\eta \sqrt{g}\, \mathbf{a}^2 \cdot \mathbf{F}_l^T \right) + \frac{\partial}{\partial \zeta} \left(\gamma_\zeta \sqrt{g}\, \mathbf{a}^3 \cdot \mathbf{F}_l^T \right)$$

$$- T_l \sum_{i=2}^{i_{max}} \left\{ \Delta s_{il}^{np} \left(\frac{\partial}{\partial \xi} \left(\gamma_\xi \sqrt{g}\, \mathbf{a}^1 \cdot \mathbf{F}_{il}^C \right) + \frac{\partial}{\partial \eta} \left(\gamma_\eta \sqrt{g}\, \mathbf{a}^2 \cdot \mathbf{F}_{il}^C \right) + \frac{\partial}{\partial \zeta} \left(\gamma_\zeta \sqrt{g}\, \mathbf{a}^3 \cdot \mathbf{F}_{il}^C \right) \right) \right\}$$

$$= \gamma_v \sqrt{g} \left[DT_l^N - T_l \sum_{i=2}^{i_{max}} \Delta s_{il}^{np} \left(\mu_{il} - C_{il} \mu_l \right) \right]. \tag{10.65}$$

The flux notation of the particles number density equation is

$$\frac{\partial}{\partial \tau} \left(n_l \sqrt{g}\, \gamma_v \right) + \frac{\partial}{\partial \xi} \left[\sqrt{g}\, \gamma_\xi n_l \mathbf{a}^1 \cdot (\mathbf{V}_l - \mathbf{V}_{cs}) \right] + \frac{\partial}{\partial \eta} \left[\sqrt{g}\, \gamma_\eta n_l \mathbf{a}^2 \cdot (\mathbf{V}_l - \mathbf{V}_{cs}) \right]$$

$$+ \frac{\partial}{\partial \zeta} \left[\sqrt{g}\, \gamma_\zeta n_l \mathbf{a}^3 \cdot (\mathbf{V}_l - \mathbf{V}_{cs}) \right]$$

$$+ \frac{\partial}{\partial \eta} \left(\sqrt{g}\, \gamma_\eta \mathbf{F}_l^n \right) + \frac{\partial}{\partial \xi} \left(\sqrt{g}\, \gamma_\xi \mathbf{F}_l^n \right) + \frac{\partial}{\partial \zeta} \left(\sqrt{g}\, \gamma_\zeta \mathbf{F}_l^n \right)$$

$$= \gamma_v \sqrt{g} \left(\dot{n}_{l,kin} - \dot{n}_{l,coal} + \dot{n}_{l,sp} \right), \tag{10.66}$$

where the turbulent diffusion flux of particles is defined as follows

$$\mathbf{F}_l^n = -\frac{v_l^t}{Sc^t} \left(\mathbf{a}^1 \frac{\partial n_l}{\partial \xi} + \mathbf{a}^2 \frac{\partial n_l}{\partial \eta} + \mathbf{a}^3 \frac{\partial n_l}{\partial \zeta} \right). \tag{10.67}$$

10.12.4 Momentum conservation in the *x*-direction

The flux notation of the *x*-component of the momentum equation is

$$\sqrt{g}\left[\alpha_l\rho_l\gamma_v\frac{\partial u_l}{\partial \tau}+\left(\gamma_\xi \mathbf{a}^1\frac{\partial u_l}{\partial \xi}+\gamma_\eta \mathbf{a}^2\frac{\partial u_l}{\partial \eta}+\gamma_\zeta \mathbf{a}^3\frac{\partial u_l}{\partial \zeta}\right)\cdot \mathbf{G}_l\right]$$

$$+\frac{\partial}{\partial \xi}\left(\gamma_\xi\sqrt{g}\,\mathbf{a}^1\cdot \mathbf{F}_l^u\right)+\frac{\partial}{\partial \eta}\left(\gamma_\xi\sqrt{g}\,\mathbf{a}^2\cdot \mathbf{F}_l^u\right)+\frac{\partial}{\partial \zeta}\left(\gamma_\xi\sqrt{g}\,\mathbf{a}^3\cdot \mathbf{F}_l^u\right)$$

$$+\sqrt{g}\alpha_l\left(a^{11}\gamma_\xi\frac{\partial p}{\partial \xi}+a^{21}\gamma_\eta\frac{\partial p}{\partial \eta}+a^{31}\gamma_\zeta\frac{\partial p}{\partial \zeta}\right)$$

$$-\sqrt{g}\gamma_v\left[\sum_{\substack{m=1\\m\ne l}}^{3}\overline{c}_{ml}^d|\Delta \mathbf{V}_{ml}|\Delta u_{ml}-\overline{c}_{wl}^d|\Delta \mathbf{V}_{wl}|\Delta u_{wl}\right]$$

$$-\gamma_v\sqrt{g}\sum_{\substack{m=1\\m\ne l}}^{l_{\max}}\overline{c}_{ml}^{L}\left[\left(\mathbf{V}_l-\mathbf{V}_m\right)\times\left(\mathbf{a}^1\times\frac{\partial \mathbf{V}_m}{\partial \xi}+\mathbf{a}^2\times\frac{\partial \mathbf{V}_m}{\partial \eta}+\mathbf{a}^3\times\frac{\partial \mathbf{V}_m}{\partial \zeta}\right)\right]_x$$

$$-\gamma_v\sqrt{g}\overline{c}_{wl}^{L}\left[\left(\mathbf{V}_l-\mathbf{V}_{cs}\right)\times\left(\mathbf{a}^1\times\frac{\partial \mathbf{V}_{cs}}{\partial \xi}+\mathbf{a}^2\times\frac{\partial \mathbf{V}_{cs}}{\partial \eta}+\mathbf{a}^3\times\frac{\partial \mathbf{V}_{cs}}{\partial \zeta}\right)\right]_x$$

$$-\gamma_v\sqrt{g}\sum_{\substack{m=1\\m\ne l}}^{3}\overline{c}_{ml}^{vm}\left[\begin{array}{l}\dfrac{\partial \Delta u_{ml}}{\partial \tau}+\mathbf{a}^1\cdot\left(\mathbf{V}_l-\mathbf{V}_{cs}\right)\dfrac{\partial \Delta u_{ml}}{\partial \xi}+\mathbf{a}^2\cdot\left(\mathbf{V}_l-\mathbf{V}_{cs}\right)\dfrac{\partial \Delta u_{ml}}{\partial \eta}\\ +\mathbf{a}^3\cdot\left(\mathbf{V}_l-\mathbf{V}_{cs}\right)\dfrac{\partial \Delta u_{ml}}{\partial \zeta}\end{array}\right]$$

$$-\gamma_v\sqrt{g}\overline{c}_{wl}^{vm}\left[\begin{array}{l}\dfrac{\partial \Delta u_{csl}}{\partial \tau}+\mathbf{a}^1\cdot\left(\mathbf{V}_l-\mathbf{V}_{cs}\right)\dfrac{\partial \Delta u_{csl}}{\partial \xi}+\mathbf{a}^2\cdot\left(\mathbf{V}_l-\mathbf{V}_{cs}\right)\dfrac{\partial \Delta u_{csl}}{\partial \eta}\\ +\mathbf{a}^3\cdot\left(\mathbf{V}_l-\mathbf{V}_{cs}\right)\dfrac{\partial \Delta u_{csl}}{\partial \zeta}\end{array}\right]$$

$$=\gamma_v\sqrt{g}\left[-\alpha_l\rho_l g_x+\sum_{m=1}^{3,w}\left[\mu_{ml}\left(u_m-u_l\right)\right]+\mu_{lw}\left(u_{lw}-u_l\right)\right], \qquad (10.68)$$

where

$$\mathbf{F}_l^u = \mathbf{F}_l^{uv} + \mathbf{F}_l^{uvb} + \mathbf{F}_l^{uvT} , \qquad (10.69)$$

is the diffusion momentum flux in the *x*-direction with components

$$\mathbf{F}_l^{uv} = -\alpha_l \rho_l v_l^* \left(\mathbf{a}^1 \frac{\partial u_l}{\partial \xi} + \mathbf{a}^2 \frac{\partial u_l}{\partial \eta} + \mathbf{a}^3 \frac{\partial u_l}{\partial \zeta} \right), \qquad (10.70)$$

$$\mathbf{F}_l^{uvb} = \alpha_l^e \rho_l v_l^* \frac{2}{3} (\nabla \cdot \mathbf{V}_l) \mathbf{i} , \qquad (10.71)$$

$$\mathbf{F}_l^{uvT} = -\alpha_l^e \rho_l v_l^* \frac{\partial \mathbf{V}_l}{\partial x} . \qquad (10.72)$$

10.12.5 Momentum conservation in the *y*-direction

The flux notation of the *y*-component of the momentum equation is

$$\sqrt{g} \left[\alpha_l \rho_l \gamma_v \frac{\partial v_l}{\partial \tau} + \left(\gamma_\xi \mathbf{a}^1 \frac{\partial v_l}{\partial \xi} + \gamma_\eta \mathbf{a}^2 \frac{\partial v_l}{\partial \eta} + \gamma_\zeta \mathbf{a}^3 \frac{\partial v_l}{\partial \zeta} \right) \cdot \mathbf{G}_l \right]$$

$$+ \frac{\partial}{\partial \xi} \left(\gamma_\xi \sqrt{g} \; \mathbf{a}^1 \cdot \mathbf{F}_l^v \right) + \frac{\partial}{\partial \eta} \left(\gamma_\xi \sqrt{g} \; \mathbf{a}^2 \cdot \mathbf{F}_l^v \right) + \frac{\partial}{\partial \zeta} \left(\gamma_\xi \sqrt{g} \; \mathbf{a}^3 \cdot \mathbf{F}_l^v \right)$$

$$+ \sqrt{g} \alpha_l \left(a^{12} \gamma_\xi \frac{\partial p}{\partial \xi} + a^{22} \gamma_\eta \frac{\partial p}{\partial \eta} + a^{32} \gamma_\zeta \frac{\partial p}{\partial \zeta} \right)$$

$$- \sqrt{g} \gamma_v \left[\sum_{\substack{m=1 \\ m \neq l}}^{3} \overline{c}_{ml}^d \left| \Delta \mathbf{V}_{ml} \right| \Delta v_{ml} - \overline{c}_{wl}^d \left| \Delta \mathbf{V}_{wl} \right| \Delta v_{wl} \right]$$

$$- \gamma_v \sqrt{g} \sum_{\substack{m=1 \\ m \neq l}}^{l_{\max}} \overline{c}_{ml}^L \left[(\mathbf{V}_l - \mathbf{V}_m) \times \left(\mathbf{a}^1 \times \frac{\partial \mathbf{V}_m}{\partial \xi} + \mathbf{a}^2 \times \frac{\partial \mathbf{V}_m}{\partial \eta} + \mathbf{a}^3 \times \frac{\partial \mathbf{V}_m}{\partial \zeta} \right) \right]_y$$

$$-\gamma_v\sqrt{g}\overline{c}_{wl}^L\left[\left(\mathbf{V}_l-\mathbf{V}_{cs}\right)\times\left(\mathbf{a}^1\times\frac{\partial\mathbf{V}_{cs}}{\partial\xi}+\mathbf{a}^2\times\frac{\partial\mathbf{V}_{cs}}{\partial\eta}+\mathbf{a}^3\times\frac{\partial\mathbf{V}_{cs}}{\partial\zeta}\right)\right]_y$$

$$-\gamma_v\sqrt{g}\sum_{\substack{m=1\\m\neq l}}^{3}\overline{c}_{ml}^{vm}\left[\begin{array}{c}\dfrac{\partial\Delta v_{ml}}{\partial\tau}+\mathbf{a}^1\cdot\left(\mathbf{V}_l-\mathbf{V}_{cs}\right)\dfrac{\partial\Delta v_{ml}}{\partial\xi}+\mathbf{a}^2\cdot\left(\mathbf{V}_l-\mathbf{V}_{cs}\right)\dfrac{\partial\Delta v_{ml}}{\partial\eta}\\ +\mathbf{a}^3\cdot\left(\mathbf{V}_l-\mathbf{V}_{cs}\right)\dfrac{\partial\Delta v_{ml}}{\partial\zeta}\end{array}\right]$$

$$-\gamma_v\sqrt{g}\,\overline{c}_{wl}^{vm}\left[\begin{array}{c}\dfrac{\partial\Delta v_{csl}}{\partial\tau}+\mathbf{a}^1\cdot\left(\mathbf{V}_l-\mathbf{V}_{cs}\right)\dfrac{\partial\Delta v_{csl}}{\partial\xi}+\mathbf{a}^2\cdot\left(\mathbf{V}_l-\mathbf{V}_{cs}\right)\dfrac{\partial\Delta v_{csl}}{\partial\eta}\\ +\mathbf{a}^3\cdot\left(\mathbf{V}_l-\mathbf{V}_{cs}\right)\dfrac{\partial\Delta v_{csl}}{\partial\zeta}\end{array}\right]$$

$$=\gamma_v\sqrt{g}\left[-\alpha_l\rho_l g_y+\sum_{m=1}^{3,w}\left[\mu_{ml}\left(v_m-v_l\right)\right]+\mu_{lw}\left(v_{lw}-v_l\right)\right], \qquad (10.73)$$

where

$$\mathbf{F}_l^v=\mathbf{F}_l^{vv}+\mathbf{F}_l^{vvb}+\mathbf{F}_l^{vvT} \qquad (10.74)$$

is the diffusion momentum flux in the y-direction with components

$$\mathbf{F}_l^{vv}=-\alpha_l\rho_l v_l^*\left(\mathbf{a}^1\frac{\partial v_l}{\partial\xi}+\mathbf{a}^2\frac{\partial v_l}{\partial\eta}+\mathbf{a}^3\frac{\partial v_l}{\partial\zeta}\right), \qquad (10.75)$$

$$\mathbf{F}_l^{vvb}=\alpha_l^e\rho_l v_l^*\frac{2}{3}\left(\nabla\cdot\mathbf{V}_l\right)\mathbf{j}, \qquad (10.76)$$

$$\mathbf{F}_l^{vvT}=-\alpha_l^e\rho_l v_l^*\frac{\partial\mathbf{V}_l}{\partial y}.$$

10.12.6 Momentum conservation in the z-direction

The flux notation of the z-component of the momentum equation is

$$\sqrt{g}\left[\alpha_l\rho_l\gamma_v\frac{\partial w_l}{\partial\tau}+\left(\gamma_\xi\mathbf{a}^1\frac{\partial w_l}{\partial\xi}+\gamma_\eta\mathbf{a}^2\frac{\partial w_l}{\partial\eta}+\gamma_\zeta\mathbf{a}^3\frac{\partial w_l}{\partial\zeta}\right)\cdot\mathbf{G}_l\right]$$

$$+\frac{\partial}{\partial\xi}\left(\gamma_\xi\sqrt{g}\,\mathbf{a}^1\cdot\mathbf{F}_l^w\right)+\frac{\partial}{\partial\eta}\left(\gamma_\xi\sqrt{g}\,\mathbf{a}^2\cdot\mathbf{F}_l^w\right)+\frac{\partial}{\partial\zeta}\left(\gamma_\xi\sqrt{g}\,\mathbf{a}^3\cdot\mathbf{F}_l^w\right)$$

$$+\sqrt{g}\alpha_l\left(a^{13}\gamma_\xi\frac{\partial p}{\partial\xi}+a^{23}\gamma_\eta\frac{\partial p}{\partial\eta}+a^{33}\gamma_\zeta\frac{\partial p}{\partial\zeta}\right)$$

$$-\sqrt{g}\gamma_v\left[\sum_{\substack{m=1\\m\neq l}}^{3}\overline{c}_{ml}^d|\Delta\mathbf{V}_{ml}|\Delta w_{ml}-\overline{c}_{wl}^d|\Delta\mathbf{V}_{wl}|\Delta w_{wl}\right]$$

$$-\gamma_v\sqrt{g}\sum_{\substack{m=1\\m\neq l}}^{l_{\max}}\overline{c}_{ml}^L\left[\left(\mathbf{V}_l-\mathbf{V}_m\right)\times\left(\mathbf{a}^1\times\frac{\partial\mathbf{V}_m}{\partial\xi}+\mathbf{a}^2\times\frac{\partial\mathbf{V}_m}{\partial\eta}+\mathbf{a}^3\times\frac{\partial\mathbf{V}_m}{\partial\zeta}\right)\right]_z$$

$$-\gamma_v\sqrt{g}\,\overline{c}_{wl}^L\left[\left(\mathbf{V}_l-\mathbf{V}_{cs}\right)\times\left(\mathbf{a}^1\times\frac{\partial\mathbf{V}_{cs}}{\partial\xi}+\mathbf{a}^2\times\frac{\partial\mathbf{V}_{cs}}{\partial\eta}+\mathbf{a}^3\times\frac{\partial\mathbf{V}_{cs}}{\partial\zeta}\right)\right]_z$$

$$-\gamma_v\sqrt{g}\sum_{\substack{m=1\\m\neq l}}^{3}\overline{c}_{ml}^{vm}\left[\begin{array}{l}\dfrac{\partial\Delta w_{ml}}{\partial\tau}+\mathbf{a}^1\cdot\left(\mathbf{V}_l-\mathbf{V}_{cs}\right)\dfrac{\partial\Delta w_{ml}}{\partial\xi}+\mathbf{a}^2\cdot\left(\mathbf{V}_l-\mathbf{V}_{cs}\right)\dfrac{\partial\Delta w_{ml}}{\partial\eta}\\[6pt]+\mathbf{a}^3\cdot\left(\mathbf{V}_l-\mathbf{V}_{cs}\right)\dfrac{\partial\Delta w_{ml}}{\partial\zeta}\end{array}\right]$$

$$-\gamma_v\sqrt{g}\,\overline{c}_{wl}^{vm}\left[\begin{array}{l}\dfrac{\partial\Delta w_{csl}}{\partial\tau}+\mathbf{a}^1\cdot\left(\mathbf{V}_l-\mathbf{V}_{cs}\right)\dfrac{\partial\Delta w_{csl}}{\partial\xi}+\mathbf{a}^2\cdot\left(\mathbf{V}_l-\mathbf{V}_{cs}\right)\dfrac{\partial\Delta w_{csl}}{\partial\eta}\\[6pt]+\mathbf{a}^3\cdot\left(\mathbf{V}_l-\mathbf{V}_{cs}\right)\dfrac{\partial\Delta w_{csl}}{\partial\zeta}\end{array}\right]$$

$$=\gamma_v\sqrt{g}\left[-\alpha_l\rho_l g_z+\sum_{m=1}^{3,w}\left[\mu_{ml}\left(w_m-w_l\right)\right]+\mu_{lw}\left(w_{lw}-w_l\right)\right], \qquad (10.77)$$

where

$$\mathbf{F}_l^w = \mathbf{F}_l^{wv} + \mathbf{F}_l^{wvb} + \mathbf{F}_l^{wvT} \qquad (10.78)$$

is the diffusion momentum flux in the z-direction with components

$$\mathbf{F}_l^{wv} = -\alpha_l \rho_l v_l^* \left(\mathbf{a}^1 \frac{\partial w_l}{\partial \xi} + \mathbf{a}^2 \frac{\partial w_l}{\partial \eta} + \mathbf{a}^3 \frac{\partial w_l}{\partial \zeta} \right), \qquad (10.79)$$

$$\mathbf{F}_l^{wvb} = \alpha_l^e \rho_l v_l^* \frac{2}{3} (\nabla \cdot \mathbf{V}_l) \mathbf{k}, \qquad (10.80)$$

$$\mathbf{F}_l^{w\theta T} = -\alpha_l^e \rho_l v_l^* \frac{\partial \mathbf{V}_l}{\partial z}. \qquad (10.81)$$

10.13 Concluding remarks

The equations derived in this Chapter may be of interest not only for scientists and engineers developing computational models but also for those using intensively computational models. One may compare the set of equation presented in this Chapter with the sets solved by different commercial providers and reveal the physical phenomena which are still not taken into account by them. This is helpful to learn the limitations of the existing products.

References

1. Kolev NI (1986) Transiente Dreiphasen Dreikomponenten Strömung, Teil 3: 3D-Dreifluid-Diffusionsmodell, KfK 4080.
2. Kolev NI (August 1987) A three field-diffusion model of three-phase, three-component flow for the transient 3D-computer code IVA2/01, Nuclear Technology, vol 78 pp 95-131
3. Kolev NI (1991) Derivatives for the equation of state of multi-component mixtures for universal multi-component flow models, Nuclear Science and Engineering: vol 108 p 74-87
4. Kolev NI (1994) The code IVA4: Modeling of mass conservation in multi-phase multi component flows in heterogeneous porous media, Kerntechnik, vol 59 no 4-5 pp 226-237
5. Kolev NI (1994) The code IVA4: Modeling of momentum conservation in multi phase flows in heterogeneous porous media, Kerntechnik, vol 59 no 6 pp 249-258
6. Kolev NI (1995) The code IVA4: Second law of thermodynamics for multi phase flows in heterogeneous porous media, Kerntechnik, vol 60 no 1 pp 1-39
7. Kolev NI (1997) Comments on the entropy concept, Kerntechnik, vol 62 no 1 pp 67-70
8. Kolev NI (1998) On the variety of notation of the energy conservation principle for single phase flow, Kerntechnik, vol 63 no 3 pp 145-156

9. Kolev NI (1999) Applied multi-phase flow analysis and its relation to constitutive physics, Proc. of the 8th International Symposium on Computational Fluid Dynamics, ISCFD '99 5 - 10 September 1999 Bremen, Germany. Japan Journal for Computational Fluid Dynamics, 2000
10. Kolev NI (2001) Conservation equations for multi-phase multi-component multi-velocity fields in general curvilinear coordinate systems, Keynote lecture, Proceedings of ASME FEDSM'01, ASME 2001 Fluids Engineering Division Summer Meeting, New Orleans, Louisiana, May 29 - June 1, 2001
11. Peyret R, editor: (1996) Handbook of computational fluid mechanics, Academic Press, London
12. Thompson J F, Warsi ZUA, Wayne MC (1985) Numerical grid generation, North-Holand, New York-Amsterdam-Oxford
13. Vinokur M (1974) Conservation equations of gas dynamics in curvilinear coordinate systems, J. Comput. Phys., vol 14 pp 105-125
14. Vivand H (1974) Formes conservatives des equations de la dynamique des gas, Conservative forms of gas dynamics equations, La Recherche Aerospatiale, no 1 (Janvier-Fevrier), p 65 a 68.

11 Type of the system of PDEs

Understanding the type of the partial differential equation systems is an important prerequisite for building successful numerical methods. This chapter gives the definition equations of eigenvalues and eigenvectors of systems of PDEs with constant coefficients. In addition the way to transform the initial system into canonical form is given after determining the type of the system. Then the following questions are answered: What relation exists between the eigenvalues and (a) the propagation velocity of perturbations of the flow parameters, (b) the propagation velocity of harmonic oscillations of the flow parameters, and (c) the critical flows?

This section is an English translation of the slightly modified Chap. 4 of [1]. The interested reader will find more information about the method of the characteristics in [1].

11.1 Eigenvalues, eigenvectors, canonical form

The general form of the conservation laws presents a *semi-linear non-homogeneous system of partial differential equations*

$$\frac{\partial \mathbf{X}}{\partial \tau} + \frac{\partial \mathbf{Y}}{\partial z} = \mathbf{D}. \tag{11.1}$$

This form is frequently called the *primary* or *conservative* form. Let us assume a flow that can be completely described by the following vector of dependent variables: $\mathbf{U} = \mathbf{U}(\tau, z)$. Using the differential forms of the state variables it is possible to rewrite Eq. (11.1) in other forms, for instance in the so called *semi-conservative* forms

$$\mathbf{J}_1 \frac{\partial \mathbf{U}}{\partial \tau} + \frac{\partial \mathbf{Y}}{\partial z} = \mathbf{D} \tag{11.2}$$

or

$$\frac{\partial \mathbf{U}}{\partial \tau} + \mathbf{F} \frac{\partial \mathbf{Y}}{\partial z} = \mathbf{C} \tag{11.3}$$

where $\mathbf{J}_1 = \partial \mathbf{X}/\partial \mathbf{U}$, $\mathbf{F} = \mathbf{J}_1^{-1}$, $\mathbf{C} = \mathbf{J}_1^{-1} \mathbf{D}$, or in the *non-conservative* forms

$$\mathbf{J}_1 \frac{\partial \mathbf{U}}{\partial \tau} + \mathbf{J}_2 \frac{\partial \mathbf{U}}{\partial z} = \mathbf{D} \text{ or } \frac{\partial \mathbf{U}}{\partial \tau} + \mathbf{A} \frac{\partial \mathbf{U}}{\partial z} = \mathbf{C}, \qquad (11.4, 5)$$

where $\mathbf{J}_2 = \partial \mathbf{Y}/\partial \mathbf{U}$, $\mathbf{A} = \mathbf{J}_1^{-1}\mathbf{J}_2$. Writing Eq. (11.5) in component form yields two equations

$$L_1 = \frac{\partial u_1}{\partial \tau} + a_{11} \frac{\partial u_1}{\partial z} + a_{12} \frac{\partial u_2}{\partial z} - c_1 = 0, \qquad (11.6)$$

$$L_2 = \frac{\partial u_2}{\partial \tau} + a_{21} \frac{\partial u_1}{\partial z} + a_{22} \frac{\partial u_2}{\partial z} - c_2 = 0. \qquad (11.7)$$

Using a vector $\mathbf{h}_i$ we build a linear combination of the two equations

$$L_i = h_{i1}\left(\frac{\partial u_1}{\partial \tau} + \underbrace{\frac{a_{11}h_{i1} + a_{21}h_{i2}}{h_{i1}}}_{\lambda_i}\frac{\partial u_1}{\partial z}\right) + h_{i2}\left(\frac{\partial u_2}{\partial \tau} + \underbrace{\frac{a_{12}h_{i1} + a_{22}h_{i2}}{h_{i2}}}_{\lambda_i}\frac{\partial u_2}{\partial z}\right) - h_{i1}c_1 - h_{i2}c_2 = 0.$$

(11.8)

If the components of the vector $\mathbf{h}_i$ are computed so that the following conditions are fulfilled

$$a_{11}h_{i1} + a_{21}h_{i2} = \lambda_i h_{i1}, \qquad (11.9)$$

$$a_{12}h_{i1} + a_{22}h_{i2} = \lambda_i h_{i2}, \qquad (11.10)$$

or in matrix notation

$$\left(\mathbf{A}^T - \lambda_i \mathbf{I}\right)\mathbf{h}_i = 0, \qquad (11.11)$$

the two Eqs. (11.8) for each $\mathbf{h}_i$ can be written in a remarkably simple form as we will see later. Equation (11.11) is the definition equation for *eigenvectors* of the matrix $\mathbf{A}^T$. Because Eq. (11.11) is homogeneous, the condition to have linear independent solutions for $\mathbf{h}_i$ is that the determinant of the coefficient matrix is equal to zero

$$\left|\mathbf{A}^T - \lambda_i \mathbf{I}\right| = 0. \qquad (11.12)$$

Actually, this is the definition equation for the *eigenvalues* of the matrix $\mathbf{A}^T$ called the *characteristic equation*. With respect to the unknown eigenvalues it is a polynomial of the n-th degree, where n is the rank of the matrix $\mathbf{A}^T$. Let us assume we have already obtained n real solutions for the eigenvalues such that at least two of them differ from each other. With them we can obtain at least n independent solutions for the eigenvectors. In our particular case of Eq. (11.8) we have

$$h_{11}\left(\frac{\partial u_1}{\partial \tau} + \lambda_1 \frac{\partial u_1}{\partial z}\right) + h_{12}\left(\frac{\partial u_2}{\partial \tau} + \lambda_1 \frac{\partial u_2}{\partial z}\right) = h_{11}c_1 + h_{12}c_2, \tag{11.13}$$

$$h_{21}\left(\frac{\partial u_1}{\partial \tau} + \lambda_2 \frac{\partial u_1}{\partial z}\right) + h_{22}\left(\frac{\partial u_2}{\partial \tau} + \lambda_2 \frac{\partial u_2}{\partial z}\right) = h_{21}c_1 + h_{22}c_2, \tag{11.14}$$

or rewritten as a scalar product

$$\mathbf{h}_i\left(\frac{\partial \mathbf{U}}{\partial \tau} + \lambda_i \frac{\partial \mathbf{U}}{\partial z}\right) = \mathbf{h}_i \mathbf{C}, \ i=1,n, \tag{11.15}$$

or in matrix notation

$$\mathbf{H}\frac{\partial \mathbf{U}}{\partial \tau} + \mathbf{\Lambda I H}\frac{\partial \mathbf{U}}{\partial z} = \mathbf{HC}. \tag{11.16}$$

Here $\mathbf{H}$ is a matrix whose rows are the eigenvectors $\mathbf{h}_i$. $\mathbf{I}$ is the unit matrix. Equations (11.15) or the system (11.16) is called the *pseudo-canonical* form of the original system.

The total time derivative of $\mathbf{U}$ along the curve defined by

$$dz/d\tau = \lambda_i \tag{11.17}$$

is

$$\frac{d\mathbf{U}}{d\tau} = \frac{\partial \mathbf{U}}{\partial \tau} + \lambda_i \frac{\partial \mathbf{U}}{\partial z}. \tag{11.18}$$

With this notation we reach the final, *canonical form* of the original system,

$$\mathbf{h}_i \frac{d\mathbf{U}}{d\tau} = \mathbf{h}_i \mathbf{C}, \ i=1,n. \tag{11.19}$$

The curves whose inclination in the space-time domain is defined by $dz/d\tau = \lambda_i$ are called *characteristic curves*. Note that each equation in Eq. (11.19) is valid only along the corresponding characteristic curve. The eigenvalues and eigenvectors determine the type of the system – see *Courant* and *Hilbert* [2] (1962):

   a) If all the solutions of the characteristic equation are *imaginer or complex* then the system is *elliptic* and from the type of the *potential equation*.
   b) If all the solutions are *real* and *equal,* then the system is *parabolic* and from the type of the *heat conduction equation*.
   c) If the characteristic equation *n different real solutions* or it has *n-real and at least two different solutions,* and the system $\mathbf{A}^T$ has *n-linear independent eigenvectors,* then the system is *hyperbolic* and from the type of the *wave equation*.

11.2 Physical interpretation

The eigenvalues contain very important physical information. This has to be well understood for several reasons. The first one the need to distinguish between different types of systems of partial differential equations with different properties. So, the proper defined multi-phase flow equations are from the type of the wave equation. If due to erroneous formulation the type of the resulting system turns to be other than hyperbolic the model is a priori not acceptable. Second, the formulation of different numerical methods has to rely on the basic mathematical properties of the system in order not to lose them during the numerical integration.

11.2.1 Eigenvalues and propagation velocity of perturbations

First of all observe that the eigenvalue λ_i has dimensions of velocity. Attaching a coordinate system with one axis tangential to the characteristic line and moving it with the *characteristic velocity* λ_i allows the initial system of partial differential equations to be transforming to the simpler system of ordinary differential equations. This simply means that (a) in the corresponding canonical equation the change of one variable is connected with the change of the other variables appearing with their derivatives, and (b) the propagation velocity of a perturbation of the participating variables is equal to the characteristic velocity λ_i. This is remarkable. The absence of real eigenvalues in a wrong formulated model means that there is no real propagation velocity of the variable or variables describing the particular physical phenomenon. If the experimental observations show a real and finite propagation velocity such models are a priori in conflict with reality and therefore not acceptable. Using the language of the mathematicians:

Non-hyperbolic models cannot adequately describe flow phenomena.

11.2.2 Eigenvalues and propagation velocity of harmonic oscillations

Consider a small perturbation $\Delta \mathbf{U}$ of the vector of the time-averaged dependent variables $\overline{\mathbf{U}}$,

$$\mathbf{U} = \overline{\mathbf{U}} + \Delta \mathbf{U} \tag{11.20}$$

in a flow described by

$$\frac{\partial \mathbf{U}}{\partial \tau} + \mathbf{A}\frac{\partial \mathbf{U}}{\partial z} = \mathbf{C}, \tag{11.21}$$

resulting in

$$\frac{\partial \Delta \mathbf{U}}{\partial \tau} + \mathbf{A}\frac{\partial \Delta \mathbf{U}}{\partial z} + \frac{\partial \overline{\mathbf{U}}}{\partial \tau} + \mathbf{A}\frac{\partial \overline{\mathbf{U}}}{\partial z} = \mathbf{C}. \tag{11.22}$$

Assuming that

$$\frac{\partial \overline{\mathbf{U}}}{\partial \tau} + \mathbf{A}\frac{\partial \overline{\mathbf{U}}}{\partial z} \approx \mathbf{C} \tag{11.23}$$

results in

$$\frac{\partial \Delta \mathbf{U}}{\partial \tau} + \mathbf{A}\frac{\partial \Delta \mathbf{U}}{\partial z} = 0. \tag{11.25}$$

Consider harmonic oscillations of magnitude $\Delta \overline{\mathbf{U}}$, the frequency of the undumped oscillation f and the wave number k defined by

$$\mathbf{U} = \Delta \overline{\mathbf{U}} e^{i(f\tau - kz)}. \tag{11.26}$$

Substituting in the perturbation equation we obtain

$$-ik\left(\mathbf{A} - \mathbf{I}f/k\right)\Delta \overline{\mathbf{U}} e^{i(f\tau - kz)} = 0 \tag{11.27}$$

or

$$-ik\left(\mathbf{A} - \mathbf{I}f/k\right) = 0. \tag{11.28}$$

We see that $f/k \triangleq \lambda_i$ defined by

$$|\mathbf{A} - \mathbf{I} f/k| = 0 \qquad (11.29)$$

is the propagation velocity of the harmonic oscillations.

> Therefore, if the eigenvalues exist and are real and finite they are equal to the propagation velocity of the harmonic oscillations inside the flow.

11.2.3 Eigenvalues and critical flow

What happens if the propagation velocity of a disturbance is equal to zero, $\lambda_i = 0$? It follows from Eq. (11.29) that for this particular case

$$|\mathbf{A}| = 0. \qquad (11.30)$$

If for instance pressure and velocity are coupled in one equation of the canonical form, the condition $\lambda_i = 0$ means that a disturbance of the pressure and the corresponding velocity disturbance cannot propagate in space. Let us take another look at this situation. Consider a steady state flow described by

$$\mathbf{A} \frac{d\mathbf{U}}{dz} = \mathbf{C}, \qquad (11.31)$$

where $\mathbf{U}$ contains also the pressure. Solving with respect to the derivatives we obtain

$$\frac{dp}{dz} = -\frac{...}{|\mathbf{A}|}. \qquad (11.32)$$

Condition (11.30) means therefore

$$\frac{dp}{dz} = -\infty. \qquad (11.33)$$

But this is the well known condition for critical flow. So we win another physical interpretation of the eigenvalues.

> If one of the eigenvalues is equal to zero the flow is stagnant or critical. For the case of non-zero velocities, the changes of any parameters downwards the critical cross section does not influence the steady state mass flow rate. The critical steady state mass flow rate is only a function of the flow parameters upwards the flow.

Critical flow is a phenomenon that is desirable if it is wished to limit the mass flow rate in a facility and undesirable if for instance due to the critical mass flow rate the coolability of a system is endangered or the productivity of a facility is reduced.

References

1. Kolev NI (1986) Transiente Zweiphasenströmung, Springer, Berlin Heidelberg New York
2. Courant RU and Hilbert D (1962) Methods of Mathematical Physics, Wiley-Interscience, New York, vol 2

12 Numerical solution methods for multi-phase flow problems

A class of implicit numerical solution methods for multi-phase flow problems based on the entropy concept is presented in this work. First order donor-cell finite difference approximation is used for the time derivatives and the convection terms. The diffusion terms are discretized using a second order finite difference approximation. A staggered grid is used for discretization of the momentum equations. The entropy concept permits analytical reduction of the algebraic problem to give pressure or pressure correction equations for each computational cell. Analytical backward substitution closes the iteration cycle. Three different methods applied in IVA computer codes are presented and the experience gained with these methods is discussed. The class of methods presented has been applied with success to a variety of practical problems in the field of nuclear engineering. High order upwinding and the constrained interpolation profile method are also presented.

This Chapter is an extended version of the theoretical part of the publication [33]:

12.1 Introduction

Numerical methods for transient single-phase flow analysis are available and in widespread use for practical applications. Methods for cost-effective solution of multi-phase flow problems are still in their infancy. There is a significant experience base available for two-phase flows, e.g. gas/liquid and dispersed solid particles/gas flow for the special case of small concentrations in the dispersed phase. To my knowledge, there is no universal method for integrating systems of partial differential equations describing multi-phase flows. The purpose of this work is to present the methods derived for the computer codes IVA2 to IVA6 and to give a short description of the experience gained with these methods.

12.2 Formulation of the mathematical problem

Consider the following mathematical problem: A multi-phase flow is described by the following vector of dependent variables

$$\mathbf{U}^T = (\alpha_m, T_1, s_2, s_3, C_{il}, n_l, p, u_l, v_l, w_l), \text{ where } l = 1, 2, 3, \qquad (12.1)$$

which is a function of the three space coordinates (r,θ,z), and of the time τ,

$$\mathbf{U} = \mathbf{U}(r,\theta,z,\tau). \qquad (12.2)$$

The relationship $\mathbf{U} = \mathbf{U}(r,\theta,z,\tau)$ is defined by the volume-averaged and successively time-averaged mass, momentum and energy conservation equations as well as by initial conditions, boundary conditions, and geometry. As shown in Chapters 1, 2 and 5 [27, 28, 29], the conservation principles lead to the following system of 21 non-linear, non-homogeneous partial differential equations with variable coefficients

$$\frac{\partial}{\partial \tau}(\alpha_l \rho_l \gamma_v) + \nabla \cdot (\alpha_l \rho_l \mathbf{V}_l \gamma) = \gamma_v \mu_l \qquad (12.3)$$

$$\alpha_l \rho_l \left[\gamma_v \frac{\partial C_{il}}{\partial \tau} + (\mathbf{V}_l \gamma) C_{il} \right] - \nabla \cdot \left(\alpha_l \rho_l D_{il}^* \gamma \nabla C_{il} \right) + \gamma_v \mu_l^+ C_{il} = \gamma_v DC_{il} \qquad (12.4)$$

$$\frac{\partial}{\partial \tau}(n_l \gamma_v) + \nabla \cdot (n_l \mathbf{V}_l \gamma) = \gamma_v \left(\dot{n}_{lkin} - \dot{n}_{lcol} + \dot{n}_{lsp} \right) \quad \text{for} \quad \alpha_l > 0, \qquad (12.5)$$

$$\frac{\partial}{\partial \tau}(\alpha_l \rho_l \mathbf{V}_l \gamma_v) + \nabla \cdot \left[(\alpha_l \rho_l \mathbf{V}_l \mathbf{V}_l + \overline{\overline{\tau}}_l) \gamma \right] + \alpha_l \gamma \nabla p + (\alpha_l \rho_l \mathbf{g} + \mathbf{f}_l) \gamma_v$$

$$= \gamma_v \left[\mu_{wl} \mathbf{V}_{wl} - \mu_{lw} \mathbf{V}_{lw} + \sum_{m=1}^{l_{max}} (\mu_{ml} \mathbf{V}_m - \mu_{lm} \mathbf{V}_l) \right], \qquad (12.6)$$

$$\rho_l c_{pl} \left[\alpha_l \gamma_v \frac{\partial T_l}{\partial \tau} + (\alpha_l^e \mathbf{V}_l \gamma \cdot \nabla) T_l \right] - \nabla \cdot (\alpha_l^e \lambda_l^* \gamma \nabla T_l)$$

$$- \left[1 - \rho_l \left(\frac{\partial h_l}{\partial p} \right)_{T_l, all\_C's} \right] \left[\alpha_l \gamma_v \frac{\partial p}{\partial \tau} + (\alpha_l^e \mathbf{V}_l \gamma \cdot \nabla) p \right] + T_l \sum_{i=2}^{i_{max}} \Delta s_{il}^{np} \nabla (\alpha_l^e \rho_l D_{il}^* \gamma \nabla C_{il})$$

$$= \gamma_v \left[DT_l^N - T_l \sum_{i=2}^{i_{max}} \Delta s_{il}^{np} (\mu_{il} - C_{il} \mu_l) \right], \quad \text{for } l = 1, \qquad (12.7a)$$

and

$$\alpha_l \rho_l \left[\gamma_v \frac{\partial s_l}{\partial \tau} + (\mathbf{V}_l \gamma \nabla) s_l \right] - \frac{1}{T_l} \nabla \cdot \left(\alpha_l^e \lambda_l^* \gamma \nabla T \right) + \gamma_v \mu_l^+ s_l = \gamma_v D s_l \quad \text{for } l = 2,3$$
(12.7b)

This system is defined in the three-dimensional domain **R**. The initial conditions of $\mathbf{U}(\tau = 0) = \mathbf{U}_a$ in **R** and the boundary conditions acting at the interface separating the integration space from its environment are given. The solution required is for conditions after the time interval $\Delta \tau$ has elapsed. The previous time variables are assigned the index a. The time variables not denoted with a are either in the new time plane, or are the best available guesses for the new time plane.

In order to enable modeling of flows with arbitrary obstacles and inclusions in the integration space as is usually expected for technical applications, surface permeabilities are defined

$$(\gamma_r, \gamma_\theta, \gamma_z) = \text{functions of } (r, \theta, z, \tau),$$
(12.8)

at the virtual surfaces that separate each computational cell from its environment. By definition, the surface permeabilities have values between one and zero,

$$0 \leq \text{each of all } (\gamma_r, \gamma_\theta, \gamma_z)\text{'s} \leq 1.$$
(12.9)

A volumetric porosity

$$\gamma_v = \gamma_v(r, \theta, z, \tau)$$
(12.10)

is assigned to each computational cell, with

$$0 < \gamma_v \leq 1.$$
(12.11)

The surface permeabilities and the volume porosities are not expected to be smooth functions of the space coordinates in the region **R** and of time. For this reason, one constructs a frame of geometrical flow obstacles which are functions of space and time. This permits a large number of extremely interesting technical applications to be done with this type of approach.

In order to construct useful numerical solutions it is essential that an appropriate set of constitutive relations be available: state equations, thermodynamic derivatives given in Chapter 3, equations for estimation of the transport properties, correlations modeling the heat, mass and momentum transport across the surfaces dividing the separate velocity fields given in Volume II, etc. These relationships together are called closure equations – see e.g. in [22,24-26,30-32].

12.3 Space discretization and location of the discrete variables

The flow is defined in a domain to rectangular coordinate system; either Cartesian or cylindrical (see Fig. 12.1).

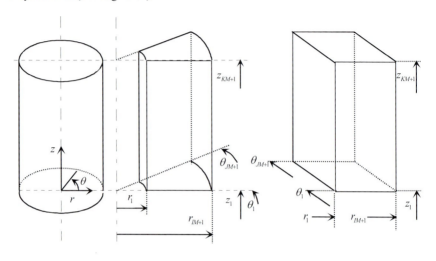

Fig. 12.1. Flow domain definition

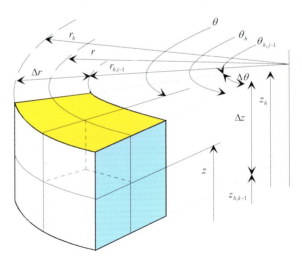

Fig. 12.2. Geometrical sizes of the mass and energy conservation computational cell referred to as a non-staggered cell

12.3 Space discretization and location of the discrete variables   509

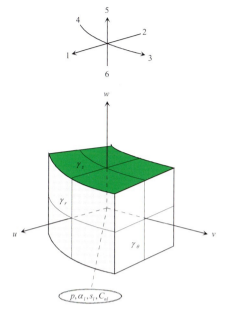

Fig. 12.3. Location of the dependent variables and of the surface permeabilities

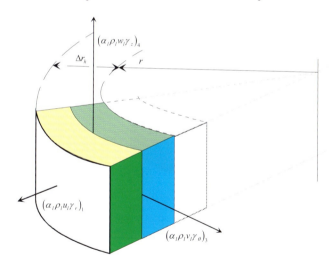

Fig. 12.4. Staggered control volume for integration of the radial momentum conservation equations

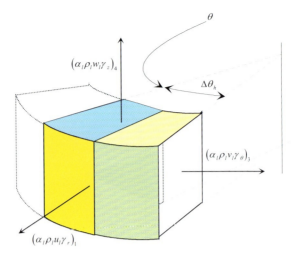

Fig. 12.5. Staggered control volume for integration of the angular momentum conservation equations

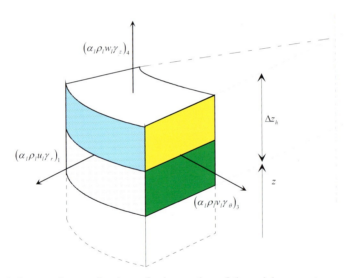

Fig. 12.6. Staggered control volume for integration of the axial momentum conservation equations

The domain is divided into computational cells as shown in Fig. 12.2, these having a center with the volume coordinates (r,θ,z). The sizes of each cell $(\Delta r, \Delta\theta, \Delta z)$ are a result of non-uniform spatial discretization in each direction. The cell boundaries are defined by (r_h, θ_h, z_h). The distances between two adjacent cell centers are $(\Delta r_h, \Delta\theta_h, \Delta z_h)$, respectively. The dependent variables

$(\alpha_m, s_l, C_{il}, n_l, p)$, the state and the transport properties $(T_l, \rho_l, \upsilon_l, \lambda_l, etc)$, and the volumetric porosity are located in the cell centers as shown in Fig. 12.3. The surface permeabilities $(\gamma_r, \gamma_\theta, \gamma_z)$, and the velocity components (u_l, v_l, w_l) are located in the cell interfaces as shown in Fig. 12.3. Thus, all dependent field variables are located at the cell center except the flow variables which are located at the cell surfaces. Fig. 12.3 also shows the control volume for integration of the field and field-component mass conservation equations, of the particle density conservation equations, and of the entropy conservation equations. Figures 12.4, 12.5 and 12.6 show the control volumes for integration of the momentum conservation equations in each separate direction. The momentum cells have the mass cell centers at their surfaces. This forms the so-called staggered grid system - see *Harlow* and *Amsden* [16] (1971). Such systems are employed extensively in fluid mechanics to avoid non-physical oscillations by allowing the use of first order discretization methods - see *Issa* [17] (1983), for example. The staggered grid system is not necessarily needed if high order discretization methods are used.

Each mass and energy cell is identified by the integer indices (i, j, k) corresponding to the three spatial directions (see Fig. 12.7). The integer indices take values i from 1 to $IM + 2$, j from 1 to $JM + 2$, and k from 1 to $KM + 2$. To achieve a code architecture that allows treatment of boundary cells in the same manner as non-boundary cells, I use a layer of fictitious cells surrounding the real cells as shown in Fig. 12.7. These cells have common indices $i = 1$ or $i = IM + 2, j = 1$, or $j = JM + 2$, $k = 1$, or $k = KM + 2$, respectively. The boundary conditions in the form of prescribed time functions are likewise applied within these auxiliary cells.

The flow field variables for the front walls of the cell, just like the velocities, are assigned the same indices as the cell itself. Thus, the back walls of the cells have the cell indices minus one for each particular direction.

It is convenient to denote with $m = 1$ through 6 the cells $i+1$, i-1, $j+1$, j-1, $k+1$ and k-1, respectively that surround the mass and energy conservation cell (i,j,k) (see Fig. 12.3). Similarly the corresponding walls attached to the (i,j,k) cell, and the flow properties defined at this surfaces are assigned designations $m = 1$ through 6. For notation convenience I define the outwards directed normal velocities to the surfaces of each computational cell, $\mathbf{V}_{lm}^n$, as follows

$$\left(\mathbf{V}_{lm}^n\right)^T = (u_l, -u_{l,i-1}, v_l, -v_{l,j-1}, w_l, -w_{l,k-1}) \qquad (12.12)$$

The donor- cell concept is used. This concept is based on the intuitive assumption that the material leaving the cell has the same properties as the cell. This concept stems from *Courant* et al. [9] (1952).

All field variables have three indices, i, j and k. For simplicity, I omit the indices except where they are not i, j, k. For example $p_{i,j,k}$ is replaced by p.

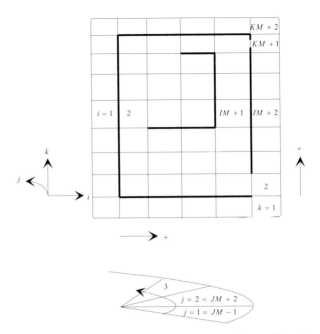

Fig. 12.7. Three-dimensional mesh construction, boundary layer cells, cell numbering

12.4 Discretization of the mass conservation equations

The first order donor cell discretized mass conservation equation (12.3) for each velocity field is given in *Appendix 12.1*. Introducing a number of convenient abbreviations also given in *Appendix 12.1* the following is obtained

$$(\alpha_l \rho_l \gamma_v - \alpha_{la} \rho_{la} \gamma_{va})/\Delta\tau + \sum_{m=1}^{6}\left(b_{lm+}\alpha_l \rho_l - b_{lm-}\alpha_{lm}\rho_{lm}\right) - \gamma_v \mu_l = 0. \quad (12.13)$$

Introducing the velocity normal to each surface of the discretization volume, as defined by Eq. (12.12), the *b* coefficients from *Appendix 12.1* can be conveniently written as

$$b_{lm+} = \beta_m \xi_{lm+} \mathbf{V}_{lm}^n \geq 0, \quad (12.14)$$

$$b_{lm-} = -\beta_m \xi_{lm-} \mathbf{V}_{lm}^n \geq 0, \quad (12.15)$$

$$\xi_{lm+} = \frac{1}{2}\left[1 + sign\left(\mathbf{V}_{lm}^n\right)\right], \qquad (12.16)$$

$$\xi_{lm-} = 1 - \xi_{lm+}. \qquad (12.17)$$

It is advisable to compute the geometry coefficients,

$$\beta_1 = \frac{r_h^\kappa \gamma_r}{r^\kappa \Delta r}, \quad \beta_2 = \frac{(r_h^\kappa \gamma_r)_{i-1}}{r^\kappa \Delta r}, \quad \beta_3 = \frac{\gamma_\theta}{r^\kappa \Delta \theta}, \quad \beta_4 = \frac{\gamma_{\theta,j-1}}{r^\kappa \Delta \theta}, \quad \beta_5 = \frac{\gamma_z}{\Delta z}, \quad \beta_5 = \frac{\gamma_{z,k-1}}{\Delta z},$$

once at the beginning of the integration and perform corrections only for those computational cells where there is a change of the geometry during the time considered. Normally there are only a few such cells relative to the total number of cells. Note that for the first order finite difference approximation

$$\nabla \gamma \approx \sum_{m=1}^{6} \beta_m. \qquad (12.18)$$

At this point the method used for computation of the field volumetric fractions in the computer codes IVA2 [20-22, 25], and IVA3 [23, 26] will be described. The method exploits the point *Gauss-Seidel* iteration assuming known velocity fields and thermal properties.

In the donor cell concept the term

$$B_{lm-} = b_{lm-}\alpha_{l,m}\rho_{l,m} \geq 0$$

plays an important role. B_{lm-} is in fact the mass flow entering the cell from the face m divided by the volume of the cell – it is in fact a non-zero volumetric mass source. Once computed for the mass conservation equation it is stored and used subsequently in all other conservation equations.

Consider the field variables $\alpha_l \rho_l$ in the convective terms associated with the output flow in the new time plane, and $\alpha_{lm}\rho_{lm}$ in the neighboring cells m as the best available guesses for the new time plane. Solving Eq. (12.13) with respect to $\alpha_l \rho_l$ gives

$$\overline{\alpha}_l \overline{\rho}_l = \left[\gamma_{va}\left(\mu_l + \frac{\alpha_{la}\rho_{la}}{\Delta \tau}\right) + \sum_{m=1}^{6} B_{lm-}\right] / \left(\frac{\gamma_v}{\Delta \tau} + \sum_{m=1}^{6} b_{lm+}\right). \qquad (12.19)$$

Here $\dfrac{\gamma_v}{\Delta \tau} + \sum_{m=1}^{6} b_{lm+} > 0$ is ensured besides by Eq. (12.15) by Eq. (12.11) that does not allow γ_v to be zero. For just originating field we have

$$\overline{\alpha}_l = \dfrac{\Delta \tau}{\overline{\rho}_l} \dfrac{\gamma_{va}\mu_l + \sum\limits_{m=1}^{6} B_{lm-}}{\gamma_v + \Delta \tau \sum\limits_{m=1}^{6} b_{lm+}}.$$

Obviously the field can originate due to convection, $\sum_{m=1}^{6} B_{lm-} > 0$, or due to in-cell mass source, $\mu_l > 0$, or due to simultaneous appearance of the both phenomena. In case of origination caused by in-cell mass source terms it is important to define the initial density, $\overline{\rho}_l$, in order to compute $\overline{\alpha}_l = \Delta \tau \dfrac{\mu_l}{\overline{\rho}_l}$.

The best mass conservation in such procedures is ensured if the following sequence is used for computation of the volume fractions:

$$\alpha_2 = \overline{\alpha}_2 \overline{\rho}_2 / \rho_2, \quad \alpha_3 = \overline{\alpha}_3 \overline{\rho}_3 / \rho_3, \quad \alpha_1 = 1 - \alpha_2 - \alpha_3.$$

12.5 First order donor-cell finite difference approximations

Each velocity field is characterized by properties such as specific entropies, concentrations of the inert components, etc. denoted by Φ. These properties can be transported (a) by convection, driven by the mass flow F and (b) by diffusion, controlled by the diffusion constant Γ. Differentiating the convective terms in the conservation equation and comparing these with the corresponding field mass conservation equation multiplied by Φ leads to a considerably simpler semi-conservative form that contains the following connective terms

$$...F\dfrac{\partial \Phi}{\partial z} - \dfrac{\partial}{\partial z}\left(\Gamma \dfrac{\partial \Phi}{\partial z}\right)... \qquad (12.20)$$

To allow use of the values for the mass flows at the boundary of the computational cell where these mass flows are originally defined, I discretize the following equivalent form

$$...\dfrac{\partial}{\partial z}\left(F\Phi - \Gamma \dfrac{\partial \Phi}{\partial z}\right) - \Phi \dfrac{\partial F}{\partial z}... \qquad (12.21)$$

12.5 First order donor-cell finite difference approximations

There are a number of methods for discretizing this type of equation for single-phase flows, see *Chow* et al. [7] (1984), *Patel* et al. [54] (1985), *Patel* et al. [56] (1986). Some of them take the form of an analytical solution of the simple convection-diffusion equation as an approximation formula for the profile $\Phi = \Phi(z)$. For multi-phase flows I use the simplest first order donor cell method for discretization of the convective terms and a second order central difference method for discretization of the diffusion term. The result

$$F\frac{\partial \Phi}{\partial z} - \frac{\partial}{\partial z}\left(\Gamma\frac{\partial \Phi}{\partial z}\right) = \frac{\partial}{\partial z}\left(F\Phi - \Gamma\frac{\partial \Phi}{\partial z}\right) - \Phi\frac{\partial F}{\partial z}$$

$$\approx \frac{1}{\Delta z}\left[F\left\{\frac{1}{2}[1+sign(F)]\Phi + \frac{1}{2}[1-sign(F)]\Phi_{k+1}\right\} - \frac{\Gamma}{\Delta z_h}(\Phi_{k+1} - \Phi)\right]$$

$$-\frac{1}{\Delta z}\left[F_{k-1}\left\{\frac{1}{2}[1+sign(F_{k-1})]\Phi_{k-1} + \frac{1}{2}[1-sign(F_{k-1})]\Phi\right\} - \frac{\Gamma_{k-1}}{\Delta z_{h,k-1}}(\Phi - \Phi_{k-1})\right]$$

$$-\frac{1}{\Delta z}\Phi(F - F_{k-1}), \tag{12.22}$$

can be further simplified taking into account

$$\frac{1}{2}[1+sign(F)] - 1 = -\frac{1}{2}[1-sign(F)], \tag{12.23}$$

$$\frac{1}{2}[1-sign(F_{k-1})] - 1 = -\frac{1}{2}[1+sign(F_{k-1})], \tag{12.24}$$

as follows

$$-\frac{1}{\Delta z}\left\{-F\frac{1}{2}[1-sign(F)] + \frac{\Gamma}{\Delta z_h}\right\}(\Phi_{k+1} - \Phi)$$

$$-\frac{1}{\Delta z}\left\{F_{k-1}\frac{1}{2}[1+sign(F_{k-1})] + \frac{\Gamma_{k-1}}{\Delta z_{h,k-1}}\right\}(\Phi_{k-1} - \Phi)$$

$$= -\frac{1}{\Delta z}\left(-F\xi_{15-} + \frac{\Gamma}{\Delta z_h}\right)(\Phi_{k+1} - \Phi) - \frac{1}{\Delta z}\left(F_{k-1}\xi_{16-} + \frac{\Gamma_{k-1}}{\Delta z_{h,k-1}}\right)(\Phi_{k-1} - \Phi)$$

$$= \left(b_{k+1} + b_{k-1}\right)\Phi - b_{k-1}\Phi_{k-1} - b_{k+1}\Phi_{k+1}. \tag{12.25}$$

An important property of the linearized coefficients b_{k+1}, b_{k-1} should be noted, namely that they are not negative $b_{k+1} \geq 0$, $b_{k-1} \geq 0$. Two consequences of this property are:

1) An increase in $\Phi_{k\pm 1}$ in the neighboring locations to k leads to an increase in Φ and vice versa;

2) If the coefficients $b_{k\pm 1}$ are equal to zero (e.g. due to the fact that $\gamma_{z,k\pm 1} = 0$), the difference $\Phi_{k\pm 1} - \Phi$ cannot influence the value of Φ, so that there is real decoupling.

3) If the diffusion is neglected, the flow leaving the cell due to convection does not influence the specific properties Φ of the velocity field in the cell. Inlet flows can influence the specific properties of the field in the cell only if they have specific properties that differ from those of the field in the cell considered.

The coefficient b contains information about the propagation speed of a disturbance Φ, namely $\Delta z / \Delta \tau \approx$ convection + diffusion velocity. The limitation of the time step is associated with the material velocity of the quality Φ. This method is very appropriate for predominant convection

$$Pe = F\Delta z_h / \Gamma < 2, \tag{12.26}$$

see Patankar [47] (1980). The local grid *Peclet* number Pe is the ratio of the amounts of the property Φ transported by convection and diffusion, respectively. Large values of Pe, e.g. $|Pe| > 10$, mean that convection is predominant and small values that diffusion is predominant. For one-dimensional processes without sources of Φ and with predominant convection, linearization of the profile leads to overestimation of the diffusion component of the flow. This is characteristic for coarse meshes, leading to $|Pe| > 10$. This consideration causes some investigators to look for a more realistic profile of the function $\Phi = \Phi(z)$ as a basis for construction of discretization schemes without a strong upper limitation on mesh size.

12.6 Discretization of the concentration equations

Equation (12.4) is discretized following the procedure already described in Section 12.5. The result is given in *Appendix 12.2*. The abbreviated notation is

$$\alpha_{la}\rho_{la}\gamma_{va}\frac{C_{il}-C_{ila}}{\Delta\tau}-\sum_{m=1}^{6}b_{lm}(C_{ilm}-C_{il})+\frac{1}{2}(\gamma_{v}+\gamma_{va})\mu_{l}^{+}C_{il}=\frac{1}{2}(\gamma_{v}+\gamma_{va})DC_{il}$$

(12.27)

where

$$b_{lm}=B_{lm-}+\beta_{m}D_{ilm}^{*}/\Delta L_{hm}.$$

(12.28)

Solving with respect to the unknown concentration we obtain

$$C_{il}=\frac{\alpha_{la}\rho_{la}\gamma_{va}C_{ila}+\Delta\tau\left[\frac{1}{2}(\gamma_{v}+\gamma_{va})DC_{il}+\sum_{m=1}^{6}b_{lm}C_{ilm}\right]}{\alpha_{la}\rho_{la}\gamma_{va}+\Delta\tau\left[\sum_{m=1}^{6}b_{lm}+\frac{1}{2}(\gamma_{v}+\gamma_{va})\mu_{l}^{+}\right]}.$$

For the case of just an originating velocity field, $\alpha_{la}=0$ and

$$\sum_{m=1}^{6}b_{lm}+\frac{1}{2}(\gamma_{v}+\gamma_{va})\mu_{l}^{+}>0,$$

we have

$$C_{il}=\frac{\frac{1}{2}(\gamma_{v}+\gamma_{va})DC_{il}+\sum_{m=1}^{6}b_{lm}C_{ilm}}{\frac{1}{2}(\gamma_{v}+\gamma_{va})\mu_{l}^{+}+\sum_{m=1}^{6}b_{lm}}.$$

The diffusion coefficients D_{ilm}^{*} are defined at the surfaces m of the mass cell such that the diffusion flows through these surfaces are continuous. Because the diffusion coefficients D_{il}^{*} are located by definition in the center of the cell, with the coefficients acting at the cell surfaces required, it is necessary to compute these by some kind of averaging procedure. Harmonic averaging, described in *Appendix 12.3*, is the natural choice for diffusion processes. *Patankar* [46] (1978) shows that this averaging procedure gives better results than simple arithmetic averaging for steady-state heat conduction problems in multi-composite materials and for diffusion in single-phase flows. This type of averaging procedure has the following interesting properties:

(a) If one of the diffusion coefficients D_{il}^{*} or $D_{il,i+1}^{*}$ is equal to zero, D_{il1}^{*} is likewise equal to zero, meaning that no diffusion can take place because one of the two elementary cells hinders it, this matching behavior in reality.

(b) The above effect also takes place if in one of the two neighboring cells the velocity field l is does not exists.

The superficial mass flow rate G_{lm} of the velocity field l through the surfaces $m = 1$ through 6, computed by means of the donor-cell concept is also given in *Appendix 12.2*. In this case the absolute value of the *Peclet* number is

$$|Pe| = |G_{lm}| \Delta L_{hm} / D^*_{ilm}, \quad D^*_{ilm} > 0 \tag{12.29}$$

$$|Pe| \to +\infty \quad D^*_{ilm} = 0, \tag{12.30}$$

where ΔL_{hm} is the distance between the centers of the elementary cells belonging to the surface m. The mass flows G_{lm} are used simultaneously in several different conservation equations. It is advisable to compute these once at the beginning of the cycle. Because different dependent variables have different diffusion coefficients, the *diffusion* part is specific to each equation.

12.7 Discretization of the entropy equation

Equation (12.7) is discretized following the procedure already described in Section 12.5. The result is

$$\alpha_{la} \rho_{la} \gamma_{va} \frac{s_l - s_{la}}{\Delta \tau} - \sum_{m=1}^{6} B_{lm-}(s_{lm} - s_l) + \frac{1}{2}(\gamma_v + \gamma_{va})\mu_l^+ s_l = \frac{1}{2}(\gamma_v + \gamma_{va}) Ds_l^*$$

$$\tag{12.31}$$

where

$$Ds_l^* = Ds_l + \frac{1}{\gamma_v T_l} \sum_{m=1}^{6} \beta_m \frac{D_{lm}^T}{\Delta L_{hm}}(T_{lm} - T_l). \tag{12.32}$$

The computation of the harmonic averaged thermal conductivity coefficients is given in *Appendix 12.3*. Solving with respect to the unknown specific entropy we obtain

$$s_l = \frac{\alpha_{la} \rho_{la} \gamma_{va} s_{la} + \Delta \tau \left[\frac{1}{2}(\gamma_v + \gamma_{va}) Ds_l^* + \sum_{m=1}^{6} B_{lm-} s_{lm} \right]}{\alpha_{la} \rho_{la} \gamma_{va} + \Delta \tau \left[\sum_{m=1}^{6} B_{lm-} + \frac{1}{2}(\gamma_v + \gamma_{va})\mu_l^+ \right]}.$$

For the case of just an originating velocity field, $\alpha_{la} = 0$ and

$$\sum_{m=1}^{6} B_{lm-} + \frac{1}{2}(\gamma_v + \gamma_{va})\mu_l^+ > 0,$$

we have

$$s_l = \frac{\frac{1}{2}(\gamma_v + \gamma_{va})Ds_l^* + \sum_{m=1}^{6} B_{lm-}s_{lm}}{\frac{1}{2}(\gamma_v + \gamma_{va})\mu_l^+ + \sum_{m=1}^{6} B_{lm-}}.$$

12.8 Discretization of the temperature equation

In place of the entropy equation, the temperature equation can also be used. The temperature equation

$$\rho_l c_{pl} \left[\alpha_l \gamma_v \frac{\partial T_l}{\partial \tau} + (\alpha_l^e \mathbf{V}_l \gamma \cdot \nabla) T_l \right] - \nabla \cdot (\alpha_l^e \lambda_l^* \nabla T)$$

$$- \left[1 - \rho_l \left(\frac{\partial h_l}{\partial p}\right)_{T_l, all\_C's}\right] \left[\alpha_l \gamma_v \frac{\partial p}{\partial \tau} + (\alpha_l^e \mathbf{V}_l \gamma \cdot \nabla) p \right]$$

$$= \gamma_v \left[DT_l^N - T_l \sum_{i=2}^{i_{\max}} \Delta s_{il}^{np} (\mu_{il} - C_{il}\mu_l) \right] \quad (12.33)$$

is discretized following the procedure already described in Section 12.5. The result is

$$\alpha_{la}\rho_{la}c_{pa}\gamma_{va} \frac{T_l - T_{la}}{\Delta \tau} - \sum_{m=1}^{6} \left(B_{lm-}c_{pm} + \beta_m \frac{\alpha_{lm}^* \lambda_{lm}}{\Delta L_{hm}} \right)(T_{lm} - T_l)$$

$$= \frac{1}{2}(\gamma_v + \gamma_{va})\left[DT_l^N - T_l \sum_{i=2}^{i_{\max}} \Delta s_{il}^{np} (\mu_{il} - C_{il}\mu_l) \right]$$

$$+ \left[1 - \rho_l \left(\frac{\partial h_l}{\partial p}\right)_{T_l, all\_C's}\right]_a \left[\alpha_{la}\gamma_{va} \frac{p - p_a}{\Delta \tau} - \sum_{m=1}^{6} \frac{B_{lm-}}{\rho_{lm}}(p_m - p) \right]. \quad (12.34)$$

Computation of the harmonic averaged thermal conductivity coefficients is shown in *Appendix 12.3*. Bearing in mind that the linearized source terms can be rewritten as linear functions of the temperatures

$$DT_l^N = c_l^T - \sum_{k=1}^{3} a_{lk}^{*T} T_k ,\qquad(12.35)$$

we obtain the following form for the above equation

$$\begin{aligned}&\left[\frac{\alpha_{la}\rho_{la}c_{pa}}{\Delta\tau} + a_{ll}^{*T} + \frac{1}{\gamma_{va}}\sum_{m=1}^{6}\left(B_{lm-}c_{pm} + \beta_m \frac{\alpha_{lm}^* \lambda_{lm}}{\Delta L_{hm}}\right)\right.\\ &\left.+\frac{1}{2}\left(1+\frac{\gamma_v}{\gamma_{va}}\right)\sum_{i=2}^{i_{max}}\Delta s_{il}^{np}\left(\mu_{il}-C_{il}\mu_l\right)\right]T_l + \frac{1}{2}\left(1+\frac{\gamma_v}{\gamma_{va}}\right)\sum_{k\ne l}a_{lk}^{*T}T_k\\ &= \frac{\alpha_{la}\rho_{la}c_{pa}}{\Delta\tau}T_{la} + \frac{1}{2}\left(1+\frac{\gamma_v}{\gamma_{va}}\right)c_l^T + \frac{1}{\gamma_{va}}\sum_{m=1}^{6}\left(B_{lm-}c_{pm}+\beta_m\frac{\alpha_{lm}^*\lambda_{lm}}{\Delta L_{hm}}\right)T_{lm}\\ &+\left[1-\rho_l\left(\frac{\partial h_l}{\partial p}\right)_{T_l,all\_C's}\right]_a\left[\alpha_{la}\frac{p-p_a}{\Delta\tau} - \frac{1}{\gamma_{va}}\sum_{m=1}^{6}\frac{B_{lm-}}{\rho_{lm}}(p_m-p)\right],\end{aligned}\qquad(12.36)$$

or in a short notation

$$a_{ll}^T T_l + \sum_{k\ne l} a_{lk}^T T_k = b_l ,\qquad(12.37)$$

or in a matrix notation for all velocity fields

$$\mathbf{AT = B}\qquad(12.38)$$

where

$$Dp_l = \left[1-\rho_l\left(\frac{\partial h_l}{\partial p}\right)_{T_l,all\_C's}\right]_a\left[\alpha_{la}\frac{p-p_a}{\Delta\tau} - \frac{1}{\gamma_{va}}\sum_{m=1}^{6}\frac{B_{lm-}}{\rho_{lm}}(p_m-p)\right],\qquad(12.39)$$

$$b_{lm-}^* = B_{lm-}c_{pm} + \beta_m\frac{\alpha_{lm}^*\lambda_{lm}}{\Delta L_{hm}},\qquad(12.40)$$

$$a_{ll}^T = \frac{\alpha_{la}\rho_{la}c_{pa}}{\Delta\tau} + \frac{1}{2}\left(1+\frac{\gamma_v}{\gamma_{va}}\right)\left[a_{ll}^{*T} + \sum_{i=2}^{i_{max}}\Delta s_{il}^{np}\left(\mu_{il}-C_{il}\mu_l\right)\right] + \frac{1}{\gamma_{va}}\sum_{m=1}^{6}b_{lm-}^*,$$
(12.41)

$$a_{lm}^T = \frac{1}{2}\left(1+\frac{\gamma_v}{\gamma_{va}}\right)a_{lm}^{*T} \quad \text{for } l \neq m,$$
(12.42)

$$b_l = \frac{\alpha_{la}\rho_{la}c_{pa}}{\Delta\tau}T_{la} + \frac{1}{2}\left(1+\frac{\gamma_v}{\gamma_{va}}\right)c_l^T + \frac{1}{\gamma_{va}}\sum_{m=1}^{6}b_{lm-}^*T_{lm} + Dp_l.$$
(12.43)

This algebraic system can be solved for the temperatures for

$$\det \mathbf{A} = a_{11}a_{22}a_{33} + a_{12}a_{23}a_{31} + a_{21}a_{32}a_{13} - a_{31}a_{22}a_{13} - a_{32}a_{23}a_{11} - a_{12}a_{21}a_{33} \neq 0.$$
(12.44)

The result is

$$\mathbf{T} = \mathbf{A}^{-1}\mathbf{B}$$
(12.45)

with components

$$T_l = \left(\sum_{m=1}^{3} b_m \bar{a}_{lm}\right)/\det \mathbf{A},$$
(12.46)

The $\bar{a}$ values are given in *Appendix 12.5*.

An important property of the diagonal elements of the matrix $\mathbf{A}$

$$a_{ll}^T = \frac{\alpha_{la}\rho_{la}c_{pa}}{\Delta\tau} + \frac{1}{2}\left(1+\frac{\gamma_v}{\gamma_{va}}\right)\left[a_{ll}^{*T} + \sum_{i=2}^{i_{max}}\Delta s_{il}^{np}\left(\mu_{il}-C_{il}\mu_l\right)\right] + \frac{1}{\gamma_{va}}\sum_{m=1}^{6}b_{lm-}^* \quad (12.47)$$

should be noted: The m-th diagonal element equal to zero indicates that at that time the velocity field does not exist, and will not originate in the next time step. For this reason, the rank of the matrix is reduced by one. Even if the field does not exist but will originate in the next time step either by convection or by mass transfer from the neighboring field, or from other mass sources, or by an arbitrary combination of these three processes, the diagonal element is not zero and the initial temperature is induced properly. It is obvious that, if one neglects all convective terms and mass sources, no initial value for a non-existent field can be defined. For numerical computations I recommend normalization of the diagonal elements $a_{ll,norm}^T = \left|a_{ll}^T\Delta\tau/(\rho_{la}c_{pla})\right|$ and their comparison with e.g. $\varepsilon = 0.001$. If

$a_{ll,norm}^T < \varepsilon$, there is no velocity field l, and the velocity field l will not originate in the next time step.

The three equations obtained in this manner are used to construct the pressure-temperature coupling. I use the term partial decoupling of the temperature equations from each other (PDTE) to describe this decoupling procedure. This step is extremely important for creation of a stable numerical algorithm even when using first order donor cell discretization. The coupling coefficients between the velocity fields correspond to each of the flow patterns modeled. For a number of these the coupling is strong e.g. bubble-liquid, for others not so strong, e.g. large diameter droplets-gas. Coupling is non-linear in all cases and must be resolved by iteration.

12.9. Physical significance of the necessary convergence condition

Writing for each cell (i,j,k) a single algebraic equation (12.27), I obtain a system of algebraic equations for all C_{il} values. This system has a specific seven diagonal structure. The coefficient matrix is defined positive. Suppose that all other variables except the C_{il} values are known. A possible method for the solution of this system for all C_{il} values is the *Gauss-Seidel* iterative method in one of its several variants. This method is not the most effective but is frequently used because of its simplicity. The necessary condition for convergence of this method is the predominance of the elements on the main diagonal compared to the other elements

$$|b_l| \ge \sum_{m=1}^{6} |b_{lm}| \quad \text{for all equations,} \tag{12.48}$$

$$|b_l| > \sum_{m=1}^{6} |b_{lm}| \quad \text{at least for one equation.} \tag{12.49}$$

This is the well-known *Scarborough* criterion, see [63] (1958). Because this is only a necessary condition, convergence is possible even if this criterion is violated. But satisfaction of this condition gives the confidence that the algebraic system can be solved at least with one iteration method, *Patankar* [47] (1980). Bearing in mind that all of the elements of the sum

$$b = \gamma_v (\alpha_{la} \rho_{la} / \Delta\tau + \mu_l^+) + \sum_{m=1}^{6} b_{lm} \tag{12.50}$$

are non-negative, the *Scarborough* criterion reduces to

12.9. Physical significance of the necessary convergence condition

$$\gamma_v (\alpha_{la}\rho_{la}/\Delta\tau + \mu_l^+) \geq 0 \quad \text{for all equations} \tag{12.51}$$

$$\gamma_v (\alpha_{la}\rho_{la}/\Delta\tau + \mu_l^+) > 0 \quad \text{at least for one equation.} \tag{12.52}$$

If $\alpha_{la} > 0$ at least for one elementary cell, the above conditions are always satisfied. If the velocity field l is missing from the integration domain, $\alpha_{la} = 0$ for all cells, satisfaction of $\mu_l^+ > 0$ is necessary for at least one single point.

The *Scarbourough* criterion will now be considered from a different point of view. Assuming that the properties associated with the flow leaving the cell are known at the new point in time and that the properties associated with the flow entering the cell are the best guesses for the new point in time, and solving Eq. (12.27) with respect to C_{il}, the following is obtained

$$C_{il} = \frac{\alpha_{la}\rho_{la}\gamma_{va}C_{ila} + \Delta\tau\left[\dfrac{1}{2}(\gamma_v + \gamma_{va})DC_{il} + \sum_{m=1}^{6}b_{lm}C_{ilm}\right]}{\alpha_{la}\rho_{la}\gamma_{va} + \Delta\tau\left[\sum_{m=1}^{6}b_{lm} + \dfrac{1}{2}(\gamma_v + \gamma_{va})\mu_l^+\right]}. \tag{12.53}$$

This resembles the use of the point *Jacobi* method for the solution of the above equation. The method consists in a successive visiting of all cells in the definition domain as many times as necessary until the improvement in the solution from iteration to iteration falls below a defined small value. Where space velocity distribution is known, this method works without any problems. Even though this method has the lowest convergence rate compared to other existing methods, this illustrates an important feature, namely the computation of the initial values for field properties when the field is in the process of origination within the current time step, $\alpha_{la} = 0$, i.e.

$$C_{il} = \frac{\dfrac{1}{2}(\gamma_v + \gamma_{va})DC_{il} + \sum_{m=1}^{6}b_{lm}C_{ilm}}{\dfrac{1}{2}(\gamma_v + \gamma_{va})\mu_l^+ + \sum_{m=1}^{6}b_{lm}}. \tag{12.54}$$

Note that initial value for C_{ila} is not required in this case. Only the following is required

$$\dfrac{1}{2}(\gamma_v + \gamma_{va})\mu_l^+ + \sum_{m=1}^{6}b_{lm} > 0. \tag{12.55}$$

The velocity field can originate in several possible ways. Two of these are as follows:

a) no convection and diffusion takes place but the source terms differ from zero

$$C_{il} = DC_{il} / \mu_l^+ ; \qquad (12.56)$$

b) source terms in the cell are zero but convection or diffusion takes place

$$C_{il} = \left(\sum_{m=1}^{6} b_{lm} C_{ilm} \right) / \sum_{m=1}^{6} b_{lm} . \qquad (12.57)$$

The *Scarborough* criterion is not satisfied if

$$\alpha_{la} \rho_{la} \gamma_{va} + \Delta \tau \left[\sum_{m=1}^{6} b_{lm} + \frac{1}{2} (\gamma_v + \gamma_{va}) \mu_l^+ \right] = 0 \qquad (12.58)$$

for all cells. In this case C_{il} is, however, not defined in accordance with Eq. (12.53). In other words, this criterion simply states that if the field (a) does not exist in the entire integration domain and (b) will not originate at the next moment, its specific properties, such as concentrations, entropies, etc. are not defined.

Note: In creation of the numerical method, it is necessary to set the *initial values* for the field properties once before starting simulation so as to avoid multiplication with undefined numbers $\alpha_{la} \rho_{la} C_{ila} / \Delta \tau$. If the field disappears during the transient, the field will retain its specific properties from the last time step. On origination of a field, its specific properties are obtained by averaging the specific properties of the flows entering it with weighting coefficients equal to the corresponding mass flow divided by the net mass flow into the cell, as obtained say using Eq. (12.54). In this case the previous value for the specific properties, such as C_{ila}, etc. do not influence the result.

12.10. Implicit discretization of momentum equations

Figure 12.4 shows the control volume for discretization of the momentum equation in the r direction. Compared to the mass conservation cell at the centre, for which pressure is defined, the volume r is displaced by $\Delta r / 2$ in the r direction. Indices 1 through 6 are used to denote the front and back surfaces of the elementary cell in r, θ, and z directions, respectively, similar to the manner shown in Fig. 12.3.

The pressure difference $p_{i+1} - p$ is the driving force for the velocity change in the r direction. As already mentioned, this type of construction for a staggered mesh of elementary cells is really only necessary in the case of schemes with the first order of accuracy for the space discretization of the convective terms in the momentum equations. The reasons behind this are as follows: If the mesh is not staggered for the case

$$p_{i+1} = p_{i-1} \tag{12.59}$$

and

$$p \neq p_{i\pm 1}, \tag{12.60}$$

the pressure gradient

$$(p_{i+1} - p_{i-1}) / (\Delta r_h + \Delta r_{h,i-1}) \tag{12.61}$$

is equal to zero which means that in spite of the non-uniform pressure field the velocity component u is then not influenced in any way by the pressure force, which does not correspond to reality. Such schemes are numerically unstable.

By definition, the permeabilities, the porosities, mass flow rates, densities, and volumetric fractions are not defined at those points where they are needed for discretization of the momentum equation. As a result, surface properties for the staggered cells are derived by surface weighting of the surface properties for the adjacent non-staggered cells. Volumetric properties of the staggered cells are computed by volumetric weighting of the corresponding volumetric properties for the adjacent non-staggered cells. For velocities transporting material between two adjacent non-staggered cells the donor cell principle is used to compute the convected property.

This calculation entails additional computational effort. Despite this, this method has been in widespread use for single-phase flows for the last three decades. A number of examples of the way in which cell mixing properties are computed will now be given.

Instead of direct discretization of the momentum equations, I derive the discretized working form through the following steps:

1. Implicit discretization of the momentum equation.

2. Implicit discretization of the mass conservation equation for the same velocity field.
3. Subtraction of the mass conservation equation obtained in this way multiplication by u_l from the discretized momentum equation.

The result is given in *Appendix 12.4*. Note that the centrifugal force gives the effective force in the r-direction which results from fluid motion in the θ direction. For computation of the centrifugal force in the relatively large elementary cell both components for surfaces 3 and 4 are added, as was performed during the derivation of the momentum equation itself. For a very small $\Delta\theta$ this yields the following

$$\lim_{\Delta\theta \to 0} \frac{\sin(\Delta\theta/2)}{\Delta\theta} = \frac{1}{2} \tag{12.62}$$

which reduces the expression for the centrifugal force to just this expression in the momentum equation. The viscous components that counteract the centrifugal force are computed in an analogous manner.

The momentum equation is now rewritten in the following compact form

$$\left[\frac{\alpha_{lua}\rho_{lua}}{\Delta\tau} - \sum_{\substack{m=1 \\ m \neq l}}^{3} a_{lm} + c_{lu}|u_l| + a_{l,conv} + \mu_{wl} - \mu_{lw} \right] u_l + \sum_{\substack{m=1 \\ m \neq l}}^{3} a_{lm} u_m$$

$$= b_{l,conv} + \alpha_{lua}\rho_{lua}(u_{la}/\Delta\tau - g_u) - \sum_{\substack{m=1 \\ m \neq l}}^{3} b_{lm} + \mu_{wl}u_{wl} - \mu_{lw}u_{lw}$$

$$- \alpha_{lua} \frac{1}{\Delta r_h} \frac{\gamma_{rua}}{\gamma_{vu}} (p_{i+1} - p), \tag{12.63}$$

where

$$a_{l,conv} = -\frac{1}{\gamma_{vu}} \sum_{m=1}^{6} bu_{lm}, \tag{12.64}$$

$$b_{l,conv} = -\frac{1}{\gamma_{vu}} \left(\bar{R}_{ul} + \sum_{m=1}^{6} bu_{lm}u_{lm} \right), \tag{12.65}$$

represents the discretized convective term.

$$\bar{R}_{ul} = -\frac{\sin(\Delta\theta/2)}{(r+\Delta r_h/2)^\kappa \Delta\theta}\{(\alpha_l\rho_l v_l\gamma_\theta)_3 v_{l3} + (\alpha_l\rho_l v_l\gamma_\theta)_4 v_{l4}\}$$

$$-\frac{2}{(r+\Delta r_h/2)^\kappa}\left\{(\alpha\rho v)_{l3}\left[\left(\frac{\partial v_l}{\partial\theta}\right)_3 + u_{l3}\right]\gamma_3 + (\alpha\rho v)_{l4}\left[\left(\frac{\partial v_l}{\partial\theta}\right)_4 + u_{l4}\right]\gamma_4\right\}.$$
(12.66)

All velocities u_l are taken as being in the new time plane, a so called implicit formulation. This process is repeated for the chosen number of the velocity fields. The result is a system of algebraic equations with respect to the velocities

$$\mathbf{Au} = \mathbf{B} - \mathbf{a}(p_{i+1} - p),$$
(12.67)

where the elements of the matrix $\mathbf{A}$ are

$$a_{lm} = a_{lm}^* - \mu_{ml},$$
(12.68)

$$a_{lm}^* = a_{ml}^* = -\bar{c}_{ml}^d - \bar{c}_{lm}^d - (\bar{c}_{ml}^{vm} + \bar{c}_{lm}^{vm})/\Delta\tau \quad \text{for} \quad m \neq l,$$
(12.69)

$$a_{ll} = \frac{\alpha_{lua}\rho_{lua}}{\Delta\tau} - \sum_{\substack{m=1\\m\neq l}}^{3} a_{lm} + c_{lu}|u_l| + a_{l,conv} + \mu_{wl} - \mu_{lw},$$
(12.70)

the elements of the algebraic vector $\mathbf{B}$ are

$$b_l = b_{l,conv} + \alpha_{lua}\rho_{lua}(u_{la}/\Delta\tau - g_u) - \sum_{\substack{m=1\\m\neq l}}^{3} b_{lm} + \mu_{wl}u_{wl} - \mu_{lw}u_{lw},$$
(12.71)

where

$$b_{lm} = -b_{ml} = (\bar{c}_{ml}^{vm} + \bar{c}_{lm}^{vm})(u_{ma} - u_{la})/\Delta\tau.$$
(12.72)

Note that by definition, if one velocity field does not exist, $\alpha_l = 0$, so that the coefficients describing its coupling with the other fields are then equal to zero, $\bar{c}_{ml}^{vm} = 0$, $\bar{c}_{lm}^{vm} = 0$. This algebraic system can be solved to derive the l velocities provided that

$$\det\mathbf{A} = a_{11}a_{22}a_{33} + a_{12}a_{23}a_{31} + a_{21}a_{32}a_{13} - a_{31}a_{22}a_{13} - a_{32}a_{23}a_{11} - a_{12}a_{21}a_{33} \neq 0.$$
(12.73)

The result is

$$\boxed{\mathbf{u} = \mathbf{du} - \mathbf{RU}\left(p_{i+1} - p\right),} \qquad (12.74)$$

where

$$\mathbf{du} = \mathbf{A}^{-1}\mathbf{B} \qquad (12.75)$$

with components

$$du_l = \left(\sum_{m=1}^{3} b_m \bar{a}_{lm}\right) / \det \mathbf{A}, \qquad (12.76)$$

and

$$\mathbf{RU} = \mathbf{A}^{-1}\mathbf{a}, \qquad (12.77)$$

with components

$$RU_l = \frac{1}{\Delta r_h} \frac{\gamma_{rua}}{\gamma_{vu}} \left(\sum_{m=1}^{3} \alpha_{mua} \bar{a}_{lm}\right) / \det \mathbf{A}, \qquad (12.78)$$

and the $\bar{a}$ values given in *Appendix 12.5*. Equation (12.74) shows that applying a spatial pressure difference to a multi-field mixture with different field densities results in relative motion between the fields. This relative motion between two adjacent fields results in forces that act at the interfaces.

An important property of the diagonal elements

$$\frac{\alpha_{lua}\rho_{lua}}{\Delta\tau} - \sum_{\substack{m=1 \\ m\neq l}}^{3} a_{lm} + c_{lu}|u_l| + a_{l,conv} + \mu_{wl} - \mu_{lw} \qquad (12.79)$$

of the matrix **A** should be noted: The *m*-th diagonal element equal to zero indicates that at that time the velocity field does not exist, and will not originate in the next time step. For this reason, the rank of the matrix is reduced by one. Even if the field does not exist, but will originate in the next time step either by convection, or by mass transfer from the neighboring field, or from other mass sources, or by an arbitrary combination of these three processes, the diagonal element is not zero and the initial velocity is induced properly.

It is obvious that, if one neglects all convective terms and mass sources, no initial value for a non-existent field can be defined. For numerical computations we

12.10. Implicit discretization of momentum equations

recommend normalization of the diagonal elements $a_{ll,norm} = |a_{ll}\Delta\tau/\rho_l|$ and their comparison with e.g. $\varepsilon = 0.001$. If $a_{ll,norm} < \varepsilon$, there is no velocity field l, and the velocity field l will not originate in the next time step.

If the convection, diffusion and mass source terms are disregarded, the matrix **A** is symmetric and the expressions for the relative velocities become very simple.

The three equations obtained in this manner are used to construct the pressure-velocity coupling. I use the term partial decoupling of the momentum equations from each other (PDME) to describe this decoupling procedure. Examples of the usefulness of this method are given in [24]. Note the difference between this procedure and the decoupling procedure used in COBRA-TF [18], where decoupling is performed with a lower degree of implicitness by solving for the directional mass flow rates instead of for the velocities. This step is extremely important for creation of a stable numerical algorithm even when using first order donor cell discretization. The coupling coefficients between the velocity fields correspond to each of the flow patterns modeled. For a number of these the coupling is strong, e.g. bubble-liquid, for others not so strong, e.g. large diameter droplets-gas. Coupling is non-linear in all cases and must be resolved by iteration.

I derive the discretized working form of the momentum equations in the other two directions analogously to Eq. (12.63). The corresponding control volumes are shown in Figs. 12.5 and 12.6. The result is given in *Appendixes 12.6* and *12.7*. The abbreviated notations are

$$\left[\frac{\alpha_{lva}\rho_{lva}}{\Delta\tau} - \sum_{\substack{m=1 \\ m\neq l}}^{3} a_{lm} + c_{lv}|v_l| + a_{l,conv} + \mu_{wl} - \mu_{lw} \right] v_l + \sum_{\substack{m=1 \\ m\neq l}}^{3} a_{lm} v_m$$

$$= b_{l,conv} + \alpha_{lva}\rho_{lva}\left(v_{la}/\Delta\tau - g_v\right) - \sum_{\substack{m=1 \\ m\neq l}}^{3} b_{lm} + \mu_{wl}v_{wl} - \mu_{lw}v_{lw}$$

$$-\alpha_{lva}\frac{1}{r^\kappa \Delta\theta_h}\frac{\gamma_{\theta va}}{\gamma_{vv}}\left(p_{j+1} - p\right), \qquad (12.80)$$

where

$$a_{l,conv} = -\frac{1}{\gamma_{vv}}\sum_{m=1}^{6} bv_{lm}, \qquad (12.81)$$

$$b_{l,conv} = -\frac{1}{\gamma_{vv}}\left(\overline{R}_{vl} + \sum_{m=1}^{6} bv_{lm} v_{lm}\right), \qquad (12.82)$$

or

$$\mathbf{v} = \mathbf{dv} - \mathbf{RV}(p_{j+1} - p), \qquad (12.83)$$

and

$$\left[\frac{\alpha_{lwa}\rho_{lwa}}{\Delta\tau} - \sum_{\substack{m=1 \\ m\neq l}}^{3} a_{lm} \mid c_{lw}\mid w_l\mid + a_{l,conv} + \mu_{wl} - \mu_{lw}\right] w_l + \sum_{\substack{m=1 \\ m\neq l}}^{3} a_{lm} w_m$$

$$= b_{l,conv} + \alpha_{lwa}\rho_{lwa}\left(w_{la}/\Delta\tau - g_w\right) - \sum_{\substack{m=1 \\ m\neq l}}^{3} b_{lm} + \mu_{wl} w_{wl} - \mu_{lw} w_{lw}$$

$$-\alpha_{lwa}\frac{1}{\Delta z_h}\frac{\gamma_{zwa}}{\gamma_{vw}}(p_{k+1} - p), \qquad (12.84)$$

where

$$a_{l,conv} = -\frac{1}{\gamma_{vw}}\sum_{m=1}^{6} bw_{lm}, \qquad (12.85)$$

$$b_{l,conv} = -\frac{1}{\gamma_{vw}}\left(\overline{R}_{wl} + \sum_{m=1}^{6} bw_{lm} w_{lm}\right), \qquad (12.86)$$

or

$$\mathbf{w} = \mathbf{dw} - \mathbf{RW}(p_{k+1} - p). \qquad (12.87)$$

The a_l and b_l terms reflect the actions of drag and of added mass forces. The $a_{l,conv}$ and $b_{l,conv}$ terms reflect the actions of the spatial inertia and viscous forces. This general structure of Eqs. (12.63), (12.80), and (12.84) proved its worth during testing of the code, permitting effects to be introduced step by step.

Note that a non-slip boundary condition at the wall is easily introduced by computing the wall friction force resisting the flow as follows

Wall friction force =

$$\alpha_l^* \rho_l v \left[\frac{1}{(r+\Delta r/2)^\kappa r_h^\kappa \Delta \theta} \left(\frac{1-\gamma_3}{\Delta \theta_h} + \frac{1-\gamma_4}{\Delta \theta_{h,j-1}} \right) + \frac{1}{\Delta z} \left(\frac{1-\gamma_5}{\Delta z_h} + \frac{1-\gamma_6}{\Delta z_{h,k-1}} \right) \right] 2u_l$$

$$-\alpha_l^* \rho_l v \frac{2\sin(\Delta \theta/2)}{(r+\Delta r/2)^{2\kappa} \Delta \theta} \left(\frac{1-\gamma_3}{\Delta \theta_h} + \frac{1-\gamma_4}{\Delta \theta_{h,j-1}} \right) 2v_l \qquad (12.88)$$

where $\alpha_l^* = 1$ for l = continuum and l = disperse.

The forms of the momentum equations obtained in this way are

$$V_{lm}^n = dV_{lm}^n - RVel_{lm}(p_m - p), \qquad (12.89)$$

$$dV_l^{n=1,6} = \left(du_l, -du_{l,i-1}, dv_l, -dv_{l,j-1}, dw_l, -dw_{l,k-1} \right), \qquad (12.90)$$

$$RVel_l^{n=1,6} = \left(RU_l, RU_{l,i-1}, RV_l, RV_{l,j-1}, RW_l, RW_{l,k-1} \right). \qquad (12.91)$$

Remember that the normal velocities, Eq. (12.12), are defined as positive if directed from the control volume to the environment on each of the six surfaces m of the computational cell.

12.11 Pressure equations for IVA2 and IVA3 computer codes

The mixture volume conservation equation (5.188) derived in Chapter 5 and [29] is

$$\frac{\gamma_v}{\rho a^2} \frac{\partial p}{\partial \tau} + \sum_{l=1}^{3} \frac{\alpha_l}{\rho_l a_l^2} \mathbf{V}_l \gamma . \nabla p + \nabla . \left(\sum_{l=1}^{3} \alpha_l \mathbf{V}_l \gamma \right) = \sum_{l=1}^{3} D\alpha_l - \frac{\partial \gamma_v}{\partial \tau}. \qquad (12.92)$$

Here a is the sonic velocity in a homogeneous multi-phase mixture defined as follows

$$\frac{1}{\rho a^2} = \sum_{l=1}^{3} \frac{\alpha_l}{\rho_l a_l^2}, \tag{12.93}$$

and

$$\rho = \sum_{l=1}^{3} \alpha_l \rho_l \tag{12.94}$$

is the mixture density. The mixture volume conservation equation is discretized directly using the donor cell concept in the IVA2 computer code. The result is Eq. (38) in [21],

$$\gamma_{va} \sum_{l=1}^{3} \frac{\alpha_{al}}{\rho_{la} a_{la}^2} \frac{p - p_a}{\Delta \tau} + \sum_{m=1}^{6} \beta_m \sum_{l=1}^{3} \alpha^*_{lm} V^n_{lm} = \sum_{l=1}^{3} D\alpha_l - \frac{\gamma_v - \gamma_{va}}{\Delta \tau} \tag{12.95}$$

where

$$D\alpha_l = \frac{1}{2}(\gamma_v + \gamma_{va}) \frac{1}{\rho_{la}} \left\{ \mu_l - \frac{1}{\rho_{la}} \left[\left(\frac{\partial \rho_l}{\partial s_l}\right)_a Ds_l^N + \sum_{i=2}^{i_{max}} \left(\frac{\partial \rho_l}{\partial C_{il}}\right)_a DC_{il}^N \right] \right\}$$

$$\approx \frac{1}{2}(\gamma_v + \gamma_{va}) \frac{1}{\rho_{la}} \left\{ \mu_l - \frac{1}{\rho_{la}} \left[\left(\frac{\partial \rho_l}{\partial T_l}\right)_{p,all\_C's} \frac{1}{c_{pl}} \left[DT_l^N - T_l \sum_{i=2}^{i_{max}} \Delta s_{il}^{np} (\mu_{il} - C_{il}\mu_l) \right] \right. \right.$$

$$\left. \left. + \sum_{i=2}^{i_{max}} \left(\frac{\partial \rho_l}{\partial C_{il}}\right)_{p,T_l,all\_C's\_except\_C_{il}} (\mu_{il} - C_{il}\mu_l) \right] \right\},$$

(12.96)

$$\alpha^*_{lm} = \xi_{lm+} \alpha_l + \xi_{lm-} \alpha_{lm} \left[1 + \frac{1}{\rho_{ma} a_{ma}^2} (p_m - p) \right], \tag{12.97}$$

$$Ds_l^N = Ds_l - \mu_l^+ s_l, \tag{12.98}$$

$$DC_{il}^N = DC_{il} - \mu_l^+ s_{il}. \tag{12.99}$$

Substituting the momentum equation (12.89) for the normal velocities

$$V^n_{lm} = dV^n_{lm} - RVel_{lm}(p_m - p),$$

one finally obtains the pressure equation used in IVA2:

$$c p + \sum_{m=1}^{6} c_m p_m = d, \quad (12.100)$$

where

$$c = \frac{\gamma_{va}}{\Delta \tau} \sum_{l=1}^{3} \frac{\alpha_{al}}{\rho_{la} a_{la}^2} - \sum_{m=1}^{6} c_m, \quad (12.101)$$

$$c_m = -\beta_m \sum_{l=1}^{3} \alpha_{lm}^* RVel_{lm}, \quad (12.102)$$

$$d = \sum_{l=1}^{3} D\alpha_l - \frac{\gamma_v - \gamma_{va}}{\Delta \tau} + p_a \frac{\gamma_{va}}{\Delta \tau} \sum_{l=1}^{3} \frac{\alpha_{al}}{\rho_{la} a_{la}^2} - \sum_{m=1}^{6} \beta_m \sum_{l=1}^{3} \alpha_{lm}^* V_{lm}^n. \quad (12.103)$$

The mixture volume conservation equation used in IVA3 was derived in the same way but starting with mass conservation equations that had already been discretized instead of starting with the analytical one. This ensures full compatibility of the pressure equation obtained in this way with the discretized mass conservation equations.

$$\gamma_v \sum_{l=1}^{3} \frac{\alpha_{la}}{\rho_{la}} (p_l - p_{la}) / \Delta \tau + \sum_{l=1}^{3} \frac{1}{\rho_{la}} \left\{ \sum_{m=1}^{6} \beta_m \left[\xi_{lm+} \alpha_l \rho_l + \xi_{lm-} (\alpha_l \rho_l)_m \right] V_{lm}^n \right\}$$

$$= \frac{\gamma_v}{\rho_{la}} \sum_{l=1}^{3} \mu_l - (\gamma_v - \gamma_{va}) / \Delta \tau. \quad (12.104)$$

Replacing the linearized state equation (3.137)

$$\rho_l - \rho_{la} = \frac{1}{a_{la}^2}(p - p_a) + \left(\frac{\partial \rho_l}{\partial s_l}\right)_{p, all\_C's} (s_l - s_{la})$$

$$+ \sum_{i=2}^{i_{max}} \left(\frac{\partial \rho_l}{\partial C_{i,l}}\right)_{p, s_l, all\_C's\_except\_C_{i,l}} (C_{i,l} - C_{i,la})$$

from [23] I obtain

$$(p-p_a)\frac{\gamma_v}{\Delta\tau}\sum_{l=1}^{3}\frac{\alpha_{al}}{\rho_{la}a_{la}^2}+\sum_{l=1}^{3}\frac{1}{\rho_{la}}\left\{\sum_{m=1}^{6}\beta_m\left[\xi_{lm+}\alpha_l\rho_l+\xi_{lm-}(\alpha_l\rho_l)_m\right]V_{lm}^n\right\}$$

$$=\sum_{l=1}^{3}D\alpha_l-\frac{\gamma_v-\gamma_{va}}{\Delta\tau}, \qquad (12.105)$$

where

$$D\alpha_l=\frac{\gamma_v}{\rho_{la}}\left\{\mu_l-\frac{\alpha_{la}}{\Delta\tau}\left[\left(\frac{\partial\rho_l}{\partial s_l}\right)_a(s_l-s_{la})+\sum_{i=2}^{i_{max}}\left(\frac{\partial\rho_l}{\partial C_{il}}\right)_a(C_{il}-C_{ila})\right]\right\}$$

$$\approx\frac{\gamma_v}{\rho_{la}}\left\{\mu_l-\alpha_{la}\sum_{i=2}^{i_{max}}\left[\left(\frac{\partial\rho_l}{\partial C_{il}}\right)_{p,T_l,all\_C's\_except\_C_{il}}-\left(\frac{\partial\rho_l}{\partial T_l}\right)_{p,all\_C's}\frac{T_l}{c_{pl}}\Delta s_{il}^{np}\right]\frac{C_{il}-C_{ila}}{\Delta\tau}\right\}.$$

$$(12.106)$$

Substituting the momentum equation (12.89) for the normal velocities V_{lm}^n, one finally obtains the pressure equation used in IVA3:

$$cp+\sum_{m=1}^{6}c_m p_m=d, \qquad (12.107)$$

where

$$c=\frac{\gamma_v}{\Delta\tau}\left(\sum_{l=1}^{3}\frac{\alpha_{la}}{\rho_{la}a_{la}^2}\right)-\sum_{m=1}^{6}c_m, \qquad (12.108)$$

$$c_m=-\beta_m\sum_{l=1}^{3}\frac{1}{\rho_{la}}\left[\xi_{lm+}\alpha_l\rho_l+\xi_{lm-}(\alpha_l\rho_l)_m\right]RVel_{lm}, \qquad (12.109)$$

$$d=\frac{\gamma_v}{\Delta\tau}\left(\sum_{l=1}^{3}\frac{\alpha_{la}}{\rho_{la}a_{la}^2}\right)p_a-\frac{\gamma_v-\gamma_{va}}{\Delta\tau}$$

$$+\sum_{l=1}^{3}\left\{D\alpha_l-\frac{1}{\rho_{la}}\sum_{m=1}^{6}\beta_m\left[\xi_{lm+}\alpha_l\rho_l+\xi_{lm-}(\alpha_l\rho_l)_m\right]dV_{lm}^n\right\}. \qquad (12.110)$$

Writing this equation for each particular cell I obtain a system of $IM\times JM\times KM$ algebraic equations with respect to the pressures in the new time plane. The coefficient matrix has the expected 7 diagonal symmetrical structure with the guaranteed diagonal dominance, see Eq. (12.108). The system coefficients are continuous non-linear functions of the solutions of the system. Therefore the system is non-linear. It has to be solved by iterations. There are two iteration cycles one called outer and another called inner. Inside the inner cycle the coefficients of the system are considered constant. Even in this case for large size of the problems an iterative procedure of solving the system of algebraic equations is necessary – see for brief introduction Appendix 12.9. I solve this system using one of the 4 successive relaxation methods built into the IVA3 computer code. The relaxation coefficient used here is unity. The four methods in IVA3 differ in the computational effort associated with direct inversion during the iterations. The first three methods solve directly the pressure equation plane by plane, for rectangle, cylinder, circle. The fourth method implements strong coupling between pressure and velocity along one line. This method is called the line-by-line solution method. The user can select the appropriate one of the four methods for the geometry of the problem which has to be simulated.

12.12 A *Newton*-type iteration method for multi-phase flows

The methods used in IVA2 and IVA3 are found to be converging and numerically stable. The reason for introduction of a new method is based on the following observations:

(a) In spite of the fact that the mixture volume conservation residuals are reduced to the values of the computational zeros, there are limitations on the reduction of the mass residuals for all mass conservation equations;

(b) Although strict convergence has been demonstrated in hundreds of numerical experiments, it has never been proven analytically.

The method presented below resolves the above two dilemmas: It (i) simultaneously leads to reduction of all residuals to strict computer zeros, and (ii) proof of convergence is derived from the principle of the Newton-type iteration method. Numerical experiments show that this method is 20% faster than the previous two methods, in spite of the fact that a single iteration step takes more time. This method does, however, require a preconditioning step. For the first iteration step I normally use the IVA3 method. All preceding methods are of course retained in IVA4.

A variety of *Newtonian* iteration methods for single- and two-phase flows were first proposed by *Patankar* and *Spalding* [48] (1972), *Spalding* and *Patankar* [47, 64-70] (1972-1981), [4, 45, 49-52] (1967-1981) and their school [6, 7, 8, 11, 15, 34, 35, 44, 53, 55, 60] (1973-1986), with widespread use in a variety of applications, e.g., TRAC development [1-3, 10, 18, 19, 39-43, 57, 59, 62, 73, 75] (1981-1985), *Vanka* [74] (1985). The IVA4 method described here can be considered as a generalization of this family of methods for multi-phase flow that use the highly efficient entropy concept and allowing analytical derivation of the pressure correction equations in contrast with all preceding *Newtonian* iteration methods. The new method will be described in this Section.

Consider the following system of algebraic non-linear equations

$$f_l^\alpha \equiv \left[\alpha_l \rho_l \gamma_v - (\alpha_l \rho_l \gamma_v)_a\right]/\Delta\tau + \sum_{m=1}^{6}[b_{lm+}\alpha_l \rho_l - b_{lm-}(\alpha_l \rho_l)_m] - \frac{1}{2}(\gamma_v + \gamma_{va})\mu_l = 0 \quad (12.111)$$

$$f_l^s \equiv \alpha_a \rho_{la} \gamma_{va} \frac{s_l - s_{la}}{\Delta\tau} - \sum_{m=1}^{6}[b_{lm}\alpha_{lm}\rho_{lm}(s_{lm} - s_l)] + \frac{1}{2}(\gamma_v + \gamma_{va})(\mu_l^+ s_l - Ds_l^*) = 0 \quad (12.112)$$

$$f_{il}^C \equiv \alpha_{la}\rho_{la}\gamma_{va}\frac{C_{il} - C_{ila}}{\Delta\tau} - \sum_{m=1}^{6}\left\{\left[b_{lm-}(\alpha_l \rho_l)_m + \beta_m \frac{D_{ilm}^*}{\Delta L_{h,m}} A(|Pe_m|)\right](C_{il,m} - C_{il})\right\}$$

$$+ \frac{1}{2}(\gamma_v + \gamma_{va})(\mu_l^+ C_{il} - DC_{il}) = 0 \quad (12.113)$$

$$f_l^n \equiv \left[n_l\gamma_v - (n_l\gamma_v)_a\right]/\Delta\tau + \sum_{m=1}^{6}(b_{lm+}n_l - b_{lm-}n_{lm}) - \frac{1}{2}(\gamma_v + \gamma_{va})Dn_l = 0 \quad (12.114)$$

$$V_{lm}^n = dV_{lm}^n - RVel_{lm}(p_m - p), \quad (12.115)$$

resulting from the first order donor-cell discretization of the 21 defining equations describing the multi-phase, multi-component flow and the condition

$$\sum_{l=1}^{l_{max}} \alpha_l - 1 = 0. \quad (12.116)$$

For simplicity, diffusion terms in the concentration and entropy equations are included within the Ds_l and DC_{il} terms. The system can be written in the following compact form

$$\mathbf{F}(\mathbf{U}) = 0, \quad (12.117)$$

where

$$\mathbf{U}^T = (\alpha_l, s_l, C_{il}, n_l, p, p_m). \quad (12.118)$$

In order to solve the system of Eqs. (12.111-12.115) I proceeded as follows. The sum of the terms on the left hand side should be equal to zero if the current values of all variables satisfy the equations. The mass, entropy, concentration and particle density equations will generally not be satisfied when the new velocities computed from the momentum equations are used to compute the convective terms in Eqs. (12.111) through (12.114). There will be some residual error in each equation as a result of the new velocities given by

$$f_l^\alpha = \frac{\partial f_l^\alpha}{\partial \alpha_l}(\alpha_l - \alpha_{l0}) \to 0 \qquad (12.119)$$

$$f_l^s = \frac{\partial f_l^s}{\partial s_l}(s_l - s_{l0}) \to 0 \qquad (12.120)$$

$$f_{il}^C = \frac{\partial f_{il}^C}{\partial C_{il}}(C_{il} - C_{il0}) \to 0 \qquad (12.121)$$

$$f_l^n = \frac{\partial f_l^n}{\partial n_l}(n_l - n_{l0}) \to 0 \qquad (12.122)$$

$$f_\Sigma = \sum_{l=1}^{l_{max}} \alpha_l - 1 \to 0 \qquad (12.123)$$

where

$$\alpha_{l0} = \left[\gamma_{va}\alpha_{la}\rho_{la}/\Delta\tau + \frac{1}{2}(\gamma_v + \gamma_{va})\mu_l + \sum_{m=1}^{6} b_{lm-}\alpha_{lm}\rho_{lm}\right]/\frac{\partial f_l^\alpha}{\partial \alpha_l}, \qquad (12.124)$$

$$s_{l0} = \left[\gamma_{va}\alpha_{la}\rho_{la}s_{la}/\Delta\tau + \frac{1}{2}(\gamma_v + \gamma_{va})Ds_l^* + \sum_{m=1}^{6} b_{lm-}\alpha_{lm}\rho_{lm}s_{lm}\right]/\frac{\partial f_l^s}{\partial s_l}, \qquad (12.125)$$

$$C_{il0} = \left[\begin{array}{c}\gamma_{va}\alpha_{la}\rho_{la}C_{ila}/\Delta\tau + \frac{1}{2}(\gamma_v + \gamma_{va})DC_{il} \\ + \sum_{m=1}^{6}\left[b_{lm-}(\alpha_l\rho_l)_m + \beta_m \frac{D_{ilm}^*}{\Delta L_{h,m}}A(|Pe_m|)\right]C_{ilm}\end{array}\right]/\frac{\partial f_{il}^C}{\partial C_{il}}, \qquad (12.126)$$

$$n_{l0} = \left[\gamma_{va} n_{la} / \Delta\tau + \frac{1}{2}(\gamma_v + \gamma_{va}) Dn_l + \sum_{m=1}^{6} b_{lm_-} n_{lm} \right] / \frac{\partial f_l^n}{\partial n_l}. \tag{12.127}$$

All terms are computed using currently known values for each of the variables not marked with a. The equations are simultaneously satisfied when $f_l^\alpha, f_l^s, f_{il}^C, f_l^n$ and f_Σ simultaneously approach zero for all cells in the mesh. The variation of the each of the dependent variables required to bring the residual errors to zero can be obtained by using the block *Newton-Raphson* method for two-phase flow see for example *Liles* and *Reed* [39] (1978), *Thurgood* et al. [73] (1983). This is implemented by linearizing the equations with respect to the dependent variables $\alpha_l, s_l, C_{il}, n_l, p$ and p_m to obtain the following equations for each cell.

$$\frac{\partial f_l^\alpha}{\partial \alpha_l}\delta\alpha_l + \frac{\partial f_l^\alpha}{\partial s_l}\delta s_l + \sum_{i=2}^{i_{max}}\frac{\partial f_l^\alpha}{\partial C_{il}}\delta C_{il} + \frac{\partial f_l^\alpha}{\partial p}\delta p + \sum_{m=1}^{6}\frac{\partial f_l^\alpha}{\partial p_m}\delta p_m = -f_l^\alpha, \tag{12.128}$$

$$\frac{\partial f_l^s}{\partial s_l}\delta s_l + \frac{\partial f_l^s}{\partial p}\delta p + \sum_{m=1}^{6}\frac{\partial f_l^s}{\partial p_m}\delta p_m = -f_l^s, \tag{12.129}$$

$$\frac{\partial f_{il}^C}{\partial C_{il}}\delta C_{il} + \frac{\partial f_{il}^C}{\partial p}\delta p + \sum_{m=1}^{6}\frac{\partial f_{il}^C}{\partial p_m}\delta p_m = -f_{il}^C, \tag{12.130}$$

$$\frac{\partial f_l^n}{\partial n_l}\delta n_l + \frac{\partial f_l^n}{\partial p}\delta p + \sum_{m=1}^{6}\frac{\partial f_l^n}{\partial p_m}\delta p_m = -f_l^n, \tag{12.131}$$

$$\sum_{l=1}^{l_{max}}\delta\alpha_l = -f_\Sigma. \tag{12.132}$$

This equation has the form

$$\frac{\partial \mathbf{f}}{\partial \mathbf{U}}\delta\mathbf{U} = -\mathbf{f}. \tag{12.133}$$

$Det(\partial \mathbf{f}/\partial \mathbf{U})$ is the *Jacobian* of the system of equations evaluated for the set of dependent variables given by the vector U. $\delta\mathbf{U}$ is the solution vector containing the linear variation of the dependent variables, and $-\mathbf{f}$ is a vector containing the negative of the residual errors required to bring the error for each equation to zero. The matrix $\partial \mathbf{f}/\partial \mathbf{U}$ is composed of analytical derivatives of each of the increments of the residuals with respect to the dependent variables. The derivatives needed to construct the *Jacobian* are easily obtained and summarized in *Appendix 12.8*. The velocities are assumed to exhibit linear dependence on the pressures, this allowing one to obtain the derivatives of the velocities with respect to pressure directly from the momentum equations (12.89). The result is given also in *Appendix 12.8*. The linear variation of the velocity with respect to pressure is given by

$$\delta V_{lm}^n = -RVel_{lm}(\delta p_m - \delta p). \tag{12.134}$$

> Note that this form of the velocity increment equation is applicable only after an explicit pass of the explicit parts of the momentum equations.

After all of the derivatives for the above equations have been calculated, the system of equations (12.133) is solved analytically by elimination for the unknown increments. In order to eliminate the volume fraction increments we divide the mass conservation equation by $\partial f_l^\alpha / \partial \alpha_l$ and sum the resulting equations. Replacing $\sum_{l=1}^{l_{max}} \delta \alpha_l$ by $-f_\Sigma$ I eliminate the $\delta \alpha_l$ values

$$\sum_{l=1}^{l_{max}} \left(\frac{\partial f_l^\alpha / \partial s_l}{\partial f_l^\alpha / \partial \alpha_l} \delta s_l + \sum_{i=2}^{i_{max}} \frac{\partial f_l^\alpha / \partial C_{il}}{\partial f_l^\alpha / \partial \alpha_l} \delta C_{il} + \frac{\partial f_l^\alpha / \partial p}{\partial f_l^\alpha / \partial \alpha_l} \delta p + \sum_{m=1}^{6} \frac{\partial f_l^\alpha / \partial p_m}{\partial f_l^\alpha / \partial \alpha_l} \delta p_m \right)$$

$$= f_\Sigma - \sum_{l=1}^{l_{max}} \frac{f_l^\alpha}{\partial f_l^\alpha / \partial \alpha_l} = f_\Sigma - \sum_{l=1}^{l_{max}} (\alpha_l - \alpha_{l0}) = \sum_{l=1}^{l_{max}} \alpha_{l0} - 1. \tag{12.135}$$

The next step is to solve the entropy, concentration and particle density equations for the entropy, concentration and particle density increments. Thus for $\partial f_l^s / \partial s_l > 0$,

$$\delta s_l = -\frac{f_l^s}{\partial f_l^s / \partial s_l} - \frac{\partial f_l^s / \partial p}{\partial f_l^s / \partial s_l} \delta p - \sum_{m=1}^{6} \frac{\partial f_l^s / \partial p_m}{\partial f_l^s / \partial s_l} \delta p_m \tag{12.136}$$

$$\delta C_{il} = -\frac{f_{il}^C}{\partial f_{il}^C / \partial C_{il}} - \frac{\partial f_{il}^C / \partial p}{\partial f_{il}^C / \partial C_{il}} \delta p - \sum_{m=1}^{6} \frac{\partial f_{il}^C / \partial p_m}{\partial f_{il}^C / \partial C_{il}} \delta p_m \tag{12.137}$$

For $\partial f_l^s / \partial s_l = 0$,

$$\delta s_l = 0, \tag{12.138}$$

$$\delta C_{il} = 0. \tag{12.139}$$

Similarly for $\partial f_l^n / \partial n_l > 0$

$$\delta n_l = -\frac{f_l^n}{\partial f_l^n / \partial n_l} - \frac{\partial f_l^n / \partial p}{\partial f_l^n / \partial n_l}\delta p - \sum_{m=1}^{6}\frac{\partial f_l^n / \partial p_m}{\partial f_l^n / \partial n_l}\delta p_m, \qquad (12.140)$$

and for $\partial f_l^n / \partial n_l = 0$,

$$\delta n_l = 0. \qquad (12.141)$$

Note that computation of the increments is only permissible if $\partial f_l^s / \partial s_l > 0$. If $\partial f_l^s / \partial s_l = 0$, the velocity field l did not exist, and will not originate in the time step considered. In this case, the properties of the velocity field l are not defined and not required for the computation of the pressure field as $\partial f_l^\alpha / \partial s_l = 0$, $\partial f_l^\alpha / \partial C_{il} = 0$.

If the velocity field l did not exist at the old time level but has just originated, this means that

$$\alpha_{la} \leq 0 + \varepsilon \text{ and } \partial f_l^s / \partial s_l > 0, \qquad (12.142)$$

I then obtain from Eqs. (12.125) through (12.127) the exact initial values of the entropies, of the concentrations, and of the particle number densities

$$s_l = s_{l0} > 0, \qquad (12.143)$$

$$C_{il} = C_{il0}, \quad 0 \leq C_{il} \leq 1, \qquad (12.144)$$

$$n_l = n_{l0} \geq 0, \qquad (12.145)$$

which obviously do not depend on the non-defined old values s_{la}, C_{ila} and n_{la}, respectively. The necessary temperature inversion is performed by the iterations using Eq. (3.129)

$$T_l = T_l^- \exp\left\{\left[s_l - s_l^-\left(T_l^-, p, C_{il,i=1,i_{\max}}\right)\right]/c_{pl}\right\}. \qquad (12.146)$$

Here the superscript "-" denotes the previous iteration value. The initial density is computed using the state equation of the new velocity field

$$\rho_l = \rho_l\left(T_l, p, C_{il,i=2,i_{\max}}\right) \qquad (12.147)$$

in this case only. In all other cases, the temperature and the density are calculated using the linearized equation of state for each velocity field (3.106) and (3.137):

$$T_l = T_{la} + \frac{\partial T_l}{\partial p}\delta p + \frac{\partial T_l}{\partial s_l}\delta s_l + \sum_{i=2}^{i_{max}} \frac{\partial T_l}{\partial C_{il}}\delta C_{il}, \qquad (12.148)$$

$$\rho_l = \rho_{la} + \frac{\partial \rho_l}{\partial p}\delta p + \frac{\partial \rho_l}{\partial s_l}\delta s_l + \sum_{i=2}^{i_{max}} \frac{\partial \rho_l}{\partial C_{il}}\delta C_{il}. \qquad (12.149)$$

The above demonstrates the remarkable property of the semi-conservative form of the discretized entropy and concentration equations that they do not depend on the field volume fractions in the actual computational cell. We therefore replace the entropy and the concentration increments in Eq. (12.135), substituting for $\partial f_l^s / \partial s_l > 0$

$$\overline{\alpha s_l} = \frac{\partial f_l^\alpha / \partial s_l}{\partial f_l^\alpha / \partial \alpha_l} / \frac{\partial f_l^s}{\partial s_l} = \frac{\alpha_l}{\rho_l}\frac{\partial \rho_l}{\partial s_l} / \frac{\partial f_l^s}{\partial s_l}, \qquad (12.150)$$

$$\overline{\alpha C_{il}} = \frac{\partial f_l^\alpha / \partial C_{il}}{\partial f_l^\alpha / \partial \alpha_l} / \frac{\partial f_l^c}{\partial C_{il}} = \frac{\alpha_l}{\rho_l}\frac{\partial \rho_l}{\partial C_{il}} / \frac{\partial f_l^s}{\partial C_{il}}, \qquad (12.151)$$

and for $\partial f_l^s / \partial s_l = 0$

$$\overline{\alpha s_l} = 0, \qquad (12.152)$$

$$\overline{\alpha C_{il}} = 0, \qquad (12.153)$$

and finally obtain the so called pressure correction equation which is the discrete analog of the *Poisson*-type equation for multi-phase flows

$$c\delta p + \sum_{m=1}^{6} c_m \delta p_m = D^*, \qquad (12.154)$$

where

$$c = \sum_{l=1}^{l_{max}}\left(\frac{\partial f_l^\alpha / \partial p}{\partial f_l^\alpha / \partial \alpha_l} - \overline{\alpha s_l}\frac{\partial f_l^s}{\partial p} - \sum_{i=2}^{i_{max}}\overline{\alpha C_{il}}\frac{\partial f_{il}^c}{\partial p}\right)$$

$$= \sum_{l=1}^{l_{max}} \left[\frac{\partial f_l^\alpha / \partial p}{\partial f_l^\alpha / \partial \alpha_l} - \frac{\alpha_l}{\rho_l} \frac{1}{\partial f_l^s / \partial s_l} \left(\frac{\partial \rho_l}{\partial s_l} \frac{\partial f_l^s}{\partial p} + \sum_{i=2}^{i_{max}} \frac{\partial \rho_l}{\partial C_{il}} \frac{\partial f_{il}^C}{\partial p} \right) \right]$$

$$= \sum_{l=1}^{l_{max}} \frac{1}{\rho_l \frac{\partial f_l^n}{\partial n_l}} \left[\partial f_l^\alpha / \partial p - \left(\frac{\partial f_l^s}{\partial p} - \sum_{i=2}^{i_{max}} \frac{\partial s_l}{\partial C_{il}} \frac{\partial f_{il}^C}{\partial p} + \sum_{i=2}^{i_{max}} c_{pl} \Delta \overline{R}_{il} \frac{\partial f_{il}^C}{\partial p} \right) \right]$$

$$= \frac{1}{\rho a^2} + \sum_{m=1}^{6} c_m^*, \qquad (12.155)$$

and

$$c_m^* = \beta_m \sum_{l=1}^{l_{max}} \frac{RVel_{lm}}{\rho_l \frac{\partial f_l^n}{\partial n_l}} \left[\xi_{lm+} \alpha_l \rho_l + \xi_{lm-} (\alpha_l \rho_l)_m \right] \left[\begin{array}{c} 1 - (s_{lm} - s_l) + \sum_{i=2}^{i_{max}} \frac{\partial s_l}{\partial C_{il}} (C_{ilm} - C_{il}) \\ - \sum_{i=2}^{i_{max}} c_{pl} \Delta \overline{R}_{il} (C_{ilm} - C_{il}) \end{array} \right],$$

$$(12.156)$$

$$c_m = \sum_{l=1}^{l_{max}} \left(\frac{\partial f_l^\alpha / \partial p_m}{\partial f_l^\alpha / \partial \alpha_l} - \overline{\alpha s_l} \frac{\partial f_l^s}{\partial p_m} - \sum_{i=2}^{i_{max}} \overline{\alpha C_{il}} \frac{\partial f_{il}^C}{\partial p_m} \right)$$

$$= \sum_{l=1}^{l_{max}} \left[\frac{\partial f_l^\alpha / \partial p_m}{\partial f_l^\alpha / \partial \alpha_l} - \frac{\alpha_l}{\rho_l} \frac{1}{\partial f_l^s / \partial s_l} \left(\frac{\partial \rho_l}{\partial s_l} \frac{\partial f_l^s}{\partial p_m} + \sum_{i=2}^{i_{max}} \frac{\partial \rho_l}{\partial C_{il}} \frac{\partial f_{il}^C}{\partial p_m} \right) \right]$$

$$= \sum_{l=1}^{l_{max}} \frac{1}{\rho_l \frac{\partial f_l^n}{\partial n_l}} \left[\partial f_l^\alpha / \partial p_m - \left(\frac{\partial f_l^s}{\partial p_m} - \sum_{i=2}^{i_{max}} \frac{\partial s_l}{\partial C_{il}} \frac{\partial f_{il}^C}{\partial p_m} + \sum_{i=2}^{i_{max}} c_{pl} \Delta \overline{R}_{il} \frac{\partial f_{il}^C}{\partial p_m} \right) \right]$$

$$= -\beta_m \sum_{l=1}^{l_{max}} \frac{1}{\rho_l \frac{\partial f_l^n}{\partial n_l}} \left[\begin{array}{c} \xi_{lm+} \alpha_l \rho_l RVel_{lm} + \xi_{lm-} (\alpha_l \rho_l)_m \left(RVel_{lm} - \frac{V_{lm}^n}{\rho_{lm} a_{lm}^2} \right) \\ \left(1 - (s_{lm} - s_l) + \sum_{i=2}^{i_{max}} \frac{\partial s_l}{\partial C_{il}} (C_{ilm} - C_{il}) - \sum_{i=2}^{i_{max}} c_{pl} \Delta \overline{R}_{il} (C_{ilm} - C_{il}) \right) \end{array} \right]$$

12.12 A Newton-type iteration method for multi-phase flows

$$= \beta_m \sum_{l=1}^{l_{max}} \frac{1}{\rho_l \frac{\partial f_l^n}{\partial n_l}} \xi_{lm-} V_{lm}^n \frac{\alpha_{lm}}{a_{lm}^2} - c_m^*, \qquad (12.157)$$

$$D^* = f_\Sigma - \sum_{l=1}^{l_{max}} \left(\frac{f_l^\alpha}{\partial f_l^\alpha / \partial \alpha_l} - \overline{\alpha s_l} f_l^s - \sum_{i=2}^{i_{max}} \overline{\alpha C_{il}} f_{il}^C \right)$$

$$= f_\Sigma - \sum_{l=1}^{l_{max}} \left[\frac{f_l^\alpha}{\partial f_l^\alpha / \partial \alpha_l} - \frac{\alpha_l}{\rho_l} \frac{1}{\partial f_l^s / \partial s_l} \left(\frac{\partial \rho_l}{\partial s_l} f_l^s + \sum_{i=2}^{i_{max}} \frac{\partial \rho_l}{\partial C_{il}} f_{il}^C \right) \right]$$

$$= f_\Sigma - \sum_{l=1}^{l_{max}} \frac{1}{\rho_l \frac{\partial f_l^n}{\partial n_l}} \left[f_l^\alpha - \frac{1}{\frac{\partial \rho_l}{\partial s_l}} \left(\frac{\partial \rho_l}{\partial s_l} f_l^s + \sum_{i=2}^{i_{max}} \frac{\partial \rho_l}{\partial C_{il}} f_{il}^C \right) \right]$$

$$= f_\Sigma - \sum_{l=1}^{l_{max}} \frac{1}{\rho_l \frac{\partial f_l^n}{\partial n_l}} \left[f_l^\alpha - \left(f_l^s - \sum_{i=2}^{i_{max}} \frac{\partial s_l}{\partial C_{il}} f_{il}^C + \sum_{i=2}^{i_{max}} c_{pl} \Delta \overline{R}_{il} f_{il}^C \right) \right], \qquad (12.158)$$

for $\partial f_l^s / \partial s_l > 0$. Using Eqs. (3.135) and (3.136), see also in [22], we obtain for the ratio

$$\frac{\partial \rho_l / \partial C_{il}}{\partial \rho_l / \partial s_l} = \frac{c_{pl}}{T_l} \frac{\left(\frac{\partial \rho_l}{\partial C_{il}}\right)_{p,T_l,\text{all}\_C's\_\text{except}\_C_{il}}}{\left(\frac{\partial \rho_l}{\partial T_l}\right)_{p,\text{all}\_C's}} - \left(\frac{\partial s_l}{\partial C_{il}}\right)_{p,T_l,\text{all}\_C's\_\text{except}\_C_{il}}$$

$$= c_{pl} \Delta \overline{R}_{il} - \frac{\partial s_l}{\partial C_{il}}, \qquad (12.159)$$

bearing in mind Eqs. (3.57) and (3.58), see also in [22],

$$\Delta \overline{R}_{il} = \frac{\left(\dfrac{\partial \rho_l}{\partial C_{il}}\right)_{p,T_l,all\_C's\_except\_C_{il}}}{T_l \left(\dfrac{\partial \rho_l}{\partial T_l}\right)_{p,all\_C's}} = \frac{\dfrac{1}{T_l}\sum_{i=2}^{i_{max}}\left[\dfrac{1}{\left(\partial \rho_{il}/\partial p_{il}\right)_{T_l}} - \dfrac{1}{\left(\partial \rho_{1l}/\partial p_{1l}\right)_{T_l}}\right]}{-\dfrac{1}{\rho_l}\sum_{i=1}^{i_{max}}\left(\dfrac{\partial \rho_{il}}{\partial T_l}\right)_{p_{il}} / \left(\dfrac{\partial \rho_{il}}{\partial p_{il}}\right)_{T_l}}$$

(12.160)

the term $\Delta \overline{R}_{il}$ can be interpreted as the relative deviation of the pseudo gas constant for the component l from the pseudo gas constant of component 1. For ideal gas mixtures this term is simply $\Delta \overline{R}_{il} = (R_{il} - R_{1l})/R_l$.

Obviously if $\Delta \tau \to 0$, $c_m \to 0$, and $c \to \dfrac{1}{\rho a^2}$ because

$$\frac{\partial f_l^n}{\partial n_l} = \frac{\gamma_v}{\Delta \tau} + \sum_{m=1}^{6} \beta_m \xi_{lm+} V_{lm}^n > 0.$$

Note that the coefficients in this equation are analytically defined. I emphasize this point as in the current state of the art for modelling of multi-dimensional, multi-phase flows this step is performed numerically e.g. *Liles* and *Reed* [75] (1978), *Thurgood* et al. [73] (1983), *Kolev* [20] (1986), *Bohl* et al. [5] (1990), this being a much more time consuming approach.

The linear variation of the pressure in cell (i,j,k) as a function of the surrounding cell pressures is given by Eq. (12.154). A similar equation can be derived for each cell in the mesh. This set of equations for the pressure variation in each mesh cell must be simultaneously satisfied. The solution to this equation set can be obtained by direct inversion for problems involving only a few mesh cells, or using the *Gauss-Seidel* iteration technique for problems involving a large number of mesh cells. In IVA4 the pressure corrector method exploits the algebraic solvers already discussed in Section 10. For brief introduction see Appendix 12.9.

The computer time required to solve Eq. (12.133) can be greatly reduced if the non-linear coefficients dV_{lm}^n and $RVel_{lm}$ are assumed to remain constant during a time step $\Delta \tau$, with the solution then obtained only for the linearized system (12.133). Checks are made on the value of each of the new time variables to assure that the variations of the new time variables from the old variables lie within reasonable limits. If the new time variables have non-physical values. e.g., void fractions less than zero or greater than one, or if the variation of the new time variable from the old variable is implausibly large, then the solution is run back to the beginning of the time step, the variables are set to their old time value, the time step is reduced and the computation repeated. This is implemented to ensure that the linearized equations are sufficiently representative of the non-linear equations to provide an acceptable level of calculation accuracy. The time step size is controlled as a function of the rate of change of the independent variables for the same reason.

12.13 Integration procedure: implicit method

The integration procedure is a logical sequence of the steps needed to obtain a set of dependent variables for each computational cell, these variables satisfying the conservation equations for each time step under the simplifying assumptions introduced and the working hypothesis for any given class of initial and boundary conditions.

The following procedure was found to lead to unconditionally stable solutions.

1. Read the information defining the problem and the required integration accuracy:

- Logical control information;
- Geometry;
- Initial conditions;
- Boundary conditions;
- External sources;
- Variable permeabilities;
- Heat structure definitions.

Perform as many time steps as required to reach the prescribed process time. A single time cycle consists of the following steps:

2. Perform computations before starting the outer iterations:

Numerics:

- Impose actual geometry;
- Store old time level information;
- Impose boundary conditions;
- Impose structure heat sources;

Cell by cell constitutive relations:

- Estimate the equations of state for each component of the velocity field (thermo-physical and transport properties for simple constituents):
- Estimate the equations of state for the mixtures of which each field consists;
- Identify the flow pattern;

- Estimate constitutive relationships dependent on flow pattern;
- Compute energy and mass source terms for each particular cell for the flow pattern identified;
- Compute interfacial drag and virtual mass coefficients;
- Compute wall-fluid interaction drag coefficients;

Numerics:

- Compute coefficients of the discretized momentum equations;
- Compute the linearized coefficients for solving the local momentum equation with respect to the local field velocities for each direction;
- Estimate velocities based on the old time level pressures;
- Impose cyclic boundary conditions in the case of θ closed cylindrical geometry.

3. Perform outer iterations:

- Estimate new $\approx s_l, \approx C_{nl}, \approx \alpha_l$;
- Repeat this step as many times as necessary to satisfy with prescribed accuracy all entropy and concentration equations and the appropriate two of the three mass conservation equations;
- Compute the coefficients of the pressure equation;
- Solve the pressure equation for $\tau + \Delta\tau$;
- Perform convergence, accuracy and time step control;
- Compute velocities for $\tau + \Delta\tau$;
- Impose cyclic boundary conditions if required;
- Compute s_l, C_{nl}, α_l for $\tau + \Delta\tau$;
- Control convergence;
- Check against general accuracy requirements, if not fulfilled perform the next outer iteration; if no convergence is achieved reduce time step, recover the old time level situation and repeat the outer iterations until convergence is achieved;
- Perform the next outer iterations until all general accuracy requirements are satisfied;

4. Perform computations after successful time step:

- Perform temperature inversion;
- Optimize time step for the next integration step;

5. Write restart information for prescribed step frequency before specified CPU time has elapsed, and at the end of the simulation.

6. Write information for post processing of the results.

12.14 Time step and accuracy control

The time step limitation dictated by the linear stability analysis for implicit donor-cell methods is

$$\Delta \tau_{CFL} < \max\left(\frac{\Delta r_h}{u_l}, \frac{r_h \Delta \theta_h}{v_l}, \frac{\Delta z_h}{w_l}\right) \qquad (12.161a)$$

for all cells. This is the so called material *Courant, Friedrichs* and *Levi* (MCFL) criterion. Numerous numerical experiments have shown that this method can work properly in many cases with larger time steps. The MCFL criterion is nevertheless retained to ensure convergence in all cases. In addition to this limitation, there are two reasons leading to further time step limits: (a) linearization limits, (b) definition limits for the dependent variables:

(a) Linearization limits: The change of the dependent variables within a time step in each computational cell should not exceed a prescribed value. This condition is associated with the linearization of the strongly non-linear system of 21 PDEs and the state equations for each time step, which is not considered in the classical *von Neumann* linear stability analysis of 1D numerical scheme for differential equations with constant coefficients.

(b) Definition limits for the dependent variables: We illustrate this problem by the following example. The velocity field mass is non-negative

$$\Delta \tau_{\max}^{\alpha} < (\alpha_l^* \rho_l \gamma_v - \alpha_{la} \rho_{la} \gamma_{va}) / \left\{\gamma_v \mu_l - \sum_m [b_{lm+} \alpha_l \rho_l - b_{lm-}(\alpha_l \rho_l)_m]\right\}, \quad (12.161b)$$

where $\alpha_l^* = 0$ for a decreasing volumetric fraction, that is for $\bar{\alpha}_l < \alpha_{la}$. For an increasing volumetric fraction, $\bar{\alpha}_l > \alpha_{la}$, the volumetric fraction of the velocity field cannot exceed the value of one by definition, $\alpha_l^* = 1$.

The outer iterations are considered as successfully completed if pressure and velocity increments from iteration to iteration and the relative mass conservation error reach values smaller than those prescribed. Additional time step optimization is imposed in order to keep the time step such as to have only a prescribed number of outer iterations, e.g. 6.

High order methods for discretization of the time derivatives require storing information for past time steps. This is the limitation that forces most of the authors to use in the multiphase flows first order explicit or implicit Euler methods. Such schemes can be easily extended to second order by using the trapezoid rule for integration over the time resulting in the popular *Crank-Nicolson* method

$$\Phi - \Phi_a = \frac{1}{2}\Delta\tau\left[f(\tau,\Phi_a) + f(\tau+\Delta\tau,\Phi)\right],$$

where f contains the remaining part of the discretized equation. Usually this method is using as a two step predictor corrector method

$$\Phi^* = \Phi_a + \frac{1}{2}\Delta\tau f(\tau,\Phi_a),$$

$$\Phi = \Phi^* + \frac{1}{2}\Delta\tau f(\tau+\Delta\tau,\Phi^*).$$

The information required is stored in any case for the old and for the new time level. The corrector step can be repeated until the solution does not change any more within prescribed limits. Alternatively the method of *Adams-Bashforth*, see in [80] p.139, can also be used

$$\Phi - \Phi_a = \frac{1}{2}\Delta\tau\left[3f(\tau,\Phi_a) - f(\tau+\Delta\tau,\Phi)\right].$$

12.15 High order discretization schemes for convection-diffusion terms

12.15.1 Space exponential scheme

Patankar and *Spalding* [49] (1972) propose the Φ profile between the locations $(k, k+1)$ and $(k-1, k)$ to be approximated by means of the following functions

$$\frac{\Phi^* - \Phi}{\Phi_{k+1} - \Phi} = \frac{\exp\left(Pe\frac{z^* - z}{\Delta z_h}\right) - 1}{\exp(Pe) - 1}, \quad Pe = \frac{F\Delta z_h}{\Gamma}, \quad z \le z^* \le z_{k+1} \quad (12.162)$$

$$\frac{\Phi^* - \Phi_{k-1}}{\Phi - \Phi_{k-1}} = \frac{\exp\left(Pe_{k-1}\frac{z^* - z_{k-1}}{\Delta z_{h,k-1}}\right) - 1}{\exp(Pe_{k-1}) - 1}, \quad Pe_{k-1} = \frac{F_{k-1}\Delta z_{h,k-1}}{\Gamma_{k-1}}, \quad z_{k-1} \le z^* \le z.$$

$$(12.163)$$

For more details see *Patankar* [47] (1980). Some reason for the choice even of this functional form is the solution of the equation

$$\frac{d}{dz}\left(F\Phi - \Gamma \frac{d\Phi}{dz}\right) = 0 ,\qquad(12.164)$$

by the following boundary conditions

$$z^* = z, \qquad \Phi^* = \Phi, \qquad(12.165)$$

$$z^* = z_{k+1}, \qquad \Phi^* = \Phi_{k+1} \qquad(12.166)$$

for the first equation and

$$z^* = z_{k-1}, \qquad \Phi^* = \Phi_{k-1}, \qquad(12.167)$$

$$z^* = z, \qquad \Phi^* = \Phi \qquad(12.168)$$

for the second. *Tomiyama* [72] (1990) found that the analytical solution of a transient convection-diffusion equation without source terms is also fully consistent with the functions (12.162) and (12.163). These functions have interesting properties, namely, the total convection-diffusion flow on the boundaries 5 and 6

$$J_{k-1} = \left(F\Phi^* - \Gamma \frac{d\Phi^*}{dz}\right)_{k-1} = F_{k-1}\left[\Phi_{k-1} + \frac{\Phi_{k-1} - \Phi}{\exp(Pe_{k-1}) - 1}\right] \qquad(12.169)$$

$$J = \left(F\Phi^* - \Gamma \frac{d\Phi^*}{dz}\right) = F\left[\Phi + \frac{\Phi - \Phi_{k+1}}{\exp(Pe) - 1}\right] \qquad(12.170)$$

depends on z only by the averaging method for the computation of Γ and Γ_{k-1}, but *not on z alone*. Using the so obtained flows on the boundaries of the computational cell we obtain the finite difference analog of

$$\frac{d}{dz}\left(F\Phi - \Gamma \frac{d\Phi}{dz}\right) - \Phi\frac{dF}{dz} = 0 \qquad(12.171)$$

namely

$$-\underbrace{\frac{F}{\Delta z}\frac{\Gamma}{\Delta z_h}\frac{1}{\exp(Pe)-1}}_{b_{k+1}}(\Phi_{k+1} - \Phi) - \underbrace{\frac{F_{k-1}}{\Delta z}\frac{\Gamma_{k-1}}{\Delta z_{h,k-1}}\frac{Pe_{k-1}\exp(Pe_{k-1})}{\exp(Pe_{k-1})-1}}_{b_{k-1}}(\Phi_{k-1} - \Phi)$$

$$= (b_{k+1} + b_{k-1}) - b_{k+1}\Phi_{k+1} - b_{k-1}\Phi_{k-1}, \tag{12.172}$$

where

$$b_{k+1} = \frac{F}{\Delta z} \frac{\Gamma}{\Delta z_h} \frac{1}{\exp(Pe) - 1}, \tag{12.173}$$

$$b_{k-1} = \frac{F_{k-1}}{\Delta z} \frac{\Gamma_{k-1}}{\Delta z_{h,k-1}} \frac{Pe_{k-1} \exp(Pe_{k-1})}{\exp(Pe_{k-1}) - 1}. \tag{12.174}$$

The computation of exponents is in general *time consuming*. Patankar [52] (1981) proposed a very accurate *approximation* of the coefficients

$$b_{k+1} = \frac{1}{\Delta z}\left\{-F\frac{1}{2}[1 - \text{sign}(F)] + \frac{\Gamma}{\Delta z_h} A(|Pe|)\right\} \tag{12.175}$$

$$b_{k-1} = \frac{1}{\Delta z}\left\{-F_{k-1}\frac{1}{2}[1 + \text{sign}(F_{k-1})] + \frac{\Gamma_{k-1}}{\Delta z_{h,k-1}} A(|Pe_{k-1}|)\right\}, \tag{12.176}$$

where

$$A(|Pe|) = \max\left[0, (1 - 0.1|Pe|)^5\right]. \tag{12.177}$$

Comparing these coefficients with the coefficients obtained with the linear profile, we see that for values of the local grid *Peclet* numbers $|Pe|$ greater than about 6 the diffusion flow is reduced to zero and the both schemes are nearly equivalent. By construction of a numerical algorithm two methods can be easily used simultaneously, programming the second one and setting optionally $A = 1$ if the use of the first method is desired.

Using such profile approximations for the discretization poses some *unanswered questions*:

- The profile of Φ in the case of no *source terms* (of Φ) has to be not the same as in the case of source terms different from zero;

- The *convection in the other two directions* in the three-dimensional flow will influence the profile in the first direction and vice versa, which is not taken into account into the above discretization schemes;

- During *strong transients* it is difficult to guess an appropriate profile of Φ;

- For most cases of practical interest, especially in three dimensions where very fine meshes are out of the question, the actual component grid *Peclet* numbers are likely to be orders of magnitude larger than 2 or 6 throughout the bulk of the flow domain. This means that the donor-cell upwind and the space exponential schemes are operating as first order upwind almost everywhere in the flow field except for a very small fraction of grid points near boundaries and stagnation regions where the convecting velocities are small. Thus instead of solving a high-convection problem, these methods simulate a low convection problem in which the effective local component grid *Peclet* number can never be greater than 2 *Leonard* and *Mokhtari* [37] (1990).

There are attempts to answer the first two questions. *Chow* and *Tien* [7, p. 91] (1978), successfully take into account some influence of the source term in their profile approximation. The resulting scheme is more accurate than the above described two schemes and possesses unconditional stability. *Pakash* [53] (1984) takes into account the influence of the flow in the other two directions in his profile approximation. *Patel* et al. [54] (1985), comparing the qualities of 7 discretization schemes with analytical one- and two-dimensional solutions, emphasize the advantages of the above discussed local analytical discretization scheme, and of the *Leonard* super upwind scheme [37, 36] (1978). In a later publication, *Patel* and *Markatos* [56] (1986) compare 8 new schemes and emphasize the advantages of the quadratic-upstream-extended-revised scheme proposed by *Pollard* and *Ali* [58] (1982).

Guenther [13] (1988) comparing the qualities of 7 discretization schemes, emphasizes the advantages of the local-analytical-method of second order of accuracy in space (donor-cell) proposed in [12] (1988) using the approximation formulas from [14] (1974).

The general tendency in the development of such schemes is to use further the donor-cell concept and increase the accuracy of the space approximations using schemes with an order of accuracy higher than one.

12.15.2 High order upwinding

The main idea behind the high order differencing methods is the approximation of the Φ function by using information from more than two adjacent cells. Usually the *Lagrange* polynomial parabolic approximation m-th order is used

$$\Phi(z_h) = \sum_m \frac{\sum_{l \neq m}(z_k - z_l)}{\sum_{l \neq m}(z_m - z_l)} \Phi(z_m) \qquad (12.178)$$

where, e.g.

$$m = k - 2, k - 1, k, k + 1, k + 2.$$

Numerical schemes for convection problems are usually constructed non-symmetrically, which means using more information for the approximation of the actual Φ from the upwind direction than from the downwind direction. Compared with symmetrical schemes the non-symmetrical schemes are more stable. Let us illustrate the main idea by the example of the successful third-order upwind scheme QUICK - see *Leonard* and *Mokharti* [37] (1990):

$$\Phi_5 = \text{Eq. (12.178)} \qquad (12.179)$$

where

$$m = k - 1, k, k + 1 \quad \text{for} \quad w \geq 0,$$

$$m = k, k + 1, k + 2 \quad \text{for} \quad w < 0,$$

and

$$\Phi_6 = \text{Eq. (12.178)} \qquad (12.180)$$

where

$$m = k - 2, k - 1, k \quad \text{for} \quad w \geq 0,$$

$$m = k - 1, k, k + 1 \quad \text{for} \quad w_{k-1} < 0.$$

For approximation of the diffusion terms QUICK uses the usual second order symmetric approximation. The disadvantage of the high order upwind schemes is the so called undershoot and overshoot if one simulates sharp discontinuities in the integration region. What is helpful to resolve this problem is a simple limiting strategy which maintains the good resolution of higher order upwinding while eliminating undershoots and overshoots without introducing artificial diffusion or destroying conservation. *Leonard* and *Mokhtari* [37] propose an extremely simple technique which can be extended to arbitrary high order accuracy. Next we explain the main idea of this proposal. In the case of positive direction of the velocity, the character of the functional form of Φ_m, $m = k - 1, k, k + 1$ is checked. One distinguishes between locally monotonic and locally non-monotonic behavior:

1) Locally monotonic behavior is defined if

$$\Phi_{k+1} > \Phi > \Phi_{k-1} \qquad (12.181)$$

or

$$\Phi_{k-1} < \Phi < \Phi_{k+1}. \tag{12.182}$$

In this case the convected face-value, Φ_5, computed by the high order differencing should lie between adjacent node values

$$\Phi < \Phi_5 < \Phi_{k+1}, \tag{12.183}$$

or

$$\Phi > \Phi_5 > \Phi_{k+1}, \tag{12.184}$$

respectively. Otherwise Φ_5 is limited by one of the adjacent node values. If

$$\Phi_{k-1} = \Phi \tag{12.185}$$

Leonard and *Mokhtari* use simply the upwind value

$$\Phi_5 = \Phi \quad (\equiv \Phi_{k-1}). \tag{12.186}$$

2) Locally non-monotonic behavior is defined if

$$\Phi > \Phi_{k+1} \quad \text{and} \quad \Phi > \Phi_{k-1}, \tag{12.187}$$

or

$$\Phi < \Phi_{k+1} \quad \text{and} \quad \Phi < \Phi_{k-1}. \tag{12.188}$$

In this case one can use the linear extrapolation of the values of the two adjacent upwind points, namely

$$\Phi_5 = \Phi + \frac{z_h - z}{z - z_{k-1}}(\Phi - \Phi_{k-1}), \tag{12.189}$$

which is simply a reduction of order of the approximation to second order upwind.

One should be careful in applying the high order differencing in a region divided by permeabilities which can be zeros. If e.g. in the above discussed cases $\gamma_{z,k-1} = 0$ we should use simply the first order upwind value

$$\Phi_5 = \Phi_k. \tag{12.190}$$

The existing codes using the classical donor-cell upwind or local analytical schemes can be easily upgraded to incorporate cost-effective high order non-oscillatory methods in order to use them in cases of coarse grids where the local grid *Peclet* number is greater than two or six respectively.

12.15.3 Constrained interpolation profile (CIP) method

Yabe and his coworkers [79] published a family of methods with the remarkable quality of resolving discontinuities with much smaller number of cells then any other known methods – see in [71] p. 588, Fig. 19. The basic idea of this group of methods is to use the primitive dependent variables and their derivatives as a set of dependent variables. For the first group the conservation equations are used and for the second group the corresponding derivatives of these equations are used. Instead of reviewing here all the methods, we give few examples of the so called exactly conservative scheme for the CIP method. The remarkable quality of this scheme is the enforced conservation of given properties even by using a non-conservative form of the conservation equations.

12.15.3.1 Exactly conservative scheme in a non-conservative form

Tanaka et al. [71] proposed in year 2000 to approximate the *f* profile between the locations $z_{h,k-1} \le z^* \le z_h$ by means of the *fourth order* polynomial spline function

$$f^* - f_{k-1} = a(z^* - z_{h,k-1})^4 + b(z^* - z_{h,k-1})^3 + c(z^* - z_{h,k-1})^2 + g_{k-1}(z^* - z_{h,k-1}), \tag{12.191}$$

which already satisfies the continuity and the smoothness conditions at the left boundary $z^* = z_{h,k-1}$,

$$f^* = f_{k-1}, \quad \frac{df}{dz^*} = g_{k-1}. \tag{12.192}$$

It has to satisfy the continuity and the smoothness conditions also at the right boundary $z^* = z_h$

$$f^* = f, \quad \frac{df}{dz^*} = g. \tag{12.193}$$

This leads to the additional two equations for computing the unknown coefficients *a*, *b*, and *c*

$$\Delta z_{k-1}^4 a + \Delta z_{k-1}^3 b + \Delta z_{k-1}^2 c = f - f_{k-1} - g_{k-1}\Delta z_{k-1}, \tag{12.194}$$

$$4\Delta z_{k-1}^3 a + 3\Delta z_{k-1}^2 b + 2\Delta z_{k-1} c = g - g_{k-1}. \tag{12.195}$$

In addition the authors enforce the so called continuity constraint. We use here the integral as proposed by the authors in [71] divided by the distance between two points $\Delta z_{k-1} = z_h - z_{h,k-1}$ which is the weighted averaged over this distance,

$$f_{av} = \frac{1}{\Delta z_{k-1}} \int_{z_{k-1}}^{z_h} f^* dz^*, \tag{12.196}$$

resulting in the third equation

$$\frac{1}{5}\Delta z_{k-1}^4 a + \frac{1}{4}\Delta z_{k-1}^3 b + \frac{1}{3}\Delta z_{k-1}^2 c = f_{av} - g_{k-1}\frac{1}{2}\Delta z_{k-1} - f_{k-1}, \tag{12.197}$$

required to compute the three unknown constants. Solving with respect the coefficients one obtains

$$a = \frac{30}{\Delta z_{k-1}^4}\left[f_{av} - \frac{1}{2}(f+f_{k-1}) + (g-g_{k-1})\frac{1}{12}\Delta z_{k-1}\right], \tag{12.198}$$

$$b = \frac{4}{\Delta z_{k-1}^3}\left[-15f_{av} + 7f + 8f_{k-1} - \left(g - \frac{3}{2}g_{k-1}\right)\Delta z_{k-1}\right], \tag{12.199}$$

$$c_{k-1} = \frac{3}{\Delta z_{k-1}^2}\left[10f_{av} - 4f - 6f_{k-1} + (g-3g_{k-1})\frac{1}{2}\Delta z_{k-1}\right]. \tag{12.200}$$

The CIP method is illustrated on integrating the following simple equation

$$\frac{\partial f}{\partial \tau} + w\frac{\partial f}{\partial z} = 0, \tag{12.201}$$

written in canonical form $df/d\tau = 0$ along the characteristic line defined by $dz/d\tau = w$. The solution for the point $k - 1$ is

$$f_{k-1}^{n+1}(\tau + \Delta\tau, z_{h,k-1}) = f_{k-1}^*(\tau, z_{h,k-1} - w\Delta\tau), \tag{12.202}$$

or in terms of the already find approximations for $w > 0$

$$f_{k-1}^{n+1} = a_{k-1}\xi^4 + b_{k-1}\xi^3 + c_{k-1}\xi^2 + g_{k-2}\xi + f_{k-2},\qquad(12.203)$$

where $\xi = z_h - z_{h,k-1} - w\Delta\tau = \Delta z_{k-1}(1 - w\Delta\tau/\Delta z_{k-1})$. We will need in a moment also the derivative

$$g_{k-1}^{n+1} = 4a_{k-1}\xi^3 + 3b_{k-1}\xi^2 + 2c_{k-1}\xi + g_{k-2},\qquad(12.204)$$

and the flux

$$F = \int_0^{\Delta\tau} fw d\tau = \int_\xi^{\Delta z_{k-1}} f_{k-1}^*(z)dz = \frac{1}{5}a_{k-1}\left(\Delta z_{k-1}^5 - \xi^5\right) + \frac{1}{4}b_{k-1}\left(\Delta z_{k-1}^4 - \xi^4\right)$$

$$+\frac{1}{3}c_{k-1}\left(\Delta z_{k-1}^3 - \xi^3\right) + \frac{1}{2}g_{k-2}\left(\Delta z_{k-1}^2 - \xi^2\right) + f_{k-2}\left(\Delta z_{k-1} - \xi\right).\qquad(12.205)$$

For $w < 0$ we have

$$f_{k-1}^{n+1} = a\xi^4 + b\xi^3 + c\xi^3 + g_{k-1}\xi + f_{k-1}.\qquad(12.206)$$

where $\xi = w\Delta\tau$. Again the derivative is

$$g_{k-1}^{n+1} = 4a\xi^3 + 3b\xi^2 + 3c\xi^2 + g_{k-1},\qquad(12.207)$$

and the flux is

$$F = \int_0^{\Delta\tau} fw d\tau = \int_\xi^0 f(z)dz = -\left(\frac{1}{5}a\xi^5 + \frac{1}{4}b\xi^4 + \frac{1}{3}c\xi^3 + \frac{1}{2}g_{k-1}\xi^2 + f_{k-1}\xi\right).$$

$$(12.208)$$

For solving the non homogeneous convection equation

$$\frac{\partial f}{\partial\tau} + \frac{\partial}{\partial z}(wf) = 0\qquad(12.209)$$

let us apply the splitting procedure as applied by *Xiao* [81]. The solution of the homogeneous convection $\frac{\partial f}{\partial\tau} + w\frac{\partial f}{\partial z} = 0$ is done by the pseudo-characteristic method as already discussed. It gives f^{*n+1}. Then the second step of the splitting provides f^{*n+1} as follows:

$$f^{n+1} = f^{*n+1} - f^{*n+1} \int_\tau^{\tau+\Delta\tau} \frac{\partial w}{\partial z} d\tau'. \tag{12.210}$$

Xiao [81] uses the donor-cell concept for computing the integral term as follows

$$\int_\tau^{\tau+\Delta\tau} \frac{\partial w}{\partial z} d\tau' = \frac{\Delta\tau}{\Delta z} \frac{1}{2} \left(w - w_{k-1} + w^{*n+1} - w_{k-1}^{*n+1} \right) \quad \text{for} \quad w > 0, \tag{12.211}$$

$$\int_\tau^{\tau+\Delta\tau} \frac{\partial w}{\partial z} d\tau' = \frac{\Delta\tau}{\Delta z_{k+1}} \frac{1}{2} \left(w_{k+1} - w + w_{k+1}^{*n+1} - w^{*n+1} \right) \quad \text{for} \quad w \le 0. \tag{12.212}$$

12.15.3.2 Computing the weighted averages in the new time plane

The non homogeneous convection defined by

$$\frac{\partial f}{\partial \tau} + \frac{\partial}{\partial z}(wf) = 0 \tag{12.213}$$

can be rewritten in integral form for $w > 0$

$$\int_0^{\Delta z} f^{*n+1} dz - \int_0^{\Delta z} f^* dz = \frac{1}{\Delta z} \left(\int_\tau^{\tau+\Delta\tau} f^*_{k-1} w_{k-1} d\tau' - \int_\tau^{\tau+\Delta\tau} f^* w d\tau' \right). \tag{12.214}$$

Substituting $wd\tau = dz$ in the left hand site integrals and setting as a lower and upper boundaries the departure and the end point of the corresponding characteristic lines at the both ends of the cell we obtain

$$\int_0^{\Delta z} f^{*n+1} dz - \int_0^{\Delta z} f^* dz = \frac{1}{\Delta z} \left(\int_{\Delta z_{k-1} - w_{k-1}\Delta\tau}^{\Delta z_{k-1}} f^*_{k-1} w_{k-1} d\tau' - \int_{\Delta z - w\Delta\tau}^{\Delta z} f^* w d\tau' \right). \tag{12.215}$$

The general notation of this result is

$$f_{av}^{n+1} = \frac{1}{\Delta z}(F_{k-1} + f_{av}\Delta z - F) = f_{av} + \frac{1}{\Delta z}(F_{k-1} - F). \tag{12.215}$$

where the F's are the fluxes already defined with Eqs. (12.205) and (12.208) that are depending on the velocity direction. With other words, the profile in the new time plane between $z_{h,k-1}$ and z_h is the transported profile from the old time plane within the boundaries $z_{h,k-1} - w_{k-1}\Delta\tau$ and $z_h - w\Delta\tau$.

12.15.3.3 Choice of the gradients

In the ordinary cubic-spline interpolation, continuity of the value, the first derivative, and the second derivative is required to generate the piecewise cubic polynomials from the data given at some discrete points. *Yabe* and *Takewaki* show in [76] that this procedure is not suitable for the present problem because the profile generated by the classical spline method is not consistent with the physical nature of the governing equations. The CIP method relaxes the construction of the spline based on gradients computed with the values at the discrete points. Instead, it requires gradients physically based on the governing equations *which is the main source of success of this method*. To estimate the gradient the authors differentiate the model equation with respect to the spatial coordinate,

$$\frac{\partial g}{\partial \tau} + w \frac{\partial g}{\partial z} = \frac{dg}{d\tau} = 0. \qquad (12.216)$$

Transforming the equation in the orthogonal coordinate system attached with the one coordinate to the characteristic line defined by $dz/d\tau = w$, gives

$$\frac{dg}{d\tau} = 0. \qquad (12.217)$$

For the solution of this particular case we have Eqs. (12.204) and (12.207).

12.15.3.4 Multi-phase flow example

Now let us consider the model equation

$$\frac{\partial f}{\partial \tau} + \frac{\partial}{\partial z}(fw) = \mu, \qquad (12.218)$$

in which μ and w are known functions of τ and z. We rewrite it in pseudo-characteristic form $\frac{\partial f}{\partial \tau} + w \frac{\partial f}{\partial z} = \mu - f \frac{\partial w}{\partial z}$ or

$$\frac{df}{d\tau} = \mu - f \frac{\partial w}{\partial z} \qquad (12.219)$$

along the line defined by $dz/d\tau = w$. The approximate solution for $w > 0$ is then

$$f_{k-1}^{*n+1} = a_{k-1}\xi^4 + b_{k-1}\xi^3 + c_{k-1}\xi^2 + g_{k-2}\xi + f_{k-2}^*, \qquad (12.220)$$

$$f_{k-1}^{n+1} = f_{k-1}^{*n+1} + \int_0^{\Delta\tau} \mu_{k-1} d\tau - f_{k-1}^{*n+1} \int_0^{\Delta\tau} \frac{\partial w}{\partial z} d\tau. \quad (12.221)$$

For $w < 0$ we have

$$f_{k-1}^{*n+1} = a\xi^4 + b\xi^3 + c\xi^3 + g_{k-1}\xi + f_{k-1}. \quad (12.222)$$

$$f_{k-1}^{n+1} = f_{k-1}^{*n+1} + \int_0^{\Delta\tau} \mu d\tau - f_{k-1}^{*n+1} \int_0^{\Delta\tau} \frac{\partial w}{\partial z} d\tau, \quad (12.223)$$

The integrals are computed by Eq. (12.211) and (12.212), Xiao [81], using the donor-cell concept.

The additional equation for the gradient is obtained by differentiating the model equation with respect to the spatial coordinate

$$\frac{\partial g}{\partial \tau} + w\frac{\partial g}{\partial z} = \frac{\partial \mu}{\partial z} - f\frac{\partial}{\partial z}\frac{\partial w}{\partial z} - 2g\frac{\partial w}{\partial z}. \quad (12.224)$$

For the construction of the approximate solution we use again the canonical form

$$\frac{dg}{d\tau} = \frac{\partial \mu}{\partial z} - f\frac{\partial}{\partial z}\frac{\partial w}{\partial z} - 2g\frac{\partial w}{\partial z}, \quad (12.225)$$

which is valid along the line defined by $dz/d\tau = w$ and obtain

$$g_{k-1}^{*n+1} = 4a_{k-1}\xi^3 + 3b_{k-1}\xi^2 + 2c_{k-1}\xi + g_{k-2}, \quad (12.226)$$

$$g_{k-1}^{n+1} = g_{k-1}^{*n+1} + \int_0^{\Delta\tau} \frac{\partial \mu_{k-1}}{\partial z} d\tau - f_{k-1}^{*n+1}\int_0^{\Delta\tau} \frac{\partial}{\partial z}\left(\frac{\partial w}{\partial z}\right) d\tau - 2g_{k-1}^{*n+1}\int_0^{\Delta\tau} \frac{\partial w}{\partial z} d\tau, \quad (12.227)$$

$$\int_0^{\Delta\tau} \frac{\partial}{\partial z}\left(\frac{\partial w}{\partial z}\right) d\tau = \frac{\Delta\tau}{\Delta z_{h,k-1}}\frac{1}{2}\left(\frac{w - w_{k-1}}{\Delta z} - \frac{w_{k-1} - w_{k-2}}{\Delta z_{k-1}} + \frac{w^* - w_{k-1}^*}{\Delta z} - \frac{w_{k-1}^* - w_{k-2}^*}{\Delta z_{k-1}}\right) \quad (12.228)$$

$$\int_0^{\Delta\tau} \frac{\partial w}{\partial z} d\tau = \frac{\Delta\tau}{\Delta z}\frac{1}{2}\left(w - w_{k-1} + w^* - w_{k-1}^*\right), \quad (12.229)$$

for $w > 0$, and

$$g_{k-1}^{*n+1} = 4a\xi^3 + 3b\xi^2 + 3c\xi^2 + g_{k-1} \tag{12.230}$$

$$g_{k-1}^{n+1} = g_{k-1}^{*n+1} + \int_0^{\Delta\tau} \frac{\partial \mu}{\partial z} d\tau - f_{k-1}^{*n+1} \int_0^{\Delta\tau} \frac{\partial}{\partial z}\left(\frac{\partial w}{\partial z}\right) d\tau - 2 g_{k-1}^{*n+1} \int_0^{\Delta\tau} \frac{\partial w}{\partial z} d\tau, \tag{12.231}$$

$$\int_0^{\Delta\tau} \frac{\partial w}{\partial z} d\tau = \frac{\Delta\tau}{\Delta z_{k+1}} \frac{1}{2}\left(w_{k+1} - w + w_{k+1}^* - w^*\right), \tag{12.232}$$

for $w < 0$.

12.15.3.5 Phase discontinuity treated with CIP

For propagation of a volumetric fraction described by the equation

$$\frac{\partial \alpha}{\partial \tau} + w \frac{\partial \alpha}{\partial z} = 0, \tag{12.231}$$

which may be monotonic or may possess sharp discontinuities, and at the same time has always to have values between 0 and 1, Yabe and coworkers [78, 77] proposed to use the following transformation

$$\Phi = \tan\left[(1-\varepsilon)\pi(\alpha - 0.5)\right], \tag{12.232}$$

$$\alpha = 0.5 + \frac{\arctan \Phi}{(1-\varepsilon)\pi}, \tag{12.233}$$

where ε is a small positive constant. The factor $(1-\varepsilon)$ enables one to get for $\Phi \to -\infty$ $\alpha \to 0$ and $\Phi \to \infty$ $\alpha \to 1$, and to tune the sharpness of the layer to be resolved. Differentiating and replacing in the original equation results in

$$\frac{1}{(1-\varepsilon)\pi(1+\Phi^2)}\left(\frac{\partial \Phi}{\partial \tau} + w \frac{\partial \Phi}{\partial z}\right) = 0, \tag{12.234}$$

or

$$\frac{\partial \Phi}{\partial \tau} + w \frac{\partial \Phi}{\partial z} = 0, \tag{12.235}$$

which again is solved by the CIP method. The merit of this transformation is that, although Φ may be slightly diffusive and may have undulation when the discretized equation is solved, the inversely transformed value of Φ is always limited to a range between 0 and 1. The method is appropriate even for the description of the boundary of a rigid body moving in space. The method is not recommended in combination with a low order solution method like upwind etc. For more discussion see [71] p. 567.

12.16 Pipe networks: some basic definitions

The methods for integration of the systems of partial differential equations governing the multi phase flows in 3D presented in the previous Sections can be used also for describing the flow in pipe networks. The system takes a very simple form. Usually the z-direction components of all the equations are necessary in differential and in finite difference form with some small modifications for the change of the axis angle. In addition coupling through the so called knots has to be mathematically defined. We call the pipe network flow 1.5-dimensional flow rather then one dimensional, because of the cross connections among the pipes. It is very important before attempting to design a new computational model for flows in pipe networks to formally describe the pipe network in quantitative characteristics. This is the subject of this section.

12.16.1 Pipes

A *pipe* is a one-dimensional flow channel, of which the axis runs arbitrarily through three-dimensional space – see Fig. 12.8.

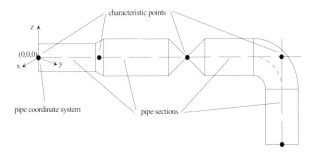

Fig. 12.8. Pipe definition: coordinate system, characteristic points, pipe sections

At specified *characteristic points* the pipe may contain

a) elbows,
b) reductions and expansions,
c) components such as valves, pumps etc.

or may experience sudden changes in

a) the pipe inner diameter defining the flow cross section, or in
b) the pipe hydraulic diameter,
c) the pipe material, or in
d) the wall thickness or in
e) the roughness,

or

d) may be interconnected with the beginning or the end of other pipes and form a knot.

The coordinates of all these *characteristic points* define the axis of the pipe in the space.

The characteristic points divide the pipe into straight *sections*, where some specific pipe attributes like

a) inner diameter,
b) hydraulic diameter,
c) wall thickness,
d) the roughness and
e) the material

of the pipe are constant – see Fig. 12.9.

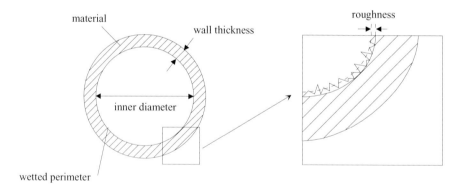

$$\text{hydraulic diameter} = \frac{4 \cdot \text{flow cross section}}{\text{wetted perimeter}}$$

Fig. 12.9. Definitions of the pipe attributes: inner diameter defining the flow cross section, hydraulic diameter, material, wall thickness, roughness

Pipes are identified by integer numbers ranging from 1 to the total number of pipes inside the network and for convenience by a text identifier - *name*. This name has to be unique throughout the network. Sometimes it may happen that pipes have exactly the same geometry and run parallel in space, i.e. they are of the same type. It is not necessary to model each of them separately. The resulting flow is instead computed by multiplying the single pipe flow by the number of parallel pipes.

In addition, *each* pipe has its own *normalizing diameter* defining the normalizing flow area. All pipe flow cross section areas are divided by the normalizing cross section for the pipe. The results are the so called flow permeabilities defining the flow cross section. Describing the portion of the cross section available for the flow is the technique used in the IVA code series and is applied in the three-dimensional analysis as well as in the one-dimensional network analysis.

12.16.2 Axis in the space

Before a pipe can be inserted into the network, its geometry has to be defined. The pipe is usually defined by an identifier name and the coordinates of its characteristic points.

$$\left(x_p, y_p, z_p\right)$$

The pipe is defined in its own rectangular left oriented coordinate system (Fig. 12.8). The coordinate system is usually attached to the beginning of the pipe. That means every pipe starts at the point $(0,0,0)$ in its own coordinate system. The coordinates of every point of interest called characteristic points, such as elbows, components, area changes or changes in material, are defined with respect to this point (*relative coordinates*). This gives an opportunity to create libraries with standardized pipes. Defining pipes in *absolute coordinates* is also possible. In this case the coordinate system is not necessarily connected to the pipe start point.

The positive orientation of the pipe axis is defined through the order of the characteristic points from the start point to the pipe end. This direction corresponds to the increasing cell indices of the discrete control volumes created after the pipe definition for computational analysis.

Two characteristic angles are specified (see Fig. 12.10) for use internally in computer codes. The first one,

$$\theta = \arccos\left[\left(\mathbf{r}_{k-1,k} \cdot \mathbf{r}_{k+1,k}\right) / \left(\left|\mathbf{r}_{k-1,k}\right|\left|\mathbf{r}_{k+1,k}\right|\right)\right],$$

is called the *deviation angle*. Here *k-1*, *k* and *k+1* define three seqential characteristic points and **r**'s are the vectors beiween them. This is the angle of a pipe section defined as the deviation from the positive oriented axis of the previous pipe section. The second one,

$$\varphi = \arccos\left[\left(z_{k+1} - z_k\right)/\left|\mathbf{r}_{k+1,k}\right|\right],$$

is called *inclination angle (polar angle)*.

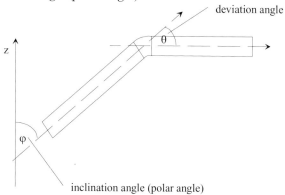

Fig. 12.10. Definition of deviation and inclination angles

Here k and $k + 1$ are the two end ponts of the segment. The inclination angle is defined as the deviation of the positive oriented section axis with respect to the upwards oriented vertical direction (the negative gravity direction).

The length of each pipe section,

$$\left|\mathbf{r}_{k+1,k}\right| = \sqrt{\left(x_{k+1} - x_k\right) + \left(y_{k+1} - y_k\right) + \left(z_{k+1} - z_k\right)},$$

and its characteristic angles can then be computed automatically with the already specified information. Both angles and the section length are basic geometrical inputs for the definition of the mathematical flow description problem.

12.16.3 Diameters of pipe sections

A *pipe section* is a part of the pipe being between two neighboring characteristic points, see Fig. 12.8. The pipe section is per definition a *straight* piece of a pipe. For every pipe section

a) the pipe inner diameter defining the flow cross section,
b) the pipe hydraulic diameter,
c) the pipe material,
d) the wall thickness and
e) the roughness.

have to be defined, see Fig. 12.9.

The default for the inner diameter is the normalizing diameter of the pipe.

The hydraulic diameter of a flow channel is defined as 4 times the flow cross section divided by the wetted perimeter corresponding to this cross section. By default the hydraulic diameter is equal to the inner diameter. That means the pipes are assumed by default to be circular tubes.

Examples for default definitions are: the material is stainless steel, the wall thickness is 0.1 times the inner diameter, and the roughness is 0.00004 m. These values can then be changed by per input.

12.16.4 Reductions

A smooth change in the pipe inner diameter or hydraulic diameter is expressed through a form piece called *reduction* – see Fig. 12.11. The reduction has a specified length and is centered on the corresponding characteristic point. At its ends it has the diameters of the adjacent pipe sections and the diameter varies linearly between the two ends. On the contrary, an *abrupt area change* is expressed exactly through the characteristic point itself.

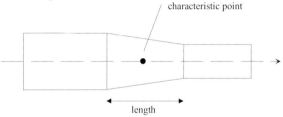

Fig. 12.11. Definition of reductions

12.16.5 Elbows

Elbows are associated with points, where the pipe segment axis changes its direction in space – see Fig. 12. The flow axis coincides before and after the bend with the pipe segment axis. Note that the cross point of these axes is the characteristic point with which an elbow is associated.

The flow axis of an elbow possesses a bend radius which is in fact the curvature radius of the flow axis. Example for default value for the bend radius is 1.5 times the inner diameter of the preceding section. This value can be then changed by input. For example characteristic points which are not associated with an elbow, may receive a very large default value of the bend radius – 100 m.

Therefore a pipe section possesses an elbow on one of its ends, if the deviation angle is greater than zero and the bend radius of the corresponding characteristic point is less than the specified default.

Note that in accordance with the above definition the start point or the end point of a pipe can never be an elbow.

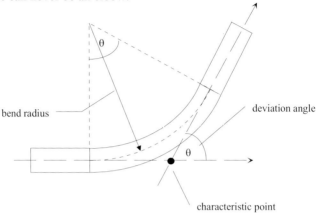

Fig. 12.12. Definition of elbows

12.16.6 Creating a library of pipes

After the pipe data have been specified correctly, the pipe definition has to be automatically saved in a file. A *pipe library* is a file system containing an arbitrary number of pipe defining files. Creating a library of files "pipes" allows one later to simply interconnect them and use some of them repeatedly for different problems.

12.16.7 Sub system network

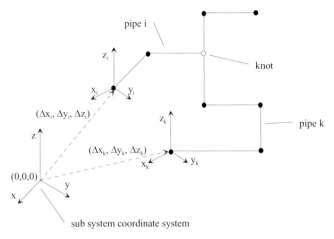

Fig. 12.13. Definition of sub system network

A *sub system network* consists of a number of *pipes* already defined in the pipe library which are linked together through *knots*. The sub system network is defined in its own coordinate system having default coordinates

$$(0,0,0).$$

The sub system network definition contains also

a) a list of all involved pipes and

b) the shift of the particular pipe attached coordinate system in the new sub system coordinate system.

Note: a sub system network can also consist of only one pipe.

Data of any pipe in the sub system network list should be allowed to be changed if necessary. Earlier defined pipes can be loaded into the list. Pipes can also be removed from the list.

Once the list of pipes necessary to form the network its complete, the pipes can be linked together by editing coordinates of the starting points.

$$(x_0, y_0, z_0)$$

They act as an offset to the particular pipe attached coordinate system. The pipe internal coordinate system itself is not affected.

12.16.8 Discretization of pipes

For the numerical computation of the system of differential equations, we use usually a finite volume technique to discretize the system. Therefore each pipe is divided into finite control volumes, at whose centers the flow properties like pressure, temperature, mass etc. are defined. We call these the real cells of the pipe. They are numbered with increasing cell indices e.g. i. The increasing cell indices define explicitly the positive flow direction as illustrated in Fig. 12.14. The staggered grid method implies a second set of cells, called the momentum cells, e.g. k. These are used for the discretization of the momentum equations and are located at the upper boundary of the corresponding real cells. The velocity is defined for these momentum cells. In addition, two auxiliary cells with zero length are introduced: one representing the pipe inlet and the other representing the pipe outlet. These two are needed to set proper boundary conditions at the pipe edges. The relation between the real cell numbering and the computational cell numbering is simply $k = i + 2(n - 1) + 1$, where n is the consecutive number of pipes. It is wise to store the computational cell numbers for the entrance and exit cells of each pipe. Then the cycles can be organized by visiting in each pipe the cells from the entrance to the exit.

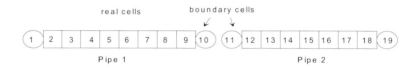

Fig. 12.14. Pipes

Equidistant discretization is reccomendet but not always possible in technical systems for which the exact positioning of bents, valves etc. is important. A remidy is to try to dicretize each segment equidistantly.

12.16.9 Knots

Several pipes can be linked together through knots to design a pipe network. The knots are arbitrarily numbered with indices ranging from 1 to the total number of knots inside the network. The flow passes through the auxiliary cells of the pipes starting or ending at the knot. The pipes starting out of the knot are called knot outputs; the pipes ending into the knot are called knot inputs. The auxiliary cells represent the openings of the knot into the pipe. They are assumed identical with the knot cell and therefore have the same cell properties. A good knot model provides a momentum transfer from one pipe into another. Therefore the angles between the pipes have to be specified. The inclination angles of the pipes connected through the knot are defined as the deviation from the positive direction of the knot cell, regardless of whether the corresponding pipe is a knot input or a knot output (see also Fig. 12.15).

How to recognize *knots*?

A knot must be a characteristic point belonging to one of the pipes included in the sub system satisfying the following conditions:

If one of the ends of at least one other pipe coincides with a characteristic point of the pipe, this characteristic point is identified also as a knot. The same is valid, if one of the ends of the pipe itself coincides with a characteristic point of the pipe different from this end.

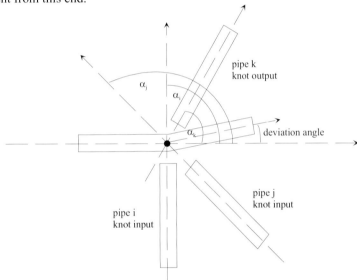

Fig. 12.15. Definition of knots

Thus, inspecting all of the characteristic points of one pipe and checking whether the first or the last points of other pipes - or of the same pipe, if the characteristic point is different from this end - coincide with it, we identify whether knots are present in this pipe or not. If during this search pipe ends of other pipes are identified to belong to a knot, they are marked as "*already belongs to knot*".

The procedure is repeated for all pipes successively in order of their appearance in the network list. The pipe ends already having the mark "*already belongs to knot*" are excluded from the further checks. So all participating pipes are visited. Finally the total number of knots is identified. The knots are numbered from one to their total number in the order in which they are determined.

Additional information is derived from the above specified knot data for the computational analysis by answering the following questions: How many pipes enter the knot? Which pipes are these? How many pipes exit the knot? Which pipes are these? What are the inclination angles of the entering pipes with respect to the positive oriented axis pointing to the knot of the pipe which the knot belongs to? What are the inclination angles of the exiting pipes with respect to the positive oriented axis pointing to the knot of the pipe which the knot belongs to? The pipe to which the knot belongs is excluded from this questionnaire.

The inclination angles of the entering and exiting pipes with respect to the positive oriented axis pointing to the knot of the pipe which the knot belongs to are computed automatically. If the characteristic point identified as a knot belongs to a change in direction, the angle is computed with respect to the section before the characteristic point. In this case the direction change is not considered as an elbow, but as a sharp kink (see Fig. 12.14).

So, the network definition can be then saved. A prior defined network definition file can be also loaded for modification or check if required. The strategy described in Section 12.16 was programmed by *Iris Roloff-Bock* in the graphical preprocessing system NETGEN and later transformed into SONIA graphical preprocessing module by *Tony Chen*. With both systems a convenient way is provided for defining a network flow problems for the IVA computer code system.

Appendix 12.1 Definitions applicable to discretization of the mass conservation equations

Following Section 12.5 the first order donor-cell discretized mass conservation equation (12.3) for each velocity field is

$$(\alpha_l \rho_l \gamma_v - \alpha_{la} \rho_{la} \gamma_{va})/\Delta\tau$$

$$+ \frac{r_h^\kappa \gamma_r}{r^\kappa \Delta r} u_l \left\{ \frac{1}{2}\left[1+\text{sign}(u_l)\right]\alpha_l\rho_l + \frac{1}{2}\left[1-\text{sign}(u_l)\right](\alpha_l\rho_l)_{i+1} \right\}$$

$$- \frac{(r_h^\kappa \gamma_r)_{i-1}}{r^\kappa \Delta r} u_{l,i-1} \left\{ \frac{1}{2}\left[1+\text{sign}(u_{l,i-1})\right](\alpha_l\rho_l)_{i-1} + \frac{1}{2}\left[1-\text{sign}(u_{l,i-1})\right]\alpha_l\rho_l \right\}$$

$$+\frac{\gamma_\theta}{r^\kappa \Delta\theta} v_l \left\{ \frac{1}{2}\left[1+sign(v_l)\right]\alpha_l\rho_l + \frac{1}{2}\left[1-sign(v_l)\right](\alpha_l\rho_l)_{j+1} \right\}$$

$$-\frac{\gamma_{\theta,j-1}}{r^\kappa \Delta\theta} v_{l,j-1} \left\{ \frac{1}{2}\left[1+sign(v_{l,j-1})\right](\alpha_l\rho_l)_{j-1} + \frac{1}{2}\left[1-sign(v_{l,j-1})\right]\alpha_l\rho_l \right\}$$

$$+\frac{\gamma_z}{\Delta z} w_l \left\{ \frac{1}{2}\left[1+sign(w_l)\right]\alpha_l\rho_l + \frac{1}{2}\left[1-sign(w_l)\right](\alpha_l\rho_l)_{k+1} \right\}$$

$$-\frac{\gamma_{z,k-1}}{\Delta z} w_{l,k-1} \left\{ \frac{1}{2}\left[1+sign(w_{l,k-1})\right](\alpha_l\rho_l)_{k-1} + \frac{1}{2}\left[1-sign(w_{l,k-1})\right]\alpha_l\rho_l \right\}$$

$$-\frac{1}{2}(\gamma_v + \gamma_{va})\mu_l = 0$$

Introducing the velocity normal to each surface of the discretization volume, Eq. (12.12), the b coefficients can be conveniently written as follows

$$b_{l1+} = +\beta_1 \frac{1}{2}\left[1+sign(u_l)\right]u_l = +\beta_1 \xi_{l1+} u_l,$$

$$b_{l2+} = -\beta_2 \frac{1}{2}\left[1-sign(u_{l,i-1})\right]u_{l,i-1} = -\beta_2 \xi_{l2+} u_{l,i-1},$$

$$b_{l3+} = +\beta_3 \frac{1}{2}\left[1+sign(v_l)\right]v_l = +\beta_3 \xi_{l3+} v_l,$$

$$b_{l4+} = -\beta_4 \frac{1}{2}\left[1-sign(v_{l,j-1})\right]v_{l,j-1} = -\beta_4 \xi_{l4+} v_{l,j-1},$$

$$b_{l5+} = +\beta_5 \frac{1}{2}\left[1+sign(w_l)\right]w_l = +\beta_5 \xi_{l5+} w_l,$$

$$b_{l6+} = -\beta_6 \frac{1}{2}\left[1-sign(w_{l,k-1})\right]w_{l,k-1} = -\beta_6 \xi_{l6+} w_{l,k-1},$$

or compactly written

$$b_{lm+} = \beta_m \xi_{lm+} V_{lm}^n,$$

and

$$b_{l1-} = -\beta_1 \frac{1}{2}\left[1 - sign(u_l)\right]u_l = -\beta_1 \xi_{l1-} u_l,$$

$$b_{l2-} = +\beta_2 \frac{1}{2}\left[1 + sign(u_{l,i-1})\right]u_{l,i-1} = +\beta_2 \xi_{l2-} u_{l,i-1},$$

$$b_{l3-} = -\beta_3 \frac{1}{2}\left[1 - sign(v_l)\right]v_l = -\beta_3 \xi_{l3-} v_l,$$

$$b_{l4-} = +\beta_4 \frac{1}{2}\left[1 + sign(v_{l,j-1})\right]v_{l,j-1} = +\beta_4 \xi_{l4-} v_{l,j-1},$$

$$b_{l5-} = -\beta_5 \frac{1}{2}\left[1 - sign(w_l)\right]w_l = -\beta_5 \xi_{l5-} w_l,$$

$$b_{l6-} = +\beta_6 \frac{1}{2}\left[1 + sign(w_{l,k-1})\right]w_{l,k-1} = +\beta_6 \xi_{l6-} w_{l,k-1},$$

or compactly written

$$b_{lm-} = -\beta_m \xi_{lm-} V_{lm}^n,$$

where the signed integers are

$$\xi_{l1+} = \frac{1}{2}\left[1 + sign(u_l)\right],$$

$$\xi_{l2+} = \frac{1}{2}\left[1 - sign(u_{l,i-1})\right],$$

$$\xi_{l3+} = \frac{1}{2}\left[1 + sign(v_l)\right],$$

$$\xi_{l4+} = \frac{1}{2}\left[1 - sign(v_{l,j-1})\right],$$

$$\xi_{l5+} = \frac{1}{2}\left[1 + sign(w_l)\right],$$

$$\xi_{l6+} = \frac{1}{2}\left[1 - sign(w_{l,k-1})\right],$$

and

$$\xi_{l1-} = \frac{1}{2}\left[1 - sign(u_l)\right],$$

$$\xi_{l2-} = \frac{1}{2}\left[1 + sign(u_{l,i-1})\right],$$

$$\xi_{l3-} = \frac{1}{2}\left[1 - sign(v_l)\right],$$

$$\xi_{l4-} = \frac{1}{2}\left[1 + sign(v_{l,j-1})\right],$$

$$\xi_{l5-} = \frac{1}{2}\left[1 - sign(w_l)\right],$$

$$\xi_{l6-} = \frac{1}{2}\left[1 + sign(w_{l,k-1})\right].$$

Obviously

$$\xi_{lm-} = 1 - \xi_{lm+}.$$

Note that the b_{lm+} coefficients are volume flows leaving the cell through the face m divided by the cell volume. b_{lm-} coefficients are volume flows entering the cell through the face m divided by the cell volume. Therefore the b coefficients have the physical meaning of volumetric mass sources and sinks in the cell due to convection.

Appendix 12.2 Discretization of the concentration equations

The concentration Eq. (12.4) will now be discretized following the procedure already described. The result is

$$\alpha_{la} \rho_{la} \gamma_v (C_{il} - C_{ila}) / \Delta \tau$$

$$-\frac{r_h^\kappa \gamma_r}{r^\kappa \Delta r}\left\{-u_l \frac{1}{2}\left[1-sign(u_l)\right](\alpha_l \rho_l)_{i+1} + \frac{D_{il1}^*}{\Delta r_h} A(|Pe_1|)\right\}(C_{il,i+1} - C_{il})$$

$$-\frac{(r_h^\kappa \gamma_r)_{i-1}}{r^\kappa \Delta r}\left\{u_{l,i-1} \frac{1}{2}\left[1+sign(u_{l,i-1})\right](\alpha_l \rho_l)_{i-1} + \frac{D_{il2}^*}{\Delta r_{h,i-1}} A(|Pe_2|)\right\}(C_{il,i-1} - C_{il})$$

$$-\frac{\gamma_\theta}{r^\kappa \Delta \theta}\left\{-v_l \frac{1}{2}\left[1-sign(v_l)\right](\alpha_l \rho_l)_{j+1} + \frac{D_{il3}^*}{r^\kappa \Delta \theta_h} A(|Pe_3|)\right\}(C_{il,j+1} - C_{il})$$

$$-\frac{\gamma_{\theta,j-1}}{r^\kappa \Delta \theta}\left\{v_{l,j-1} \frac{1}{2}\left[1+sign(v_{l,j-1})\right](\alpha_l \rho_l)_{j-1} + \frac{D_{il4}^*}{r^\kappa \Delta \theta_{h,j-1}} A(|Pe_4|)\right\}(C_{il,j-1} - C_{il})$$

$$-\frac{\gamma_z}{\Delta z}\left\{-w_l \frac{1}{2}\left[1-sign(w_l)\right](\alpha_l \rho_l)_{k+1} + \frac{D_{il5}^*}{\Delta z_h} A(|Pe_5|)\right\}(C_{il,k+1} - C_{il})$$

$$-\frac{\gamma_{z,k-1}}{\Delta z}\left\{w_{l,k-1} \frac{1}{2}\left[1+sign(w_{l,k-1})\right](\alpha_l \rho_l)_{k-1} + \frac{D_{il6}^*}{\Delta z_{h,k-1}} A(|Pe_6|)\right\}(C_{il,k-1} - C_{il})$$

$$+\frac{1}{2}(\gamma_v + \gamma_{va}) \mu_l^+ C_{il} = \frac{1}{2}(\gamma_v + \gamma_{va}) DC_{il},$$

where

$$A(|Pe_m|) = 1,$$

or

$$\alpha_{la} \rho_{la} \gamma_{va}(C_{il} - C_{ila})/\Delta \tau$$

$$-\left\{b_{l1-}(\alpha_l \rho_l)_{i+1} + \beta_1 \frac{D_{il1}^*}{\Delta r_h} A(|Pe_1|)\right\}(C_{il,i+1} - C_{il})$$

$$-\left\{b_{l2-}(\alpha_l \rho_l)_{i-1} + \beta_2 \frac{D_{il2}^*}{\Delta r_{h,i-1}} A(|Pe_2|)\right\}(C_{il,i-1} - C_{il})$$

$$-\left\{b_{l3-}(\alpha_l\rho_l)_{j+1}+\beta_3\frac{D^*_{il3}}{r^\kappa\Delta\theta_h}A(|Pe_3|)\right\}(C_{il,j+1}-C_{il})$$

$$-\left\{b_{l4-}(\alpha_l\rho_l)_{j-1}+\beta_4\frac{D^*_{il4}}{r^\kappa\Delta\theta_{h,j-1}}A(|Pe_4|)\right\}(C_{il,j-1}-C_{il})$$

$$-\left\{b_{l5-}(\alpha_l\rho_l)_{k+1}+\beta_5\frac{D^*_{il5}}{\Delta z_h}A(|Pe_5|)\right\}(C_{il,k+1}-C_{il})$$

$$-\left\{b_{l6-}(\alpha_l\rho_l)_{k-1}+\beta_6\frac{D^*_{il6}}{\Delta z_{h,k-1}}A(|Pe_6|)\right\}(C_{il,k-1}-C_{il})$$

$$+\frac{1}{2}(\gamma_v+\gamma_{va})\mu_l^+ C_{il}=\frac{1}{2}(\gamma_v+\gamma_{va})DC_{il},$$

or compactly written

$$\alpha_{la}\rho_{la}\gamma_{va}(C_{il}-C_{ila})/\Delta\tau-\sum_{m=1}^6\left[\underbrace{B_{lm-}+\beta_m\frac{D^*_{ilm}}{\Delta L_{h,m}}A(|Pe_m|)}_{b_{lm}}\right](C_{il,m}-C_{il})$$

$$+\frac{1}{2}(\gamma_v+\gamma_{va})\mu_l^+ C_{il}=\frac{1}{2}(\gamma_v+\gamma_{va})DC_{il}.$$

The harmonic averaged diffusion coefficients are given in Appendix 12.3. The l-field mass flow rate across the m-face is computed by means of the donor-cell concept as follows

$$G_{l1}=(\alpha_l\rho_l u_l)_1=\left[\xi_{l1+}\alpha_l\rho_l+\xi_{l1-}(\alpha_l\rho_l)_{i+1}\right]u_l,$$

$$G_{l2}=(\alpha_l\rho_l u_l)_2=\left[\xi_{l2-}(\alpha_l\rho_l)_{i-1}+\xi_{l2+}\alpha_l\rho_l\right]u_{l,i-1},$$

$$G_{l3}=(\alpha_l\rho_l v_l)_3=\left[\xi_{l3+}\alpha_l\rho_l+\xi_{l3-}(\alpha_l\rho_l)_{j+1}\right]v_l,$$

$$G_{l4}=(\alpha_l\rho_l v_l)_4=\left[\xi_{l4-}(\alpha_l\rho_l)_{j-1}+\xi_{l4+}\alpha_l\rho_l\right]v_{l,j-1},$$

$$G_{15} = (\alpha_l \rho_l w_l)_5 = \left[\xi_{15+} \alpha_l \rho_l + \xi_{15-} (\alpha_l \rho_l)_{k+1} \right] w_l,$$

$$G_{16} = (\alpha_l \rho_l w_l)_6 = \left[\xi_{16-} (\alpha_l \rho_l)_{k-1} + \xi_{16+} \alpha_l \rho_l \right] w_{l,k-1}.$$

Note that the donor-cell concept takes into account the velocity directions. Computation of the harmonic averaged thermal conductivity coefficients is shown also in *Appendix 12.3*.

Appendix 12.3 Harmonic averaged diffusion coefficients

A natural averaging of the coefficients describing diffusion across the face m, having surface cross section S_m is then the harmonic averaging

$$\frac{D^\Phi_{l,m}}{\Delta L_{h,m}} = \left(\frac{\Phi_l}{\Delta V} \right)_m \quad S_m = S_m \frac{2(\Phi_l)(\Phi_l)_m}{\Delta V_m (\Phi_l) + \Delta V (\Phi_l)_m}$$

where in the right hand side m = 1, 2, 3, 4, 5, 6 is equivalent to i + 1, i - 1, j + 1, j - 1, k + 1, k - 1, respectively regarding the properties inside a control volumes. ΔV is the non-staggered cell volume, and ΔV_m is the volume of the cell at the other side of face m. It guaranties that if the field in one of the neighboring cell is missing the diffusion coefficient is zero. This property is derived from the solution of the steady state one-dimensional diffusion equations.

For computation of

$$\frac{D^*_{il,m}}{\Delta L_{h,m}} = \left(\frac{\alpha_l \rho_l D^*_{il}}{\Delta V} \right)_m \quad S_m = S_m \frac{2(\alpha_l \rho_l D^*_{il})(\alpha_l \rho_l D^*_{il})_m}{\Delta V_m (\alpha_l \rho_l D^*_{il}) + \Delta V (\alpha_l \rho_l D^*_{il})_m}$$

we simply set $\Phi_l = \alpha_l \rho_l D^*_{il}$.

For computation of $\dfrac{D^T_{l,m}}{\Delta L_{h,m}}$ we simply set $\Phi_l = \alpha_l \lambda_l$.

For computation of $\dfrac{D^{sC}_{il,m}}{\Delta L_{h,m}}$ we simply set $\Phi_l = \alpha_l \rho_l D^*_{il} (s_{il} - s_{1l})$. Note that

$$\frac{D^C_{il,m}}{\Delta L_{h,m}} = 0$$

for $s_{il} = s_{1l}$ or $s_{il,m} = s_{1l,m}$.

For computation of the turbulent particle diffusion coefficient $\dfrac{D^n_{l,m}}{\Delta L_{h,m}}$ we simply set $\Phi_l = \dfrac{v^t_l}{Sc^t}$.

For computation of $\dfrac{D^V_{l,m}}{\Delta L_{h,m}}$ we simply set $\Phi_l = \alpha_l \rho_l v^*_l$.

In case of cylindrical or Cartesian coordinate systems we have zero off-diagonal diffusion terms and

$$\frac{D^\Phi_{l1}}{\Delta r_h} = \frac{2(\Phi_l)(\Phi_l)_{i+1}}{\Delta r (\Phi_l)_{i+1} + \Delta r_{i+1} \Phi_l},$$

$$\frac{D^\Phi_{l2}}{\Delta r_{h,i-1}} = \frac{2(\Phi_l)(\Phi_l)_{i-1}}{\Delta r (\Phi_l)_{i-1} + \Delta r_{i-1} \Phi_l},$$

$$\frac{D^\Phi_{l3}}{r^\kappa \Delta \theta_h} = \frac{2(\Phi_l)(\Phi_l)_{j+1}}{r^\kappa \left[\Delta \theta (\Phi_l)_{j+1} + \Delta \theta_{j+1} \Phi_l \right]},$$

$$\frac{D^\Phi_{l4}}{r^\kappa \Delta \theta_{h,j-1}} = \frac{2(\Phi_l)(\Phi_l)_{j-1}}{r^\kappa \left[\Delta \theta (\Phi_l)_{j-1} + \Delta \theta_{j-1} \Phi_l \right]},$$

$$\frac{D^\Phi_{l5}}{\Delta z_h} = \frac{2(\Phi_l)(\Phi_l)_{k+1}}{\Delta z (\Phi_l)_{k+1} + \Delta z_{k+1} \Phi_l},$$

$$\frac{D^\Phi_{l6}}{\Delta z_{h,k-1}} = \frac{2(\Phi_l)(\Phi_l)_{k-1}}{\Delta z (\Phi_l)_{k-1} + \Delta z_{k-1} \Phi_l}.$$

Appendix 12.4. Discretized radial momentum equation

The *r*-momentum equation discretized by using the donor-cell concept is

$$\gamma_{vu}\alpha_{lua}\rho_{lua}(u_l - u_{la})/\Delta\tau$$

$$-\frac{r_{i+1}^{\kappa}}{(r+\Delta r_h/2)^{\kappa}\Delta r_h}\left\{-\min\left[0,(\alpha_l\rho_l u_l\gamma_r)_1\right]+(\alpha_l\rho_l v_l\gamma_r)_1\frac{1}{\Delta r_{i+1}}\right\}(u_{li+1}-u_l)$$

$$-\frac{r^{\kappa}}{(r+\Delta r_h/2)^{\kappa}\Delta r_h}\left\{\max\left[0,(\alpha_l\rho_l u_l\gamma_r)_2\right]+(\alpha_l\rho_l v_l\gamma_r)_2\frac{1}{\Delta r}\right\}(u_{li-1}-u_l)$$

$$-\frac{1}{(r+\Delta r_h/2)^{\kappa}\Delta\theta}\left\{-\min\left[0,(\alpha_l\rho_l v_l\gamma_\theta)_3\right]+(\alpha_l\rho_l v_l\gamma_\theta)_3\frac{1}{r_h^{\kappa}\Delta\theta_h}\right\}(u_{lj+1}-u_l)$$

$$-\frac{1}{(r+\Delta r_h/2)^{\kappa}\Delta\theta}\left\{\max\left[0,(\alpha_l\rho_l v_l\gamma_\theta)_4\right]+(\alpha_l\rho_l v_l\gamma_\theta)_4\frac{1}{r_h^{\kappa}\Delta\theta_{h,j-1}}\right\}(u_{lj-1}-u_l)$$

$$-\frac{1}{\Delta z}\left\{-\min\left[0,(\alpha_l\rho_l w_l\gamma_z)_5\right]+(\alpha_l\rho_l v_l\gamma_z)_5\frac{1}{\Delta z_h}\right\}(u_{lk+1}-u_l)$$

$$-\frac{1}{\Delta z}\left\{\max\left[0,(\alpha_l\rho_l w_l\gamma_z)_6\right]+(\alpha_l\rho_l v_l\gamma_z)_6\frac{1}{\Delta z_{h,k-1}}\right\}(u_{lk-1}-u_l)$$

$$-\frac{\sin(\Delta\theta/2)}{(r+\Delta r_h/2)^K \Delta\theta} \left\{ \begin{array}{l} (\alpha_l \rho_l v_l \gamma_\theta)_3 v_{l3} + (\alpha_l \rho_l v_l \gamma_\theta)_4 v_{l4} \\ \\ -\frac{2}{(r+\Delta r_h/2)^K} \left\{ \begin{array}{l} (\alpha\rho v)_{l3} \left[\left(\frac{\partial v_l}{\partial\theta}\right)_3 + u_{l3} \right] \gamma_3 \\ \\ +(\alpha\rho v)_{l4} \left[\left(\frac{\partial v_l}{\partial\theta}\right)_4 + u_{l4} \right] \gamma_4 \end{array} \right. \end{array} \right.$$

$$+\alpha_{lua} \frac{\gamma_{rua}}{\Delta r_h}(p_{i+1} - p)$$

$$+\gamma_{vu}\left[\alpha_{lua}\rho_{lua}g_r + c_{wlu}|u_l|u_l - \sum_{\substack{m=1\\m\neq l}}^{3}\left[\bar{c}_{ml}^d(u_m - u_l) + \bar{c}_{ml}^{vm}\frac{\partial}{\partial\tau}(u_m - u_l)\right]\right]$$

$$= \gamma_{vu}\left[\mu_{wlu}(u_{wl} - u_l) - \mu_{lwu}(u_{lw} - u_l) + \sum_{m\neq l}\mu_{mlu}(u_m - u_l)\right] + f_{vlu}.$$

Now we introduce the volume weighting coefficient

$$C^* = \frac{V_{i+1}}{V + V_{i+1}}.$$

This coefficient allows us easily to compute the mass flows entering the staggered cell through each face divided by the staggered cell volume by using the B_{lm-} flux densities already computed for the non-staggered cells

$$bu_{l,m} = -C^*\left(B_{lm-} + \beta_m \frac{D_{l,m}^v}{\Delta L_{h,m}}\right) - (1-C^*)\left(B_{lm-} + \beta_m \frac{D_{l,m}^v}{\Delta L_{h,m}}\right)_{i+1}.$$

With this abbreviation we have

$$\gamma_{vu}\alpha_{lua}\rho_{lua}\left(u_l - u_{la}\right)/\Delta\tau + \sum_{m=1}^{6} bu_{l,m}\left(u_{l,m} - u_l\right)\ldots$$

Comparison with simple test problems for pure radial flow, pure rotational flow and superposition of both gives rise to improvements that will be described in a moment.

The first improvement is to use the following expression for the centrifugal force and its viscous counterpart

$$-\frac{1}{2}\frac{1}{\left(r+\Delta r_h/2\right)^\kappa}\left\{\left(\alpha_l\rho_l v_l\gamma_\theta\right)_3 v_{l3} + \left(\alpha_l\rho_l v_l\gamma_\theta\right)_4 v_{l4}\right.$$

$$\left.-\frac{2}{\left(r+\Delta r_h/2\right)^\kappa}\left\{\left(\alpha\rho v\right)_{l3}\left[\left(\frac{\partial v_l}{\partial\theta}\right)_3 + u_{l3}\right]\gamma_3 + \left(\alpha\rho v\right)_{l4}\left[\left(\frac{\partial v_l}{\partial\theta}\right)_4 + u_{l4}\right]\gamma_4\right\}\right\},$$

instead of the original form. The method of computing the centrifugal force can be understood as first computing the average flow rotation frequency and then the corresponding force. The analogy to the single-phase rigid body steady rotation, Appendix 2.3,

$$p_{i+1} - p = \frac{1}{2}\rho\omega^2\left(r_{i+1}^2 - r^2\right)$$

or

$$\frac{p_{i+1} - p}{\Delta r_h} = \rho\omega^2\left(r + \Delta r_h/2\right),$$

is obvious.

The second improvement regards the convective term in the r direction. The exact solution of the single-phase steady state mass and momentum equation for radial flow gives the well-known *Bernulli* equation

$$\frac{1}{\Delta r_h}\left(p_{i+1} - p\right) = -\frac{1}{2}\rho\frac{1}{\Delta r_h}\left(u^{*2}_{i+1} - u^{*2}\right),$$

see Appendix 2.3. In this equation the velocities are defined at the same places where the pressures are defined. This is not the case with our finite difference equation. In order to make the finite difference form equivalent to *Bernulli* form we use the single-phase case to derive a corrector multiplier as follows. We still use the donor-cell idea to compute the displaced velocities

$u > 0$ $\qquad\qquad u \le 0$

$$u*_{i+1} = u \frac{r_h}{r_{i+1}} \qquad\qquad u*_{i+1} = u_{i+1} \frac{r_{h,i+1}}{r_{i+1}}$$

$$u* = u_{i-1} \frac{r_{h,i-1}}{r} \qquad\qquad u* = u \frac{r_h}{r}$$

Replacing in the *Bernulli* equation results in

$$\rho \frac{1}{\Delta r_h} \frac{1}{2} \left(u \frac{r_h}{r_{i+1}} + u_{i-1} \frac{r_{h,i-1}}{r} \right) \left(u \frac{r_h}{r_{i+1}} - u_{i-1} \frac{r_{h,i-1}}{r} \right) + \frac{1}{\Delta r_h} (p_{i+1} - p) = 0 \quad \text{for} \quad u > 0$$

and

$$\rho \frac{1}{\Delta r_h} \frac{1}{2} \left(u_{i+1} \frac{r_{h,i+1}}{r_{i+1}} + u \frac{r_h}{r} \right) \left(u_{i+1} \frac{r_{h,i+1}}{r_{i+1}} - u \frac{r_h}{r} \right) + \frac{1}{\Delta r_h} (p_{i+1} - p) = 0 \quad \text{for} \quad u \le 0,$$

respectively. We multiply and divide the convective term by $(u + u_{i-1})(u - u_{i-1})$ and use outside this product

$$u_{i-1} = u \frac{r_h}{r_{h,i-1}} .$$

The result is

$$\rho \frac{1}{\Delta r_h} \frac{\frac{1}{r_{i+1}^2} - \frac{1}{r^2}}{\frac{1}{r_h^2} - \frac{1}{r_{h,i-1}^2}} \frac{1}{2} (u + u_{i-1})(u - u_{i-1}) + \frac{1}{\Delta r_h} (p_{i+1} - p) = 0, \qquad\qquad u > 0.$$

For the singularity $r_{h,i-1} = 0$ we have

$$\rho \frac{1}{\Delta r_h} \left(\frac{r_h}{r_{i+1}} \right)^2 \frac{1}{2} (u + u_{i-1})(u - u_{i-1}) + \frac{1}{\Delta r_h} (p_{i+1} - p) = 0 .$$

Similarly for negative velocity between the two pressures we have

$$u_{i+1} = u \frac{r_h}{r_{h,i+1}}$$

and therefore

$$\rho \frac{1}{\Delta r_h} \frac{\frac{1}{r_{i+1}^2} - \frac{1}{r^2}}{\frac{1}{r_{h,i+1}^2} - \frac{1}{r_h^2}} \frac{1}{2}(u_{i+1} + u)(u_{i+1} - u) + \frac{1}{\Delta r_h}(p_{i+1} - p) = 0, \qquad u \leq 0.$$

Thus the final form of the multiphase convective term is written as

$$-\frac{1}{\Delta r_h} \frac{\frac{1}{r_{i+1}^2} - \frac{1}{r^2}}{\frac{1}{r_h^2} - \frac{1}{r_{h,i-1}^2}} \left\{ -\min\left[0, \, (\alpha_l \rho_l u_l \gamma_r)_1\right] + (\alpha_l \rho_l v_l \gamma_r)_1 \frac{1}{\Delta r_{i+1}} \right\} (u_{li+1} - u_l)$$

$$-\frac{1}{\Delta r_h} \frac{\frac{1}{r_{i+1}^2} - \frac{1}{r^2}}{\frac{1}{r_{h,i+1}^2} - \frac{1}{r_h^2}} \left\{ \max\left[0, \, (\alpha_l \rho_l u_l \gamma_r)_2\right] + (\alpha_l \rho_l v_l \gamma_r)_2 \frac{1}{\Delta r} \right\} (u_{li-1} - u_l).$$

It is convenient to compute the geometry coefficients at the start

$$\beta_{r1} = \frac{1}{\Delta r_h}, \; \beta_{r2} = \beta_{r1}, \; \beta_{r3} = \frac{1}{(r + \Delta r_h / 2)^\kappa \Delta \theta}, \; \beta_{r4} = \beta_{r3}, \; \beta_{r5} = \frac{1}{\Delta z}, \; \beta_{r6} = \beta_{r5}.$$

The bu coefficients are defined as follows:

$$bu_{l1} = -\beta_{r1} \frac{\frac{1}{r_{i+1}^2} - \frac{1}{r^2}}{\frac{1}{r_h^2} - \frac{1}{r_{h,i-1}^2}} \left\{ -\min\left[0, (\alpha_l \rho_l u_l \gamma_r)_1\right] + (\alpha_l \rho_l v_l \gamma_r)_1 \frac{1}{\Delta r_{i+1}} \right\},$$

$$bu_{l2} = -\beta_{r2} \frac{\frac{1}{r_{i+1}^2} - \frac{1}{r^2}}{\frac{1}{r_{h,i+1}^2} - \frac{1}{r_h^2}} \left\{ \max\left[0, (\alpha_l \rho_l u_l \gamma_r)_2\right] + (\alpha_l \rho_l v_l \gamma_r)_2 \frac{1}{\Delta r} \right\},$$

$$bu_{l3} = -\beta_{r3} \left\{ -\min\left[0, (\alpha_l \rho_l v_l \gamma_\theta)_3\right] + (\alpha_l \rho_l v_l \gamma_\theta)_3 \frac{1}{r_h^K \Delta \theta_h} \right\},$$

$$bu_{l4} = -\beta_{r4} \left\{ \max\left[0, (\alpha_l \rho_l v_l \gamma_\theta)_4\right] + (\alpha_l \rho_l v_l \gamma_\theta)_4 \frac{1}{r_h^K \Delta \theta_{h,j-1}} \right\},$$

$$bu_{l5} = -\beta_{r5} \left\{ -\min\left[0, (\alpha_l \rho_l w_l \gamma_z)_5\right] + (\alpha_l \rho_l v_l \gamma_z)_5 \frac{1}{\Delta z_h} \right\},$$

$$bu_{l6} = -\beta_{r6} \left\{ \max\left[0, (\alpha_l \rho_l w_l \gamma_z)_6\right] + (\alpha_l \rho_l v_l \gamma_z)_6 \frac{1}{\Delta z_{h,k-1}} \right\}.$$

Now we introduce the volume weighting coefficient

$$C^* = \frac{V_{i+1}}{V + V_{i+1}}.$$

This coefficient allows us easily to compute the mass flows entering the staggered cell through each face divided by the staggered cell volume by using the B_{lm-} flux densities already computed for the non-staggered cells

$$bu_{l,m} = -C^* \left(B_{lm-} + \beta_m \frac{D_{l,m}^u}{\Delta L_{h,m}} \right) - (1 - C^*) \left(B_{lm-} + \beta_m \frac{D_{l,m}^u}{\Delta L_{h,m}} \right)_{i+1}.$$

With this abbreviation we have

$$\gamma_{vu} \alpha_{lua} \rho_{lua} (u_l - u_{la})/\Delta \tau + \sum_{m=1}^{6} bu_{l,m}(u_{l,m} - u_l) \ldots$$

Appendix 12.5 The $\bar{a}$ coefficients for Eq. (12.46)

$$\bar{a}_{11} = a_{22}a_{33} - a_{32}a_{23} \quad \bar{a}_{12} = a_{32}a_{13} - a_{12}a_{33} \quad \bar{a}_{13} = a_{12}a_{23} - a_{22}a_{13}$$

$$\bar{a}_{21} = a_{23}a_{31} - a_{21}a_{33} \quad \bar{a}_{22} = a_{11}a_{33} - a_{31}a_{13} \quad \bar{a}_{23} = a_{21}a_{13} - a_{23}a_{11}$$

$$\bar{a}_{31} = a_{21}a_{32} - a_{31}a_{22} \quad \bar{a}_{32} = a_{12}a_{31} - a_{32}a_{11} \quad \bar{a}_{33} = a_{11}a_{22} - a_{21}a_{12}$$

Appendix 12.6 Discretization of the angular momentum equation

The θ momentum equation discretized by using the donor cell concept is

$$\gamma_{vv}\alpha_{lva}\rho_{lva}(v_l - v_{la})/\Delta\tau$$

$$-\frac{r_h^\kappa}{r^\kappa \Delta r}\left\{-\min\left[0,\left(\alpha_l\rho_l u_l \gamma_r\right)_1\right] + \left(\alpha_l\rho_l v_l \gamma_r\right)_1 \frac{1}{\Delta r_h}\right\}(v_{li+1} - v_l)$$

$$-\frac{r_{h,i-1}^\kappa}{r^\kappa \Delta r}\left\{\max\left[0,\left(\alpha_l\rho_l u_l \gamma_r\right)_2\right] + \left(\alpha_l\rho_l v_l \gamma_r\right)_2 \frac{1}{\Delta r_{h,i-1}}\right\}(v_{li-1} - v_l)$$

$$-\frac{1}{r^\kappa \Delta\theta_h}\left\{-\min\left[0,\left(\alpha_l\rho_l v_l \gamma_\theta\right)_3\right] + \left(\alpha_l\rho_l v_l \gamma_\theta\right)_3 \frac{1}{r^\kappa \Delta\theta_{j+1}}\right\}(v_{lj+1} - v_l)$$

$$-\frac{1}{r^\kappa \Delta\theta_h}\left\{\max\left[0,\left(\alpha_l\rho_l v_l \gamma_\theta\right)_4\right] + \left(\alpha_l\rho_l v_l \gamma_\theta\right)_4 \frac{1}{r^\kappa \Delta\theta}\right\}(v_{lj-1} - v_l)$$

$$-\frac{1}{\Delta z}\left\{-\min\left[0,\left(\alpha_l\rho_l w_l \gamma_z\right)_5\right] + \left(\alpha_l\rho_l v_l \gamma_z\right)_5 \frac{1}{\Delta z_h}\right\}(v_{lk+1} - v_l)$$

$$-\frac{1}{\Delta z}\left\{\max\left[0,\left(\alpha_l\rho_l w_l \gamma_z\right)_6\right] + \left(\alpha_l\rho_l v_l \gamma_z\right)_6 \frac{1}{\Delta z_{h,k-1}}\right\}(v_{lk-1} - v_l)$$

$$+\frac{\sin(\Delta\theta/2)}{r^\kappa \Delta\theta_h}$$

$$\times \left\{ (\alpha_l \rho_l v_l \gamma_\theta)_3 u_{l3} + (\alpha_l \rho_l v_l \gamma_\theta)_4 u_{l4} - \left\{ \begin{array}{l} (\alpha \rho v)_{lj+1} \gamma_{\theta 3} \left[r \dfrac{\partial}{\partial r}\left(\dfrac{v_l}{r}\right) + \dfrac{1}{r}\dfrac{\partial u_l}{\partial \theta} \right]_3 \\ \\ + (\alpha \rho v)_l \gamma_{\theta 4} \left[r \dfrac{\partial}{\partial r}\left(\dfrac{v_l}{r}\right) + \dfrac{1}{r}\dfrac{\partial u_l}{\partial \theta} \right]_4 \end{array} \right\} \right.$$

$$+ \alpha_{lva} \frac{\gamma_\theta}{r^\kappa \Delta \theta_h}(p_{j+1} - p)$$

$$+ \gamma_{vv} \left[\alpha_{lva}\rho_{lva} g_\theta + c_{wlv}|v_l|v_l - \sum_{\substack{m=1 \\ m \neq l}}^{3} \left[\overline{c}_{ml}^d (v_m - v_l) + \overline{c}_{ml}^{vm} \frac{\partial}{\partial \tau}(v_m - v_l) \right] \right]$$

$$= \gamma_{vv} \left[\mu_{wlv}(v_{wl} - v_l) - \mu_{lwv}(v_{lw} - v_l) + \sum_{\substack{m=1 \\ m \neq l}}^{3} \mu_{mlv}(v_m - v_l) \right] + f_{vlv}$$

Introducing the geometry coefficients

$$\beta_{\theta 1} = \frac{r_h^\kappa}{r^\kappa \Delta r},\ \beta_{\theta 2} = \frac{r_{h,i-1}^\kappa}{r^\kappa \Delta r},\ \beta_{\theta 3} = \frac{1}{r^\kappa \Delta \theta_h},\ \beta_{\theta 4} = \beta_{\theta 3},\ \beta_{\theta 5} = \frac{1}{\Delta z},\ \beta_{\theta 6} = \beta_{\theta 5},$$

the bv coefficients are defined as follows

$$bv_{l1} = -\beta_{\theta 1} \left\{ -\min\left[0,(\alpha_l \rho_l u_l \gamma_r)_1\right] + (\alpha_l \rho_l v_l \gamma_r)_1 \frac{1}{\Delta r_h} \right\},$$

$$bv_{l2} = -\beta_{\theta 2} \left\{ \max\left[0,(\alpha_l \rho_l u_l \gamma_r)_2\right] + (\alpha_l \rho_l v_l \gamma_r)_2 \frac{1}{\Delta r_{h,i-1}} \right\},$$

$$bv_{l3} = -\beta_{\theta 3} \left\{ -\min\left[0,(\alpha_l \rho_l v_l \gamma_\theta)_3\right] + (\alpha_l \rho_l v_l \gamma_\theta)_3 \frac{1}{r^\kappa \Delta \theta_{j+1}} \right\},$$

$$bv_{l4} = -\beta_{\theta 4} \left\{ \max\left[0,(\alpha_l \rho_l v_l \gamma_\theta)_4\right] + (\alpha_l \rho_l v_l \gamma_\theta)_4 \frac{1}{r^\kappa \Delta \theta} \right\},$$

$$bv_{l5} = -\beta_{05}\left\{-\min\left[0,(\alpha_l\rho_l w_l\gamma_z)_5\right]+(\alpha_l\rho_l v_l\gamma_z)_5\frac{1}{\Delta z_h}\right\},$$

$$bv_{l6} = -\beta_{06}\left\{\max\left[0,(\alpha_l\rho_l w_l\gamma_z)_6\right]+(\alpha_l\rho_l v_l\gamma_z)_6\frac{1}{\Delta z_{h,k-1}}\right\}.$$

Now we introduce the volume weighting coefficient

$$C^* = \frac{V_{j+1}}{V+V_{j+1}}.$$

This coefficient allows us easily to compute the mass flows entering the staggered cell through each face divided by the staggered cell volume by using the B_{lm-} flux densities already computed for the non-staggered cells

$$bv_{l,m} = -C^*\left(B_{lm-}+\beta_m\frac{D^v_{l,m}}{\Delta L_{h,m}}\right)-(1-C^*)\left(B_{lm-}+\beta_m\frac{D^v_{l,m}}{\Delta L_{h,m}}\right)_{j+1}.$$

With this abbreviation we have

$$\gamma_{vv}\alpha_{lva}\rho_{lva}(v_l-v_{la})/\Delta\tau + \sum_{m=1}^{6}bv_{l,m}(v_{l,m}-v_l)\ldots$$

Appendix 12.7 Discretization of the axial momentum equation

The z momentum equation discretized by using the donor cell concept is

$$\gamma_{vw}\alpha_{lwa}\rho_{lwa}(w_l-w_{la})/\Delta\tau$$

$$-\frac{r_h^\kappa}{r^\kappa\Delta r}\left\{-\min\left[0,(\alpha_l\rho_l u_l\gamma_r)_1\right]+(\alpha_l\rho_l v_l\gamma_r)_1\frac{1}{\Delta r_h}\right\}(w_{li+1}-w_l)$$

$$-\frac{r_{h,i-1}^\kappa}{r^\kappa\Delta r}\left\{\max\left[0,(\alpha_l\rho_l u_l\gamma_r)_2\right]+(\alpha_l\rho_l v_l\gamma_r)_2\frac{1}{\Delta r_{h,i-1}}\right\}(w_{li-1}-w_l)$$

$$-\frac{1}{r^\kappa\Delta\theta}\left\{-\min\left[0,(\alpha_l\rho_l v_l\gamma_\theta)_3\right]+(\alpha_l\rho_l v_l\gamma_\theta)_3\frac{1}{r^\kappa\Delta\theta_h}\right\}(w_{lj+1}-w_l)$$

$$-\frac{1}{r^{\kappa}\Delta\theta}\left\{\max\left[0,(\alpha_l\rho_l v_l\gamma_\theta)_4\right]+(\alpha_l\rho_l v_l\gamma_\theta)_4\frac{1}{r^{\kappa}\Delta\theta_{h,j-1}}\right\}(w_{lj-1}-w_l)$$

$$-\frac{1}{\Delta z_h}\left\{-\min\left[0,(\alpha_l\rho_l w_l\gamma_z)_5\right]+(\alpha_l\rho_l v_l\gamma_z)_5\frac{1}{\Delta z_{k+1}}\right\}(w_{lk+1}-w_l)$$

$$-\frac{1}{\Delta z_h}\left\{\max\left[0,(\alpha_l\rho_l w_l\gamma_z)_6\right]+(\alpha_l\rho_l v_l\gamma_z)_6\frac{1}{\Delta z}\right\}(w_{lk-1}-w_l)+\alpha_{lwa}\frac{\gamma_z}{\Delta z}(p_{k+1}-p)$$

$$+\gamma_{vw}\left[\alpha_{lwa}\rho_{lwa}g_z+c_{wlw}|w_l|w_l-\sum_{\substack{m=1\\m\neq l}}^{3}\left[\overline{c}_{ml}^{d}(w_m-w_l)+\overline{c}_{ml}^{vm}\frac{\partial}{\partial\tau}(w_m-w_l)\right]\right]$$

$$=\gamma_{vw}\left[\mu_{wlw}(w_{wl}-w_l)-\mu_{lww}(w_{lw}-w_l)+\sum_{\substack{m=1\\m\neq l}}^{3}\mu_{mlw}(w_m-w_l)\right]+f_{vlw}$$

Introducing the geometry coefficients

$$\beta_{z1}=\frac{r_h^{\kappa}}{r^{\kappa}\Delta r},\ \beta_{z2}=\frac{r_{h,i-1}^{\kappa}}{r^{\kappa}\Delta r},\ \beta_{z3}=\frac{1}{r^{\kappa}\Delta\theta},\ \beta_{z4}=\beta_{z3},\ \beta_{z5}=\frac{1}{\Delta z_h},\ \beta_{z6}=\beta_{z5}$$

the *bw* coefficients are defined as follows

$$bw_{l1}=-\beta_{z1}\left\{-\min\left[0,(\alpha_l\rho_l u_l\gamma_r)_1\right]+(\alpha_l\rho_l v_l\gamma_r)_1\frac{1}{\Delta r_h}\right\},$$

$$bw_{l2}=-\beta_{z2}\left\{\max\left[0,(\alpha_l\rho_l u_l\gamma_r)_2\right]+(\alpha_l\rho_l v_l\gamma_r)_2\frac{1}{\Delta r_{h,i-1}}\right\},$$

$$bw_{l3}=-\beta_{z3}\left\{-\min\left[0,(\alpha_l\rho_l v_l\gamma_\theta)_3\right]+(\alpha_l\rho_l v_l\gamma_\theta)_3\frac{1}{r^{\kappa}\Delta\theta_h}\right\},$$

$$bw_{l4}=-\beta_{z4}\left\{\max\left[0,(\alpha_l\rho_l v_l\gamma_\theta)_4\right]+(\alpha_l\rho_l v_l\gamma_\theta)_4\frac{1}{r^{\kappa}\Delta\theta_{h,j-1}}\right\},$$

$$bw_{15} = -\beta_{z5}\left\{-\min\left[0,\left(\alpha_l\rho_l w_l\gamma_z\right)_5\right]+\left(\alpha_l\rho_l v_l\gamma_z\right)_5\frac{1}{\Delta z_{k+1}}\right\},$$

$$bw_{16} = -\beta_{z6}\left\{\max\left[0,\left(\alpha_l\rho_l w_l\gamma_z\right)_6\right]+\left(\alpha_l\rho_l v_l\gamma_z\right)_6\frac{1}{\Delta z}\right\}.$$

Now we introduce the volume weighting coefficient

$$C^* = \frac{V_{k+1}}{V + V_{k+1}}.$$

This coefficient allows us easily to compute the mass flows entering the staggered cell through each face divided by the staggered cell volume by using the B_{lm-} flux densities already computed for the non-staggered cells

$$bw_{l,m} = -C^*\left(B_{lm-}+\beta_m\frac{D^v_{l,m}}{\Delta L_{h,m}}\right)-\left(1-C^*\right)\left(B_{lm-}+\beta_m\frac{D^v_{l,m}}{\Delta L_{h,m}}\right)_{k+1}.$$

With this abbreviation we have

$$\gamma_{vw}\alpha_{lwa}\rho_{lwa}\left(w_l - w_{la}\right)/\Delta\tau + \sum_{m=1}^{6}bw_{l,m}\left(w_{l,m}-w_l\right)\ldots$$

Appendix 12.8 Analytical derivatives for the residual error of each equation with respect to the dependent variables

The velocities are assumed to be linearly dependent on the pressures, with the result that derivatives for the velocities with respect to pressure can be obtained directly from the momentum equations

$$V^n_{lm} = dV^n_{lm} - RVel_{lm}(p_m - p).$$

The result is

$$\frac{\partial V^n_{lm}}{\partial p} = RVel_{lm}, \quad \frac{\partial V^n_{lm}}{\partial p_m} = -RVel_{lm}, \quad \frac{\partial V^n_{lm}}{\partial p_x} = 0 \quad \text{for} \quad x \neq m,$$

and therefore

$$\frac{\partial b_{lm+}}{\partial p} = \beta_m \xi_{lm+} \frac{\partial V_{lm}^n}{\partial p} = \beta_m \xi_{lm+} RVel_{lm},$$

$$\frac{\partial b_{lm+}}{\partial p_m} = \beta_m \xi_{lm+} \frac{\partial V_{lm}^n}{\partial p_m} = -\beta_m \xi_{lm+} RVel_{lm},$$

$$\frac{\partial b_{lm+}}{\partial p_x} = 0 \quad \text{for} \quad x \neq m,$$

$$\frac{\partial b_{lm-}}{\partial p} = -\beta_m \xi_{lm-} \frac{\partial V_{lm}^n}{\partial p} = -\beta_m \xi_{lm-} RVel_{lm}, \quad \frac{\partial b_{lm}}{\partial p} = \frac{\partial b_{lm-}}{\partial p},$$

$$\frac{\partial b_{lm-}}{\partial p_m} = -\beta_m \xi_{lm-} \frac{\partial V_{lm}^n}{\partial p_m} = \beta_m \xi_{lm-} RVel_{lm}, \quad \frac{\partial b_{lm}}{\partial p_m} = \frac{\partial b_{lm-}}{\partial p_m},$$

$$\frac{\partial b_{lm-}}{\partial p_x} = 0 \quad \text{for} \quad x \neq m, \qquad \frac{\partial b_{lm}}{\partial p_x} = 0 \quad \text{for} \quad x \neq m.$$

Particle number density derivatives: Taking the derivatives of the LHS of the discretized particle number density conservation equation

$$f_l^n \equiv \left[n_l \gamma_v - (n_l \gamma_v)_a \right] / \Delta \tau + \sum_{m=1}^{6} \left(b_{lm+} n_l - b_{lm-} n_{lm} \right) - \frac{1}{2} (\gamma_v + \gamma_{va}) Dn_l$$

$$= \frac{\partial f_l^n}{\partial n_l} (n_l - n_{l0}) = 0$$

we obtain:

$$\frac{\partial f_l^n}{\partial n_l} = \frac{\gamma_v}{\Delta \tau} + \sum_{m=1}^{6} b_{lm+} > 0,$$

$$\frac{\partial f_l^n}{\partial p_m} = \sum_{x=1}^{6} \left(n_l \frac{\partial b_{lm+}}{\partial p_x} - n_{lm} \frac{\partial b_{lm-}}{\partial p_x} \right) = -\beta_m \left(\xi_{lm+} n_l + \xi_{lm-} n_{lm} \right) RVel_{lm},$$

for $m = 1, 6,$

$$\frac{\partial f_l^n}{\partial p} = \sum_{m=1}^{6}\left(n_l\frac{\partial b_{lm+}}{\partial p}-n_{lm}\frac{\partial b_{lm-}}{\partial p}\right) = \sum_{m=1}^{6}\left[\beta_m\left(\xi_{lm+}n_l+\xi_{lm-}n_{lm}\right)RVel_{lm}\right] = -\sum_{m=1}^{6}\frac{\partial f_l^n}{\partial p_m}.$$

Mass conservation derivatives: Taking the derivatives of the LHS of the discretized mass conservation equation

$$f_l^{\alpha} \equiv \left[\alpha_l\rho_l\gamma_v - \left(\alpha_l\rho_l\gamma_v\right)_a\right]/\Delta\tau + \sum_{m=1}^{6}[b_{lm+}\alpha_l\rho_l - b_{lm-}(\alpha_l\rho_l)_m] - \frac{1}{2}\left(\gamma_v+\gamma_{va}\right)\mu_l$$

$$= \frac{\partial f_l^{\alpha}}{\partial\alpha_l}(\alpha_l - \alpha_{l0}) = 0$$

we obtain

$$\frac{\partial f_l^{\alpha}}{\partial \alpha_l} = \rho_l\left(\frac{\gamma_v}{\Delta\tau}+\sum_{m=1}^{6}b_{lm+}\right) \equiv \rho_l\frac{\partial f_l^n}{\partial n_l} \succ 0,$$

$$\frac{\partial f_l^{\alpha}}{\partial s_l} = \alpha_l\frac{\partial\rho_l}{\partial s_l}\left(\frac{\gamma_v}{\Delta\tau}+\sum_{m=1}^{6}b_{lm+}\right) \equiv \frac{\alpha_l}{\rho_l}\frac{\partial\rho_l}{\partial s_l}\frac{\partial f_l^{\alpha}}{\partial\alpha_l} \equiv \alpha_l\frac{\partial\rho_l}{\partial s_l}\frac{\partial f_l^n}{\partial n_l},$$

$$\frac{\partial f_l^{\alpha}}{\partial C_{il}} = \alpha_l\frac{\partial\rho_l}{\partial C_{il}}\left(\frac{\gamma_v}{\Delta\tau}+\sum_{m=1}^{6}b_{lm+}\right) \equiv \frac{\alpha_l}{\rho_l}\frac{\partial\rho_l}{\partial C_{il}}\frac{\partial f_l^{\alpha}}{\partial\alpha_l} \equiv \alpha_l\frac{\partial\rho_l}{\partial C_{il}}\frac{\partial f_l^n}{\partial n_l},$$

$$\frac{\partial f_l^{\alpha}}{\partial p} = \alpha_l\frac{\partial\rho_l}{\partial p}\left(\frac{\gamma_v}{\Delta\tau}+\sum_{m=1}^{6}b_{lm+}\right) + \sum_{m=1}^{6}\left[\alpha_l\rho_l\frac{\partial b_{lm+}}{\partial p}-(\alpha_l\rho_l)_m\frac{\partial b_{lm-}}{\partial p}\right]$$

$$= \frac{\alpha_l}{\rho_l a_l^2}\frac{\partial f_l^{\alpha}}{\partial\alpha_l} + \sum_{m=1}^{6}\left\{\beta_m\left[\xi_{lm+}\alpha_l\rho_l+\xi_{lm-}(\alpha_l\rho_l)_m\right]RVel_{lm}\right\},$$

$$\frac{\partial f_l^{\alpha}}{\partial p_m} = \alpha_l\rho_l\sum_{x=1}^{6}\frac{\partial b_{lx+}}{\partial p_m} - \sum_{x=1}^{6}\left[(\alpha_l\rho_l)_x\frac{\partial b_{lx-}}{\partial p_m}\right] - \alpha_{lm}b_{lm-}\frac{\partial\rho_{lm}}{\partial p_m}$$

$$= -\beta_m\left[\xi_{lm+}\alpha_l\rho_l+\xi_{lm-}(\alpha_l\rho_l)_m\right]RVel_{lm} + \beta_m\alpha_{lm}\xi_{lm-}V_{lm}^n\frac{\partial\rho_{lm}}{\partial p_m}$$

$$= -\beta_m \left[\xi_{lm+} \alpha_l \rho_l RVel_{lm} + \xi_{lm-} (\alpha_l \rho_l)_m \left(RVel_{lm} - \frac{V_{lm}^n}{\rho_{lm}} \frac{\partial \rho_{lm}}{\partial p_m} \right) \right] \quad \text{for} \quad m = 1, 6.$$

Specific entropy derivatives: Taking the derivatives of the LHS of the discretized entropy conservation equation

$$f_l^s \equiv \alpha_a \rho_{la} \gamma_{va} \frac{s_l - s_{la}}{\Delta \tau} - \sum_{m=1}^{6} [b_{lm-} \alpha_{lm} \rho_{lm} (s_{lm} - s_l)] + \frac{1}{2} (\gamma_v + \gamma_{va}) (\mu_l^+ s_l - Ds_l^*)$$

$$= \frac{\partial f_l^s}{\partial s_l} (s_l - s_{l0}) = 0$$

we obtain:

$$\frac{\partial f_l^s}{\partial s_l} = \frac{\alpha_{la} \rho_{la} \gamma_{va}}{\Delta \tau} + \frac{1}{2} (\gamma_v + \gamma_{va}) \mu_l^+ + \sum_{m=1}^{6} b_{lm-} (\alpha_l \rho_l)_m \geq 0 ,$$

$$\frac{\partial f_l^s}{\partial p} = -\sum_{m=1}^{6} (\alpha_l \rho_l)_m (s_{lm} - s_l) \frac{\partial b_{lm-}}{\partial p} = \sum_{m=1}^{6} (\alpha_l \rho_l)_m (s_{lm} - s_l) \beta_m \xi_{lm-} RVel_{lm} ,$$

$$\frac{\partial f_l^s}{\partial p_m} = -\sum_{x=1}^{6} (\alpha_l \rho_l)_x (s_{lx} - s_l) \frac{\partial b_{lx-}}{\partial p_m} - b_{lm-} \alpha_{lm} (s_{lm} - s_l) \frac{\partial \rho_{lm}}{\partial p_m}$$

$$= -\beta_m \xi_{lm-} \alpha_{lm} (s_{lm} - s_l) \left(\rho_{lm} RVel_{lm} - V_{lm}^n \frac{\partial \rho_{lm}}{\partial p_m} \right) \quad \text{for} \quad m = 1, 6.$$

Inert mass conservation equation derivatives: Taking the derivatives of the LHS of the discretized inert mass conservation equation

$$f_{il}^C \equiv \alpha_{la} \rho_{la} \gamma_{va} \frac{C_{il} - C_{ila}}{\Delta \tau} - \sum_{m=1}^{6} \left\{ \left[b_{lm-} (\alpha_l \rho_l)_m + \beta_m \frac{D_{ilm}^*}{\Delta L_{h,m}} A(|Pe_m|) \right] (C_{il,m} - C_{il}) \right\}$$

$$+ \frac{1}{2} (\gamma_v + \gamma_{va}) (\mu_l^+ C_{il} - DC_{il}) = \frac{\partial f_{il}^C}{\partial C_{il}} (C_{il} - C_{il0}) = 0$$

we obtain

$$\frac{\partial f_{il}^{C}}{\partial C_{il}} = \frac{\gamma_{va}\alpha_{la}\rho_{la}}{\Delta\tau} + \frac{1}{2}(\gamma_{v} + \gamma_{va})\mu_{l}^{+} + \sum_{m=1}^{6}\left[b_{lm-}(\alpha_{l}\rho_{l})_{m} + \beta_{m}\frac{D_{ilm}^{*}}{\Delta L_{h,m}}A(|Pe_{m}|)\right] \geq 0,$$

$$\frac{\partial f_{il}^{C}}{\partial p} = -\sum_{m=1}^{6}(\alpha_{l}\rho_{l})_{m}(C_{ilm} - C_{il})\frac{\partial b_{lm}}{\partial p} = \sum_{m=1}^{6}(\alpha_{l}\rho_{l})_{m}(C_{ilm} - C_{il})\beta_{m}\xi_{lm-}RVel_{lm},$$

$$\frac{\partial f_{il}^{C}}{\partial p_{m}} = -\sum_{x=1}^{6}(\alpha_{l}\rho_{l})_{x}(C_{ilx} - C_{il})\frac{\partial b_{lx-}}{\partial p_{m}} - b_{lm-}\alpha_{lm}(C_{ilm} - C_{il})\frac{\partial \rho_{lm}}{\partial p_{m}}$$

$$= -\alpha_{lm}(C_{ilm} - C_{il})\left(\rho_{lm}\frac{\partial b_{lm-}}{\partial p_{m}} + b_{lm-}\frac{\partial \rho_{lm}}{\partial p_{m}}\right)$$

$$= -\beta_{m}\xi_{lm-}\alpha_{lm}(C_{ilm} - C_{il})\left(\rho_{lm}RVel_{lm} - V_{lm}^{n}\frac{\partial \rho_{lm}}{\partial p_{m}}\right) \quad \text{for} \quad m=1,6.$$

Appendix 12.9 Simple introduction to iterative methods for solution of algebraic systems

Most of the iterative methods for solving the linear system of equations

$$\mathbf{Ax} = \mathbf{b}$$

are based on the regular splitting of the coefficient matrix

$$\mathbf{A} = \mathbf{B} - \mathbf{R}.$$

All three matrices above are of the same range. The iteration method is then

$$\mathbf{Bx}^{n+1} = \mathbf{Rx}^{n} + \mathbf{b}$$

or after eliminating $\mathbf{R}$

$$\mathbf{Bx}^{n+1} = \mathbf{Bx}^{n} + \mathbf{b} - \mathbf{Ax}^{n}.$$

Introducing the residual error vector at the previous solution

$$\mathbf{r}^{n} = \mathbf{b} - \mathbf{Ax}^{n},$$

we have a method for computing the next approximation

$$\mathbf{x}^{n+1} = \mathbf{x}^{n} + \mathbf{B}^{-1}\mathbf{r}^{n},$$

by adding to the previous estimate for the solution a correction vector $\Delta \mathbf{x}^n$,

$$\mathbf{x}^{n+1} = \mathbf{x}^n + \Delta \mathbf{x}^n.$$

$\Delta \mathbf{x}^n$ is in fact the solution of the algebraic system

$$\mathbf{B}\Delta \mathbf{x}^n = \mathbf{r}^n.$$

The matrix **B** has to be selected so as to enable non-expensive solution of the above system in terms of computer time and storage. Selecting

$$\mathbf{B} = \text{diag}(\mathbf{A})$$

results in the *Jacobi* method. Selection of **B** to be the lower off diagonal part including the diagonal leads to the *Gauss-Seidel* method. Selecting **B** to consists of non-expensively invertable blocks of **A** leads to the so called *Block-Gauss-Seidel* method. The secret of creating a powerful method is in appropriate selection of **B**. The reader will find valuable information on this subject in [61].

References

1. Addessio FL et al. TRAC-PF1 (February 1984) An advanced best-estimate computer program for pressurised water reactor analysis, NUREG/CR-3567, LA-9844-MS
2. Addessio FL et al. (July-August 1985) TRAC-PF1/MOD1 Computer Code and Developmental Assessment, Nuclear Safety, vol 26 no 4 pp 438-454
3. Andersen JGM, Schang JC (1984) A predictor- corrector method for the BWR version of the TRAC computer code, AICHE Symposium Series, Heat Transfer - Niagara Falls, Farukhi NM (ed) 236.80, pp 275-280
4. Amsden AA (1985) KIVA: A computer program for two- and three-dimensional fluid flows with chemical reactions and fuel sprays, LA-10245-MS
5. Bohl WR et al. (1988) Multi-phase flow in the advanced fluid dynamics model, ANS Proc. 1988, Nat. Heat Transfer Conf., July 24-July 27, Houston, Texas. HTC-3 pp 61-70
6. Caretto LS, Gosman AD, Patankar SV, Spalding DB (1973) Two calculation procedures for steady, three dimensional flow with recirculation, Proc. 3rd Int. Conf. on Numerical Methods in Fluid Mechanics, Springer, Lecture Notes in Physics, vol 11 no 19 pp 60-68
7. Chow LC, Tien CL (1978) An examination of four differencing schemes for some elliptic-type convection equations, Numerical Heat Transfer, vol 1 pp 87-100
8. Connell SD, Stow P (1986) A discussion and comparison of numerical techniques used to solve the Navier-Stokes and Euler equations, Int. J. for Num. Math. in Eng., vol 6 pp 155-163

9. Courant R, Isacson E, Rees M (1952) On the solution of non-linear hyperbolic differential equations by finite differentials, Commun. Pure Appl. Math., vol 5 p 243
10. Dearing JF (1985) A four-fluid model of PWR degraded cores, Third Int. Top. Meeting on Reactor Thermal Hydraulics, Newport, Rhode Island, Oct.15-18 1985, LA-UR- 85-947/ CONF-85/007--3
11. van Doormaal JP, Raithby GD (1984) Enhancement of the SIMPLE method for prediction of incompressible fluid flows, Numerical Heat Transfer, vol 7 pp 147-163
12. Günther C (1988) Monotone Upwind-Verfahren 2.Ordnung zur Loesung der Konvektions-Diffusionsgleichungen, ZAMM. Z. angew. Math. Mech. vol 68 no 5, T383-T384
13. Günther C (August 1988) Vergleich verschiedener Differenzenverfahren zur numerischen Loesungen der 2-d Konvektions-Diffusionsgleichungen anhand eines Beispieles mit bekannter exakter Loesungen, Kernforschungszentrum Karlsruhe, KfK 4439
14. Gushchin VA, Shchennikov VV (1974) A monotonic difference scheme of second order accuracy, Zh. Vychisl. Mat. Mat. Fiz., vol 14 no 3 pp 789-792
15. Haaland SE (1984) Calculation of entrainment rate, initial values, and transverse velocity for the Patankar-Spalding method, Numerical Heat Transfer, vol 7 pp 39-57
16. Harlow FH, Amsden A A (1971) A numerical fluid dynamics calculations method for all flow speeds, J. Comp. Physics, vol 8 pp 197-213
17. Issa RI (1983) Numerical Methods for Two- and Tree-Dimensional Recirculating Flows, in Comp. Methods for Turbulent Transonic and Viscous Flow, Ed. Essers, J.A.: Hemisphere, Springer, pp 183
18. Kelly JM, Kohrt JR (1983) COBRA-TF: Flow blockage heat transfer program, Proc. Eleventh Water Reactor Safety Research Information Meeting, Gaithersburg - Maryland, NUREG/CP-0048, 1 (Oct.24-28 1983) pp 209-232
19. Knight TD Ed. (1984) TRAC-PD2 Independent Assessment, NUREG/CR-3866, LA-10166
20. Kolev NI (October 1986) IVA-2 A Computer Code for Modeling of Three Dimensional, Three- Phase, Three-Component Flow by Means of Three Velocity Fields in Cylindrical Geometry with Arbitrary Internal Structure Including Nuclear Reactor Core, Proc. of the Int. Seminar "Thermal physics 86" held in Rostock, German Democratic Republic, (in Russian)
21. Kolev NI (Aug. 1987) A three field-diffusion model of three-phase, threecomponent flow for the transient 3D-computer code IVA2/001, Nuclear Technology, vol 78 pp 95-131
22. Kolev NI (1990) Derivatives for the state equations of multi-component mixtures for universal multi-component flow models, Nuclear Science and Engineering, vol 108 pp 74-87
23. Kolev N I (Sept. 1991) A three-field model of transient 3D multi-phase, three-component flow for the computer code IVA3, Part 2: Models for the interfacial transport phenomena. Code validation. KfK 4949 Kernforschungszentrum Karlsruhe
24. Kolev NI , Tomiyama A, Sakaguchi T (Sept. 1991) Modeling of the mechanical interaction between the velocity fields in three phaseflow, Experimental Thermal and Fluid Science, vol 4 no 5 pp 525-545
25. Kolev NI (1993) Fragmentation and coalescence dynamics in multi-phase flows, Experimental Thermal and Fluid Science, vol 6 pp 211-251
26. Kolev NI (Apr. 23-27 1994) IVA4 computer code: dynamic fragmentation model for liquid and its aplication to melt water interaction, Proc. ICONE-3, Third International Conf. on Nucl. Engineering, "Nuclear Power and Energy Future", Kyoto, Japan. Pre-

sented at the „Workshop zur Kühlmittel/ Schmelze - Wechselwirkung", 14-15 Nov. 1994, Cologne, Germany
27. Kolev NI (1994) The code IVA4: Modeling of mass conservation in multi-phase multi-component flows in heterogeneous porous media, Kerntechnik, vol 59 no 4-5 pp 226-237
28. Kolev NI (1994) The code IVA4: Modeling of momentum conservation in multi-phase flows in heterogeneous porous media, Kerntechnik, vol 59 no 6 pp 249-258
29. Kolev NI (1995) The code IVA4: Second Law of Thermodynamics for Multi-Phase Multi-Component Flows in Heterogeneous Porous Media, Kerntechnik vol 60 no 1 pp 28-39
30. Kolev NI (1995) How accurate can we predict nucleate boiling, Experimental Thermal and Fluid Science, vol 10 pp 370-378
31. Kolev NI (1995) The code IVA4: Nucleation and flashing model, Kerntechnik vol 60 (1995) no 6, pp 157-164. Also in: Proc. Second Int. Conf. On Multiphase Flow, Apr. 3-7, 1995, Kyoto; 1995 ASME & JSME Fluid Engineering Conference International Symposium on Validation of System Transient Analysis Codes, Aug. 13-18, 1995, Hilton Head (SC) USA; Int. Symposium on Two-Phase Flow Modeling and Experimentation, ERGIFE Place Hotel, Rome, Italy, October 9-11, 1995
32. Kolev NI (1995) IVA4 computer code: The model for film boiling on a sphere in subcooled, saturated and superheated water, Proc. Second Int. Conference On Multiphase Flow, Apr. 3-7 1995, Kyoto, Japan. Presented At The „Workshop zur Kühlmittel/Schmelze - Wechselwirkung", 14-15 Nov. 1994, Cologne, Germany
33. Kolev NI (1996) Three fluid modeling with dynamic fragmentation and coalescence fiction or daily practice? 7th FARO Experts Group Meeting Ispra, October 15-16, 1996; Proceedings of OECD/CSNI Workshop on Transient thermal-hydraulic and neutronic codes requirements, Annapolis, Md, U.S.A., 5th-8th November 1996; 4th World Conference on Experimental Heat Transfer, Fluid Mechanics and Thermodynamics, ExHFT 4, Brussels, June 2-6, 1997; ASME Fluids Engineering Conference & Exhibition, The Hyatt Regency Vancouver, Vancouver, British Columbia, CANADA June 22-26, 1997, Invited Paper; Proceedings of 1997 International Seminar on Vapor Explosions and Explosive Eruptions (AMIGO-IMI), May 22-24, Aoba Kinen Kaikan of Tohoku University, Sendai-City, Japan
34. Köller A (1980) Anwendung numerischer Lösungsverfahren der Navier-Stokes-Gleichung zur Vermeidung von Wirbeln, Siemens Forsch.-u. Entwickl.-Ber., vol 9 no 2 pp 99-104
35. Latimaer BR, Pollard A (1985) Comparison of pressure-velocity coupling solution algorithms, Num. Heat Transfer, vol 8 pp 635-652
36. Leonard BP (1978) Third-order finite-difference method for steady two- dimensional convection, Num. Meth. in Laminar and Turbulent Flow, pp 807-819
37. Leonard BP, Mokhtari S (1990) Beyond first - order up winding: The ultra - sharp alternative for non - oscillatory Steady - state simulation of convection. Int. J. for Numerical Methods in Engineering, vol 30 pp 729-766
38. Leonard BP (1990) New flash: Upstream parabolic interpolation, Proc. 2nd GAMM Conference on Num. Meth. in Fluid Mechanics, Köln, Germany
39. Liles D, Reed WR (1978) A semi-implicit method for two-phase fluid dynamics, J. of Comp. Physics, vol 26 pp 390-407

40. Liles D et al (April 1981) TRAC-FD2 An advanced best-estimate computer program for pressurised water reactor loss-of-coolant accident analysis, NUREG/CR-2054, LA-8709 MS
41. Liles D, Mahaffy JM (1984) An approach to fluid mechanics calculations on serial and parallel computer architectures in large scale scientific computation, Parter SV (ed) Academic Press, Inc., Orlando, pp 141-159
42. Mahaffy JH, Liles D (April 1979) Applications of implicit numerical methods to problems in two phase flow, NUREG/CR-0763, LA-7770-MS
43. Mahaffy JH (1979) A stability-enhancing two-phase method for fluid flow calculations, NUREG/CR-0971, LA-7951-MS or J. of Comp. Physics, vol 46 no 3 (June 1983)
44. Neuberger AW (March 1984) Optimierung eines numerischen Verfahrens zur Berechnung elliptischer Strömungen, DFVLR- FB84-16
45. Patankar SV, Spalding DB (1967) A finite-difference procedure for solving the equations of the two dimensional boundary layer, Int. J. Heat Mass Transfer vol 10 pp 1389-1411
46. Patankar SV (1978) A numerical method for conduction in composite materials, Flow in Irregular Geometries and Conjugate Heat Transfer, Proc. 6th Int. Heat Transfer Conf., Toronto, vol 3 p 297
47. Patankar SV (1980) Numerical heat transfer and fluid flow, Hemisphere, New York
48. Patankar SV, Spalding DB (1972) A calculation procedure for heat, mass and momentum transfer in three dimensional parabolic flows, Int. J. Heat Mass Transfer, vol 15 pp 1787-1806
49. Patankar SV, Rafiinejad D, Spalding DB (1975) Calculation of the three-dimensional boundary layer with solution of all three momentum equations, Computer Methods in Applied Mechanics and Engineering, North-Holland Publ. Company, vol 6 pp 283-292
50. Patankar SV (1975) Numerical prediction of three dimensional flows, in Ed. Lauder B E, Studies in convection, theory, measurement and application, Academic Press, London, vol 1
51. Patankar SV, Basn DK, Alpay S (December 1977) A prediction of the three-dimensional velocity field of a deflected turbulent jet, Transactions of the ASME, Journal of Fluids Engineering, pp 758-767
52. Patankar S V (1981) A calculation procedure for two-dimensional elliptic situations, Numerical Heat Transfer, vol 4 pp 409-425
53. Prakash C (1984) Application of the locally analytic differencing scheme to some test problems for the convection- diffusion equation, Numerical Heat Transfer vol 7 pp 165-182
54. Patel MK, Markatos NC, Cross MS (1985) A critical evaluation of seven discretization schemes for convection-diffusion equations, Int. J. Num. Methods in Fluids, pp 225-244
55. Patel MK, Markatos NC (1986) An evaluation of eight discretization schemes for two-dimensional convection- diffusion equations, Int. J. for Numerical Methods in Fluids, vol 6 pp 129-154
56. Patel MK, Markatos NC (1986) An evaluation of eight discretization schemes for two-dimensional convection-Diffusion equations, Int. J. for Numerical Methods in Fluids, vol 6 pp 129-154
57. Prior RJ (June 1979) Computational methods in thermal reactor safety, NUREG/CR-0851, LA-7856-MS

58. Pollard A, Aiu LWA (1982) The calculation of some laminar flows using various discretization schemes, Comp. Math. Appl. Mesh. Eng., vol 35 pp 293-313
59. Rohatgi US (April 1985) Assessment of TRAC codes with dartmouth college countercurrent flow tests, Nuclear Technology vol 69 pp 100-106
60. Roscoe DF (1976) The solution of the three-dimensional Navier-Stokes equations using a new finite difference approach, Int. J. for Num. Math. in Eng., vol 10 pp 1299-1308
61. Saad Y (1996) Iterative methods for sparse linear systems, PWS Publishing Company, Boston,
62. Sargis DA, Chan PC (1984) An implicit fractional step method for two-phase flow. Basic Aspects of Two Phase Flow and Heat Transfer, HTD-34 pp 127-136
63. Scarborough JB (1958) Numerical mathematical analysis, 4th ed., John Hopkins Press, Baltimore
64. Spalding DB (1976) The Calculation of Free-Convection Phenomena in Gas-Liquid Mixtures, ICHMT Seminar Dubrovnik (1976), in Turbulent Buoyant Convection. Eds. Afgan N, Spalding DB, Hemisphere, Washington pp 569- 586, or HTS Repport 76/11 (1977)
65. Spalding DB (March 1979) Multi-phase flow prediction in power system equipment and components, EPRI Workshop on Basic Two-Phase Flow Modelling in Reactor Safety and Performance, Tampa, Florida
66. Spalding DB (January 1980) Mathematical modeling of fluid mechanics, Heat-Transfer and Chemical - Reaction Processes, A lecture course, Imperial College of Science and Technology, Mech. Eng. Dep., London. HTS Report 80/1
67. Spalding DB (1980) Numerical computation of multi-phase fluid flow and heat transfer, in Recent Advances in Numerical Methods in Fluids, Eds. Taylor C, Morgan K, pp 139-167
68. Spalding DB (June 1981) Development in the IPSA procedure for numerical computation of multi-phase, slip, unequal temperature, etc, Imperial College of Science and Technology, Mech. Eng. Dep., London. HTS Report 81/11
69. Spalding DB (1981) A general purpose computer program for multi-dimensional one- and two-phase flow, Mathematics and Computers in Simulation vol XXIII pp 267-276
70. Spalding DB (1981) Numerical computation of multiphase flows, A course of 12 lectures with GENMIX 2P, Listing and 5 Appendices, HTS Report 81/8
71. Tanaka R, Nakamura T, Yabe T (2000) Constructing an exactly conservative scheme in a non conservative form, Comput. Phys. Commun., vol 126 pp 232
72. Tomiyama A (1990) Study on finite difference methods for fluid flow using difference operators and iteration matrices, PhD Thesis, Tokyo Institute of Technology
73. Thurgood MJ, Cuta JM, Koontz AS, Kelly JM (1983) COBRA/TRAC - A thermal hydraulic code for transient analysis of nuclear reactor vessels and primary coolant systems, NUREG/CR-3046 1-5
74. Vanka SP (1985) Block-implicit calculation of steady turbulent recirculating flows, Int. J. Heat Mass Transfer, vol 28 no 11 pp 2093-2103
75. Williams KA, Liles DR (1984) Development and assessment of a numerical fluid dynamics model for non-equilibrium steam-water flows with entrained droplets, AICHE Symposium Series, Heat Transfer - Niagara Falls 1984, ed. by Farukhi NM, 236. 80 pp 416-425
76. Yabe T, Takewaki H (Dezember 1986) CIP, A new numerical solver for general non linear hyperbolic equations in multi-dimension, KfK 4154, Kernforschungszentrum Karlsruhe

77. Yabe T, Xiao F (1993) Description of complex sharp interface during shock wave interaction with liquid drop, Journal of Computational Physics of Japan, vol 62, no 8 pp 2537-2540
78. Yabe T, Xiao F, Mochizuki H (1995) Simulation technique for dynamic evaporation processes, Nuclear Engineering and Design, vol 155 pp 45-53
79. Yabe T, Xiao F, Utsumi T (2001) The constrained interpolation profile method for multiphase analysis, Journal of Computational Physics, vol 169 pp 556-593
80. Ferzinger JH and Paric (2002) Computational methods for fluid dynamics, Springer, 3rd Edition, Berlin
81. Xiao F, Ikebata A and Hasegawa T (2004) Multi-fluid simulation by multi integrated moment method, to be published in Computers & Structures

13 Numerical methods for multi-phase flow in curvilinear coordinate systems

This chapter presents a numerical solution method for multi-phase flow analysis based on local volume and time averaged conservation equations. The emphasis of this development was to create a computer code architecture that absorb all the constitutive physics and functionality from the past 25years development of the three fluid multi-component IVA-entropy concept for multi-phase flows into a boundary fitted orthogonal coordinate framework. Collocated discretization for the momentum equations is used followed by weighted averaging for the staggered grids resulting in analytical expressions for the normal velocities. Using the entropy concept analytical reduction to a pressure-velocity coupling is found. The performance of the method is demonstrated by comparison of two cases for which experimental results and numerical solution with the previous method are available. The agreement demonstrates the success of this development.

13.1 Introduction

We extend now the method described in the previous chapter to more arbitrary geometry. Instead of considering Cartesian or cylindrical geometry only, we will consider an integration space called a *block* in which the computational finite volumes fit inside the block so that the outermost faces of the external layer of the finite volumes create the face of the block. Similarly, bodies immersed into this space have external faces identical with the faces of the environmental computational cells. Such blocks can be inter-connected. With this technology multi-phase flows in arbitrary interfaces can be conveniently handled.

For understanding the material presented in this section I strongly recommend going over Appendixes 1 and 2 before continuing reading.

Before starting with the description of the new method let us summarize briefly the state of the art in this field.

In the last ten years the numerical modeling of single-phase flow in boundary fitted coordinates is becoming standard in the industry. This is not the case with the numerical modeling of multiphase flows. There are some providers of single-phase-computer codes claming that their codes can simulate multi-phase flows.

Taking close looks of the solution methods of these codes reveals that existing single phase solvers are used and a provision is given to the user to add an other velocity field and define explicit the interfacial interaction physics. This strategy does not account for the feed back of the strong interfacial interactions on the mathematical solution methods - see the discussion in *Miettinen* and *Schmidt* [17]. Multi-phase flow simulations require specific solution methods accounting for this specific physics - see for instance the discussion by *Antal* et al. [1].

There are groups of methods that are solving single phase conservation equations with surface tracking, see the state of the art part of *Tryggavson* et al. [29] work. This is in fact a direct numerical simulation that is outside of the scope of this chapter. To mention few of them: In Japan a powerful family of cubic-interpolation methods (CIP) is developed based on the pseudo-characteristic method of lines [25, 26, 18, 40-42, 32-39]. In USA particle tracking and level-set surface tracking methods are very popular; see for instance [23, 20, 24, 29]. The third group of DNS method with surface tracking is the lattice-*Boltzman* family, see [6, 19] and the references given there. To the family of emerging methods the so called diffuse interface methods based on high order thermodynamics can be mentioned, see *Verschueren* [30], *Jamet* et al. [7]. Let as emphasize once again, that unlike those methods, our work is concentrated on methods solving the local volume and time averaged multiphase flow equations which are much different then the single phase equations.

In Europe two developments for solving two-fluid conservation equations in unstructured grids are known to me. *Staedke* et al. [22] developed a solution method based on the method of characteristics using unstructured grid in a single domain. The authors added artificial terms to enforce hyperbolicy in the initially incomplete system of partial differential equations that contain derivatives which do not have any physical meaning. *Toumi* et al. [28] started again from the incomplete system for two-fluid two phase flows without interaction terms and included them later for a specific class of processes; see *Kumbaro* et al. [15]. These authors extended the approximate *Riemann* solver originally developed for single phase flows by *Roe* to two-fluid flows. One application example of the method is demonstrated in a single space domain in [28, 15]. No industrial applications of these two methods have been reported so far. One should note that it is well know that if proper local volume averaging is applied the originating interfacial interaction terms provide naturally hyperbolicy of the system of PDS and there is no need for artificial terms without any physical meaning. An example for the resolution of this problem is given by *van Wijngaarden* [31] in 1976 among many others.

In USA *Lahey* and *Drew* [16] demonstrated clearly in 1999 haw by careful elaboration of the constitutive relationships starting from first principles variety of steady state processes including *frequency dependent acoustics* can be successfully simulated. Actually, the idea by *Lahey* and *Drew* [16] is a further development of the proposal made by *Harlow* and *Amsden* [4] in 1975 where liquid (1) in vapor (2) and vapor (3) in liquid (4) are grouped in two velocity fields 1 + 2 and 3 + 4. The treatment of *Lahey* and *Drew* [16] is based on four velocity fields. *Antal* et al. [1] started developing the NPHASE multi-domain multi-phase flows code based on the single phase *Rhie* and *Chow* numerical method extended to multi-phase flows. Application example is given for T-pipe bubble flows with 10 groups of bubble diameters. Two works that can be considered as a subset of this approach are reported in [27, 3]. Another direction of development in USA that can be observed is the use of the volume of fluid method with computing the surface tension force as a volumetric force [5, 14, 2, 21].

13.2 Nodes, grids, meshes, topology - some basic definitions

Database concept: Let as consider the data base concept. The data volume is made up of points, or *nodes*, which themselves define in their neighborhood a volume element. We use hexahedrons (Fig. 4 in Appendix 1). A hexahedron is a 3D volume element with six sides and eight vertices. The vertices are connected in an order that mimics the way nodes are numbered in the data volume. The nodes in the data volume are numbered by beginning with 1 at the data volume's origin. Node numbering increases, with x changing fastest. This means node numbering increases along the x axis first, the y axis second, and the z axis last, until all nodes are numbered. The numbering of the vertices of the volume elements follows the same rule. It starts with the vertex being closest to the data volume's origin, moves along x, then y, and then z.

Grid: A *grid* is a set of locations in a 3D data volume defined with x, y, and z coordinates. The locations are called *nodes*, which are connected in a specific order to create the topology of the grid. A grid can be regular or irregular depending on how its nodes are represented as points.

A *regular grid*'s nodes are evenly spaced in x, y, and z directions, respectively. A *regular grid*'s nodes are specified with x, y, and z offsets from the data volume's origin. A regular grid may have equidistant or non-equidistant spacing. If equidistant spacing is used all areas of the data volumes have the same resolution. This suits data with regular sample intervals.

An *irregular grid*'s nodes need not be evenly spaced or in a rectangular configuration. This suits data with a specific area of interest that require finer sampling. Because nodes may not be evenly spaced, each node's *xyz* coordinates must be explicitly listed.

Topology: A *topology* defines an array of elements by specifying the connectivity of the element's vertices or nodes. It builds a volume from separate elements by specifying how they are connected together. The elements can be 3D volume elements or 2D surface elements. A topology is either regular or irregular, depending upon what types of elements it defines and how they are structured.

A *regular topology* defines a data volume's node connectivity. We assume a hexahedron volume element type. A regular topology can be used for regular or irregular grids.

An *irregular topology* defines the node connectivity of either a data volume or geometric surface elements. An irregular topology data volume can be composed of either hexahedron or tetrahedron volume elements. Geometry objects are composed of points, lines, or polygons. An element list has to explicitly specify how the nodes connect to form these elements.

Volume elements: The volume elements are the smallest building blocks of a data volume topology.

Mesh: A *mesh* is a grid combined with specific topology for the volume of data. We distinguish the following mesh types: regular meshes, irregular or structured meshes, unstructured meshes, and geometry meshes.

A *regular mesh* consists of grid having regular spacing and regular topology that consists of simple, rectangular array of volume elements.

An *irregular mesh* explicitly specifies the *xyz* coordinates of each node in a node list. As in the regular mesh, the topology is regular, although individual elements are formed by explicit *xyz* node locations. The grid may be irregular or rectilinear.

An unstructured mesh explicitly defines the topology. Each topology element is explicitly defined by its node connectivity in an element list. The grid may be regular or irregular.

In this sense we are dealing in this Chapter with *multi-blocks* each of them consisting of

- irregular grid's nodes, irregular meshes,
- regular topology with hexahedron volume element type.

The integration space is built by a specified number of interconnected blocks.

13.3 Formulation of the mathematical problem

Consider the following mathematical problem: A multi-phase flow is described by the following vector of dependent variables

$$\mathbf{U}^T = (\alpha_m, T_l, s_2, s_3, C_{il}, n_l, p, u_l, v_l, w_l),$$

where

$$l = 1, 2, 3, \; i1 = 1...n1, \; i2 = 1...n2, \; i3 = 1...n3$$

which is a function of the three space coordinates (x, y, z), and of the time τ,

$$\mathbf{U} = \mathbf{U}(x, y, z, \tau).$$

The relationship $\mathbf{U} = \mathbf{U}(x, y, z, \tau)$ is defined by the volume-averaged and successively time-averaged mass, momentum and energy conservation equations derived in Chapters 1, 2, 5 [8-12] as well as by initial conditions, boundary conditions, and geometry. The conservation equations are transformed in a curvilinear coordinate system ξ, η, ζ as shown in Chapter 11, [13]. The flux form of these equations is given in Chapter 11. As shown in Chapter 11 [13], the conservation principles lead to a system of 19+n1+n2+n3 non-linear, non-homogeneous partial differential equations with variable coefficients. This system is defined in the three-dimensional domain $\mathbf{R}$. The initial conditions of $\mathbf{U}(\tau = 0) = \mathbf{U}_a$ in $\mathbf{R}$ and the boundary conditions acting at the interface separating the integration space from its environment are given. The solution required is for conditions after the time interval $\Delta\tau$ has elapsed. The previous time variables are assigned the index a. The time variables not denoted with a are either in the new time plane, or are the best available guesses for the new time plane.

In order to enable modeling of flows with arbitrary obstacles and inclusions in the integration space as is usually expected for technical applications, surface permeabilities are defined

$$(\gamma_\xi, \gamma_\eta, \gamma_\zeta) = \text{functions of } (\xi, \eta, \zeta, \tau),$$

at the virtual surfaces that separate each computational cell from its environment. By definition, the surface permeabilities have values between one and zero,

$$0 \leq \text{each of all } (\gamma_\xi, \gamma_\eta, \gamma_\zeta)\text{'s} \leq 1.$$

A volumetric porosity

$$\gamma_v = \gamma_v(\xi, \eta, \zeta, \tau)$$

is assigned to each computational cell, with

$$0 < \gamma_v \leq 1.$$

The surface permeabilities and the volume porosities are not expected to be smooth functions of the space coordinates in the region **R** and of time. For this reason, one constructs a frame of geometrical flow obstacles which are functions of space and time. This permits a large number of extremely interesting technical applications of this type of approach.

In order to construct useful numerical solutions it is essential that an appropriate set of constitutive relations be available: state equations, thermodynamic derivatives, equations for estimation of the transport properties, correlations modeling the heat, mass and momentum transport across the surfaces dividing the separate velocity fields, etc. These relationships together are called closure equations. This very complex problem will not be discussed in Volume II. Only the numerics will be addressed here.

13.4 Discretization of the mass conservation equations

13.4.1 Integration over a finite time step and finite control volume

We start with the conservation equation (10.56) for the species i inside the velocity field l in the curvilinear coordinate system

$$\frac{\partial}{\partial \tau}\left(\alpha_l \rho_l C_{il} \sqrt{g}\, \gamma_v \right) + \frac{\partial}{\partial \xi}\left(\gamma_\xi \sqrt{g}\, \mathbf{a}^1 \cdot \mathbf{G}_{il}\right) + \frac{\partial}{\partial \eta}\left(\gamma_\eta \sqrt{g}\, \mathbf{a}^2 \cdot \mathbf{G}_{il}\right)$$

$$+ \frac{\partial}{\partial \zeta}\left(\gamma_\zeta \sqrt{g}\, \mathbf{a}^3 \cdot \mathbf{G}_{il}\right) = \gamma_v \sqrt{g}\, \mu_{il}, \qquad (13.1)$$

where the species mass flow rate vector is defined as follows

$$\mathbf{G}_{il} = \alpha_l \rho_l \left[C_{il}\left(\mathbf{V}_l - \mathbf{V}_{cs}\right) - D_{il}^* \left(\mathbf{a}^1 \frac{\partial C_{il}}{\partial \xi} + \mathbf{a}^2 \frac{\partial C_{il}}{\partial \eta} + \mathbf{a}^3 \frac{\partial C_{il}}{\partial \zeta} \right) \right]. \qquad (13.2)$$

Note that for $C_{il} = 1$ we have the mass flow vector of the velocity field

$$\mathbf{G}_l = \alpha_l \rho_l \left(\mathbf{V}_l - \mathbf{V}_{cs}\right). \qquad (13.3)$$

Next we will use the following basic relationships from Appendix 2 between the surface vectors and the contravariant vectors, and between the *Jacobian* determinant and the infinitesimal spatial and volume increments

13.4 Discretization of the mass conservation equations

$$\sqrt{g}\mathbf{a}^1 = \frac{\mathbf{S}^1}{\partial \eta \partial \zeta}, \quad \sqrt{g}\mathbf{a}^2 = \frac{\mathbf{S}^2}{\partial \xi \partial \zeta}, \quad \sqrt{g}\mathbf{a}^3 = \frac{\mathbf{S}^3}{\partial \xi \partial \eta}, \quad \sqrt{g} = \frac{dV}{d\xi d\eta d\zeta}. \quad (13.4\text{-}7)$$

We will integrate both sides of the equation over the time, and spatial intervals $\partial \tau$, $\partial \xi$, $\partial \eta$, and $\partial \zeta$ respectively. We start with the first term

$$\iiint_{\Delta V}\left[\int_0^{\Delta \tau} \frac{\partial}{\partial \tau}\left(\alpha_l \rho_l C_{il} \sqrt{g}\, \gamma_v\right) \partial \tau\right] \partial \xi \partial \eta \partial \zeta = \left[\int_0^{\Delta \tau} \frac{\partial}{\partial \tau}\left(\alpha_l \rho_l C_{il} \gamma_v\right) \partial \tau\right] \iiint_{\Delta V} dV$$

$$= \left[\left(\alpha_l \rho_l C_{il} \gamma_v\right) - \left(\alpha_l \rho_l C_{il} \gamma_v\right)_a\right] \Delta V. \quad (13.8)$$

This result is obtained under the assumption that there is no spatial variation of the properties $\left(\alpha_l \rho_l C_{il} \gamma_v\right)$ inside the cell. The integration of the other terms gives the following results:

$$\int_0^{\Delta \tau} \iiint_{\Delta V} \frac{\partial}{\partial \xi}\left(\gamma_\xi \sqrt{g}\, \mathbf{a}^1 \cdot \mathbf{G}_{il}\right) \partial \xi \partial \eta \partial \zeta \partial \tau = \Delta \tau \int_0^{\Delta \xi} \frac{\partial}{\partial \xi}\left(\gamma_\xi \mathbf{S}^1 \cdot \mathbf{G}_{il}\right) \partial \xi$$

$$= \Delta \tau \left[\left(\gamma_\xi \mathbf{S}^1 \cdot \mathbf{G}_{il}\right) - \left(\gamma_\xi \mathbf{S}^1 \cdot \mathbf{G}_{il}\right)_{i-1}\right]$$

$$= \Delta \tau \Delta V \left[\gamma_\xi \frac{S_1}{\Delta V}\left(\mathbf{e}^1 \cdot \mathbf{G}_{il}\right) - \gamma_{\xi,i-1} \frac{S_2}{\Delta V}\left(\mathbf{e}^1 \cdot \mathbf{G}_{il}\right)_{i-1}\right], \quad (13.9)$$

$$\int_0^{\Delta \tau} \iiint_{\Delta V} \frac{\partial}{\partial \eta}\left(\gamma_\eta \sqrt{g}\, \mathbf{a}^2 \cdot \mathbf{G}_{il}\right) \partial \xi \partial \eta \partial \zeta \partial \tau = \Delta \tau \int_0^{\Delta \eta} \frac{\partial}{\partial \eta}\left(\gamma_\eta \mathbf{S}^2 \cdot \mathbf{G}_{il}\right) \partial \eta$$

$$= \Delta \tau \left[\left(\gamma_\eta \mathbf{S}^2 \cdot \mathbf{G}_{il}\right) - \left(\gamma_\eta \mathbf{S}^2 \cdot \mathbf{G}_{il}\right)_{j-1}\right]$$

$$= \Delta \tau \Delta V \left[\gamma_\eta \frac{S_3}{\Delta V}\left(\mathbf{e}^2 \cdot \mathbf{G}_{il}\right) - \gamma_{\eta,j-1} \frac{S_4}{\Delta V}\left(\mathbf{e}^2 \cdot \mathbf{G}_{il}\right)_{j-1}\right], \quad (13.10)$$

$$\int_0^{\Delta \tau} \iiint_{\Delta V} \frac{\partial}{\partial \zeta}\left(\gamma_\zeta \sqrt{g}\, \mathbf{a}^3 \cdot \mathbf{G}_{il}\right) \partial \xi \partial \eta \partial \zeta \partial \tau = \Delta \tau \int_0^{\Delta \zeta} \frac{\partial}{\partial \zeta}\left(\gamma_\zeta \mathbf{S}^3 \cdot \mathbf{G}_{il}\right) \partial \zeta$$

$$= \Delta \tau \left[\left(\gamma_\zeta \mathbf{S}^3 \cdot \mathbf{G}_{il}\right) - \left(\gamma_\zeta \mathbf{S}^3 \cdot \mathbf{G}_{il}\right)_{k-1}\right]$$

$$= \Delta \tau \Delta V \left[\gamma_\zeta \frac{S_5}{\Delta V} \left(\mathbf{e}^3 \cdot \mathbf{G}_{il} \right) - \gamma_{\zeta,k-1} \frac{S_6}{\Delta V} \left(\mathbf{e}^3 \cdot \mathbf{G}_{il} \right)_{k-1} \right], \quad (13.11)$$

$$\int_0^{\Delta \tau} \iiint_{\Delta V} \gamma_v \sqrt{g}\, \mu_{il}\, \partial \xi \partial \eta \partial \zeta \partial \tau = \Delta \tau \iiint_{\Delta V} \gamma_v \mu_{il} dV = \Delta \tau \gamma_v \mu_{il} \Delta V. \quad (13.12)$$

It is convenient to introduce the numbering at the surfaces of the control volumes 1 to 6 corresponding to high-i, low-i, high-j, low-j, high-k and low-k, respectively. We first define the unit surface vector $(\mathbf{e})^m$ at each surface m as outwards directed:

$$(\mathbf{e})^1 = \mathbf{e}^1,\ (\mathbf{e})^2 = -\mathbf{e}^1_{i-1},\ (\mathbf{e})^3 = \mathbf{e}^2,\ (\mathbf{e})^4 = -\mathbf{e}^2_{j-1},\ (\mathbf{e})^5 = \mathbf{e}^3,\ (\mathbf{e})^6 = -\mathbf{e}^3_{k-1}. \quad (13.13\text{-}18)$$

With this we have a short notation of the corresponding discretized concentration conservation equation

$$\alpha_l \rho_l C_{il} \gamma_v - \alpha_{la} \rho_{la} C_{ila} \gamma_{va} + \Delta \tau \sum_{m=1}^{6} \beta_m (\mathbf{e})^m \cdot \mathbf{G}_{il,m} = \Delta \tau \gamma_v \mu_{il}. \quad (13.19)$$

We immediately recognize that it is effective to compute once the geometry coefficients

$$\beta_1 = \gamma_\xi \frac{S_1}{\Delta V},\ \beta_2 = \gamma_{\xi,i-1} \frac{S_2}{\Delta V},\ \beta_3 = \gamma_\eta \frac{S_3}{\Delta V},\ \beta_4 = \gamma_{\eta,j-1} \frac{S_4}{\Delta V},$$

$$\beta_5 = \gamma_\zeta \frac{S_5}{\Delta V},\ \beta_6 = \gamma_{\zeta,k-1} \frac{S_6}{\Delta V}, \quad (13.20\text{-}25)$$

before the process simulation, to store them, and to update only those that change during the computation. Secondly, we see that these coefficients contain exact physical geometry information. Note that for cylindrical coordinate systems we have

$$\beta_1 = \frac{r_h^\kappa \gamma_r}{r^\kappa \Delta r},\ \beta_2 = \frac{(r_h^\kappa \gamma_r)_{i-1}}{r^\kappa \Delta r},\ \beta_3 = \frac{\gamma_\theta}{r^\kappa \Delta \theta},\ \beta_4 = \frac{\gamma_{\theta,j-1}}{r^\kappa \Delta \theta},$$

$$\beta_5 = \frac{\gamma_z}{\Delta z},\ \beta_6 = \frac{\gamma_{z,k-1}}{\Delta z}, \quad (13.26\text{-}31)$$

and for Cartesian setting $\kappa = 0$ and $r = x$, $\theta = y$,

$$\beta_1 = \frac{\gamma_x}{\Delta x}, \ \beta_2 = \frac{\gamma_{x,i-1}}{\Delta x}, \ \beta_3 = \frac{\gamma_y}{\Delta y}, \ \beta_4 = \frac{\gamma_{y,j-1}}{\Delta y}, \ \beta_5 = \frac{\gamma_z}{\Delta z}, \ \beta_6 = \frac{\gamma_{z,k-1}}{\Delta z}, \quad (13.32\text{-}39)$$

compare with Section 11.4. Setting $C_{il} = 1$ in Eq. (13.19) we obtain the discretized mass conservation equation of each velocity field

$$\alpha_l \rho_l \gamma_v - \alpha_{la} \rho_{la} \gamma_{va} + \Delta \tau \sum_{m=1}^{6} \beta_m (\mathbf{e})^m \cdot \mathbf{G}_{l,m} = \Delta \tau \gamma_v \mu_l \ . \qquad (13.40)$$

Next we derive the useful non-conservative form of the concentration equations. We multiply Eq. (13.40) by the concentration at the new time plane and subtract the resulting equation from Eq. (13.19). Then the field mass source term is split in two non-negative parts $\mu_l = \mu_l^+ + \mu_l^-$. The result is

$$\alpha_{la} \rho_{la} (C_{il} - C_{ila}) \gamma_{va} + \Delta \tau \sum_{m=1}^{6} \beta_m (\mathbf{e})^m \cdot (\mathbf{G}_{il,m} - C_{il} \mathbf{G}_{l,m}) + \Delta \tau \gamma_v \mu_l^+ C_{il} = \Delta \tau \gamma_v DC_{il} \ .$$
$$(13.41)$$

where $DC_{il} = \mu_{il} + \mu_l^- C_{il}$. Note that

$$\mathbf{G}_{il,m} - C_{il} \mathbf{G}_{l,m}$$

$$= \left[\alpha_l \rho_l (\mathbf{V}_l - \mathbf{V}_{cs}) \right]_m (C_{il,m} - C_{il}) - \left[\alpha_l \rho_l D_{il}^* \left(\mathbf{a}^1 \frac{\partial C_{il}}{\partial \xi} + \mathbf{a}^2 \frac{\partial C_{il}}{\partial \eta} + \mathbf{a}^3 \frac{\partial C_{il}}{\partial \zeta} \right) \right]_m \ .$$
$$(13.42)$$

Up to this point of the derivation we did not made any assumption about the computation of the properties at the surfaces of the control volume.

13.4.2 The donor-cell concept

The concept of the so called *donor-cell* for the convective terms is now introduced. Flow of given scalars takes the values of the scalars at the cell where the flow is coming from. Mathematically it is expressed as follows. First we define velocity normal to the cell surfaces and outwards directed

$$V_{l,m}^n = (\mathbf{e})^m \cdot (\mathbf{V}_l - \mathbf{V}_{cs})_m, \qquad (13.43)$$

then the switch functions (to store them use *signet integers* in computer codes, it saves memory)

$$\xi_{lm+} = \frac{1}{2}\left[1 + sign\left(V_{lm}^{n}\right)\right], \tag{13.44}$$

$$\xi_{lm-} = 1 - \xi_{lm+}, \tag{13.45}$$

and then the *b* coefficients as follows

$$b_{lm+} = \beta_{m}\xi_{lm+}V_{lm}^{n} \geq 0, \tag{13.46}$$

$$b_{lm-} = -\beta_{m}\xi_{lm-}V_{lm}^{n} \geq 0. \tag{13.47}$$

If the normal outwards directed velocity is positive the +b coefficients are unity and the –b coefficients are zero and vice versa. In this case the normal mass flow rate at the surfaces is

$$(\mathbf{e})^{m} \cdot \mathbf{G}_{l,m} = \left[\alpha_{l}\rho_{l}\mathbf{e}\cdot(\mathbf{V}_{l} - \mathbf{V}_{cs})\right]_{m} = \left(\xi_{lm+}\alpha_{l}\rho_{l} + \xi_{lm-}\alpha_{l,m}\rho_{l,m}\right)V_{lm}^{n}. \tag{13.48}$$

Using the above result the mass conservation equations for each field result is

$$\alpha_{l}\rho_{l}\gamma_{v} - \alpha_{la}\rho_{la}\gamma_{va} + \Delta\tau\sum_{m=1}^{6}\left(b_{lm+}\alpha_{l}\rho_{l} - B_{lm-}\right) - \Delta\tau\gamma_{v}\mu_{l} = 0. \tag{13.49}$$

In the donor-cell concept the term

$$B_{lm-} = b_{lm-}\alpha_{l,m}\rho_{l,m} \tag{13.50}$$

plays an important role. B_{lm-} is in fact the mass flow entering the cell from the face *m* divided by the volume of the cell. Once computed for the mass conservation equation it is stored and used subsequently in all other conservation equations.

At this point the method used for computation of the field volumetric fractions by iteration using the point *Gauss-Seidel* method for known velocity vectors and thermal properties will be described.

Consider the field variables $\alpha_{l}\rho_{l}$ in the convective terms associated with the output flow in the new time plane, and $\alpha_{lm}\rho_{lm}$ in the neighboring cells *m* as the best available guesses for the new time plane. Solving Eq. (13.49) with respect to $\alpha_{l}\rho_{l}$ gives

13.4 Discretization of the mass conservation equations

$$\bar{\alpha}_l \bar{\rho}_l = \left[\gamma_{va} \left(\mu_l + \frac{\alpha_{la} \rho_{la}}{\Delta \tau} \right) + \sum_{m=1}^{6} B_{lm-} \right] \bigg/ \left(\frac{\gamma_v}{\Delta \tau} + \sum_{m=1}^{6} b_{lm+} \right). \tag{13.51}$$

Here

$$\frac{\gamma_v}{\Delta \tau} + \sum_{m=1}^{6} b_{lm+} > 0 \tag{13.52}$$

is ensured because γ_v is not allowed to be zero. For a field that is just originating we have

$$\bar{\alpha}_l = \frac{\Delta \tau}{\bar{\rho}_l} \cdot \frac{\gamma_{va} \mu_l + \sum_{m=1}^{6} B_{lm-}}{\gamma_v + \Delta \tau \sum_{m=1}^{6} b_{lm+}}. \tag{13.53}$$

Obviously the field can originate due to convection, $\sum_{m=1}^{6} B_{lm-} > 0$, or due to an in-cell mass source, $\mu_l > 0$, or due to the simultaneous appearance of both phenomena. In case of origination caused by in-cell mass source terms it is important to define the initial density, $\bar{\rho}_l$, in order to compute $\bar{\alpha}_l = \Delta \tau \mu_l / \bar{\rho}_l$.

The best mass conservation in such procedures is ensured if the following sequence is used for computation of the volume fractions:

$$\alpha_2 = \bar{\alpha}_2 \bar{\rho}_2 / \rho_2, \quad \alpha_3 = \bar{\alpha}_3 \bar{\rho}_3 / \rho_3, \quad \alpha_1 = 1 - \alpha_2 - \alpha_3. \tag{13.54-56}$$

For designing the pressure-velocity coupling the form of the discretized mass conservation is required that explicitly contains the normal velocities,

$$(\alpha_l \rho_l \gamma_v - \alpha_{la} \rho_{la} \gamma_{va}) / \Delta \tau + \sum_{m=1}^{6} \beta_m \left(\xi_{lm+} \alpha_l \rho_l + \xi_{lm-} \alpha_{lm} \rho_{lm} \right) V_{lm}^n - \gamma_v \mu_l = 0. \tag{13.57}$$

The mass flow rate of the species i inside the field l at the cell surface m is then

$$(\mathbf{e})^m \cdot \mathbf{G}_{il,m} = \left[\alpha_l \rho_l C_{il} \mathbf{e} \cdot (\mathbf{V}_l - \mathbf{V}_{cs}) \right]_m - \left[\alpha_l \rho_l D_{il}^* \mathbf{e} \cdot \left(\mathbf{a}^1 \frac{\partial C_{il}}{\partial \xi} + \mathbf{a}^2 \frac{\partial C_{il}}{\partial \eta} + \mathbf{a}^3 \frac{\partial C_{il}}{\partial \zeta} \right) \right]_m$$

$$= \left(\xi_{lm+} \alpha_l \rho_l C_{il} + \xi_{lm-} \alpha_{l,m} \rho_{l,m} C_{il,m} \right) V_{lm}^n - \left[\alpha_l \rho_l D_{il}^* \mathbf{e} \cdot \left(\mathbf{a}^1 \frac{\partial C_{il}}{\partial \xi} + \mathbf{a}^2 \frac{\partial C_{il}}{\partial \eta} + \mathbf{a}^3 \frac{\partial C_{il}}{\partial \zeta} \right) \right]_m, \tag{13.58}$$

and consequently

$$(\mathbf{e})^m \cdot (\mathbf{G}_{il,m} - C_{il}\mathbf{G}_{l,m})$$

$$= \xi_{lm-} \alpha_{l,m} \rho_{l,m} V_{lm}^n (C_{il,m} - C_{il}) - \left[\alpha_l \rho_l D_{il}^* \mathbf{e} \cdot \left(\mathbf{a}^1 \frac{\partial C_{il}}{\partial \xi} + \mathbf{a}^2 \frac{\partial C_{il}}{\partial \eta} + \mathbf{a}^3 \frac{\partial C_{il}}{\partial \zeta} \right) \right]_m.$$

(13.59)

Thus Eq. (13.41) takes the intermediate form

$$\alpha_{la}\rho_{la}(C_{il} - C_{ila})\gamma_{va} - \Delta\tau \sum_{m=1}^{6} \left\{ \begin{array}{c} B_{lm-}(C_{il,m} - C_{il}) \\ \\ + B_m \left[\alpha_l \rho_l D_{il}^* \mathbf{e} \cdot \left(\mathbf{a}^1 \frac{\partial C_{il}}{\partial \xi} + \mathbf{a}^2 \frac{\partial C_{il}}{\partial \eta} + \mathbf{a}^3 \frac{\partial C_{il}}{\partial \zeta} \right) \right]_m \end{array} \right\}$$

$$= \Delta\tau\gamma_v (\mu_{il} - C_{il}\mu_l).$$

(13.60)

13.4.3 Two methods for computing the finite difference approximations of the contravariant vectors at the cell center

The contravariant vectors for each particular surface can be expressed by

$$\mathbf{a}^1 = \frac{\mathbf{S}^1}{dV} \partial\xi, \quad \mathbf{a}^2 = \frac{\mathbf{S}^2}{dV} \partial\eta, \quad \mathbf{a}^3 = \frac{\mathbf{S}^3}{dV} \partial\zeta.$$

(13.61-63)

Note that the contravariant vectors normal to each control volume surface are conveniently computed for **equidistant discretization in the computational space** as follows

$$(\mathbf{a}^1)_1 = \frac{\mathbf{S}^1}{\Delta V_1} = \frac{(\mathbf{S})_1}{\Delta V_1} = \frac{S_1}{\Delta V_1}(\mathbf{e})^1, \quad (\mathbf{a}^1)_2 = \frac{\mathbf{S}^1}{\Delta V_2} = -\frac{(\mathbf{S})_2}{\Delta V_2} = -\frac{S_2}{\Delta V_2}(\mathbf{e})^2, \quad (13.64\text{-}65)$$

$$(\mathbf{a}^2)_3 = \frac{\mathbf{S}^2}{\Delta V_3} = \frac{(\mathbf{S})_3}{\Delta V_3} = \frac{S_3}{\Delta V_3}(\mathbf{e})^3, \quad (\mathbf{a}^2)_4 = \frac{\mathbf{S}^2}{\Delta V_4} = -\frac{(\mathbf{S})_4}{\Delta V_4} = -\frac{S_4}{\Delta V_4}(\mathbf{e})^4, \quad (13.66\text{-}67)$$

$$(\mathbf{a}^3)_5 = \frac{\mathbf{S}^3}{\Delta V_5} = \frac{(\mathbf{S})_5}{\Delta V_5} = \frac{S_5}{\Delta V_5}(\mathbf{e})^5, \quad (\mathbf{a}^3)_6 = \frac{\mathbf{S}^3}{\Delta V_6} = -\frac{(\mathbf{S})_6}{\Delta V_6} = -\frac{S_6}{\Delta V_6}(\mathbf{e})^6, \quad (13.68\text{-}69)$$

where the volume associated with these vectors is

$$\overline{\Delta V}_m = \frac{1}{2}\left(\Delta V + \Delta V_m\right). \tag{13.70}$$

The finite volume method: There are two practicable methods for approximation of the contravariant vectors at the cell center. The first one makes use of the already computed normal interface vectors in the following way:

$$\mathbf{a}_c^1 = \frac{1}{2}\left[\left(\mathbf{a}^1\right)_1 + \left(\mathbf{a}^1\right)_2\right], \quad \mathbf{a}_c^2 = \frac{1}{2}\left[\left(\mathbf{a}^2\right)_3 + \left(\mathbf{a}^2\right)_4\right], \quad \mathbf{a}_c^3 = \frac{1}{2}\left[\left(\mathbf{a}^3\right)_5 + \left(\mathbf{a}^3\right)_6\right] \tag{13.71-73}$$

The finite difference method: The second method uses the coordinates of the vertices of the control volume directly. First we define the position at the cell surfaces that will be used to compute the transformation metrics as follows:

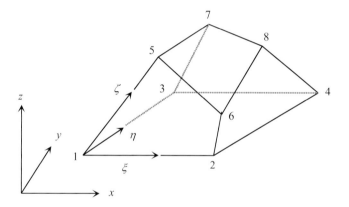

Fig. 12.1. Numbering of the vertices

$$\mathbf{r}_{S1} = \frac{1}{4}\left(\mathbf{r}_2 + \mathbf{r}_4 + \mathbf{r}_8 + \mathbf{r}_6\right), \qquad \mathbf{r}_{S2} = \frac{1}{4}\left(\mathbf{r}_1 + \mathbf{r}_3 + \mathbf{r}_7 + \mathbf{r}_5\right), \tag{13.74-75}$$

$$\mathbf{r}_{S3} = \frac{1}{4}\left(\mathbf{r}_3 + \mathbf{r}_4 + \mathbf{r}_8 + \mathbf{r}_7\right), \qquad \mathbf{r}_{S4} = \frac{1}{4}\left(\mathbf{r}_1 + \mathbf{r}_2 + \mathbf{r}_6 + \mathbf{r}_5\right), \tag{13.76-77}$$

$$\mathbf{r}_{S5} = \frac{1}{4}\left(\mathbf{r}_5 + \mathbf{r}_6 + \mathbf{r}_8 + \mathbf{r}_7\right), \qquad \mathbf{r}_{S6} = \frac{1}{4}\left(\mathbf{r}_1 + \mathbf{r}_2 + \mathbf{r}_4 + \mathbf{r}_3\right). \tag{13.78-79}$$

Then we compute the *inverse metrics* of the coordinate transformation for **equidistant discretization in the transformed space**

$$\begin{pmatrix} \dfrac{\partial x}{\partial \xi} & \dfrac{\partial x}{\partial \eta} & \dfrac{\partial x}{\partial \zeta} \\ \dfrac{\partial y}{\partial \xi} & \dfrac{\partial y}{\partial \eta} & \dfrac{\partial y}{\partial \zeta} \\ \dfrac{\partial z}{\partial \xi} & \dfrac{\partial z}{\partial \eta} & \dfrac{\partial z}{\partial \zeta} \end{pmatrix} = \begin{pmatrix} x_{s1} - x_{s2} & x_{s3} - x_{s4} & x_{s5} - x_{s6} \\ y_{s1} - y_{s2} & y_{s3} - y_{s4} & y_{s5} - y_{s6} \\ z_{s1} - z_{s2} & z_{s3} - z_{s4} & z_{s5} - z_{s6} \end{pmatrix}. \tag{13.80}$$

Then we compute the *Jacobian determinant* and the *metrics* of the coordinate transformation for equidistant discretization in the transformed space.

As already mentioned all this information belongs to the center of the cell. However, the off-diagonal geometry information is required at the cell surfaces. For both cases we use the two corresponding neighbor vectors to compute the contravariant vectors at the cell surfaces as follows

$$\left(\mathbf{a}^2\right)_1 = \frac{1}{2}\left(\mathbf{a}_c^2 + \mathbf{a}_{c,i+1}^2\right), \left(\mathbf{a}^3\right)_1 = \frac{1}{2}\left(\mathbf{a}_c^3 + \mathbf{a}_{c,i+1}^3\right),$$

$$\left(\mathbf{a}^2\right)_2 = \frac{1}{2}\left(\mathbf{a}_c^2 + \mathbf{a}_{c,i-1}^2\right), \left(\mathbf{a}^3\right)_2 = \frac{1}{2}\left(\mathbf{a}_c^3 + \mathbf{a}_{c,i-1}^3\right),$$

$$\left(\mathbf{a}^1\right)_3 = \frac{1}{2}\left(\mathbf{a}_c^1 + \mathbf{a}_{c,j+1}^1\right), \left(\mathbf{a}^3\right)_3 = \frac{1}{2}\left(\mathbf{a}_c^3 + \mathbf{a}_{c,j+1}^3\right),$$

$$\left(\mathbf{a}^1\right)_4 = \frac{1}{2}\left(\mathbf{a}_c^1 + \mathbf{a}_{c,j-1}^1\right), \left(\mathbf{a}^3\right)_4 = \frac{1}{2}\left(\mathbf{a}_c^3 + \mathbf{a}_{c,j-1}^3\right),$$

$$\left(\mathbf{a}^1\right)_5 = \frac{1}{2}\left(\mathbf{a}_c^1 + \mathbf{a}_{c,k+1}^1\right), \left(\mathbf{a}^2\right)_5 = \frac{1}{2}\left(\mathbf{a}_c^2 + \mathbf{a}_{c,k+1}^2\right),$$

$$\left(\mathbf{a}^1\right)_6 = \frac{1}{2}\left(\mathbf{a}_c^1 + \mathbf{a}_{c,k-1}^1\right), \left(\mathbf{a}^2\right)_6 = \frac{1}{2}\left(\mathbf{a}_c^2 + \mathbf{a}_{c,k-1}^2\right). \tag{13.81-94}$$

13.4.4 Discretization of the diffusion terms

13.4.4.1 General

Our next task is to find appropriate finite difference approximation for the six diffusion terms

13.4 Discretization of the mass conservation equations

$$\left[\alpha_l \rho_l D_{il}^* \mathbf{e} \cdot \left(\mathbf{a}^1 \frac{\partial C_{il}}{\partial \xi} + \mathbf{a}^2 \frac{\partial C_{il}}{\partial \eta} + \mathbf{a}^3 \frac{\partial C_{il}}{\partial \zeta}\right)\right]_m,$$

The geometric properties computed by using the control volume approach in the previous section are used to transform the diagonal diffusion terms as direct finite differences

$$\sum_{m=1}^{6} \beta_m \left[\alpha_l \rho_l D_{il}^* \mathbf{e} \cdot \left(\mathbf{a}^1 \frac{\partial C_{il}}{\partial \xi} + \mathbf{a}^2 \frac{\partial C_{il}}{\partial \eta} + \mathbf{a}^3 \frac{\partial C_{il}}{\partial \zeta}\right)\right]_m$$

$$= \beta_1 \left(\frac{\alpha_l \rho_l D_{il}^*}{\Delta V}\right)_1 S_1 \left[C_{il,i+1} - C_{il} + \frac{\overline{\Delta V_1}}{S_1}\left(\mathbf{e} \cdot \mathbf{a}^2 \frac{\partial C_{il}}{\partial \eta} + \mathbf{e} \cdot \mathbf{a}^3 \frac{\partial C_{il}}{\partial \zeta}\right)_1\right]$$

$$+ \beta_2 \left(\frac{\alpha_l \rho_l D_{il}^*}{\Delta V}\right)_2 S_2 \left[C_{il,i-1} - C_{il} + \frac{\overline{\Delta V_2}}{S_2}\left(\mathbf{e} \cdot \mathbf{a}^2 \frac{\partial C_{il}}{\partial \eta} + \mathbf{e} \cdot \mathbf{a}^3 \frac{\partial C_{il}}{\partial \zeta}\right)_2\right]$$

$$+ \beta_3 \left(\frac{\alpha_l \rho_l D_{il}^*}{\Delta V}\right)_3 S_3 \left[C_{il,j+1} - C_{il} + \frac{\overline{\Delta V_3}}{S_3}\left(\mathbf{e} \cdot \mathbf{a}^1 \frac{\partial C_{il}}{\partial \xi} + \mathbf{e} \cdot \mathbf{a}^3 \frac{\partial C_{il}}{\partial \zeta}\right)_3\right]$$

$$+ \beta_4 \left(\frac{\alpha_l \rho_l D_{il}^*}{\Delta V}\right)_4 S_4 \left[C_{il,j-1} - C_{il} + \frac{\overline{\Delta V_4}}{S_4}\left(\mathbf{e} \cdot \mathbf{a}^1 \frac{\partial C_{il}}{\partial \xi} + \mathbf{e} \cdot \mathbf{a}^3 \frac{\partial C_{il}}{\partial \zeta}\right)_4\right]$$

$$+ \beta_5 \left(\frac{\alpha_l \rho_l D_{il}^*}{\Delta V}\right)_5 S_5 \left[C_{il,k+1} - C_{il} + \frac{\overline{\Delta V_5}}{S_5}\left(\mathbf{e} \cdot \mathbf{a}^1 \frac{\partial C_{il}}{\partial \xi} + \mathbf{e} \cdot \mathbf{a}^2 \frac{\partial C_{il}}{\partial \eta}\right)_5\right]$$

$$+ \beta_6 \left(\frac{\alpha_l \rho_l D_{il}^*}{\Delta V}\right)_6 S_6 \left[C_{il,k-1} - C_{il} + \frac{\overline{\Delta V_6}}{S_6}\left(\mathbf{e} \cdot \mathbf{a}^1 \frac{\partial C_{il}}{\partial \xi} + \mathbf{e} \cdot \mathbf{a}^2 \frac{\partial C_{il}}{\partial \eta}\right)_6\right]. \quad (13.95)$$

A natural averaging of the diffusion coefficients is then the harmonic averaging as given in Appendix 12.1

$$\frac{D_{il,m}^C}{\Delta L_{h,m}} = \left(\frac{\alpha_l \rho_l D_{il}^*}{\Delta V}\right)_m S_m = S_m \frac{2\left(\alpha_l \rho_l D_{il}^*\right)\left(\alpha_l \rho_l D_{il}^*\right)_m}{\Delta V_m \left(\alpha_l \rho_l D_{il}^*\right) + \Delta V \left(\alpha_l \rho_l D_{il}^*\right)_m}, \quad (13.96)$$

where in the right hand side $m = 1, 2, 3, 4, 5, 6$ is equivalent to $i + 1, i - 1, j + 1, j - 1, k + 1, k - 1$, respectively regarding the properties inside a control volumes. It guaranties that if the field in one of the neighboring cells is missing the diffusion coefficient is zero.

13.4.4.2 Orthogonal coordinate systems

In the case of orthogonal coordinate systems we see that:

- the off-diagonal diffusion terms are equal to zero,
- the finite volume approximations of the diagonal terms are obtained without the need to know anything about the contravariant vectors.

This illustrates the advantage of using orthogonal coordinate systems. This is valid for any diffusion terms in the conservation equations, e.g. the thermal heat diffusion terms in the energy conservation equations, the viscous diffusion terms in the momentum equations etc.

13.4.4.3 Off-diagonal diffusion terms in the general case

The geometric coefficients of the off-diagonal diffusion terms can then be computed as follows

$$d_{12} = (\mathbf{e})^1 \cdot (\mathbf{a}^2)_1 = (e)^{11}(a^{21})_1 + (e)^{12}(a^{22})_1 + (e)^{13}(a^{23})_1, \quad (13.97)$$

$$d_{13} = (\mathbf{e})^1 \cdot (\mathbf{a}^3)_1 = (e)^{11}(a^{31})_1 + (e)^{12}(a^{32})_1 + (e)^{13}(a^{33})_1, \quad (13.98)$$

$$d_{22} = (\mathbf{e})^2 \cdot (\mathbf{a}^2)_2 = (e)^{21}(a^{21})_2 + (e)^{22}(a^{22})_2 + (e)^{23}(a^{23})_2 = -(d_{12})_{i-1}, \quad (13.99)$$

$$d_{23} = (\mathbf{e})^2 \cdot (\mathbf{a}^3)_2 = (e)^{21}(a^{31})_2 + (e)^{22}(a^{32})_2 + (e)^{23}(a^{33})_2 = -(d_{13})_{i-1}, \quad (13.100)$$

$$d_{31} = (\mathbf{e})^3 \cdot (\mathbf{a}^1)_3 = (e)^{31}(a^{11})_3 + (e)^{32}(a^{12})_3 + (e)^{33}(a^{13})_3, \quad (13.101)$$

$$d_{33} = (\mathbf{e})^3 \cdot (\mathbf{a}^3)_3 = (e)^{31}(a^{31})_3 + (e)^{32}(a^{32})_3 + (e)^{33}(a^{33})_3, \quad (13.102)$$

$$d_{41} = (\mathbf{e})^4 \cdot (\mathbf{a}^1)_4 = (e)^{41}(a^{11})_4 + (e)^{42}(a^{12})_4 + (e)^{43}(a^{13})_4 = -(d_{31})_{j-1}, \quad (13.103)$$

$$d_{43} = (\mathbf{e})^4 \cdot (\mathbf{a}^3)_4 = (e)^{41} (a^{31})_4 + (e)^{42} (a^{32})_4 + (e)^{43} (a^{33})_4 = -(d_{33})_{j-1}, \quad (13.104)$$

$$d_{51} = (\mathbf{e})^5 \cdot (\mathbf{a}^1)_5 = (e)^{51} (a^{11})_5 + (e)^{52} (a^{12})_5 + (e)^{53} (a^{13})_5, \quad (13.105)$$

$$d_{52} = (\mathbf{e})^5 \cdot (\mathbf{a}^2)_5 = (e)^{51} (a^{21})_5 + (e)^{52} (a^{22})_5 + (e)^{53} (a^{23})_5, \quad (13.106)$$

$$d_{61} = (\mathbf{e})^6 \cdot (\mathbf{a}^1)_6 = (e)^{61} (a^{11})_6 + (e)^{62} (a^{12})_6 + (e)^{63} (a^{13})_6 = -(d_{51})_{k-1}, \quad (13.107)$$

$$d_{62} = (\mathbf{e})^6 \cdot (\mathbf{a}^2)_6 = (e)^{61} (a^{21})_6 + (e)^{62} (a^{22})_6 + (e)^{63} (a^{23})_6 = -(d_{52})_{k-1}. \quad (13.108)$$

With this notation the diffusion term takes the form

$$\sum_{m=1}^{6} \beta_m \left[\alpha_l \rho_l D_{il}^* \mathbf{e} \cdot \left(\mathbf{a}^1 \frac{\partial C_{il}}{\partial \xi} + \mathbf{a}^2 \frac{\partial C_{il}}{\partial \eta} + \mathbf{a}^3 \frac{\partial C_{il}}{\partial \zeta} \right) \right]_m$$

$$= \sum_{m=1}^{6} \beta_m \frac{D_{il,m}^C}{\Delta L_{h,m}} \left(C_{il,m} - C_{il} + \frac{\overline{\Delta V}_m}{S_m} DI\_C_{il,m} \right) \quad (13.109)$$

where

$$DI\_C_{il,1} = d_{12} \left.\frac{\partial C_{il}}{\partial \eta}\right|_1 + d_{13} \left.\frac{\partial C_{il}}{\partial \zeta}\right|_1, \qquad DI\_C_{il,2} = d_{22} \left.\frac{\partial C_{il}}{\partial \eta}\right|_2 + d_{23} \left.\frac{\partial C_{il}}{\partial \zeta}\right|_2,$$

(13.110-111)

$$DI\_C_{il,3} = d_{31} \left.\frac{\partial C_{il}}{\partial \xi}\right|_3 + d_{33} \left.\frac{\partial C_{il}}{\partial \zeta}\right|_3, \qquad DI\_C_{il,4} = d_{41} \left.\frac{\partial C_{il}}{\partial \xi}\right|_4 + d_{43} \left.\frac{\partial C_{il}}{\partial \zeta}\right|_4,$$

(13.112-113)

$$DI\_C_{il,5} = d_{51} \left.\frac{\partial C_{il}}{\partial \xi}\right|_5 + d_{52} \left.\frac{\partial C_{il}}{\partial \eta}\right|_5, \qquad DI\_C_{il,6} = d_{61} \left.\frac{\partial C_{il}}{\partial \xi}\right|_6 + d_{62} \left.\frac{\partial C_{il}}{\partial \eta}\right|_6.$$

(13.114-115)

The twelve concentration derivatives are computed as follows

$$\left.\frac{\partial C_{il}}{\partial \eta}\right|_1 = \frac{1}{4}\left(C_{il,j+1} + C_{il,i+1,j+1} - C_{il,j-1} - C_{il,i+1,j-1}\right), \tag{13.116}$$

$$\left.\frac{\partial C_{il}}{\partial \zeta}\right|_1 = \frac{1}{4}\left(C_{il,k+1} + C_{il,i+1,k+1} - C_{il,k-1} - C_{il,i+1,k-1}\right), \tag{13.117}$$

$$\left.\frac{\partial C_{il}}{\partial \eta}\right|_2 = \frac{1}{4}\left(C_{il,i-1,j+1} + C_{il,j+1} - C_{il,i-1,j-1} - C_{il,j-1}\right), \tag{13.118}$$

$$\left.\frac{\partial C_{il}}{\partial \zeta}\right|_2 = \frac{1}{4}\left(C_{il,k+1} + C_{il,i-1,k+1} - C_{il,k-1} - C_{il,i-1,k-1}\right), \tag{13.119}$$

$$\left.\frac{\partial C_{il}}{\partial \xi}\right|_3 = \frac{1}{4}\left(C_{il,i+1} + C_{il,i+1,j+1} - C_{il,i-1} - C_{il,i-1,j+1}\right), \tag{13.120}$$

$$\left.\frac{\partial C_{il}}{\partial \zeta}\right|_3 = \frac{1}{4}\left(C_{il,k+1} + C_{il,j+1,k+1} - C_{il,k-1} - C_{il,j+1,k-1}\right), \tag{13.121}$$

$$\left.\frac{\partial C_{il}}{\partial \xi}\right|_4 = \frac{1}{4}\left(C_{il,i+1} + C_{il,i+1,j-1} - C_{il,i-1} - C_{il,i-1,j-1}\right), \tag{13.122}$$

$$\left.\frac{\partial C_{il}}{\partial \zeta}\right|_4 = \frac{1}{4}\left(C_{il,k+1} + C_{il,j-1,k+1} - C_{il,k-1} - C_{il,j-1,k-1}\right), \tag{13.123}$$

$$\left.\frac{\partial C_{il}}{\partial \xi}\right|_5 = \frac{1}{4}\left(C_{il,i+1,k+1} + C_{il,i+1} - C_{il,i-1,k+1} - C_{il,i-1}\right), \tag{13.124}$$

$$\left.\frac{\partial C_{il}}{\partial \eta}\right|_5 = \frac{1}{4}\left(C_{il,j+1} + C_{il,j+1,k+1} - C_{il,j-1} - C_{il,j-1,k+1}\right), \tag{13.125}$$

$$\left.\frac{\partial C_{il}}{\partial \xi}\right|_6 = \frac{1}{4}\left(C_{il,i+1} + C_{il,i+1,k-1} - C_{il,i-1} - C_{il,i-1,k-1}\right), \qquad (13.126)$$

$$\left.\frac{\partial C_{il}}{\partial \eta}\right|_6 = \frac{1}{4}\left(C_{il,j+1} + C_{il,j+1,k-1} - C_{il,j-1} - C_{il,j-1,k-1}\right). \qquad (13.127)$$

13.4.4.4 Final form of the finite volume concentration equation

Thus the final form of the discretized concentration equation (13.1) is

$$\alpha_{la}\rho_{la}\left(C_{il} - C_{ila}\right)\gamma_{va} - \Delta\tau\sum_{m=1}^{6}\left(B_{lm-} + \beta_m \frac{D_{il,m}^C}{\Delta L_{h,m}}\right)\left(C_{il,m} - C_{il}\right) + \Delta\tau\gamma_v\mu_l^+ C_{il}$$

$$= \Delta\tau\gamma_v DC_{il} + \Delta\tau\sum_{m=1}^{6}\beta_m \frac{D_{il,m}^C}{\Delta L_{h,m}}\frac{\overline{\Delta V}_m}{S_m} DI\_C_{il,m}, \qquad (13.128)$$

Solving with respect to the unknown concentration we obtain

$$C_{il} = \frac{\alpha_{la}\rho_{la}\gamma_{va}C_{ila} + \Delta\tau\left\{\gamma_v DC_{il} + \sum_{m=1}^{6}\left[\begin{array}{c}B_{lm-}C_{il,m}\\ +\beta_m \dfrac{D_{il,m}^C}{\Delta L_{h,m}}\left(C_{il,m} + \dfrac{\overline{\Delta V}_m}{S_m} DI\_C_{il,m}\right)\end{array}\right]\right\}}{\alpha_{la}\gamma_{va}\rho_{la} + \Delta\tau\left[\gamma_v\mu_l^+ + \sum_{m=1}^{6}\left(B_{lm-} + \beta_m \dfrac{D_{il,m}^C}{\Delta L_{h,m}}\right)\right]}.$$

(13.129)

For the case of a just originating velocity field, $\alpha_{la} = 0$ and

$$\gamma_v\mu_l^+ + \sum_{m=1}^{6}\left(b_{lm-}\alpha_{l,m}\rho_{l,m} + \beta_m \frac{D_{il,m}^C}{\Delta L_{h,m}}\right) > 0, \qquad (13.130)$$

we have

$$C_{il} = \frac{\gamma_v DC_{il} + \sum_{m=1}^{6} \gamma_v DC_{il} + \sum_{m=1}^{6} \left[B_{lm-} C_{il,m} + \beta_m \frac{D_{l,m}^C}{\Delta L_{h,m}} \left(C_{il,m} + \frac{\overline{\Delta V}_m}{S_m} DI\_C_{il,m} \right) \right]}{\gamma_v \mu_l^+ + \sum_{m=1}^{6} \left(B_{lm-} + \beta_m \frac{D_{l,m}^C}{\Delta L_{h,m}} \right)}.$$

(13.131)

13.5 Discretization of the entropy equation

The entropy equation (10.62) is discretized following the procedure already described in the previous section. The result is

$$\alpha_{la} \rho_{la} \left(s_l - s_{la} \right) \gamma_{va} - \Delta \tau \sum_{m=1}^{6} B_{lm-} \left(s_{l,m} - s_l \right) + \Delta \tau \gamma_v \mu_l^+ s_l = \Delta \tau \gamma_v Ds_l^*, \quad (13.132)$$

where

$$Ds_l^* = Ds_l + \frac{1}{\gamma_v} \sum_{m=1}^{6} \beta_m \left[\frac{1}{T_l} \frac{D_{l,m}^T}{\Delta L_{h,m}} \left(T_{l,m} - T_l + \frac{\overline{\Delta V}_m}{S_m} DI\_T_{l,m} \right) + \frac{D_{l,m}^{sC}}{\Delta L_{h,m}} \sum_{i=2}^{i_{max}} \left(C_{il,m} - C_{il} + \frac{\overline{\Delta V}_m}{S_m} DI\_C_{il} \right) \right]. \quad (13.133)$$

The term $DI\_T_{l,m}$ is computed by replacing the concentrations in Eqs. (11.110-127) with the corresponding temperatures. The computation of the harmonic averaged thermal conductivity coefficients is given in Appendix 12.1. Solving with respect to the unknown specific entropy we obtain

$$s_l = \frac{\alpha_{la} \rho_{la} \gamma_{va} s_{la} + \Delta \tau \left(\gamma_v Ds_l^* + \sum_{m=1}^{6} B_{lm-} s_{lm} \right)}{\alpha_{la} \rho_{la} \gamma_{va} + \Delta \tau \left(\sum_{m=1}^{6} B_{lm-} + \gamma_v \mu_l^+ \right)}. \quad (13.134)$$

For the case of a just originating velocity field, $\alpha_{la} = 0$ and $\sum_{m=1}^{6} B_{lm-} + \gamma_v \mu_l^+ > 0$, we have

$$S_l = \frac{\gamma_v Ds_l^* + \sum_{m=1}^{6} B_{lm-} s_{lm}}{\sum_{m=1}^{6} B_{lm-} + \gamma_v \mu_l^+}.$$ (13.135)

13.6 Discretization of the temperature equation

The temperature equation (10.65) is discretized following the procedure already described in the previous section. The result is

$$\alpha_{la} \rho_{la} c_{pla} (T_l - T_{la}) \gamma_{va} - \Delta\tau \sum_{m=1}^{6} \left(B_{lm-} c_{pl,m} + \beta_m \frac{D_{l,m}^T}{\Delta L_{h,m}} \right) (T_{l,m} - T_l)$$

$$\left[1 - \rho_l \left(\frac{\partial h_l}{\partial p} \right)_{T_l, all\_C's} \right] \left[\alpha_{la} (p - p_a) \gamma_{va} - \Delta\tau \sum_{m=1}^{6} \frac{B_{lm-}}{\rho_{lm}} (p_m - p) \right]$$

$$= \Delta\tau\gamma_v \left[DT_l^N - T_l \sum_{i=2}^{i_{max}} \Delta s_{il}^{np} (\mu_{il} - C_{il} \mu_l) \right]$$

$$- \Delta\tau \sum_{m=1}^{6} \beta_m \left[\frac{D_{l,m}^C}{\Delta L_{h,m}} T_l \sum_{i=2}^{i_{max}} \Delta s_{il}^{np} \left(C_{il,m} - C_{il} + \frac{\overline{\Delta V}_m}{S_m} DI\_C_{il} \right) - \frac{D_{l,m}^T}{\Delta L_{h,m}} \frac{\overline{\Delta V}_m}{S_m} DI\_T_l \right].$$ (13.136)

13.7 Discretization of the particle number density equation

The particle number density equation (10.66) is discretized following the procedure already described in the previous section. The result is

$$n_l \gamma_v - n_{la} \gamma_{va} + \Delta\tau \sum_{m=1}^{6} \beta_m \left[(\mathbf{e})^m \cdot (\mathbf{V}_{l,m} - \mathbf{V}_{cs}) n_{l,m} - \frac{D_{l,m}^n}{\Delta L_{h,m}} (n_{l,m} - n_l) \right]$$

$$= \Delta\tau\gamma_v (\dot{n}_{l,kin} - \dot{n}_{l,coal} + \dot{n}_{l,sp}) + \Delta\tau \sum_{m=1}^{6} \beta_m \frac{D_{l,m}^n}{\Delta L_{h,m}} \frac{\overline{\Delta V}_m}{S_m} DI\_n_{l,m}.$$ (13.137)

The turbulent diffusion coefficient is again a result of harmonic volume averaging – Appendix 12.1.

$$\frac{D_{l,m}^n}{\Delta L_{h,m}} = \left(\frac{\frac{v_l^t}{Sc^t}}{\Delta V}\right)_m S_m = S_m \frac{2\left(\frac{v_l^t}{Sc^t}\right)\left(\frac{v_l^t}{Sc^t}\right)_m}{\Delta V_m \left(\frac{v_l^t}{Sc^t}\right) + \Delta V \left(\frac{v_l^t}{Sc^t}\right)_m}. \tag{13.138}$$

13.8 Discretization of the *x* momentum equation

The *x* momentum equation (10.68) is discretized as already discussed in the previous Section. The result is

$$\alpha_{la}\rho_{la}\left(u_l - u_{la}\right)\gamma_{va} - \Delta\tau\sum_{m=1}^{6}\left\{B_{lm-} + \beta_m \frac{D_{l,m}^v}{\Delta L_{h,m}}\left[1 + \frac{1}{3}(e)^{m1}(e)^{m1}\right]\right\}\left(u_{l,m} - u_l\right)$$

$$+\Delta\tau\alpha_l\left(a^{11}\gamma_\xi \frac{\partial p}{\partial \xi} + a^{21}\gamma_\eta \frac{\partial p}{\partial \eta} + a^{31}\gamma_\zeta \frac{\partial p}{\partial \zeta}\right)$$

$$-\Delta\tau\gamma_v\left[\sum_{\substack{m=1\\m\neq l}}^{3}\overline{c}_{ml}^d|\Delta\mathbf{V}_{ml}|(u_m - u_l) + \overline{c}_{wl}^d|\Delta\mathbf{V}_{wl}|(u_{cs} - u_l)\right]$$

$$-\Delta\tau\gamma_v\sum_{\substack{m=1\\m\neq l}}^{l_{max}}\overline{c}_{ml}^L\left[(v_l - v_m)b_{m,3} - (w_l - w_m)b_{m,2}\right]$$

$$-\Delta\tau\gamma_v\overline{c}_{wl}^L\left[(v_l - v_{cs})b_{w,3} - (w_l - w_{cs})b_{w,2}\right]$$

$$-\Delta\tau\gamma_v\sum_{\substack{m=1\\m\neq l}}^{3}\overline{c}_{ml}^{vm}\left(\frac{\partial \Delta u_{ml}}{\partial \tau} + \overline{V}^1\frac{\partial \Delta u_{ml}}{\partial \xi} + \overline{V}^2\frac{\partial \Delta u_{ml}}{\partial \eta} + \overline{V}^3\frac{\partial \Delta u_{ml}}{\partial \zeta}\right)$$

$$-\gamma_v\Delta\tau\overline{c}_{wl}^{vm}\left(\frac{\partial \Delta u_{csl}}{\partial \tau} + \overline{V}^1\frac{\partial \Delta u_{csl}}{\partial \xi} + \overline{V}^2\frac{\partial \Delta u_{csl}}{\partial \eta} + \overline{V}^3\frac{\partial \Delta u_{csl}}{\partial \zeta}\right)$$

$$= \Delta\tau\gamma_v\left[-\alpha_l\rho_l g_x + \sum_{m=1}^{3,w}\left[\mu_{ml}(u_m - u_l)\right] + \mu_{lw}(u_{lw} - u_l)\right]$$

$$+\Delta\tau\sum_{m=1}^{6}\beta_m\frac{D_{l,m}^v}{\Delta L_{h,m}}\left(\begin{array}{c}DI\_u_{l,m}-\dfrac{2}{3}(e)^{m1}DI\_u_{l,m}^b+DI\_vis_{lm}^{uT}\\ \\ +\dfrac{1}{3}(e)^{m1}\left[(e)^{m2}\left(v_{l,m}-v_l\right)+(e)^{m3}\left(w_{l,m}-w_l\right)\right]\end{array}\right). \quad (13.139)$$

A natural averaging of the diffusion coefficients is the harmonic averaging – Appendix 12.1

$$\frac{D_{il,m}^v}{\Delta L_{h,m}}=\left(\frac{\alpha_l^e\rho_l v_l^*}{\Delta V}\right)_m \quad S_m=S_m\frac{2\left(\alpha_l^e\rho_l v_l^*\right)\left(\alpha_l^e\rho_l v_l^*\right)_m}{\Delta V_m\left(\alpha_l^e\rho_l v_l^*\right)+\Delta V\left(\alpha_l^e\rho_l v_l^*\right)_m}. \quad (13.140)$$

It is valid for all momentum equations. Note how we arrive to the integral form of the pressure term:

$$\frac{1}{\Delta V}\int_0^{\Delta\tau}\iiint_{\Delta V}\sqrt{g}\alpha_l\left(a^{11}\gamma_\xi\frac{\partial p}{\partial\xi}+a^{21}\gamma_\eta\frac{\partial p}{\partial\eta}+a^{31}\gamma_\zeta\frac{\partial p}{\partial\zeta}\right)\partial\xi\partial\eta\partial\zeta\partial\tau$$

$$=\Delta\tau\alpha_l\left(a^{11}\gamma_\xi\frac{\partial p}{\partial\xi}+a^{21}\gamma_\eta\frac{\partial p}{\partial\eta}+a^{31}\gamma_\zeta\frac{\partial p}{\partial\zeta}\right). \quad (13.141)$$

The b coefficients of in the lift force expressions result from the Cartesian component decomposition:

$$b_{m,1}=a^{12}\frac{\partial w_m}{\partial\xi}-a^{13}\frac{\partial v_m}{\partial\xi}+a^{22}\frac{\partial w_m}{\partial\eta}-a^{23}\frac{\partial v_m}{\partial\eta}+a^{32}\frac{\partial w_m}{\partial\zeta}-a^{33}\frac{\partial v_m}{\partial\zeta}, \quad (13.142)$$

$$b_{m,2}=a^{13}\frac{\partial u_m}{\partial\xi}-a^{11}\frac{\partial w_m}{\partial\xi}+a^{23}\frac{\partial u_m}{\partial\eta}-a^{21}\frac{\partial w_m}{\partial\eta}+a^{33}\frac{\partial u_m}{\partial\zeta}-a^{31}\frac{\partial w_m}{\partial\zeta}, \quad (13.143)$$

$$b_{m,3}=a^{11}\frac{\partial v_m}{\partial\xi}-a^{12}\frac{\partial u_m}{\partial\xi}+a^{21}\frac{\partial v_m}{\partial\eta}-a^{22}\frac{\partial u_m}{\partial\eta}+a^{31}\frac{\partial v_m}{\partial\zeta}-a^{32}\frac{\partial u_m}{\partial\zeta}, \quad (13.144)$$

and

$$b_{w,1}=a^{12}\frac{\partial w_{cs}}{\partial\xi}-a^{13}\frac{\partial v_{cs}}{\partial\xi}+a^{22}\frac{\partial w_{cs}}{\partial\eta}-a^{23}\frac{\partial v_{cs}}{\partial\eta}+a^{32}\frac{\partial w_{cs}}{\partial\zeta}-a^{33}\frac{\partial v_{cs}}{\partial\zeta}, \quad (13.145)$$

$$b_{w,2} = a^{13}\frac{\partial u_{cs}}{\partial \xi} - a^{11}\frac{\partial w_{cs}}{\partial \xi} + a^{23}\frac{\partial u_{cs}}{\partial \eta} - a^{21}\frac{\partial w_{cs}}{\partial \eta} + a^{33}\frac{\partial u_{cs}}{\partial \zeta} - a^{31}\frac{\partial w_{cs}}{\partial \zeta}, \quad (13.146)$$

$$b_{w,3} = a^{11}\frac{\partial v_{cs}}{\partial \xi} - a^{12}\frac{\partial u_{cs}}{\partial \xi} + a^{21}\frac{\partial v_{cs}}{\partial \eta} - a^{22}\frac{\partial u_{cs}}{\partial \eta} + a^{31}\frac{\partial v_{cs}}{\partial \zeta} - a^{32}\frac{\partial u_{cs}}{\partial \zeta}. \quad (13.147)$$

We proceed in a similar way for the other momentum equations.

13.9 Discretization of the *y* momentum equation

The result of the discretization of the *y* momentum equation (10.73) is

$$\alpha_{la}\rho_{la}\left(v_l - v_{la}\right)\gamma_{va} - \Delta\tau\sum_{m=1}^{6}\left\{B_{lm-} + \beta_m\frac{D_{l,m}^v}{\Delta L_{h,m}}\left[1 + \frac{1}{3}(e)^{m2}(e)^{m2}\right]\right\}(v_{l,m} - v_l)$$

$$+\Delta\tau\alpha_l\left(a^{12}\gamma_{\xi}\frac{\partial p}{\partial \xi} + a^{22}\gamma_{\eta}\frac{\partial p}{\partial \eta} + a^{32}\gamma_{\zeta}\frac{\partial p}{\partial \zeta}\right)$$

$$-\Delta\tau\gamma_v\left[\sum_{\substack{m=1\\m\ne l}}^{3}\overline{c}_{ml}^d\left|\Delta\mathbf{V}_{ml}\right|(v_m - v_l) + \overline{c}_{wl}^d\left|\Delta\mathbf{V}_{wl}\right|(v_{cs} - v_l)\right]$$

$$-\Delta\tau\gamma_v\sum_{\substack{m=1\\m\ne l}}^{l_{max}}\overline{c}_{ml}^L\left[(w_l - w_m)b_{m,1} - (u_l - u_m)b_{m,3}\right]$$

$$-\Delta\tau\gamma_v\overline{c}_{wl}^L\left[(w_l - w_{cs})b_{w,1} - \Delta u_{lw}(u_l - u_{cs})b_{w,3}\right]$$

$$-\Delta\tau\gamma_v\sum_{\substack{m=1\\m\ne l}}^{3}\overline{c}_{ml}^{vm}\left(\frac{\partial \Delta v_{ml}}{\partial \tau} + \overline{V}^1\frac{\partial \Delta v_{ml}}{\partial \xi} + \overline{V}^2\frac{\partial \Delta v_{ml}}{\partial \eta} + \overline{V}^3\frac{\partial \Delta v_{ml}}{\partial \zeta}\right)$$

$$-\gamma_v\Delta\tau\overline{c}_{wl}^{vm}\left(\frac{\partial \Delta v_{csl}}{\partial \tau} + \overline{V}^1\frac{\partial \Delta v_{csl}}{\partial \xi} + \overline{V}^2\frac{\partial \Delta v_{csl}}{\partial \eta} + \overline{V}^3\frac{\partial \Delta v_{csl}}{\partial \zeta}\right)$$

$$= \Delta\tau\gamma_v\left[-\alpha_l\rho_l g_y + \sum_{m=1}^{3,w}\left[\mu_{ml}(v_m - v_l)\right] + \mu_{lw}(v_{lw} - v_l)\right]$$

$$+\Delta\tau\sum_{m=1}^{6}\beta_{m}\frac{D_{l,m}^{v}}{\Delta L_{h,m}}\left(\begin{array}{c}DI\_v_{l,m}-\frac{2}{3}(e)^{m2}DI\_u_{l,m}^{b}+DI\_vis_{lm}^{vT}\\ +\frac{1}{3}(e)^{m2}\left[(e)^{m1}(u_{l,m}-u_{l})+(e)^{m3}(w_{l,m}-w_{l})\right]\end{array}\right). \quad (13.148)$$

13.10. Discretization of the z momentum equation

The result of the discretization of the z momentum equation (10.77) is

$$\alpha_{la}\rho_{la}\left(w_{l}-w_{la}\right)\gamma_{va}-\Delta\tau\sum_{m=1}^{6}\left\{B_{lm-}+\beta_{m}\frac{D_{l,m}^{v}}{\Delta L_{h,m}}\left[1+\frac{1}{3}(e)^{m3}(e)^{m3}\right]\right\}(w_{l,m}-w_{l})$$

$$+\Delta\tau\alpha_{l}\left(a^{13}\gamma_{\xi}\frac{\partial p}{\partial\xi}+a^{23}\gamma_{\eta}\frac{\partial p}{\partial\eta}+a^{33}\gamma_{\zeta}\frac{\partial p}{\partial\zeta}\right)$$

$$-\Delta\tau\gamma_{v}\left[\sum_{\substack{m=1\\m\neq l}}^{3}\overline{c}_{ml}^{d}|\Delta\mathbf{V}_{ml}|(w_{m}-w_{l})+\overline{c}_{wl}^{d}|\Delta\mathbf{V}_{wl}|(w_{cs}-w_{l})\right]$$

$$-\Delta\tau\gamma_{v}\sum_{\substack{m=1\\m\neq l}}^{l_{max}}\overline{c}_{ml}^{L}\left[(u_{l}-u_{m})b_{m,2}-(v_{l}-v_{m})b_{m,1}\right]$$

$$-\Delta\tau\gamma_{v}\overline{c}_{wl}^{L}\left[(u_{l}-u_{cs})b_{w,2}-(v_{l}-v_{cs})b_{w,1}\right]$$

$$-\Delta\tau\gamma_{v}\sum_{\substack{m=1\\m\neq l}}^{3}\overline{c}_{ml}^{vm}\left(\frac{\partial\Delta w_{ml}}{\partial\tau}+\overline{V}^{1}\frac{\partial\Delta w_{ml}}{\partial\xi}+\overline{V}^{2}\frac{\partial\Delta w_{ml}}{\partial\eta}+\overline{V}^{3}\frac{\partial\Delta w_{ml}}{\partial\zeta}\right)$$

$$-\gamma_{v}\Delta\tau\overline{c}_{wl}^{vm}\left(\frac{\partial\Delta w_{csl}}{\partial\tau}+\overline{V}^{1}\frac{\partial\Delta w_{csl}}{\partial\xi}+\overline{V}^{2}\frac{\partial\Delta w_{csl}}{\partial\eta}+\overline{V}^{3}\frac{\partial\Delta w_{csl}}{\partial\zeta}\right)$$

$$=\Delta\tau\gamma_{v}\left[-\alpha_{l}\rho_{l}g_{z}+\sum_{m=1}^{3,w}\left[\mu_{ml}(w_{m}-w_{l})\right]+\mu_{lw}(w_{lw}-w_{l})\right]$$

$$+\Delta\tau\sum_{m=1}^{6}\beta_m\frac{D^v_{l,m}}{\Delta L_{h,m}}\left(\begin{array}{c}DI\_w_{l,m}-\frac{2}{3}(e)^{m3}DI\_u^b_{l,m}+DI\_vis^{wT}_{lm}\\ +\frac{1}{3}(e)^{m3}\left[(e)^{m1}(u_{l,m}-u_l)+(e)^{m2}(v_{l,m}-v_l)\right]\end{array}\right). \qquad (13.149)$$

13.11 Pressure-velocity coupling

The IVA3 method: An important target of the numerical methods is to guarantee a strict mass conservation in the sense of the overall mass balance as for the single cell as well for the sum of the cells inside the physical domain of interest. We use the discretized mass conservation equations of each field in a special way to construct the so called pressure-velocity coupling keeping in mind the above requirement. First we note that the difference resulting from the time derivative divided by the new time level density can be rearranged as follows

$$\frac{1}{\rho_l}(\alpha_l\rho_l\gamma_v-\alpha_{la}\rho_{la}\gamma_{va})=(\alpha_l-\alpha_{la})\gamma_v+\alpha_{la}(\gamma_v-\gamma_{va})+\frac{\alpha_{la}}{\rho_l}(\rho_l-\rho_{la})\gamma_{va}. \qquad (13.150)$$

Then, we divide each of the discretized field mass conservation equations by the corresponding new time level density. Having in mind Eq. (13.150) we obtain

$$(\alpha_l-\alpha_{la})\gamma_v+\frac{\alpha_{la}}{\rho_l}(\rho_l-\rho_{la})\gamma_{va}+\Delta\tau\frac{1}{\rho_l}\sum_{m=1}^{6}\beta_m\left(\xi_{lm+}\alpha_l\rho_l+\xi_{lm-}\alpha_{lm}\rho_{lm}\right)V^n_{lm}$$

$$=\frac{1}{\rho_l}\Delta\tau\gamma_v\mu_l-\alpha_{la}(\gamma_v-\gamma_{va}). \qquad (13.151)$$

We sum all of the l_{max} mass conservation equations. The first term disappears because the sum of all volume fractions is equal to unity. In the resulting equation the temporal density difference is replaced by the linearized form of the equation of state, Eq. (3.173),

$$\rho_l-\rho_{la}=\frac{1}{a_{la}^2}(p-p_a)+\left(\frac{\partial\rho_l}{\partial s_l}\right)_a(s_l-s_{la})+\sum_{i=2}^{i_{max}}\left(\frac{\partial\rho_l}{\partial C_{i,l}}\right)_a(C_{i,l}-C_{i,la}). \qquad (13.152)$$

The result is

$$(p-p_a)\gamma_{va}\sum_{l=1}^{l_{max}}\frac{\alpha_{la}}{\rho_l a_{la}^2}+\Delta\tau\sum_{l=1}^{l_{max}}\frac{1}{\rho_l}\sum_{m=1}^{6}\beta_m\left(\xi_{lm+}\alpha_l\rho_l+\xi_{lm-}\alpha_{lm}\rho_{lm}\right)V_{lm}^n=\Delta\tau\sum_{l=1}^{l_{max}}D\alpha_l,$$
(13.153)

where

$$\Delta\tau D\alpha_l = \frac{1}{\rho_l}\Delta\tau\gamma_v\mu_l - \alpha_{la}\left(\gamma_v - \gamma_{va}\right)$$

$$-\gamma_{va}\frac{\alpha_{la}}{\rho_l}\left[\left(\frac{\partial\rho_l}{\partial s_l}\right)_a (s_l - s_{la}) + \sum_{i=2}^{i_{max}}\left(\frac{\partial\rho_l}{\partial C_{i,l}}\right)_a (C_{i,l} - C_{i,la})\right].$$
(13.154)

This equation is equivalent exactly to the sum of the discretized mass conservation equations divided by the corresponding densities. It takes into account the influence of the variation of the density with the time on the pressure change. The spatial variation of the density in the second term is still not resolved. With the next step we will derive a approximated approach to change also the influence of the spatial variation of the density on the pressure change.

Writing the discretized momentum equation in the linearized form

$$V_{lm}^n = dV_{lm}^n - \left(\mathbf{e}\right)^m \cdot \mathbf{V}_{cs,m} - RVel_{lm}(p_m - p),$$
(13.155)

and replacing we finally obtain the so called pressure equation

$$p\gamma_{va}\sum_{l=1}^{l_{max}}\frac{\alpha_{la}}{\rho_l a_{la}^2} - \Delta\tau\sum_{m=1}^{6}\beta_m\left[\sum_{l=1}^{l_{max}}\left(\xi_{lm+}\alpha_l+\xi_{lm-}\alpha_{lm}\frac{\rho_{lm}}{\rho_l}\right)RVel_{lm}\right](p_m - p)$$

$$= p_a\gamma_{va}\sum_{l=1}^{l_{max}}\frac{\alpha_{la}}{\rho_l a_{la}^2} - \Delta\tau\sum_{l=1}^{l_{max}}\frac{1}{\rho_l}\sum_{m=1}^{6}\beta_m\left(\xi_{lm+}\alpha_l\rho_l+\xi_{lm-}\alpha_{lm}\rho_{lm}\right)\left[dV_{lm}^n - \left(\mathbf{e}\right)^m \cdot \mathbf{V}_{cs,m}\right]$$

$$+\Delta\tau\sum_{l=1}^{l_{max}}D\alpha_l.$$
(13.156)

Defining the coefficients

$$c_m = -\Delta\tau\beta_m\sum_{l=1}^{l_{max}}\left(\xi_{lm+}\alpha_l+\xi_{lm-}\alpha_{lm}\frac{\rho_{lm}}{\rho_l}\right)RVel_{lm},$$
(13.157)

$$c = p\gamma_{va}\sum_{l=1}^{l_{max}}\frac{\alpha_{la}}{\rho_l a_{la}^2} - \sum_{m=1}^{6}c_m,$$
(13.158)

$$d = p_a \gamma_{va} \sum_{l=1}^{l_{max}} \frac{\alpha_{la}}{\rho_l a_{la}^2} + \Delta\tau \sum_{l=1}^{l_{max}} D\alpha_l$$

$$-\Delta\tau \sum_{m=1}^{6} \beta_m \sum_{l=1}^{l_{max}} \left(\xi_{lm+}\alpha_l + \xi_{lm-}\alpha_{lm} \frac{\rho_{lm}}{\rho_l} \right) \left[dV_{lm}^n - (e)^m \cdot V_{cs,m} \right], \qquad (13.159)$$

we obtain the pressure equation

$$cp + \sum_{m=1}^{6} c_m p_m = d, \qquad (13.160)$$

connecting each cell pressure with the pressure of the surrounding cells. We see that the system of algebraic equations *has positive diagonal elements*, is *symmetric* and *strictly diagonally dominant* because

$$|c| > \sum_{m=1}^{6} |c_m|. \qquad (13.161)$$

These are very important properties.

The IVA2 method: The spatial deviation of the density of the surrounding cells from the density of the cell considered can be introduced into Eq. (13.153) as follows

$$\beta_m \left(\xi_{lm+}\alpha_l \rho_l + \xi_{lm-}\alpha_{lm}\rho_{lm} \right) = \beta_m \left(\xi_{lm+}\alpha_l + \xi_{lm-}\alpha_{lm} \right) \rho_l + \beta_m \xi_{lm-}\alpha_{lm} \left(\rho_{lm} - \rho_l \right). \qquad (13.161)$$

The result is

$$(p - p_a) \gamma_{va} \sum_{l=1}^{l_{max}} \frac{\alpha_{la}}{\rho_l a_{la}^2} + \Delta\tau \sum_{l=1}^{l_{max}} \sum_{m=1}^{6} \beta_m \left(\xi_{lm+}\alpha_l + \xi_{lm-}\alpha_{lm} \right) V_{lm}^n$$

$$+ \Delta\tau \sum_{l=1}^{l_{max}} \frac{1}{\rho_l} \sum_{m=1}^{6} \beta_m \xi_{lm-}\alpha_{lm} \left(\rho_{lm} - \rho_l \right) V_{lm}^n$$

$$= \sum_{l=1}^{l_{\max}} \begin{bmatrix} \dfrac{1}{\rho_l} \Delta\tau\gamma_v\mu_l - \alpha_{la}(\gamma_v - \gamma_{va}) \\ -\gamma_{va}\dfrac{\alpha_{la}}{\rho_l}\left[\left(\dfrac{\partial\rho_l}{\partial s_l}\right)_a (s_l - s_{la}) + \sum_{i=2}^{i_{\max}}\left(\dfrac{\partial\rho_l}{\partial C_{i,l}}\right)_a (C_{i,l} - C_{i,la})\right] \end{bmatrix}. \qquad (13.163)$$

The spatial density variation can again be expressed as follows, Eq. (3.173),

$$\rho_{lm} - \rho_l = \dfrac{1}{a_{la}^2}(p_m - p) + \left(\dfrac{\partial\rho_l}{\partial s_l}\right)_a (s_{lm} - s_l) + \sum_{i=2}^{i_{\max}}\left(\dfrac{\partial\rho_l}{\partial C_{i,l}}\right)_a (C_{i,lm} - C_{i,l}). \qquad (13.164)$$

With this we obtain

$$(p - p_a)\gamma_{va}\sum_{l=1}^{l_{\max}}\dfrac{\alpha_{la}}{\rho_l a_{la}^2} + \Delta\tau\sum_{l=1}^{l_{\max}}\sum_{m=1}^{6}\beta_m\left[\xi_{lm+}\alpha_l + \xi_{lm-}\alpha_{lm}\left(1+\dfrac{p_m - p}{\rho_l a_{la}^2}\right)\right]V_{lm}^n$$

$$= \Delta\tau\sum_{l=1}^{l_{\max}} D\alpha_l, \qquad (13.165)$$

where

$$\rho_l\Delta\tau D\alpha_l = \Delta\tau\gamma_v\mu_l - \alpha_{la}\rho_l(\gamma_v - \gamma_{va}) - \left(\dfrac{\partial\rho_l}{\partial s_l}\right)_a \begin{bmatrix} \gamma_{va}\alpha_{la}(s_l - s_{la}) \\ -\Delta\tau\sum_{m=1}^{6}b_{lm-}\alpha_{lm}(s_{lm} - s_l) \end{bmatrix}$$

$$- \sum_{i=2}^{i_{\max}}\left(\dfrac{\partial\rho_l}{\partial C_{i,l}}\right)_a \left[\gamma_{va}\alpha_{la}(C_{i,l} - C_{i,la}) - \Delta\tau\sum_{m=1}^{6}b_{lm-}\alpha_{lm}(C_{i,lm} - C_{i,l})\right], \qquad (13.166)$$

Equation (13.165) is equivalent exactly to the sum of the discretized mass conservation equations. The discretized concentration equation divided by the old time level density is

$$\gamma_{va}\alpha_{la}(C_{il} - C_{ila}) - \Delta\tau\sum_{m=1}^{6}b_{lm-}\alpha_{l,m}\dfrac{\rho_{l,m}}{\rho_{la}}(C_{il,m} - C_{il}) = \dfrac{\Delta\tau}{\rho_{la}}DC_{il}^N, \qquad (13.167)$$

where

$$DC_{il}^N = \gamma_v \left(DC_{il} - \mu_l^+ C_{il} \right) + \sum_{m=1}^{6} \beta_m \frac{D_{il,m}^*}{\Delta L_{h,m}} \left(C_{il,m} - C_{il} + DI\_C_{il,m} \right), \quad (13.168)$$

does not contain time derivatives and convection terms. Even these terms are the most strongly varying in transient processes during a single time step. In the case of negligible diffusion DC_{il}^N contains only source terms.

We realize that the expression on the right hand side is very similar to the left hand side of the concentration equation divided by the old time level density. A very useful approximation is then

$$\gamma_{va} \alpha_{la} \left(C_{i,l} - C_{i,la} \right) \, \Delta\tau \sum_{m=1}^{6} b_{lm\_} \alpha_{lm} \left(C_{i,lm} - C_{i,l} \right)$$

$$\approx \alpha_{la} \left(C_{il} - C_{ila} \right) \gamma_{va} - \Delta\tau \sum_{m=1}^{6} b_{lm\_} \alpha_{l,m} \frac{\rho_{l,m}}{\rho_{la}} \left(C_{il,m} - C_{il} \right) = \frac{\Delta\tau}{\rho_{la}} DC_{il}^N \quad (13.169)$$

and

$$\gamma_{va} \alpha_{la} \left(s_l - s_{la} \right) - \Delta\tau \sum_{m=1}^{6} b_{lm\_} \alpha_{lm} \left(s_{lm} - s_l \right)$$

$$\approx \alpha_{la} \left(s_l - s_{la} \right) \gamma_{va} - \Delta\tau \sum_{m=1}^{6} b_{lm\_} \alpha_{l,m} \frac{\rho_{l,m}}{\rho_{la}} \left(s_{l,m} - s_l \right) = \frac{\Delta\tau}{\rho_{la}} Ds_l^N, \quad (13.170)$$

Thus $D\alpha_l$ can be approximated as follows

$$D\alpha_l = \gamma_v \frac{\mu_l}{\rho_l} - \alpha_{la} \left(\frac{\gamma_v - \gamma_{va}}{\Delta\tau} \right) - \frac{1}{\rho_l \rho_{la}} \left\{ \left(\frac{\partial \rho_l}{\partial s_l} \right)_a Ds_l^N + \sum_{i=2}^{i_{max}} \left(\frac{\partial \rho_l}{\partial C_{i,l}} \right)_a DC_{il}^N \right\}.$$
$$(13.171)$$

Replacing with the normal velocities computed from the discretized momentum equation in linearized form we finally obtain the so called pressure equation

$$p\gamma_{va} \sum_{l=1}^{l_{max}} \frac{\alpha_{la}}{\rho_l a_{la}^2} - \Delta\tau \sum_{m=1}^{6} \beta_m \sum_{l=1}^{l_{max}} \left[\xi_{lm+} \alpha_l + \xi_{lm-} \alpha_{lm} \left(1 + \frac{p_m - p}{\rho_l a_{la}^2} \right) \right] RVel_{lm} (p_m - p)$$

$$= p_a \gamma_{va} \sum_{l=1}^{l_{max}} \frac{\alpha_{la}}{\rho_l a_{la}^2} + \Delta\tau \sum_{l=1}^{l_{max}} D\alpha_l$$

$$-\Delta\tau \sum_{m=1}^{6} \beta_m \sum_{l=1}^{l_{max}} \left[\xi_{lm+}\alpha_l + \xi_{lm-}\alpha_{lm}\left(1 + \frac{p_m - p}{\rho_l a_{la}^2}\right)\right]\left[dV_{lm}^n - (\mathbf{e})^m \cdot \mathbf{V}_{cs,m}\right] \quad (13.172)$$

or

$$cp + \sum_{m=1}^{6} c_m p_m = d, \quad (13.173)$$

where

$$c_m = -\Delta\tau \beta_m \sum_{l=1}^{l_{max}} \left[\xi_{lm+}\alpha_l + \xi_{lm-}\alpha_{lm}\left(1 + \frac{p_m - p}{\rho_l a_{la}^2}\right)\right] RVel_{lm}, \quad (13.174)$$

$$c = p\gamma_{va} \sum_{l=1}^{l_{max}} \frac{\alpha_{la}}{\rho_l a_{la}^2} - \sum_{m=1}^{6} c_m, \quad (13.175)$$

$$d = p_a \gamma_{va} \sum_{l=1}^{l_{max}} \frac{\alpha_{la}}{\rho_l a_{la}^2} + \Delta\tau \sum_{l=1}^{l_{max}} D\alpha_l$$

$$-\Delta\tau \sum_{m=1}^{6} \beta_m \sum_{l=1}^{l_{max}} \left[\xi_{lm+}\alpha_l + \xi_{lm-}\alpha_{lm}\left(1 + \frac{p_m - p}{\rho_l a_{la}^2}\right)\right]\left[dV_{lm}^n - (\mathbf{e})^m \cdot \mathbf{V}_{cs,m}\right]. \quad (13.176)$$

The advantage of Eq. (13.172) for the very first outer iteration step is that it takes the influence of all sources on the pressure change which is not the case in Eq. (13.156). The advantage of Eq. (13.156) for all subsequent outer iterations is that it reduces the residuals to very low value which is not the case with Eq. (13.172) because of the approximations (13.169) and (13.170). An additional source of numerical error is that the new density is usually computed within the outer iteration by using Eq. (13.159) and not Eq. (13.164). Combined, both equations result in a useful algorithm. As a predictor step use Eq. (13.172) and for all other iterations use Eq. (13.156).

13.12 Staggered *x* momentum equation

Two families of methods are known in the literature for solving partial differential equations with low order methods, the so called co-located and staggered grid methods. In the co-located methods all dependent variables are defined at the center of the mass control volume. In these methods unless the staggered grid method is used, discretization of order higher then the first order is required to create a stable numerical method. In the staggered grid method all dependent variables are defined in the center of the mass control volume except the velocities which are defined at the faces of the volume. In both cases the velocities are required for the center as well for the faces, so that the one group of velocities is usually computed by interpolation from the known other group. The control volume for the staggered grid methods consists of the half of the volumes belonging to each face. Strictly speaking the required geometrical information that has to be stored for these methods is four times those for the co-located methods. A compromise between low order methods using low storage and stability is to derive the discretized form of the momentum equation in the staggered cell from already discretized momentum equations in the two neighboring cells. This is possible for the following reason. Momentum equations are force balances per unit mixture volume and therefore they can be volumetrically averaged over the staggered grids. In this section we will use this idea. As already mentioned the staggered computational cell in the ξ direction consists of the half of the mass control volumes belonging to the both sites of the ξ face. We will discretize the three components of the momentum equation in this staggered cell. Then we will use the dot product of the so discretized vector momentum equation with the unit face vector to obtain the normal velocity at the cell face. In doing this, we will try to keep the computational effort small by finding common coefficients for all three equations. This approach leads to a pressure gradient component normal to the face instead of a pressure gradient to each of the Cartesian directions, which is simply numerically treated. This is the key for designing implicit or semi-implicit methods.

Time derivatives: We start with the term $\alpha_{la}\rho_{la}(u_l - u_{la})\gamma_{va}$, perform volume averaging

$$\overline{\alpha_{la}\rho_{la}(u_l - u_{la})\gamma_{va}}$$

$$= \alpha_{la}\rho_{la}(u_l - u_{la})\frac{\gamma_{va}\Delta V}{\Delta V + \Delta V_{i+1}} + \alpha_{la,i+1}\rho_{la,i+1}(u_{l,i+1} - u_{la,i+1})\frac{\gamma_{va,i+1}\Delta V_{i+1}}{\Delta V + \Delta V_{i+1}} \quad (13.177)$$

and approximate the average with

$$\overline{\alpha_{la}\rho_{la}(u_l - u_{la})\gamma_{va}} \approx (\alpha_{la}\rho_{la}\gamma_{va})_u (u_l^u - u_{la}^u), \quad (13.178)$$

where

13.12 Staggered x momentum equation

$$(\alpha_{la}\rho_{la}\gamma_{va})_u = \alpha_{la}\rho_{la}\frac{\gamma_{va}\Delta V}{\Delta V + \Delta V_{i+1}} + \alpha_{la,i+1}\rho_{la,i+1}\frac{\gamma_{va,i+1}\Delta V_{i+1}}{\Delta V + \Delta V_{i+1}}. \tag{13.179}$$

Note that this procedure of averaging does not give

$$u_l^u = \frac{1}{2}\left(u_l + u_{l,i+1}\right) \tag{13.180}$$

in the general case. Similarly we have for the other directions momentum equations

$$\overline{\alpha_{la}\rho_{la}\left(v_l - v_{la}\right)\gamma_{va}} \approx \left(\alpha_{la}\rho_{la}\gamma_{va}\right)_u \left(v_l^u - v_{la}^u\right), \tag{13.181}$$

$$\overline{\alpha_{la}\rho_{la}\left(w_l - w_{la}\right)\gamma_{va}} \approx \left(\alpha_{la}\rho_{la}\gamma_{va}\right)_u \left(w_l^u - w_{la}^u\right). \tag{13.182}$$

We realize that in this way of approximation the component velocity differences for all three Cartesian directions possess a common coefficient.

Convective terms: The following approximation for the convective terms is proposed

$$\overline{\sum_{m=1}^{6} b_{lm-}\alpha_{l,m}\rho_{l,m}\left(u_{l,m} - u_l\right)}$$

$$= \frac{1}{\Delta V + \Delta V_{i+1}}\left\{\Delta V \sum_{m=1}^{6} B_{lm-}\left(u_{l,m} - u_l\right) + \Delta V_{i+1}\sum_{m=1}^{6}\left[B_{lm-}\left(u_{l,m} - u_l\right)\right]_{i+1}\right\}$$

$$\approx \sum_{m=1}^{6} b_{lm-}^u \alpha_{l,m}^u \rho_{l,m}^u \left(u_{l,m}^u - u_l^u\right), \tag{13.183}$$

where

$$b_{lm-}^u \alpha_{l,m}^u \rho_{l,m}^u = C^* B_{lm-} + \left(1 - C^*\right)\left(B_{lm-}\right)_{i+1}, \tag{13.184}$$

is the m-th face mass flow into the staggered cell divided by its volume and

$$C^* = \frac{\Delta V}{\Delta V + \Delta V_{i+1}}. \tag{13.185}$$

As in the case of the time derivatives we realize that in this way of approximation the component velocity differences for all three Cartesian directions possess a common coefficient.

Diagonal diffusion terms: We apply a similar procedure to the diagonal diffusion terms

$$\sum_{m=1}^{6} \beta_m \frac{D_{l,m}^V}{\Delta L_{h,m}} \left[1 + \frac{1}{3}(e)^{m1}(e)^{m1} \right] (u_{l,m} - u_l) =$$

$$= C^* \sum_{m=1}^{6} \beta_m \frac{D_{l,m}^V}{\Delta L_{h,m}} \left[1 + \frac{1}{3}(e)^{m1}(e)^{m1} \right] (u_{l,m} - u_l)$$

$$+ (1 - C^*) \sum_{m=1}^{6} \left\{ \beta_m \frac{D_{l,m}^V}{\Delta L_{h,m}} \left[1 + \frac{1}{3}(e)^{m1}(e)^{m1} \right] (u_{l,m} - u_l) \right\}_{i+1}$$

$$\approx \sum_{m=1}^{6} \left(\begin{array}{c} C^* \beta_m \dfrac{D_{l,m}^V}{\Delta L_{h,m}} \left[1 + \dfrac{1}{3}(e)^{m1}(e)^{m1} \right] \\ + \left\{ (1 - C^*) \beta_m \dfrac{D_{l,m}^V}{\Delta L_{h,m}} \left[1 + \dfrac{1}{3}(e)^{m1}(e)^{m1} \right] \right\}_{i+1} \end{array} \right) (u_{l,m}^u - u_l^u). \quad (13.186)$$

Thus the combined convection-diffusion terms are finally approximated as follows

$$-\sum_{m=1}^{6} \left\{ B_{lm-} + \beta_m \frac{D_{l,m}^V}{\Delta L_{h,m}} \left[1 + \frac{1}{3}(e)^{m1}(e)^{m1} \right] \right\} (u_{l,m} - u_l)$$

$$\approx \sum_{m=1}^{6} a_{lm,cd} (u_{l,m}^u - u_l^u) + \sum_{m=1}^{6} a_{lm,u\_dif} (u_{l,m}^u - u_l^u), \quad (13.187)$$

where

$$a_{lm,cd} = -C^* \left(B_{lm-} + \beta_m \frac{D_{l,m}^V}{\Delta L_{h,m}} \right) - (1 - C^*) \left(B_{lm-} + \beta_m \frac{D_{l,m}^V}{\Delta L_{h,m}} \right)_{i+1}, \quad (13.188)$$

$$a_{lm,u\_dif} = -\frac{1}{3}\left\{C^*\beta_m \frac{D_{l,m}^V}{\Delta L_{h,m}}(e)^{m1}(e)^{m1} + (1-C^*)\left[\beta_m \frac{D_{l,m}^V}{\Delta L_{h,m}}(e)^{m1}(e)^{m1}\right]_{i+1}\right\}. \tag{13.189}$$

Similarly we have

$$-\sum_{m=1}^{6}\left\{B_{lm-} + \beta_m \frac{D_{l,m}^V}{\Delta L_{h,m}}\left[1+\frac{1}{3}(e)^{m2}(e)^{m2}\right]\right\}(v_{l,m} - v_l)$$

$$\approx \sum_{m=1}^{6} a_{lm,cd}\left(v_{l,m}^u - v_l^u\right) + \sum_{m=1}^{6} a_{lm,v\_dif}\left(v_{l,m}^u - v_l^u\right), \tag{13.190}$$

$$a_{lm,v\_dif} = -\frac{1}{3}\left\{C^*\beta_m \frac{D_{l,m}^V}{\Delta L_{h,m}}(e)^{m2}(e)^{m2} + (1-C^*)\left[\beta_m \frac{D_{l,m}^V}{\Delta L_{h,m}}(e)^{m2}(e)^{m2}\right]_{i+1}\right\}. \tag{13.191}$$

$$-\sum_{m=1}^{6}\left\{B_{lm-} + \beta_m \frac{D_{l,m}^V}{\Delta L_{h,m}}\left[1+\frac{1}{3}(e)^{m3}(e)^{m3}\right]\right\}(w_{l,m} - w_l)$$

$$\approx \sum_{m=1}^{6} a_{lm,cd}\left(w_{l,m}^u - w_l^u\right) + \sum_{m=1}^{6} a_{lm,w\_dif}\left(w_{l,m}^u - w_l^u\right), \tag{13.192}$$

$$a_{lm,w\_dif} = -\frac{1}{3}\left\{C^*\beta_m \frac{D_{l,m}^V}{\Delta L_{h,m}}(e)^{m3}(e)^{m3} + (1-C^*)\left[\beta_m \frac{D_{l,m}^V}{\Delta L_{h,m}}(e)^{m3}(e)^{m3}\right]_{i+1}\right\}. \tag{13.193}$$

We realize again that the coefficients $a_{lm,cd}$ are common for all the momentum equations in the staggered cell.

Drag force terms: The following approximation contains in fact computation of the volume averages of the linearized drag coefficients.

$$\gamma_v\left[\sum_{\substack{m=1\\m\neq l}}^{3}\overline{c}_{ml}^d\left|\Delta\mathbf{V}_{ml}\right|(u_m - u_l) + \overline{c}_{wl}^d\left|\Delta\mathbf{V}_{wl}\right|(u_{cs} - u_l)\right]$$

$$\approx \sum_{\substack{m=1 \\ m \neq l}}^{3} \left(\gamma_v c_{ml}^d \right)_u \left(u_m^u - u_l^u \right) + \left(\gamma_v c_{wl}^d \right)_u \left(u_{cs}^u - u_l^u \right), \tag{13.194}$$

where

$$\left(\gamma_v c_{ml}^d \right)_u = \frac{1}{\Delta V + \Delta V_{i+1}} \left(\gamma_v \Delta V \overline{c}_{ml}^d \left| \Delta \mathbf{V}_{ml} \right| + \gamma_{v,i+1} \Delta V_{i+1} \overline{c}_{ml,i+1}^d \left| \Delta \mathbf{V}_{ml} \right|_{i+1} \right), \tag{13.195}$$

$$\left(\gamma_v c_{wl}^d \right)_u = \frac{1}{\Delta V + \Delta V_{i+1}} \left(\gamma_v \Delta V \overline{c}_{wl}^d \left| \Delta \mathbf{V}_{csl} \right| + \gamma_{v,i+1} \Delta V_{i+1} \overline{c}_{wl,i+1}^d \left| \Delta \mathbf{V}_{csl} \right|_{i+1} \right). \tag{13.196}$$

Similarly we have

$$\gamma_v \left[\sum_{\substack{m=1 \\ m \neq l}}^{3} \overline{c}_{ml}^d \left| \Delta \mathbf{V}_{ml} \right| \left(v_m - v_l \right) + \overline{c}_{wl}^d \left| \Delta \mathbf{V}_{wl} \right| \left(v_{cs} - v_l \right) \right]$$

$$\approx \sum_{\substack{m=1 \\ m \neq l}}^{3} \left(\gamma_v c_{ml}^d \right)_u \left(v_m^u - v_l^u \right) + \left(\gamma_v c_{wl}^d \right)_u \left(v_{cs}^u - v_l^u \right), \tag{13.197}$$

$$\gamma_v \left[\sum_{\substack{m=1 \\ m \neq l}}^{3} \overline{c}_{ml}^d \left| \Delta \mathbf{V}_{ml} \right| \left(w_m - w_l \right) + \overline{c}_{wl}^d \left| \Delta \mathbf{V}_{wl} \right| \left(w_{cs} - w_l \right) \right]$$

$$\approx \sum_{\substack{m=1 \\ m \neq l}}^{3} \left(\gamma_v c_{ml}^d \right)_u \left(w_m^u - w_l^u \right) + \left(\gamma_v c_{wl}^d \right)_u \left(w_{cs}^u - w_l^u \right). \tag{13.198}$$

Again the drag term coefficients for the momentum equations in the staggered cell are common.

Gravitational force: The volume averaging for the gravitational force gives

$$\overline{\alpha_l \rho_l g_x \gamma_v} = g_x \left(\alpha_{la} \rho_{la} \gamma_{va} \right)_u, \tag{13.199}$$

$$\overline{\alpha_l \rho_l g_y \gamma_v} = g_y \left(\alpha_{la} \rho_{la} \gamma_{va} \right)_u, \tag{13.200}$$

$$\overline{\alpha_l \rho_l g_z \gamma_v} = g_z \left(\alpha_{la} \rho_{la} \gamma_{va} \right)_u. \tag{13.201}$$

Interfacial momentum transfer due to mass transfer: The interfacial momentum transfer is again approximated by first volume averaging the volume mass source terms due to interfacial mass transfer. The source terms due to external injection or suction are computed exactly because it is easy to prescribe the velocities corresponding to the sources at the mass cell center.

$$-\left[\sum_{m=1}^{3,w}\left[\mu_{ml}\gamma_v\left(u_m-u_l\right)\right]+\mu_{lw}\gamma_v\left(u_{lw}-u_l\right)\right]$$

$$\approx \sum_{m=1}^{3}\left(\gamma_v\mu_{ml}\right)_u\left(u_m^u-u_l^u\right)+\left(\gamma_v\mu_{wl}\right)_u\left(u_{wl}^u-u_l^u\right)-\left(\gamma_v\mu_{lw}\right)_u\left(u_{lw}^u-u_l^u\right) \quad (13.202)$$

$$\left(\gamma_v\mu_{ml}\right)_u = \frac{1}{\Delta V + \Delta V_{i+1}}\left(\gamma_v \Delta V \mu_{ml} + \gamma_{v,i+1}\Delta V_{i+1}\mu_{ml,i+1}\right), \quad (13.203)$$

$$\left(\gamma_v\mu_{wl}\right)_u = \frac{1}{\Delta V + \Delta V_{i+1}}\left(\gamma_v \Delta V \mu_{wl} + \gamma_{v,i+1}\Delta V_{i+1}\mu_{wl,i+1}\right), \quad (13.204)$$

$$\left(\gamma_v\mu_{lw}\right)_u = \frac{1}{\Delta V + \Delta V_{i+1}}\left(\gamma_v \Delta V \mu_{lw} + \gamma_{v,i+1}\Delta V_{i+1}\mu_{lw,i+1}\right), \quad (13.205)$$

$$\left(\gamma_v\mu_{wl}\right)_u u_{wl} = \frac{1}{\Delta V + \Delta V_{i+1}}\left(\gamma_v \Delta V \mu_{wl} u_{wl} + \gamma_{v,i+1}\Delta V_{i+1}\mu_{wl,i+1}u_{wl,i+1}\right), \quad (13.206)$$

$$\left(\gamma_v\mu_{lw}\right)_u u_{lw} = \frac{1}{\Delta V + \Delta V_{i+1}}\left(\gamma_v \Delta V \mu_{lw} u_{lw} + \gamma_{v,i+1}\Delta V_{i+1}\mu_{lw,i+1}u_{lw,i+1}\right), \quad (13.207)$$

Similarly we have

$$-\left[\sum_{m=1}^{3,w}\left[\mu_{ml}\gamma_v\left(v_m-v_l\right)\right]+\mu_{lw}\gamma_v\left(v_{lw}-v_l\right)\right]$$

$$\approx \sum_{m=1}^{3}\left(\gamma_v\mu_{ml}\right)_u\left(v_m^u-v_l^u\right)+\left(\gamma_v\mu_{wl}\right)_u\left(v_{wl}^u-v_l^u\right)-\left(\gamma_v\mu_{lw}\right)_u\left(v_{lw}^u-v_l^u\right), \quad (13.208)$$

$$-\overline{\left[\sum_{m=1}^{3,w}\left[\mu_{ml}\gamma_v\left(w_m-w_l\right)\right]+\mu_{lw}\gamma_v\left(w_{lw}-w_l\right)\right]}$$

$$\approx \sum_{m=1}^{3}\left(\gamma_v\mu_{ml}\right)_u\left(w_m^u-w_l^u\right)+\left(\gamma_v\mu_{wl}\right)_u\left(w_{wl}^u-w_l^u\right)-\left(\gamma_v\mu_{lw}\right)_u\left(w_{lw}^u-w_l^u\right). \quad (13.209)$$

Lift force, off-diagonal viscous forces: The lift force and the off-diagonal viscous forces are explicitly computed by strict volume averaging.

Pressure gradient:

$$\overline{\alpha_l\gamma_\xi\left(\nabla p\right)\cdot\mathbf{i}}=\alpha_{lu}\gamma_\xi\left(\nabla p\right)\cdot\mathbf{i}, \quad (13.210)$$

where

$$\alpha_{lu}=\frac{\gamma_v\Delta V\alpha_l+\gamma_{v,i+1}\Delta V_{i+1}\alpha_{l,i+1}}{\gamma_v\Delta V+\gamma_{v,i+1}\Delta V_{i+1}}. \quad (13.211)$$

Similarly we have

$$\overline{\alpha_l\gamma_\xi\left(\nabla p\right)\cdot\mathbf{j}}=\alpha_{lu}\gamma_\xi\left(\nabla p\right)\cdot\mathbf{j}, \quad (13.212)$$

$$\overline{\alpha_l\gamma_\xi\left(\nabla p\right)\cdot\mathbf{k}}=\alpha_{lu}\gamma_\xi\left(\nabla p\right)\cdot\mathbf{k}. \quad (13.213)$$

Virtual mass force:

$$\gamma_v\sum_{\substack{m=1\\m\neq l}}^{3,cs}\overline{c}_{ml}^{vm}\left(\frac{\partial\Delta u_{ml}}{\partial\tau}+\overline{V}^1\frac{\partial\Delta u_{ml}}{\partial\xi}+\overline{V}^2\frac{\partial\Delta u_{ml}}{\partial\eta}+\overline{V}^3\frac{\partial\Delta u_{ml}}{\partial\zeta}\right)$$

$$=\sum_{\substack{m=1\\m\neq l}}^{3,cs}\left(\gamma_v\overline{c}_{ml}^{vm}\right)_u\left(\begin{array}{c}\dfrac{u_m^u-u_l^u}{\Delta\tau}-\dfrac{u_{ma}^u-u_{la}^u}{\Delta\tau}+\left(\overline{V}^1\right)_1\left(u_{m,i+1}-u_{l,i+1}-u_m+u_l\right)\\+\left(\overline{V}^2\right)_1\left(\left.\dfrac{\partial u_m}{\partial\eta}\right|_1-\left.\dfrac{\partial u_l}{\partial\eta}\right|_1\right)+\left(\overline{V}^3\right)_1\left(\left.\dfrac{\partial u_m}{\partial\zeta}\right|_1-\left.\dfrac{\partial u_l}{\partial\zeta}\right|_1\right)\end{array}\right). \quad (13.214)$$

Similarly we have

$$\gamma_v \sum_{\substack{m=1\\m\neq l}}^{3,cs} \overline{c}_{ml}^{vm} \left(\frac{\partial \Delta v_{ml}}{\partial \tau} + \overline{V}^1 \frac{\partial \Delta v_{ml}}{\partial \xi} + \overline{V}^2 \frac{\partial \Delta v_{ml}}{\partial \eta} + \overline{V}^3 \frac{\partial \Delta v_{ml}}{\partial \zeta} \right)$$

$$= \sum_{\substack{m=1\\m\neq l}}^{3,cs} \left(\gamma_v \overline{c}_{ml}^{vm} \right)_u \left(\begin{array}{c} \dfrac{v_m^u - v_l^u}{\Delta \tau} - \dfrac{v_{ma}^u - v_{la}^u}{\Delta \tau} + \left(\overline{V}^1\right)_1 \left(v_{m,i+1} - v_{l,i+1} - v_m + v_l\right) \\ + \left(\overline{V}^2\right)_1 \left(\left.\dfrac{\partial v_m}{\partial \eta}\right|_1 - \left.\dfrac{\partial v_l}{\partial \eta}\right|_1 \right) + \left(\overline{V}^3\right)_1 \left(\left.\dfrac{\partial v_m}{\partial \zeta}\right|_1 - \left.\dfrac{\partial v_l}{\partial \zeta}\right|_1 \right) \end{array} \right), \quad (13.215)$$

$$\gamma_v \sum_{\substack{m=1\\m\neq l}}^{3,cs} \overline{c}_{ml}^{vm} \left(\frac{\partial \Delta w_{ml}}{\partial \tau} + \overline{V}^1 \frac{\partial \Delta w_{ml}}{\partial \xi} + \overline{V}^2 \frac{\partial \Delta w_{ml}}{\partial \eta} + \overline{V}^3 \frac{\partial \Delta w_{ml}}{\partial \zeta} \right)$$

$$= \sum_{\substack{m=1\\m\neq l}}^{3,cs} \left(\gamma_v \overline{c}_{ml}^{vm} \right)_u \left(\begin{array}{c} \dfrac{w_m^u - w_l^u}{\Delta \tau} - \dfrac{w_{ma}^u - w_{la}^u}{\Delta \tau} + \left(\overline{V}^1\right)_1 \left(w_{m,i+1} - w_{l,i+1} - w_m + w_l\right) \\ + \left(\overline{V}^2\right)_1 \left(\left.\dfrac{\partial w_m}{\partial \eta}\right|_1 - \left.\dfrac{\partial w_l}{\partial \eta}\right|_1 \right) + \left(\overline{V}^3\right)_1 \left(\left.\dfrac{\partial w_m}{\partial \zeta}\right|_1 - \left.\dfrac{\partial w_l}{\partial \zeta}\right|_1 \right) \end{array} \right).$$

(13.216)

Implicit treatment of the interfacial interaction:

$$\left(\alpha_{la}\rho_{la}\gamma_{va}\right)_u \frac{u_l^u - u_{la}^u}{\Delta\tau} + \sum_{m=1}^{6} a_{lm,cd}\left(u_{l,m}^u - u_l^u\right) + \alpha_{l,u}\gamma_\xi\left(\nabla p\right)\cdot\mathbf{i}$$

$$-\sum_{\substack{m=1\\m\neq l}}^{3,cs} \left(\gamma_v \overline{c}_{ml}^{vm} \right)_u \left(\begin{array}{c} \dfrac{u_m^u - u_l^u}{\Delta \tau} - \dfrac{u_{ma}^u - u_{la}^u}{\Delta \tau} + \left(\overline{V}^1\right)_1 \left(u_{m,i+1} - u_{l,i+1} - u_m + u_l\right) \\ + \left(\overline{V}^2\right)_1 \left(\left.\dfrac{\partial u_m}{\partial \eta}\right|_1 - \left.\dfrac{\partial u_l}{\partial \eta}\right|_1 \right) + \left(\overline{V}^3\right)_1 \left(\left.\dfrac{\partial u_m}{\partial \zeta}\right|_1 - \left.\dfrac{\partial u_l}{\partial \zeta}\right|_1 \right) \end{array} \right) - \left(\gamma_v f_l^L\right)_u$$

$$= -\sum_{m=1}^{6} a_{lm,u\_dif}\left(u_{l,m}^u - u_l^u\right) - g_x\left(\alpha_{la}\rho_{la}\gamma_{va}\right)_u$$

$$+\sum_{m=1}^{3}\left(\gamma_v\mu_{ml}\right)_u\left(u_m^u-u_l^u\right)+\left(\gamma_v\mu_{wl}\right)_u\left(u_{wl}^u-u_l^u\right)-\left(\gamma_v\mu_{lw}\right)_u\left(u_{lw}^u-u_l^u\right)$$

$$+\sum_{\substack{m=1\\m\neq l}}^{3}\left(\gamma_v c_{ml}^d\right)_u\left(u_m^u-u_l^u\right)+\left(\gamma_v c_{wl}^d\right)_u\left(u_{cs}^u-u_l^u\right)+Vis_l^u. \tag{13.217}$$

For all three velocity fields we have a system of algebraic equations with respect to the corresponding field velocity components in the x direction

$$\left[\left(\alpha_{la}\rho_{la}\gamma_{va}\right)_u\frac{1}{\Delta\tau}-\sum_{\substack{m=1\\m\neq l}}^{3}a_{lm}+a_{l,cd}+\left(\gamma_v\bar{c}_{wl}^{vm}\right)_u\frac{1}{\Delta\tau}+\left(\gamma_v c_{wl}^d\right)_u+\left(\gamma_v\mu_{wl}\right)_u-\left(\gamma_v\mu_{lw}\right)_u\right]u_l^u$$

$$+\sum_{\substack{m=1\\m\neq l}}^{3}a_{lm}u_m^u=b_l-\alpha_{l,u}\gamma_\xi\left(\nabla p\right)\cdot\mathbf{i} \tag{13.218}$$

or

$$\mathbf{A}^u\mathbf{u}^u=\mathbf{b}\mathbf{u}^u-\mathbf{a}^u\gamma_\xi\left(\nabla p\right)\cdot\mathbf{i}. \tag{13.219}$$

The non-diagonal and the diagonal elements of the $\mathbf{A}^u$ matrix are

$$a_{ml}=-\left[\left(\gamma_v\bar{c}_{ml}^{vm}\right)_u\frac{1}{\Delta\tau}+\left(\gamma_v\mu_{ml}\right)_u+\left(\gamma_v c_{ml}^d\right)_u\right], \tag{13.220}$$

and

$$a_{ll}=\left(\alpha_{la}\rho_{la}\gamma_{va}\right)_u\frac{1}{\Delta\tau}-\sum_{\substack{m=1\\m\neq l}}^{3}a_{lm}+a_{l,cd}+\left(\gamma_v\bar{c}_{wl}^{vm}\right)_u\frac{1}{\Delta\tau}+\left(\gamma_v c_{wl}^d\right)_u+\left(\gamma_v\mu_{wl}\right)_u-\left(\gamma_v\mu_{lw}\right)_u, \tag{13.221}$$

respectively, where

$$a_{l,cd}=-\sum_{m=1}^{6}a_{lm,cd}. \tag{13.222}$$

The elements of the algebraic vector $\mathbf{b}\mathbf{u}^u$ are

$$bu_l=bu_{l,conv}+Vis_l^u+\left(\alpha_{la}\rho_{la}\gamma_{va}\right)_u\left(\frac{u_{la}^u}{\Delta\tau}-g_x\right)-\sum_{\substack{m=1\\m\neq l}}^{3,cs}bu_{lm}+\left(\gamma_v f_l^L\right)_u$$

$$+\left[\left(\gamma_v \overline{c}_{csl}^{vm}\right)_u \frac{1}{\Delta \tau}+\left(\gamma_v c_{wl}^d\right)_u\right]u_{cs}^u+\left[\left(\gamma_v \mu_{wl}\right)_u-\left(\gamma_v \mu_{lw}\right)_u\right]u_{lw}^u \qquad (13.223)$$

where

$$bu_{ml}=-bu_{lm}=\left(\gamma_v \overline{c}_{ml}^{vm}\right)_u \begin{bmatrix} -\dfrac{u_{ma}^u-u_{la}^u}{\Delta \tau}+\left(\overline{V}^1\right)_1\left(u_{m,i+1}-u_{l,i+1}-u_m+u_l\right) \\ +\left(\overline{V}^2\right)_1\left(\left.\dfrac{\partial u_m}{\partial \eta}\right|_1-\left.\dfrac{\partial u_l}{\partial \eta}\right|_1\right)+\left(\overline{V}^3\right)_1\left(\left.\dfrac{\partial u_m}{\partial \zeta}\right|_1-\left.\dfrac{\partial u_l}{\partial \zeta}\right|_1\right) \end{bmatrix}$$
$$(13.224)$$

and

$$bu_{l,conv}=-\sum_{m=1}^{6}\left[\left(a_{lm,cd}+a_{lm,u\_dif}\right)u_{l,m}^u-a_{lm,u\_dif}u_l^u\right]. \qquad (13.225)$$

The elements of the algebraic vector $\mathbf{a}^u$ are $\alpha_{l,u}$, where $l = 1,2,3$. Note that by definition, if one velocity field does not exist, $\alpha_l = 0$, the coefficients describing its coupling with the other fields are then equal to zero. Similarly we can discretize the momentum equations in the same staggered control volume for the other Cartesian components. The result in component form is then

$$\mathbf{A}^u \mathbf{u}^u = \mathbf{b}\mathbf{u}^u - \mathbf{a}^u \gamma_\xi \left(\nabla p\right) \cdot \mathbf{i}, \qquad (13.226)$$

$$\mathbf{A}^u \mathbf{v}^u = \mathbf{b}\mathbf{v}^u - \mathbf{a}^u \gamma_\xi \left(\nabla p\right) \cdot \mathbf{j}, \qquad (13.227)$$

$$\mathbf{A}^u \mathbf{w}^u = \mathbf{b}\mathbf{w}^u - \mathbf{a}^u \gamma_\xi \left(\nabla p\right) \cdot \mathbf{k}. \qquad (13.228)$$

It is remarkable that the $\mathbf{A}$ matrix and the coefficients of the pressure gradient are common for all the three systems of equations. If we take the dot product of each u-v-w equation with the unit normal vector at the control volume face we then obtain

$$\mathbf{A}^u \left[\left(e^{11}\right)_1 \mathbf{u}^u + \left(e^{12}\right)_1 \mathbf{v}^u + \left(e^{13}\right)_1 \mathbf{w}^u\right]$$
$$= \left(e^{11}\right)_1 \mathbf{b}\mathbf{u}^u + \left(e^{12}\right)_1 \mathbf{b}\mathbf{v}^u + \left(e^{13}\right)_1 \mathbf{b}\mathbf{w}^u - \mathbf{a}^u \gamma_\xi \left[\left(\mathbf{e}^1\right)_1 \cdot \left(\nabla p\right)\right]. \qquad (13.229)$$

Having in mind that the outwards pointing normal face velocity is

$$\overline{\mathbf{V}}^n = (\mathbf{e})^1 \cdot \mathbf{V}^u, \tag{13.230}$$

and

$$\frac{\partial p}{\partial \xi} = (\mathbf{e})^1 \cdot (\nabla p), \tag{13.231}$$

we obtain finally

$$\mathbf{A}^u \overline{\mathbf{V}}^n = \mathbf{b}^u - \mathbf{a}^u \gamma_\xi \frac{\partial p}{\partial \xi}, \tag{13.232}$$

where

$$\mathbf{b}^u = \left(e^{11}\right)_1 \mathbf{b}u^u + \left(e^{12}\right)_1 \mathbf{b}v^u + \left(e^{13}\right)_1 \mathbf{b}w^u. \tag{13.233}$$

This algebraic system can be solved with respect to each field velocity provided that

$$\det \mathbf{A}^u = a_{11}a_{22}a_{33} + a_{12}a_{23}a_{31} + a_{21}a_{32}a_{13} - a_{31}a_{22}a_{13} - a_{32}a_{23}a_{11} - a_{12}a_{21}a_{33} \neq 0. \tag{13.234}$$

The result is

$$\overline{\mathbf{V}}^n = \mathbf{dV}_\xi - \mathbf{RV}_\xi \gamma_\xi (p_{i+1} - p), \tag{13.235}$$

where

$$\mathbf{dV}_\xi = \left(\mathbf{A}^u\right)^{-1} \mathbf{b}^u \tag{13.236}$$

with components

$$d\overline{V}_{\xi,l} = \left(\sum_{m=1}^{3} b_m \overline{a}_{lm}\right) / \det \mathbf{A}^u, \tag{13.237}$$

and

$$\mathbf{RV}_\xi = \left(\mathbf{A}^u\right)^{-1} \mathbf{a}_u, \tag{13.238}$$

with components

$$RU_l = \left(\sum_{m=1}^{3} \alpha_{ma,u} \bar{a}_{lm}\right) / \det \mathbf{A}_u,\qquad (13.239)$$

and the $\bar{a}$ values are

$$\bar{a}_{11} = a_{22}a_{33} - a_{32}a_{23}, \quad \bar{a}_{12} = a_{32}a_{13} - a_{12}a_{33}, \quad \bar{a}_{13} = a_{12}a_{23} - a_{22}a_{13},$$

$$\bar{a}_{21} = a_{23}a_{31} - a_{21}a_{33}, \quad \bar{a}_{22} = a_{11}a_{33} - a_{31}a_{13}, \quad \bar{a}_{23} = a_{21}a_{13} - a_{23}a_{11},$$

$$\bar{a}_{31} = a_{21}a_{32} - a_{31}a_{22}, \quad \bar{a}_{32} = a_{12}a_{31} - a_{32}a_{11}, \quad \bar{a}_{33} = a_{11}a_{22} - a_{21}a_{12}. \qquad (13.240\text{-}248)$$

Actually, not the absolute but the relative normal face velocity is required to construct the pressure equation which is readily obtained

$$\mathbf{V}^n = \overline{\mathbf{V}}^n - (\mathbf{e})^1 \cdot \mathbf{V}_{cs}. \qquad (13.249)$$

Appendix 13.1 Harmonic averaged diffusion coefficients

A natural averaging of the coefficients describing diffusion across the face m, having surface cross section S_m is then the harmonic averaging

$$\frac{D_{l,m}^{\Phi}}{\Delta L_{h,m}} = \left(\frac{\Phi_l}{\Delta V}\right)_m S_m = S_m \frac{2(\Phi_l)(\Phi_l)_m}{\Delta V_m (\Phi_l) + \Delta V (\Phi_l)_m}$$

where on the right hand side m = 1, 2, 3, 4, 5, 6 is equivalent to $i + 1, i - 1, j + 1, j - 1, k + 1, k - 1$, respectively regarding the properties inside a control volumes. ΔV is the non-staggered cell volume, and ΔV_m is the volume of the cell at the other side of face m. It guaranties that if the field in one of the neighboring cells is missing the diffusion coefficient is zero. This property is derived from the solution of the steady state one-dimensional diffusion equations.

For computation of

$$\frac{D_{il,m}^*}{\Delta L_{h,m}} = \left(\frac{\alpha_l \rho_l D_{il}^*}{\Delta V}\right)_m S_m = S_m \frac{2(\alpha_l \rho_l D_{il}^*)(\alpha_l \rho_l D_{il}^*)_m}{\Delta V_m (\alpha_l \rho_l D_{il}^*) + \Delta V (\alpha_l \rho_l D_{il}^*)_m}$$

we simply set $\Phi_l = \alpha_l \rho_l D_{il}^*$.

For computation of $\dfrac{D_{l,m}^T}{\Delta L_{h,m}}$ we simply set $\Phi_l = \alpha_l \lambda_l$.

For computation of $\dfrac{D_{il,m}^{sC}}{\Delta L_{h,m}}$ we simply set $\Phi_l = \alpha_l \rho_l D_{il}^* (s_{il} - s_{1l})$. Note that

$$\dfrac{D_{il,m}^C}{\Delta L_{h,m}} = 0$$

for $s_{il} = s_{1l}$ or $s_{il,m} = s_{1l,m}$.

For computation of the turbulent particle diffusion coefficient $\dfrac{D_{l,m}^n}{\Delta L_{h,m}}$ we simply set $\Phi_l = \dfrac{\nu_l^t}{Sc^t}$.

For computation of $\dfrac{D_{l,m}^V}{\Delta L_{h,m}}$ we simply set $\Phi_l = \alpha_l \rho_l \nu_l^*$.

In the case of cylindrical or Cartesian coordinate systems we have zero off-diagonal diffusion terms and

$$\dfrac{D_{l1}^\Phi}{\Delta r_h} = \dfrac{2(\Phi_l)(\Phi_l)_{i+1}}{\Delta r (\Phi_l)_{i+1} + \Delta r_{i+1} \Phi_l},$$

$$\dfrac{D_{l2}^\Phi}{\Delta r_{h,i-1}} = \dfrac{2(\Phi_l)(\Phi_l)_{i-1}}{\Delta r (\Phi_l)_{i-1} + \Delta r_{i-1} \Phi_l},$$

$$\dfrac{D_{l3}^\Phi}{r^\kappa \Delta \theta_h} = \dfrac{2(\Phi_l)(\Phi_l)_{j+1}}{r^\kappa \left[\Delta \theta (\Phi_l)_{j+1} + \Delta \theta_{j+1} \Phi_l \right]},$$

$$\dfrac{D_{l4}^\Phi}{r^\kappa \Delta \theta_{h,j-1}} = \dfrac{2(\Phi_l)(\Phi_l)_{j-1}}{r^\kappa \left[\Delta \theta (\Phi_l)_{j-1} + \Delta \theta_{j-1} \Phi_l \right]},$$

$$\frac{D_{l5}^{\Phi}}{\Delta z_h} = \frac{2(\Phi_l)(\Phi_l)_{k+1}}{\Delta z (\Phi_l)_{k+1} + \Delta z_{k+1} \Phi_l},$$

$$\frac{D_{l6}^{\Phi}}{\Delta z_{h,k-1}} = \frac{2(\Phi_l)(\Phi_l)_{k-1}}{\Delta z (\Phi_l)_{k-1} + \Delta z_{k-1} \Phi_l}.$$

Appendix 13.2 Off-diagonal viscous diffusion terms of the *x* momentum equation

The off-diagonal viscous diffusion terms in the *x* momentum equation are

$$\sum_{m=1}^{6} \beta_m \frac{D_{l,m}^V}{\Delta L_{h,m}} \left(DI\_u_{l,m} - \frac{2}{3}(e)^{m1} DI\_u_{l,m}^b + DI\_vis_{lm}^{uT} \right)$$

$$= \beta_1^* \left\{ q_{x1,12} \frac{\partial u_l}{\partial \eta}\bigg|_1 + q_{x1,13} \frac{\partial u_l}{\partial \zeta}\bigg|_1 + q_{x1,22} \frac{\partial v_l}{\partial \eta}\bigg|_1 + q_{x1,23} \frac{\partial v_l}{\partial \zeta}\bigg|_1 + q_{x1,32} \frac{\partial w_l}{\partial \eta}\bigg|_1 + q_{x1,33} \frac{\partial w_l}{\partial \zeta}\bigg|_1 \right\}$$

$$+ \beta_2^* \left\{ q_{x2,12} \frac{\partial u_l}{\partial \eta}\bigg|_2 + q_{x2,13} \frac{u_l}{\partial \zeta}\bigg|_2 q_{x2,22} \frac{\partial v_l}{\partial \eta}\bigg|_2 + q_{x2,23} \frac{\partial v_l}{\partial \zeta}\bigg|_2 + q_{x2,32} \frac{\partial w_l}{\partial \eta}\bigg|_2 + q_{x2,33} \frac{\partial w_l}{\partial \zeta}\bigg|_2 \right\}$$

$$+ \beta_3^* \left\{ q_{x3,11} \frac{\partial u_l}{\partial \xi}\bigg|_3 + q_{x3,13} \frac{\partial u_l}{\partial \zeta}\bigg|_3 + q_{x3,21} \frac{\partial v_l}{\partial \xi}\bigg|_3 + q_{x3,23} \frac{\partial v_l}{\partial \zeta}\bigg|_3 + q_{x3,31} \frac{\partial w_l}{\partial \xi}\bigg|_3 + q_{x3,33} \frac{\partial w_l}{\partial \zeta}\bigg|_3 \right\}$$

$$+ \beta_4^* \left\{ q_{x4,11} \frac{\partial u_l}{\partial \xi}\bigg|_4 + q_{x4,13} \frac{\partial u_l}{\partial \zeta}\bigg|_4 + q_{x4,21} \frac{\partial v_l}{\partial \xi}\bigg|_4 + q_{x4,23} \frac{\partial v_l}{\partial \zeta}\bigg|_4 + q_{x4,31} \frac{\partial w_l}{\partial \xi}\bigg|_4 + q_{x4,33} \frac{\partial w_l}{\partial \zeta}\bigg|_4 \right\}$$

$$+ \beta_5^* \left\{ q_{x5,11} \frac{\partial u_l}{\partial \xi}\bigg|_5 + q_{x5,12} \frac{\partial u_l}{\partial \eta}\bigg|_5 + q_{x5,21} \frac{\partial v_l}{\partial \xi}\bigg|_5 + q_{x5,22} \frac{\partial v_l}{\partial \eta}\bigg|_5 + q_{x5,31} \frac{\partial w_l}{\partial \xi}\bigg|_5 + q_{x5,32} \frac{\partial w_l}{\partial \eta}\bigg|_5 \right\}$$

$$+ \beta_6^* \left\{ q_{x6,11} \frac{\partial u_l}{\partial \xi}\bigg|_6 + q_{x6,12} \frac{\partial u_l}{\partial \eta}\bigg|_6 + q_{x6,21} \frac{\partial v_l}{\partial \xi}\bigg|_6 + q_{x6,22} \frac{\partial v_l}{\partial \eta}\bigg|_6 + q_{x6,31} \frac{\partial w_l}{\partial \xi}\bigg|_6 + q_{x6,32} \frac{\partial w_l}{\partial \eta}\bigg|_6 \right\}.$$

Here the coefficients

$$\beta_m^* = \beta_m = \frac{D_{l,m}^V}{\Delta L_{h,m}} \frac{\overline{\Delta V}_m}{S_m}$$

are used also in the other momentum equations. The following 36 coefficients are functions of the geometry only.

$$q_{x1,12} = \frac{4}{3}(e)^{11}(a^{21})_1 + (e)^{12}(a^{22})_1 + (e)^{13}(a^{23})_1 = d_{12} + \frac{1}{3}(e)^{11}(a^{21})_1,$$

$$q_{x1,13} = \frac{4}{3}(e)^{11}(a^{31})_1 + (e)^{12}(a^{32})_1 + (e)^{13}(a^{33})_1 = d_{13} + \frac{1}{3}(e)^{11}(a^{31})_1,$$

$$q_{x1,22} = (e)^{12}(a^{21})_1 - \frac{2}{3}(e)^{11}(a^{22})_1,$$

$$q_{x1,23} = (e)^{12}(a^{31})_1 - \frac{2}{3}(e)^{11}(a^{32})_1,$$

$$q_{x1,32} = (e)^{13}(a^{21})_1 - \frac{2}{3}(e)^{11}(a^{23})_1,$$

$$q_{x1,33} = (e)^{13}(a^{31})_1 - \frac{2}{3}(e)^{11}(a^{33})_1,$$

$$q_{x2,12} = \frac{4}{3}(e)^{21}(a^{21})_2 + (e)^{22}(a^{22})_2 + (e)^{23}(a^{23})_2 = d_{22} + \frac{1}{3}(e)^{21}(a^{21})_2$$

$$= -(q_{x1,12})_{i-1},$$

$$q_{x2,13} = \frac{4}{3}(e)^{21}(a^{31})_2 + (e)^{22}(a^{32})_2 + (e)^{23}(a^{33})_2 = d_{23} + \frac{1}{3}(e)^{21}(a^{31})_2$$

$$= -(q_{x1,13})_{i-1},$$

$$q_{x2,22} = (e)^{22}(a^{21})_2 - \frac{2}{3}(e)^{21}(a^{22})_2 = -(q_{x1,22})_{i-1},$$

$$q_{x2,23} = (e)^{22}(a^{31})_2 - \frac{2}{3}(e)^{21}(a^{32})_2 = -(q_{x1,23})_{i-1},$$

$$q_{x2,32} = (e)^{23}(a^{21})_2 - \frac{2}{3}(e)^{21}(a^{23})_2 = -(q_{x1,32})_{i-1},$$

$$q_{x2,33} = (e)^{23}(a^{31})_2 - \frac{2}{3}(e)^{21}(a^{33})_2 = -(q_{x1,33})_{i-1},$$

$$q_{x3,11} = \frac{4}{3}(e)^{31}(a^{11})_3 + (e)^{32}(a^{12})_3 + (e)^{33}(a^{13})_3 = d_{31} + \frac{1}{3}(e)^{31}(a^{11})_3,$$

$$q_{x3,13} = \frac{4}{3}(e)^{31}(a^{31})_3 + (e)^{32}(a^{32})_3 + (e)^{33}(a^{33})_3 = d_{33} + \frac{1}{3}(e)^{31}(a^{31})_3,$$

$$q_{x3,21} = (e)^{32}(a^{11})_3 - \frac{2}{3}(e)^{31}(a^{12})_3,$$

$$q_{x3,23} = (e)^{32}(a^{31})_3 - \frac{2}{3}(e)^{31}(a^{32})_3,$$

$$q_{x3,31} = (e)^{33}(a^{11})_3 - \frac{2}{3}(e)^{31}(a^{13})_3,$$

$$q_{x3,33} = (e)^{33}(a^{31})_3 - \frac{2}{3}(e)^{31}(a^{33})_3,$$

$$q_{x4,11} = \frac{4}{3}(e)^{41}(a^{11})_4 + (e)^{42}(a^{12})_4 + (e)^{43}(a^{13})_4 = d_{41} + \frac{1}{3}(e)^{41}(a^{11})_4$$

$$= -(q_{x3,11})_{j-1},$$

$$q_{x4,13} = \frac{4}{3}(e)^{41}(a^{31})_4 + (e)^{42}(a^{32})_4 + (e)^{43}(a^{33})_4 = d_{43} + \frac{1}{3}(e)^{41}(a^{31})_4$$

$$= -(q_{x3,13})_{j-1},$$

$$q_{x4,21} = (e)^{42}(a^{11})_4 - \frac{2}{3}(e)^{41}(a^{12})_4 = -(q_{x3,21})_{j-1},$$

$$q_{x4,23} = (e)^{42}(a^{31})_4 - \frac{2}{3}(e)^{41}(a^{32})_4 = -(q_{x3,23})_{j-1},$$

$$q_{x4,31} = (e)^{43}(a^{11})_4 - \frac{2}{3}(e)^{41}(a^{13})_4 = -(q_{x3,31})_{j-1},$$

$$q_{x4,33} = (e)^{43}(a^{31})_4 - \frac{2}{3}(e)^{41}(a^{33})_4 = -(q_{x3,33})_{j-1},$$

$$q_{x5,11} = \frac{4}{3}(e)^{51}(a^{11})_5 + (e)^{52}(a^{12})_5 + (e)^{53}(a^{13})_5 = d_{51} + \frac{1}{3}(e)^{51}(a^{11})_5,$$

$$q_{x5,12} = \frac{4}{3}(e)^{51}(a^{21})_5 + (e)^{52}(a^{22})_5 + (e)^{53}(a^{23})_5 = d_{52} + \frac{1}{3}(e)^{51}(a^{21})_5,$$

$$q_{x5,21} = (e)^{52}(a^{11})_5 - \frac{2}{3}(e)^{51}(a^{12})_5,$$

$$q_{x5,22} = (e)^{52}(a^{21})_5 - \frac{2}{3}(e)^{51}(a^{22})_5,$$

$$q_{x5,31} = (e)^{53}(a^{11})_5 - \frac{2}{3}(e)^{51}(a^{13})_5,$$

$$q_{x5,32} = (e)^{53}(a^{21})_5 - \frac{2}{3}(e)^{51}(a^{23})_5,$$

$$q_{x6,11} = \frac{4}{3}(e)^{61}(a^{11})_6 + (e)^{62}(a^{12})_6 + (e)^{63}(a^{13})_6 = d_{61} + \frac{1}{3}(e)^{61}(a^{11})_6$$

$$= -(q_{x5,11})_{k-1},$$

$$q_{x6,12} = \frac{4}{3}(e)^{61}(a^{21})_6 + (e)^{62}(a^{22})_6 + (e)^{63}(a^{23})_6 = d_{62} + \frac{1}{3}(e)^{61}(a^{21})_6$$

$$= -(q_{x5,12})_{k-1},$$

$$q_{x6,21} = (e)^{62}(a^{11})_6 - \frac{2}{3}(e)^{61}(a^{12})_6 = -(q_{x5,21})_{k-1},$$

$$q_{x6,22} = (e)^{62}(a^{21})_6 - \frac{2}{3}(e)^{61}(a^{22})_6 = -(q_{x5,22})_{k-1},$$

$$q_{x6,31} = (e)^{63}(a^{11})_6 - \frac{2}{3}(e)^{61}(a^{13})_6 = -(q_{x5,31})_{k-1},$$

$$q_{x6,32} = (e)^{63}(a^{21})_6 - \frac{2}{3}(e)^{61}(a^{23})_6 = -(q_{x5,32})_{k-1}.$$

Appendix 13.3 Off-diagonal viscous diffusion terms of the *y* momentum equation

The off-diagonal viscous diffusion terms in the *y* momentum equation

$$\sum_{m=1}^{6} \beta_m \frac{D_{l,m}^V}{\Delta L_{h,m}} \left(DI\_v_{l,m} - \frac{2}{3}(e)^{m2} DI\_u_{l,m}^b + DI\_vis_{lm}^{vT} \right)$$

are computed using the same procedure as those for *x* equation replacing simply the subscript *x* with *y* and using the following geometry coefficients.

$$q_{y1,12} = (e)^{11}(a^{22})_1 - \frac{2}{3}(e)^{12}(a^{21})_1,$$

$$q_{y1,13} = (e)^{11}(a^{32})_1 - \frac{2}{3}(e)^{12}(a^{31})_1,$$

$$q_{y1,22} = (e)^{11}(a^{21})_1 + \frac{4}{3}(e)^{12}(a^{22})_1 + (e)^{13}(a^{23})_1,$$

$$q_{y1,23} = (e)^{11}(a^{31})_1 + \frac{4}{3}(e)^{12}(a^{32})_1 + (e)^{13}(a^{33})_1,$$

$$q_{y1,32} = (e)^{13}(a^{22})_1 - \frac{2}{3}(e)^{12}(a^{23})_1,$$

$$q_{y1,33} = (e)^{13}\left(a^{32}\right)_1 - \frac{2}{3}(e)^{12}\left(a^{33}\right)_1,$$

$$q_{y2,12} = (e)^{21}\left(a^{22}\right)_2 - \frac{2}{3}(e)^{22}\left(a^{21}\right)_2 = -\left(q_{y1,12}\right)_{i-1},$$

$$q_{y2,13} = (e)^{21}\left(a^{32}\right)_2 - \frac{2}{3}(e)^{22}\left(a^{31}\right)_2 = -\left(q_{y1,13}\right)_{i-1},$$

$$q_{y2,22} = (e)^{21}\left(a^{21}\right)_2 + \frac{4}{3}(e)^{22}\left(a^{22}\right)_2 + (e)^{23}\left(a^{23}\right)_2 = -\left(q_{y1,22}\right)_{i-1},$$

$$q_{y2,23} = (e)^{21}\left(a^{31}\right)_2 + \frac{4}{3}(e)^{22}\left(a^{32}\right)_2 + (e)^{23}\left(a^{33}\right)_2 = -\left(q_{y1,23}\right)_{i-1},$$

$$q_{y2,32} = (e)^{23}\left(a^{22}\right)_2 - \frac{2}{3}(e)^{22}\left(a^{23}\right)_2 = -\left(q_{y1,32}\right)_{i-1},$$

$$q_{y2,33} = (e)^{23}\left(a^{32}\right)_2 - \frac{2}{3}(e)^{22}\left(a^{33}\right)_2 = -\left(q_{y1,33}\right)_{i-1},$$

$$q_{y3,11} = (e)^{31}\left(a^{12}\right)_3 - \frac{2}{3}(e)^{32}\left(a^{11}\right)_3,$$

$$q_{y3,13} = (e)^{31}\left(a^{32}\right)_3 - \frac{2}{3}(e)^{32}\left(a^{31}\right)_3,$$

$$q_{y3,21} = (e)^{31}\left(a^{11}\right)_3 + \frac{4}{3}(e)^{32}\left(a^{12}\right)_3 + (e)^{33}\left(a^{13}\right)_3,$$

$$q_{y3,23} = (e)^{31}\left(a^{31}\right)_3 + \frac{4}{3}(e)^{32}\left(a^{32}\right)_3 + (e)^{33}\left(a^{33}\right)_3,$$

$$q_{y3,31} = (e)^{33}\left(a^{12}\right)_3 - \frac{2}{3}(e)^{32}\left(a^{13}\right)_3,$$

$$q_{y3,33} = (e)^{33}\left(a^{32}\right)_3 - \frac{2}{3}(e)^{32}\left(a^{33}\right)_3,$$

Appendix 13.3 Off-diagonal viscous diffusion terms of the y momentum equation

$$q_{y4,11} = (e)^{41} (a^{12})_4 - \frac{2}{3}(e)^{42} (a^{11})_4 = -(q_{y3,11})_{j-1},$$

$$q_{y4,13} = (e)^{41} (a^{32})_4 - \frac{2}{3}(e)^{42} (a^{31})_4 = -(q_{y3,13})_{j-1},$$

$$q_{y4,21} = (e)^{41} (a^{11})_4 + \frac{4}{3}(e)^{42} (a^{12})_4 + (e)^{43} (a^{13})_4 = -(q_{y3,21})_{j-1},$$

$$q_{y4,23} = (e)^{41} (a^{31})_4 + \frac{4}{3}(e)^{42} (a^{32})_4 + (e)^{43} (a^{33})_4 = -(q_{y3,23})_{j-1},$$

$$q_{y4,31} = (e)^{43} (a^{12})_4 - \frac{2}{3}(e)^{42} (a^{13})_4 = -(q_{y3,31})_{j-1},$$

$$q_{y4,33} = (e)^{43} (a^{32})_4 - \frac{2}{3}(e)^{42} (a^{33})_4 = -(q_{y3,33})_{j-1},$$

$$q_{y5,11} = (e)^{51} (a^{12})_5 - \frac{2}{3}(e)^{52} (a^{11})_5,$$

$$q_{y5,12} = (e)^{51} (a^{22})_5 - \frac{2}{3}(e)^{52} (a^{21})_5,$$

$$q_{y5,21} = (e)^{51} (a^{11})_5 + \frac{4}{3}(e)^{52} (a^{12})_5 + (e)^{53} (a^{13})_5,$$

$$q_{y5,22} = (e)^{51} (a^{21})_5 + \frac{4}{3}(e)^{52} (a^{22})_5 + (e)^{53} (a^{23})_5,$$

$$q_{y5,31} = (e)^{53} (a^{12})_5 - \frac{2}{3}(e)^{52} (a^{13})_5,$$

$$q_{y5,32} = (e)^{53} (a^{22})_5 - \frac{2}{3}(e)^{52} (a^{23})_5,$$

$$q_{y6,11} = (e)^{61}(a^{12})_6 - \frac{2}{3}(e)^{62}(a^{11})_6 = -(q_{y5,11})_{k-1},$$

$$q_{y6,12} = (e)^{61}(a^{22})_6 - \frac{2}{3}(e)^{62}(a^{21})_6 = -(q_{y5,12})_{k-1},$$

$$q_{y6,21} = (e)^{61}(a^{11})_6 + \frac{4}{3}(e)^{62}(a^{12})_6 + (e)^{63}(a^{13})_6 = -(q_{y5,21})_{k-1},$$

$$q_{y6,22} = (e)^{61}(a^{21})_6 + \frac{4}{3}(e)^{62}(a^{22})_6 + (e)^{63}(a^{23})_6 = -(q_{y5,22})_{k-1},$$

$$q_{y6,31} = (e)^{63}(a^{12})_6 - \frac{2}{3}(e)^{62}(a^{13})_6 = -(q_{y5,31})_{k-1},$$

$$q_{y6,32} = (e)^{63}(a^{22})_6 - \frac{2}{3}(e)^{62}(a^{23})_6 = -(q_{y5,32})_{k-1}.$$

Appendix 13.4 Off-diagonal viscous diffusion terms of the z momentum equation

The off-diagonal viscous diffusion terms in the z momentum equation

$$\sum_{m=1}^{6} \beta_m \frac{D_{l,m}^v}{\Delta L_{h,m}} \left(DI\_w_{l,m} - \frac{2}{3}(e)^{m3} DI\_u_{l,m}^b + DI\_vis_{lm}^{wT} \right)$$

are computed using the same procedure as those for x equation replacing simply the subscript x with z and using the following geometry coefficients.

$$q_{z1,12} = (e)^{11}(a^{23})_1 - \frac{2}{3}(e)^{13}(a^{21})_1,$$

$$q_{z1,13} = (e)^{11}(a^{33})_1 - \frac{2}{3}(e)^{13}(a^{31})_1,$$

$$q_{z1,22} = (e)^{12}(a^{23})_1 - \frac{2}{3}(e)^{13}(a^{22})_1,$$

Appendix 13.4 Off-diagonal viscous diffusion terms of the z momentum equation

$$q_{z1,23} = (e)^{12}(a^{33})_1 - \frac{2}{3}(e)^{13}(a^{32})_1,$$

$$q_{z1,32} = (e)^{11}(a^{21})_1 + (e)^{12}(a^{22})_1 + \frac{4}{3}(e)^{13}(a^{23})_1,$$

$$q_{z1,33} = (e)^{11}(a^{31})_1 + (e)^{12}(a^{32})_1 + \frac{4}{3}(e)^{13}(a^{33})_1,$$

$$q_{z2,12} = (e)^{21}(a^{23})_2 - \frac{2}{3}(e)^{23}(a^{21})_2 = -(q_{z1,12})_{i-1},$$

$$q_{z2,13} = (e)^{21}(a^{33})_2 - \frac{2}{3}(e)^{23}(a^{31})_2 = -(q_{z1,13})_{i-1},$$

$$q_{z2,22} = (e)^{22}(a^{23})_2 - \frac{2}{3}(e)^{23}(a^{22})_2 = -(q_{z1,22})_{i-1},$$

$$q_{z2,23} = (e)^{22}(a^{33})_2 - \frac{2}{3}(e)^{23}(a^{32})_2 = -(q_{z1,23})_{i-1},$$

$$q_{z2,32} = (e)^{21}(a^{21})_2 + (e)^{22}(a^{22})_2 + \frac{4}{3}(e)^{23}(a^{23})_2 = -(q_{z1,32})_{i-1},$$

$$q_{z2,33} = (e)^{21}(a^{31})_2 + (e)^{22}(a^{32})_2 + \frac{4}{3}(e)^{23}(a^{33})_2 = -(q_{z1,33})_{i-1},$$

$$q_{z3,11} = (e)^{31}(a^{13})_3 - \frac{2}{3}(e)^{33}(a^{11})_3,$$

$$q_{z3,13} = (e)^{31}(a^{33})_3 - \frac{2}{3}(e)^{33}(a^{31})_3,$$

$$q_{z3,21} = (e)^{32}(a^{13})_3 - \frac{2}{3}(e)^{33}(a^{12})_3,$$

$$q_{z3,23} = (e)^{32}(a^{33})_3 - \frac{2}{3}(e)^{33}(a^{32})_3,$$

$$q_{z3,31} = (e)^{31}(a^{11})_3 + (e)^{32}(a^{12})_3 + \frac{4}{3}(e)^{33}(a^{13})_3,$$

$$q_{z3,33} = (e)^{31}(a^{31})_3 + (e)^{32}(a^{32})_3 + \frac{4}{3}(e)^{33}(a^{33})_3,$$

$$q_{z4,11} = (e)^{41}(a^{13})_4 - \frac{2}{3}(e)^{43}(a^{11})_4 = -(q_{z3,11})_{j-1},$$

$$q_{z4,13} = (e)^{41}(a^{33})_4 - \frac{2}{3}(e)^{43}(a^{31})_4 = -(q_{z3,13})_{j-1},$$

$$q_{z4,21} = (e)^{42}(a^{13})_4 - \frac{2}{3}(e)^{43}(a^{12})_4 = -(q_{z3,21})_{j-1},$$

$$q_{z4,23} = (e)^{42}(a^{33})_4 - \frac{2}{3}(e)^{43}(a^{32})_4 = -(q_{z3,23})_{j-1},$$

$$q_{z4,31} = (e)^{41}(a^{11})_4 + (e)^{42}(a^{12})_4 + \frac{4}{3}(e)^{43}(a^{13})_4 = -(q_{z3,31})_{j-1},$$

$$q_{z4,33} = (e)^{41}(a^{31})_4 + (e)^{42}(a^{32})_4 + \frac{4}{3}(e)^{43}(a^{33})_4 = -(q_{z3,33})_{j-1},$$

$$q_{z5,11} = (e)^{51}(a^{13})_5 - \frac{2}{3}(e)^{53}(a^{11})_5,$$

$$q_{z5,12} = (e)^{51}(a^{23})_5 - \frac{2}{3}(e)^{53}(a^{21})_5,$$

$$q_{z5,21} = (e)^{52}(a^{13})_5 - \frac{2}{3}(e)^{53}(a^{12})_5,$$

$$q_{z5,22} = (e)^{52}(a^{23})_5 - \frac{2}{3}(e)^{53}(a^{22})_5,$$

$$q_{z5,31} = (e)^{51}(a^{11})_5 + (e)^{52}(a^{12})_5 + \frac{4}{3}(e)^{53}(a^{13})_5,$$

$$q_{z5,32} = (e)^{51}(a^{21})_5 + (e)^{52}(a^{22})_5 + \frac{4}{3}(e)^{53}(a^{23})_5,$$

$$q_{z6,11} = (e)^{61}(a^{13})_6 - \frac{2}{3}(e)^{63}(a^{11})_6 = -(q_{z5,11})_{k-1},$$

$$q_{z6,12} = (e)^{61}(a^{23})_6 - \frac{2}{3}(e)^{63}(a^{21})_6 = -(q_{z5,12})_{k-1},$$

$$q_{z6,21} = (e)^{62}(a^{13})_6 - \frac{2}{3}(e)^{63}(a^{12})_6 = -(q_{z5,21})_{k-1},$$

$$q_{z6,22} = (e)^{62}(a^{23})_6 - \frac{2}{3}(e)^{63}(a^{22})_6 = -(q_{z5,22})_{k-1},$$

$$q_{z6,31} = (e)^{61}(a^{11})_6 + (e)^{62}(a^{12})_6 + \frac{4}{3}(e)^{63}(a^{13})_6 = -(q_{z5,31})_{k-1},$$

$$q_{z6,32} = (e)^{61}(a^{21})_6 + (e)^{62}(a^{22})_6 + \frac{4}{3}(e)^{63}(a^{23})_6 = -(q_{z5,32})_{k-1}.$$

References

1. Antal SP et al (2000) Development of a next generation computer code for the prediction of multi-component multiphase flows, Int. Meeting on Trends in Numerical and Physical Modeling for Industrial Multiphase Flow, Cargese, France
2. Brackbill JU, Kothe DB and Zeinach C (1992) A continuum method for modelling surface tension, Journal of Computational Physics, vol 100, p 335
3. Gregor C, Petelin S and Tiselj I (2000) Upgrade of the VOF method for the simulation of the dispersed flow, Proc. of ASME 2000 Fluids Engineering Division Summer Meeting, Boston, Massachusetts, June 11-15
4. Harlow FH and Amsden AA (1975) Numerical calculation of multiphase flow, Journal of Computational Physics vol 17 pp 19-52
5. Hirt CW (1993) Volume-fraction techniques: powerful tools for wind engineering, J. Wind Engineering and Industrial Aerodynamics, vol 46-47, p 327
6. Hou S, Zou Q, Chen S, Doolen G and Cogley AC (1995) Simulation of cavity flow by the lattice Boltzmann method, J. Comput. Phys., vol 118, p 329

7. Jamet D, Lebaigue O, Courtis N and Delhaye (2001) The second gradient method for the direct numerical simulation of liquid-vapor flows with phase change, Jornal of Comp. Physics vol 169 pp 624-651
8. Kolev NI (1994) IVA4: Modeling of mass conservation in multi-phase multi-component flows in heterogeneous porous media. Kerntechnik, vol 59 no 4-5 pp 226-237
9. Kolev NI (1994) The code IVA4: Modelling of momentum conservation in multi-phase multi-component flows in heterogeneous porous media, Kerntechnik, vol 59 no 6 pp 249-258
10. Kolev NI (1995) The code IVA4: Second law of thermodynamics for multi phase flows in heterogeneous porous media, Kerntechnik, vol 60 no 1, pp 1-39
11. Kolev NI (1997) Comments on the entropy concept, Kerntechnik, vol 62 no 1 pp 67-70
12. Kolev NI (1998) On the variety of notation of the energy conservation principle for single phase flow, Kerntechnik, vol 63 no 3 pp 145-156
13. Kolev NI (2001) Conservation equations for multi-phase multi-component multi-velocity fields in general curvilinear coordinate systems, Keynote lecture, Proceedings of ASME FEDSM'01, ASME 2001 Fluids Engineering Division Summer Meeting, New Orleans, Louisiana, May 29 - June 1, 2001
14. Kothe DB, Rider WJ, Mosso SJ, Brock JS and Hochstein JI (1996) Volume tracking of Interfaces having surface tension in two and three dimensions. AIAA 96-0859
15. Kumbaro A, Toumi I and Seignole V (April 14-18, 2002) Numerical modeling of two-phase flows using advanced two fluid system, Proc. of ICONE10, 10th Int. Conf. On Nuclear Engineering, Arlington, VA, USA
16. Lahey RT Jr and Drew D (October 3-8, 1999) The analysis of two-phase flow and heat transfer using a multidimensional, four field, two-fluid model, Ninth Int. Top. Meeting on Nuclear Reactor Thermal-Hydraulics (NURETH-9), San Francisco, California
17. Miettinen J and Schmidt H (April 14-18, 2002) CFD analyses for water-air flow with the Euler-Euler two-phase model in the FLUENT4 CFD code, Proc. of ICONE10, 10th Int. Conf. On Nuclear Engineering, Arlington, VA, USA
18. Nakamura T, Tanaka R, Yabe T and Takizawa K (2001) Exactly conservative semi-Lagrangian scheme for multi-dimensional hyperbolic equations with directional splitting technique, J. Comput. Phys., vol 147 p 171
19. Nourgaliev RR, Dinh TN and Sehgal BR (2002) On lattice Boltzmann modelling of phase transition in isothermal non-ideal fluid, Nuclear Engineering and Design, vol 211 pp 153-171
20. Osher S and Fredkiw R (2003) Level set methods and dynamic implicit surfaces, Springer-Verlag New York, Inc.
21. Rider WJ and Kothe DB (1998) Reconstructing volume tracking. Journal of Computational Physics, vol 141 p 112
22. Staedke H, Franchello G and Worth B (June 8-12, 1998) Towards a high resolution numerical simulation of transient two-phase flow, Third International Conference on Multi-Phase, ICMF'98
23. Sussman M, Smereka P and Oslier S (1994) A level set approach for computing solutions to incompressible two-phase flow, Journal of Computational Physics, vol 114 p 146
24. Swthian JA (1996) Level set methods, Cambridge University Press
25. Takewaki H, Nishiguchi A and Yabe T, (1985) The Cubic-Interpolated Pseudo Particle (CIP) Method for Solving Hyperbolic-Type Equations, J. Comput. Phys., vol 61 p 261

26. Takewaki H and Yabe Y (1987) Cubic-Interpolated Pseudo Particle (CIP) Method - Application to Nonlinear Multi-Dimensional Problems, J. Cornput. Phys., vol 70 p 355
27. Tomiyama A et al. (2000) (N+2)-Field modelling of dispersed multiphase flow, Proc. of ASME 2000 Fluids Engineering Division Summer Meeting, Boston, Massachusetts, June 11-15
28. Toumi I at al (April 2-6, 2000) Development of a multi-dimensional upwind solver for two-phase water/steam flows, Proc. of ICONE 8, 8th Int. Conf. On Nuclear Engineering, Baltimore, MD USA
29. Tryggavson G at al. (2001) A front tracking method for the computations of multiphase flows, Journal of Comp. Physics vol 169 pp 708-759
30. Verschueren M (1999) A difuse-interface model for structure development in flow, PhD Thesis, Technische Universitteit Eindhoven
31. Wijngaarden L (1976) Hydrodynamic interaction between gas bubbles in liquid, J. Fluid Mech., vol 77 pp 27-44
32. Yabe T and Takei E (1988) A New Higher-Order Godunov Method for General Hyperbolic Equations. J.Phys. Soc. Japan, vol 57 p 2598
33. Xiao F and Yabe T (2001) Completely conservative and oscillation-less semi-Lagrangian schemes for advection transportation, J. Comput. Phys., vol 170 p 498
34. Xiao F, Yabe T, Peng X and Kobayashi H, (2002) Conservation and oscillation-less transport schemes based an rational functions, J. Geophys. Res., vol 107, p 4609
35. Xiao F (2003) Profile-modifiable conservative transport schemes and a simple multi integrated moment formulation for hydrodynamics, in Computational Fluid Dynamics 2002, Amfield S, Morgan P and Srinivas K, eds, Springer, p 106
36. Xiao F and Ikebata A (2003) An efficient method for capturing free boundary in multi-fluid simulations, Int. J. Numer. Method in Fluid. vol 42 pp 187-210
37. Yabe T and Wang PY (1991) Unified Numerical Procedure for Compressible and Incompressible Fluid, J. Phys. Soc. Japan vol 60 pp 2105-2108
38. Yabe T and Aoki A (1991) A Universal Solver for Hyperbolic-Equations by Cubic Polynomial Interpolation. I. One-Dimensional Solver, Comput. Phys. Commun., vol 66 p 219
39. Yabe T, Ishikawa T, Wang PY, Aoki T, Kadota Y and Ikeda F (1991) A universal solver for hyperbolic-equations by cubic-polynomial interpolation. II. 2-dimensional and 3-dimensional solvers. Comput. Phys. Commun., vol 66 p 233
40. Yabe T, Xiao F and Utsumi T (2001) Constrained Interpolation Profile Method for Multiphase Analysis, J. Comput. Phys., vol 169 pp 556-593
41. Yabe T, Tanaka R, Nakamura T and Xiao F (2001) Exactly conservative semi-Lagrangian scheme (CIP-CSL) in one dimension. Mon. Wea. Rev., vol. 129 p 332
42. Yabe T, Xiao F and Utsumi T (2001) The constrained interpolation profile method for multiphase analysis, J. Comput. Phys., vol 169 p 556

Appendix 1 Brief introduction to vector analysis

Before starting to study the theory of multi-phase flows it is advisable to refresh your knowledge on vector analysis. My favorite choice is the book "Calculus and Analytic Geometry" by *Thomas* et al. [5]. Of course you may use any text book to this topic also. Here only a brief summary is given in order to assist in the understanding of the next chapters of this book.

Right handed Cartesian coordinate system: To locate points in space, we use three mutually perpendicular coordinate axes (x, y, z) as proposed by *Descartes*. When you hold your right hand so that the fingers curl from the positive *x*-axis toward the positive *y*-axis and your thumb points along the positive *z*-axis the coordinate systems is *right handed*. We use right handed coordinate systems.

Vector (The Gibbs - Heaviside concept from 1870): A *vector* in a space is a directed line segment. Two vectors are *equal* or *the same* if they have the same length and direction.

The vector between two points: The vector from point $P_1(x_1, y_1, z_1)$ to point $P_2(x_2, y_2, z_2)$ is

$$\overrightarrow{P_1P_2} = (x_2 - x_1)\mathbf{i} + (y_2 - y_1)\mathbf{j} + (z_2 - z_1)\mathbf{k}. \tag{1}$$

Midpoints: The position vector of the midpoint M of the line segment joining points $P_1(x_1, y_1, z_1)$ and $P_2(x_1, y_1, z_1)$ is

$$\mathbf{r}_M = \frac{x_2 + x_1}{2}\mathbf{i} + \frac{y_2 + y_1}{2}\mathbf{j} + \frac{z_2 + z_1}{2}\mathbf{k}. \tag{2}$$

Centroid of a triangle: The position vector of the center of mass of a triangle defined by the points $P_0(x_0, y_0, z_0)$, $P_1(x_1, y_1, z_1)$ and $P_2(x_2, y_2, z_2)$ is

$$\mathbf{r}_{cm} = \frac{x_2 + x_1 + x_0}{3}\mathbf{i} + \frac{y_2 + y_1 + y_0}{3}\mathbf{j} + \frac{z_2 + z_1 + z_0}{3}\mathbf{k}. \tag{3}$$

Centroid of the union of non-overlapping plane regions: The Alexandrian Greek *Pappus* knew in the third century that a centroid of the union of two non-overlapping plane regions lies on the line segment joining their individual centroids. More specifically, suppose m_1 and m_2 are the masses of thin plates P_1 and P_2 that cover non overlapping regions in the xy plane. Let $\mathbf{c}_1$ and $\mathbf{c}_2$ be the vectors from the origin to the respective centers of mass of P_1 and P_2. Then the center of mass of the union $P_1 \cup P_2$ of the two plates is determined by the vector

$$\mathbf{r}_{cm} = \frac{m_1 \mathbf{c}_1 + m_2 \mathbf{c}_2}{m_1 + m_2}. \qquad (4)$$

This equation is known as the *Pappus*'s formula. For more than two non-overlapping plates, as long a their number is finite, the formula generalizes to

$$\mathbf{r}_{cm} = \frac{\sum_m m_m \mathbf{c}_m}{\sum_m m_m}. \qquad (5)$$

This formula is especially useful for finding the centroid of a plate of irregular shape that is made up of pieces of constant density whose centroids we know from geometry. We find the centroid of each piece and apply the above equation to find the centroid of the plane.

Magnitude: The *magnitude (length)* of the vector $\mathbf{A} = a_1\mathbf{i} + a_2\mathbf{j} + a_3\mathbf{k}$ is

$$|\mathbf{A}| = |a_1\mathbf{i} + a_2\mathbf{j} + a_3\mathbf{k}| = \sqrt{a_1^2 + a_2^2 + a_3^2}. \qquad (6)$$

Vectors with magnitude equal to one are called *unit vectors*. Unit vectors are built from direction cosines. Vectors with magnitude zero are called *zero vectors* and are denoted with **0**.

Lines and line segments in space: The vector equation for the line through $P_0(x_0, y_0, z_0)$ parallel to a vector $\mathbf{v} = A\mathbf{i} + B\mathbf{j} + C\mathbf{k}$ is

$$\overrightarrow{P_0P} = t\mathbf{v}, \qquad -\infty < t < \infty \qquad (7)$$

or

$$(x - x_0)\mathbf{i} + (y - y_0)\mathbf{j} + (z - z_0)\mathbf{k} = t(A\mathbf{i} + B\mathbf{j} + C\mathbf{k}), \quad -\infty < t < \infty, \qquad (8)$$

where all points P lie on the line and t is a parameter. For line segments t is bounded, e.g. $a < t < b$.

Appendix 1 Brief introduction to vector analysis

Standard parametrization of the line through $P_0(x_0, y_0, z_0)$ **parallel to** $\mathbf{v} = A\mathbf{i} + B\mathbf{j} + C\mathbf{k}$ **:** The standard parametrization of the line through $P_0(x_0, y_0, z_0)$ parallel to the vector $\mathbf{v} = A\mathbf{i} + B\mathbf{j} + C\mathbf{k}$ is

$$x = x_0 + tA, \quad y = y_0 + tB, \quad z = z_0 + tC, \quad -\infty < t < \infty. \tag{9}$$

In other words the points $(x, y, z) = (x_0 + tA, y_0 + tB, z_0 + tC)$ in the interval $-\infty < t < \infty$ make up the line.

Position vector defining a line: The vector

$$\mathbf{r}(x, y, z) = \mathbf{r}(t) = (x_0 + tA)\mathbf{i} + (y_0 + tB)\mathbf{j} + (z_0 + tC)\mathbf{k}, \quad -\infty < t < \infty \tag{10}$$

from the origin to the point $P(x, y, z)$ is the *position vector*. The position vector defining a line crossing the points $P_1(x_1, y_1, z_1)$ and $P_2(x_1, y_1, z_1)$ is

$$\mathbf{r}(x, y, z) = \mathbf{r}(t) = \left[x_1 + t(x_2 - x_1)\right]\mathbf{i} + \left[y_1 + t(y_2 - y_1)\right]\mathbf{j} + \left[z_1 + t(z_2 - z_1)\right]\mathbf{k},$$

$$0 < t < 1. \tag{11}$$

Derivative of a position vector defining a line: At any point t the derivative of the position vector is

$$\frac{d\mathbf{r}}{dt} = \lim_{\Delta t \to 0} \frac{\mathbf{r}(t + \Delta t) - \mathbf{r}(t)}{\Delta t} = A\mathbf{i} + B\mathbf{j} + C\mathbf{k}, \tag{12}$$

which is parallel to the line.

Position vector defining a curve: The vector

$$\mathbf{r}(x, y, z) = \mathbf{r}(t) = f(t)\mathbf{i} + g(t)\mathbf{j} + h(t)\mathbf{k}, \quad -\infty < t < \infty \tag{13}$$

from the origin to the point $P(x, y, z)$ that belongs to the curve is the *position vector* of the curve.

Derivative of a position vector defining a curve: At any point t the derivative of the position vector is the vector

$$\frac{d\mathbf{r}}{dt} = \lim_{\Delta t \to 0} \frac{\mathbf{r}(t+\Delta t) - \mathbf{r}(t)}{\Delta t} = \frac{\partial f}{\partial t}\mathbf{i} + \frac{\partial g}{\partial t}\mathbf{j} + \frac{\partial h}{\partial t}\mathbf{k}, \qquad (14)$$

- see Fig. A1.1.

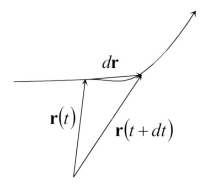

Fig. A1.1. Infinitesimal change of the position vector

The curve traced by **r** is smooth if $d\mathbf{r}/dt$ is continuous and never **0**, i.e. if f, g, and h have continuous first derivatives that are not simultaneously 0. The vector $d\mathbf{r}/dt$, when different from **0**, is also a vector *tangent* to the curve. The tangent line through the point $\left[f(t_0), g(t_0), h(t_0)\right]$ is defined to be a line through the point parallel to $d\mathbf{r}/dt$ at $t = t_0$. The magnitude of the derivative of the position vector defining the curve is

$$\left|\frac{d\mathbf{r}}{dt}\right| = \sqrt{\left(\frac{df}{dt}\right)^2 + \left(\frac{dg}{dt}\right)^2 + \left(\frac{dh}{dt}\right)^2} = \sqrt{\left(\frac{dx}{dt}\right)^2 + \left(\frac{dy}{dt}\right)^2 + \left(\frac{dz}{dt}\right)^2}. \qquad (15)$$

For the case where the position vector defines a line between two points $P_1(x_1, y_1, z_1)$ and $P_2(x_1, y_1, z_1)$ the magnitude of the derivative of the position vector is

$$\left|\frac{d\mathbf{r}}{dt}\right| = \sqrt{(x_2 - x_1)^2 + (y_2 - y_1)^2 + (z_2 - z_1)^2} \qquad (16)$$

the distance between the two points.

Arc length: The length of a smooth curve $\mathbf{r}(t) = f(t)\mathbf{i} + g(t)\mathbf{j} + h(t)\mathbf{k}$, $a < t < b$, that is traced exactly once as t increases from $t = a$ to $t = b$ is

$$L = \int_a^b \sqrt{\left(\frac{df}{dt}\right)^2 + \left(\frac{dg}{dt}\right)^2 + \left(\frac{dh}{dt}\right)^2}\, dt = \int_a^b \sqrt{\left(\frac{dx}{dt}\right)^2 + \left(\frac{dy}{dt}\right)^2 + \left(\frac{dz}{dt}\right)^2}\, dt = \int_a^b \left|\frac{d\mathbf{r}}{dt}\right| dt.$$

(17)

For the case where the position vector defines a line between two points $P_1(x_1, y_1, z_1)$ and $P_2(x_2, y_2, z_2)$ the length of the line segment between the two points is

$$L = \int_0^1 \sqrt{(x_2 - x_1)^2 + (y_2 - y_1)^2 + (z_2 - z_1)^2}\, dt = \sqrt{(x_2 - x_1)^2 + (y_2 - y_1)^2 + (z_2 - z_1)^2}.$$

(18)

Arc length parameter base point $P_0(x_0, y_0, z_0)$: If we choose $P_0(t_0)$ on a smooth curve C parametrized by t, each value of t determines a point $P[x(t), y(t), z(t)]$ on C and a "directed distance"

$$s(t) = \int_{t_0}^t \left|\frac{d\mathbf{r}}{dt}\right| d\tau,$$

(19)

measured along C from the base point. If $t > t_0$, $s(t)$ is the distance from $P_0(t_0)$ to $P(t)$. If $t < t_0$, $s(t)$ is the negative of the distance. Each value of s determines a point on C and parametrizes C with respect to s. We call s an *arc length parameter* for the curve. The parameter's value increases in direction of increasing t. For a smooth curve the derivatives beneath the radical are continuous and the fundamental theorem of calculus tells us that s is a differentiable function of t with derivative

$$\frac{ds}{dt} = \left|\frac{d\mathbf{r}}{dt}\right|.$$

(20)

Note that, while the base point $P_0(t_0)$ plays a role in defining s, it plays no role in defining ds/dt. Note also that $ds/dt > 0$ since the curve is smooth and s increases with increasing t.

Time derivatives, velocity, speed, acceleration, direction of motion If

$$\mathbf{r}(x, y, z) = \mathbf{r}(\tau) = f(\tau)\mathbf{i} + g(\tau)\mathbf{j} + h(\tau)\mathbf{k}$$

(21)

is a position vector of a particle moving along a smooth curve in space, and the components are smooth functions of time τ, then

$$\mathbf{v}(t) = \frac{d\mathbf{r}}{dt} \tag{22}$$

is the particle's *velocity vector*, tangent to a curve. At any time τ, the direction of $\mathbf{v}$ is the *direction of motion*. The magnitude of $\mathbf{v}$ is the particle's speed.

$$\text{Speed} = |\mathbf{v}(\tau)| = \left|\frac{d\mathbf{r}}{d\tau}\right|. \tag{23}$$

The derivative

$$\mathbf{a} = \frac{d\mathbf{v}}{d\tau} = \frac{d^2\mathbf{r}}{d\tau^2}, \tag{24}$$

when it exists, is the particle's *acceleration vector*.

Differentiation rules for time derivatives: Because the derivatives of vector functions may be computed component by component, the rules for differentiating vector functions have the same form as the rules for differentiating scalar functions.

$$\text{Constant Function Rule:} \quad \frac{d\mathbf{C}}{d\tau} = \mathbf{0} \quad \text{(any constant vector } \mathbf{C}\text{).} \tag{25}$$

$$\text{Scalar Multiple Rules:} \quad \frac{d}{d\tau}(c\mathbf{u}) = c\frac{d\mathbf{u}}{d\tau} \quad \text{(any number } c\text{)} \tag{26}$$

$$\frac{d}{d\tau}(f\mathbf{u}) = f\frac{d\mathbf{u}}{d\tau} + \mathbf{u}\frac{df}{d\tau} \quad \text{(any differentiable scalar function).} \tag{27}$$

If $\mathbf{u}$ and $\mathbf{v}$ are differentiable vector functions of t, then

$$\text{Sum Rule:} \quad \frac{d}{d\tau}(\mathbf{u}+\mathbf{v}) = \frac{d\mathbf{u}}{d\tau} + \frac{d\mathbf{v}}{d\tau}. \tag{28}$$

$$\text{Difference Rule:} \quad \frac{d}{d\tau}(\mathbf{u}-\mathbf{v}) = \frac{d\mathbf{u}}{d\tau} - \frac{d\mathbf{v}}{d\tau}. \tag{29}$$

$$\text{Dot Product Rule:} \quad \frac{d}{d\tau}(\mathbf{u}\cdot\mathbf{v}) = \frac{d\mathbf{u}}{d\tau}\cdot\frac{d\mathbf{v}}{d\tau}. \tag{30}$$

Cross Product Rule: $\dfrac{d}{d\tau}(\mathbf{u}\times\mathbf{v}) = \dfrac{d\mathbf{u}}{d\tau}\times\mathbf{v} + \mathbf{u}\times\dfrac{d\mathbf{v}}{d\tau}$ (in that order). (31)

Chain Rule (Short Form): If $\mathbf{r}$ is a differentiable function of τ and τ is a differentiable function of s, then

$$\frac{d\mathbf{r}}{ds} = \frac{d\mathbf{r}}{d\tau}\frac{d\tau}{ds}. \tag{32}$$

Unit tangent vector T: The *unit tangent vector* of a differentiable curve $\mathbf{r}(t)$ is

$$\mathbf{T} = \frac{d\mathbf{r}}{ds} = \frac{d\mathbf{r}/dt}{ds/dt} = \frac{d\mathbf{r}/dt}{|d\mathbf{r}/dt|}, \tag{33}$$

or

$$\mathbf{T} = \frac{d\mathbf{r}}{ds} = \left(\frac{\partial x}{\partial s}\mathbf{i} + \frac{\partial y}{\partial s}\mathbf{j} + \frac{\partial z}{\partial s}\mathbf{k}\right). \tag{34}$$

For the case where the position vector defines a line between two points $P_1(x_1, y_1, z_1)$ and $P_2(x_2, y_2, z_2)$ the unit tangent vector is

$$\mathbf{T} = \frac{d\mathbf{r}}{ds} = \frac{d\mathbf{r}/dt}{ds/dt} = \frac{d\mathbf{r}/dt}{|d\mathbf{r}/dt|} = \frac{(x_2-x_1)\mathbf{i} + (y_2-y_1)\mathbf{j} + (z_2-z_1)\mathbf{k}}{\sqrt{(x_2-x_1)^2 + (y_2-y_1)^2 + (z_2-z_1)^2}}. \tag{35}$$

The unit tangent vector plays an important role in the theory of curvilinear coordinate transformation.

Curvature of space curve: As a particle moves along a smooth curve in the space, $\mathbf{T} = d\mathbf{r}/ds$ turns as the curve bends. Since $\mathbf{T}$ is a unit vector, its length remains constant and only its direction changes as the particle moves along the curve. The rate at which $\mathbf{T}$ turns per unit of the length along the curve is called *curvature*. The traditional symbol for the curvature is the Greek letter kappa, κ. If $\mathbf{T}$ is the tangent vector of a smooth curve, the curvature function of the curve is

$$\kappa = \left|\frac{d\mathbf{T}}{ds}\right|. \tag{36}$$

The principal unit normal vector for space curve: Since $\mathbf{T}$ has a constant length, the vector $d\mathbf{T}/ds$ is orthogonal to $\mathbf{T}$. Therefore, if we divide $d\mathbf{T}/ds$ by its magnitude κ, we obtain a unit vector orthogonal to **T**. At a point where $\kappa \neq 0$, the principal unit normal vector for a curve in the space is

$$\mathbf{N} = \frac{d\mathbf{T}/ds}{|d\mathbf{T}/ds|} = \frac{1}{\kappa}\frac{d\mathbf{T}}{ds}. \tag{37}$$

The vector $d\mathbf{T}/ds$ points in the direction in which **T** turns as the curve bends. Therefore, if we face the direction of increasing length, the vector $d\mathbf{T}/ds$ points toward the right if **T** turns clockwise and towards the left if **T** turns counterclockwise. Because the arc length parameter for a smooth curve is defined with ds/dt positive, $ds/dt = |ds/dt|$, the chain rule gives

$$\mathbf{N} = \frac{d\mathbf{T}/ds}{|d\mathbf{T}/ds|} = \frac{(d\mathbf{T}/dt)(dt/ds)}{|d\mathbf{T}/dt||dt/ds|} = \frac{d\mathbf{T}/dt}{|d\mathbf{T}/dt|}. \tag{38}$$

Scalar (dot) product: The *scalar product* (*dot product*) $\mathbf{A} \cdot \mathbf{B}$ ("A dot B") of vectors

$$\mathbf{A} = a_1\mathbf{i} + a_2\mathbf{j} + a_3\mathbf{k}, \tag{39}$$

and

$$\mathbf{B} = b_1\mathbf{i} + b_2\mathbf{j} + b_3\mathbf{k} \tag{40}$$

is the number

$$\mathbf{A} \cdot \mathbf{B} = |\mathbf{A}||\mathbf{B}|\cos\theta \tag{41}$$

where θ is the angle between **A** and **B**. In words, it is the length of **A** times the length **B** times the cosine of the angle between **A** and **B**. The law of cosines for the triangle whose sides represent **A**, **B** and **C** is $|\mathbf{C}|^2 = |\mathbf{A}|^2 + |\mathbf{B}|^2 - 2|\mathbf{A}||\mathbf{B}|\cos\theta$ is used to obtain

$$\mathbf{A} \cdot \mathbf{B} = a_1b_1 + a_2b_2 + a_3b_3. \tag{42}$$

To find the scalar product of two vectors we multiply their corresponding components and add the results.

$$\mathbf{A} \cdot \mathbf{A} = a_1^2 + a_2^2 + a_3^2 = |\mathbf{A}|^2 \tag{43}$$

is called the *Euclidian norm*.

Splitting a vector into components parallel and orthogonal to another vector: The vector **B** can be split into a component which is parallel to **A**

$$proj_A \mathbf{B} = \left(\frac{\mathbf{B} \cdot \mathbf{A}}{\mathbf{A} \cdot \mathbf{A}} \right) \mathbf{A} \tag{44}$$

and a component orthogonal to **A**, $\mathbf{B} - proj_A \mathbf{B}$,

$$\mathbf{B} = \left(\frac{\mathbf{B} \cdot \mathbf{A}}{\mathbf{A} \cdot \mathbf{A}} \right) \mathbf{A} + \left[\mathbf{B} - \left(\frac{\mathbf{B} \cdot \mathbf{A}}{\mathbf{A} \cdot \mathbf{A}} \right) \mathbf{A} \right]. \tag{45}$$

Equations for a plane in space: Suppose a plane M passes through a point $P_0(x_0, y_0, z_0)$ and is normal (perpendicular) to the non-zero vector

$$\mathbf{n} = A\mathbf{i} + B\mathbf{j} + C\mathbf{k}. \tag{46}$$

Then M is the set of all points $P(x, y, z)$ for which $\overrightarrow{P_0P}$ is orthogonal to **n**. That is, P lies on M if and only if

$$\mathbf{n} \cdot \overrightarrow{P_0P} = 0. \tag{47}$$

This equation is equivalent to

$$(A\mathbf{i} + B\mathbf{j} + C\mathbf{k}) \cdot \left[(x - x_0)\mathbf{i} + (y - y_0)\mathbf{j} + (z - z_0)\mathbf{k} \right] = 0 \tag{48}$$

or

$$A(x - x_0) + B(y - y_0) + C(z - z_0) = 0. \tag{49}$$

Note that the angle between two intersecting planes is defined to be the (acute) angle determined by their normal vectors.

A plane determined by three points: Consider a plane determined by the three points $P_0(x_0, y_0, z_0)$, $P_1(x_1, y_1, z_1)$ and $P_2(x_2, y_2, z_2)$. The vector connecting point P_0 with point P_1 is then

$$\overrightarrow{P_0P_1} = (x_1 - x_0)\mathbf{i} + (y_1 - y_0)\mathbf{j} + (z_1 - z_0)\mathbf{k}. \tag{50}$$

The vector connecting point P_0 with point P_2 is then

$$\vec{P_0P_2} = (x_2 - x_0)\mathbf{i} + (y_2 - y_0)\mathbf{j} + (z_2 - z_0)\mathbf{k} . \tag{51}$$

The vector normal to the plane is defined by

$$\vec{P_0P_1} \times \vec{P_0P_2} = \begin{vmatrix} \mathbf{i} & \mathbf{j} & \mathbf{k} \\ x_1 - x_0 & y_1 - y_0 & z_1 - z_0 \\ x_2 - x_0 & y_2 - y_0 & z_2 - z_0 \end{vmatrix} . \tag{52}$$

The plane is then defined by

$$\begin{vmatrix} \mathbf{i} & \mathbf{j} & \mathbf{k} \\ x_1 - x_0 & y_1 - y_0 & z_1 - z_0 \\ x_2 - x_0 & y_2 - y_0 & z_2 - z_0 \end{vmatrix} \cdot \left[(x - x_0)\mathbf{i} + (y - y_0)\mathbf{j} + (z - z_0)\mathbf{k} \right] = 0, \tag{53}$$

or

$$\begin{vmatrix} x_0 - x & y_0 - y & z_0 - z \\ x_1 - x & y_1 - y & z_1 - z \\ x_2 - x & y_2 - y & z_2 - z \end{vmatrix} = 0 . \tag{54}$$

Laws of the dot product: The dot product is commutative

$$\mathbf{A} \cdot \mathbf{B} = \mathbf{B} \cdot \mathbf{A} . \tag{55}$$

If c is any number (or scalar), then

$$(c\mathbf{A}) \cdot \mathbf{B} = \mathbf{A} \cdot (c\mathbf{B}) = c(\mathbf{A} \cdot \mathbf{B}) . \tag{56}$$

If $\mathbf{C} = c_1\mathbf{i} + c_2\mathbf{j} + c_3\mathbf{k}$ is any third vector, then

$$\mathbf{A} \cdot (\mathbf{B} + \mathbf{C}) = \mathbf{A} \cdot \mathbf{B} + \mathbf{A} \cdot \mathbf{C} , \tag{57}$$

that is the dot product obeys the *distributive low*. Combined with the commutative law it is also evident that

$$(\mathbf{A} + \mathbf{B}) \cdot \mathbf{C} = \mathbf{A} \cdot \mathbf{C} + \mathbf{B} \cdot \mathbf{C} . \tag{58}$$

The last two equations permit us to multiply sums of vectors by the familiar laws of algebra. For example

$$(\mathbf{A}+\mathbf{B})\cdot(\mathbf{C}+\mathbf{D}) = \mathbf{A}\cdot\mathbf{C} + \mathbf{A}\cdot\mathbf{D} + \mathbf{B}\cdot\mathbf{C} + \mathbf{B}\cdot\mathbf{D}. \tag{59}$$

Angle between two non-zero vectors: The angle between two non-zero vectors $\mathbf{A}$ and $\mathbf{B}$ is

$$\theta = \arccos\left(\frac{\mathbf{A}\cdot\mathbf{B}}{|\mathbf{A}||\mathbf{B}|}\right). \tag{60}$$

Perpendicular (orthogonal) vectors: Non-zero vectors $\mathbf{A}$ and $\mathbf{B}$ are *perpendicular* (*orthogonal*) if and only if

$$\mathbf{A}\cdot\mathbf{B} = 0. \tag{61}$$

The gradient vector (gradient): The *gradient vector* (*gradient*) of the differentiable function $f(x,y,z)$ at point $P_0(x_0, y_0, z_0)$ is the vector

$$\nabla f = \frac{\partial f}{\partial x}\mathbf{i} + \frac{\partial f}{\partial y}\mathbf{j} + \frac{\partial f}{\partial z}\mathbf{k} \tag{62}$$

obtained by evaluating the partial derivatives of f at $P_0(x_0, y_0, z_0)$. The notation ∇f is read "grad f" as well as "gradient of f" and "del f". The symbol ∇ by itself is read "del". Another notation for the gradient is grad f, read the way it is written.

Equations for tangent planes and normal lines: If

$$\mathbf{r}(x,y,z) = \mathbf{r}(t) = g(t)\mathbf{i} + h(t)\mathbf{j} + k(t)\mathbf{k}$$

is a smooth curve on the level surface $f(x,y,z)=c$ of a differentiable function f, then $f[g(t), h(t), k(t)] = c$. Differentiating both sides of this equation with respect to t leads to

$$\frac{d}{dt}f[g(t), h(t), k(t)] = \frac{dc}{dt} \tag{63}$$

$$\frac{df}{dt} = \frac{\partial f}{\partial x}\frac{dg}{dt} + \frac{\partial f}{\partial y}\frac{dh}{dt} + \frac{\partial f}{\partial z}\frac{dk}{dt} = \left(\frac{\partial f}{\partial x}\mathbf{i} + \frac{\partial f}{\partial y}\mathbf{j} + \frac{\partial f}{\partial z}\mathbf{k}\right)\cdot\left(\frac{\partial g}{\partial t}\mathbf{i} + \frac{\partial h}{\partial t}\mathbf{j} + \frac{\partial k}{\partial t}\mathbf{k}\right)$$

$$= \nabla f \cdot \frac{d\mathbf{r}}{dt} = 0. \tag{64}$$

At every point along the curve, ∇f is orthogonal to the curve's velocity vector $\dfrac{d\mathbf{r}}{dt}$, and therefore orthogonal to the surface $f(x,y,z) = c$. The unit normal vector to this surface is

$$\mathbf{n} = \frac{\nabla f}{|\nabla f|}. \tag{65}$$

If the curve passes through the point $P_0(x_0, y_0, z_0)$:

The *tangent plane* at the point P_0 on the level surface $f(x,y,z) = c$ is the plane through P_0 normal to $\nabla f\big|_{P_0}$,

$$\left(\frac{\partial f}{\partial x}\mathbf{i} + \frac{\partial f}{\partial y}\mathbf{j} + \frac{\partial f}{\partial z}\mathbf{k} \right) \cdot \left[(x-x_0)\mathbf{i} + (y-y_0)\mathbf{j} + (z-z_0)\mathbf{k} \right] = 0; \tag{66}$$

The *normal line* of the surface at P_0 is the line through P_0 parallel to $\nabla f\big|_{P_0}$.

Gradient of a plane defined by three points: Consider a plane determined by the three points $P_0(x_0, y_0, z_0)$, $P_1(x_1, y_1, z_1)$ and $P_2(x_2, y_2, z_2)$

$$f(x,y,z) = \begin{vmatrix} x_0 - x & y_0 - y & z_0 - z \\ x_1 - x & y_1 - y & z_1 - z \\ x_2 - x & y_2 - y & z_2 - z \end{vmatrix}$$

$$= (x_0 - x)(y_1 - y)(z_2 - z) + (x_2 - x)(y_0 - y)(z_1 - z) + (x_1 - x)(y_2 - y)(z_0 - z)$$

$$- (x_2 - x)(y_1 - y)(z_0 - z) - (x_1 - x)(y_0 - y)(z_2 - z) - (x_0 - x)(y_2 - y)(z_1 - z) = 0 \tag{67}$$

The *gradient vector* (gradient) of the differentiable function $f(x,y,z)$ at point $P(x,y,z)$ is the vector

$$\nabla f = \frac{\partial f}{\partial x}\mathbf{i} + \frac{\partial f}{\partial y}\mathbf{j} + \frac{\partial f}{\partial z}\mathbf{k}, \tag{68}$$

with components

$$\frac{\partial f}{\partial x} = (y_1 - y_2)(z_0 - z) + (y_2 - y_0)(z_1 - z) + (y_0 - y_1)(z_2 - z), \tag{69}$$

$$\frac{\partial f}{\partial y} = (z_1 - z_2)(x_0 - x) + (z_2 - z_0)(x_1 - x) + (z_0 - z_1)(x_2 - x), \tag{70}$$

$$\frac{\partial f}{\partial z} = (x_1 - x_2)(y_0 - y) + (x_2 - x_0)(y_1 - y) + (x_0 - x_1)(y_2 - y). \tag{71}$$

Note that the vectors $\overrightarrow{P_0P_1}$, $\overrightarrow{P_0P_2}$, and the vector normal to the plane forms a left oriented vector system. The gradient vector is therefore parallel to the surface normal vector.

Curvature of a surface: Consider the smooth curves

$$\mathbf{r}_1(x, y, z) = \mathbf{r}_1(t) = g_1(t)\mathbf{i} + h_1(t)\mathbf{j} + k_1(t)\mathbf{k}, \tag{72}$$

$$\mathbf{r}_2(x, y, z) = \mathbf{r}_2(t) = g_2(t)\mathbf{i} + h_2(t)\mathbf{j} + k_2(t)\mathbf{k}, \tag{73}$$

on the level surface $f(x, y, z) = c$ of a differentiable function f. In this case $f[g_1(t), h_1(t), k_1(t)] = c$ and $f[g_2(t), h_2(t), k_2(t)] = c$ along the curves. The curves crosses the point $P_0(x_0, y_0, z_0)$. At this point they are orthogonal. The tangents at P_0 of these curves are $\mathbf{T}_1$ and $\mathbf{T}_2$. s_1 and s_2 are the arc distances counted from P_0. As a particle moves along the first curve in the space, $\mathbf{T}_1 = d\mathbf{r}_1 / ds_1$ turns as the curve bends. Since $\mathbf{T}_1$ is a unit vector, its length remains constant and only its direction changes as the particle moves along the curve. The rate at which $\mathbf{T}_1$ turns per unit of the length along the curve is called the *curvature* κ_1. As $\mathbf{T}_1$ is the tangent vector of a smooth curve, the curvature function of the curve is

$$\kappa_1 = \left| \frac{d\mathbf{T}_1}{ds_1} \right| = \left| (\mathbf{T}_1 \cdot \nabla) \mathbf{T}_1 \right|. \tag{74}$$

Similarly, the curvature function of the second curve is

$$\kappa_2 = \left|\frac{d\mathbf{T}_2}{ds_2}\right| = \left|(\mathbf{T}_2 \cdot \nabla)\mathbf{T}_2\right|. \tag{75}$$

e

Both curvature functions are called *principle curvatures*. Their reciprocals are called *principal radii*. The *curvature function of the level surface* $f(x,y,z) = c$ is defined as

$$\kappa = \kappa_1 + \kappa_2. \tag{76}$$

The gradient along a direction normal and tangential to a surface: Consider the level surface $f(x,y,z) = c$ of a differentiable function f with a normal unit vector

$$\mathbf{n} = \frac{\nabla f}{|\nabla f|}. \tag{77}$$

The gradient of any differentiable scalar function in space can then be split into a component parallel to the normal vector and a component tangential to the surface:

$$\nabla = \left(\frac{\nabla \cdot \mathbf{n}}{\mathbf{n} \cdot \mathbf{n}}\right)\mathbf{n} + \left[\nabla - \left(\frac{\nabla \cdot \mathbf{n}}{\mathbf{n} \cdot \mathbf{n}}\right)\mathbf{n}\right] = \nabla_n + \nabla_t, \tag{78}$$

where

$$\nabla_n = (\nabla \cdot \mathbf{n})\mathbf{n} \tag{79}$$

and

$$\nabla_t = \nabla - \nabla_n. \tag{80}$$

Curvature of a surface defined by the gradient of its unit normal vector: Consider a smooth surface in space defined by its normal vector

$$\mathbf{n}(x,y,z) = g(x,y,z)\mathbf{i} + h(x,y,z)\mathbf{j} + k(x,y,z)\mathbf{k}. \tag{81}$$

In this case there is a convenient method reported by *Brackbill* et al. [2] for computation of the curvature of this surface

$$\kappa = -(\nabla \cdot \mathbf{n}). \tag{82}$$

Let us estimate as an example the curvature of a liquid layer in stratified flow between two horizontal planes in the plane $y = const$. The interface is described by the curve $z = \delta_2(x)$ where $\delta_2(x)$ is the film thickness. Thus the level surface function is $f(x,y,z) \equiv \delta_2(x) - z = 0$. The gradient of the level surface function is

$$\nabla f(x, y = const, z) = \frac{\partial f(x,z)}{\partial x}\mathbf{i} + \frac{\partial f(x,z)}{\partial z}\mathbf{k} = \frac{\partial \delta_2(x)}{\partial x}\mathbf{i} - \mathbf{k}, \qquad (83)$$

and the magnitude of the gradient

$$|\nabla f| = \left\{1 + \left[\frac{\partial \delta_2(x)}{\partial x}\right]^2\right\}^{1/2}. \qquad (84)$$

The normal vector is

$$\mathbf{n}_2 = -\frac{\nabla f}{|\nabla f|} = -\left(\frac{\frac{\partial \delta_2(x)}{\partial x}}{\left\{1 + \left[\frac{\partial \delta_2(x)}{\partial x}\right]^2\right\}^{1/2}}\mathbf{i} - \frac{1}{\left\{1 + \left[\frac{\partial \delta_2(x)}{\partial x}\right]^2\right\}^{1/2}}\mathbf{k}\right). \qquad (85)$$

Note that for decreasing film thickness with x the normal vector points outside the liquid. This is the result of selecting $f(x,y,z) \equiv \delta_2(x) - z = 0$ instead of $f(x,y,z) \equiv z - \delta_2(x) = 0$ which is also possible. The curvature is then

$$\kappa_l = -(\nabla \cdot \mathbf{n}_l) = \nabla \cdot \left[\frac{\nabla f}{|\nabla f|}\right] = \nabla \cdot \left(\frac{\frac{\partial \delta_2(x)}{\partial x}}{\left\{1 + \left[\frac{\partial \delta_2(x)}{\partial x}\right]^2\right\}^{1/2}}\mathbf{i} - \frac{1}{\left\{1 + \left[\frac{\partial \delta_2(x)}{\partial x}\right]^2\right\}^{1/2}}\mathbf{k}\right)$$

$$= \frac{\partial}{\partial x}\frac{\frac{\partial \delta_2(x)}{\partial x}}{\left\{1 + \left[\frac{\partial \alpha_2(x)}{\partial x}\right]^2\right\}^{1/2}} - \frac{\partial}{\partial z}\frac{1}{\left\{1 + \left[\frac{\partial \delta_2(x)}{\partial x}\right]^2\right\}^{1/2}}. \qquad (86)$$

Having in mind that

$$\frac{\partial}{\partial z}\frac{1}{\left\{1+\left[\frac{\partial \delta_2(x)}{\partial x}\right]^2\right\}^{1/2}} = 0 \qquad (87)$$

one finally obtain the well-known expression

$$\kappa_2 = \frac{\partial^2 \delta_2(x)}{\partial x^2} \bigg/ \left\{1+\left[\frac{\partial \delta_2(x)}{\partial x}\right]^2\right\}^{3/2}. \qquad (88)$$

Speed of displacement of geometrical surface (Truesdell and Toupin [7] 1960): Consider a set of geometrical surfaces defined by the equation

$$\mathbf{r} = \mathbf{r}(t_1, t_2, \tau), \qquad (89)$$

where t_1 and t_2 are the coordinates of a point on this surface and τ is the time. The velocity of the surface point (t_1, t_2) is defined by

$$\mathbf{V}_\sigma = \left(\frac{\partial \mathbf{r}}{\partial \tau}\right)_{t_1, t_2}. \qquad (90)$$

Expressing the surface equation with

$$f(x, y, z, \tau) = 0 \qquad (91)$$

we have

$$\frac{\partial f}{\partial \tau} + \mathbf{V}_\sigma \cdot \nabla f = 0. \qquad (92)$$

Knowing that the outwards directed unit surface vector is

$$\mathbf{n} = \frac{\nabla f}{|\nabla f|} \qquad (93)$$

the surface velocity can be expressed as

$$\mathbf{V}_\sigma \cdot \mathbf{n} = -\frac{\partial f/\partial \tau}{|\nabla f|}.\qquad(94)$$

Potential function: If the vector **F** is defined on D and

$$\mathbf{F} = \nabla f \qquad(95)$$

for some scalar function on D then f is called a *potential function* for **F**. The important property of the potential functions is derived from the following consideration. Suppose A and B are two points in D and that the curve

$$\mathbf{r}(x,y,z) = \mathbf{r}(t) = f(t)\mathbf{i} + g(t)\mathbf{j} + h(t)\mathbf{k}, \qquad a \le t \le b, \qquad(96)$$

is smooth on D joining A and B. Along the curve, f is a differentiable function of t and

$$\frac{df}{dt} = \frac{\partial f}{\partial x}\frac{dx}{dt} + \frac{\partial f}{\partial y}\frac{dy}{dt} + \frac{\partial f}{\partial z}\frac{dz}{dt} = \left(\frac{\partial f}{\partial x}\mathbf{i} + \frac{\partial f}{\partial y}\mathbf{j} + \frac{\partial f}{\partial z}\mathbf{k}\right) \cdot \left(\frac{\partial x}{\partial t}\mathbf{i} + \frac{\partial y}{\partial t}\mathbf{j} + \frac{\partial z}{\partial t}\mathbf{k}\right)$$

$$= \nabla f \cdot \frac{d\mathbf{r}}{dt} = \mathbf{F} \cdot \frac{d\mathbf{r}}{dt}.\qquad(97)$$

Therefore,

$$\int_C \mathbf{F} \cdot d\mathbf{r} = \int_a^b \mathbf{F} \cdot \frac{d\mathbf{r}}{dt} dt = \int_a^b \frac{df}{dt} dt = f(A) - f(B),\qquad(98)$$

depends only of the values of f at A and B and not on the path between. In other words if

$$\mathbf{F} = M(x,y,z)\mathbf{i} + N(x,y,z)\mathbf{j} + P(x,y,z)\mathbf{k} \qquad(99)$$

is a field whose component functions have continuous first partial derivatives and there is a solution of the equation

$$M(x,y,z)\mathbf{i} + N(x,y,z)\mathbf{j} + P(x,y,z)\mathbf{k} = \frac{\partial f}{\partial x}\mathbf{i} + \frac{\partial f}{\partial y}\mathbf{j} + \frac{\partial f}{\partial z}\mathbf{k},\qquad(100)$$

which means solution of the system

$$\frac{\partial f}{\partial x} = M(x,y,z),\qquad(101)$$

$$\frac{\partial f}{\partial y} = N(x,y,z), \qquad (102)$$

$$\frac{\partial f}{\partial z} = P(x,y,z), \qquad (103)$$

the vector **F** possesses a potential function f and is called *conservative*.

Differential form, exact differential form: The form

$$M(x,y,z)dx + N(x,y,z)dy + P(x,y,z)dz \qquad (104)$$

is called the *differential form*. A differential form is *exact* on a domain D in space if

$$M(x,y,z)dx + N(x,y,z)dy + P(x,y,z)dz = \frac{\partial f}{\partial x}dx + \frac{\partial f}{\partial y}dy + \frac{\partial f}{\partial z}dz \qquad (105)$$

for some scalar function f through D. In this case f is a smooth function of x, y, and z. Obviously the test for the differential form being exact is the same as the test for **F**'s being conservative that is the test for the existence of f. Another way for testing whether the differential form is exact is *Euler's* relations saying that if the differential form is exact the mutual cross derivatives are equal, that is

$$\frac{\partial}{\partial y}\left(\frac{\partial f}{\partial x}\right) = \frac{\partial}{\partial x}\left(\frac{\partial f}{\partial y}\right), \quad \frac{\partial}{\partial z}\left(\frac{\partial f}{\partial x}\right) = \frac{\partial}{\partial x}\left(\frac{\partial f}{\partial z}\right), \text{ and } \frac{\partial}{\partial y}\left(\frac{\partial f}{\partial z}\right) = \frac{\partial}{\partial z}\left(\frac{\partial f}{\partial y}\right). \qquad (106)$$

Directional derivatives: Suppose that the function $f(x,y,z)$ is defined through the region R in the xyz space, that $P_0(x_0, y_0, z_0)$ is a point in R, and that $\mathbf{u} = u_1\mathbf{i} + u_2\mathbf{j} + u_3\mathbf{k}$ is a unit vector. Then the equations

$$x = x_0 + su_1, \quad y = y_0 + su_2, \quad z = z_0 + su_3, \quad -\infty < s < \infty, \qquad (107)$$

parametrize the line through $P_0(x_0, y_0, z_0)$ parallel to **u**. The parameter s measures the arc length from P_0 in the direction **u**. We find the rate of change of f at P_0 in the direction **u** by calculating df/ds at P_0. The derivative of f at P_0 in the direction of the unit vector $\mathbf{u} = u_1\mathbf{i} + u_2\mathbf{j} + u_3\mathbf{k}$ is the number

$$\left(\frac{df}{ds}\right)_{\mathbf{u},P_0} = \lim_{s \to 0} \frac{f(x_0 + su_1, y_0 + su_2, z_0 + su_3) - f(x_0, y_0, z_0)}{s} \qquad (108)$$

provided the limit exists. Useful relations for practical calculations are

$$\left(\frac{df}{ds}\right)_{\mathbf{u},P_0} = \left[\left(\frac{df}{dx}\right)_{P_0}\mathbf{i} + \left(\frac{df}{dy}\right)_{P_0}\mathbf{j} + \left(\frac{df}{dz}\right)_{P_0}\mathbf{k}\right] \cdot (u_1\mathbf{i} + u_2\mathbf{j} + u_3\mathbf{k}) = (\nabla f)_{P_0} \cdot \mathbf{u}$$

$$= |\nabla f| \cdot |\mathbf{u}| \cos\theta . \qquad (109)$$

At any given point, f increases most rapidly in the direction of ∇f and decreases most rapidly in the direction of $-\nabla f$. In any direction orthogonal to ∇f, the derivative is zero.

Derivatives of a function along the unit tangent vector: Suppose that the function $f(x,y,z)$ is defined through the region R in the xyz space, and is differentiable, and that

$$\mathbf{T} = \frac{d\mathbf{r}}{ds} = \left(\frac{\partial x}{\partial s}\mathbf{i} + \frac{\partial y}{\partial s}\mathbf{j} + \frac{\partial z}{\partial s}\mathbf{k}\right) \qquad (110)$$

is a unit tangent vector to the differentiable curve $\mathbf{r}(t)$ through the region R. $P_0(x_0, y_0, z_0)$ is a point at $\mathbf{r}(t)$ in R. Then the *derivative* of f at P_0 along the unit tangent vector is the number

$$\frac{\partial f}{\partial s} = \frac{\partial f}{\partial x}\frac{\partial x}{\partial s} + \frac{\partial f}{\partial y}\frac{\partial y}{\partial s} + \frac{\partial f}{\partial z}\frac{\partial z}{\partial s} = \left(\frac{\partial f}{\partial x}\mathbf{i} + \frac{\partial f}{\partial y}\mathbf{j} + \frac{\partial f}{\partial z}\mathbf{k}\right) \cdot \left(\frac{\partial x}{\partial s}\mathbf{i} + \frac{\partial y}{\partial s}\mathbf{j} + \frac{\partial z}{\partial s}\mathbf{k}\right)$$

$$= (\nabla f) \cdot \frac{\partial \mathbf{r}}{\partial s} = (\nabla f) \cdot \mathbf{T} . \qquad (111)$$

The vector (cross) products: If $\mathbf{A}$ and $\mathbf{B}$ are two non-zero vectors in space which are not parallel, they determine a plane. We select a unit vector $\mathbf{n}$ perpendicular to the plane by the *right-handed rule*. This means we choose $\mathbf{n}$ to be unit (normal) vector that points the way your right thumb points when your fingers curl through the angle θ from $\mathbf{A}$ to $\mathbf{B}$. We then define the *vector product* $\mathbf{A} \times \mathbf{B}$ ("$\mathbf{A}$ cross $\mathbf{B}$") to be the vector

$$\mathbf{A} \times \mathbf{B} = (|\mathbf{A}||\mathbf{B}|\sin\theta)\mathbf{n} . \qquad (112)$$

The vector $\mathbf{A}\times\mathbf{B}$ is orthogonal to both $\mathbf{A}$ and $\mathbf{B}$ because it is a scalar multiple of $\mathbf{n}$. The vector product of $\mathbf{A}$ and $\mathbf{B}$ is often called the cross product of $\mathbf{A}$ and $\mathbf{B}$ because of the cross in the notation $\mathbf{A}\times\mathbf{B}$.

Since the sines of 0 and π are both zero in the definition equation, it makes sense to define the cross product of two parallel non-zero vectors to be $\mathbf{0}$.

If one or both of the vectors $\mathbf{A}$ and $\mathbf{B}$ are zero, we also define $\mathbf{A}\times\mathbf{B}$ to be zero. This way, the cross product of two vectors $\mathbf{A}$ and $\mathbf{B}$ is zero if and only if $\mathbf{A}$ and $\mathbf{B}$ are parallel or one or both of them are zero.

Note that

$$\mathbf{A}\times\mathbf{B} = -(\mathbf{B}\times\mathbf{A}). \tag{113}$$

The determination formula for $\mathbf{A}\times\mathbf{B}$: If $\mathbf{A} = a_1\mathbf{i} + a_2\mathbf{j} + a_3\mathbf{k}$, and $\mathbf{B} = b_1\mathbf{i} + b_2\mathbf{j} + b_3\mathbf{k}$, then

$$\mathbf{A}\times\mathbf{B} = \begin{vmatrix} \mathbf{i} & \mathbf{j} & \mathbf{k} \\ a_1 & a_2 & a_3 \\ b_1 & b_2 & b_3 \end{vmatrix}. \tag{114}$$

As an example consider the expression appearing in the multi-phase fluid mechanics for computation of the lift forces $(\mathbf{V}_l - \mathbf{V}_m)\times(\nabla\times\mathbf{V}_m)$ as a function of the velocity vectors. The result is given in two steps

$$(\nabla\times\mathbf{V}_m) = \begin{vmatrix} \mathbf{i} & \mathbf{j} & \mathbf{k} \\ \dfrac{\partial}{\partial x} & \dfrac{\partial}{\partial y} & \dfrac{\partial}{\partial z} \\ u_m & v_m & w_m \end{vmatrix} = \left(\dfrac{\partial w_m}{\partial y} - \dfrac{\partial v_m}{\partial z}\right)\mathbf{i} - \left(\dfrac{\partial w_m}{\partial x} - \dfrac{\partial u_m}{\partial z}\right)\mathbf{j} + \left(\dfrac{\partial v_m}{\partial x} - \dfrac{\partial u_m}{\partial y}\right)\mathbf{k},$$

$$\tag{115}$$

$$(\mathbf{V}_l - \mathbf{V}_m)\times(\nabla\times\mathbf{V}_m) =$$

$$= \left[(v_l - v_m)\left(\dfrac{\partial v_m}{\partial x} - \dfrac{\partial u_m}{\partial y}\right) + (w_l - w_m)\left(\dfrac{\partial w_m}{\partial x} - \dfrac{\partial u_m}{\partial z}\right)\right]\mathbf{i}$$

$$- \left[(u_l - u_m)\left(\dfrac{\partial v_m}{\partial x} - \dfrac{\partial u_m}{\partial y}\right) - (w_l - w_m)\left(\dfrac{\partial w_m}{\partial y} - \dfrac{\partial v_m}{\partial z}\right)\right]\mathbf{j}$$

$$+\left[-\left(u_l - u_m\right)\left(\frac{\partial w_m}{\partial x} - \frac{\partial u_m}{\partial z}\right) - \left(v_l - v_m\right)\left(\frac{\partial w_m}{\partial y} - \frac{\partial v_m}{\partial z}\right)\right]\mathbf{k}. \tag{116}$$

Parallel vectors: Non-zero vectors **A** and **B** are parallel if and only if

$$\mathbf{A} \times \mathbf{B} = \mathbf{0}. \tag{117}$$

The associative and distributive lows for the cross product: The *scalar distributive law* is

$$(r\mathbf{A}) \times (s\mathbf{B}) = (rs)(\mathbf{A} \times \mathbf{B}), \tag{118}$$

with the special case

$$(-\mathbf{A}) \times \mathbf{B} = \mathbf{A} \times (-\mathbf{B}) = -(\mathbf{A} \times \mathbf{B}). \tag{119}$$

The vector distributive laws are

$$\mathbf{A} \times (\mathbf{B} + \mathbf{C}) = \mathbf{A} \times \mathbf{B} + \mathbf{A} \times \mathbf{C}, \tag{120}$$

$$(\mathbf{B} + \mathbf{C}) \times \mathbf{A} = \mathbf{B} \times \mathbf{A} + \mathbf{C} \times \mathbf{A}. \tag{121}$$

Note the interesting vector identities [6], p. 100 and 101,

$$(\mathbf{A} \times \mathbf{B}) \cdot (\mathbf{C} \times \mathbf{D}) = (\mathbf{A} \cdot \mathbf{C})(\mathbf{B} \cdot \mathbf{D}) - (\mathbf{A} \cdot \mathbf{D})(\mathbf{B} \cdot \mathbf{C}), \tag{122}$$

and

$$\mathbf{A} \times (\mathbf{B} \times \mathbf{C}) = (\mathbf{A} \cdot \mathbf{C})\mathbf{B} - (\mathbf{A} \cdot \mathbf{B})\mathbf{C}. \tag{123}$$

The area of a parallelogram: Because **n** is a unit vector, the magnitude of $\mathbf{A} \times \mathbf{B}$ is

$$|\mathbf{A} \times \mathbf{B}| = |\mathbf{A}||\mathbf{B}||\sin\theta||\mathbf{n}| = |\mathbf{A}||\mathbf{B}||\sin\theta|. \tag{124}$$

This is the area of the parallelogram determined by **A** and **B**, $|\mathbf{A}|$ being the base of the parallelogram and $|\mathbf{B}||\sin\theta|$ the height.

Area of union of two triangle plane regions: Consider four vertices points defined by the position vector

$$\mathbf{r}_i = x_i\mathbf{i} + y_i\mathbf{j} + z_i\mathbf{k},\tag{125}$$

where $i = 5, 6, 8, 7$, see Fig.A1.2.

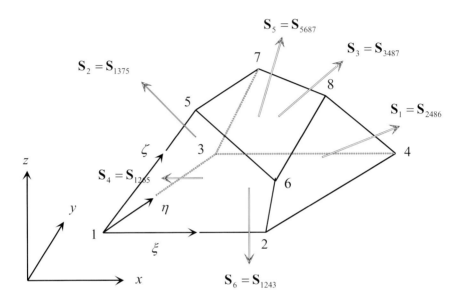

Fig. A1.2. Tetrahedron defined by joint triangles

In general the four points must not belong to the same plane. The surface vector defined by these points is computed as follows. First we compute the surface vectors of the triangles 587 and 568

$$\mathbf{S}_{587} = \frac{1}{2}(\mathbf{r}_5 - \mathbf{r}_7) \times (\mathbf{r}_8 - \mathbf{r}_7),\tag{126}$$

$$\mathbf{S}_{568} = \frac{1}{2}(\mathbf{r}_6 - \mathbf{r}_8) \times (\mathbf{r}_5 - \mathbf{r}_6).\tag{127}$$

The vector sum of the above two vectors yields an expression for the vector surface

$$\mathbf{S}_{5687} = \frac{1}{2}(\mathbf{r}_6 - \mathbf{r}_5) \times (\mathbf{r}_8 - \mathbf{r}_5),\tag{128}$$

Kordulla and *Vinokur* [3]. In fact this is the surface defined by the vector product of the vectors joining the couple of the opposite vertices, respectively. One should note that the first, the second and the resulting vector form a right handed coordinate systems in order to specify the outward direction of the surface if it is part of the control volume surface. Thus, for the hexadron presented in Fig. A1.2 we have

$$S_1 = S_{2486} = \frac{1}{2}(r_4 - r_6) \times (r_8 - r_2),\qquad(129)$$

$$S_2 = S_{1375} = \frac{1}{2}(r_3 - r_5) \times (r_1 - r_7),\qquad(130)$$

$$S_3 = S_{3487} = \frac{1}{2}(r_4 - r_7) \times (r_3 - r_8),\qquad(131)$$

$$S_4 = S_{1265} = \frac{1}{2}(r_6 - r_1) \times (r_5 - r_2),\qquad(132)$$

$$S_5 = S_{5687} = \frac{1}{2}(r_8 - r_5) \times (r_7 - r_6),\qquad(133)$$

$$S_6 = S_{1243} = \frac{1}{2}(r_4 - r_1) \times (r_2 - r_3).\qquad(134)$$

The triple scalar or box product: The product $(A \times B) \cdot C$ is called the triple scalar product of **A**, **B**, and **C** (in that order). As you can see from the formula

$$|(A \times B) \cdot C| = |A \times B||C|\cos\theta,\qquad(135)$$

the absolute value of the product is the volume of the parallelepiped (parallelogram-sided box) determined by **A**, **B**, and **C**. The number $|A \times B|$ is the area of the parallelogram. The number $|C|\cos\theta$ is the parallelogram's height. Because of the geometry $(A \times B) \cdot C$ is called the *box product* of **A**, **B**, and **C**.

By treating the planes of **B** and **C** and of **C** and **A** as the base planes of the parallelepiped determined by **A**, **B**, and **C**, we see that

$$(A \times B) \cdot C = (B \times C) \cdot A = (C \times A) \cdot B.\qquad(136)$$

Since the dot product is commutative, the above equation gives

$$(\mathbf{A} \times \mathbf{B}) \cdot \mathbf{C} = \mathbf{A} \cdot (\mathbf{B} \times \mathbf{C}). \tag{137}$$

The triple scalar product can be evaluated as a determinant

$$(\mathbf{A} \times \mathbf{B}) \cdot \mathbf{C} = \begin{vmatrix} a_1 & a_2 & a_3 \\ b_1 & b_2 & b_3 \\ c_1 & c_2 & c_3 \end{vmatrix}. \tag{138}$$

Volume of a tetrahedron: Consider a tetrahedron defined by the four vertices numbered as shown in Fig. A1.3.

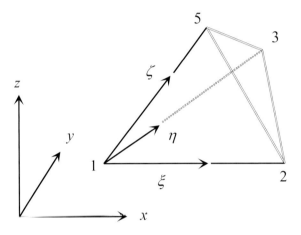

Fig. A1.3. Definition of the vertices numbering of the tetrahedron

The volume of the tetrahedron is then

$$V = \frac{1}{6}(\mathbf{r}_5 - \mathbf{r}_1) \cdot \left[(\mathbf{r}_2 - \mathbf{r}_1) \times (\mathbf{r}_3 - \mathbf{r}_1) \right]. \tag{139}$$

Volume of a hexahedron: Consider a hexahedron defined by the eight vertices numbered starting with base in the counterclockwise direction as shown in Fig. A1.4.

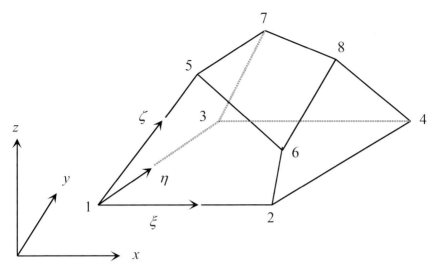

Fig. A1.4. Definition of the vertices numbering of the hexahedron

An alternative approach for computation of the volume of the hexahedron is to compute the volumes of the tetrahedra making up the hexahedron. *Kordulla* and *Vinokur* [3] used symmetric partitioning of the faces decomposing the hexahedron into four tetrahedra and obtained

$$V_{12435687} = -\frac{1}{3}(\mathbf{r}_8 - \mathbf{r}_1) \cdot (\mathbf{S}_{1243} + \mathbf{S}_{1265} + \mathbf{S}_{1375}) = -\frac{1}{3}(\mathbf{r}_8 - \mathbf{r}_1) \cdot (\mathbf{S}_2 + \mathbf{S}_4 + \mathbf{S}_6).$$
(140)

Here the surfaces $\mathbf{S}_{1243}$, $\mathbf{S}_{1265}$, and $\mathbf{S}_{1375}$ belongs to the same vertices and $\mathbf{r}_8 - \mathbf{r}_1$ is the diagonal joining these vertices with the opposite ones.

Dyadic product of two vectors: Consider the vectors $\mathbf{A} = a_1\mathbf{i} + a_2\mathbf{j} + a_3\mathbf{k}$, $\mathbf{B} = b_1\mathbf{i} + b_2\mathbf{j} + b_3\mathbf{k}$. The dyadic product of the two vectors is written as $\mathbf{AB}$ without any sign between them, and is defined as the following second order tensor

$$\mathbf{AB} = \begin{pmatrix} a_1b_1 & a_1b_2 & a_1b_3 \\ a_2b_1 & a_2b_2 & a_2b_3 \\ a_3b_1 & a_3b_2 & a_3b_3 \end{pmatrix}.$$
(141)

Note that the dyadic product of $\mathbf{AB}$ is not equivalent to $\mathbf{BA}$. As examples we give some dyadic products between the vectors $\nabla \cdot = \frac{\partial}{\partial x}\mathbf{i} + \frac{\partial}{\partial y}\mathbf{j} + \frac{\partial}{\partial z}\mathbf{k}$ and

$\mathbf{V} = u\mathbf{i} + v\mathbf{j} + v\mathbf{k}$, which are used in fluid mechanics for definition of the friction forces in flows:

$$\nabla \mathbf{V} = \begin{pmatrix} \frac{\partial u}{\partial x} & \frac{\partial v}{\partial x} & \frac{\partial w}{\partial x} \\ \frac{\partial u}{\partial y} & \frac{\partial v}{\partial y} & \frac{\partial w}{\partial y} \\ \frac{\partial u}{\partial z} & \frac{\partial v}{\partial z} & \frac{\partial w}{\partial z} \end{pmatrix}, \tag{142}$$

$$\left(\nabla \mathbf{V}\right)^T = \begin{pmatrix} \frac{\partial u}{\partial x} & \frac{\partial u}{\partial y} & \frac{\partial u}{\partial z} \\ \frac{\partial v}{\partial x} & \frac{\partial v}{\partial y} & \frac{\partial v}{\partial z} \\ \frac{\partial w}{\partial x} & \frac{\partial w}{\partial y} & \frac{\partial w}{\partial z} \end{pmatrix}. \tag{144}$$

Another dyadic product of the velocity vector

$$\mathbf{V}\mathbf{V} = \begin{pmatrix} uu & uv & uw \\ vu & vv & vw \\ wu & wv & ww \end{pmatrix}, \tag{145}$$

is used to describe the convective momentum transport.

Eigenvalues: The eigenvalues λ_i of a matrix $\mathbf{A}$ are the solution of the characteristic polynomial

$$\left|\mathbf{A} - \lambda\mathbf{I}\right| = \det\left(\mathbf{A} - \lambda\mathbf{I}\right) = 0, \tag{146}$$

where $\mathbf{I}$ is the identity matrix.

Eigenvectors: The right eigenvector of a matrix $\mathbf{A}$ corresponding to a an eigenvalue λ_i of $\mathbf{A}$ is a vector $\mathbf{K}^{(i)} = \left[k_1^{(i)}, k_2^{(i)}, ..., k_m^{(i)}\right]^T$ satisfying

$$\mathbf{A}\mathbf{K}^{(i)} = \lambda_i \mathbf{K}^{(i)}. \tag{147}$$

Similarly, a left eigenvector of a matrix **A** corresponding to a an eigenvalue λ_i of **A** is a vector $\mathbf{L}^{(i)} = \left[k_1^{(i)}, k_2^{(i)}, ..., k_m^{(i)} \right]^T$ satisfying

$$\mathbf{A}\mathbf{L}^{(i)} = \lambda_i \mathbf{L}^{(i)}. \tag{148}$$

Diagonalizable matrix: A matrix **A** is said to be diagonalizable if **A** can be expressed as

$$\mathbf{A} = \mathbf{K}\mathbf{H}\mathbf{K}^{-1}, \tag{149}$$

in terms of a diagonal matrix **H** and a matrix **K**. The diagonal element of **H** are the eigenvalues λ_i of **A** and the columns of **K** are the right eigenvectors of **A** corresponding to the eigenvalues λ_i, that is

$$\mathbf{H} = \begin{bmatrix} \lambda_1 & 0 & . & 0 \\ 0 & \lambda_2 & . & 0 \\ . & . & . & . \\ 0 & 0 & . & \lambda_m \end{bmatrix}, \; \mathbf{K} = \left[\mathbf{K}^{(i)}, \mathbf{K}^{(i)}, ..., \mathbf{K}^{(i)} \right]^T, \; \mathbf{A}\mathbf{K}^{(i)} = \lambda_i \mathbf{K}^{(i)}. \tag{150}$$

Rotation around axes: Given a point (x_0, y_0, z_0). A new position of this point (x, y, z) obtained after rotation by angle φ with respect to the x, y or z axes respectively is

$$\begin{pmatrix} x \\ y \\ z \end{pmatrix} = \mathbf{R}_i \begin{pmatrix} x_0 \\ y_0 \\ z_0 \end{pmatrix}, \tag{151}$$

where

$$\mathbf{R}_1 = \begin{pmatrix} 1 & 0 & 0 \\ 0 & \cos\varphi & -\sin\varphi \\ 0 & \sin\varphi & \cos\varphi \end{pmatrix}, \mathbf{R}_2 = \begin{pmatrix} \cos\varphi & 0 & \sin\varphi \\ 0 & 1 & 0 \\ -\sin\varphi & 0 & \cos\varphi \end{pmatrix}, \mathbf{R}_3 = \begin{pmatrix} \cos\varphi & -\sin\varphi & 0 \\ \sin\varphi & \cos\varphi & 0 \\ 0 & 0 & 1 \end{pmatrix}.$$
$$(152, 153, 154)$$

Scaling: Given a point (x_0, y_0, z_0). A new position of this point (x, y, z) obtained after scaling by

$$\mathbf{S}_1 = \begin{pmatrix} s_x & 0 & 0 \\ 0 & s_y & 0 \\ 0 & 0 & s_z \end{pmatrix}, \tag{155}$$

is

$$\begin{pmatrix} x \\ y \\ z \end{pmatrix} = \mathbf{S}_i \begin{pmatrix} x_0 \\ y_0 \\ z_0 \end{pmatrix}. \tag{156}$$

Translation: Graphical systems make use of the general notation of transformation operation defining by multiplying the coordinates with a 4 by 4 matrix

$$\begin{pmatrix} x \\ y \\ z \\ w \end{pmatrix} = \mathbf{T} \begin{pmatrix} x_0 \\ y_0 \\ z_0 \\ 1 \end{pmatrix},$$

called translation matrix

$$\mathbf{T} = \begin{pmatrix} 1 & 0 & 0 & \Delta x \\ 0 & 1 & 0 & \Delta y \\ 0 & 0 & 1 & \Delta z \\ 0 & 0 & 0 & 1 \end{pmatrix}.$$

w is used for hardware scaling in graphical representations. Replacing the upper 3 by 3 elements with the matrices defining rotation or scaling results in general notation of transformation. Then multiple transformations on objects are simply sequences of matrices multiplications. Note that the round off errors may damage the objects coordinates after many multiplications. Therefore one should design systems by starting the transformations always from a initial state.

References

1. Anderson JD (1995) Computational fluid dynamics, McGraw-Hill, Inc., New York
2. Brackbill JU, Kothe DB, Zemach C (1992) A continuum method for modeling surface tension, Journal of Computational Physics, vol 100 pp 335-354
3. Kordulla W, Vinokur M (1983) AIAA J., vol 21 pp 917-918
4. Peyret R editor (1996) Handbook of computational fluid mechanics, Academic Press, London
5. Thomas GB Jr, Finney RL, Weir MD (1998) Calculus and analytic geometry, 9th Edition, Addison-Wesley Publishing Company, Reading, Massachusetts etc.

6. Thompson JF, Warsi ZUA, Mastin CW (1985) Numerical grid generation, North-Holand, New York-Amsterdam-Oxford
7. Truesdell CA, Toupin RA (1960) The classical field theories, Encyclopedia of Physics, S. Fluge Ed., III/1, Principles of Classical Mechanics and Field Theory, Springer, pp 226-858
8. Vivand H (1974) Conservative forms of gas dynamics equations, Rech. Aerosp., no 1971-1, pp 65-68
9. Vinokur M (1974) Conservation equations of gas dynamics in curvilinear coordinate systems, J. Comput. Phys., vol 14, pp 105-125

Appendix 2 Basics of the coordinate transformation theory

Cartesian coordinate systems are usually the first choice of the scientist and the engineer. Unfortunately in nature and technology the boundaries of a flow may have a very complicated form. Prescribing physically adequate boundary condition at such surfaces is frequently not a solvable problem. The next choice of course is one of the well-known curvilinear orthogonal coordinate systems, polar, or bipolar, or cylindrical.

The purpose of this Appendix is to collect from the literature the most important basics on which the general coordinate transformation theory relies. The material is ordered in a generic way. Each statement follows from the already defined statement or statements. For those who would like to perfect his knowledge in this field there are two important books that are my favorite choices are *Thomas* et al. [5] - about the calculus and analytic geometry, and *Thompson* et al. [7] - about numerical grid generation. For a complete collection of basic definitions, rules and useful formula see [10].

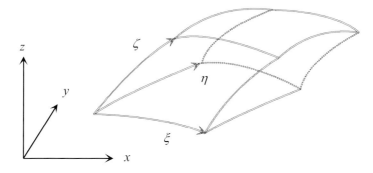

Fig. A2.1. Cartesian and curvilinear coordinate systems

Consider three non identical and non-parallel curves in the space ξ, η, ζ having only one common point designated as an origin and presented in Fig. A2.1. The curves are smooth (al least one times differentiable). We call these curves *coordinate lines* or *curvilinear coordinates*. The curvilinear coordinate lines of a three-dimensional system can be considered also as space curves formed by the intersection of surfaces on which one of the coordinates is hold constant. One coordinate varies along a coordinate line, while the other two are constant thereon. Thus, we have the coordinate transformation defined by

$$x = f(\xi,\eta,\zeta),\tag{1}$$

$$y = g(\xi,\eta,\zeta),\tag{2}$$

$$z = h(\xi,\eta,\zeta).\tag{3}$$

The surfaces defined by $\xi = const$, $\eta = const$, $\zeta = const$, are called coordinate surfaces. We talk then of physical and transformed (computational) space. In the *physical space* the coordinates (x,y,z) become independent variables. In the *transformed space* the transformed computational coordinates (ξ,η,ζ) become independent variables.

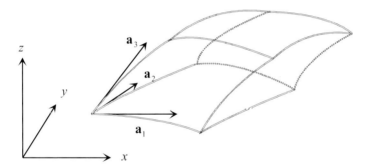

Fig. A2.2. The covariant tangent vectors to the coordinate lines form the so called covariant base vectors of the curvilinear coordinate system

The tangent vectors to the coordinate lines form the so called *base vectors of the coordinate system*. These base vectors are called *covariant base vectors*, Fig. A2.2. The normal vectors to the coordinate surfaces form the so called *contravariant base vectors*. The two types of the base vectors are illustrated in Fig. A2.3, showing an element of volume with six sides, each of which lies on some coordinate surface.

Another frequently used notation is x_i $(i = 1,2,3)$: x_1, x_2, x_3 and ξ^i $(i = 1,2,3)$: ξ^1, ξ^2, ξ^3. In the latter case the superscripts serve only as labels and not as powers.

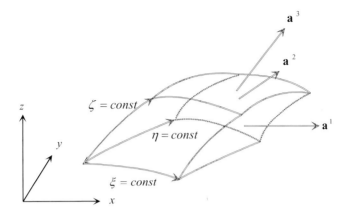

Fig. A2.3. The contravariant normal vectors to the coordinate surfaces form the so called contravariant base vectors of the curvilinear coordinate system

Metrics and inverse metrics: The differential increments in the transformed coordinate system as a function of the differential increments in the physical coordinate system can be computed as follows

$$\begin{pmatrix} d\xi \\ d\eta \\ d\zeta \end{pmatrix} = \begin{pmatrix} \frac{\partial \xi}{\partial x} & \frac{\partial \xi}{\partial y} & \frac{\partial \xi}{\partial z} \\ \frac{\partial \eta}{\partial x} & \frac{\partial \eta}{\partial y} & \frac{\partial \eta}{\partial z} \\ \frac{\partial \zeta}{\partial x} & \frac{\partial \zeta}{\partial y} & \frac{\partial \zeta}{\partial z} \end{pmatrix} \begin{pmatrix} dx \\ dy \\ dz \end{pmatrix}. \qquad (4)$$

The elements of the matrix in Eq. (4) are called *metrics* of the coordinate transformation [1] p.178. An alternative notation of the metrics is a^{ij}. The differential increments in the physical coordinate system are a function of the differential increments in the transformed coordinate system

$$\begin{pmatrix} dx \\ dy \\ dz \end{pmatrix} = \begin{pmatrix} \frac{\partial x}{\partial \xi} & \frac{\partial x}{\partial \eta} & \frac{\partial x}{\partial \zeta} \\ \frac{\partial y}{\partial \xi} & \frac{\partial y}{\partial \eta} & \frac{\partial y}{\partial \zeta} \\ \frac{\partial z}{\partial \xi} & \frac{\partial z}{\partial \eta} & \frac{\partial z}{\partial \zeta} \end{pmatrix} \begin{pmatrix} d\xi \\ d\eta \\ d\zeta \end{pmatrix} \quad \text{or} \quad \begin{pmatrix} dx \\ dy \\ dz \end{pmatrix} = \mathbf{J} \begin{pmatrix} d\xi \\ d\eta \\ d\zeta \end{pmatrix}. \qquad (5)$$

where

$$\mathbf{J}(\xi,\eta,\zeta) = \begin{pmatrix} \dfrac{\partial x}{\partial \xi} & \dfrac{\partial x}{\partial \eta} & \dfrac{\partial x}{\partial \zeta} \\ \dfrac{\partial y}{\partial \xi} & \dfrac{\partial y}{\partial \eta} & \dfrac{\partial y}{\partial \zeta} \\ \dfrac{\partial z}{\partial \xi} & \dfrac{\partial z}{\partial \eta} & \dfrac{\partial z}{\partial \zeta} \end{pmatrix} \qquad (6)$$

is the so called *Jacobian matrix* of the coordinate transformation $x = f(\xi,\eta,\zeta)$, $y = g(\xi,\eta,\zeta)$, $z = h(\xi,\eta,\zeta)$. The elements of the *Jacobian matrix* are called *inverse metrics* of the coordinate transformation [1]. An alternative notation of the inverse metrics is a_{ij}^T. The relation between the so called *metrics* and the inverse metrics is obvious by noting that the transformation is invertable

$$\begin{pmatrix} d\xi \\ d\eta \\ d\zeta \end{pmatrix} = \begin{pmatrix} \dfrac{\partial x}{\partial \xi} & \dfrac{\partial x}{\partial \eta} & \dfrac{\partial x}{\partial \zeta} \\ \dfrac{\partial y}{\partial \xi} & \dfrac{\partial y}{\partial \eta} & \dfrac{\partial y}{\partial \zeta} \\ \dfrac{\partial z}{\partial \xi} & \dfrac{\partial z}{\partial \eta} & \dfrac{\partial z}{\partial \zeta} \end{pmatrix}^{-1} \begin{pmatrix} dx \\ dy \\ dz \end{pmatrix} = \begin{pmatrix} \dfrac{\partial \xi}{\partial x} & \dfrac{\partial \xi}{\partial y} & \dfrac{\partial \xi}{\partial z} \\ \dfrac{\partial \eta}{\partial x} & \dfrac{\partial \eta}{\partial y} & \dfrac{\partial \eta}{\partial z} \\ \dfrac{\partial \zeta}{\partial x} & \dfrac{\partial \zeta}{\partial y} & \dfrac{\partial \zeta}{\partial z} \end{pmatrix} \begin{pmatrix} dx \\ dy \\ dz \end{pmatrix}, \qquad (7)$$

and therefore

$$\mathbf{J}^{-1} = \begin{pmatrix} \dfrac{\partial \xi}{\partial x} & \dfrac{\partial \xi}{\partial y} & \dfrac{\partial \xi}{\partial z} \\ \dfrac{\partial \eta}{\partial x} & \dfrac{\partial \eta}{\partial y} & \dfrac{\partial \eta}{\partial z} \\ \dfrac{\partial \zeta}{\partial x} & \dfrac{\partial \zeta}{\partial y} & \dfrac{\partial \zeta}{\partial z} \end{pmatrix} = \begin{pmatrix} \dfrac{\partial x}{\partial \xi} & \dfrac{\partial x}{\partial \eta} & \dfrac{\partial x}{\partial \zeta} \\ \dfrac{\partial y}{\partial \xi} & \dfrac{\partial y}{\partial \eta} & \dfrac{\partial y}{\partial \zeta} \\ \dfrac{\partial z}{\partial \xi} & \dfrac{\partial z}{\partial \eta} & \dfrac{\partial z}{\partial \zeta} \end{pmatrix}^{-1}$$

$$= \dfrac{1}{\sqrt{g}} \begin{pmatrix} \dfrac{\partial y}{\partial \eta}\dfrac{\partial z}{\partial \zeta} - \dfrac{\partial y}{\partial \zeta}\dfrac{\partial z}{\partial \eta} & \dfrac{\partial x}{\partial \zeta}\dfrac{\partial z}{\partial \eta} - \dfrac{\partial x}{\partial \eta}\dfrac{\partial z}{\partial \zeta} & \dfrac{\partial x}{\partial \eta}\dfrac{\partial y}{\partial \zeta} - \dfrac{\partial x}{\partial \zeta}\dfrac{\partial y}{\partial \eta} \\ \dfrac{\partial y}{\partial \zeta}\dfrac{\partial z}{\partial \xi} - \dfrac{\partial y}{\partial \xi}\dfrac{\partial z}{\partial \zeta} & \dfrac{\partial x}{\partial \xi}\dfrac{\partial z}{\partial \zeta} - \dfrac{\partial x}{\partial \zeta}\dfrac{\partial z}{\partial \xi} & \dfrac{\partial x}{\partial \zeta}\dfrac{\partial y}{\partial \xi} - \dfrac{\partial x}{\partial \xi}\dfrac{\partial y}{\partial \zeta} \\ \dfrac{\partial y}{\partial \xi}\dfrac{\partial z}{\partial \eta} - \dfrac{\partial y}{\partial \eta}\dfrac{\partial z}{\partial \xi} & \dfrac{\partial x}{\partial \eta}\dfrac{\partial z}{\partial \xi} - \dfrac{\partial x}{\partial \xi}\dfrac{\partial z}{\partial \eta} & \dfrac{\partial x}{\partial \xi}\dfrac{\partial y}{\partial \eta} - \dfrac{\partial x}{\partial \eta}\dfrac{\partial y}{\partial \xi} \end{pmatrix}, \qquad (8)$$

where

$$\sqrt{g} = |\mathbf{J}(\xi,\eta,\zeta)|$$

$$= \frac{\partial x}{\partial \xi}\left(\frac{\partial y}{\partial \eta}\frac{\partial z}{\partial \zeta} - \frac{\partial y}{\partial \zeta}\frac{\partial z}{\partial \eta}\right) - \frac{\partial x}{\partial \eta}\left(\frac{\partial y}{\partial \xi}\frac{\partial z}{\partial \zeta} - \frac{\partial y}{\partial \zeta}\frac{\partial z}{\partial \xi}\right) + \frac{\partial x}{\partial \zeta}\left(\frac{\partial y}{\partial \xi}\frac{\partial z}{\partial \eta} - \frac{\partial y}{\partial \eta}\frac{\partial z}{\partial \xi}\right),$$
(9)

is called the *Jacobian determinant* or *Jacobian* of the coordinate transformation. An alternative notation of the metrics is a^{ij}. We see that

$$\mathbf{J} = a_{ij}^T = \left(a^{ij}\right)^{-1},$$
(10)

$$a^{ij} = \mathbf{J}^{-1} = \left(a_{ij}^T\right)^{-1},$$
(11)

$$a_{ij} = \left[\left(a^{ij}\right)^{-1}\right]^T = \left[\left(a^{ij}\right)^T\right]^{-1}.$$
(12)

Covariant base vectors: The *tangent vectors* $(\mathbf{a}_1, \mathbf{a}_2, \mathbf{a}_3)$ to the three curvilinear coordinate lines represented by (ξ, η, ζ) are called the three *covariant* base vectors of the curvilinear coordinate system, and are designated with

$$\mathbf{a}_1: \quad \frac{\partial \mathbf{r}}{\partial \xi} = \frac{\partial x}{\partial \xi}\mathbf{i} + \frac{\partial y}{\partial \xi}\mathbf{j} + \frac{\partial z}{\partial \xi}\mathbf{k},$$
(13)

$$\mathbf{a}_2: \quad \frac{\partial \mathbf{r}}{\partial \eta} = \frac{\partial x}{\partial \eta}\mathbf{i} + \frac{\partial y}{\partial \eta}\mathbf{j} + \frac{\partial z}{\partial \eta}\mathbf{k},$$
(14)

$$\mathbf{a}_3: \quad \frac{\partial \mathbf{r}}{\partial \zeta} = \frac{\partial x}{\partial \zeta}\mathbf{i} + \frac{\partial y}{\partial \zeta}\mathbf{j} + \frac{\partial z}{\partial \zeta}\mathbf{k}.$$
(15)

The components of the three *covariant* base vectors of the curvilinear coordinate system are the columns of the matrix of the inverse metrics (Jacobian matrix – Eq. 6). They are *not* unit vectors. The corresponding unit vectors are

$$\mathbf{e}_1 = \frac{\mathbf{a}_1}{|\mathbf{a}_1|}, \quad \mathbf{e}_2 = \frac{\mathbf{a}_2}{|\mathbf{a}_2|}, \quad \mathbf{e}_3 = \frac{\mathbf{a}_3}{|\mathbf{a}_3|}.$$
(16,17,18)

Contravariant base vectors: The *normal vectors* to a coordinate surface on which the coordinates ξ, η and ζ are constant, respectively are given by

$$\mathbf{a}^1: \quad \nabla \xi = \frac{\partial \xi}{\partial x}\mathbf{i} + \frac{\partial \xi}{\partial y}\mathbf{j} + \frac{\partial \xi}{\partial z}\mathbf{k}, \tag{19}$$

$$\mathbf{a}^2: \quad \nabla \eta = \frac{\partial \eta}{\partial x}\mathbf{i} + \frac{\partial \eta}{\partial y}\mathbf{j} + \frac{\partial \eta}{\partial z}\mathbf{k}, \tag{20}$$

$$\mathbf{a}^3: \quad \nabla \zeta = \frac{\partial \zeta}{\partial x}\mathbf{i} + \frac{\partial \zeta}{\partial y}\mathbf{j} + \frac{\partial \zeta}{\partial z}\mathbf{k}. \tag{21}$$

The components of the three *contravariant* base vectors of the curvilinear coordinate system are the rows of the matrix of the metrics – Eq. (4). They are *not* unit vectors. The corresponding unit vectors are

$$\xi = const, \quad \mathbf{e}^1 = \frac{\mathbf{a}^1}{|\mathbf{a}^1|}, \tag{22}$$

$$\eta = const, \quad \mathbf{e}^2 = \frac{\mathbf{a}^2}{|\mathbf{a}^2|}, \tag{23}$$

$$\zeta = const, \quad \mathbf{e}^3 = \frac{\mathbf{a}^3}{|\mathbf{a}^3|}. \tag{24}$$

At any given point on the surface defined with $\xi = const$, ξ increases most rapidly in the direction of $\nabla \xi$ and decreases most rapidly in the direction of $-\nabla \xi$. In any direction orthogonal to $\nabla \xi$, the derivative is zero. The corresponding statements are valid for the remaining surfaces $\eta = const$ and $\zeta = const$.

These normal vectors to the three coordinate surfaces are called the three *contravariant* base vectors of the curvilinear coordinate system.

Cartesian vector in co- and contravariant coordinate systems: Any Cartesian vector $\mathbf{V} = u\mathbf{i} + v\mathbf{j} + w\mathbf{k}$ has two sets of distinct components with respect to the frames of the covariant and contravariant base vectors

$$\mathbf{V} = V'^1 \mathbf{a}_1 + V'^2 \mathbf{a}_2 + V'^3 \mathbf{a}_3, \tag{25}$$

$$\mathbf{V} = V'_1\mathbf{a}^1 + V'_2\mathbf{a}^2 + V'_3\mathbf{a}^3 . \tag{26}$$

The components V'^i and V'_i are called *contravariant* and *covariant components* of **V** respectively. The components are easily computed by equalizing the Cartesian components as follows

$$u\mathbf{i} + v\mathbf{j} + w\mathbf{k} = V'^1\mathbf{a}_1 + V'^2\mathbf{a}_2 + V'^3\mathbf{a}_3$$

$$= \left(V'^1 a_{11} + V'^2 a_{21} + V'^3 a_{31}\right)\mathbf{i} + \left(V'^1 a_{12} + V'^2 a_{22} + V'^3 a_{32}\right)\mathbf{j}$$

$$+ \left(V'^1 a_{13} + V'^2 a_{23} + V'^3 a_{33}\right)\mathbf{k} , \tag{27}$$

or

$$\begin{pmatrix} a_{11} & a_{21} & a_{31} \\ a_{12} & a_{22} & a_{32} \\ a_{13} & a_{23} & a_{33} \end{pmatrix} \begin{pmatrix} V'^1 \\ V'^2 \\ V'^3 \end{pmatrix} = \begin{pmatrix} u \\ v \\ w \end{pmatrix} \quad \text{or} \quad \mathbf{J} \begin{pmatrix} V'^1 \\ V'^2 \\ V'^3 \end{pmatrix} = \begin{pmatrix} u \\ v \\ w \end{pmatrix} \tag{28}$$

or after comparing with Eq. (9) we have

$$\begin{pmatrix} V'^1 \\ V'^2 \\ V'^3 \end{pmatrix} = \begin{pmatrix} a_{11} & a_{21} & a_{31} \\ a_{12} & a_{22} & a_{32} \\ a_{13} & a_{23} & a_{33} \end{pmatrix}^{-1} \begin{pmatrix} u \\ v \\ w \end{pmatrix} = \mathbf{J}^{-1} \cdot \mathbf{V} . \tag{29}$$

Therefore

$$V'^1 = a^{11}u + a^{12}v + a^{13}w = \mathbf{a}^1 \cdot \mathbf{V} , \tag{30}$$

$$V'^2 = a^{21}u + a^{22}v + a^{23}w = \mathbf{a}^2 \cdot \mathbf{V} , \tag{31}$$

$$V'^3 = a^{31}u + a^{32}v + a^{33}w = \mathbf{a}^3 \cdot \mathbf{V} , \tag{32}$$

and Eq. (25) can be rewritten as follows

$$\mathbf{V} = \left(\mathbf{a}^1 \cdot \mathbf{V}\right)\mathbf{a}_1 + \left(\mathbf{a}^2 \cdot \mathbf{V}\right)\mathbf{a}_2 + \left(\mathbf{a}^3 \cdot \mathbf{V}\right)\mathbf{a}_3 , \tag{33}$$

compare with Eq. (35) in [7] p. 109. Similarly we have

$$\begin{pmatrix} V_1' \\ V_2' \\ V_3' \end{pmatrix} = \begin{pmatrix} a^{11} & a^{21} & a^{31} \\ a^{12} & a^{22} & a^{32} \\ a^{13} & a^{23} & a^{33} \end{pmatrix}^{-1} \begin{pmatrix} u \\ v \\ w \end{pmatrix} = \left[\left(a^{ij} \right)^T \right]^{-1} = \left[\left(a^{ij} \right)^{-1} \right]^T = a_{ij}^T = \mathbf{J} \cdot \mathbf{V}, \quad (34)$$

and

$$\mathbf{V} = (\mathbf{a}_1 \cdot \mathbf{V})\mathbf{a}^1 + (\mathbf{a}_2 \cdot \mathbf{V})\mathbf{a}^2 + (\mathbf{a}_3 \cdot \mathbf{V})\mathbf{a}^3, \quad (35)$$

compare with Eq. (36) in [7] p. 109. Remember again that the components $\mathbf{a}^i \cdot \mathbf{V}$ and $\mathbf{a}_i \cdot \mathbf{V}$ are called contravariant and covariant components of $\mathbf{V}$ respectively.

Increment of the position vector: The infinitesimal increment of the position vector is

$$d\mathbf{r} = \mathbf{a}_1 d\xi + \mathbf{a}_2 d\eta + \mathbf{a}_3 d\zeta. \quad (36)$$

Increment of the arc length: The *infinitesimal increment of the arc length* along a general space curve can be approximated with the magnitude if the infinitesimal increment of the position vector

$$ds = |d\mathbf{r}| = \sqrt{d\mathbf{r} \cdot d\mathbf{r}} \quad (37)$$

or

$$|ds|^2 = d\mathbf{r} \cdot d\mathbf{r} = (\mathbf{a}_1 d\xi + \mathbf{a}_2 d\eta + \mathbf{a}_3 d\zeta) \cdot (\mathbf{a}_1 d\xi + \mathbf{a}_2 d\eta + \mathbf{a}_3 d\zeta)$$

$$= \begin{pmatrix} (\mathbf{a}_1 \cdot \mathbf{a}_1) d\xi d\xi + (\mathbf{a}_2 \cdot \mathbf{a}_1) d\xi d\eta + (\mathbf{a}_3 \cdot \mathbf{a}_1) d\xi d\zeta \\ + (\mathbf{a}_1 \cdot \mathbf{a}_2) d\eta d\xi + (\mathbf{a}_2 \cdot \mathbf{a}_2) d\eta d\eta + (\mathbf{a}_3 \cdot \mathbf{a}_2) d\eta d\zeta \\ + (\mathbf{a}_1 \cdot \mathbf{a}_3) d\zeta d\xi + (\mathbf{a}_2 \cdot \mathbf{a}_3) d\zeta d\eta + (\mathbf{a}_3 \cdot \mathbf{a}_3) d\zeta d\zeta \end{pmatrix}. \quad (38)$$

The tensor built by the dot products is obviously symmetric, and is called the *co-variant tensor*.

Covariant metric tensor: The *covariant metric tensor* is defined as

$$g_{ij} = \begin{pmatrix} \mathbf{a}_1 \cdot \mathbf{a}_1 & \mathbf{a}_1 \cdot \mathbf{a}_2 & \mathbf{a}_1 \cdot \mathbf{a}_3 \\ \mathbf{a}_2 \cdot \mathbf{a}_1 & \mathbf{a}_2 \cdot \mathbf{a}_2 & \mathbf{a}_2 \cdot \mathbf{a}_3 \\ \mathbf{a}_3 \cdot \mathbf{a}_1 & \mathbf{a}_3 \cdot \mathbf{a}_2 & \mathbf{a}_3 \cdot \mathbf{a}_3 \end{pmatrix}, \tag{39}$$

which is a symmetric tensor.

Angles between the unit covariant vectors: The angles between the covariant unit vectors can be expressed in terms of the elements of the covariant metric tensor as follows

$$\theta_{12} = \arccos\left(\frac{\mathbf{a}_1 \cdot \mathbf{a}_2}{|\mathbf{a}_1||\mathbf{a}_2|}\right) = \arccos\left(\frac{g_{12}}{\sqrt{g_{11}g_{22}}}\right), \tag{40}$$

$$\theta_{13} = \arccos\left(\frac{\mathbf{a}_1 \cdot \mathbf{a}_3}{|\mathbf{a}_1||\mathbf{a}_3|}\right) = \arccos\left(\frac{g_{13}}{\sqrt{g_{11}g_{33}}}\right), \tag{41}$$

$$\theta_{23} = \arccos\left(\frac{\mathbf{a}_2 \cdot \mathbf{a}_3}{|\mathbf{a}_2||\mathbf{a}_3|}\right) = \arccos\left(\frac{g_{23}}{\sqrt{g_{22}g_{33}}}\right). \tag{42}$$

It is obvious that only in a specially designed curvilinear coordinate systems can the covariant vectors be mutually perpendicular. Such systems are called *orthogonal*. In this case the transformed region *is rectangular*.

General curvilinear coordinate systems are *not orthogonal* in many applications. In such systems the transformed region where the curvilinear coordinates are independent variables *can be thought of as being rectangular*, and can be treated as such from a coding standpoint by formation of the finite difference equations and in the solution thereof. The problem is thus much simpler in the transformed field, *since the boundaries here are all thought of as rectangular*.

Orthogonality: In an orthogonal coordinate system: a) the two types of base vectors are parallel and b) three base vectors on each type are mutually perpendicular. The consequence is that the off diagonal terms of covariant metric tensor are zeros.

Surface area increment: Consider a parallelepiped with finite sizes $\mathbf{a}_1 d\xi$, $\mathbf{a}_2 d\eta$, and $\mathbf{a}_3 d\zeta$. The areas of the surfaces defined with constant curvilinear coordinates are

$$\xi = const, \quad d\mathbf{S}^1 = (\mathbf{a}_2 \times \mathbf{a}_3) d\eta d\zeta, \tag{43}$$

$$\eta = const, \quad d\mathbf{S}^2 = (\mathbf{a}_3 \times \mathbf{a}_1) d\xi d\zeta, \tag{44}$$

$$\zeta = const, \quad d\mathbf{S}^3 = (\mathbf{a}_1 \times \mathbf{a}_2) d\xi d\eta, \tag{45}$$

respectively. Using the vector identity

$$(\mathbf{A} \times \mathbf{B}) \cdot (\mathbf{C} \times \mathbf{D}) = (\mathbf{A} \cdot \mathbf{C})(\mathbf{B} \cdot \mathbf{D}) - (\mathbf{A} \cdot \mathbf{D})(\mathbf{B} \cdot \mathbf{C}) \tag{46}$$

for **C** = **A** and **D** = **B**

$$(\mathbf{A} \times \mathbf{B}) \cdot (\mathbf{A} \times \mathbf{B}) = (\mathbf{A} \cdot \mathbf{A})(\mathbf{B} \cdot \mathbf{B}) - (\mathbf{A} \cdot \mathbf{B})^2, \tag{47}$$

or

$$|\mathbf{A} \times \mathbf{B}| = \sqrt{(\mathbf{A} \cdot \mathbf{A})(\mathbf{B} \cdot \mathbf{B}) - (\mathbf{A} \cdot \mathbf{B})^2} \tag{48}$$

the magnitude of the increment of the surface area is

$\xi = const$,

$$dS^1 = |\mathbf{a}_2 \times \mathbf{a}_3| d\eta d\zeta = \sqrt{(\mathbf{a}_2 \cdot \mathbf{a}_2)(\mathbf{a}_3 \cdot \mathbf{a}_3) - (\mathbf{a}_2 \cdot \mathbf{a}_3)^2} \, d\eta d\zeta = \sqrt{g_{22} g_{33} - g_{23}^2} \, d\eta d\zeta, \tag{49}$$

$\eta = const$,

$$dS^2 = |\mathbf{a}_1 \times \mathbf{a}_3| d\xi d\zeta = \sqrt{(\mathbf{a}_1 \cdot \mathbf{a}_1)(\mathbf{a}_3 \cdot \mathbf{a}_3) - (\mathbf{a}_3 \cdot \mathbf{a}_1)^2} \, d\xi d\zeta = \sqrt{g_{11} g_{33} - g_{31}^2} \, d\xi d\zeta, \tag{50}$$

$\zeta = const$,

$$dS^3 = |\mathbf{a}_1 \times \mathbf{a}_2| d\xi d\eta = \sqrt{(\mathbf{a}_1 \cdot \mathbf{a}_1)(\mathbf{a}_2 \cdot \mathbf{a}_2) - (\mathbf{a}_1 \cdot \mathbf{a}_2)^2} \, d\xi d\eta = \sqrt{g_{11} g_{22} - g_{12}^2} \, d\xi d\eta. \tag{51}$$

Infinite volume in curvilinear coordinate systems: An infinitesimal parallelepiped formed by the vectors $\mathbf{a}_1 d\xi$, $\mathbf{a}_2 d\eta$, and $\mathbf{a}_3 d\zeta$ has a infinitesimal volume computed using the box product rule

$$dV = (\mathbf{a}_1 d\xi \times \mathbf{a}_2 d\eta) \cdot \mathbf{a}_3 d\zeta = (\mathbf{a}_1 \times \mathbf{a}_2) \cdot \mathbf{a}_3 d\xi d\eta d\zeta$$

$$= \begin{vmatrix} a_{11} & a_{12} & a_{13} \\ a_{21} & a_{22} & a_{23} \\ a_{31} & a_{32} & a_{33} \end{vmatrix} d\xi d\eta d\zeta = \begin{vmatrix} \frac{\partial x}{\partial \xi} & \frac{\partial x}{\partial \eta} & \frac{\partial x}{\partial \zeta} \\ \frac{\partial y}{\partial \xi} & \frac{\partial y}{\partial \eta} & \frac{\partial y}{\partial \zeta} \\ \frac{\partial z}{\partial \xi} & \frac{\partial z}{\partial \eta} & \frac{\partial z}{\partial \zeta} \end{vmatrix} d\xi d\eta d\zeta , \quad (52)$$

or using the *Jacobian determinant* or *Jacobian* of the coordinate transformation $x = f(\xi,\eta,\zeta)$, $y = g(\xi,\eta,\zeta)$, $z = h(\xi,\eta,\zeta)$ we have

$$dV = |\mathbf{J}(\xi,\eta,\zeta)| d\xi d\eta d\zeta . \quad (53)$$

Another notation of the *Jacobian* is

$$\sqrt{g} = |\mathbf{J}(\xi,\eta,\zeta)| = (\mathbf{a}_1 \times \mathbf{a}_2) \cdot \mathbf{a}_3 = (\mathbf{a}_2 \times \mathbf{a}_3) \cdot \mathbf{a}_1 = (\mathbf{a}_3 \times \mathbf{a}_1) \cdot \mathbf{a}_2 , \quad (54)$$

so that the volume increment can also be written as

$$dV = \sqrt{g}\, d\xi d\eta d\zeta . \quad (55)$$

Note the cyclic permutation of the subscripts inside the brackets in Eq. (54). In finite difference form we have

$$\Delta V = \sqrt{g}\, \Delta\xi \Delta\eta \Delta\zeta . \quad (56)$$

Selecting strict equidistant discretization into the transformed space with $\Delta\xi = 1$, $\Delta\eta = 1$, $\Delta\zeta = 1$ we have a visualization of the meaning of the *Jacobian*, namely

$$\sqrt{g} = \Delta V . \quad (57)$$

Some authors are using this approach to compute the *Jacobian* from the volume of the computational cell provided the discretization of the computational domain is strictly equidistant with steps of unity in all directions. The real volume of the cell is computed in this case from the Cartesian coordinates of the vertices of the real cell as will be shown later.

The Jacobian of the coordinate transformation in terms of dot products of the covariant vectors: The square of the Jacobian

$$g = \left| \mathbf{J}(\xi, \eta, \zeta) \right|^2 = \left[\mathbf{a}_1 \cdot (\mathbf{a}_2 \times \mathbf{a}_3) \right]^2 \tag{58}$$

can be rewritten as a function of the dot products of the components of the *Jacobian* determinant as follows. Using the vector identities

$$(\mathbf{A} \times \mathbf{B}) \cdot (\mathbf{C} \times \mathbf{D}) = (\mathbf{A} \cdot \mathbf{C})(\mathbf{B} \cdot \mathbf{D}) - (\mathbf{A} \cdot \mathbf{D})(\mathbf{B} \cdot \mathbf{C}) \tag{59}$$

for $\mathbf{C} = \mathbf{A}$ and $\mathbf{D} = \mathbf{B}$ results in

$$(\mathbf{A} \times \mathbf{B}) \cdot (\mathbf{A} \times \mathbf{B}) = (\mathbf{A} \cdot \mathbf{A})(\mathbf{B} \cdot \mathbf{B}) - (\mathbf{A} \cdot \mathbf{B})^2 \tag{60}$$

or

$$(\mathbf{A} \cdot \mathbf{B})^2 = (\mathbf{A} \cdot \mathbf{A})(\mathbf{B} \cdot \mathbf{B}) - (\mathbf{A} \times \mathbf{B}) \cdot (\mathbf{A} \times \mathbf{B}). \tag{61}$$

With this result we have

$$\left[\mathbf{a}_1 \cdot (\mathbf{a}_2 \times \mathbf{a}_3) \right]^2 = (\mathbf{a}_1 \cdot \mathbf{a}_1)(\mathbf{a}_2 \times \mathbf{a}_3) \cdot (\mathbf{a}_2 \times \mathbf{a}_3) - \left[\mathbf{a}_1 \times (\mathbf{a}_2 \times \mathbf{a}_3) \right] \cdot \left[\mathbf{a}_1 \times (\mathbf{a}_2 \times \mathbf{a}_3) \right]. \tag{62}$$

Using the same identity Eq. (59)

$$(\mathbf{a}_2 \times \mathbf{a}_3) \cdot (\mathbf{a}_2 \times \mathbf{a}_3) = (\mathbf{a}_2 \cdot \mathbf{a}_2)(\mathbf{a}_3 \cdot \mathbf{a}_3) - (\mathbf{a}_2 \cdot \mathbf{a}_3)^2 \tag{63}$$

and by the vector identity

$$\mathbf{A} \times (\mathbf{B} \times \mathbf{C}) = (\mathbf{A} \cdot \mathbf{C})\mathbf{B} - (\mathbf{A} \cdot \mathbf{B})\mathbf{C} \tag{64}$$

we have

$$\mathbf{a}_1 \times (\mathbf{a}_2 \times \mathbf{a}_3) = (\mathbf{a}_1 \cdot \mathbf{a}_3)\mathbf{a}_2 - (\mathbf{a}_1 \cdot \mathbf{a}_2)\mathbf{a}_3. \tag{65}$$

Finally we obtain

$$\left[\mathbf{a}_1 \cdot (\mathbf{a}_2 \times \mathbf{a}_3) \right]^2 = (\mathbf{a}_1 \cdot \mathbf{a}_1)\left[(\mathbf{a}_2 \cdot \mathbf{a}_2)(\mathbf{a}_3 \cdot \mathbf{a}_3) - (\mathbf{a}_2 \cdot \mathbf{a}_3)^2 \right] - \left| (\mathbf{a}_1 \cdot \mathbf{a}_3)\mathbf{a}_2 - (\mathbf{a}_1 \cdot \mathbf{a}_2)\mathbf{a}_3 \right|^2$$

$$= (\mathbf{a}_1 \cdot \mathbf{a}_1)(\mathbf{a}_2 \cdot \mathbf{a}_2)(\mathbf{a}_3 \cdot \mathbf{a}_3) - (\mathbf{a}_1 \cdot \mathbf{a}_1)(\mathbf{a}_2 \cdot \mathbf{a}_3)^2 - (\mathbf{a}_2 \cdot \mathbf{a}_2)(\mathbf{a}_1 \cdot \mathbf{a}_3)^2 - (\mathbf{a}_3 \cdot \mathbf{a}_3)(\mathbf{a}_1 \cdot \mathbf{a}_2)^2$$

$$+ 2(\mathbf{a}_1 \cdot \mathbf{a}_2)(\mathbf{a}_3 \cdot \mathbf{a}_2)(\mathbf{a}_1 \cdot \mathbf{a}_3)$$

$$= \begin{vmatrix} \mathbf{a}_1 \cdot \mathbf{a}_1 & \mathbf{a}_1 \cdot \mathbf{a}_2 & \mathbf{a}_1 \cdot \mathbf{a}_3 \\ \mathbf{a}_2 \cdot \mathbf{a}_1 & \mathbf{a}_2 \cdot \mathbf{a}_2 & \mathbf{a}_2 \cdot \mathbf{a}_3 \\ \mathbf{a}_3 \cdot \mathbf{a}_1 & \mathbf{a}_3 \cdot \mathbf{a}_2 & \mathbf{a}_3 \cdot \mathbf{a}_3 \end{vmatrix} = |g_{ij}|$$

$$= g_{11}\left(g_{22}g_{33} - g_{23}^2\right) - g_{12}\left(g_{21}g_{33} - g_{23}g_{31}\right) + g_{13}\left(g_{21}g_{32} - g_{22}g_{31}\right), \quad (66)$$

which is the determinant of the covariant symmetric tensor.

Relation between the partial derivatives with respect to Cartesian and curvilinear coordinates: Partial derivatives with respect to *Cartesian* coordinates are related to partial derivatives with respect to the curvilinear coordinates by the chain rule. If $\varphi(x, y, z)$ is a scalar-valued function of the main variables x, y, z, through the three intermediate variables ξ, η, ζ, then

$$\frac{\partial \varphi}{\partial x} = \frac{\partial \varphi}{\partial \xi}\frac{\partial \xi}{\partial x} + \frac{\partial \varphi}{\partial \eta}\frac{\partial \eta}{\partial x} + \frac{\partial \varphi}{\partial \zeta}\frac{\partial \zeta}{\partial x} = \frac{\partial \varphi}{\partial \xi}a^{11} + \frac{\partial \varphi}{\partial \eta}a^{21} + \frac{\partial \varphi}{\partial \zeta}a^{31}, \quad (67)$$

$$\frac{\partial \varphi}{\partial y} = \frac{\partial \varphi}{\partial \xi}\frac{\partial \xi}{\partial y} + \frac{\partial \varphi}{\partial \eta}\frac{\partial \eta}{\partial y} + \frac{\partial \varphi}{\partial \zeta}\frac{\partial \zeta}{\partial y} = \frac{\partial \varphi}{\partial \xi}a^{12} + \frac{\partial \varphi}{\partial \eta}a^{22} + \frac{\partial \varphi}{\partial \zeta}a^{32}, \quad (68)$$

$$\frac{\partial \varphi}{\partial z} = \frac{\partial \varphi}{\partial \xi}\frac{\partial \xi}{\partial z} + \frac{\partial \varphi}{\partial \eta}\frac{\partial \eta}{\partial z} + \frac{\partial \varphi}{\partial \zeta}\frac{\partial \zeta}{\partial z} = \frac{\partial \varphi}{\partial \xi}a^{13} + \frac{\partial \varphi}{\partial \eta}a^{23} + \frac{\partial \varphi}{\partial \zeta}a^{33}, \quad (69)$$

with the second superscript indicating the Cartesian component of the contravariant vectors. If

$$\mathbf{V}(x, y, z) = u(x, y, z)\mathbf{i} + v(x, y, z)\mathbf{j} + v(x, y, z)\mathbf{k}$$

is a vector-valued function of the main variables x, y, z, through the three intermediate variables ξ, η, ζ, then

$$\frac{\partial \mathbf{V}}{\partial x} = \left(\frac{\partial u}{\partial \xi}a^{11} + \frac{\partial u}{\partial \eta}a^{21} + \frac{\partial u}{\partial \zeta}a^{31}\right)\mathbf{i} + \left(\frac{\partial v}{\partial \xi}a^{11} + \frac{\partial v}{\partial \eta}a^{21} + \frac{\partial v}{\partial \zeta}a^{31}\right)\mathbf{j}$$

$$+\left(\frac{\partial w}{\partial \xi}a^{11}+\frac{\partial w}{\partial \eta}a^{21}+\frac{\partial w}{\partial \zeta}a^{31}\right)\mathbf{k}, \tag{70}$$

$$\frac{\partial \mathbf{V}}{\partial y}=\left(\frac{\partial u}{\partial \xi}a^{12}+\frac{\partial u}{\partial \eta}a^{22}+\frac{\partial u}{\partial \zeta}a^{32}\right)\mathbf{i}+\left(\frac{\partial v}{\partial \xi}a^{12}+\frac{\partial v}{\partial \eta}a^{22}+\frac{\partial v}{\partial \zeta}a^{32}\right)\mathbf{j}$$

$$+\left(\frac{\partial w}{\partial \xi}a^{12}+\frac{\partial w}{\partial \eta}a^{22}+\frac{\partial w}{\partial \zeta}a^{32}\right)\mathbf{k}, \tag{71}$$

$$\frac{\partial \mathbf{V}}{\partial z}=\left(\frac{\partial u}{\partial \xi}a^{13}+\frac{\partial u}{\partial \eta}a^{23}+\frac{\partial u}{\partial \zeta}a^{33}\right)\mathbf{i}+\left(\frac{\partial v}{\partial \xi}a^{13}+\frac{\partial v}{\partial \eta}a^{23}+\frac{\partial v}{\partial \zeta}a^{33}\right)\mathbf{j}$$

$$+\left(\frac{\partial w}{\partial \xi}a^{13}+\frac{\partial w}{\partial \eta}a^{23}+\frac{\partial w}{\partial \zeta}a^{33}\right)\mathbf{k}. \tag{72}$$

The divergence theorem (Gauss-Ostrogradskii): The divergence theorem says that under suitable conditions the outward flux of a vector field across a closed surface (oriented outward) equals the triple integral of the divergence of the field over the region enclosed by the surface. The flux of vector $\mathbf{F} = M\mathbf{i} + N\mathbf{j} + P\mathbf{k}$ across a closed oriented surface S in the direction of the surface's outward unit normal field $\mathbf{n}$ equals the integral of $\nabla \cdot \mathbf{F}$ over the region D enclosed by the surface:

$$\iint_S \mathbf{F} \cdot \mathbf{n} d\sigma = \iiint_D \nabla \cdot \mathbf{F} dV. \tag{73}$$

Volume of space enveloped by a closed surface: Consider in a space closed surface S described by the position vector $\mathbf{r}(x,y,z)$. The volume inside the surface can be expressed in terms of the area integral over its boundary using the divergence theorem

$$\iint_S \mathbf{r} \cdot \mathbf{n} d\sigma = \iiint_D \nabla \cdot \mathbf{r} dV = \iiint_D \left(\frac{\partial}{\partial x}\mathbf{i}+\frac{\partial}{\partial y}\mathbf{j}+\frac{\partial}{\partial z}\mathbf{k}\right) \cdot (x\mathbf{i}+y\mathbf{j}+z\mathbf{k}) dV$$

$$= \iiint_D \left(\frac{\partial x}{\partial x}+\frac{\partial y}{\partial y}+\frac{\partial z}{\partial z}\right) dV = 3\iiint_D dV = 3D, \tag{74}$$

or

$$D = \frac{1}{3} \iint_S \mathbf{r} \cdot \mathbf{n} d\sigma . \tag{75}$$

A practical application of this relation is the computation of the volume enclosed in plane elements with area S_m, outwards unit normal vectors $\mathbf{n}_m$, and position vector of the centroids of the faces $\mathbf{r}_m$,

$$D = \frac{1}{3} \sum_m S_m \mathbf{n}_m \cdot \mathbf{r}_m . \tag{76}$$

The divergence theorem for a curvilinear coordinate system: Consider a differential element of volume D bounded by six faces lying on coordinate surfaces, as shown in Fig. A2.3.

Divergence: Applying the *Divergence Theorem* we obtain

$$\iiint_D (\nabla \cdot \mathbf{F}) \sqrt{g} \, d\xi d\eta d\zeta$$

$$= \iint_{S_1} \mathbf{F} \cdot (\mathbf{a}_2 \times \mathbf{a}_3) d\eta d\zeta - \iint_{S_2} \mathbf{F} \cdot (\mathbf{a}_2 \times \mathbf{a}_3) d\eta d\zeta$$

$$+ \iint_{S_3} \mathbf{F} \cdot (\mathbf{a}_3 \times \mathbf{a}_1) d\xi d\zeta - \iint_{S_4} \mathbf{F} \cdot (\mathbf{a}_3 \times \mathbf{a}_1) d\xi d\zeta$$

$$+ \iint_{S_5} \mathbf{F} \cdot (\mathbf{a}_1 \times \mathbf{a}_2) d\xi d\eta - \iint_{S_6} \mathbf{F} \cdot (\mathbf{a}_1 \times \mathbf{a}_2) d\xi d\eta . \tag{77}$$

Dividing by D and letting D approach zero we obtain the so called *conservative* expression for the divergence

$$\nabla \cdot \mathbf{F} = \frac{1}{\sqrt{g}} \left\{ \frac{\partial}{\partial \xi} \left[(\mathbf{a}_2 \times \mathbf{a}_3) \cdot \mathbf{F} \right] + \frac{\partial}{\partial \eta} \left[(\mathbf{a}_3 \times \mathbf{a}_1) \cdot \mathbf{F} \right] + \frac{\partial}{\partial \zeta} \left[(\mathbf{a}_1 \times \mathbf{a}_2) \cdot \mathbf{F} \right] \right\} . \tag{78}$$

Having in mind that $\mathbf{a}_2$ and $\mathbf{a}_3$ are independent of ξ, and $\mathbf{a}_1$ and $\mathbf{a}_3$ are independent of η, and $\mathbf{a}_1$ and $\mathbf{a}_2$ are independent of ζ results in the so called *first fundamental metric identity* [4]

$$\frac{\partial}{\partial \xi}(\mathbf{a}_2 \times \mathbf{a}_3) + \frac{\partial}{\partial \eta}(\mathbf{a}_3 \times \mathbf{a}_1) + \frac{\partial}{\partial \zeta}(\mathbf{a}_1 \times \mathbf{a}_2) = 0 . \tag{79}$$

The so called *non-conservative* expression for the divergence is then

$$\nabla \cdot \mathbf{F} = \frac{1}{\sqrt{g}}\left[\left(\mathbf{a}_2 \times \mathbf{a}_3\right)\cdot\frac{\partial \mathbf{F}}{\partial \xi} + \left(\mathbf{a}_3 \times \mathbf{a}_1\right)\cdot\frac{\partial \mathbf{F}}{\partial \eta} + \left(\mathbf{a}_1 \times \mathbf{a}_2\right)\cdot\frac{\partial \mathbf{F}}{\partial \zeta}\right]. \tag{80}$$

Curl: The Divergence Theorem is valid if the dot product is replaced by a cross product. The conservative form of the *curl* is

$$\nabla \times \mathbf{F} = \frac{1}{\sqrt{g}}\left\{\frac{\partial}{\partial \xi}\left[\left(\mathbf{a}_2 \times \mathbf{a}_3\right)\times \mathbf{F}\right] + \frac{\partial}{\partial \eta}\left[\left(\mathbf{a}_3 \times \mathbf{a}_1\right)\times \mathbf{F}\right] + \frac{\partial}{\partial \zeta}\left[\left(\mathbf{a}_1 \times \mathbf{a}_2\right)\times \mathbf{F}\right]\right\}, \tag{81}$$

and the non-conservative form

$$\nabla \times \mathbf{F} = \frac{1}{\sqrt{g}}\left[\left(\mathbf{a}_2 \times \mathbf{a}_3\right)\times\frac{\partial \mathbf{F}}{\partial \xi} + \left(\mathbf{a}_3 \times \mathbf{a}_1\right)\times\frac{\partial \mathbf{F}}{\partial \eta} + \left(\mathbf{a}_1 \times \mathbf{a}_2\right)\times\frac{\partial \mathbf{F}}{\partial \zeta}\right]. \tag{82}$$

Gradient: The Divergence Theorem is valid if the tensor or the vector **F** is replaced by a scalar φ. The conservative form of the *gradient vector* (*gradient*) of the differentiable function φ is

$$\nabla \varphi = \frac{1}{\sqrt{g}}\left\{\frac{\partial}{\partial \xi}\left[\left(\mathbf{a}_2 \times \mathbf{a}_3\right)\varphi\right] + \frac{\partial}{\partial \eta}\left[\left(\mathbf{a}_3 \times \mathbf{a}_1\right)\varphi\right] + \frac{\partial}{\partial \zeta}\left[\left(\mathbf{a}_1 \times \mathbf{a}_2\right)\varphi\right]\right\}, \tag{83}$$

and the non-conservative form

$$\nabla \varphi = \frac{1}{\sqrt{g}}\left[\left(\mathbf{a}_2 \times \mathbf{a}_3\right)\frac{\partial \varphi}{\partial \xi} + \left(\mathbf{a}_3 \times \mathbf{a}_1\right)\frac{\partial \varphi}{\partial \eta} + \left(\mathbf{a}_1 \times \mathbf{a}_2\right)\frac{\partial \varphi}{\partial \zeta}\right]. \tag{84}$$

Laplacian: The expression for the *Laplacian* follows from the expression for the divergence and gradient as follows: The conservative form is

$$\nabla^2 \varphi = \nabla \cdot \left(\nabla \varphi\right)$$

$$
\begin{aligned}
= &\frac{1}{\sqrt{g}}\frac{\partial}{\partial \xi}\left[(\mathbf{a}_2 \times \mathbf{a}_3) \cdot \frac{1}{\sqrt{g}}\left\{\begin{array}{l}\dfrac{\partial}{\partial \xi}\left[(\mathbf{a}_2 \times \mathbf{a}_3)\varphi\right]+\dfrac{\partial}{\partial \eta}\left[(\mathbf{a}_3 \times \mathbf{a}_1)\varphi\right] \\ +\dfrac{\partial}{\partial \zeta}\left[(\mathbf{a}_1 \times \mathbf{a}_2)\varphi\right]\end{array}\right\}\right] \\
&+\frac{1}{\sqrt{g}}\frac{\partial}{\partial \eta}\left[(\mathbf{a}_1 \times \mathbf{a}_3) \cdot \frac{1}{\sqrt{g}}\left\{\begin{array}{l}\dfrac{\partial}{\partial \xi}\left[(\mathbf{a}_2 \times \mathbf{a}_3)\varphi\right]+\dfrac{\partial}{\partial \eta}\left[(\mathbf{a}_3 \times \mathbf{a}_1)\varphi\right] \\ +\dfrac{\partial}{\partial \zeta}\left[(\mathbf{a}_1 \times \mathbf{a}_2)\varphi\right]\end{array}\right\}\right] \\
&+\frac{1}{\sqrt{g}}\frac{\partial}{\partial \zeta}\left[(\mathbf{a}_1 \times \mathbf{a}_2) \cdot \frac{1}{\sqrt{g}}\left\{\begin{array}{l}\dfrac{\partial}{\partial \xi}\left[(\mathbf{a}_2 \times \mathbf{a}_3)\varphi\right]+\dfrac{\partial}{\partial \eta}\left[(\mathbf{a}_3 \times \mathbf{a}_1)\varphi\right] \\ +\dfrac{\partial}{\partial \zeta}\left[(\mathbf{a}_1 \times \mathbf{a}_2)\varphi\right]\end{array}\right\}\right].
\end{aligned} \qquad (85)
$$

The non-conservative form is

$$\nabla^2 \varphi = \nabla \cdot (\nabla \varphi)$$

$$
= \frac{1}{\sqrt{g}}\left\{\begin{array}{l}
\dfrac{\partial}{\partial \xi}\left[(\mathbf{a}_2 \times \mathbf{a}_3)\cdot\dfrac{1}{\sqrt{g}}\left[(\mathbf{a}_2\times\mathbf{a}_3)\dfrac{\partial \varphi}{\partial \xi}+(\mathbf{a}_3\times\mathbf{a}_1)\dfrac{\partial \varphi}{\partial \eta}+(\mathbf{a}_1\times\mathbf{a}_2)\dfrac{\partial \varphi}{\partial \zeta}\right]\right] \\
+\dfrac{\partial}{\partial \eta}\left[(\mathbf{a}_1 \times \mathbf{a}_3)\cdot\dfrac{1}{\sqrt{g}}\left[(\mathbf{a}_2\times\mathbf{a}_3)\dfrac{\partial \varphi}{\partial \xi}+(\mathbf{a}_3\times\mathbf{a}_1)\dfrac{\partial \varphi}{\partial \xi}+(\mathbf{a}_1\times\mathbf{a}_2)\dfrac{\partial \varphi}{\partial \zeta}\right]\right] \\
+\dfrac{\partial}{\partial \zeta}\left[(\mathbf{a}_1 \times \mathbf{a}_2)\cdot\dfrac{1}{\sqrt{g}}\left[(\mathbf{a}_2\times\mathbf{a}_3)\dfrac{\partial \varphi}{\partial \xi}+(\mathbf{a}_3\times\mathbf{a}_1)\dfrac{\partial \varphi}{\partial \xi}+(\mathbf{a}_1\times\mathbf{a}_2)\dfrac{\partial \varphi}{\partial \zeta}\right]\right]
\end{array}\right\}.
$$

(86)

The above relations were used for the first time to derive conservative transformed equations in the gas dynamics by *Vivand* [8] and *Vinokur* [9] in 1974.

Contravariant metric tensor: It is very interesting to note that in the *Laplacian* the nine scalar products form a symmetric tensor

$$g^{ij} = \begin{pmatrix} \mathbf{a}^1 \cdot \mathbf{a}^1 & \mathbf{a}^1 \cdot \mathbf{a}^2 & \mathbf{a}^1 \cdot \mathbf{a}^3 \\ \mathbf{a}^2 \cdot \mathbf{a}^1 & \mathbf{a}^2 \cdot \mathbf{a}^2 & \mathbf{a}^2 \cdot \mathbf{a}^3 \\ \mathbf{a}^3 \cdot \mathbf{a}^1 & \mathbf{a}^3 \cdot \mathbf{a}^2 & \mathbf{a}^3 \cdot \mathbf{a}^3 \end{pmatrix}, \tag{87}$$

called the *contravariant metric tensor*. Using the identity

$$(\mathbf{A} \times \mathbf{B}) \cdot (\mathbf{C} \times \mathbf{D}) = (\mathbf{A} \cdot \mathbf{C})(\mathbf{B} \cdot \mathbf{D}) - (\mathbf{A} \cdot \mathbf{D})(\mathbf{B} \cdot \mathbf{C}) \tag{88}$$

we compute the six contravariant metric numbers as follows

$$\mathbf{a}^1 \cdot \mathbf{a}^1 = \frac{1}{g}(\mathbf{a}_2 \times \mathbf{a}_3) \cdot (\mathbf{a}_2 \times \mathbf{a}_3) = \frac{1}{g}\left[(\mathbf{a}_2 \cdot \mathbf{a}_2)(\mathbf{a}_3 \cdot \mathbf{a}_3) - (\mathbf{a}_2 \cdot \mathbf{a}_3)^2\right]$$

$$= \frac{1}{g}(g_{22}g_{33} - g_{23}^2), \tag{89}$$

$$\mathbf{a}^2 \cdot \mathbf{a}^2 = \frac{1}{g}(\mathbf{a}_3 \times \mathbf{a}_1) \cdot (\mathbf{a}_3 \times \mathbf{a}_1) = \frac{1}{g}\left[(\mathbf{a}_3 \cdot \mathbf{a}_3)(\mathbf{a}_1 \cdot \mathbf{a}_1) - (\mathbf{a}_3 \cdot \mathbf{a}_1)^2\right]$$

$$= \frac{1}{g}(g_{33}g_{11} - g_{31}^2), \tag{90}$$

$$\mathbf{a}^3 \cdot \mathbf{a}^3 = \frac{1}{g}(\mathbf{a}_1 \times \mathbf{a}_2) \cdot (\mathbf{a}_1 \times \mathbf{a}_2) = \frac{1}{g}\left[(\mathbf{a}_1 \cdot \mathbf{a}_1)(\mathbf{a}_2 \cdot \mathbf{a}_2) - (\mathbf{a}_1 \cdot \mathbf{a}_2)^2\right]$$

$$= \frac{1}{g}(g_{11}g_{22} - g_{12}^2), \tag{91}$$

$$\mathbf{a}^1 \cdot \mathbf{a}^2 = \frac{1}{g}(\mathbf{a}_2 \times \mathbf{a}_3) \cdot (\mathbf{a}_3 \times \mathbf{a}_1) = \frac{1}{g}\left[(\mathbf{a}_2 \cdot \mathbf{a}_3)(\mathbf{a}_3 \cdot \mathbf{a}_1) - (\mathbf{a}_2 \cdot \mathbf{a}_1)(\mathbf{a}_3 \cdot \mathbf{a}_3)\right]$$

$$= \frac{1}{g}(g_{23}g_{31} - g_{21}g_{33}), \tag{92}$$

$$\mathbf{a}^1 \cdot \mathbf{a}^3 = \frac{1}{g}(\mathbf{a}_2 \times \mathbf{a}_3) \cdot (\mathbf{a}_1 \times \mathbf{a}_2) = \frac{1}{g}\left[(\mathbf{a}_2 \cdot \mathbf{a}_1)(\mathbf{a}_3 \cdot \mathbf{a}_2) - (\mathbf{a}_2 \cdot \mathbf{a}_2)(\mathbf{a}_3 \cdot \mathbf{a}_1)\right]$$

$$= \frac{1}{g}\left(g_{21}g_{32} - g_{22}g_{31}\right), \tag{93}$$

$$\mathbf{a}^2 \cdot \mathbf{a}^3 = \frac{1}{g}(\mathbf{a}_3 \times \mathbf{a}_1) \cdot (\mathbf{a}_1 \times \mathbf{a}_2) = \frac{1}{g}\left[(\mathbf{a}_3 \cdot \mathbf{a}_1)(\mathbf{a}_1 \cdot \mathbf{a}_2) - (\mathbf{a}_3 \cdot \mathbf{a}_2)(\mathbf{a}_1 \cdot \mathbf{a}_1)\right]$$

$$= \frac{1}{g}\left(g_{31}g_{12} - g_{32}g_{11}\right). \tag{94}$$

Computing the determinant of the contravariant metric tensor we realize that

$$\begin{vmatrix} \mathbf{a}^1 \cdot \mathbf{a}^1 & \mathbf{a}^1 \cdot \mathbf{a}^2 & \mathbf{a}^1 \cdot \mathbf{a}^3 \\ \mathbf{a}^2 \cdot \mathbf{a}^1 & \mathbf{a}^2 \cdot \mathbf{a}^2 & \mathbf{a}^2 \cdot \mathbf{a}^3 \\ \mathbf{a}^3 \cdot \mathbf{a}^1 & \mathbf{a}^3 \cdot \mathbf{a}^2 & \mathbf{a}^3 \cdot \mathbf{a}^3 \end{vmatrix} = \frac{1}{g}. \tag{95}$$

Relation between the contravariant and the covariant base vectors, dual vectors: The non-conservative form of the *gradient vector* of the differentiable function φ

$$\nabla \varphi = \frac{1}{\sqrt{g}}\left[(\mathbf{a}_2 \times \mathbf{a}_3)\frac{\partial \varphi}{\partial \xi} + (\mathbf{a}_3 \times \mathbf{a}_1)\frac{\partial \varphi}{\partial \eta} + (\mathbf{a}_1 \times \mathbf{a}_2)\frac{\partial \varphi}{\partial \zeta}\right] \tag{96}$$

is applied for all of the coordinates ξ, η, ζ. Since the three coordinates are independent of each other their derivatives with respect to each other are zero. With this in mind we obtain

$$\nabla \xi = \frac{1}{\sqrt{g}}\left[(\mathbf{a}_2 \times \mathbf{a}_3)\frac{\partial \xi}{\partial \xi} + (\mathbf{a}_3 \times \mathbf{a}_1)\frac{\partial \xi}{\partial \eta} + (\mathbf{a}_1 \times \mathbf{a}_2)\frac{\partial \xi}{\partial \zeta}\right] = \frac{1}{\sqrt{g}}(\mathbf{a}_2 \times \mathbf{a}_3), \tag{97}$$

$$\nabla \eta = \frac{1}{\sqrt{g}}\left[(\mathbf{a}_2 \times \mathbf{a}_3)\frac{\partial \eta}{\partial \xi} + (\mathbf{a}_3 \times \mathbf{a}_1)\frac{\partial \eta}{\partial \eta} + (\mathbf{a}_1 \times \mathbf{a}_2)\frac{\partial \eta}{\partial \zeta}\right] = \frac{1}{\sqrt{g}}(\mathbf{a}_3 \times \mathbf{a}_1), \tag{98}$$

$$\nabla \zeta = \frac{1}{\sqrt{g}}\left[(\mathbf{a}_2 \times \mathbf{a}_3)\frac{\partial \zeta}{\partial \xi} + (\mathbf{a}_3 \times \mathbf{a}_1)\frac{\partial \zeta}{\partial \eta} + (\mathbf{a}_1 \times \mathbf{a}_2)\frac{\partial \zeta}{\partial \zeta}\right] = \frac{1}{\sqrt{g}}(\mathbf{a}_1 \times \mathbf{a}_2), \tag{99}$$

or simply

$$\mathbf{a}^1 = \frac{1}{\sqrt{g}}(\mathbf{a}_2 \times \mathbf{a}_3), \qquad (100)$$

$$\mathbf{a}^2 = \frac{1}{\sqrt{g}}(\mathbf{a}_3 \times \mathbf{a}_1), \qquad (101)$$

$$\mathbf{a}^3 = \frac{1}{\sqrt{g}}(\mathbf{a}_1 \times \mathbf{a}_2). \qquad (102)$$

The contravariant vectors are not unit vectors. This result allows us to simplify the expressions for the gradient, divergence, curl, *Laplacian*, etc. Note that with these results the *first fundamental metric identity*,

$$\frac{\partial}{\partial \eta}(\mathbf{a}_2 \times \mathbf{a}_3) + \frac{\partial}{\partial \xi}(\mathbf{a}_3 \times \mathbf{a}_1) + \frac{\partial}{\partial \zeta}(\mathbf{a}_1 \times \mathbf{a}_2) = 0, \qquad (103)$$

can be written as

$$\frac{\partial}{\partial \xi}\left(\sqrt{g}\,\mathbf{a}^1\right) + \frac{\partial}{\partial \eta}\left(\sqrt{g}\,\mathbf{a}^2\right) + \frac{\partial}{\partial \zeta}\left(\sqrt{g}\,\mathbf{a}^3\right) = 0. \qquad (104)$$

It is very important that the grid generated for numerical integration fulfills the above identity strictly. Otherwise spurious numerical errors are introduced in the computational results.

Note the useful relations between the covariant and the contravariant vectors

$$\mathbf{a}_1 \cdot \mathbf{a}^1 = \frac{1}{\sqrt{g}}\mathbf{a}_1 \cdot (\mathbf{a}_2 \times \mathbf{a}_3) = 1, \quad \mathbf{a}_2 \cdot \mathbf{a}^1 = 0, \quad \mathbf{a}_3 \cdot \mathbf{a}^1 = 0, \qquad (105\text{-}107)$$

$$\mathbf{a}_1 \cdot \mathbf{a}^2 = 0, \quad \mathbf{a}_2 \cdot \mathbf{a}^2 = \frac{1}{\sqrt{g}}\mathbf{a}_2 \cdot (\mathbf{a}_3 \times \mathbf{a}_1) = 1, \quad \mathbf{a}_3 \cdot \mathbf{a}^2 = 0, \qquad (108\text{-}110)$$

$$\mathbf{a}_1 \cdot \mathbf{a}^3 = 0, \quad \mathbf{a}_2 \cdot \mathbf{a}^3 = 0, \quad \mathbf{a}_3 \cdot \mathbf{a}^3 = \frac{1}{\sqrt{g}}\mathbf{a}_3 \cdot (\mathbf{a}_1 \times \mathbf{a}_2) = 1. \qquad (111\text{-}113)$$

These properties distinguish the couple of vectors $\mathbf{a}_i$ and $\mathbf{a}^i$. Such vectors are called *dual*. Using these properties Eqs. (33) and (35) can be easily proved.

Approximating the contravariant base vectors: Consider a parallelepiped with finite sizes $\mathbf{a}_1 \Delta \xi$, $\mathbf{a}_2 \Delta \eta$, and $\mathbf{a}_3 \Delta \zeta$. The areas of the surfaces defined with constant curvilinear coordinates are

$$\xi = const, \quad \mathbf{S}^1 = (\mathbf{a}_2 \times \mathbf{a}_3) \Delta \eta \Delta \zeta = \sqrt{g} \mathbf{a}^1 \Delta \eta \Delta \zeta, \tag{114}$$

$$\eta = const, \quad \mathbf{S}^2 = (\mathbf{a}_3 \times \mathbf{a}_1) \Delta \xi \Delta \zeta = \sqrt{g} \mathbf{a}^2 \Delta \xi \Delta \zeta, \tag{115}$$

$$\zeta = const, \quad \mathbf{S}^3 = (\mathbf{a}_1 \times \mathbf{a}_2) \Delta \xi \Delta \eta = \sqrt{g} \mathbf{a}^3 \Delta \xi \Delta \eta, \tag{116}$$

respectively. Solving with respect to the contravariant vectors gives

$$\mathbf{a}^1 = \frac{\mathbf{S}^1}{\sqrt{g} \Delta \eta \Delta \zeta}, \tag{117}$$

$$\mathbf{a}^2 = \frac{\mathbf{S}^2}{\sqrt{g} \Delta \xi \Delta \zeta}, \tag{118}$$

$$\mathbf{a}^3 = \frac{\mathbf{S}^3}{\sqrt{g} \Delta \xi \Delta \eta}. \tag{119}$$

If one selects $\Delta \xi = 1$, $\Delta \eta = 1$, and $\Delta \zeta = 1$, the contravariant vectors are then

$$\mathbf{a}^1 = \frac{\mathbf{S}^1}{\sqrt{g}}, \tag{120}$$

$$\mathbf{a}^2 = \frac{\mathbf{S}^2}{\sqrt{g}}, \tag{121}$$

$$\mathbf{a}^3 = \frac{\mathbf{S}^3}{\sqrt{g}}. \tag{122}$$

The contravariant metric tensor is in this case

$$g^{ij} = \mathbf{a}^i \cdot \mathbf{a}^j = \frac{\mathbf{S}^i \cdot \mathbf{S}^j}{g}. \tag{123}$$

In a sense of the numerical construction of the computational grid the above formulas can be used for computation of the contravariant vectors for equidistant grids with steps of unity in all directions in the transformed space.

Relations between the inverse metrics and the metrics of the coordinate transformation: Usually in practical problems the inverse metrics are computed analytically or numerically and then the metrics are computed using the following procedure

$$\mathbf{a}^1 := \frac{\partial \xi}{\partial x}\mathbf{i} + \frac{\partial \xi}{\partial y}\mathbf{j} + \frac{\partial \xi}{\partial z}\mathbf{k} = \frac{1}{\sqrt{g}}(\mathbf{a}_2 \times \mathbf{a}_3) = \frac{1}{\sqrt{g}}\begin{vmatrix} \mathbf{i} & \mathbf{j} & \mathbf{k} \\ \frac{\partial x}{\partial \eta} & \frac{\partial y}{\partial \eta} & \frac{\partial z}{\partial \eta} \\ \frac{\partial x}{\partial \zeta} & \frac{\partial y}{\partial \zeta} & \frac{\partial z}{\partial \zeta} \end{vmatrix}, \quad (124)$$

where the Cartesian components are

$$\frac{\partial \xi}{\partial x} = \frac{1}{\sqrt{g}}\left(\frac{\partial y}{\partial \eta}\frac{\partial z}{\partial \zeta} - \frac{\partial z}{\partial \eta}\frac{\partial y}{\partial \zeta}\right), \quad (125)$$

$$\frac{\partial \xi}{\partial y} = -\frac{1}{\sqrt{g}}\left(\frac{\partial x}{\partial \eta}\frac{\partial z}{\partial \zeta} - \frac{\partial z}{\partial \eta}\frac{\partial x}{\partial \zeta}\right), \quad (126)$$

$$\frac{\partial \xi}{\partial z} = \frac{1}{\sqrt{g}}\left(\frac{\partial x}{\partial \eta}\frac{\partial y}{\partial \zeta} - \frac{\partial y}{\partial \eta}\frac{\partial x}{\partial \zeta}\right). \quad (127)$$

Similarly we have for the other vectors

$$\mathbf{a}^2 := \frac{\partial \eta}{\partial x}\mathbf{i} + \frac{\partial \eta}{\partial y}\mathbf{j} + \frac{\partial \eta}{\partial z}\mathbf{k} = \frac{1}{\sqrt{g}}(\mathbf{a}_3 \times \mathbf{a}_1) = \frac{1}{\sqrt{g}}\begin{vmatrix} \mathbf{i} & \mathbf{j} & \mathbf{k} \\ \frac{\partial x}{\partial \zeta} & \frac{\partial y}{\partial \zeta} & \frac{\partial z}{\partial \zeta} \\ \frac{\partial x}{\partial \xi} & \frac{\partial y}{\partial \xi} & \frac{\partial z}{\partial \xi} \end{vmatrix} \quad (128)$$

resulting in

$$\frac{\partial \eta}{\partial x} = \frac{1}{\sqrt{g}}\left(\frac{\partial y}{\partial \zeta}\frac{\partial z}{\partial \xi} - \frac{\partial z}{\partial \zeta}\frac{\partial y}{\partial \xi}\right), \quad (129)$$

$$\frac{\partial \eta}{\partial y} = -\frac{1}{\sqrt{g}} \left(\frac{\partial x}{\partial \zeta} \frac{\partial z}{\partial \xi} - \frac{\partial z}{\partial \zeta} \frac{\partial x}{\partial \xi} \right), \tag{130}$$

$$\frac{\partial \eta}{\partial z} = \frac{1}{\sqrt{g}} \left(\frac{\partial x}{\partial \zeta} \frac{\partial y}{\partial \xi} - \frac{\partial y}{\partial \zeta} \frac{\partial x}{\partial \xi} \right), \tag{131}$$

and

$$\mathbf{a}^3 := \frac{\partial \zeta}{\partial x} \mathbf{i} + \frac{\partial \zeta}{\partial y} \mathbf{j} + \frac{\partial \zeta}{\partial z} \mathbf{k} = \frac{1}{\sqrt{g}} (\mathbf{a}_1 \times \mathbf{a}_2) = \frac{1}{\sqrt{g}} \begin{vmatrix} \mathbf{i} & \mathbf{j} & \mathbf{k} \\ \frac{\partial x}{\partial \xi} & \frac{\partial y}{\partial \xi} & \frac{\partial z}{\partial \xi} \\ \frac{\partial x}{\partial \eta} & \frac{\partial y}{\partial \eta} & \frac{\partial z}{\partial \eta} \end{vmatrix} \tag{132}$$

resulting in

$$\frac{\partial \zeta}{\partial x} = \frac{1}{\sqrt{g}} \left(\frac{\partial y}{\partial \xi} \frac{\partial z}{\partial \eta} - \frac{\partial z}{\partial \xi} \frac{\partial y}{\partial \eta} \right), \tag{133}$$

$$\frac{\partial \zeta}{\partial y} = -\frac{1}{\sqrt{g}} \left(\frac{\partial x}{\partial \xi} \frac{\partial z}{\partial \eta} - \frac{\partial z}{\partial \xi} \frac{\partial x}{\partial \eta} \right), \tag{134}$$

$$\frac{\partial \zeta}{\partial z} = \frac{1}{\sqrt{g}} \left(\frac{\partial x}{\partial \xi} \frac{\partial y}{\partial \eta} - \frac{\partial y}{\partial \xi} \frac{\partial x}{\partial \eta} \right). \tag{135}$$

A plausibility proof of this computation is obtained by comparing the partial derivatives with the components of Eq. (8).

Restatement of the conservative derivative operations:

In what follows the *covariant* differential operators in conservative form are given.

Divergence:

$$\nabla \cdot \mathbf{F} = \frac{1}{\sqrt{g}} \left[\frac{\partial}{\partial \xi} \left(\sqrt{g}\, \mathbf{a}^1 \cdot \mathbf{F} \right) + \frac{\partial}{\partial \eta} \left(\sqrt{g}\, \mathbf{a}^2 \cdot \mathbf{F} \right) + \frac{\partial}{\partial \zeta} \left(\sqrt{g}\, \mathbf{a}^3 \cdot \mathbf{F} \right) \right]. \tag{136}$$

Curl:

$$\nabla \times \mathbf{F} = \frac{1}{\sqrt{g}} \left[\frac{\partial}{\partial \xi}\left(\sqrt{g}\, \mathbf{a}^1 \times \mathbf{F}\right) + \frac{\partial}{\partial \eta}\left(\sqrt{g}\, \mathbf{a}^2 \times \mathbf{F}\right) + \frac{\partial}{\partial \zeta}\left(\sqrt{g}\, \mathbf{a}^3 \times \mathbf{F}\right) \right]. \quad (137)$$

Gradient:

$$\nabla \varphi = \frac{1}{\sqrt{g}} \left[\frac{\partial}{\partial \xi}\left(\sqrt{g}\, \mathbf{a}^1 \varphi\right) + \frac{\partial}{\partial \eta}\left(\sqrt{g}\, \mathbf{a}^2 \varphi\right) + \frac{\partial}{\partial \zeta}\left(\sqrt{g}\, \mathbf{a}^3 \varphi\right) \right]. \quad (138)$$

From this expression the conservative expressions for the first derivatives result

$$\frac{\partial \varphi}{\partial x} = \frac{1}{\sqrt{g}} \left[\frac{\partial}{\partial \xi}\left(\sqrt{g}\, a^{11} \varphi\right) + \frac{\partial}{\partial \eta}\left(\sqrt{g}\, a^{12} \varphi\right) + \frac{\partial}{\partial \zeta}\left(\sqrt{g}\, a^{13} \varphi\right) \right], \quad (139)$$

$$\frac{\partial \varphi}{\partial y} = \frac{1}{\sqrt{g}} \left[\frac{\partial}{\partial \xi}\left(\sqrt{g}\, a^{21} \varphi\right) + \frac{\partial}{\partial \eta}\left(\sqrt{g}\, a^{22} \varphi\right) + \frac{\partial}{\partial \zeta}\left(\sqrt{g}\, a^{33} \varphi\right) \right], \quad (140)$$

$$\frac{\partial \varphi}{\partial z} = \frac{1}{\sqrt{g}} \left[\frac{\partial}{\partial \xi}\left(\sqrt{g}\, a^{31} \varphi\right) + \frac{\partial}{\partial \eta}\left(\sqrt{g}\, a^{32} \varphi\right) + \frac{\partial}{\partial \zeta}\left(\sqrt{g}\, a^{33} \varphi\right) \right]. \quad (141)$$

Using the first fundamental metric identity the above expressions reduce to the simple non-conservative form, which are also immediately obtained by the chain rule. If conservative integration schemes are designed the use of the conservative form is strongly recommended.

Laplacian:

$$\nabla^2 \varphi = \nabla \cdot (\nabla \varphi)$$

$$= \frac{1}{\sqrt{g}} \begin{cases} \dfrac{\partial}{\partial \xi}\left\{ \mathbf{a}^1 \cdot \left[\dfrac{\partial}{\partial \xi}\left(\sqrt{g}\, \mathbf{a}^1 \varphi\right) + \dfrac{\partial}{\partial \eta}\left(\sqrt{g}\, \mathbf{a}^2 \varphi\right) + \dfrac{\partial}{\partial \zeta}\left(\sqrt{g}\, \mathbf{a}^3 \varphi\right) \right] \right\} \\ + \dfrac{\partial}{\partial \eta}\left\{ \mathbf{a}^2 \cdot \left[\dfrac{\partial}{\partial \xi}\left(\sqrt{g}\, \mathbf{a}^1 \varphi\right) + \dfrac{\partial}{\partial \eta}\left(\sqrt{g}\, \mathbf{a}^2 \varphi\right) + \dfrac{\partial}{\partial \zeta}\left(\sqrt{g}\, \mathbf{a}^3 \varphi\right) \right] \right\} \\ + \dfrac{\partial}{\partial \zeta}\left\{ \mathbf{a}^3 \cdot \left[\dfrac{\partial}{\partial \xi}\left(\sqrt{g}\, \mathbf{a}^1 \varphi\right) + \dfrac{\partial}{\partial \eta}\left(\sqrt{g}\, \mathbf{a}^2 \varphi\right) + \dfrac{\partial}{\partial \zeta}\left(\sqrt{g}\, \mathbf{a}^3 \varphi\right) \right] \right\} \end{cases}.$$

(142)

Diffusion Laplacian: Usually the Laplacian appears in the diffusion terms of the conservation equations in the form

$$\nabla \cdot (\lambda \nabla \varphi) = \frac{1}{\sqrt{g}} \begin{cases} \dfrac{\partial}{\partial \xi}\left\{ \mathbf{a}^1 \cdot \lambda \left[\dfrac{\partial}{\partial \xi}\left(\sqrt{g}\, \mathbf{a}^1 \varphi\right) + \dfrac{\partial}{\partial \eta}\left(\sqrt{g}\, \mathbf{a}^2 \varphi\right) + \dfrac{\partial}{\partial \zeta}\left(\sqrt{g}\, \mathbf{a}^3 \varphi\right) \right] \right\} \\ + \dfrac{\partial}{\partial \eta}\left\{ \mathbf{a}^2 \cdot \lambda \left[\dfrac{\partial}{\partial \xi}\left(\sqrt{g}\, \mathbf{a}^1 \varphi\right) + \dfrac{\partial}{\partial \eta}\left(\sqrt{g}\, \mathbf{a}^2 \varphi\right) + \dfrac{\partial}{\partial \zeta}\left(\sqrt{g}\, \mathbf{a}^3 \varphi\right) \right] \right\} \\ + \dfrac{\partial}{\partial \zeta}\left\{ \mathbf{a}^3 \cdot \lambda \left[\dfrac{\partial}{\partial \xi}\left(\sqrt{g}\, \mathbf{a}^1 \varphi\right) + \dfrac{\partial}{\partial \eta}\left(\sqrt{g}\, \mathbf{a}^2 \varphi\right) + \dfrac{\partial}{\partial \zeta}\left(\sqrt{g}\, \mathbf{a}^3 \varphi\right) \right] \right\} \end{cases},$$

(143)

where λ is a scalar-valued function of the local flow parameters.

Restatement of the non-conservative derivative operations:

In what follows the *covariant* differential operators in non-conservative form are given.

Divergence: $\nabla \cdot \mathbf{F} = \mathbf{a}^1 \cdot \dfrac{\partial \mathbf{F}}{\partial \xi} + \mathbf{a}^2 \cdot \dfrac{\partial \mathbf{F}}{\partial \eta} + \mathbf{a}^3 \cdot \dfrac{\partial \mathbf{F}}{\partial \zeta}.$ (144)

Curl: $\nabla \times \mathbf{F} = \mathbf{a}^1 \times \dfrac{\partial \mathbf{F}}{\partial \xi} + \mathbf{a}^2 \times \dfrac{\partial \mathbf{F}}{\partial \eta} + \mathbf{a}^3 \times \dfrac{\partial \mathbf{F}}{\partial \zeta}.$ (145)

Gradient: $\nabla \varphi = \mathbf{a}^1 \dfrac{\partial \varphi}{\partial \xi} + \mathbf{a}^2 \dfrac{\partial \varphi}{\partial \eta} + \mathbf{a}^3 \dfrac{\partial \varphi}{\partial \zeta}$. (146)

From this expression the non-conservative expressions for the first derivatives result

$$\dfrac{\partial \varphi}{\partial x} = a^{11} \dfrac{\partial \varphi}{\partial \xi} + a^{21} \dfrac{\partial \varphi}{\partial \eta} + a^{31} \dfrac{\partial \varphi}{\partial \zeta}, \qquad (147)$$

$$\dfrac{\partial \varphi}{\partial y} = a^{12} \dfrac{\partial \varphi}{\partial \xi} + a^{22} \dfrac{\partial \varphi}{\partial \eta} + a^{32} \dfrac{\partial \varphi}{\partial \zeta}, \qquad (148)$$

$$\dfrac{\partial \varphi}{\partial z} = a^{13} \dfrac{\partial \varphi}{\partial \xi} + a^{23} \dfrac{\partial \varphi}{\partial \eta} + a^{33} \dfrac{\partial \varphi}{\partial \zeta}, \qquad (149)$$

with the second superscript indicating the Cartesian components of the contravariant vectors.

Laplacian:

$$\nabla^2 \varphi = \nabla \cdot (\nabla \varphi) = \dfrac{1}{\sqrt{g}} \left\{ \begin{array}{l} \dfrac{\partial}{\partial \xi}\left[\sqrt{g}\, \mathbf{a}^1 \cdot \left(\mathbf{a}^1 \dfrac{\partial \varphi}{\partial \xi} + \mathbf{a}^2 \dfrac{\partial \varphi}{\partial \eta} + \mathbf{a}^3 \dfrac{\partial \varphi}{\partial \zeta}\right)\right] \\[1em] + \dfrac{\partial}{\partial \eta}\left[\sqrt{g}\, \mathbf{a}^2 \cdot \left(\mathbf{a}^1 \dfrac{\partial \varphi}{\partial \xi} + \mathbf{a}^2 \dfrac{\partial \varphi}{\partial \eta} + \mathbf{a}^3 \dfrac{\partial \varphi}{\partial \zeta}\right)\right] \\[1em] + \dfrac{\partial}{\partial \zeta}\left[\sqrt{g}\, \mathbf{a}^3 \cdot \left(\mathbf{a}^1 \dfrac{\partial \varphi}{\partial \xi} + \mathbf{a}^2 \dfrac{\partial \varphi}{\partial \eta} + \mathbf{a}^3 \dfrac{\partial \varphi}{\partial \zeta}\right)\right] \end{array} \right\}$$

$$= \dfrac{1}{\sqrt{g}} \left\{ \begin{array}{l} \dfrac{\partial}{\partial \xi}\left[\sqrt{g}\left(g^{11} \dfrac{\partial \varphi}{\partial \xi} + g^{12} \dfrac{\partial \varphi}{\partial \eta} + g^{13} \dfrac{\partial \varphi}{\partial \zeta}\right)\right] \\[1em] + \dfrac{\partial}{\partial \eta}\left[\sqrt{g}\left(g^{21} \dfrac{\partial \varphi}{\partial \xi} + g^{22} \dfrac{\partial \varphi}{\partial \eta} + g^{23} \dfrac{\partial \varphi}{\partial \zeta}\right)\right] \\[1em] + \dfrac{\partial}{\partial \zeta}\left[\sqrt{g}\left(g^{31} \dfrac{\partial \varphi}{\partial \xi} + g^{32} \dfrac{\partial \varphi}{\partial \eta} + g^{33} \dfrac{\partial \varphi}{\partial \zeta}\right)\right] \end{array} \right\}. \qquad (150)$$

Diffusion Laplacian:

$$\nabla \cdot (\lambda \nabla \varphi) = \frac{1}{\sqrt{g}} \left\{ \begin{array}{l} \dfrac{\partial}{\partial \xi} \left[\sqrt{g}\, \mathbf{a}^1 \cdot \lambda \left(\mathbf{a}^1 \dfrac{\partial \varphi}{\partial \xi} + \mathbf{a}^2 \dfrac{\partial \varphi}{\partial \eta} + \mathbf{a}^3 \dfrac{\partial \varphi}{\partial \zeta} \right) \right] \\[2ex] + \dfrac{\partial}{\partial \eta} \left[\sqrt{g}\, \mathbf{a}^2 \cdot \lambda \left(\mathbf{a}^1 \dfrac{\partial \varphi}{\partial \xi} + \mathbf{a}^2 \dfrac{\partial \varphi}{\partial \eta} + \mathbf{a}^3 \dfrac{\partial \varphi}{\partial \zeta} \right) \right] \\[2ex] + \dfrac{\partial}{\partial \zeta} \left[\sqrt{g}\, \mathbf{a}^3 \cdot \lambda \left(\mathbf{a}^1 \dfrac{\partial \varphi}{\partial \xi} + \mathbf{a}^2 \dfrac{\partial \varphi}{\partial \eta} + \mathbf{a}^3 \dfrac{\partial \varphi}{\partial \zeta} \right) \right] \end{array} \right\}$$

$$= \frac{1}{\sqrt{g}} \left\{ \begin{array}{l} \dfrac{\partial}{\partial \xi} \left[\sqrt{g}\, \lambda \left(g^{11} \dfrac{\partial \varphi}{\partial \xi} + g^{12} \dfrac{\partial \varphi}{\partial \eta} + g^{13} \dfrac{\partial \varphi}{\partial \zeta} \right) \right] \\[2ex] + \dfrac{\partial}{\partial \eta} \left[\sqrt{g}\, \lambda \left(g^{21} \dfrac{\partial \varphi}{\partial \xi} + g^{22} \dfrac{\partial \varphi}{\partial \eta} + g^{33} \dfrac{\partial \varphi}{\partial \zeta} \right) \right] \\[2ex] + \dfrac{\partial}{\partial \zeta} \left[\sqrt{g}\, \lambda \left(g^{31} \dfrac{\partial \varphi}{\partial \xi} + g^{32} \dfrac{\partial \varphi}{\partial \eta} + g^{33} \dfrac{\partial \varphi}{\partial \zeta} \right) \right] \end{array} \right\}. \quad (151)$$

Divergence of a tensor: Consider the dyadic product of the vector **V**, **VV**, which is a second order tensor. The divergence of this tensor is then

$$\nabla \cdot (\mathbf{VV}) = \left[\nabla \cdot (\mathbf{V}u) \right] \mathbf{i} + \left[\nabla \cdot (\mathbf{V}v) \right] \mathbf{j} + \left[\nabla \cdot (\mathbf{V}w) \right] \mathbf{k}. \quad (152)$$

In this case the already derived expressions for divergence of vectors can be used. Another example of the divergence of a tensor is the expression

$$\nabla \cdot (\eta \nabla \mathbf{V})^T = \left[\nabla \cdot \left(\eta \frac{\partial \mathbf{V}}{\partial x} \right) \right] \mathbf{i} + \left[\nabla \cdot \left(\eta \frac{\partial \mathbf{V}}{\partial y} \right) \right] \mathbf{j} + \left[\nabla \cdot \left(\eta \frac{\partial \mathbf{V}}{\partial z} \right) \right] \mathbf{k}. \quad (153)$$

Also in this case the already derived expressions for divergence of vectors can be used.

Diffusion Laplacian of a tensor: Usually the diffusion Laplacian of a tensor appears in the diffusion terms of the momentum conservation equations in the form

$$\nabla \cdot (\eta \nabla \mathbf{V})$$

which is equivalent to

$$\nabla \cdot (\eta \nabla \mathbf{V}) = [\nabla \cdot (\eta \nabla u)]\mathbf{i} + [\nabla \cdot (\eta \nabla v)]\mathbf{j} + [\nabla \cdot (\eta \nabla w)]\mathbf{k} . \tag{154}$$

In this case the already derived expressions for each component are used.

Derivatives along the normal to the curvilinear coordinate surface: If $\varphi(\xi,\eta,\zeta)$ is a scalar-valued function of the main variables ξ,η,ζ, through the three intermediate variables x,y,z, then the derivatives along the normal to the curvilinear coordinate surface are

$$\xi = const, \quad \frac{\partial \varphi}{\partial \xi^*} = \frac{\mathbf{a}^1}{|\mathbf{a}^1|} \cdot \nabla \varphi = \mathbf{e}^1 \cdot \nabla \varphi, \tag{155}$$

$$\eta = const, \quad \frac{\partial \varphi}{\partial \eta^*} = \frac{\mathbf{a}^2}{|\mathbf{a}^2|} \cdot \nabla \varphi = \mathbf{e}^2 \cdot \nabla \varphi, \tag{156}$$

$$\zeta = const, \quad \frac{\partial \varphi}{\partial \zeta^*} = \frac{\mathbf{a}^3}{|\mathbf{a}^3|} \cdot \nabla \varphi = \mathbf{e}^3 \cdot \nabla \varphi. \tag{157}$$

Derivatives along the tangent of the curvilinear coordinate lines: If $\varphi(\xi,\eta,\zeta)$ is a scalar-valued function of the main variables ξ,η,ζ, through the three intermediate variables x,y,z, then

$$\frac{\partial \varphi}{\partial \xi} = \frac{\partial \varphi}{\partial x}\frac{\partial x}{\partial \xi} + \frac{\partial \varphi}{\partial y}\frac{\partial y}{\partial \xi} + \frac{\partial \varphi}{\partial z}\frac{\partial z}{\partial \xi} = \left(\frac{\partial \varphi}{\partial x}\mathbf{i} + \frac{\partial \varphi}{\partial y}\mathbf{j} + \frac{\partial \varphi}{\partial z}\mathbf{k}\right) \cdot \left(\frac{\partial x}{\partial \xi}\mathbf{i} + \frac{\partial y}{\partial \xi}\mathbf{j} + \frac{\partial z}{\partial \xi}\mathbf{k}\right)$$

$$= (\nabla \varphi) \cdot \frac{\partial \mathbf{r}}{\partial \xi} = \mathbf{a}_1 \cdot \nabla \varphi, \tag{158}$$

$$\frac{\partial \varphi}{\partial \eta} = \frac{\partial \varphi}{\partial x}\frac{\partial x}{\partial \eta} + \frac{\partial \varphi}{\partial y}\frac{\partial y}{\partial \eta} + \frac{\partial \varphi}{\partial z}\frac{\partial z}{\partial \eta} = \left(\frac{\partial \varphi}{\partial x}\mathbf{i} + \frac{\partial \varphi}{\partial y}\mathbf{j} + \frac{\partial \varphi}{\partial z}\mathbf{k}\right) \cdot \left(\frac{\partial x}{\partial \eta}\mathbf{i} + \frac{\partial y}{\partial \eta}\mathbf{j} + \frac{\partial z}{\partial \eta}\mathbf{k}\right)$$

$$=(\nabla\varphi)\cdot\frac{\partial \mathbf{r}}{\partial \eta}=\mathbf{a}_2\cdot\nabla\varphi, \qquad (159)$$

$$\frac{\partial\varphi}{\partial\zeta}=\frac{\partial\varphi}{\partial x}\frac{\partial x}{\partial\zeta}+\frac{\partial\varphi}{\partial y}\frac{\partial y}{\partial\zeta}+\frac{\partial\varphi}{\partial z}\frac{\partial z}{\partial\zeta}=\left(\frac{\partial\varphi}{\partial x}\mathbf{i}+\frac{\partial\varphi}{\partial y}\mathbf{j}+\frac{\partial\varphi}{\partial z}\mathbf{k}\right)\cdot\left(\frac{\partial x}{\partial\zeta}\mathbf{i}+\frac{\partial y}{\partial\zeta}\mathbf{j}+\frac{\partial z}{\partial\zeta}\mathbf{k}\right)$$

$$=(\nabla\varphi)\cdot\frac{\partial \mathbf{r}}{\partial \zeta}=\mathbf{a}_3\cdot\nabla\varphi. \qquad (160)$$

These are the three directional derivatives of $\varphi(x,y,z)$ along ξ,η,ζ, respectively. Another quick check is obtained by replacing the non-conservative form of the gradient in the transformed coordinate systems and using the dual properties of the co- and contravariant vectors.

Derivatives along the normal to the coordinate lines and tangent to the curvilinear coordinate surface: The vector $\mathbf{a}^i\times\mathbf{a}_i$ is normal to the coordinate line on which ξ^i varies and is also tangent to the coordinate surface on which ξ^i is constant. Using the relations between the covariant and the contravariant vectors and the vector identity

$$(\mathbf{B}\times\mathbf{C})\times\mathbf{A} = -\big[(\mathbf{A}\cdot\mathbf{C})\mathbf{B}-(\mathbf{A}\cdot\mathbf{B})\mathbf{C}\big] \qquad (161)$$

we have

$$\mathbf{a}^1\times\mathbf{a}_1 = \frac{1}{\sqrt{g}}(\mathbf{a}_2\times\mathbf{a}_3)\times\mathbf{a}_1 = -\frac{1}{\sqrt{g}}\big[(\mathbf{a}_1\cdot\mathbf{a}_3)\mathbf{a}_2-(\mathbf{a}_1\cdot\mathbf{a}_2)\mathbf{a}_3\big]$$

$$= -\frac{1}{\sqrt{g}}(g_{13}\mathbf{a}_2-g_{12}\mathbf{a}_3), \qquad (162)$$

$$\mathbf{a}^2\times\mathbf{a}_2 = \frac{1}{\sqrt{g}}(\mathbf{a}_3\times\mathbf{a}_1)\times\mathbf{a}_2 = -\frac{1}{\sqrt{g}}\big[(\mathbf{a}_2\cdot\mathbf{a}_1)\mathbf{a}_3-(\mathbf{a}_2\cdot\mathbf{a}_3)\mathbf{a}_1\big]$$

$$= -\frac{1}{\sqrt{g}}(g_{21}\mathbf{a}_3-g_{23}\mathbf{a}_1), \qquad (163)$$

$$\mathbf{a}^3\times\mathbf{a}_3 = \frac{1}{\sqrt{g}}(\mathbf{a}_1\times\mathbf{a}_2)\times\mathbf{a}_3 = -\frac{1}{\sqrt{g}}\big[(\mathbf{a}_3\cdot\mathbf{a}_2)\mathbf{a}_1-(\mathbf{a}_3\cdot\mathbf{a}_1)\mathbf{a}_2\big]$$

$$= -\frac{1}{\sqrt{g}}(g_{32}\mathbf{a}_1 - g_{31}\mathbf{a}_2). \tag{164}$$

The magnitude is then

$$|\mathbf{a}^1 \times \mathbf{a}_1|^2 = \frac{1}{g}\left(g_{13}^2 g_{22} - 2g_{12}g_{13}g_{32} + g_{12}^2 g_{33}\right)$$

$$= \frac{1}{g}\left[g_{13}\left(g_{13}g_{22} - g_{12}g_{32}\right) - g_{12}\left(g_{13}g_{32} - g_{12}g_{33}\right)\right], \tag{165}$$

$$|\mathbf{a}^2 \times \mathbf{a}_2|^2 = \frac{1}{g}\left(g_{21}^2 g_{33} - 2g_{21}g_{23}g_{31} + g_{23}^2 g_{11}\right)$$

$$= \frac{1}{g}\left[g_{21}\left(g_{21}g_{33} - g_{23}g_{31}\right) - g_{23}\left(g_{21}g_{31} - g_{23}g_{11}\right)\right], \tag{166}$$

$$|\mathbf{a}^3 \times \mathbf{a}_3|^2 = \frac{1}{g}\left(g_{32}^2 g_{11} - 2g_{31}g_{21}g_{32} + g_{31}^2 g_{22}\right)$$

$$= \frac{1}{g}\left[g_{32}\left(g_{32}g_{11} - g_{31}g_{21}\right) - g_{31}\left(g_{21}g_{32} - g_{31}g_{22}\right)\right]. \tag{167}$$

Comparing with

$$g = g_{11}\left(g_{22}g_{33} - g_{23}^2\right) - g_{12}\left(g_{21}g_{33} - g_{23}g_{31}\right) + g_{13}\left(g_{21}g_{32} - g_{22}g_{31}\right) \tag{168}$$

results in computationally cheaper expressions

$$|\mathbf{a}^1 \times \mathbf{a}_1|^2 = \frac{1}{g}g_{11}\left(g_{22}g_{33} - g_{23}^2\right) - 1, \tag{169}$$

$$|\mathbf{a}^2 \times \mathbf{a}_2|^2 = \frac{1}{g}g_{22}\left(g_{11}g_{33} - g_{31}^2\right) - 1, \tag{170}$$

$$|\mathbf{a}^3 \times \mathbf{a}_3|^2 = \frac{1}{g}g_{33}\left(g_{11}g_{22} - g_{21}^2\right) - 1. \tag{171}$$

With this result the derivatives normal to the coordinate line on which ξ^i varies and are also tangent to the coordinate surface on which ξ^i is constant are

$$\frac{\mathbf{a}^1 \times \mathbf{a}_1}{|\mathbf{a}^1 \times \mathbf{a}_1|} \cdot \nabla \varphi = -\frac{1}{\sqrt{g_{11}\left(g_{22}g_{33} - g_{23}^2\right) - g}} \left(g_{13}\mathbf{a}_2 - g_{12}\mathbf{a}_3\right) \cdot \nabla \varphi, \qquad (172)$$

$$\frac{\mathbf{a}^2 \times \mathbf{a}_2}{|\mathbf{a}^2 \times \mathbf{a}_2|} \cdot \nabla \varphi = -\frac{1}{\sqrt{g_{22}\left(g_{11}g_{33} - g_{31}^2\right) - g}} \left(g_{21}\mathbf{a}_3 - g_{23}\mathbf{a}_1\right) \cdot \nabla \varphi, \qquad (173)$$

$$\frac{\mathbf{a}^3 \times \mathbf{a}_3}{|\mathbf{a}^3 \times \mathbf{a}_3|} \cdot \nabla \varphi = -\frac{1}{\sqrt{g_{33}\left(g_{11}g_{22} - g_{21}^2\right) - g}} \left(g_{32}\mathbf{a}_1 - g_{31}\mathbf{a}_2\right) \cdot \nabla \varphi. \qquad (174)$$

Replacing the non-conservative form of the gradient in the transformed coordinate systems and using the dual properties of the co- and contravariant vectors results in

$$\frac{\mathbf{a}^1 \times \mathbf{a}_1}{|\mathbf{a}^1 \times \mathbf{a}_1|} \cdot \nabla \varphi = -\frac{g_{13}\dfrac{\partial \varphi}{\partial \eta} - g_{12}\dfrac{\partial \varphi}{\partial \zeta}}{\sqrt{g_{11}\left(g_{22}g_{33} - g_{23}^2\right) - g}}, \qquad (175)$$

$$\frac{\mathbf{a}^2 \times \mathbf{a}_2}{|\mathbf{a}^2 \times \mathbf{a}_2|} \cdot \nabla \varphi = -\frac{g_{23}\dfrac{\partial \varphi}{\partial \xi} - g_{21}\dfrac{\partial \varphi}{\partial \zeta}}{\sqrt{g_{22}\left(g_{11}g_{33} - g_{31}^2\right) - g}}, \qquad (176)$$

$$\frac{\mathbf{a}^3 \times \mathbf{a}_3}{|\mathbf{a}^3 \times \mathbf{a}_3|} \cdot \nabla \varphi = -\frac{g_{32}\dfrac{\partial \varphi}{\partial \xi} - g_{31}\dfrac{\partial \varphi}{\partial \eta}}{\sqrt{g_{33}\left(g_{11}g_{22} - g_{21}^2\right) - g}}. \qquad (177)$$

Moving coordinate system: Consider a coordinate transformation defined by

$$x = f(\xi, \eta, \zeta), \qquad (178)$$

$$y = g(\xi, \eta, \zeta), \qquad (179)$$

$$z = h(\xi, \eta, \zeta), \qquad (180)$$

which moves with time in space. The velocity

$$\mathbf{V}_{cs} = \left(\frac{\partial x}{\partial \tau}\right)_{\eta,\xi,\zeta} \mathbf{i} + \left(\frac{\partial y}{\partial \tau}\right)_{\eta,\xi,\zeta} \mathbf{j} + \left(\frac{\partial z}{\partial \tau}\right)_{\eta,\xi,\zeta} \mathbf{k}, \qquad (181)$$

is the *grid point velocity*. Note an interesting property of the time derivative of the covariant unit vectors

$$\frac{\partial \mathbf{a}_1}{\partial \tau} = \frac{\partial}{\partial \tau}\left(\frac{\partial \mathbf{r}}{\partial \xi}\right) = \frac{\partial}{\partial \xi}\left(\frac{\partial \mathbf{r}}{\partial \tau}\right) = \frac{\partial \mathbf{V}_{cs}}{\partial \xi}, \qquad (182)$$

$$\frac{\partial \mathbf{a}_2}{\partial \tau} = \frac{\partial}{\partial \tau}\left(\frac{\partial \mathbf{r}}{\partial \eta}\right) = \frac{\partial}{\partial \eta}\left(\frac{\partial \mathbf{r}}{\partial \tau}\right) = \frac{\partial \mathbf{V}_{cs}}{\partial \eta}, \qquad (183)$$

$$\frac{\partial \mathbf{a}_3}{\partial \tau} = \frac{\partial}{\partial \tau}\left(\frac{\partial \mathbf{r}}{\partial \zeta}\right) = \frac{\partial}{\partial \zeta}\left(\frac{\partial \mathbf{r}}{\partial \tau}\right) = \frac{\partial \mathbf{V}_{cs}}{\partial \zeta}. \qquad (184)$$

If $f = f(\tau, x, y, z)$ is a function of time and space using the chain rule we obtain

$$\frac{df}{d\tau} = \left(\frac{\partial f}{\partial \tau}\right)_{x,y,z}$$
$$+ \left[\left(\frac{\partial f}{\partial x}\right)_{t,y,z} \mathbf{i} + \left(\frac{\partial f}{\partial y}\right)_{t,x,z} \mathbf{j} + \left(\frac{\partial f}{\partial z}\right)_{t,x,y} \mathbf{k}\right] \cdot \left[\left(\frac{dx}{d\tau}\right)_{\xi,\eta,\zeta} \mathbf{i} + \left(\frac{dy}{d\tau}\right)_{\xi,\eta,\zeta} \mathbf{j} + \left(\frac{dz}{d\tau}\right)_{\xi,\eta,\zeta} \mathbf{k}\right]. \qquad (185)$$

Comparing the second part of the right hand side of the above equation with the definition of the gradient and of the grid point velocity we see that

$$\left(\frac{df}{d\tau}\right)_{\xi,\eta,\zeta} = \left(\frac{\partial f}{\partial \tau}\right)_{x,y,z} + (\nabla f) \cdot \mathbf{V}_{cs}, \qquad (186)$$

the absolute change with the time of a function f is the velocity with the respect to the stationary coordinate system xyz plus a component resulting from the movement of the coordinate system ξ, η, ζ. Therefore

$$\left(\frac{\partial f}{\partial \tau}\right)_{x,y,z} = \left(\frac{\partial f}{\partial \tau}\right)_{\xi,\eta,\zeta} - \mathbf{V}_{cs} \cdot \nabla f. \qquad (187)$$

Here the time derivative $(df/dt)_{\xi,\eta,\zeta}$ is understood to be at a fixed point in the transformed region. In a moving coordinate system the orientation of the unit covariant vectors will change. That is the elementary parallelepiped volume built by these vectors, the Jacobian, changes also. The derivative of the Jacobian of the coordinate transformation with time can be visualized by doting the change of the normal component to the covariant vector with the corresponding elementary coordinate surfaces formed by the other two vectors

$$\left(\frac{d\sqrt{g}}{d\tau}\right)_{\xi,\eta,\zeta} = \frac{d\mathbf{a}_1}{d\tau}\cdot(\mathbf{a}_2\times\mathbf{a}_3) + \frac{d\mathbf{a}_2}{d\tau}\cdot(\mathbf{a}_3\times\mathbf{a}_1) + \frac{d\mathbf{a}_3}{d\tau}\cdot(\mathbf{a}_1\times\mathbf{a}_2) \qquad (188)$$

or

$$\frac{d\sqrt{g}}{d\tau} = \sqrt{g}\left(\frac{d\mathbf{a}_1}{d\tau}\cdot\mathbf{a}^1 + \frac{d\mathbf{a}_2}{d\tau}\cdot\mathbf{a}^2 + \frac{d\mathbf{a}_3}{d\tau}\cdot\mathbf{a}^3\right), \qquad (189)$$

see p.130 [7]. Having in mind that

$$\frac{\partial \mathbf{a}_1}{\partial \tau} = \frac{\partial \mathbf{V}_{cs}}{\partial \xi}, \qquad (190)$$

$$\frac{\partial \mathbf{a}_2}{\partial \tau} = \frac{\partial \mathbf{V}_{cs}}{\partial \eta}, \qquad (191)$$

$$\frac{\partial \mathbf{a}_3}{\partial \tau} = \frac{\partial \mathbf{V}_{cs}}{\partial \zeta}, \qquad (192)$$

results in

$$\frac{d\sqrt{g}}{d\tau} = \sqrt{g}\left(\mathbf{a}^1\cdot\frac{\partial \mathbf{V}_{cs}}{\partial \xi} + \mathbf{a}^2\cdot\frac{\partial \mathbf{V}_{cs}}{\partial \eta} + \mathbf{a}^3\cdot\frac{\partial \mathbf{V}_{cs}}{\partial \zeta}\right). \qquad (193)$$

Note some very interesting properties of the following expression

$$\frac{\partial}{\partial \xi}\left(\sqrt{g}\,\mathbf{a}^1 f\cdot\mathbf{V}_{cs}\right) + \frac{\partial}{\partial \eta}\left(\sqrt{g}\,\mathbf{a}^2 f\cdot\mathbf{V}_{cs}\right) + \frac{\partial}{\partial \zeta}\left(\sqrt{g}\,\mathbf{a}^3 f\cdot\mathbf{V}_{cs}\right)$$

$$= f\sqrt{g}\left(\mathbf{a}^1\cdot\frac{\partial \mathbf{V}_{cs}}{\partial \xi} + \mathbf{a}^2\cdot\frac{\partial \mathbf{V}_{cs}}{\partial \eta} + \mathbf{a}^3\cdot\frac{\partial \mathbf{V}_{cs}}{\partial \zeta}\right)$$

$$+\left[\frac{\partial}{\partial \xi}\left(\sqrt{g}\,\mathbf{a}^1 f\right)+\frac{\partial}{\partial \eta}\left(\sqrt{g}\,\mathbf{a}^2 f\right)+\frac{\partial}{\partial \zeta}\left(\sqrt{g}\,\mathbf{a}^3 f\right)\right]\cdot \mathbf{V}_{cs}$$

$$= f\frac{d\sqrt{g}}{d\tau}+\left[\frac{\partial}{\partial \xi}\left(\sqrt{g}\,\mathbf{a}^1 f\right)+\frac{\partial}{\partial \eta}\left(\sqrt{g}\,\mathbf{a}^2 f\right)+\frac{\partial}{\partial \zeta}\left(\sqrt{g}\,\mathbf{a}^3 f\right)\right]\cdot \mathbf{V}_{cs} \quad (194)$$

Transformed fluid properties conservation equation: Consider $f = f(\tau, x, y, z)$ being a function of time and space which is a controlled by the following conservation equation

$$\left(\frac{\partial f}{\partial \tau}\right)_{x,y,z}+\nabla\cdot(f\mathbf{V})-\nabla\cdot(\lambda\nabla f)=\mu. \quad (195)$$

The convection-diffusion problem described by this equation is called *isotropic*. Using the expression for the time derivative, the conservative form of the divergence of a vector and gradient of a scalar in the transformed space, we obtain

$$\frac{\partial}{\partial \tau}\left(f\sqrt{g}\right)$$

$$+\frac{\partial}{\partial \xi}\left(f\sqrt{g}\,\mathbf{a}^1\cdot(\mathbf{V}-\mathbf{V}_{cs})-\mathbf{a}^1\cdot\lambda\left[\frac{\partial}{\partial \xi}\left(\sqrt{g}\,\mathbf{a}^1 f\right)+\frac{\partial}{\partial \eta}\left(\sqrt{g}\,\mathbf{a}^2 f\right)+\frac{\partial}{\partial \zeta}\left(\sqrt{g}\,\mathbf{a}^3 f\right)\right]\right)$$

$$+\frac{\partial}{\partial \eta}\left(f\sqrt{g}\,\mathbf{a}^2\cdot(\mathbf{V}-\mathbf{V}_{cs})-\mathbf{a}^2\cdot\lambda\left[\frac{\partial}{\partial \xi}\left(\sqrt{g}\,\mathbf{a}^1 f\right)+\frac{\partial}{\partial \eta}\left(\sqrt{g}\,\mathbf{a}^2 f\right)+\frac{\partial}{\partial \zeta}\left(\sqrt{g}\,\mathbf{a}^3 f\right)\right]\right)$$

$$+\frac{\partial}{\partial \zeta}\left(f\sqrt{g}\,\mathbf{a}^3\cdot(\mathbf{V}-\mathbf{V}_{cs})-\mathbf{a}^3\cdot\lambda\left[\frac{\partial}{\partial \xi}\left(\sqrt{g}\,\mathbf{a}^1 f\right)+\frac{\partial}{\partial \eta}\left(\sqrt{g}\,\mathbf{a}^2 f\right)+\frac{\partial}{\partial \zeta}\left(\sqrt{g}\,\mathbf{a}^3 f\right)\right]\right)$$

$$=\sqrt{g}\,\mu. \quad (196)$$

Using the non-conservative form of the Diffusion *Laplacian* results in a simpler form which is still conservative with respect to the spatial derivatives in the transformed coordinate system

$$\frac{\partial}{\partial \tau}\left(f\sqrt{g}\right)$$

$$+\frac{\partial}{\partial \xi}\left\{\sqrt{g}\left[f\mathbf{a}^1\cdot(\mathbf{V}-\mathbf{V}_{cs})-\lambda\left(\mathbf{a}^1\cdot\mathbf{a}^1\frac{\partial f}{\partial \xi}+\mathbf{a}^1\cdot\mathbf{a}^2\frac{\partial f}{\partial \eta}+\mathbf{a}^1\cdot\mathbf{a}^3\frac{\partial f}{\partial \zeta}\right)\right]\right\}$$

$$+\frac{\partial}{\partial \eta}\left\{\sqrt{g}\left[f\mathbf{a}^2\cdot(\mathbf{V}-\mathbf{V}_{cs})-\lambda\left(\mathbf{a}^2\cdot\mathbf{a}^1\frac{\partial f}{\partial \xi}+\mathbf{a}^2\cdot\mathbf{a}^2\frac{\partial f}{\partial \eta}+\mathbf{a}^2\cdot\mathbf{a}^3\frac{\partial f}{\partial \zeta}\right)\right]\right\}$$

$$+\frac{\partial}{\partial \zeta}\left\{\sqrt{g}\left[f\mathbf{a}^3\cdot(\mathbf{V}-\mathbf{V}_{cs})-\lambda\left(\mathbf{a}^3\cdot\mathbf{a}^1\frac{\partial f}{\partial \xi}+\mathbf{a}^3\cdot\mathbf{a}^2\frac{\partial f}{\partial \eta}+\mathbf{a}^3\cdot\mathbf{a}^3\frac{\partial f}{\partial \zeta}\right)\right]\right\}=\sqrt{g}\,\mu.$$

$$(197)$$

The equation can be rewritten in more compact form using the relative contravariant velocity components in the transformed space defined as follows

$$\overline{\mathbf{V}}=\begin{pmatrix}\mathbf{a}^1\cdot(\mathbf{V}-\mathbf{V}_{cs})\\ \mathbf{a}^2\cdot(\mathbf{V}-\mathbf{V}_{cs})\\ \mathbf{a}^3\cdot(\mathbf{V}-\mathbf{V}_{cs})\end{pmatrix}=\begin{pmatrix}\overline{V}^1\\ \overline{V}^2\\ \overline{V}^3\end{pmatrix} \qquad (198)$$

and the terms of the inverse metrics matrices

$$\frac{\partial}{\partial \tau}\left(f\sqrt{g}\right)$$

$$+\frac{\partial}{\partial \xi}\left\{\sqrt{g}\left[f\overline{V}^1-\lambda\left(g^{11}\frac{\partial f}{\partial \xi}+g^{12}\frac{\partial f}{\partial \eta}+g^{13}\frac{\partial f}{\partial \zeta}\right)\right]\right\}$$

$$+\frac{\partial}{\partial \eta}\left\{\sqrt{g}\left[f\overline{V}^2-\lambda\left(g^{21}\frac{\partial f}{\partial \xi}+g^{22}\frac{\partial f}{\partial \eta}+g^{23}\frac{\partial f}{\partial \zeta}\right)\right]\right\}$$

$$+\frac{\partial}{\partial \zeta}\left\{\sqrt{g}\left[f\overline{V}^3-\lambda\left(g^{31}\frac{\partial f}{\partial \xi}+g^{32}\frac{\partial f}{\partial \eta}+g^{33}\frac{\partial f}{\partial \zeta}\right)\right]\right\}=\sqrt{g}\,\mu. \qquad (199)$$

We see a remarkable property of this equation. In the transformed coordinate system the property $f\sqrt{g}$ having convection-diffusion flux i-components

$$\hat{I}^i = \sqrt{g}\left[f\mathbf{a}^i\cdot(\mathbf{V}-\mathbf{V}_{cs}) - \lambda\left(g^{i1}\frac{\partial f}{\partial \xi} + g^{i2}\frac{\partial f}{\partial \eta} + g^{i3}\frac{\partial f}{\partial \zeta}\right)\right] \tag{200}$$

is conserved. The convective component can be visualized as

$$\sqrt{g}\, f\mathbf{a}^i\cdot(\mathbf{V}-\mathbf{V}_{cs}) = \sqrt{g}\, f\left|\mathbf{a}^i\right|\mathbf{e}^i\cdot(\mathbf{V}-\mathbf{V}_{cs}) = \sqrt{g}\, f\left|\mathbf{a}^i\right|V^{n,i}. \tag{201}$$

Here

$$V^{n,i} = \mathbf{e}^i\cdot(\mathbf{V}-\mathbf{V}_{cs}) \tag{202}$$

is the relative velocity component normal to the surface defined by $\xi^i = const$. Remember Eqs. (117-119) giving

$$\sqrt{g}\,\mathbf{a}^i = \frac{\mathbf{S}^i}{d\xi^j d\xi^k}, \tag{203}$$

or

$$\sqrt{g}\left|\mathbf{a}^i\right| = \frac{\left|\mathbf{S}^i\right|}{d\xi^j d\xi^k}. \tag{204}$$

For equidistant discretization in the computational space we have

$$\sqrt{g}\left|\mathbf{a}^i\right| = \left|\mathbf{S}^i\right| = S^i. \tag{205}$$

Therefore the convective flux can be presented in several ways

$$\sqrt{g}\, f\mathbf{a}^i\cdot(\mathbf{V}-\mathbf{V}_{cs}) = \sqrt{g}\, f\left|\mathbf{a}^i\right|\mathbf{e}^i\cdot(\mathbf{V}-\mathbf{V}_{cs}) = S^i V^{n,i} f = \sqrt{g}\,\overline{V}^i f = \mathbf{S}^i\cdot(\mathbf{V}-\mathbf{V}_{cs})f. \tag{206}$$

With this the conservation equation can be rewritten in terms of the normal velocity components

$$\frac{\partial}{\partial \tau}\left(f\sqrt{g}\right)$$

$$+\frac{\partial}{\partial \xi}\left[S^1V^{n,1}f - \lambda\sqrt{g}\left(g^{11}\frac{\partial f}{\partial \xi} + g^{12}\frac{\partial f}{\partial \eta} + g^{13}\frac{\partial f}{\partial \zeta}\right)\right]$$

$$+\frac{\partial}{\partial \eta}\left[S^2V^{n,2}f - \lambda\sqrt{g}\left(g^{21}\frac{\partial f}{\partial \xi} + g^{22}\frac{\partial f}{\partial \eta} + g^{23}\frac{\partial f}{\partial \zeta}\right)\right]$$

$$+\frac{\partial}{\partial \zeta}\left[S^3V^{n,3}f - \lambda\sqrt{g}\left(g^{31}\frac{\partial f}{\partial \xi} + g^{32}\frac{\partial f}{\partial \eta} + g^{33}\frac{\partial f}{\partial \zeta}\right)\right] = \sqrt{g}\,\mu. \quad (207)$$

The convection-diffusion problem in the transformed space is called *anisotropic*. Thus, the isotropic convection-diffusion problem in the physical space turns out to be anisotropic in the transformed space.

Note that it is computationally more effective to store $g^{ij}\sqrt{g}$ instead of g^{ij} if $\sqrt{g}$ does not change with the time.

For an orthogonal transformed system the equation simplifies much more

$$\frac{\partial}{\partial \tau}\left(f\sqrt{g}\right) + \frac{\partial}{\partial \xi}\left[\sqrt{g}\left(f\bar{V}^1 - \lambda g^{11}\frac{\partial f}{\partial \xi}\right)\right] + \frac{\partial}{\partial \eta}\left[\sqrt{g}\left(f\bar{V}^2 - \lambda g^{22}\frac{\partial f}{\partial \eta}\right)\right]$$

$$+\frac{\partial}{\partial \zeta}\left[\sqrt{g}\left(f\bar{V}^3 - \lambda g^{33}\frac{\partial f}{\partial \zeta}\right)\right] = \sqrt{g}\,\mu. \quad (208)$$

The convection-diffusion problem in an orthogonal transformed system is called *orthotropic*.

Time dependent grid metric identity (second fundamental metric identity):
Consider a conservation equation in the transformed space for $f = const$, and $\mathbf{V} = const$ without diffusion and mass sources

$$\frac{\partial \sqrt{g}}{\partial \tau} - \frac{\partial}{\partial \xi}\left(\sqrt{g}\,\mathbf{a}^1 \cdot \mathbf{V}_{cs}\right) - \frac{\partial}{\partial \eta}\left(\sqrt{g}\,\mathbf{a}^2 \cdot \mathbf{V}_{cs}\right) - \frac{\partial}{\partial \zeta}\left(\sqrt{g}\,\mathbf{a}^3 \cdot \mathbf{V}_{cs}\right)$$

$$= -\mathbf{V}\cdot\left[\frac{\partial}{\partial \xi}\left(\sqrt{g}\,\mathbf{a}^1\right) - \frac{\partial}{\partial \eta}\left(\sqrt{g}\,\mathbf{a}^2\right) - \frac{\partial}{\partial \zeta}\left(\sqrt{g}\,\mathbf{a}^3\right)\right]. \quad (209)$$

Using the first fundamental metric identity *Thompson* et al. [7], p.159, obtained

$$\frac{\partial \sqrt{g}}{\partial \tau} - \frac{\partial}{\partial \xi}\left(\sqrt{g}\,\mathbf{a}^1 \cdot \mathbf{V}_{cs}\right) - \frac{\partial}{\partial \eta}\left(\sqrt{g}\,\mathbf{a}^2 \cdot \mathbf{V}_{cs}\right) - \frac{\partial}{\partial \zeta}\left(\sqrt{g}\,\mathbf{a}^3 \cdot \mathbf{V}_{cs}\right) = 0. \quad (210)$$

This is an important identity for the numerical analysis called sometimes the *second fundamental metric identity*. As pointed out by *Thompson* et al. [7] this identity should be used to numerically determine updated values of the *Jacobian*, $\sqrt{g}$, instead of updating it directly from the new values of the Cartesian coordinates. In the later case spurious source terms will appear [7].

Transformed total time derivative of a scalar: The total derivative of a scalar φ traveling in the Cartesian coordinate system with velocity **V** is defined with

$$\frac{d\varphi}{d\tau} = \frac{\partial \varphi}{\partial \tau} + \mathbf{V} \cdot \nabla \varphi. \quad (211)$$

Using the already derived relations we obtain

$$\frac{d\varphi}{d\tau} = \left(\frac{\partial \varphi}{\partial \tau}\right)_{x,y,z} + \mathbf{V}_l \cdot \nabla \varphi$$

$$= \left(\frac{\partial \varphi}{\partial \tau}\right)_{\xi,\eta,\zeta} + \mathbf{a}^1 \cdot (\mathbf{V}_l - \mathbf{V}_{cs})\frac{\partial \varphi}{\partial \xi} + \mathbf{a}^2 \cdot (\mathbf{V}_l - \mathbf{V}_{cs})\frac{\partial \varphi}{\partial \eta} + \mathbf{a}^3 \cdot (\mathbf{V}_l - \mathbf{V}_{cs})\frac{\partial \varphi}{\partial \zeta}$$

$$= \left(\frac{\partial \varphi}{\partial \tau}\right)_{\xi,\eta,\zeta} + \overline{V}_l^1 \frac{\partial \varphi}{\partial \xi} + \overline{V}_l^2 \frac{\partial \varphi}{\partial \eta} + \overline{V}_l^3 \frac{\partial \varphi}{\partial \zeta}. \quad (212)$$

Examples of some frequently used curvilinear coordinate transformations:

Next we apply the derivations from this Appendix to two well-known coordinate systems – the cylindrical and the spherical. If the derived formulas are applied correctly the well-known expressions will finally arise for the differential operators. It is recommended to go through these examples in order to understand better the transformation theory.

Cylindrical coordinates: The transformation is defined analytically by

$$x = \xi \cos \eta, \qquad \xi = +\sqrt{x^2 + y^2}, \qquad (213,214)$$

$$y = \xi \sin \eta, \qquad \eta = \arctan(y/x), \qquad (215,216)$$

$$z = \zeta, \qquad \zeta = z. \qquad (217,218)$$

Here ξ is the radial coordinate, η is the azimuthal coordinate and the axial coordinate ζ is equivalent with z. An equidistant transformation in the computational space is defined by

$$x(\xi_i, \eta_j) = \left[r_{min} + (r_{max} - r_{min}) \frac{\xi_i}{i_{max}} \right] \cos\left(\frac{\theta_{max} - \theta_{min}}{j_{max}} \eta_j \right),$$

$$y(\xi_i, \eta_j) = \left[r_{min} + (r_{max} - r_{min}) \frac{\xi_i}{i_{max}} \right] \sin\left(\frac{\theta_{max} - \theta_{min}}{j_{max}} \eta_j \right),$$

$$z(\zeta_k) = \frac{z_{max} - z_{min}}{k_{max}} \zeta_k,$$

where

$$\xi_i = 0,1,2,...,i_{max}, \quad \eta_j = 0,1,2,...,j_{max}, \quad \zeta_k = 0,1,2,...,k_{max}.$$

The *inverse metrics* of the coordinate transformation are then

$$\begin{pmatrix} \frac{\partial x}{\partial \xi} & \frac{\partial x}{\partial \eta} & \frac{\partial x}{\partial \zeta} \\ \frac{\partial y}{\partial \xi} & \frac{\partial y}{\partial \eta} & \frac{\partial y}{\partial \zeta} \\ \frac{\partial z}{\partial \xi} & \frac{\partial z}{\partial \eta} & \frac{\partial z}{\partial \zeta} \end{pmatrix} = \begin{pmatrix} \cos\eta & -\xi\sin\eta & 0 \\ \sin\eta & \xi\cos\eta & 0 \\ 0 & 0 & 1 \end{pmatrix}. \qquad (219)$$

The *Jacobian* is then

$$\sqrt{g} = \frac{\partial x}{\partial \xi}\left(\frac{\partial y}{\partial \eta}\frac{\partial z}{\partial \zeta} - \frac{\partial y}{\partial \zeta}\frac{\partial z}{\partial \eta} \right) - \frac{\partial x}{\partial \eta}\left(\frac{\partial y}{\partial \xi}\frac{\partial z}{\partial \zeta} - \frac{\partial y}{\partial \zeta}\frac{\partial z}{\partial \xi} \right) + \frac{\partial x}{\partial \zeta}\left(\frac{\partial y}{\partial \xi}\frac{\partial z}{\partial \eta} - \frac{\partial y}{\partial \eta}\frac{\partial z}{\partial \xi} \right)$$

$$= \xi. \qquad (220)$$

The *metrics* of the coordinate transformation are

$$
\begin{pmatrix}
\dfrac{\partial \xi}{\partial x} & \dfrac{\partial \xi}{\partial y} & \dfrac{\partial \xi}{\partial z} \\[4pt]
\dfrac{\partial \eta}{\partial x} & \dfrac{\partial \eta}{\partial y} & \dfrac{\partial \eta}{\partial z} \\[4pt]
\dfrac{\partial \zeta}{\partial x} & \dfrac{\partial \zeta}{\partial y} & \dfrac{\partial \zeta}{\partial z}
\end{pmatrix}
= \dfrac{1}{\sqrt{g}}
\begin{pmatrix}
\dfrac{\partial y}{\partial \eta}\dfrac{\partial z}{\partial \zeta} - \dfrac{\partial y}{\partial \zeta}\dfrac{\partial z}{\partial \eta} & \dfrac{\partial x}{\partial \zeta}\dfrac{\partial z}{\partial \eta} - \dfrac{\partial x}{\partial \eta}\dfrac{\partial z}{\partial \zeta} & \dfrac{\partial x}{\partial \eta}\dfrac{\partial y}{\partial \zeta} - \dfrac{\partial x}{\partial \zeta}\dfrac{\partial y}{\partial \eta} \\[4pt]
\dfrac{\partial y}{\partial \zeta}\dfrac{\partial z}{\partial \xi} - \dfrac{\partial y}{\partial \xi}\dfrac{\partial z}{\partial \zeta} & \dfrac{\partial x}{\partial \xi}\dfrac{\partial z}{\partial \zeta} - \dfrac{\partial x}{\partial \zeta}\dfrac{\partial z}{\partial \xi} & \dfrac{\partial x}{\partial \zeta}\dfrac{\partial y}{\partial \xi} - \dfrac{\partial x}{\partial \xi}\dfrac{\partial y}{\partial \zeta} \\[4pt]
\dfrac{\partial y}{\partial \xi}\dfrac{\partial z}{\partial \eta} - \dfrac{\partial y}{\partial \eta}\dfrac{\partial z}{\partial \xi} & \dfrac{\partial x}{\partial \eta}\dfrac{\partial z}{\partial \xi} - \dfrac{\partial x}{\partial \xi}\dfrac{\partial z}{\partial \eta} & \dfrac{\partial x}{\partial \xi}\dfrac{\partial y}{\partial \eta} - \dfrac{\partial x}{\partial \eta}\dfrac{\partial y}{\partial \xi}
\end{pmatrix}
$$

$$
= \begin{pmatrix}
\cos\eta & \sin\eta & 0 \\
-\dfrac{1}{\xi}\sin\eta & \dfrac{1}{\xi}\cos\eta & 0 \\
0 & 0 & 1
\end{pmatrix}. \tag{221}
$$

The components of the three *covariant* base vectors of the curvilinear coordinate system are the columns of the matrix of the inverse metrics. They are given below together with the corresponding covariant unit vectors.

$$\mathbf{a}_1 = \dfrac{\partial x}{\partial \xi}\mathbf{i} + \dfrac{\partial y}{\partial \xi}\mathbf{j} + \dfrac{\partial z}{\partial \xi}\mathbf{k} = \cos\eta\,\mathbf{i} + \sin\eta\,\mathbf{j} \qquad |\mathbf{a}_1| = 1 \qquad \mathbf{e}_1 = \cos\eta\,\mathbf{i} + \sin\eta\,\mathbf{j},$$

$$\mathbf{a}_2 = \dfrac{\partial x}{\partial \eta}\mathbf{i} + \dfrac{\partial y}{\partial \eta}\mathbf{j} + \dfrac{\partial z}{\partial \eta}\mathbf{k} = -\xi\sin\eta\,\mathbf{i} + \xi\cos\eta\,\mathbf{j} \qquad |\mathbf{a}_2| = \xi \qquad \mathbf{e}_2 = -\sin\eta\,\mathbf{i} + \cos\eta\,\mathbf{j}$$

$$\mathbf{a}_3 = \dfrac{\partial x}{\partial \zeta}\mathbf{i} + \dfrac{\partial y}{\partial \zeta}\mathbf{j} + \dfrac{\partial z}{\partial \zeta}\mathbf{k} = \mathbf{k} \qquad |\mathbf{a}_3| = 1 \qquad \mathbf{e}_3 = \mathbf{k}$$

(222-230)

The components of the three *contravariant* base vectors of the curvilinear coordinate system are the rows of the matrix of the metrics. They are given below together with the corresponding contravariant unit vectors.

$$\mathbf{a}^1 = \dfrac{\partial \xi}{\partial x}\mathbf{i} + \dfrac{\partial \xi}{\partial y}\mathbf{j} + \dfrac{\partial \xi}{\partial z}\mathbf{k} = \cos\eta\,\mathbf{i} + \sin\eta\,\mathbf{j} \qquad |\mathbf{a}^1| = 1 \qquad \mathbf{e}^1 = \cos\eta\,\mathbf{i} + \sin\eta\,\mathbf{j}$$

$$\mathbf{a}^2 = \dfrac{\partial \eta}{\partial x}\mathbf{i} + \dfrac{\partial \eta}{\partial y}\mathbf{j} + \dfrac{\partial \eta}{\partial z}\mathbf{k} = -\dfrac{1}{\xi}\sin\eta\,\mathbf{i} + \dfrac{1}{\xi}\cos\eta\,\mathbf{j} \qquad |\mathbf{a}^2| = \dfrac{1}{\xi} \qquad \mathbf{e}^2 = -\sin\eta\,\mathbf{i} + \cos\eta\,\mathbf{j}$$

$$\mathbf{a}^3 = \frac{\partial \zeta}{\partial x}\mathbf{i} + \frac{\partial \zeta}{\partial y}\mathbf{j} + \frac{\partial \zeta}{\partial z}\mathbf{k} = \mathbf{k} \qquad |\mathbf{a}^3| = 1 \quad \mathbf{e}^3 = \mathbf{k}$$

(231-239)

We see that a) the two types of base vectors are parallel and that b) three base vectors on each type are mutually perpendicular which means that the cylindrical coordinate system is orthogonal.

The gradient of the vector is then

$$\nabla \cdot \mathbf{F} = \frac{1}{\sqrt{g}}\left[\frac{\partial}{\partial \xi}\left(\sqrt{g}\,\mathbf{a}^1 \cdot \mathbf{F}\right) + \frac{\partial}{\partial \eta}\left(\sqrt{g}\,\mathbf{a}^2 \cdot \mathbf{F}\right) + \frac{\partial}{\partial \zeta}\left(\sqrt{g}\,\mathbf{a}^3 \cdot \mathbf{F}\right)\right],$$

$$= \frac{1}{\sqrt{g}}\left[\frac{\partial}{\partial \xi}\left(\sqrt{g}\,|\mathbf{a}^1|\mathbf{e}^1 \cdot \mathbf{F}\right) + \frac{\partial}{\partial \eta}\left(\sqrt{g}\,|\mathbf{a}^2|\mathbf{e}^2 \cdot \mathbf{F}\right) + \frac{\partial}{\partial \zeta}\left(\sqrt{g}\,|\mathbf{a}^1|\mathbf{e}^3 \cdot \mathbf{F}\right)\right],$$

$$= \frac{1}{\xi}\frac{\partial}{\partial \xi}(\xi F_\xi) + \frac{1}{\xi}\frac{\partial}{\partial \eta}(F_\eta) + \frac{\partial}{\partial \zeta}(F_\zeta), \qquad (240)$$

where $F_\xi = \mathbf{e}^1 \cdot \mathbf{F}$, $F_\eta = \mathbf{e}^2 \cdot \mathbf{F}$, $F_\zeta = \mathbf{e}^3 \cdot \mathbf{F}$ are the components of $\mathbf{F}$ normal to the coordinate surfaces defined by $\xi = const$, $\eta = const$, $\zeta = const$, respectively. This is the expected form for cylindrical coordinates – compare for instance [2] p. 739B.

The covariant metric tensor is

$$g_{ij} = \begin{pmatrix} \mathbf{a}_1 \cdot \mathbf{a}_1 & \mathbf{a}_1 \cdot \mathbf{a}_2 & \mathbf{a}_1 \cdot \mathbf{a}_3 \\ \mathbf{a}_2 \cdot \mathbf{a}_1 & \mathbf{a}_2 \cdot \mathbf{a}_2 & \mathbf{a}_2 \cdot \mathbf{a}_3 \\ \mathbf{a}_3 \cdot \mathbf{a}_1 & \mathbf{a}_3 \cdot \mathbf{a}_2 & \mathbf{a}_3 \cdot \mathbf{a}_3 \end{pmatrix} = \begin{pmatrix} 1 & 0 & 0 \\ 0 & \xi^2 & 0 \\ 0 & 0 & 1 \end{pmatrix}, \qquad (241)$$

which is a symmetric tensor. The angles between the covariant unit vectors are

$$\theta_{12} = \arccos\left(\frac{g_{12}}{\sqrt{g_{11}g_{22}}}\right) = \frac{\pi}{2}, \qquad (242)$$

$$\theta_{13} = \arccos\left(\frac{g_{13}}{\sqrt{g_{11}g_{33}}}\right) = \frac{\pi}{2}, \qquad (243)$$

$$\theta_{23} = \arccos\left(\frac{g_{23}}{\sqrt{g_{22}g_{33}}}\right) = \frac{\pi}{2}. \tag{244}$$

The magnitude of the increment of the surface area is

$$\xi = const, \quad dS^1 = \sqrt{g_{22}g_{33} - g_{23}^2}\, d\eta d\zeta = \xi d\eta d\zeta, \tag{245}$$

$$\eta = const, \quad dS^2 = \sqrt{g_{11}g_{33} - g_{13}^2}\, d\xi d\zeta = d\xi d\zeta, \tag{246}$$

$$\zeta = const, \quad dS^3 = \sqrt{g_{11}g_{22} - g_{12}^2}\, d\xi d\eta = \xi d\xi d\eta. \tag{247}$$

The volume increment is then

$$dV = \sqrt{g}\, d\xi d\eta d\zeta = \xi d\xi d\eta d\zeta. \tag{248}$$

The elements of the contravariant metric tensor are

$$g^{ij} = \begin{pmatrix} \mathbf{a}^1 \cdot \mathbf{a}^1 & \mathbf{a}^1 \cdot \mathbf{a}^2 & \mathbf{a}^1 \cdot \mathbf{a}^3 \\ \mathbf{a}^2 \cdot \mathbf{a}^1 & \mathbf{a}^2 \cdot \mathbf{a}^2 & \mathbf{a}^2 \cdot \mathbf{a}^3 \\ \mathbf{a}^3 \cdot \mathbf{a}^1 & \mathbf{a}^3 \cdot \mathbf{a}^2 & \mathbf{a}^3 \cdot \mathbf{a}^3 \end{pmatrix}$$

$$= \frac{1}{g}\begin{pmatrix} g_{22}g_{33} - g_{23}^2 & g_{21}g_{33} - g_{23}g_{31} & g_{21}g_{32} - g_{22}g_{31} \\ g_{21}g_{33} - g_{23}g_{31} & g_{11}g_{33} - g_{13}^2 & g_{11}g_{32} - g_{12}g_{31} \\ g_{21}g_{32} - g_{22}g_{31} & g_{11}g_{32} - g_{12}g_{31} & g_{11}g_{22} - g_{12}^2 \end{pmatrix} = \begin{pmatrix} 1 & 0 & 0 \\ 0 & \frac{1}{\xi^2} & 0 \\ 0 & 0 & 1 \end{pmatrix}. \tag{249}$$

With this result we can compute the Diffusion Laplacian of a scalar for the orthogonal system

$$\nabla \cdot (\lambda \nabla \varphi)$$

$$= \frac{1}{\sqrt{g}}\left[\frac{\partial}{\partial \xi}\left(\sqrt{g}\,\lambda g^{11}\frac{\partial \varphi}{\partial \xi}\right) + \frac{\partial}{\partial \eta}\left(\sqrt{g}\,\lambda g^{22}\frac{\partial \varphi}{\partial \eta}\right) + \frac{\partial}{\partial \zeta}\left(\sqrt{g}\,\lambda g^{33}\frac{\partial \varphi}{\partial \zeta}\right)\right]$$

$$= \frac{1}{\xi}\frac{\partial}{\partial \xi}\left(\xi \lambda \frac{\partial \varphi}{\partial \xi}\right) + \frac{1}{\xi}\frac{\partial}{\partial \eta}\left(\lambda \frac{1}{\xi}\frac{\partial \varphi}{\partial \eta}\right) + \frac{\partial}{\partial \zeta}\left(\lambda \frac{\partial \varphi}{\partial \zeta}\right). \tag{250}$$

This is the well-known form in cylindrical coordinates – compare for instance with [2] p. 739B

Spherical coordinates:

$$x = \xi \sin \eta \cos \zeta, \qquad \xi = +\sqrt{x^2 + y^2 + z^2}, \qquad (251,252)$$

$$y = \xi \sin \eta \sin \zeta, \qquad x = \arctan \frac{\sqrt{x^2 + y^2}}{z}, \qquad (253,254)$$

$$z = \xi \cos \eta, \qquad \zeta = \arctan(y/x). \qquad (255,256)$$

Here ξ is the radial coordinate, η is the azimuthal coordinate and ζ is the polar coordinate. An equidistant transformation in the computational space is defined by

$$x(\xi_i, \eta_j) = \left[r_{min} + (r_{max} - r_{min}) \frac{\xi_i}{i_{max}} \right] \sin\left(\frac{\theta_{max} - \theta_{min}}{j_{max}} \eta_j \right) \cos\left(\frac{\varphi_{max} - \varphi_{min}}{k_{max}} \zeta_j \right),$$

(257)

$$y(\xi_i, \eta_j) = \left[r_{min} + (r_{max} - r_{min}) \frac{\xi_i}{i_{max}} \right] \sin\left(\frac{\theta_{max} - \theta_{min}}{j_{max}} \eta_j \right) \sin\left(\frac{\varphi_{max} - \varphi_{min}}{k_{max}} \zeta_j \right),$$

(258)

$$z(\xi_i, \eta_j) = \left[r_{min} + (r_{max} - r_{min}) \frac{\xi_i}{i_{max}} \right] \cos\left(\frac{\theta_{max} - \theta_{min}}{j_{max}} \eta_j \right),$$

(259)

where

$$\xi_i = 0, 1, 2, \ldots, i_{max}, \quad \eta_j = 0, 1, 2, \ldots, j_{max}, \quad \zeta_k = 0, 1, 2, \ldots, k_{max}.$$

The *inverse metrics* of the coordinate transformation are then

$$\begin{pmatrix} \dfrac{\partial x}{\partial \xi} & \dfrac{\partial x}{\partial \eta} & \dfrac{\partial x}{\partial \zeta} \\ \dfrac{\partial y}{\partial \xi} & \dfrac{\partial y}{\partial \eta} & \dfrac{\partial y}{\partial \zeta} \\ \dfrac{\partial z}{\partial \xi} & \dfrac{\partial z}{\partial \eta} & \dfrac{\partial z}{\partial \zeta} \end{pmatrix} = \begin{pmatrix} \sin\eta\cos\zeta & \xi\cos\eta\cos\zeta & -\xi\sin\eta\sin\zeta \\ \sin\eta\sin\zeta & \xi\cos\eta\sin\zeta & \xi\sin\eta\cos\zeta \\ \cos\eta & -\xi\sin\eta & 0 \end{pmatrix}. \qquad (260)$$

The *Jacobian* is then

$$\sqrt{g} = \frac{\partial x}{\partial \xi}\left(\frac{\partial y}{\partial \eta}\frac{\partial z}{\partial \zeta} - \frac{\partial y}{\partial \zeta}\frac{\partial z}{\partial \eta}\right) - \frac{\partial x}{\partial \eta}\left(\frac{\partial y}{\partial \xi}\frac{\partial z}{\partial \zeta} - \frac{\partial y}{\partial \zeta}\frac{\partial z}{\partial \xi}\right) + \frac{\partial x}{\partial \zeta}\left(\frac{\partial y}{\partial \xi}\frac{\partial z}{\partial \eta} - \frac{\partial y}{\partial \eta}\frac{\partial z}{\partial \xi}\right)$$

$$= \xi^2 \sin\eta. \qquad (261)$$

The *metrics* of the coordinate transformation are

$$\begin{pmatrix} \dfrac{\partial \xi}{\partial x} & \dfrac{\partial \xi}{\partial y} & \dfrac{\partial \xi}{\partial z} \\ \dfrac{\partial \eta}{\partial x} & \dfrac{\partial \eta}{\partial y} & \dfrac{\partial \eta}{\partial z} \\ \dfrac{\partial \zeta}{\partial x} & \dfrac{\partial \zeta}{\partial y} & \dfrac{\partial \zeta}{\partial z} \end{pmatrix} = \frac{1}{\sqrt{g}}\begin{pmatrix} \dfrac{\partial y}{\partial \eta}\dfrac{\partial z}{\partial \zeta} - \dfrac{\partial y}{\partial \zeta}\dfrac{\partial z}{\partial \eta} & \dfrac{\partial x}{\partial \zeta}\dfrac{\partial z}{\partial \eta} - \dfrac{\partial x}{\partial \eta}\dfrac{\partial z}{\partial \zeta} & \dfrac{\partial x}{\partial \eta}\dfrac{\partial y}{\partial \zeta} - \dfrac{\partial x}{\partial \zeta}\dfrac{\partial y}{\partial \eta} \\ \dfrac{\partial y}{\partial \zeta}\dfrac{\partial z}{\partial \xi} - \dfrac{\partial y}{\partial \xi}\dfrac{\partial z}{\partial \zeta} & \dfrac{\partial x}{\partial \xi}\dfrac{\partial z}{\partial \zeta} - \dfrac{\partial x}{\partial \zeta}\dfrac{\partial z}{\partial \xi} & \dfrac{\partial x}{\partial \zeta}\dfrac{\partial y}{\partial \xi} - \dfrac{\partial x}{\partial \xi}\dfrac{\partial y}{\partial \zeta} \\ \dfrac{\partial y}{\partial \xi}\dfrac{\partial z}{\partial \eta} - \dfrac{\partial y}{\partial \eta}\dfrac{\partial z}{\partial \xi} & \dfrac{\partial x}{\partial \eta}\dfrac{\partial z}{\partial \xi} - \dfrac{\partial x}{\partial \xi}\dfrac{\partial z}{\partial \eta} & \dfrac{\partial x}{\partial \xi}\dfrac{\partial y}{\partial \eta} - \dfrac{\partial x}{\partial \eta}\dfrac{\partial y}{\partial \xi} \end{pmatrix}$$

$$= \begin{pmatrix} \sin\eta\cos\zeta & \sin\eta\sin\zeta & \cos\eta \\ \dfrac{1}{\xi}\cos\zeta\cos\eta & \dfrac{1}{\xi}\sin\zeta\cos\eta & -\dfrac{1}{\xi}\sin\eta \\ -\dfrac{1}{\xi}\dfrac{\sin\zeta}{\sin\eta} & \dfrac{1}{\xi}\dfrac{\cos\zeta}{\sin\eta} & 0 \end{pmatrix}. \qquad (262)$$

The components of the three *covariant* base vectors of the curvilinear coordinate system are the columns of the matrix of the inverse metrics. They are given below together with the corresponding covariant unit vectors.

$$\mathbf{a}_1 = \frac{\partial x}{\partial \xi}\mathbf{i} + \frac{\partial y}{\partial \xi}\mathbf{j} + \frac{\partial z}{\partial \xi}\mathbf{k} = \sin\eta\cos\zeta\,\mathbf{i} + \sin\eta\sin\zeta\,\mathbf{j} + \cos\eta\,\mathbf{k}, \qquad |\mathbf{a}_1| = 1,$$

$$\mathbf{e}_1 = \sin\eta\cos\zeta\,\mathbf{i} + \sin\eta\sin\zeta\,\mathbf{j} + \cos\eta\,\mathbf{k}, \qquad (263\text{-}265)$$

$$\mathbf{a}_2 = \frac{\partial x}{\partial \eta}\mathbf{i} + \frac{\partial y}{\partial \eta}\mathbf{j} + \frac{\partial z}{\partial \eta}\mathbf{k} = \xi\cos\eta\cos\zeta\,\mathbf{i} + \xi\cos\eta\sin\zeta\,\mathbf{j} - \xi\sin\eta\,\mathbf{k}\;, \quad |\mathbf{a}_2| = \xi\;,$$

$$\mathbf{e}_2 = \cos\eta\cos\zeta\,\mathbf{i} + \cos\eta\sin\zeta\,\mathbf{j} - \sin\eta\,\mathbf{k}\;, \tag{266-268}$$

$$\mathbf{a}_3 = \frac{\partial x}{\partial \zeta}\mathbf{i} + \frac{\partial y}{\partial \zeta}\mathbf{j} + \frac{\partial z}{\partial \zeta}\mathbf{k} = -\xi\sin\eta\sin\zeta\,\mathbf{i} + \xi\sin\eta\cos\zeta\,\mathbf{j}\;, \quad |\mathbf{a}_3| = \xi\sin\eta\;,$$

$$\mathbf{e}_3 = -\sin\zeta\,\mathbf{i} + \cos\zeta\,\mathbf{j}\;. \tag{269-271}$$

The components of the three *contravariant* base vectors of the curvilinear coordinate system are the rows of the matrix of the metrics. They are given below together with the corresponding contravariant unit vectors.

$$\mathbf{a}^1 = \frac{\partial \xi}{\partial x}\mathbf{i} + \frac{\partial \xi}{\partial y}\mathbf{j} + \frac{\partial \xi}{\partial z}\mathbf{k} = \sin\eta\cos\zeta\,\mathbf{i} + \sin\eta\sin\zeta\,\mathbf{j} + \cos\eta\,\mathbf{k}\;, \quad |\mathbf{a}^1| = 1\;,$$

$$\mathbf{e}^1 = \sin\eta\cos\zeta\,\mathbf{i} + \sin\eta\sin\zeta\,\mathbf{j} + \cos\eta\,\mathbf{k}\;, \tag{272-274}$$

$$\mathbf{a}^2 = \frac{\partial \eta}{\partial x}\mathbf{i} + \frac{\partial \eta}{\partial y}\mathbf{j} + \frac{\partial \eta}{\partial z}\mathbf{k} = \frac{1}{\xi}\cos\zeta\cos\eta\,\mathbf{i} + \frac{1}{\xi}\sin\zeta\cos\eta\,\mathbf{j} - \frac{1}{\xi}\sin\eta\,\mathbf{k}\;, \quad |\mathbf{a}^2| = \frac{1}{\xi}\;,$$

$$\mathbf{e}^2 = \cos\zeta\cos\eta\,\mathbf{i} + \sin\zeta\cos\eta\,\mathbf{j} - \sin\eta\,\mathbf{k}\;, \tag{275-277}$$

$$\mathbf{a}^3 = \frac{\partial \zeta}{\partial x}\mathbf{i} + \frac{\partial \zeta}{\partial y}\mathbf{j} + \frac{\partial \zeta}{\partial z}\mathbf{k} = -\frac{1}{\xi}\frac{\sin\zeta}{\sin\eta}\mathbf{i} + \frac{1}{\xi}\frac{\cos\zeta}{\sin\eta}\mathbf{j}\;, \quad |\mathbf{a}^3| = \frac{1}{\xi\sin\eta}\;,$$

$$\mathbf{e}^3 = -\sin\zeta\,\mathbf{i} + \cos\zeta\,\mathbf{j}\;. \tag{278-280}$$

We see that a) the two types of base vectors are parallel and that b) three base vectors of each type are mutually perpendicular which means that the cylindrical coordinate system is *orthogonal*.

$$\nabla\cdot\mathbf{F} = \frac{1}{\sqrt{g}}\left[\frac{\partial}{\partial \xi}\left(\sqrt{g}\,|\mathbf{a}^1|\mathbf{e}^1\cdot\mathbf{F}\right) + \frac{\partial}{\partial \eta}\left(\sqrt{g}\,|\mathbf{a}^2|\mathbf{e}^2\cdot\mathbf{F}\right) + \frac{\partial}{\partial \zeta}\left(\sqrt{g}\,|\mathbf{a}^3|\mathbf{e}^3\cdot\mathbf{F}\right)\right]$$

$$= \frac{1}{\xi^2\sin\eta}\left[\frac{\partial}{\partial \xi}\left(\xi^2\sin\eta\,F_\xi\right) + \frac{\partial}{\partial \eta}\left(\xi\sin\eta\,F_\eta\right) + \frac{\partial}{\partial \zeta}\left(\xi F_\zeta\right)\right]. \tag{281}$$

This is the expected form for spherical coordinates – compare for instance with [2] p. 739C.

The *covariant metric tensor* is

$$g_{ij} = \begin{pmatrix} \mathbf{a}_1 \cdot \mathbf{a}_1 & \mathbf{a}_1 \cdot \mathbf{a}_2 & \mathbf{a}_1 \cdot \mathbf{a}_3 \\ \mathbf{a}_2 \cdot \mathbf{a}_1 & \mathbf{a}_2 \cdot \mathbf{a}_2 & \mathbf{a}_2 \cdot \mathbf{a}_3 \\ \mathbf{a}_3 \cdot \mathbf{a}_1 & \mathbf{a}_3 \cdot \mathbf{a}_2 & \mathbf{a}_3 \cdot \mathbf{a}_3 \end{pmatrix} = \begin{pmatrix} 1 & 0 & 0 \\ 0 & \xi^2 & 0 \\ 0 & 0 & \xi^2 \sin^2 \eta \end{pmatrix}, \quad (282)$$

which is a symmetric tensor. The angles between the covariant unit vectors are

$$\theta_{12} = \arccos\left(\frac{g_{12}}{\sqrt{g_{11} g_{22}}}\right) = \frac{\pi}{2}, \quad (283)$$

$$\theta_{13} = \arccos\left(\frac{g_{13}}{\sqrt{g_{11} g_{33}}}\right) = \frac{\pi}{2}, \quad (284)$$

$$\theta_{23} = \arccos\left(\frac{g_{23}}{\sqrt{g_{22} g_{33}}}\right) = \frac{\pi}{2}. \quad (285)$$

The magnitude of the increment of the surface area is

$$\xi = const, \quad dS^1 = \sqrt{g_{22} g_{33} - g_{23}^2}\, d\eta d\zeta = \xi^2 \sin \eta\, d\eta d\zeta, \quad (286)$$

$$\eta = const, \quad dS^2 = \sqrt{g_{11} g_{33} - g_{13}^2}\, d\xi d\zeta = \xi \sin \eta\, d\xi d\zeta, \quad (287)$$

$$\zeta = const, \quad dS^3 = \sqrt{g_{11} g_{22} - g_{12}^2}\, d\xi d\eta = \xi\, d\xi d\eta. \quad (288)$$

The volume increment is then

$$dV = \sqrt{g}\, d\xi d\eta d\zeta = \xi^2 \sin \eta\, d\xi d\eta d\zeta. \quad (289)$$

The elements of the contravariant metric tensor are

$$g^{ij} = \begin{pmatrix} \mathbf{a}^1 \cdot \mathbf{a}^1 & \mathbf{a}^1 \cdot \mathbf{a}^2 & \mathbf{a}^1 \cdot \mathbf{a}^3 \\ \mathbf{a}^2 \cdot \mathbf{a}^1 & \mathbf{a}^2 \cdot \mathbf{a}^2 & \mathbf{a}^2 \cdot \mathbf{a}^3 \\ \mathbf{a}^3 \cdot \mathbf{a}^1 & \mathbf{a}^3 \cdot \mathbf{a}^2 & \mathbf{a}^3 \cdot \mathbf{a}^3 \end{pmatrix}$$

$$= \frac{1}{g}\begin{pmatrix} g_{22}g_{33} - g_{23}^2 & g_{21}g_{33} - g_{23}g_{31} & g_{21}g_{32} - g_{22}g_{31} \\ g_{21}g_{33} - g_{23}g_{31} & g_{11}g_{33} - g_{13}^2 & g_{11}g_{32} - g_{12}g_{31} \\ g_{21}g_{32} - g_{22}g_{31} & g_{11}g_{32} - g_{12}g_{31} & g_{11}g_{22} - g_{12}^2 \end{pmatrix} = \begin{pmatrix} 1 & 0 & 0 \\ 0 & \dfrac{1}{\xi^2} & 0 \\ 0 & 0 & \dfrac{1}{\xi^2 \sin^2 \eta} \end{pmatrix}.$$

(290)

With this result we can compute the Diffusion Laplacian of a scalar for the orthogonal system

$$\nabla \cdot (\lambda \nabla \varphi)$$

$$= \frac{1}{\sqrt{g}} \left[\frac{\partial}{\partial \xi} \left(\sqrt{g} \, \lambda g^{11} \frac{\partial \varphi}{\partial \xi} \right) + \frac{\partial}{\partial \eta} \left(\sqrt{g} \, \lambda g^{22} \frac{\partial \varphi}{\partial \eta} \right) + \frac{\partial}{\partial \zeta} \left(\sqrt{g} \, \lambda g^{33} \frac{\partial \varphi}{\partial \zeta} \right) \right]$$

$$= \frac{1}{\xi^2} \frac{\partial}{\partial \xi} \left(\xi^2 \lambda \frac{\partial \varphi}{\partial \xi} \right) + \frac{1}{\xi^2 \sin \eta} \frac{\partial}{\partial \eta} \left(\sin \eta \lambda \frac{\partial \varphi}{\partial \eta} \right) + \frac{1}{\xi^2 \sin^2 \eta} \frac{\partial}{\partial \zeta} \left(\lambda \frac{\partial \varphi}{\partial \zeta} \right).$$

(291)

This is the well-known form in spherical coordinates – compare for instance with [2] p. 739C.

Elliptic grid generation systems: The most popular grid generating system makes use of the *Laplace* equation. The solution of the *Laplace* equations (harmonic functions) obeys minimum and maximum values only at the boundary (maximum principle) and is smooth (derivatives of all orders exists). One dimensional stretching along the three axes correspondingly, delivers also a solution which has the similar properties as the solution of the *Laplace* equation. The later is proven to be equivalent to solving the *Poison*-like equations [7]

$$\xi_{xx} + \xi_{yy} + \xi_{zz} = P(\xi, \eta, \zeta), \qquad (292)$$

$$\eta_{xx} + \eta_{yy} + \eta_{zz} = Q(\xi, \eta, \zeta), \qquad (293)$$

$$\zeta_{xx} + \zeta_{yy} + \zeta_{zz} = R(\xi, \eta, \zeta), \qquad (294)$$

with boundary conditions prescribed at the surface containing the domain (*Dirichlet* boundary value problem). The control functions P, Q and R define the concentration of the grid lines to a prescribed lines or points. Interchanging the dependent (ξ,η,ζ) and the independent (x,y,z) variables results in

$$g^{11}\frac{\partial^2 \mathbf{r}}{\partial \xi^2}+g^{22}\frac{\partial^2 \mathbf{r}}{\partial \eta^2}+g^{33}\frac{\partial^2 \mathbf{r}}{\partial \zeta^2}+2\left(g^{12}\frac{\partial^2 \mathbf{r}}{\partial \xi \partial \eta}+g^{13}\frac{\partial^2 \mathbf{r}}{\partial \xi \partial \zeta}+g^{23}\frac{\partial^2 \mathbf{r}}{\partial \eta \partial \zeta}\right)$$

$$+P\frac{\partial \mathbf{r}}{\partial \xi}+Q\frac{\partial \mathbf{r}}{\partial \eta}+R\frac{\partial \mathbf{r}}{\partial \zeta}=0,$$

or rewritten in components form

$$g^{11}\frac{\partial^2 x}{\partial \xi^2}+g^{22}\frac{\partial^2 x}{\partial \eta^2}+g^{33}\frac{\partial^2 x}{\partial \zeta^2}+2\left(g^{12}\frac{\partial^2 x}{\partial \xi \partial \eta}+g^{13}\frac{\partial^2 x}{\partial \xi \partial \zeta}+g^{23}\frac{\partial^2 x}{\partial \eta \partial \zeta}\right)$$

$$+P\frac{\partial x}{\partial \xi}+Q\frac{\partial x}{\partial \eta}+R\frac{\partial x}{\partial \zeta}=0, \qquad (295)$$

$$g^{11}\frac{\partial^2 y}{\partial \xi^2}+g^{22}\frac{\partial^2 y}{\partial \eta^2}+g^{33}\frac{\partial^2 y}{\partial \zeta^2}+2\left(g^{12}\frac{\partial^2 y}{\partial \xi \partial \eta}+g^{13}\frac{\partial^2 y}{\partial \xi \partial \zeta}+g^{23}\frac{\partial^2 y}{\partial \eta \partial \zeta}\right)$$

$$+P\frac{\partial y}{\partial \xi}+Q\frac{\partial y}{\partial \eta}+R\frac{\partial y}{\partial \zeta}=0, \qquad (296)$$

$$g^{11}\frac{\partial^2 z}{\partial \xi^2}+g^{22}\frac{\partial^2 z}{\partial \eta^2}+g^{33}\frac{\partial^2 z}{\partial \zeta^2}+2\left(g^{12}\frac{\partial^2 z}{\partial \xi \partial \eta}+g^{13}\frac{\partial^2 z}{\partial \xi \partial \zeta}+g^{23}\frac{\partial^2 z}{\partial \eta \partial \zeta}\right)$$

$$+P\frac{\partial z}{\partial \xi}+Q\frac{\partial z}{\partial \eta}+R\frac{\partial z}{\partial \zeta}=0, \qquad (297)$$

where

$$g^{ij}=\frac{1}{g}\begin{pmatrix} g_{22}g_{33}-g_{23}^2 & g_{21}g_{33}-g_{23}g_{31} & g_{21}g_{32}-g_{22}g_{31} \\ g_{21}g_{33}-g_{23}g_{31} & g_{11}g_{33}-g_{13}^2 & g_{11}g_{32}-g_{12}g_{31} \\ g_{21}g_{32}-g_{22}g_{31} & g_{11}g_{32}-g_{12}g_{31} & g_{11}g_{22}-g_{12}^2 \end{pmatrix} \qquad (298)$$

are the elements of the contravariant metric tensor which is symmetric [7]. The elements of the contravariant metric tensor are computed as a function of the elements of the covariant metric tensor. The covariant metric tensor is defined as

$$g_{ij} = \begin{pmatrix} \mathbf{a}_1 \cdot \mathbf{a}_1 & \mathbf{a}_1 \cdot \mathbf{a}_2 & \mathbf{a}_1 \cdot \mathbf{a}_3 \\ \mathbf{a}_2 \cdot \mathbf{a}_1 & \mathbf{a}_2 \cdot \mathbf{a}_2 & \mathbf{a}_2 \cdot \mathbf{a}_3 \\ \mathbf{a}_3 \cdot \mathbf{a}_1 & \mathbf{a}_3 \cdot \mathbf{a}_2 & \mathbf{a}_3 \cdot \mathbf{a}_3 \end{pmatrix}$$

$$= \begin{pmatrix} \left(\frac{\partial x}{\partial \xi}\right)^2 + \left(\frac{\partial y}{\partial \xi}\right)^2 + \left(\frac{\partial z}{\partial \xi}\right)^2 & \frac{\partial x}{\partial \xi}\frac{\partial x}{\partial \eta} + \frac{\partial y}{\partial \xi}\frac{\partial y}{\partial \eta} + \frac{\partial z}{\partial \xi}\frac{\partial z}{\partial \eta} & \frac{\partial x}{\partial \xi}\frac{\partial x}{\partial \zeta} + \frac{\partial y}{\partial \xi}\frac{\partial y}{\partial \zeta} + \frac{\partial z}{\partial \xi}\frac{\partial z}{\partial \zeta} \\ \frac{\partial x}{\partial \xi}\frac{\partial x}{\partial \eta} + \frac{\partial y}{\partial \xi}\frac{\partial y}{\partial \eta} + \frac{\partial z}{\partial \xi}\frac{\partial z}{\partial \eta} & \left(\frac{\partial x}{\partial \eta}\right)^2 + \left(\frac{\partial y}{\partial \eta}\right)^2 + \left(\frac{\partial z}{\partial \eta}\right)^2 & \frac{\partial x}{\partial \eta}\frac{\partial x}{\partial \zeta} + \frac{\partial y}{\partial \eta}\frac{\partial y}{\partial \zeta} + \frac{\partial z}{\partial \eta}\frac{\partial z}{\partial \zeta} \\ \frac{\partial x}{\partial \xi}\frac{\partial x}{\partial \zeta} + \frac{\partial y}{\partial \xi}\frac{\partial y}{\partial \zeta} + \frac{\partial z}{\partial \xi}\frac{\partial z}{\partial \zeta} & \frac{\partial x}{\partial \eta}\frac{\partial x}{\partial \zeta} + \frac{\partial y}{\partial \eta}\frac{\partial y}{\partial \zeta} + \frac{\partial z}{\partial \eta}\frac{\partial z}{\partial \zeta} & \left(\frac{\partial x}{\partial \zeta}\right)^2 + \left(\frac{\partial y}{\partial \zeta}\right)^2 + \left(\frac{\partial z}{\partial \zeta}\right)^2 \end{pmatrix},$$
(299)

where

$$\mathbf{a}_1 = \frac{\partial x}{\partial \xi}\mathbf{i} + \frac{\partial y}{\partial \xi}\mathbf{j} + \frac{\partial z}{\partial \xi}\mathbf{k}, \tag{300}$$

$$\mathbf{a}_2 = \frac{\partial x}{\partial \eta}\mathbf{i} + \frac{\partial y}{\partial \eta}\mathbf{j} + \frac{\partial z}{\partial \eta}\mathbf{k}, \tag{301}$$

$$\mathbf{a}_3 = \frac{\partial x}{\partial \zeta}\mathbf{i} + \frac{\partial y}{\partial \zeta}\mathbf{j} + \frac{\partial z}{\partial \zeta}\mathbf{k}. \tag{302}$$

In fact all coefficients are functions of the elements in the *Jacobian matrix* of the coordinate transformation $x = f(\xi,\eta,\zeta)$, $y = g(\xi,\eta,\zeta)$, $z = h(\xi,\eta,\zeta)$,

$$\mathbf{J}(\xi,\eta,\zeta) = \begin{vmatrix} \frac{\partial x}{\partial \xi} & \frac{\partial x}{\partial \eta} & \frac{\partial x}{\partial \zeta} \\ \frac{\partial y}{\partial \xi} & \frac{\partial y}{\partial \eta} & \frac{\partial y}{\partial \zeta} \\ \frac{\partial z}{\partial \xi} & \frac{\partial z}{\partial \eta} & \frac{\partial z}{\partial \zeta} \end{vmatrix}. \tag{303}$$

The elements of the Jacobian matrix are called *inverse metrics* of the coordinate transformation. The *Jacobian determinant* or *Jacobian* of the coordinate transformation is

$$\sqrt{g} = |\mathbf{J}(\xi,\eta,\zeta)|$$

$$= \frac{\partial x}{\partial \xi}\left(\frac{\partial y}{\partial \eta}\frac{\partial z}{\partial \zeta} - \frac{\partial y}{\partial \zeta}\frac{\partial z}{\partial \eta}\right) - \frac{\partial x}{\partial \eta}\left(\frac{\partial y}{\partial \xi}\frac{\partial z}{\partial \zeta} - \frac{\partial y}{\partial \zeta}\frac{\partial z}{\partial \xi}\right) + \frac{\partial x}{\partial \zeta}\left(\frac{\partial y}{\partial \xi}\frac{\partial z}{\partial \eta} - \frac{\partial y}{\partial \eta}\frac{\partial z}{\partial \xi}\right).$$
(304)

For two dimensional case for instance for not existing *j*-direction set $\partial x_j / \xi_j = 1$, $\partial x_i / \xi_{j \neq i} = 0$. A simple prescription for computation of gg^{ij} is

$$gg^{ij} = \alpha_{ij} = \sum_{m=1}^{3} A_{mi} A_{mj},$$
(298a)

[10] where A_{mi} is the mi^{th} cofactor of the *Jacobian* matrix of the coordinate transformation, Eq. (303).

Boundary conditions: The surface containing the domain of interest is divided into 6 non-overlapping patches, Γ_i ($i = 1,2,3,4,5,6$), building 3 pairs perpendicular to each transformed directions. The boundary conditions, that is, the coordinates at the 6 patches are prescribed in advance as follows

$$\begin{pmatrix} x \\ y \\ z \end{pmatrix} = \begin{pmatrix} f_i(\xi_i, \eta, \zeta) \\ g_i(\xi_i, \eta, \zeta) \\ h_i(\xi_i, \eta, \zeta) \end{pmatrix}, \quad (\xi_i, \eta, \zeta) \in \Gamma_i, \quad (i = 1, 2),$$
(305)

$$\begin{pmatrix} x \\ y \\ z \end{pmatrix} = \begin{pmatrix} f_i(\xi, \eta_i, \zeta) \\ g_i(\xi, \eta_i, \zeta) \\ h_i(\xi, \eta_i, \zeta) \end{pmatrix}, \quad (\xi, \eta_i, \zeta) \in \Gamma_i, \quad (i = 3, 4),$$
(306)

$$\begin{pmatrix} x \\ y \\ z \end{pmatrix} = \begin{pmatrix} f_i(\xi, \eta, \zeta_i) \\ g_i(\xi, \eta, \zeta_i) \\ h_i(\xi, \eta, \zeta_i) \end{pmatrix}, \quad (\xi, \eta, \zeta_i) \in \Gamma_i, \quad (i = 5, 6).$$
(307)

Line spacing control functions: The line spacing control functions

$$P(\xi, \eta, \zeta) = -\sum_{l=1}^{n} a_{l1} \text{sign}(\xi - \xi_l) \exp(-b_{l1} T_l),$$
(308)

$$Q(\xi, \eta, \zeta) = -\sum_{l=1}^{n} a_{l2} \text{sign}(\eta - \eta_l) \exp(-b_{l2} T_l),$$
(309)

$$R(\xi,\eta,\zeta) = -\sum_{l=1}^{n} a_{l3} \mathrm{sign}(\zeta - \zeta_l)\exp(-b_{l3}T_l), \tag{310}$$

where

$$T = \sqrt{c_{l1}(\xi-\xi_l)^2 + c_{l2}(\eta-\eta_l)^2 + c_{l3}(\zeta-\zeta_l)^2}, \tag{311}$$

are recommended by *Miki* and *Takagi* [10]. The sets of the coefficients a_{lm} and c_{lm} attracts the ξ, η, and/or ζ =constant planes to the specified plane, coordinate line, or grid point. The range of the attraction effect is determined by the decay factor b_{lm}. The effect of the coefficient is demonstrated below:

Attracted planes	Planes, line, or point attracting neighboring planes
$a_{l1} \neq 0$ for ξ=const plane	$a_{l1} \neq 0$, $a_{l2} = 0$, $a_{l3} = 0$ for $\xi=\xi_l$ plane $c_{l1} \neq 0$, $c_{l2} = 0$, $c_{l3} = 0$
$a_{l2} \neq 0$ for η=const plane	$a_{l1} = 0$, $a_{l2} \neq 0$, $a_{l3} = 0$ for $\eta=\eta_l$ plane $c_{l1} = 0$, $c_{l2} \neq 0$, $c_{l3} = 0$
$a_{l3} \neq 0$ for ζ=const plane	$a_{l1} = 0$, $a_{l2} = 0$, $a_{l3} \neq 0$ for $\zeta=\zeta_l$ plane $c_{l1} = 0$, $c_{l2} = 0$, $c_{l3} \neq 0$
	For $(\eta,\zeta)=(\eta_l,\zeta_l)$ line: $a_{l1} = 0$, $a_{l2} \neq 0$, $a_{l3} \neq 0$ $c_{l1} = 0$, $c_{l2} \neq 0$, $c_{l3} \neq 0$
	For $(\xi,\zeta)=(\xi_l,\zeta_l)$ line: $a_{l1} = 0$, $a_{l2} = 0$, $a_{l3} \neq 0$ $c_{l1} = 0$, $c_{l2} = 0$, $c_{l3} \neq 0$
	For $(\xi,\eta)=(\xi_l,\eta_l)$ line: $a_{l1} \neq 0$, $a_{l2} \neq 0$, $a_{l3} = 0$ $c_{l1} \neq 0$, $c_{l2} \neq 0$, $c_{l3} = 0$
	For $(\xi,\eta,\zeta)=(\xi_l,\eta_l,\zeta_l)$ point: $a_{l1} \neq 0$, $a_{l2} \neq 0$, $a_{l3} \neq 0$ $c_{l1} \neq 0$, $c_{l2} \neq 0$, $c_{l3} \neq 0$

Simplification for 2D plane: For grid generation in 2D plane the grid generating system (ξ,η) simplifies to

$$g^{11}\frac{\partial^2 x}{\partial \xi^2}+g^{22}\frac{\partial^2 x}{\partial \eta^2}+2g^{12}\frac{\partial^2 x}{\partial \xi \partial \eta}+P\frac{\partial x}{\partial \xi}+Q\frac{\partial x}{\partial \eta}=0, \qquad (312)$$

$$g^{11}\frac{\partial^2 y}{\partial \xi^2}+g^{22}\frac{\partial^2 y}{\partial \eta^2}+2g^{12}\frac{\partial^2 y}{\partial \xi \partial \eta}+P\frac{\partial y}{\partial \xi}+Q\frac{\partial y}{\partial \eta}=0, \qquad (313)$$

where

$$g^{ij}=\frac{1}{g}\begin{pmatrix} g_{22} & g_{21} \\ g_{21} & g_{11} \end{pmatrix}=\frac{1}{g}\begin{pmatrix} \left(\frac{\partial x}{\partial \eta}\right)^2+\left(\frac{\partial y}{\partial \eta}\right)^2 & \frac{\partial x}{\partial \xi}\frac{\partial x}{\partial \eta}+\frac{\partial y}{\partial \xi}\frac{\partial y}{\partial \eta} \\ \frac{\partial x}{\partial \xi}\frac{\partial x}{\partial \eta}+\frac{\partial y}{\partial \xi}\frac{\partial y}{\partial \eta} & \left(\frac{\partial x}{\partial \xi}\right)^2+\left(\frac{\partial y}{\partial \xi}\right)^2 \end{pmatrix}, \qquad (314)$$

and

$$\sqrt{g}=\frac{\partial x}{\partial \xi}\frac{\partial y}{\partial \eta}-\frac{\partial x}{\partial \eta}\frac{\partial y}{\partial \xi}. \qquad (315)$$

Numerical solution: Central difference numerical scheme is applied for computation of the partial derivatives assuming $\Delta\xi=\Delta\eta=\Delta\zeta=1$:

$$\frac{\partial^2 x}{\partial \xi^2}=\frac{1}{4}(x_{i+1}-2x+x_{i-1}), \quad \frac{\partial x}{\partial \xi}=\frac{1}{2}(x_{i+1}-x_{i-1}), \qquad (316, 317)$$

$$\frac{\partial^2 x}{\partial \eta^2}=\frac{1}{4}(x_{j+1}-2x+x_{j-1}), \quad \frac{\partial x}{\partial \eta}=\frac{1}{2}(x_{j+1}-x_{j-1}), \qquad (318, 319)$$

$$\frac{\partial^2 x}{\partial \zeta^2}=\frac{1}{4}(x_{k+1}-2x+x_{k-1}), \quad \frac{\partial x}{\partial \zeta}=\frac{1}{2}(x_{k+1}-x_{k-1}), \qquad (320, 321)$$

$$\frac{\partial^2 y}{\partial \xi^2}=\frac{1}{4}(y_{i+1}-2y+y_{i-1}), \quad \frac{\partial x}{\partial \xi}=\frac{1}{2}(y_{i+1}-y_{i-1}), \qquad (322, 323)$$

$$\frac{\partial^2 y}{\partial \eta^2}=\frac{1}{4}(y_{j+1}-2y+y_{j-1}), \quad \frac{\partial x}{\partial \eta}=\frac{1}{2}(y_{j+1}-y_{j-1}), \qquad (324, 325)$$

$$\frac{\partial^2 y}{\partial \zeta^2} = \frac{1}{4}\left(y_{k+1} - 2y + y_{k-1}\right), \quad \frac{\partial y}{\partial \zeta} = \frac{1}{2}\left(y_{k+1} - y_{k-1}\right), \tag{326, 327}$$

$$\frac{\partial^2 z}{\partial \xi^2} = \frac{1}{4}\left(z_{i+1} - 2z + z_{i-1}\right), \quad \frac{\partial z}{\partial \xi} = \frac{1}{2}\left(z_{i+1} - z_{i-1}\right), \tag{326, 327}$$

$$\frac{\partial^2 z}{\partial \eta^2} = \frac{1}{4}\left(z_{j+1} - 2z + z_{j-1}\right), \quad \frac{\partial z}{\partial \eta} = \frac{1}{2}\left(z_{j+1} - z_{j-1}\right), \tag{328, 329}$$

$$\frac{\partial^2 z}{\partial \zeta^2} = \frac{1}{4}\left(z_{k+1} - 2z + z_{k-1}\right), \quad \frac{\partial z}{\partial \zeta} = \frac{1}{2}\left(z_{k+1} - z_{k-1}\right), \tag{330, 331}$$

$$\frac{\partial^2 x}{\partial \xi \partial \eta} = \frac{\partial}{\partial \xi}\frac{\partial x}{\partial \eta} = \frac{1}{4}\left(x_{i+1,j+1} - x_{i+1,j-1} - x_{i-1,j+1} + x_{i-1,j-1}\right), \tag{332, 333}$$

$$\frac{\partial^2 x}{\partial \xi \partial \zeta} = \frac{\partial}{\partial \xi}\frac{\partial x}{\partial \zeta} = \frac{1}{4}\left(x_{i+1,k+1} - x_{i+1,k-1} - x_{i-1,k+1} + x_{i-1,k-1}\right), \tag{334, 335}$$

$$\frac{\partial^2 x}{\partial \eta \partial \zeta} = \frac{\partial}{\partial \eta}\frac{\partial x}{\partial \zeta} = \frac{1}{4}\left(x_{j+1,k+1} - x_{j+1,k-1} - x_{j-1,k+1} + x_{j-1,k-1}\right), \tag{336, 337}$$

$$\frac{\partial^2 y}{\partial \xi \partial \eta} = \frac{\partial}{\partial \xi}\frac{\partial y}{\partial \eta} = \frac{1}{4}\left(y_{i+1,j+1} - y_{i+1,j-1} - y_{i-1,j+1} + y_{i-1,j-1}\right), \tag{339, 340}$$

$$\frac{\partial^2 y}{\partial \xi \partial \zeta} = \frac{\partial}{\partial \xi}\frac{\partial y}{\partial \zeta} = \frac{1}{4}\left(y_{i+1,k+1} - y_{i+1,k-1} - y_{i-1,k+1} + y_{i-1,k-1}\right), \tag{341, 342}$$

$$\frac{\partial^2 y}{\partial \eta \partial \zeta} = \frac{\partial}{\partial \eta}\frac{\partial y}{\partial \zeta} = \frac{1}{4}\left(y_{j+1,k+1} - y_{j+1,k-1} - y_{j-1,k+1} + y_{j-1,k-1}\right), \tag{343, 344}$$

$$\frac{\partial^2 z}{\partial \xi \partial \eta} = \frac{\partial}{\partial \xi}\frac{\partial z}{\partial \eta} = \frac{1}{4}\left(z_{i+1,j+1} - z_{i+1,j-1} - z_{i-1,j+1} + z_{i-1,j-1}\right), \tag{345, 346}$$

$$\frac{\partial^2 z}{\partial \xi \partial \zeta} = \frac{\partial}{\partial \xi}\frac{\partial z}{\partial \zeta} = \frac{1}{4}\left(z_{i+1,k+1} - z_{i+1,k-1} - z_{i-1,k+1} + z_{i-1,k-1}\right), \quad (347, 348)$$

$$\frac{\partial^2 z}{\partial \eta \partial \zeta} = \frac{\partial}{\partial \eta}\frac{\partial z}{\partial \zeta} = \frac{1}{4}\left(z_{j+1,k+1} - z_{j+1,k-1} - z_{j-1,k+1} + z_{j-1,k-1}\right). \quad (349, 350)$$

Replacing in Eq. (312) results in

$$c_{k-1}x_{k-1} + c_{j-1}x_{j-1} + c_{i-1}x_{i-1} + cx + c_{i+1}x_{i+1} + c_{j+1}x_{j+1} + c_{k+1}x_{k+1} = d_1,$$

$$c_{k-1}y_{k-1} + c_{j-1}y_{j-1} + c_{i-1}y_{i-1} + cy + c_{i+1}y_{i+1} + c_{j+1}y_{j+1} + c_{k+1}y_{k+1} = d_2,$$

$$c_{k-1}z_{k-1} + c_{j-1}z_{j-1} + c_{i-1}z_{i-1} + cz + c_{i+1}z_{i+1} + c_{j+1}z_{j+1} + c_{k+1}z_{k+1} = d_3,$$

where

$$c_{k+1} = g^{33}/2 + R,\ c_{j+1} = g^{22}/2 + Q,\ c_{i+1} = g^{11}/2 + P,$$

$$c_{k-1} = g^{33}/2 - R,\ c_{j-1} = g^{22}/2 - Q,\ c_{i-1} = g^{11}/2 - P,$$

$$c = -\left(g^{11} + g^{22} + g^{33}\right) = -\left(c_{k-1} + c_{j-1} + c_{i-1} + c_{i+1} + c_{j+1} + c_{k+1}\right),$$

$$d_1 = -g^{12}\left(x_{i+1,j+1} - x_{i+1,j-1} - x_{i-1,j+1} + x_{i-1,j-1}\right)$$

$$-g^{13}\left(x_{i+1,k+1} - x_{i+1,k-1} - x_{i-1,k+1} + x_{i-1,k-1}\right) - g^{22}\left(x_{j+1,k+1} - x_{j+1,k-1} - x_{j-1,k+1} + x_{j-1,k-1}\right),$$

$$d_2 = -g^{12}\left(y_{i+1,j+1} - y_{i+1,j-1} - y_{i-1,j+1} + y_{i-1,j-1}\right)$$

$$-g^{13}\left(y_{i+1,k+1} - y_{i+1,k-1} - y_{i-1,k+1} + y_{i-1,k-1}\right) - g^{22}\left(y_{j+1,k+1} - y_{j+1,k-1} - y_{j-1,k+1} + y_{j-1,k-1}\right),$$

$$d_3 = -g^{12}\left(z_{i+1,j+1} - z_{i+1,j-1} - z_{i-1,j+1} + z_{i-1,j-1}\right)$$

$$-g^{13}\left(z_{i+1,k+1} - z_{i+1,k-1} - z_{i-1,k+1} + z_{i-1,k-1}\right) - g^{22}\left(z_{j+1,k+1} - z_{j+1,k-1} - z_{j-1,k+1} + z_{j-1,k-1}\right),$$

The resulting system is solved using SOR methods. Note that the coefficient matrix is the same for all directions. Therefore once the matrix is inverted it can be used for the three right hand sites which safes computer time. Do not forget to treat properly the boundary conditions by putting into the right hand sites the known boundary terms.

Orthogonal systems: Numerical orthogonality is reached by setting the off-diagonal elements of the contravariant metric tensor equal to zero. This result in

$$\frac{\partial}{\partial \xi}\left(\frac{\partial x}{\partial \xi}\sqrt{\frac{g_{22}g_{33}}{g_{11}}}\right) + \frac{\partial}{\partial \eta}\left(\frac{\partial x}{\partial \eta}\sqrt{\frac{g_{11}g_{33}}{g_{22}}}\right) + \frac{\partial}{\partial \zeta}\left(\frac{\partial x}{\partial \zeta}\sqrt{\frac{g_{11}g_{22}}{g_{33}}}\right) = 0, \quad (351)$$

$$\frac{\partial}{\partial \xi}\left(\frac{\partial y}{\partial \xi}\sqrt{\frac{g_{22}g_{33}}{g_{11}}}\right) + \frac{\partial}{\partial \eta}\left(\frac{\partial y}{\partial \eta}\sqrt{\frac{g_{11}g_{33}}{g_{22}}}\right) + \frac{\partial}{\partial \zeta}\left(\frac{\partial y}{\partial \zeta}\sqrt{\frac{g_{11}g_{22}}{g_{33}}}\right) = 0, \quad (352)$$

$$\frac{\partial}{\partial \xi}\left(\frac{\partial z}{\partial \xi}\sqrt{\frac{g_{22}g_{33}}{g_{11}}}\right) + \frac{\partial}{\partial \eta}\left(\frac{\partial z}{\partial \eta}\sqrt{\frac{g_{11}g_{33}}{g_{22}}}\right) + \frac{\partial}{\partial \zeta}\left(\frac{\partial z}{\partial \zeta}\sqrt{\frac{g_{11}g_{22}}{g_{33}}}\right) = 0. \quad (353)$$

see *Thompson* and *Warsi* [6]. If, in addition to orthogonality, the condition $g_{11} = g_{22} = g_{33} = const$ is imposed, then the above three equations reduce to *Laplace* equations.

For 2D systems the equations set (351-353) reduces to

$$\frac{\partial}{\partial \xi}\left(f_r \frac{\partial x}{\partial \xi}\right) + \frac{\partial}{\partial \eta}\left(\frac{1}{f_r}\frac{\partial x}{\partial \eta}\right) = 0, \quad (354)$$

$$\frac{\partial}{\partial \xi}\left(f_r \frac{\partial y}{\partial \xi}\right) + \frac{\partial}{\partial \eta}\left(\frac{1}{f_r}\frac{\partial y}{\partial \eta}\right) = 0, \quad (355)$$

where

$$f_r = \sqrt{g_{22}/g_{11}} > 0 \quad (356)$$

is called "distortion function". It specifies the ratio of the sides of small rectangle in the *x*, *y* plane which is mapped onto a rectangle in the ξ, η plane. A reasonable upper limit for f_r is 10. The condition of orthogonality also leads to

$$\frac{\partial x}{\partial \eta} = -f_r \frac{\partial y}{\partial \xi}, \quad \frac{\partial y}{\partial \eta} = f_r \frac{\partial x}{\partial \xi}. \quad (357, 358)$$

This conditions have to be satisfied also at the boundary. For

$$f_r(\xi,\eta)=1, \tag{359}$$

Eqs. (357, 358) become the *Cauchy-Riemann* equations which are used in conformal mapping. *Tamamidis* and *Assanis* [11] recognized that non uniform f_r over the domain of interest cane be used for controlling the spacing of the nodes even generating orthogonal grids. The authors successfully examined the properties of

$$\frac{\partial^2 f_r}{\partial \xi^2}+\frac{\partial^2 f_r}{\partial \eta^2}=G(\xi,\eta) \tag{360}$$

for generating a smooth f_r's over the domain of interests. $G(\xi,\eta)$ typically involve some combination of sinusoidal, cosinusoidal and exponential functions, i.e.,

$$G(\xi,\eta)=const\, f_1(\xi)f_2(\eta). \tag{361}$$

An initial approximation can be obtained by setting $f_r = 1$. The authors generated coefficients of the algebraic systems finite difference to Eqs. (354) and (355) having the same sign by using for the first order differences: forward finite differences for x_ξ when f_ξ is positive, and backward differences when f_ξ is negative; backward finite differences for x_η when f_η is positive, and forward differences when f_η is negative.

Setting

$$f=\left(g_{11}g_{22}g_{33}\right)^{-1/2}=g^{-1/2} \tag{362}$$

the equations set (351-353) can be rewritten as

$$\frac{\partial}{\partial \xi}\left(fg_{22}g_{33}\frac{\partial x}{\partial \xi}\right)+\frac{\partial}{\partial \eta}\left(fg_{11}g_{33}\frac{\partial x}{\partial \eta}\right)+\frac{\partial}{\partial \zeta}\left(fg_{11}g_{22}\frac{\partial x}{\partial \zeta}\right)=0, \tag{363}$$

$$\frac{\partial}{\partial \xi}\left(fg_{22}g_{33}\frac{\partial y}{\partial \xi}\right)+\frac{\partial}{\partial \eta}\left(fg_{11}g_{33}\frac{\partial y}{\partial \eta}\right)+\frac{\partial}{\partial \zeta}\left(fg_{11}g_{22}\frac{\partial y}{\partial \zeta}\right)=0, \tag{364}$$

$$\frac{\partial}{\partial \xi}\left(fg_{22}g_{33}\frac{\partial z}{\partial \xi}\right) + \frac{\partial}{\partial \eta}\left(fg_{11}g_{33}\frac{\partial z}{\partial \eta}\right) + \frac{\partial}{\partial \zeta}\left(fg_{11}g_{22}\frac{\partial z}{\partial \zeta}\right) = 0. \quad (365)$$

Christov [12] proposed to use $f = f(x,y,z)$ for controlling the spacing of the grid by controlling actually the local value of the Jacobian. For 2D,

$$\frac{\partial}{\partial \xi}\left(fg_{22}\frac{\partial x}{\partial \xi}\right) + \frac{\partial}{\partial \eta}\left(fg_{11}\frac{\partial x}{\partial \eta}\right) = 0, \quad (366)$$

$$\frac{\partial}{\partial \xi}\left(fg_{22}\frac{\partial y}{\partial \xi}\right) + \frac{\partial}{\partial \eta}\left(fg_{11}\frac{\partial y}{\partial \eta}\right) = 0, \quad (367)$$

Christov proposed to impose a control function of the type

$$f = \sqrt{1 + z_x^2 + z_y^2}, \quad (368)$$

that is simply the slope of a predefined surface $z = z(x,y)$. The stronger the slope of the chosen surface at a given point the more dense the grid around this point.

The Sorenson method for enforcing orthogonality at the boundaries: Sorenson [13] oriented his development towards creating a multiple interconnected blocks. He simplified the P, Q, and R functions to facilitate attractions to the corresponding outer boundary only by specifying attenuation factors so as not to influence the opposite boundary surface e.g.

$$P(\xi,\eta,\zeta) = p_1(\eta,\zeta)\exp\left[-a(\xi-1)\right] + p_2(\eta,\zeta)\exp\left[-a(\xi_{max}-\xi)\right]$$
$$+ p_3(\xi,\zeta)\exp\left[-a(\eta-1)\right] + p_4(\xi,\zeta)\exp\left[-a(\eta_{max}-\eta)\right]$$
$$+ p_5(\xi,\eta)\exp\left[-a(\zeta-1)\right] + p_6(\xi,\eta)\exp\left[-a(\zeta_{max}-\zeta)\right]. \quad (369)$$

By appropriate selection of the p, q, r coefficients he enforced the ortogonality at the boundary. As example consider face 3. At this face the derivatives $\mathbf{r}_\xi$, $\mathbf{r}_\zeta$, $\mathbf{r}_{\xi\zeta}$, $\mathbf{r}_{\xi\xi}$, and $\mathbf{r}_{\zeta\zeta}$ are computed from the point values known at the surface once for the generation process. The derivatives $\mathbf{r}_\eta$ are found from the desired clustering and orthogonality on face 3. The desired orthogonality and clustering is specified by the following three relations $\mathbf{r}_\xi \cdot \mathbf{r}_\eta = 0$, $\mathbf{r}_\zeta \cdot \mathbf{r}_\eta = 0$, $\mathbf{r}_\eta \cdot \mathbf{r}_\eta = S^2$, where S is the height to be imposed on the cell on the boundary or in expanded form

$$x_\xi x_\eta + y_\xi y_\eta + z_\xi z_\eta = 0, \tag{370}$$

$$x_\varsigma x_\eta + y_\varsigma y_\eta + z_\varsigma z_\eta = 0, \tag{371}$$

$$x_\varsigma^2 + y_\eta^2 + z_\varsigma^2 = S^2. \tag{372}$$

Solving the first two equations with respect to x_η and y_η results in

$$x_\eta = z_\eta \left(-z_\xi y_\varsigma + z_\varsigma y_\xi\right) / \left(x_\xi y_\varsigma - x_\varsigma y_\xi\right), \tag{373}$$

$$y_\eta = z_\eta \left(-x_\xi z_\varsigma + x_\varsigma z_\xi\right) / \left(x_\xi y_\varsigma - x_\varsigma y_\xi\right), \tag{374}$$

or comparing with the cofactors of the *Jacobian* matrix of the coordinate transformation

$$x_\eta = z_\eta \left(-A_{12}\right) / \left(-A_{32}\right), \tag{375}$$

$$y_\eta = z_\eta \left(-A_{22}\right) / \left(-A_{32}\right). \tag{376}$$

Substituting in the third equation and solving with respect to z_η we obatin

$$z_\eta = \frac{SA_{32}}{\pm\sqrt{A_{12}^2 + A_{22}^2 + A_{32}^2}}, \tag{377}$$

and consequently

$$x_\eta = \frac{SA_{12}}{\pm\sqrt{A_{12}^2 + A_{22}^2 + A_{32}^2}}, \tag{378}$$

$$y_\eta = \frac{SA_{22}}{\pm\sqrt{A_{12}^2 + A_{22}^2 + A_{32}^2}}. \tag{379}$$

The positive sign is used for right-handed and the negative for left-handed coordinate systems. Now we have the derivatives $\mathbf{r}_\eta$. These derivatives can be differentiated to obtain the mixed derivatives $\mathbf{r}_{\eta\xi}$, $\mathbf{r}_{\eta\varsigma}$. All these derivatives are obtained as a function of the information stored at the surface and have to be computed once over the generation process. The only derivatives lacking is

$$\left(\mathbf{r}_{\eta\eta}\right)_{i,1,k} = \frac{-7\mathbf{r}_{i,1,k} + 8\mathbf{r}_{i,2,k} - \mathbf{r}_{i,3,k}}{2(\Delta\eta)^2} - \frac{3\mathbf{r}_{\eta,i,1,k}}{\Delta\eta}, \qquad (380)$$

which results from the *Taylor* series. Now the governing three equations written for the point $(i,1,k)$

$$\frac{\partial \mathbf{r}}{\partial \xi} p_{i,1,k} + \frac{\partial \mathbf{r}}{\partial \eta} q_{i,1,k} + \frac{\partial \mathbf{r}}{\partial \zeta} r_{i,1,k}$$

$$= -\frac{1}{g}\left[\alpha_{11}\frac{\partial^2 \mathbf{r}}{\partial \xi^2} + \alpha_{33}\frac{\partial^2 \mathbf{r}}{\partial \zeta^2} + 2\left(\alpha_{12}\frac{\partial^2 \mathbf{r}}{\partial \xi \partial \eta} + \alpha_{13}\frac{\partial^2 \mathbf{r}}{\partial \xi \partial \zeta} + \alpha_{23}\frac{\partial^2 \mathbf{r}}{\partial \eta \partial \zeta}\right)\right] - \frac{\alpha_{22}}{g}\frac{\partial^2 \mathbf{r}}{\partial \eta^2}$$

can be solved with respect to the unknown coefficients p, q and r. In each successive iteration the value of the derivatives $\left(\mathbf{r}_{\eta\eta}\right)_{i,1,k}$ and consequently the values of p, q and r are updated until convergence is reached.

Practical recommendation: Specify first the 6 non-overlapping patches. Divide the peripheral boundary on 4 patches. Generate the grid on each boundary line. Use simple 2D generating system and generate approximate 2D grid at each patch. Improve the 2D grid by using elliptic 2D generating system. Use simple algebraic system to generate initial guess for the 3D system. Generate 3D grids by using 3D elliptic generating system. If other domain has to be attached to the already generated grid specify the patch at which the new domain will be attached. Then use the same procedure to generate the 3D grid by using the already generated grid on the common patch. Plot the grid and control the result.

References

1. Anderson JD (1995) Computational fluid dynamics, McGraw-Hill, Inc., New York
2. Bird RB, Stewart WE, Lightfood E N (1960) Transport phenomena, John Wiley & Sons, New York
3. Kordulla W, Vinokur M (1983) AIAA J., vol 21 pp 917-918
4. Peyret R editor (1996) Handbook of computational fluid mechanics, Academic Press, London
5. Thomas GB Jr, Finney RL, Weir MD (1998) Calculus and analytic geometry, 9th Edition, Addison-Wesley Publishing Company, Reading, Massachusetts
6. Thompson JF, Warsi ZUA (1982) Boundary-fitted coordinate systems for numerical solutions of partial differential equations – a review, J. Comput. Phys., vol 47 pp 1-108
7. Thompson JF, Warsi ZUA, Mastin CW (1985) Numerical grid generation, North-Holand, New York-Amsterdam-Oxford

8. Vivand H (1974) Conservative forms of gas dynamics equations, Rech. Aerosp., no 1971-1 pp 65-68
9. Vinokur M (1974) Conservation equations of gas dynamics in curvilinear coordinate systems, J. Comput. Phys., vol 14 pp 105-125
10. Miki K and Takagi T, 1984, J. Comp. Phys., vol 53 pp 319-330
11. Tamamidis P and Assanis DN (1991) J. Comp. Physics, vol 94 pp 437-453
12. Christov CI (April 1982) Orthogonal coordinate meshes with manageable Jacobian, Proc. of a Symposium on the Numerical Generation of Curvilinear Coordinate Systems and their Use in the Numerical Solution of Partial Differential Equations, Nashville, Tennessee, pp 885-894, Thompson JF ed
13. Sorenson RL (July 1989) The 3DGRAPE book: theory, user's manual, examples, NASA Technical Memorandum 102224

Index

absolute coordinates 519
absolute temperature 170, 222
acoustic waves 396
added mass force 87
Al_2O_2 – water 422
anergy 295
angle between two nonzero vectors 619
angles between the unit covariant
　vectors 645
anisotropic forces 88
aperiodic 348, 351
aperiodic case 350
approximating the contravariant base
　vectors 657
arc length 612
arc length parameter base point 613
arc length vector 47
area of parallelogram 630
area of union of two triangle plane
　regions 630
associative and distributive lows for the
　cross product 629
asymptotic 349, 352
asymptotic case 351
averaging 4
axial pump 378

back pressing 342
baro-diffusion 21
b-coefficients 560
Bernulli equation 535
block Newton-Raphson method 493
Block-Gauss-Seidel method 547
boiling at the wall 268
Boussinesq hypothesis 63
bulk pressure 56

canonical form 174
Carnot coefficient 293
Carnot-cycle 300

Cartesian vector in co- and contravariant
　coordinate systems 642
Cauchy's lemma 46
center of mass velocity 20, 322
centrifugal force 481
centrifugal pump 378
centrifugal pump drive model 386
centrifugal pumps 366
centroid of triangle 609
centroid of union of non overlapping
　plane regions 610
centrufugal forces 103
Chapman-Jouquet condition 398, 405,
　408
characteristic points 517
chemical reaction in gas mixture being
　in closed volume 284
Clausius 167, 170
closest possible packing of particles 28
coalescence 30
coefficient of performance 300
cofactor of the Jacobian matrix 687
collision 37
compression work 297
concentration equations 473, 528
condensation 262
conservation equations in general
　curvilinear coordinate systems 427
conservation of energy 168
conservative form of the energy
　conservation equation 213
conservative form of the mass
　conservation equation for multi-phase
　flows 430
conservative scheme 510
constrained interpolation profile (CIP)
　method 509
continuum - sound waves 395
continuum surface force 59
contraction number 313
contravariant base vectors 642

748   Index

contravariant metric tensor 654
contravariant vectors 428, 563, 565
control volume 4, 168
convection + diffusion velocity 472
coolant entrainment ratio 419
Coriolis force 103
Courant, Friedrichs and Levi 503
covariant base vectors 641
covariant metric tensor 645
criticality mass flow rate 330
curl 652, 662
curvature 49
curvature of a surface 621
curvature of a surface defined by the gradient of its unit normal vector 622
curvature of space curve 615
curvilinear coordinate system 427, 556
Cylindrical coordinates 675

D'Alembert 168
Dalton's law 189
Dalton's law 2, 20, 117, 214
debris quenching 394
deformation 312
deformation of the mean values of the velocities 228
dependent variables vector 176
deposition 31
derivative of a position vector defining a curve 611
derivative of a position vector defining a line 611
derivatives 665
derivatives along the normal to the curvilinear coordinate surface 664
derivatives along the tangent of the curvilinear coordinate lines 664
derivatives for the equations of state 115
derivatives of a function along the unit tangent vector 627
determination formula for 628
detonation adiabatic 399
detonation wave 398
detonation waves 393
diagonal dominance 490
diagonal pump 378
diagonalizable matrix 635
diagonally dominant 580
differential form 626

differentiation rules for time derivatives 614
diffusion Laplacian 432, 661, 663
diffusion Laplacian of tensor 664
diffusion velocity 20, 215
directional derivatives 626
Dirichlet boundary value problem 685
discharge of gas from a volume 276
discontinuum - shock waves 396, 402
discretization 464
discretized concentration conservation equation 558
discretized concentration equation 570
discretized mass conservation 562
discretized momentum equation 579
dispersed flow 84
dispersed interface 82
divergence 651, 660, 661
divergence of tensor 663
divergence theorem 650
divergence theorem for curvilinear coordinate system 651
donor cell concept 487
donor cell principle 481
donor-cell 560
drag force 587
dual vectors 655
dumping time constant 347
dyadic product of two vectors 633

eddy conductivity 229
effective gas constant 274
eigen values 174
eigen vectors 174
eigenvalues 634
eigenvectors 634
elasticity modules 313
elbows 520
elliptic grid generation systems 684
energy concept 133, 139, 145
energy conservation 168
energy conservation equation 206
energy equation 203
energy principle 206
enthalphy 205
enthalphy concept 138
enthalphy diffusion coefficient 234
enthalpy concept 129, 145
enthalpy equation 216
entrainment 31
entropy 170

entropy concept 116, 133, 139, 146, 252
entropy diffusion coefficient 233
entropy equation 203, 222, 309, 453, 474, 571
equation of state 171
equations for plane in space 617
equations for tangent planes and normal lines 619
equilibrium mixture of liquid and solid phase 160
Euler 1, 168
Eulerian description 168
evaporation 262
exact differential form 626
exergetic coefficient of performance 300
exergy 291

Fe – water 422
Fick 20
Fick's law 20, 216
field entropy conservation equations 435
field mass conservation equations for each component 431
finite difference method 564
finite volume method 564
first fundamental metric identity 431, 651
first order donor cell 468
first order donor-cell finite difference approximations 470
first-order donor-cell 525
flashing 267
flow patterns 2
flux concept 451
form drag 87
forth-order-polynomial spline 510
Fourier equation 393
fragmentation 30
free turbulence 229
frequently used curvilinear coordinate transformations 675
frictional pressure drop 324
fundamental exergy equation 297

gas phase 2
Gauss-Ostrogradskii 650
Gauss-Ostrogradskii theorem 11, 54
Gauss-Seidel 478, 500
Gauss-Seidel iteration 469

Gauss-Seidel method 547, 561
Gay-Lussac 172
general transport equation 13
Gibbs equation 222
Gibbs equation for mixtures 133, 191
Gibbs potential 133
Gibbs relation 170
Gouy exergy definition 297
gradient 652, 660, 662
gradient along a direction normal and tangential to a surface 622
gradient of plane defined by three points 620
gradient vector (gradient): 619
grid 553

Hagen-Poiseuille law 246
harmonic averaging 530, 595
harmonic oscillations 347, 351
harmonic oscillations case 349
head and torque curves 377
heat input in gas being in closed volume 279
heat pump 299
Helmholtz and Stokes hypothesis 244, 437
high order discretization 504
high-order upwinding 507
homogeneous multiphase flow 401, 402
homologous curve 373
Hugoniot 398
hydraulic diameter 15
hydrogen combustion in inert atmosphere 286

impermeable fixed surfaces 16
incompressible flow 314
increment of the arc length 644
increment of the position vector 644
inert components 2
inert mass conservation equation derivatives 545
infinite volume in curvilinear coordinate systems 646
injection of inert gas in closed volume initially filled with inert gas 278
integration procedure 501
interface conditions 267
interfacial area density 15, 77
interfacial energy jump condition 211
interfacial heat transfer 328

interfacial mass transfer 329
interfacial momentum transfer due to mass transfer 588
internal energy equation 221, 231
intrinsic field average 9
intrinsic surface average 9
intrinsic surface-averaged field pressure 56
inverse metrics 565
inverted metrics tensors 428
irregular grid's 554
irregular mesh 554
irreversible bulk viscous dissipation 219
irreversible dissipated power 228
irreversible power dissipation 437
Isaac Newton 45
isentropic 176
isentropic exponent 141
isotropic convection-diffusion problem 433
iteration method 546
IVA2-method 580

Jacobi method 479, 547
Jacobian 121, 428, 494, 648
Jacobian determinant 557, 565

Kelvin-Helmholtz stability 75, 81
kg-mole concentrations 119
knots 523

Lagrangian 168
Laplace 396
Laplace equations 684
Laplacian 652, 660, 662
laws of the dot product 618
Legendre transformation 223
Leibnitz rule 13, 54, 207
lift force 90, 590
line spacing control functions 687
linear stability analysis 503
linear system of equations 546
linearized form of the equation of state 578
linearized state equation 489
lines and line segments in space 610
liquid water 2
local critical mass flow rate 326
local instantaneous entropy equations 224

local volume and time average mass conservation equation 23
local volume-average 8, 169
local volume-averaged momentum equations 46
locally monotonic behavior 508
Lomonosov 1, 168
lumped parameter volumes 268

Mach number 123
macroscopic density 9
magnitude 610
main pump equation 370, 388
Marangoni effect 50, 82
mass conservation 1, 13, 307, 489
mass conservation derivatives 544
mass conservation equation 18, 452, 468, 525
mass conservation equation for each specie inside the velocity field 308
mass conservation equation for the microscopic component 19
mass conservation equations 578
mass flow concentrations 322
mass flow vector 557
mass jump condition 26
mass source 15
melt-coolant interaction detonations 404
melt-water detonation 393
mesh 554
method of characteristics 178
metrics 428
metrics and inverse metrics 639
midpoints 609
mixture 141
mixture consisting of steam and air 160
mixture density 273, 322, 442
mixture enthalpy 190
mixture isentropic exponent 257
mixture mass flow rate 322
mixture of saturated steam and saturated liquid 159
mixture specific heat 273
mixture volume conservation equation 252, 489
mixtures consisting of solid particles and liquid 142
molar concentration 119
molecular diffusion 21
momentum conservation 45

momentum conservation equations 443
momentum equation 55, 308, 480, 481, 538
momentum equations 485, 487
momentum jump condition 46, 52
mono-dispersity 27, 32
moving coordinate system 668
multi-blocks 555
multi-component gas mixtures 117
multi-phase flow patterns 3

necessary condition for convergence 478
Newton 168
Newton method 138
Newtonian continuum 63
Newton-Raphson 119
Newton-type iteration method 491
nodes 553
non-condensibles 268
non-conservative form 326
non-conservative form of the concentration equations 559
non-conservative form of the energy equation 213
non-conservative form of the enthalpy equation 217
non-conservative form of the entropy equation 236
non-conservative form of the momentum equation 217
non-monotonic behavior 509
non-physical diffusion 16
non-slip boundary condition 486
non-structured 2
nozzle flow 328
nozzle flow with instantaneous heat exchange without mass exchange 321
nozzles frozen flow 319
nucleation rate 29
Nukiama - Tanasawa distribution 33
number density of particles 26
numerical diffusion 16
numerical methods 461
numerical orthogonality 691
numerical solution 181
numerical solution method 410

off-diagonal diffusion terms 567
off-diagonal viscous forces 590

one-dimensional three-fluid flow 307
orthogonal coordinate systems 567
orthogonal systems 691
orthogonality 642
orthotropic 674
oscillation period 348
outer iterations 502

parallel vectors 629
partial decoupling 477
partial derivatives 649
partial differential equations 462
partial pressure 117
particle number density conservation equation 434
particle number density derivatives 543
particle number density equation 308, 573
particles number density equation 454
pdv-exergy 293
perfect fluid 180
perfect gas 158, 180
perfect gases 118, 126
perpendicular (orthogonal) vectors 619
Pfaff's form 170
phase 2
phase discontinuity treated with CIP 515
physical space 638
pipe 516
pipe deformation 312
pipe library 521
pipe networks 307, 342, 516
pipe section 519
pipes 307
plane determined by three points 617
Poison-like equations 684
Poisson-type equation for multi-phase flows 497
politropic exponent 274
position vector defining a curve 611
position vector defining a line 611
potential function 625
Prandtl-Kolmogorov law 251
pressure equation 488
pressure equations 487
pressure-velocity coupling 484, 577
principal unit normal vector for space curve 616
principle of conservation of mass 13
pseudo gas constant 141

pseudo-exergy equation 291, 293
pump behavior 366
pump homologous head curves 376
pump homologous torque curves 376
pump model 366
pumps 365

Rankine 398
Rankine-Hugoniot curve 399
Ransom 79, 374
reduction 520
regular grid's 554
regular mesh 554
regular topology 554
relation between the contravariant and the covariant base vectors 655
relations between the inverse metrics and the metrics 658
relative coordinates 519
reliefe valves 342
residual error vector 546
restatement of the conservative derivative operations 659
restatement of the non conservative derivative operations 661
Reynolds averaging 17
Reynolds stresses 62
right handed Cartesian coordinate system 609
r-momentum equation 532
rotation around axes 635

Sauter mean diameter 34
scalar (dot) product 616
Scarborough criterion 478
second fundamental metric identity 431, 674
second law of the thermodynamics 251
second law of thermodynamics 252
semi-conservative form for the entropy equation 237
shock adiabatic 399
shock tube 180
shock wave discontinuity 17
signed integers 527
signet integers 560
single component flashing nozzle flow without external sources 327
single phase momentum equations 46
single-phase test cases 107
slip models 322

slip velocity ratio 322
solid particles 2
solid phase 160
sonic velocity 256, 310, 442, 487
sound velocity 396
sound wave propagation in multi-phase flows 399
space exponential scheme 504
spatial averaging theorem 10
specific enthalphy equation 233
specific enthalpy 206
specific entropy derivatives 545
specific mixture entropy of the velocity field 224
specific technical work 297
spectrum of particle size 32
spectrum of particle sizes 27
speed of displacement of geometrical surface 624
spherical coordinates 679
splitting a vector 617
stability criterion for bubbly flow 97
staggered grid 467
staggered grid methods 583
staggered x-momentum equation 583
standard parametrization of the line 611
steady sate flow 315
steam injection in steam-air mixture 280
Stokes theorem 48
stratified flow 68, 75, 311
stratified structure 38
stress tensor 244
structured 2
sub system network 522
successive relaxation methods 490
superheated steam 149
surface area increment 645
surface averaged velocity 18
surface force 47, 58
surface permeabilities 463, 556
surface permeability 7
surface tension 48
Suter-diagram 377
Suter-diagrams 378
system of algebraic equations 482
system of algebraic non-linear equations 491

temperature 204

temperature equation 247, 309, 454, 474, 572
temperature inversion 496
thermal expansion coefficient 150
thermo-diffusion 21
thermodynamic and transport properties 162
thermodynamic equilibrium 148
thermodynamic properties of water and steam 163
thermodynamic state 150
third low of the thermodynamics 153
thrusts acting on the pipes 334
time average 17
time averaged entropy equation 225
time dependent grid metric identity 674
time step limitation 503
topology 554
total time derivative 13
transformed coordinate system 428
transformed fluid properties conservation equation 670
transformed space 638
transformed total time derivative of a scalar 674
transient forces 342
triple scalar or box product 631
turbulent coefficient of thermal conductivity 229
turbulent diffusion 23, 228
turbulent pulsations 228

unit tangent vector 615
UO_2 – water 422
UO_2 – water system 412

valve characteristics 352
vdp-exergy 295
vector 609
vector (cross) products 628
vector between two points 609
vector of dependent variables 461, 555

velocity discontinuity in the shock plane 398
velocity field 2
velocity field no. 2 3
velocity field no. 3 4
velocity field no.1 2
velocity of sound 153
vena contracta 345
virtual mass force 86
virtual mass term 447
viscous diffusion 597, 601, 604
viscous dissipation 244, 245
viscous dissipation rate 246
volume conservation equation 310, 441
volume elements 554
volume fraction 6
volume of hexahedron 633
volume of space enveloped by a closed surface 650
volume of tetrahedron 632
volume porosity 463, 556
volumetric mixture flow rate 257
volumetric porosity 6

Wall friction force 486
weak waves 396
weighted average velocity 10
wet perimeters 77
Whitaker's spatial averaging theorem 54

x-component of the momentum equation 455
x-momentum equation 573, 597

y-component of the momentum equation 456
y-momentum equation 575, 601

z momentum equation 576
z-component of the momentum equation 458
z-momentum equation 540, 604

Printing: Mercedes-Druck, Berlin
Binding: Stein+Lehmann, Berlin